Microbial
Physiology

Microbial
Physiology
Unity and Diversity

Ann M. Stevens
Virginia Tech
Blacksburg, Virginia

Jayna L. Ditty
University of St. Thomas
St. Paul, Minnesota

Rebecca E. Parales
University of California, Davis
Davis, California

Susan M. Merkel
Cornell University
Ithaca, New York

ASM
PRESS
WASHINGTON, DC

WILEY

Editorial Correspondence:
ASM Press, 1752 N Street, NW, Washington, DC 20036-2904, USA

Registered Offices:
John Wiley & Sons, Inc., 111 River Street, Hoboken, NJ 07030, USA

For details of our global editorial offices, customer services, and more information
about Wiley products, visit us at www.wiley.com.

Wiley also publishes its books in a variety of electronic formats and by
print-on-demand. Some content that appears in standard print versions of this
book may not be available in other formats.

Library of Congress Cataloging-in-Publication Data

Applied for

Cover and chapter opener images: Sarah Lily Guest
Cover and interior design by: Debra Naylor, Naylor Design Inc.

SKY10070177_032124

We dedicate this book to our mentors, Abigail A. Salyers (AMS) and Caroline (Carrie) S. Harwood (JLD, REP, and SMM), pioneering women scientists in the field of microbial physiology and diversity who inspired the writing of this text. We also acknowledge the additional scientific mentors who have guided us in our career paths: David Gibson (REP), Jane Gibson (SMM, REP), Susan Golden (JLD), E. Peter Greenberg (AMS), Nick Ornston (REP), and Nadja Shoemaker (AMS), and our families for their steadfast support.

CONTENTS

Preface   *xv*
About the Authors   *xvii*
About the Companion Website   *xviii*

PART I: UNITY   3

1 Microbial Phylogeny—The Three Domains of Life   5

Introduction   6

The Three Branches of Life: *Bacteria, Archaea,* and *Eukarya*   6

The 16S/18S rRNA Gene as a Basis for Phylogenetic Comparisons   7

The Modern Molecular Phylogenetic Tree of Life   12

Phylogenetics and Earth History   14

2 Metabolic Unity—Generation of Biosynthetic Precursors   21

Making Connections   22

The Purpose of Central Metabolism   22

The 12 Essential Precursors   23

The Embden-Meyerhof-Parnas (EMP) Pathway/Glycolysis   25

Structure and Energy Exchange of Key Coenzymes   28

Controlling the Direction of Carbon Flow during Glycolysis   29

The Pentose Phosphate Pathway (PPP)   31

The Entner-Doudoroff (ED) Pathway   33

The Transition Reaction: Carbon Flow into the Tricarboxylic
Acid (TCA) Cycle   36

The Tricarboxylic Acid (TCA) Cycle   37

Anaplerotic Reactions   37

The Branched or Incomplete Tricarboxylic Acid (TCA) Pathway 41

The Glyoxylate Cycle 41

Reversing Carbon Flow from the Tricarboxylic Acid (TCA) Cycle to the Embden-Meyerhof-Parnas (EMP) Pathway 43

3 Cellular Components—What's In a Cell 51

Making Connections 52

Estimating Molecular Concentrations 52

Physiologically Relevant Protein Concentrations 54

Measuring Enzyme Activity: Basic Principles of Enzyme Assays 55

Michaelis-Menten Kinetics 58

Studying the Proteome 59

The Physiological Role and Composition of Cellular RNA 61

The Physiological Role and Composition of Cellular DNA 63

Studying the Genome and the Transcriptome 64

4 Cellular Growth 73

Making Connections 74

Methods to Monitor Bacterial Growth 74

The Phases of Bacterial Growth in Batch Culture 78

Requirements for Microbial Growth 80

Diauxic Growth 80

Exponential Growth Kinetics 81

Chemostats 83

Characteristics of Stationary-Phase Cells 84

Proteins Important for Cell Shape and Cell Division 85

Chromosome Segregation 86

5 Bioenergetics and the Proton Motive Force 95

Making Connections 96

Cellular Mechanisms for ATP Synthesis 96

Chemiosmotic Theory 98

ATP Synthase 99

The Proton Motive Force (PMF) 99

Quantifying the Proton Motive Force 99

Cellular Proton Levels 100

Environmental Impacts on the Proton Motive Force (PMF) 100

Experimentally Measuring the Proton Motive Force (PMF) 101

6 **Respiration and Fermentation** **107**

Making Connections 108

The Basic Components of an Electron Transport
Chain (ETC) 108

Electrode/Reduction Potential (E_0') 109

Brief Review of the Electron Transport Chain (ETC) in
Mitochondria 110

Q Cycle of Mitochondria 113

Bacterial Electron Transport Chains (ETCs) 113

Q Loop of Bacteria 115

Electron Donors and Acceptors in Bacteria 115

Fermentation 117

7 **Regulation—Posttranslational Control** **127**

Making Connections 128

Importance of Regulatory Processes 128

Allosteric Regulation of Enzymes 129

Allosteric Regulation of Branched Pathways 131

Covalent Modifications 134

Posttranslational Regulation in the Sugar
Phosphotransferase System (PTS) 138

8 **Gene Regulation—Transcription Initiation and
Posttranscriptional Control** **147**

Making Connections 148

Transcription Terminology 148

Bacterial Transcription Initiation and Elongation 149

Bacterial Transcription Termination 151

Regulatory *cis*- and *trans*-Acting Elements
Impacting Transcription 153

Examples of Different Promoter Structures 154

Transcriptional Regulation of the *lac* Operon 156

Activation and Repression by the Global Regulator Cra 158

Attenuation   158

Posttranscriptional Regulation   161

Methods Used to Study Gene Regulation   163

Methods to Demonstrate Protein–DNA Interactions   164

Interlude: From Unity to Diversity   177

Metabolic Diversity   178

Global Nutrient Cycles   179

Structural and Regulatory Diversity of Microbes   180

PART II: DIVERSITY   183

9 Autotrophy   185

Making Connections   186

Autotrophy   186

Calvin Cycle   187

Reductive Tricarboxylic Acid (rTCA) Cycle   191

Reductive Acetyl-CoA Pathway   193

3-Hydroxypropionate (3HP) Bi-cycle   195

3-Hydroxypropionate-4-Hydroxybutyrate (3HP-4HB) and Dicarboxylate-4-Hydroxybutyrate (DC-4HB) Cycles   195

Why So Many CO_2 Fixation Pathways?   197

10 Phototrophy   207

Making Connections   208

Phototrophy   208

Chlorophyll-Based Phototrophy   209

Cellular Structures Needed for Phototrophy: Light-Harvesting Complexes, Reaction Centers, and Unique Membrane Organizations   211

Oxygenic Photoautotrophy in the *Cyanobacteria*   215

Anaerobic Anoxygenic Phototrophy in the Phototrophic Purple Sulfur and Purple Nonsulfur Bacteria   218

Anaerobic Anoxygenic Phototrophy in the *Chlorobi* and *Chloroflexi* (Green Sulfur and Green Nonsulfur Bacteria, Respectively)   221

Anaerobic Anoxygenic Photoheterotrophy in the *Firmicutes*   224

Aerobic Anoxygenic Phototrophy   224

Retinal-Based Phototrophy   225

11 Chemotrophy in the Carbon and Sulfur Cycles   233

Making Connections   234

The Carbon Cycle   234

The Chemoorganotrophic Degradation of Polymers   236

The Chemoorganotrophic Degradation of Aromatic Acids   236

Chemoorganotrophy in *Escherichia coli*   241

Chemolithoautotrophy   246

Chemolithoautotrophy in Methanogens   248

Methylotrophy Enables Cycling of One-Carbon (C1) Compounds   251

One-Carbon (C1) Chemolithotrophy in Acetogens   253

The Sulfur Cycle   256

Chemoheterotrophy and Chemolithoautotrophy in the Sulfur Cycle: Sulfate Reducers   256

Chemolithoautotrophy in the Sulfur Cycle: Sulfur Oxidizers   259

The Anaerobic Food Web and Syntrophy   261

12 Microbial Contributions to the Nitrogen Cycle   275

Making Connections   276

Overview of the Nitrogen Cycle   276

Nitrogen Fixation   277

Biochemistry of Nitrogen Fixation   278

Regulation of Nitrogen Fixation   280

Symbiotic Plant-Microbe Interactions during Nitrogen Fixation   282

Assimilatory Nitrate Reduction   284

Ammonia Assimilation into Cellular Biomass   285

Nitrification: Ammonia Oxidation, Nitrite Oxidation, and Comammox   287

Anammox: Anaerobic Ammonia Oxidation   290

Denitrification   293

13 Structure and Function of the Cell Envelope   303

Making Connections   304

Fundamental Structure of the Cytoplasmic Membrane   304

Variation in Cytoplasmic Membranes   306

Transport across Cytoplasmic Membranes   306

Cell Wall Structures   311

Gram-Negative Outer Membrane   315

Periplasm   320

Additional Extracellular Layers   321

14 Transport and Localization of Proteins and Cell Envelope Macromolecules   333

Making Connections   334

Introduction to Cytoplasmic Membrane Protein Transport Systems   334

Secretory (Sec)-Dependent Protein Transport System   334

The Secretory (Sec)-Dependent Protein Transport Process   337

Signal Recognition Particle (SRP)-Dependent Protein Transport Process   338

Twin-Arginine Translocation (Tat) Protein Transport Process   339

Integration of Cytoplasmic Membrane Proteins   340

Gram-Negative Bacterial Outer Membrane Protein Secretion Systems   341

Secretory (Sec)- and Twin-Arginine Translocation (Tat)-Dependent Protein Secretion Systems   341

Secretory (Sec)-Independent and Mixed-Mechanism Protein Secretion Systems   343

Importance of Disulfide Bonds   347

Transport and Localization of Other Cell Envelope Components   348

15 Microbial Motility and Chemotaxis   363

Making Connections   364

Motility in Microorganisms   364

Bacterial Flagella and Swimming Motility   364

Regulation of Flagellar Synthesis in *Escherichia coli*   367

Mechanism of Swimming Motility   369

Archaeal Flagella   370

Bacterial Surface Motility   371

Chemotaxis   372

Conservation and Variation in Chemotaxis Systems
among Bacteria and Archaea   380

Methods to Study Bacterial Motility and Chemotaxis   381

16 Quorum Sensing   389

Making Connections   390

Fundamentals of Quorum Sensing   390

Quorum Sensing and Bioluminescence in the *Vibrio
fischeri*-Squid Symbiosis   391

Basic Model of Quorum Sensing in Gram-Negative
Proteobacteria   395

Basic Model of Quorum Sensing in Gram-Positive
Bacteria   398

Interspecies Communication: the LuxS System   400

Regulatory Cascade Controlling Quorum Sensing in
Vibrio cholerae   400

Quorum Quenching   402

17 Stress Responses   415

Making Connections   416

Oxidative Stress   416

Heat Shock Response   419

Sporulation   420

18 Lifestyles Involving Bacterial Differentiation   441

Making Connections   442

A Simple Model for Bacterial Cellular Differentiation:
Caulobacter crescentus   443

Differentiation in Filamentous Cyanobacterial Species   444

Life Cycle of Filamentous Spore-Forming *Streptomyces*: An Example of Bacterial Multicellularity   **447**

Life Cycle of Myxobacteria: Predatory Spore-Forming Social Bacteria   **449**

Biofilms: The Typical State of Microorganisms in the Environment   **452**

Index   *467*

PREFACE

This text focuses on key learning concepts important for undergraduate students in the ever-evolving and exciting field of microbial physiology. Our goal is to provide students with an idea of how all of the different components of a microbe work together to create a functioning living cell that has the capacity to adapt to new environments and changing environmental conditions. Thus, the text is roughly divided into two halves; the first part highlights the UNITY of essential microbial processes while the second part demonstrates the amazing DIVERSITY of the microbial world that surrounds us. This microbial physiology textbook has been designed to accompany a one-semester upper-level undergraduate course or an entry-level graduate course for students who have already taken an introductory general microbiology course. It is assumed that the reader has some prior knowledge of the processes associated with the central dogma and basic biochemistry (i.e., the four basic classes of organic molecules: carbohydrates, lipids, nucleic acids, and proteins). The American Society for Microbiology Curriculum Guidelines were used to design the desired student-learning outcomes that appear at the beginning of each chapter. In addition, supplemental active-learning materials are provided to give students the opportunity to delve deeper into the material for each chapter. We focus on building student quantitative skills by providing example calculations relevant to measuring aspects of microbial physiology. The design of the supplemental materials is also meant to provide flexibility for instructors in terms of course delivery methods (e.g., flipped classroom, hybrid).

The content of the book is designed to build upon itself from beginning to end and to examine select topics from multiple perspectives. Each chapter after the first begins with a "Making Connections" section that outlines the goals of the chapter and how the material expands upon information presented in previous chapters. Rather than cover an exhaustive list of topics and examples, this text is designed to introduce students to select specific examples. The purpose of the UNITY topics (Chapters 1–8) is to provide fundamental concepts and tools so that this knowledge can be subsequently applied more broadly to systems of interest to the reader. For example, one member of a protein family may be discussed, but with an understanding of how that protein works for the cell, much can be inferred about other family members. Many of our discussions will center on *Escherichia coli* because it is arguably the best-studied model bacterial organism. Most bacteria have similarities in their macromolecular composition and biosynthetic pathways; therefore, much of what we know about *E. coli* can be applied to other bacteria in general terms. The DIVERSITY topics (Chapters 9–18) highlight the variety of strategies used by microbes to conserve energy and acquire nutrients while adapting to the wide range of environments that they inhabit, using distinctive regulatory pathways and evolving unique physical structures.

The UNITY section of the book begins by presenting the microbes that will be discussed and used as examples in terms of phylogeny. Subsequently, the production of essential precursors during central metabolism, which are building blocks for the functional macromolecules of all microbes, is described. From this, the roles and the physiologically relevant concentrations of the macromolecules that are present in bacterial cells are further examined, leading to a discussion about how bacteria grow using these molecules. Next the fundamentals of how cells harness the proton motive force produced from an electron transport chain are presented. The unity portion of the material ends with a section on the different levels of regulation (i.e., translation, transcription, and posttranscription) that drive control of overall cellular physiology.

A brief interlude section provides key definitions in preparation for the DIVERSITY section of the book, which begins with an overview of the different types of microbial metabolism in terms of the carbon, sulfur, and nitrogen cycles. The structure of the cell envelope and mechanisms used to transport nutrients and proteins across these structures are then compared and contrasted in different microbes. The ability of bacterial cells to sense and respond to changing environmental conditions through chemotaxis and motility, quorum sensing, and stress responses is dependent upon a variety of regulatory mechanisms. Finally, topics focusing on microbial differentiation, including biofilm formation, bring together ways that complex regulatory pathways control phenotypic outputs in a variety of microbes.

We are grateful to numerous colleagues who helped us in the preparation of this textbook by reviewing and commenting on the initial outline, reading and editing specific sections of the text, answering queries, and providing material for figures. We sincerely appreciate their helpful comments and contributions. In addition, we thank the ASM Press staff for their patience and guidance throughout the writing process and our students who have informed and inspired our teaching practices over the years.

It is our hope that this microbial physiology textbook will give students greater insights and appreciation for how all of the different components of a microbial cell come together to create a complex living organism, and how different microbial species can form dynamic communities with the capacity to influence the environment on local and global scales.

ANN M. STEVENS
JAYNA L. DITTY
REBECCA E. PARALES
SUSAN M. MERKEL
2023

ABOUT THE AUTHORS

Ann M. Stevens is a Professor in the Department of Biological Sciences at Virginia Tech, Blacksburg, VA. She is a member of the Virginia Tech Academy of Teaching Excellence and has served on the American Society for Microbiology (ASM) Committee on Undergraduate Education.

Jayna L. Ditty is a Professor of Biology and an Associate Dean in the College of Arts and Sciences at the University of St. Thomas in St. Paul, MN.

Rebecca E. Parales is a Professor Emerita in the Department of Microbiology and Molecular Genetics at the University of California, Davis and has served as an instructor for the Microbial Diversity summer course at the Marine Biological Laboratory, Woods Hole, MA.

Susan M. Merkel is a Senior Lecturer emerita in the Department of Microbiology at Cornell University in Ithaca, NY. She is a past chair of the ASM Committee on Undergraduate Education and was the 2015 recipient of the ASM Carski Award for Undergraduate Education.

ABOUT THE COMPANION WEBSITE

Microbial Physiology: Unity and Diversity, is accompanied by a companion website:
www.wiley.com/go/stevens/microbialphysiology

The website includes:

- Answers to Questions for instructors
- Powerpoint files of all figures in the book

PART I: UNITY

6 Introduction

6 The Three Branches of Life: *Bacteria*, *Archaea*, and *Eukarya*

7 The 16S/18S rRNA Gene as a Basis for Phylogenetic Comparisons

12 The Modern Molecular Phylogenetic Tree of Life

14 Phylogenetics and Earth History

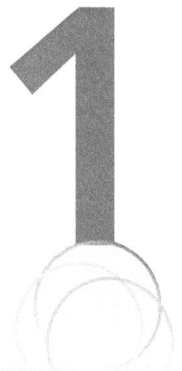

MICROBIAL PHYLOGENY—
THE THREE DOMAINS OF LIFE

After completing the material in this chapter, students should be able to . . .

1. identify distinguishing characteristics for each of the three domains of the modern molecular-based tree of life

2. discuss the strengths and limitations of using rRNA genes to generate phylogenetic trees

3. interpret a simple phylogenetic tree to determine relatedness

4. explain how endosymbiotic events and horizontal gene transfer can influence phylogenetic relationships

5. describe characteristics of two major phyla of *Archaea*

Microbial Physiology: Unity and Diversity, First Edition. Ann M. Stevens, Jayna L. Ditty, Rebecca E. Parales and Susan M. Merkel.
© 2024 American Society for Microbiology.
Companion website: www.wiley.com/go/stevens/microbialphysiology

Introduction

Microbial physiology is, in the most basic terms, the study of the structure and function of microbes. A fundamental understanding of microbial physiology is broad-reaching and has important implications across the field including the subdisciplines of environmental, medical, food, agricultural, and industrial microbiology. This first chapter begins with a background discussion on how microbes are classified, both from a historical and modern perspective, and incorporates the themes of unity and diversity that will be woven throughout this textbook. Molecular components and processes that are common across the microbial world will also be introduced.

The Three Branches of Life: *Bacteria*, *Archaea*, and *Eukarya*

Historically, microbial cells have been considered prokaryotic or eukaryotic based on their cellular substructure. Eukaryotic cells have organelles, the most obvious of which is the nucleus, but prokaryotes do not. Distinguishing between different prokaryotes is more difficult since most of these organisms look very similar to one another under a microscope. Microbes were initially classified based on their phenotypes with regard to their morphology (e.g., rods, cocci, spirilla, filaments), structures (e.g., flagella, pili), growth characteristics (e.g., aerobes versus anaerobes), and metabolic capacities (e.g., phototrophy, nitrogen fixation). However, since the 1970s it has been possible to classify microbes based on molecular **phylogeny**, which represents the evolutionary relationships between organisms. We now recognize that all organisms belong to three major groups known as **domains**. The prokaryotes consist of the domains *Bacteria* and *Archaea*, which are distinct from the third domain, *Eukarya* (Fig. 1.1). As discussed later in the text, some, but not all, of the phenotypic characteristics described above track with phylogenetic relationships.

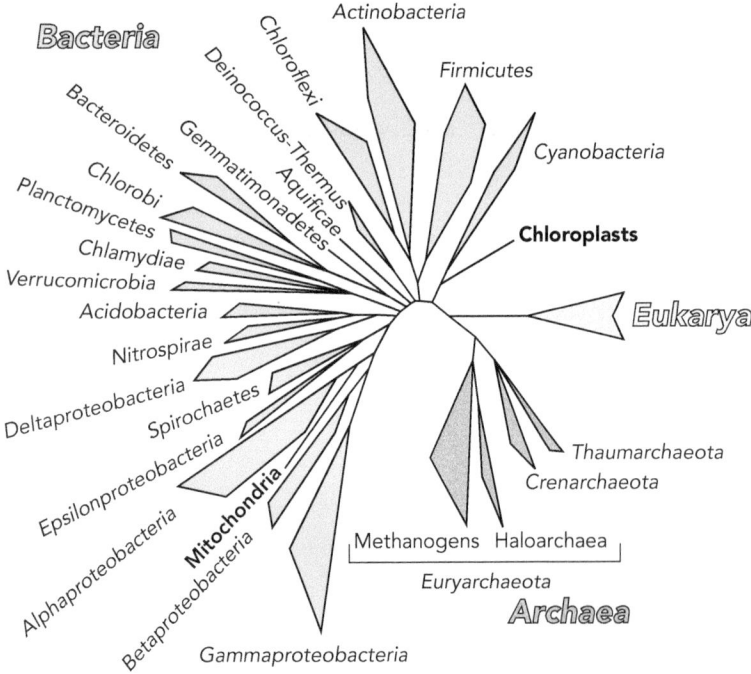

FIGURE 1.1 **Universal phylogenetic tree of life showing the three domains: *Bacteria*, *Archaea*, and *Eukarya*.** This phylogenetic tree is a model based on comparisons of small subunit rRNA (16S and 18S) sequences. Branches indicate some of the major lineages (of organisms in each domain. The major lineages represented here are based upon phyla and classes as categorized by the National Center for Biotechnology Information (NCBI) and terminology common in the literature. It should be noted that phylogenetic organization is frequently revised based on growing knowledge, technological advances, and the identification of new microbial species. As denoted by the line length of each branch, the *Bacteria* (green) and *Archaea* (pink) diverged first and the *Eukarya* (yellow) diverged from the *Archaea*. Note the close relationships of the chloroplast to the *Cyanobacteria* and the mitochondrion to the *Alphaproteobacteria*, which provide evidence for the endosymbiotic theory and the origin of eukaryotic organelles. Similar versions of this tree will be used throughout this book to illustrate the phylogeny of organisms of interest as we discuss their unique physiologies.

TABLE 1.1 **Unique attributes of the three domains of life**

Attribute	*Bacteria*	*Archaea*	*Eukarya*
Membrane-enclosed nucleus	NO	NO	YES
Peptidoglycan cell wall	YES	NO	NO
Membrane lipid structure	Ester-linked	Ether-linked	Ester-linked
Translation initiation	fMet (formylmethionyl)	Met	Met
Ribosome size	70S (5S, 16S, 23S)	70S (5S, 16S, 23S)	80S (5.8S, 18S, 28S)
Ribosome is sensitive to	Chloramphenicol, kanamycin, and streptomycin	Diphtheria toxin	Diphtheria toxin
RNA polymerase	One core with 4 essential subunits and multiple exchangeable sigma subunits	One with 12 to 14 subunits (like eukaryotic RNA polymerase II)	Three (I, II, and III)
Transcription factors required	NO	YES	YES

In addition to a nucleus, eukaryotic cells contain other membrane-bound organelles such as mitochondria and chloroplasts, which contain their own sets of genetic material. Current evidence indicates that mitochondria and chloroplasts evolved from free-living bacteria that became obligate symbionts within a host eukaryotic cell; this is the basis of the **endosymbiotic theory**. This theory, which is now strongly supported by genetic evidence, posits that a respiring free-living bacterium related to the *Alphaproteobacteria* was engulfed by another cell and evolved to form the mitochondrion. Similarly, a photosynthetic cyanobacterium that was engulfed (by a mitochondria-containing eukaryotic cell) is proposed to be the progenitor of the present-day chloroplast. Mitochondria and chloroplasts, which are roughly the same size as an average bacterium, have remnant genomes and ribosomes. Their ribosomal RNA (rRNA) gene sequences are more similar to those of bacteria than eukaryotic cells. Therefore mitochondria and chloroplasts fall within the *Bacteria* on the **phylogenetic tree** (Fig. 1.1), a diagram that depicts the evolutionary relationships among organisms. Although some *Bacteria* and *Archaea* possess subcellular structures, these are not generally considered to be organelles.

Each domain of the tree of life has distinguishing characteristics, some of which are summarized in Table 1.1. For example, there are several differences in the components of bacterial, archaeal, and eukaryotic cell envelopes (Chapter 13). In the domain *Bacteria*, cell walls are composed of **peptidoglycan**, a cross-linked polymer of *N*-acetylglucosamine and *N*-acetylmuramic acid, whereas the *Archaea* and *Eukarya* do not utilize this macromolecule. This difference is important in the treatment of human diseases caused by pathogenic bacteria, as antibiotics such as penicillin and other β-lactams target the specific cell wall structure of the *Bacteria* and thus do not harm host eukaryotic cells. A specific type of membrane lipid with ether linkages is present in only *Archaea*, whereas the *Bacteria* and *Eukarya* use ester linkages. The *Eukarya* have larger ribosomes (composed of rRNA and proteins) in comparison to the *Bacteria* and *Archaea*.

The 16S/18S rRNA Gene as a Basis for Phylogenetic Comparisons

The manner in which living organisms have been classified and compared to one another has changed multiple times over scientific history. Traditionally, taxonomic classifications were wholly based on structural and physiological traits. Here, we will utilize a modern molecular phylogenetic tree (Fig. 1.1), which is instead based on 16S and 18S rRNA gene sequences. This text focuses on the domains *Bacteria* and *Archaea*.

In the 1970s, Carl Woese, a nucleic acid biologist at the University of Illinois at Urbana-Champaign, studied phylogenetic relationships based on 16S rRNA gene sequences. The essential process of translation, whereby cells convert the information in messenger RNA (mRNA) to produce polypeptides, is universally carried out by ribosomes and transfer RNAs (tRNAs) and thus is a highly conserved process across all domains of life. The ribosome is a large complex made of rRNA and protein molecules. The rRNA molecules are classified based on their sedimentation rates when subjected to centrifugation (reported in Svedberg units, "S"). *Bacteria* and *Archaea* contain 70S ribosomes, which are composed of ~50 ribosomal proteins and three types of rRNA: 5S, 16S, and 23S. These single-stranded rRNA molecules are approximately 500, 1600, and 2300 nucleotides in length, respectively. Eukaryotes, in contrast, have larger (80S) ribosomes containing 5.8S, 18S, and 28S rRNAs plus ribosomal proteins.

Back in the 1970s, nucleic acid sequencing was in its infancy and Maxim and Gilbert sequencing was the first method developed. The 16S rRNA molecule, found in *Bacteria* and *Archaea*, was of an appropriate size to yield sufficient genetic information so that it was technically possible to obtain its full sequence, making it useable for phylogenetic comparisons. The 5S rRNA, while easier to sequence, contained less genetic information for comparisons to be made, and the 23S rRNA was at the time more difficult and costlier to sequence. Further reasons for selecting the 16S rRNA, and the corresponding 18S rRNA in eukaryotes, for phylogenetic analysis include the following. First, rRNA molecules are believed to be very ancient and therefore inherited from a common, but unknown, ancestor of all organisms (i.e., **LUCA**, the **last universal common ancestor**). This hypothesis has been supported by the discovery and characterization of small RNA molecules called ribozymes that have the ability to encode genetic information as well as perform enzymatic activities (e.g., self-splicing introns). Thus, it is thought that the evolution of RNA preceded the evolution of DNA and proteins. Second, phylogenetic relationships among organisms can only be inferred using cellular components that are homologous, since they were inherited from a common ancestor. All living cellular organisms contain rRNA, and it performs the same essential function during translation in all organisms. Third, because of the essential role of the ribosome in all cells, mutations in the rRNA gene may be lethal and can only accumulate at a slow rate that allows the molecule to maintain its function. Indeed, within the rRNA molecule, which has a complex secondary structure (Fig. 1.2), there are conserved regions that have been resilient to change over time and variable regions (V1 to V9) with more base changes. Therefore, the 16S and 18S rRNA genes were targeted as the ideal candidates for comparing sequences to determine phylogenetic relatedness.

Based on all of these characteristics, the sequence of the 16S/18S rRNA gene was used to distinguish different microorganisms from one another, and Woese proceeded to perform a comparative analysis of these sequences across a variety of microorganisms (Fig. 1.3). Current methods used to determine 16S

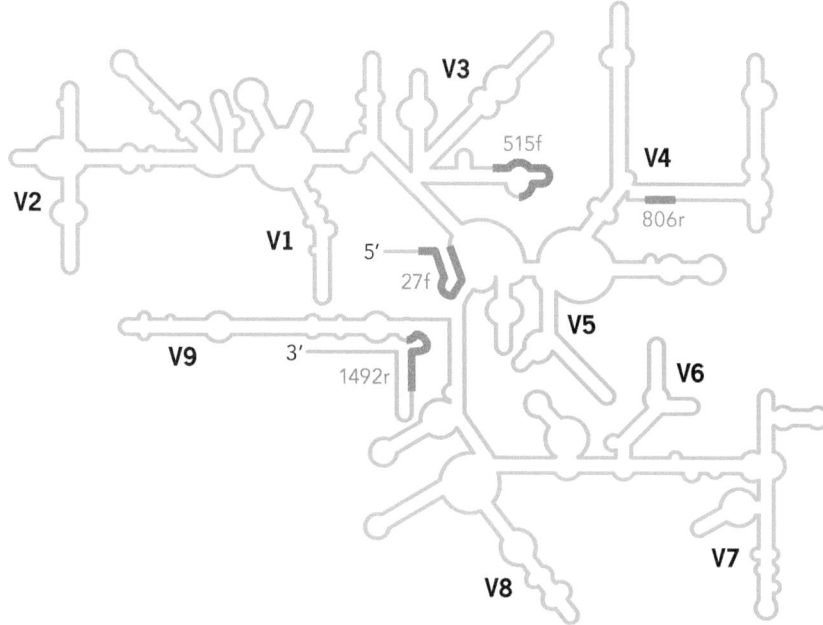

FIGURE 1.2 **Model of the secondary structure of 16S ribosomal RNA.** The conserved structure of 16S rRNA is composed of stems, which are formed by regions of the rRNA with complementary sequences that form base pairs, and loops, which are not base-paired. The molecule has both conserved and variable regions. V1 to V9 indicate the variable regions. Some conserved regions that are used as universal primer sites for PCR amplification are indicated by bold orange lines; numbers indicate nucleotide positions within the 16S rRNA gene, and r and f indicate reverse and forward directions, respectively.

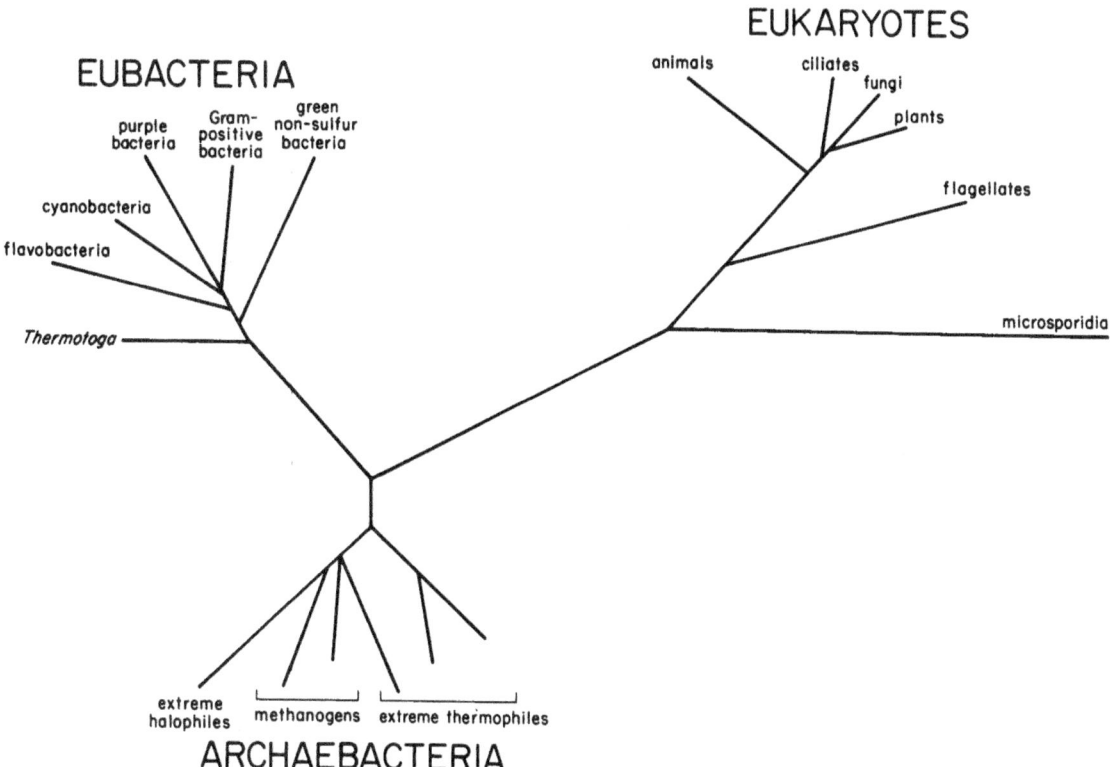

FIGURE 1.3 **An unrooted Woese phylogenetic tree.** Using rRNA sequence comparisons, Carl Woese was the first person to identify the domain *Archaea*, which he originally named the *Archaebacteria*. This is one of his early phylogenetic trees with the *Archaebacteria* broken into three groups based on physiological traits (i.e., extreme halophiles, methanogens, and extreme thermophiles). Compare this to the more recent tree shown in Fig. 1.1. Reprinted from Woese CR. 1987. *Microbiol Rev* 51(2):221–271.

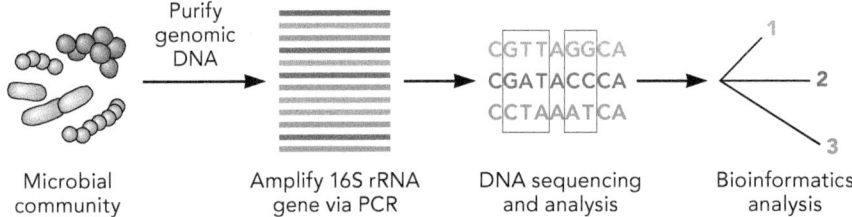

Microbial community → Purify genomic DNA → Amplify 16S rRNA gene via PCR → DNA sequencing and analysis → Bioinformatics analysis

FIGURE 1.4 **Process of generating a phylogenetic tree.** Genomic DNA is first purified from a microbial community (shown) or a pure culture. The 16S rRNA genes are amplified by PCR using universal oligonucleotide primers that bind to conserved sequences in the gene. The amplified DNA fragments are then sequenced and compared to one another using bioinformatics to generate a phylogenetic tree. The length of the line between organisms represents their phylogenetic distance from one another. This method allows for the sequencing, and potential identification, of microorganisms that cannot be grown in culture.

rRNA gene sequences (Fig. 1.4) involve first isolating the genomic DNA from the organism of interest (or from an environmental sample) and then using the **polymerase chain reaction (PCR)** to amplify specific regions of the 16S or 18S rRNA gene. Typically, this involves the use of universal primers (that anneal to conserved regions in any rRNA gene) to amplify across a variable region or across the whole rRNA gene. DNA sequencing is performed and then the obtained sequences are compared using different bioinformatics methods to generate a phylogenetic tree.

Evolutionary distance is a very simple method for comparing sequences, whereby the number of different nucleotides found between two sequences is used as the primary criterion for comparing two organisms. That is, the more different the 16S rRNA gene sequences are, the more distantly related the organisms. **Maximum parsimony** (an algorithm that assumes the smallest number of changes in all probability reflects reality) is an alternative approach that considers not only the number of differences between sequences, but also the precise position of the change in the sequence and the nature of the substitution (i.e., purine for purine, pyrimidine for pyrimidine, or purine for pyrimidine). Some methods also consider the secondary structure of the rRNA as well as the primary sequence when making comparisons. Additional computer algorithms are used to convert the resulting sequence alignments into a phylogenetic tree. The sum of the linear distance on a phylogenetic tree is typically proportional to the evolutionary distance between two organisms. The greater the length of the line, the more distantly related the two organisms are. However, because we cannot know how many times a particular base change has occurred at any position over time, it is likely that these methods underestimate evolutionary distances. It is important to note that any phylogenetic tree is a snapshot of the relationship between the organisms being analyzed at one given point in time.

While the use of 16S/18S rRNA genes has revolutionized how we understand evolutionary relationships between microorganisms, the interpretation of phylogenetic relationships can also be complicated by endosymbiotic events (i.e., mitochondria and chloroplasts). More commonly, **horizontal gene transfer** events (also known as lateral gene transfer), in which DNA is transferred between unrelated cells rather than by hereditary or vertical transmission from parent to offspring, can also complicate the interpretation of evolutionary relationships. DNA can be horizontally transferred to bacterial and archaeal cells through conjugation of plasmids or transduction by phage, and in some cases through natural transformation, which involves the uptake of DNA from the environment. These types of genetic exchange are taken into consideration in the work of Ford Doolittle and others that represent relationships in a "Web of Life" (Fig. 1.5). One common characteristic of a horizontally transferred gene is a significant deviation in its G+C content relative to that of the whole genome of an organism. Thus, it is important to base phylogenetic analyses on genes or proteins that are unlikely to have been acquired by horizontal transfer. It is generally assumed that "housekeeping genes" that encode essential functions for all cells (e.g., transcription and translation) are acquired by vertical transmission. Currently, in addition to 16S rRNA gene sequence comparisons, researchers often use comparisons of the deduced amino acid sequences of various housekeeping genes (e.g., *rpoB* or *gyrB*) to determine phylogenetic relationships. In some cases, concatenated (linked) amino acid sequences of multiple housekeeping proteins or even whole genome nucleotide sequences are used to analyze phylogeny.

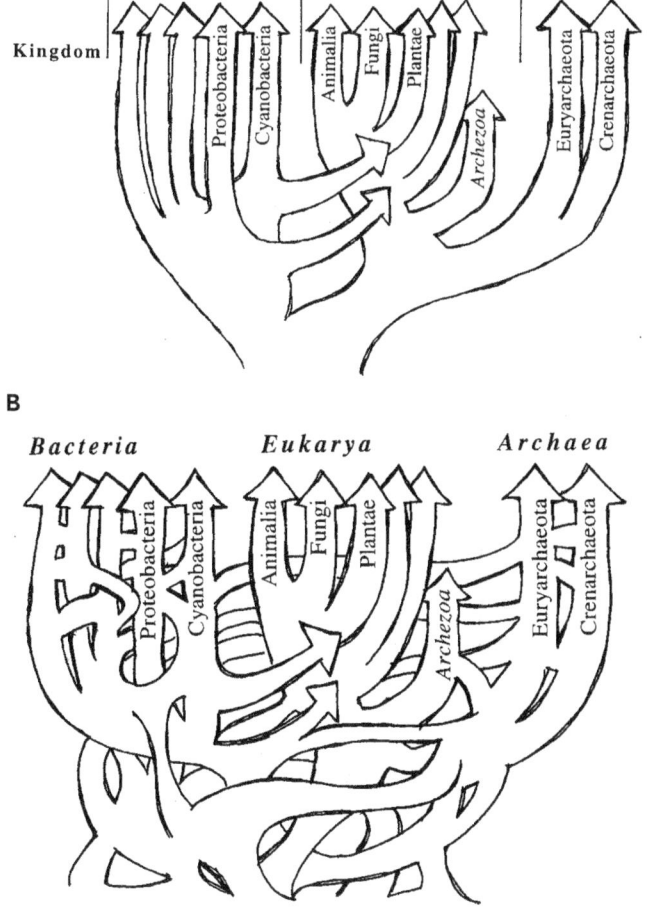

FIGURE 1.5 **A Doolittle phylogenetic tree illustrating the impact of horizontal gene transfer.** (A) An illustration of the role of horizontal gene transfer in the evolution of mitochondria and chloroplasts from free-living bacteria through endosymbiosis. (B) The impact of horizontal gene transfer through mechanisms like transformation, transduction, and conjugation is shown through this reticulated (web-like) phylogenetic tree. Reprinted from Doolittle WF. 1999. *Science* 284:2124–2128, with permission.

The Modern Molecular Phylogenetic Tree of Life

As discussed in the previous section, our modern molecular phylogenetic tree has three domains. How did Woese discover the *Archaea*? It was actually by coincidence. As the story goes, Woese asked his colleagues at the University of Illinois for samples of the various organisms that they studied in their laboratories. Ralph Wolfe was studying methanogens, which back in the 1970s were thought to be bacteria. When Woese sequenced the 16S rRNA gene of a methanogen, he discovered that the sequence was quite different from the 16S rRNA gene of comparison bacteria. Wolfe recalled this moment: "He told me methanogens weren't bacteria, and I said of course they're bacteria. They look like bacteria under the microscope." (Peterson D. 2015. The microbial man. *MCB* 9:17). When published, Woese's findings were met with great skepticism. Woese's phylogenetic tree clearly shows three domains with the *Archaea* being more closely related to the *Eukarya* than they are to the *Bacteria*. Over time Woese's theory has been validated and he was eventually awarded the Crafoord Prize in Biological Sciences, given by the Royal Swedish Academy of Sciences for accomplishments in fields not covered by the Nobel Prize.

THE *ARCHAEA*

The *Archaea* are currently divided into multiple phyla, the best studied of which are the *Crenarchaeota* and *Euryarchaeota*. Many of the microbes within the archaeal domain exhibit unique metabolic abilities. For example, the strictly anaerobic archaeal methanogens (*Euryarchaeota*) are the only microbes on the planet capable of reducing carbon dioxide fully to methane due to their repertoire of specialized coenzymes (Chapter 11). They are found in the sediments of lakes, ponds, and swamps; in sewage; and in the intestinal tracts of animals and humans. Some *Archaea* are able to thrive in extreme environments (**extremophiles**). Other members of the *Euryarchaeota* are extreme halophiles that live in environments such as salt lakes and solar evaporation ponds with high salt concentrations (e.g., 3 to 5 M; Chapter 10). While thermophiles, which grow optimally at temperatures >45°C, can be found in both the bacterial and archaeal domains, the extreme hyperthermophiles that grow at temperatures >95°C are generally *Archaea*, many of which are members of the *Crenarchaeota*. Similarly, examples of both bacterial and archaeal acidophiles and alkaliphiles that inhabit low- and high-pH environments, respectively, have been identified. Therefore, *Bacteria* and *Archaea* cannot be differentiated phylogenetically based on their ability to grow in extreme environments.

THE *BACTERIA*

Looking at a modern molecular tree of life (Fig. 1.1), it becomes readily apparent that microorganisms are the dominant and most diverse organisms on the planet. The visible world, including plants, animals, and some fungi, occupies just the tip of one branch of the tree (Fig. 1.3). This text will focus predominantly on the physiology of the *Bacteria* but will also touch upon the *Archaea*. The well-studied Gram-negative *Escherichia coli* (member of the *Gammaproteobacteria*) will be used as a model to discuss many basic physiological processes, and the well-studied Gram-positive organism *Bacillus subtilis* (member of the *Firmicutes*) will serve a similar purpose for that distinctive branch of the tree. Although almost all *Bacteria* have cell walls composed of peptidoglycan, there are two distinct types of bacterial envelope structures that can be differentiated by a specific type of cell staining known as the Gram stain (Fig. 1.6). Gram-positive bacteria have a very thick layer of peptidoglycan surrounding their cytoplasmic membrane that is associated with positively charged molecules (e.g., **teichoic acid** and lipoteichoic acid), whereas Gram-negative bacteria have a thin single layer of peptidoglycan surrounded by an outer membrane containing **lipopolysaccharide (LPS)**. The space between the two membranes

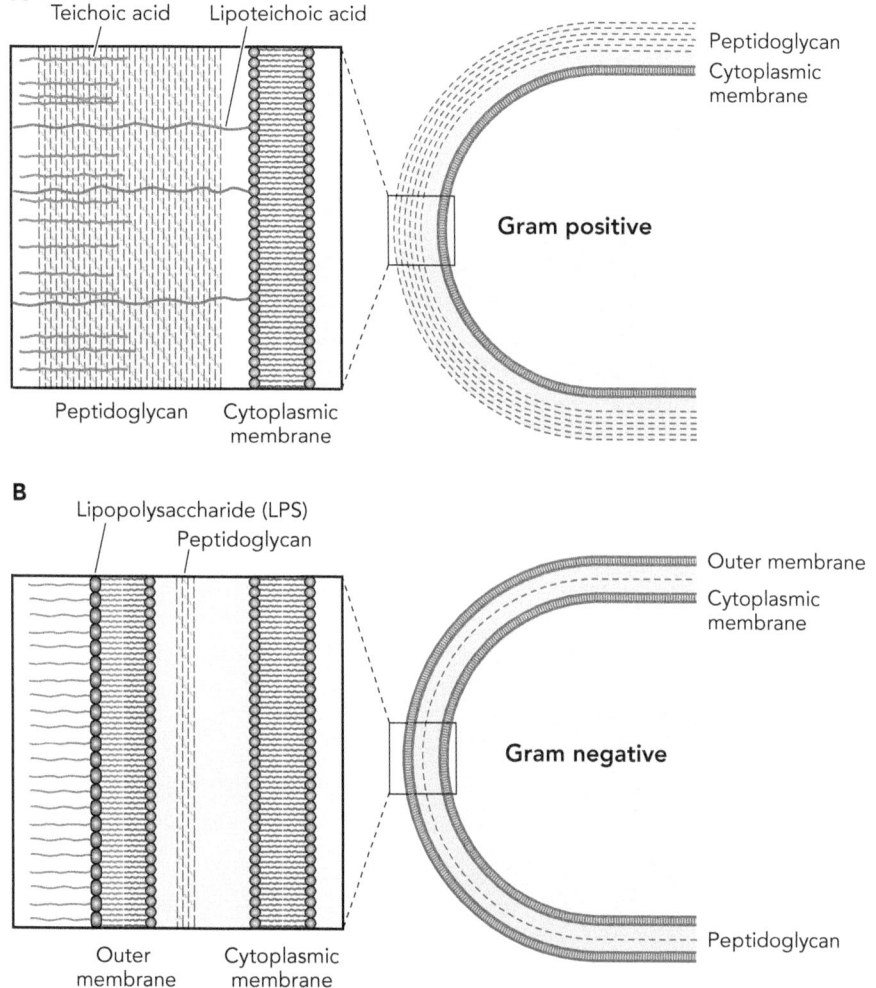

FIGURE 1.6 **Cell envelopes of Gram-positive and Gram-negative bacteria.** (A) Gram-positive bacteria have a cytoplasmic membrane and a thick peptidoglycan cell wall that is associated with other charged constituents (e.g., teichoic acid and lipoteichoic acid), whereas (B) Gram-negative bacteria have both a cytoplasmic (inner) membrane and an outer membrane that flank a thin layer of peptidoglycan. Unlike the inner membrane, which is a phospholipid bilayer, the outer membrane is asymmetrical, with the inner leaflet composed of phospholipids and the outer leaflet composed of lipopolysaccharide. Note that all inner and outer membranes also have loosely associated (peripheral) and embedded (integral) proteins (not shown).

is called the **periplasm**. This difference in cell structure has consequences for the physiology and permeability of cells that will become apparent in later chapters (Chapter 13). It is also notable that all Gram-positive bacteria (*Firmicutes* and *Actinobacteria*) are clustered together on the phylogenetic tree (Fig. 1.1) suggesting that this type of cell envelope evolved once in the last common ancestor of this group.

Although it might appear that we have a great deal of information about microbes, in actuality it is estimated that <5% of the bacteria present on Earth can currently be grown in pure cultures in the laboratory. This is likely due to the great interconnectivity that exists between microorganisms with regard to the generation and utilization of products or substrates that move between different types of cells. Many phylogenetic trees include microorganisms that have been detected in environmental samples using rRNA gene amplification and sequencing but have never been grown in the laboratory (Fig. 1.4).

Phylogenetics and Earth History

The phylogenetic tree of life can be mapped to Earth's geological history, correlating the evolutionary relationships of the domains with the geological time scale (Fig. 1.7). Based on cells and organic molecules preserved in rocks, cellular life is believed to have evolved between 3.5 billion and 4 billion years ago. However, the first original progenitor cell is unknown, as it no longer exists. Most phylogenetic trees are rooted at a point called the last universal common ancestor (LUCA). Given that the atmosphere on Earth was anaerobic and the surface temperatures were very hot, this organism likely was an anaerobic thermophile. Today, thermophiles can be found distributed in both the *Bacteria* and *Archaea* and most grow optimally in the range of 55 to 100°C in environments such as

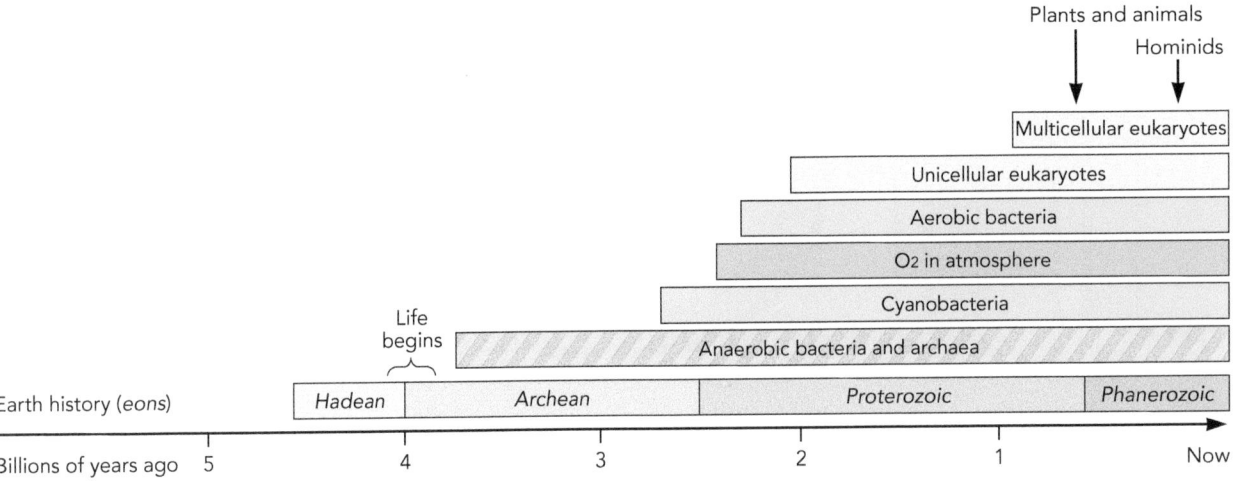

FIGURE 1.7 **Timeline of the evolution of life on Earth.** The origin of life is estimated to have occurred approximately 4 billion years ago when the planet was still hot and anoxic. *Bacteria* (green) and *Archaea* (pink) diverged early in evolution when atmospheric oxygen was completely absent. The evolution of cyanobacteria, which release oxygen as a by-product of oxygenic photosynthesis, led to the gradual oxygenation of the atmosphere. The presence of oxygen led to the evolution of aerobic forms of metabolism in *Bacteria* and *Archaea*. Early unicellular eukaryotes (yellow) with mitochondria evolved approximately 2 billion years ago, and multicellular eukaryotes began evolving less than 1 billion years ago.

hot springs, geysers, and hydrothermal vents. Some archaeal thermophiles are even capable of growth at temperatures well over 100°C. The evolution of oxygenic photosynthesis in the cyanobacteria (~3 billion years ago) resulted in the gradual accumulation of oxygen in the atmosphere, which led to the evolution of life forms capable of respiring oxygen and eventually to the first eukaryotic life (~1.5 billion to 2 billion years ago). The resulting environmental changes paved the way for the great diversification of microbial life on Earth as microbes adapted and evolved to fill all possible biological niches. As will be seen in the "Diversity" section of this book, microbial metabolism is absolutely essential to drive the Earth's carbon, nitrogen, and sulfur cycles.

Learning Outcomes: After completing the material in this chapter, students should be able to . . .

1. identify distinguishing characteristics for each of the three domains of the modern molecular-based tree of life

2. discuss the strengths and limitations of using rRNA genes to generate phylogenetic trees

3. interpret a simple phylogenetic tree to determine relatedness

4. explain how endosymbiotic events and horizontal gene transfer can influence phylogenetic relationships

5. describe characteristics of two major phyla of *Archaea*

Check Your Understanding

1. Define this terminology: phylogeny, domains, endosymbiotic theory, phylogenetic tree, peptidoglycan, LUCA, polymerase chain reaction (PCR), evolutionary distance, maximum parsimony, horizontal gene transfer, extremophiles, periplasm, teichoic acid, lipopolysaccharide (LPS)

2. Who originated the idea of a three-domain system for the phylogeny of life on Earth? What is the basis of the three-domain system? (LO2)

3. While we now know that *Archaea* and *Bacteria* are very different, before the advent of rRNA gene sequencing, *Archaea* were considered to be in the same kingdom as *Bacteria*. List two traits that caused scientists to group them together and two traits that support the hypothesis that they are very distantly related. (LO2)

4. List and describe five reasons why the 16S/18S rRNA gene is used as the basis for molecular phylogenetic comparisons. (LO2)

5. A segment of the 16S rRNA gene sequence of *Escherichia coli* is shown below. Based on the concept of evolutionary distance, which organism is *most* closely related to *E. coli*? How did you come to this conclusion? (LO3)

    > *E. coli*: T G G C A C G G T A G C
    > Organism A: T G A C A G C G T A A C
    > Organism B: T C G A C G C G T A G C
    > Organism C: A G T G A C T G T A T A

6. This phylogenetic tree was created based on 16S rRNA gene sequences. Which two organisms do you think are most closely related? On what did you base your conclusion? (LO3)

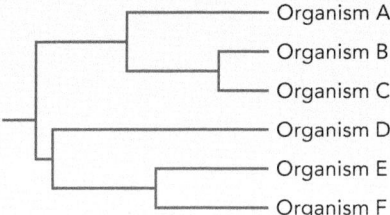

- Organism A
- Organism B
- Organism C
- Organism D
- Organism E
- Organism F

7. The purpose of completing the following information table is to help you compare and contrast important characteristics relative to the classification of *Bacteria*, *Eukarya*, and *Archaea* and the importance of two groups of *Bacteria*, the Gram-positive and Gram-negative bacteria. For each characteristic structure, describe the function it serves for the cell and

indicate if it is present (P) or absent (A) in the *Eukarya* and/or *Archaea*, and if it is present (P) or absent (A) in the Gram-positive and/or Gram-negative bacteria. (LO1)

Characteristic (Does the cell have it?)	Definition (What is it?)	Eukarya	Archaea	Bacteria	
				Gram +	Gram –
Peptidoglycan					
Teichoic acid					
Lipopolysaccharide					
16S rRNA					
18S rRNA					
Ester-linked lipids					
Ether-linked lipids					

8. Extremophiles can be found in both the *Archaea* and *Bacteria* domains. List three general categories of extremophiles and give one characteristic of each. (LO5)

9. While the use of rRNA gene sequencing has revolutionized the way in which we approach establishing evolutionary relationships among organisms, endosymbiotic events and horizontal gene transfer complicate our interpretation of these relationships. Explain how this is true. (LO4)

Dig Deeper

10. You have isolated a new bacterial species from a volcanic hot spring. Results of the many laboratory experiments you conducted are outlined below. (LO1)

Test	Result	
Growth		
Temperature	Range: 55 to 77°C	Optimal: 70°C
pH	Range: 5 to 8	Optimal: 7.5
Salt	Range: 0 to 1.5% NaCl	Optimal: 0.1%
Structures		
Lipids	Fatty acid esters	
Cell wall	Penicillin sensitive	

a) Would you classify this organism as a thermophile, acidophile, or halophile?

b) Based on the data provided, would you classify this new species as belonging to the *Bacteria*, *Archaea*, or *Eukarya*? Defend your answer by explaining your rationale for all the results provided.

11. In some cases, 16S rRNA gene sequences may not be sufficiently variable to ensure differentiation between closely related species of bacteria. An alternative approach to create a phylogenetic tree is to use the deduced amino acid sequence of a protein-encoding gene. For each entry listed below, state whether the trait would be useful to generate phylogenetic trees. Explain your reasoning. (LO2)
Suppose the gene:

- can undergo horizontal gene transfer.
- coded for a nonessential protein.
- had hypervariable regions or constant regions between species.

12. The three phylogenetic trees shown below were constructed using 16S rRNA gene sequences, the amino acid sequences of GyrB (the subunit B protein of DNA gyrase, required for DNA replication), and the amino acid sequences of NifH (a protein involved in nitrogen fixation) as labeled. Which gene was likely transferred via horizontal gene transfer and which via vertical transfer? How do you know? (LO4)

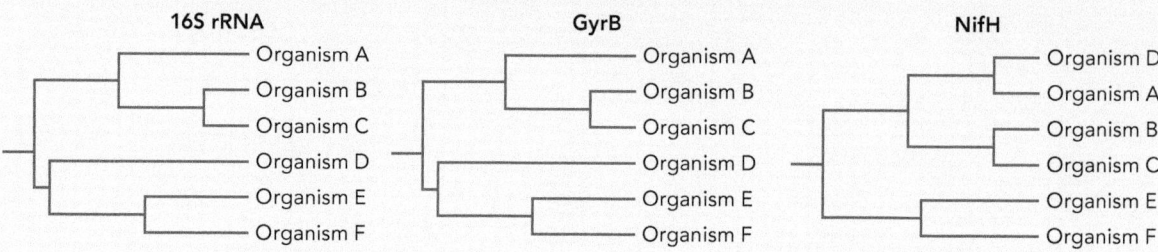

22 Making Connections

22 The Purpose of Central Metabolism

23 The 12 Essential Precursors

25 The Embden-Meyerhof-Parnas (EMP) Pathway/Glycolysis

28 Structure and Energy Exchange of Key Coenzymes

29 Controlling the Direction of Carbon Flow during Glycolysis

31 The Pentose Phosphate Pathway (PPP)

33 The Entner-Doudoroff (ED) Pathway

36 The Transition Reaction: Carbon Flow into the Tricarboxylic Acid (TCA) Cycle

37 The Tricarboxylic Acid (TCA) Cycle

37 Anaplerotic Reactions

41 The Branched or Incomplete Tricarboxylic Acid (TCA) Pathway

41 The Glyoxylate Cycle

43 Reversing Carbon Flow from the Tricarboxylic Acid (TCA) Cycle to the Embden-Meyerhof-Parnas (EMP) Pathway

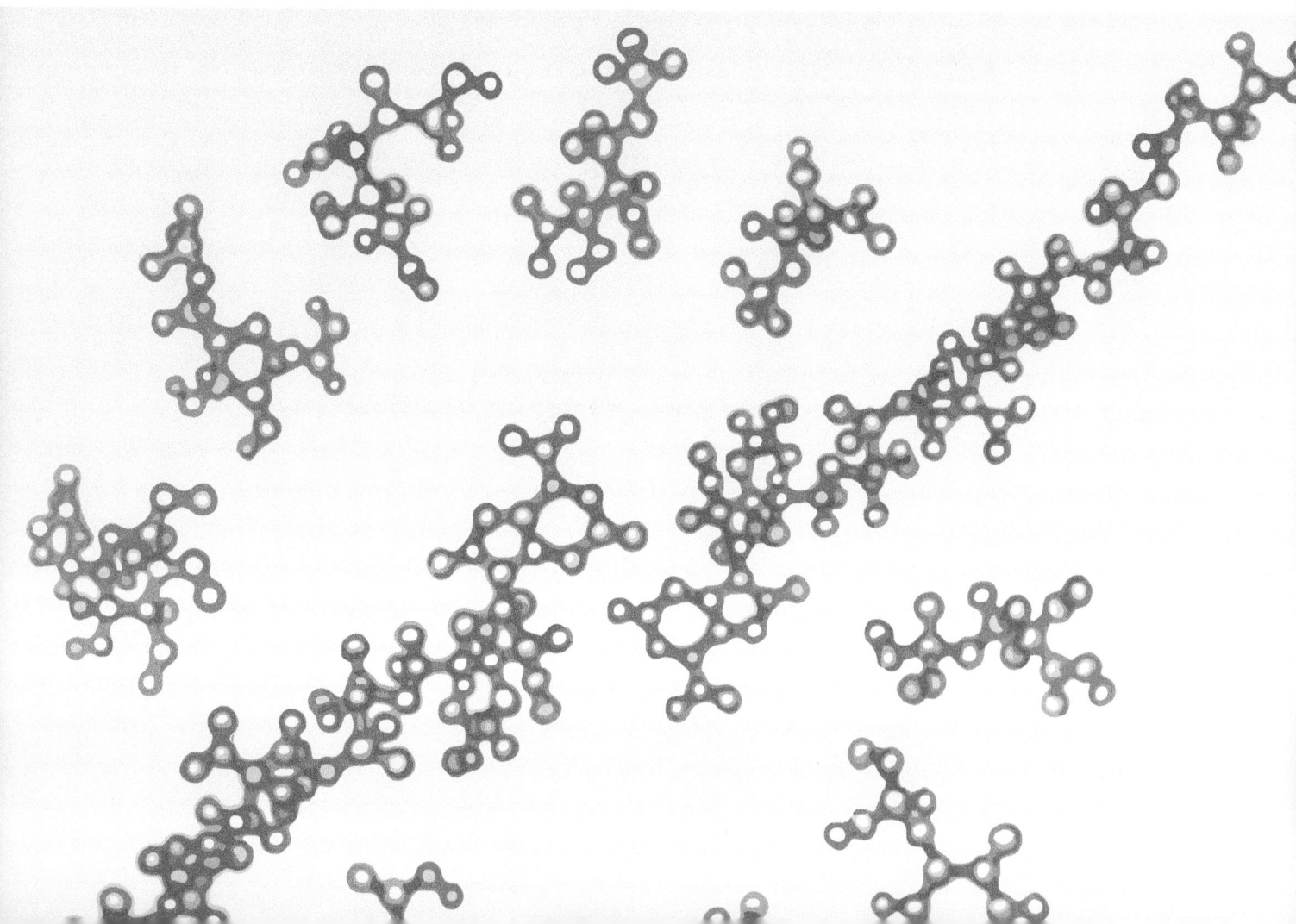

METABOLIC UNITY—
GENERATION OF BIOSYNTHETIC PRECURSORS

After completing the material in this chapter, students should be able to . . .

1. identify where in the conserved pathways of central metabolism the 12 essential precursors are produced

2. explain how cells use redox reactions and substrate-level phosphorylation to generate ATP in the Embden-Meyerhof-Parnas (EMP) pathway

3. discuss the function of the coenzymes ATP, $NAD(P)^+/NAD(P)H + H^+$, and CoA in the cell, including the importance of recycling these compounds

4. deduce the impact of effector molecules on key unidirectional enzymes in central metabolism through allosteric regulation

5. list two possible functions of the Entner-Doudoroff (ED) pathway

6. compare and contrast the role of the tricarboxylic acid (TCA) cycle in both catabolism and anabolism

Microbial Physiology: Unity and Diversity, First Edition. Ann M. Stevens, Jayna L. Ditty, Rebecca E. Parales and Susan M. Merkel.
© 2024 American Society for Microbiology.
Companion website: www.wiley.com/go/stevens/microbialphysiology

Making Connections

All organisms in the phylogenetic tree of life (Chapter 1) must be able to perform metabolism to conserve energy and produce essential chemical precursors needed for their reproduction and growth. All organisms evolved to use the reactions found in the pathways of central metabolism, which will be described in this chapter.

The Purpose of Central Metabolism

During central metabolism, cells utilize reduced organic compounds to produce the essential precursor molecules, as well as the adenosine triphosphate (ATP) and reduced coenzymes (e.g., NADH + H$^+$ and NADPH + H$^+$) needed

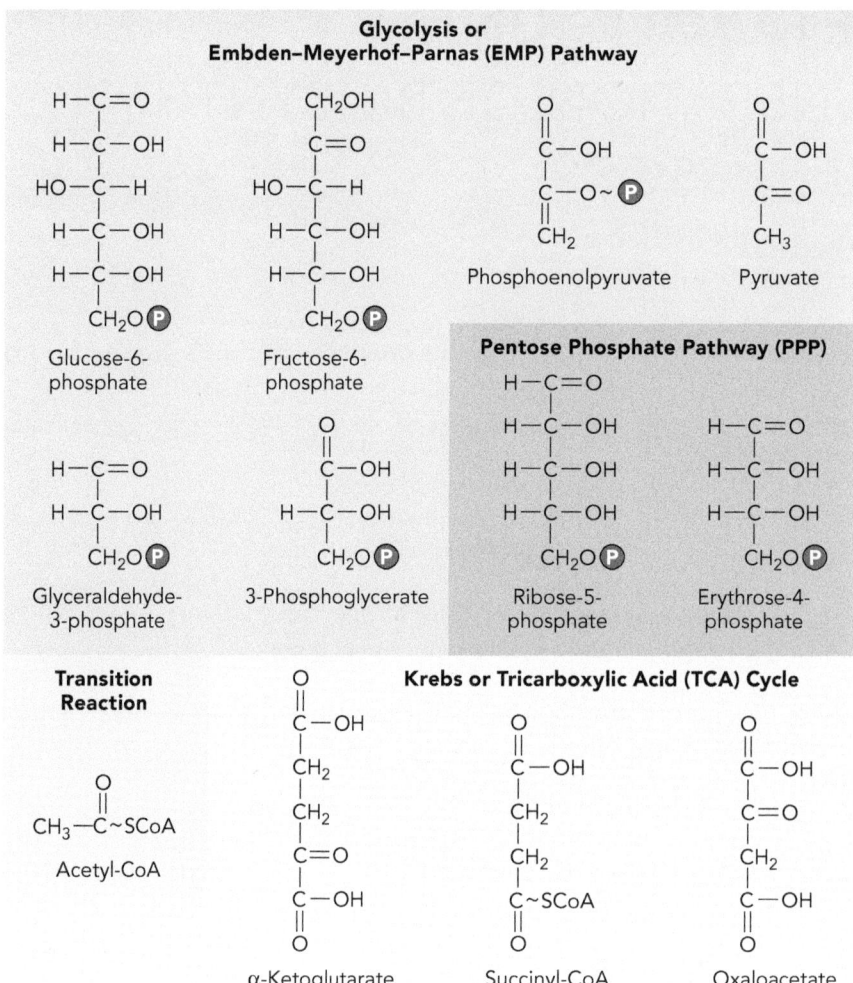

FIGURE 2.1 **The 12 essential precursors.** These universal precursor molecules are essential for life in all cells. Here they are highlighted by colored boxes to indicate the pathway of central metabolism in which each is synthesized as a metabolic intermediate during catabolism. Blue, glycolysis or Embden-Meyerhof-Parnas (EMP) pathway; pink, pentose phosphate pathway (PPP); yellow, tricarboxylic acid (TCA), Krebs, or citric acid cycle; green, acetyl-CoA, which serves as a transitional intermediate between glycolysis and the TCA cycle. The P within an orange circle indicates a phosphoryl group.

for biosynthesis. Central metabolism consists of a core set of reactions that have been maintained in almost every living organism. Essential for survival, the pathways considered to be part of central metabolism do not function separately; they work together as a web of reactions. Cells have sophisticated systems to appropriately direct the flow of metabolites through the pathways of central metabolism in order to live, grow, and successfully compete in the environment.

Overall cellular metabolism is a combination of the anabolic and catabolic reactions taking place. **Anabolism** refers to biosynthetic pathways that convert the 12 essential precursors into cellular structures. **Catabolism** refers to the degradative pathways that allow the consumption of substrates present in the external medium, the production of precursors required for growth, and the conversion of energy to useful forms. In this chapter, we will emphasize (i) the pathways used to convert any carbon source to the 12 essential precursor metabolites needed for biosynthesis and (ii) the mechanisms used to produce ATP and high-energy electrons in reduced coenzymes.

The 12 Essential Precursors

Most students are familiar with the concept that central metabolism provides energy for cells. However, an equally important concept is that central metabolism produces 12 essential precursors necessary for life (Fig. 2.1). These precursors

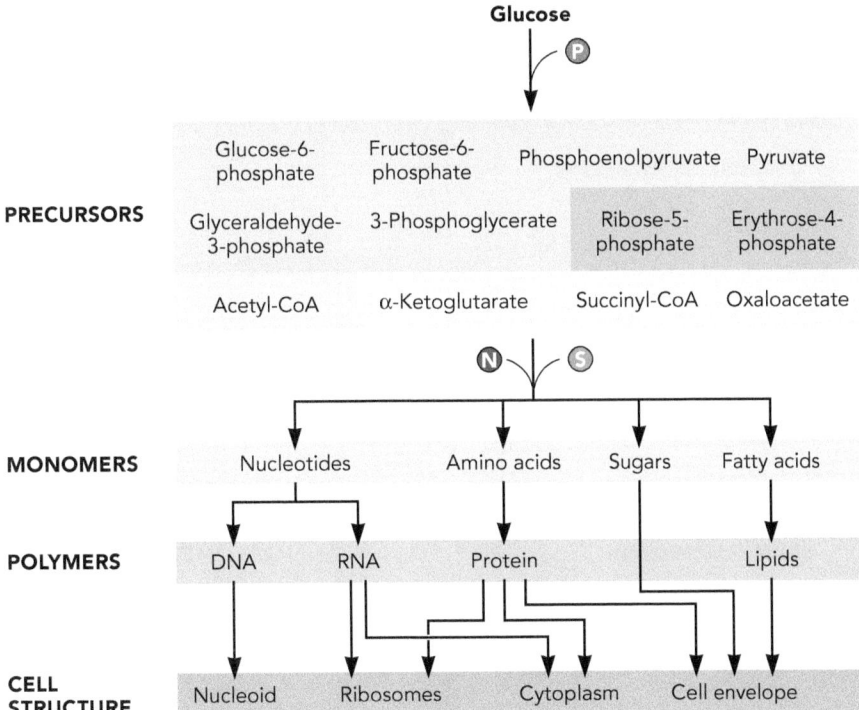

FIGURE 2.2 **The 12 essential precursors are the building blocks of cells.** The 12 essential precursors serve as the organic backbones of cellular monomers (i.e., nucleotides, amino acids, sugars, and fatty acids) during biosynthetic (anabolic) reactions. Inorganic or organic sources of phosphate, nitrogen, and sulfur are also required. The monomers are then used to form polymers and complex cellular structures (Chapter 3).

are the fundamental building blocks from which the monomers (nucleotides, amino acids, carbohydrates, and fatty acids) are synthesized, ultimately leading to the production of all the macromolecules and structures that comprise the cell (Fig. 2.2). If an organism cannot produce the monomers needed for biosynthesis from the essential precursors, they must be acquired from the environment. Every known cellular component, ranging from DNA to complex cellular structures, like the cell membrane, is synthesized from the 12 precursor metabolites, along with sources of nitrogen, sulfur, and phosphate. These biosynthetic reactions require energy in the form of the high-energy phosphate bonds in ATP and/or the high-energy electrons with reducing power in coenzymes such as nicotinamide adenine dinucleotide (NADH + H$^+$).

The 12 precursors serve as a resource or "pool" of organic molecules that link catabolism and anabolism (Fig. 2.3). While these molecules play a role in

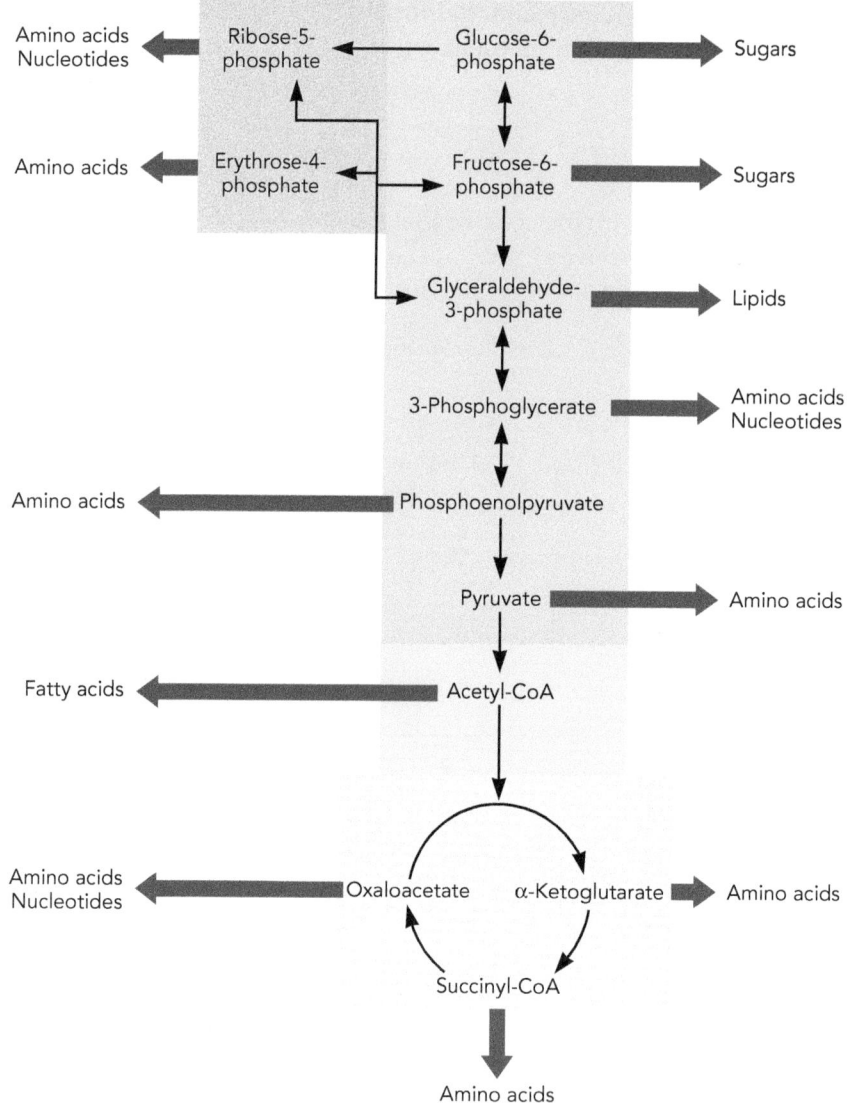

FIGURE 2.3 **Biosynthetic reactions consume the 12 essential precursors.** As nucleotides, amino acids, sugars, and fatty acids are produced during biosynthetic (anabolic) reactions, the 12 essential precursors are siphoned away (blue arrows) from central metabolism.

catabolic energy generation, eight of the 12 precursors are also substrates for the anabolic biosynthesis of amino acids. The four precursors not involved in amino acid biosynthesis are glucose-6-phosphate and fructose-6-phosphate, which are important substrates in building the carbohydrate precursors for cell wall synthesis, and glyceraldehyde-3-phosphate and acetyl coenzyme A (acetyl-CoA), which are involved in the production of membrane lipids. In addition to serving as a substrate for amino acid biosynthesis, some precursors have other roles. For example, succinyl-CoA is also used to generate the tetrapyrrole rings found in chlorophylls and cytochromes. Ribose-5-phosphate is part of the sugar-phosphate backbone of nucleotides, and oxaloacetate and 3-phosphoglycerate are used in nucleotide biosynthesis. There is a 13th precursor, sedoheptulose-7-phosphate, which is essential for the production of lipopolysaccharide in Gram-negative bacteria, but it will not be discussed here since this precursor is not universally required for all domains of life.

Organic molecules are utilized by cells through a series of pathways. The Embden-Meyerhof-Parnas (EMP) pathway (i.e., **glycolysis**), the pentose phosphate pathway (PPP), and the tricarboxylic acid (TCA) cycle (also known as the Krebs cycle or citric acid cycle), including the branched and glyoxylate pathway variations of it, are used to generate the 12 essential precursors. These three pathways are, at least in part, maintained in almost all cells. The Entner-Doudoroff (ED) pathway is an alternative pathway for glycolysis found in some prokaryotes and a few eukaryotes; this pathway offers an alternative carbon flow for the generation of precursors when there are blocks in glycolysis, or when **aldonic acids** (sugar molecules with a carboxylic acid instead of an aldehyde group) are the primary carbon source rather than glucose.

The Embden-Meyerhof-Parnas (EMP) Pathway/Glycolysis

Some steps of the EMP pathway (Fig. 2.4) are common to all living cells, as this pathway is responsible for producing six of the 12 essential precursors. The pathway can be divided into two phases: the energy input phase and the energy output phase, both of which are important for precursor metabolite production. In the energy input phase, ATP is consumed as the six-carbon (6C) glucose is ultimately converted into glyceraldehyde-3-phosphate and dihydroxyacetone phosphate (DHAP), each with three carbons (3C) in their backbones. A high-energy phosphate bond in the form of ATP is required for the conversion of glucose into glucose-6-phosphate by hexokinase. Alternatively, in some enteric bacteria, the high-energy phosphate bond in phosphoenolpyruvate (PEP) is used by the phosphotransferase transport system (PTS) for production of glucose-6-phosphate during glucose transport (Chapter 7). Glucose-6-phosphate is converted into fructose-6-phosphate by an **isomerase** (an enzyme that changes the structure of a molecule without changing its chemical composition). Then ATP is consumed a second time when fructose-6-phosphate is converted into fructose-1,6-bisphosphate by the enzyme phosphofructokinase. Thus, the energy inputs in the first phase of the EMP pathway serve to "prime the pump" for the second phase of the pathway, which yields energy for the cell. The steps in the second phase are completed twice for each molecule of glucose, because fructose-1,6-bisphosphate (6C) is cleaved by an aldolase into two molecules with

Energy Input

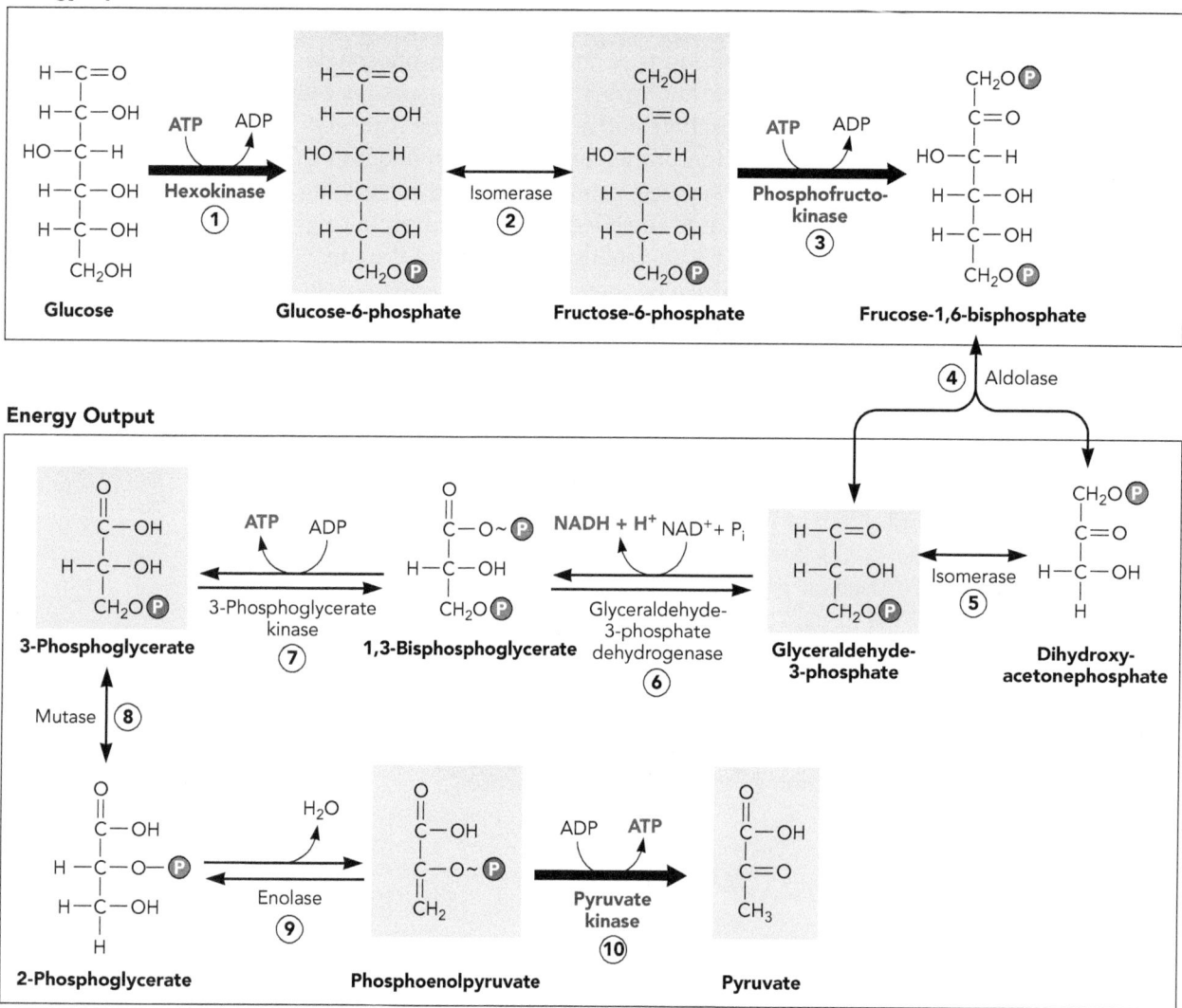

Energy Output

FIGURE 2.4 **The reactions of glycolysis or the Embden-Meyerhof-Parnas (EMP) pathway.** Essential precursors from glycolysis are highlighted by blue boxes. Energy inputs (e.g., ATP) and outputs (e.g., NADH + H⁺ and ATP) are highlighted in red. Key unidirectional enzymes are indicated in purple and the reactions they catalyze are shown with bold arrows. The specific enzymes catalyzing each reaction are (1) hexokinase, (2) phosphoglucose isomerase, (3) phosphofructokinase, (4) fructose-1,6-bisphosphate aldolase, (5) triose phosphate isomerase, (6) glyceraldehyde-3-phosphate dehydrogenase, (7) 3-phosphoglycerate kinase, (8) 3-phosphoglycerate mutase, (9) enolase, and (10) pyruvate kinase.

three carbons each, dihydroxyacetone phosphate (3C) and glyceraldehyde-3-phosphate (3C). Another isomerase converts dihydroxyacetone phosphate into glyceraldehyde-3-phosphate. Both molecules of glyceraldehyde-3-phosphate then enter the second phase of glycolysis, which yields energy.

In the energy output phase of glycolysis, glyceraldehyde-3-phosphate is converted into 1,3-bisphosphoglycerate (3C) in a **reduction/oxidation (redox) reaction** catalyzed by glyceraldehyde-3-phosphate dehydrogenase. The reduced coenzyme NADH + H⁺ (Fig. 2.5), which contains high-energy electrons as reducing power and potential energy for the electron transport chain (ETC) (Chapter 6), is also generated. ATP (Fig. 2.5) is formed by the process of **substrate-level phosphorylation**

(Fig. 2.6) in two steps of the energy output phase of glycolysis. ATP production occurs for the first time when phosphoglycerate kinase converts 1,3-bisphosphoglycerate to 3-phosphoglycerate (3C), yielding ATP. The 3-phosphoglycerate is subsequently converted into 2-phosphoglycerate, which in turn is converted into phosphoenolpyruvate. When pyruvate kinase converts phosphoenolpyruvate (3C) to pyruvate (3C), a second molecule of ATP is formed via substrate-level phosphorylation. In both reactions, a phosphoryl group is transferred to adenosine diphosphate (ADP) from a carbon-containing compound that is an intermediate in the pathway (Fig. 2.5) (as opposed to ATP formation during respiration from an ETC via ATP synthase [Fig. 2.6]; Chapter 6). Therefore, the net energy yield from glycolysis is two ATP (input of two ATP and output of four ATP) and two reduced NADH + H$^+$ coenzymes per molecule of glucose entering the pathway. In addition, six essential precursors are made: glucose-6-phosphate, fructose-6-phosphate, glyceraldehyde-3-phosphate, 3-phosphoglycerate, phosphoenolpyruvate, and pyruvate.

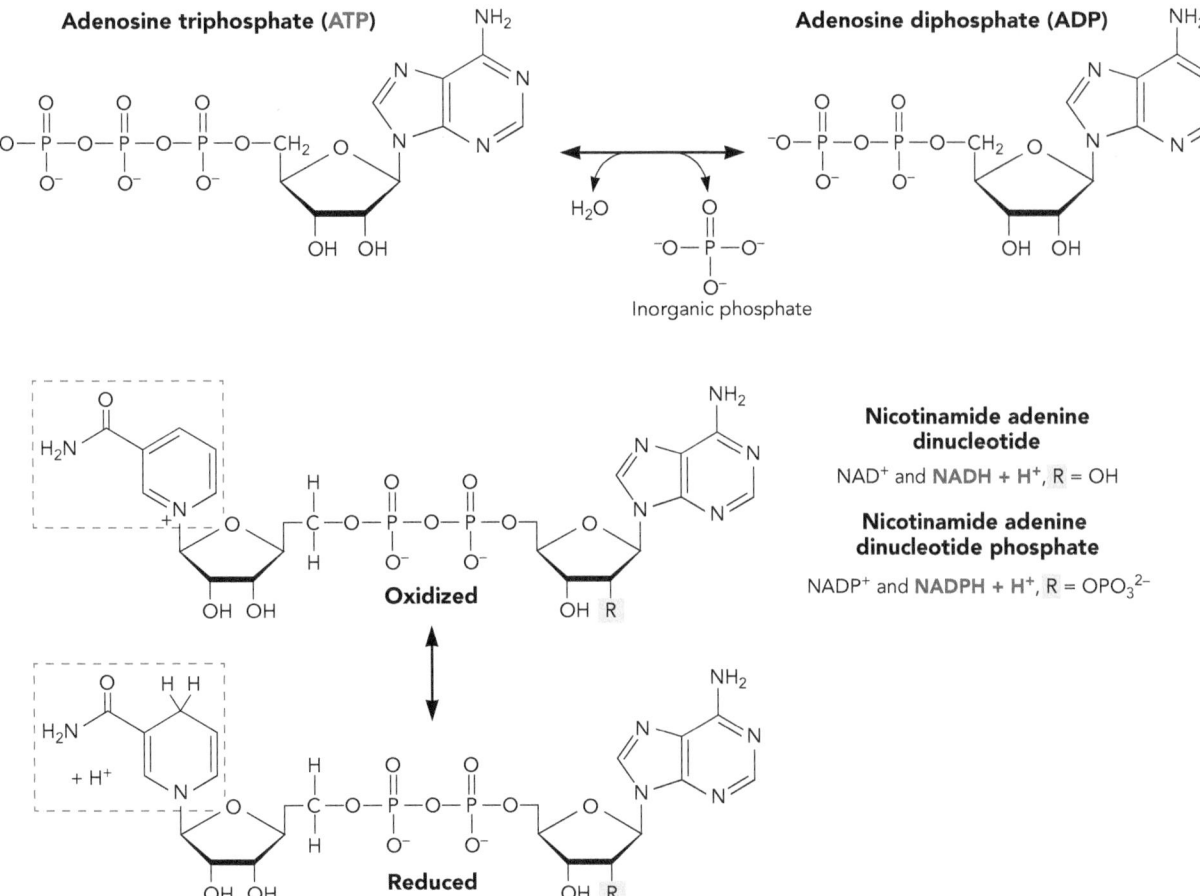

FIGURE 2.5 **Structures and energy exchange of key metabolic coenzymes associated with glycolysis.** Adenosine triphosphate (ATP) contains two high-energy phosphate anhydride bonds, whereas adenosine diphosphate (ADP) contains just one. Nicotinamide adenine dinucleotide (NAD$^+$) differs from nicotinamide adenine dinucleotide phosphate (NADP$^+$) by the absence or presence, respectively, of a phosphate group at the position of the R group highlighted in yellow. The boxed regions highlight the nicotinamide portion of the molecules associated with the oxidized (top) or reduced (bottom) state.

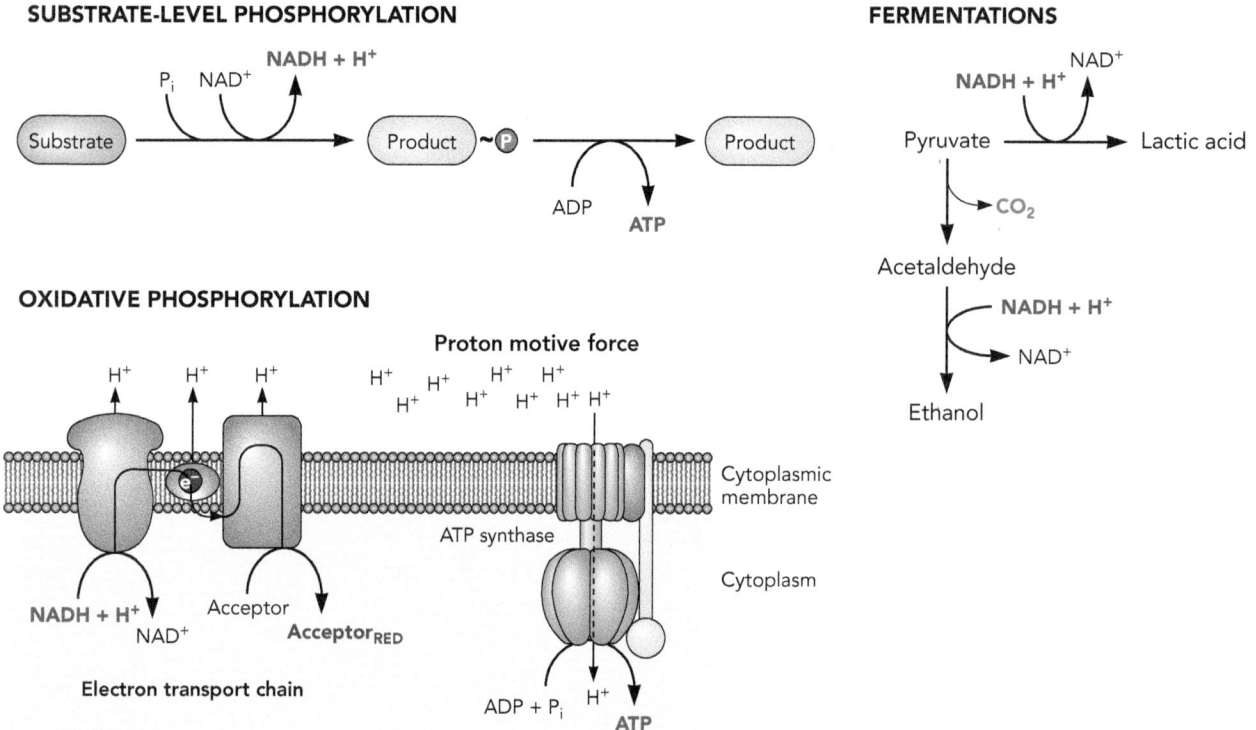

SUBSTRATE-LEVEL PHOSPHORYLATION

FERMENTATIONS

OXIDATIVE PHOSPHORYLATION

FIGURE 2.6 **ATP generation and recycling of NAD⁺.** Production of ATP through the metabolism of organic metabolic intermediates containing high-energy phosphate bonds is called substrate-level phosphorylation (upper left). The reduced NADH + H⁺ produced during substrate-level phosphorylation may be reoxidized through the process of fermentation (upper right) or oxidative phosphorylation (lower left). Oxidative phosphorylation is the term used to describe ATP production during respiration when an ETC generates a proton motive force that is utilized by ATP synthase, generating ATP.

Structure and Energy Exchange of Key Coenzymes

Substrate-level phosphorylation is used by all living cells, whether growing via fermentation or respiration, to produce ATP during glycolysis as described above. The high-energy phosphate bond in ATP serves as an energy source for many cellular reactions; thus, ATP is considered a coenzyme. ATP and ADP differ from one another due to the presence or absence, respectively, of a high-energy phosphate anhydride bond; ATP has two, whereas ADP has one (Fig. 2.5) (adenosine monophosphate [AMP] contains a phosphate group, but the phosphate bond is not highly energetic). Reduced NADH + H⁺ and reduced nicotinamide adenine dinucleotide phosphate (NADPH + H⁺) are two of the most important coenzymes for the redox reactions of cellular metabolism (Figure 2.5). Structurally, NAD⁺ and NADP⁺ (and their reduced counterparts) are identical except for the presence of an additional phosphate group on the NADP⁺ molecule. However, NAD⁺ and NADP⁺ are not interchangeably utilized in reactions; NADH + H⁺ is typically used in catabolic reactions and NADPH + H⁺ is typically used to drive biosynthetic/anabolic pathways. These two coenzymes may be enzymatically interconverted by the addition or removal of a phosphoryl group (e.g., catalyzed by a kinase or phosphatase, respectively), allowing the cell to balance its catabolic and anabolic reactions. ATP is the source of the phosphoryl group for conversion of NAD^+ to $NADP^+$. Both NAD^+ and $NADP^+$ can acquire and be reduced by two electrons. Under physiological conditions, only one proton is incorporated into the molecule and the other is liberated into the surrounding medium; hence, the reduced form is written as $NAD(P)H + H^+$.

Coenzymes such as ATP and $NAD(P)H + H^+$ are not consumed during enzymatic reactions, but rather are recycled. For example, energy is released when the reduced form of the coenzyme is reoxidized to enable "work" to occur, and the oxidized form of the coenzyme must be regenerated to permit the essential reactions of central metabolism to continue. Most cells reoxidize $NAD(P)H + H^+$ via fermentation or during either aerobic or anaerobic respiration processes.

During fermentation only substrate-level phosphorylation is used to generate ATP since there is no ETC utilized as in respiration (Fig. 2.6). Therefore, NADH $+ H^+$ is reoxidized using a variety of organic molecules, and this enables glycolysis to continue, albeit with less ATP production than via respiration (Chapter 6). One simple example of a fermentation reaction is the production of lactic acid from pyruvate (Fig. 2.6). In humans, this occurs when muscle cells become deprived of oxygen during intense exercising; the lactic acid buildup generates soreness in the muscle. Another example is the production of ethanol by yeast; pyruvate is first decarboxylated to form acetaldehyde, which is subsequently reduced by NADH $+ H^+$ to ethanol, yielding NAD^+ (Fig. 2.6). As will be discussed in Chapter 6, there are many other types of bacterial fermentations, which most commonly occur in the absence of or at low levels of oxygen, although some fermentations can occur in the presence of oxygen. For example, lactic acid bacteria are only capable of fermentative metabolism because they lack respiratory ETCs, but they can grow in the presence or absence of oxygen.

Controlling the Direction of Carbon Flow during Glycolysis

Theoretically, any enzymatic reaction is reversible since enzymes control the rate, but not the direction, of the reactions they catalyze. However, under the physiological conditions that exist in living cells, some enzymes can only function in one direction due to parameters such as substrate and product concentrations. Within glycolysis there are three key unidirectional enzymes that drive the flow of carbon from glucose to pyruvate: hexokinase, phosphofructokinase, and pyruvate kinase (Fig. 2.7). The enzymatic reactions between fructose-1,6-bisphosphate and phosphoenolpyruvate are all reversible.

The two key unidirectional enzymes phosphofructokinase and pyruvate kinase are allosterically regulated enzymes (Fig. 2.7). **Allosteric regulation** involves a small effector molecule binding to a regulatory site in the enzyme, thereby either positively or negatively regulating the activity of the enzyme to the benefit of the cell (Chapter 7). Fructose-1,6-bisphosphate is a positive effector of pyruvate kinase. It is a precursor for pyruvate synthesis, and as its levels build up in the cell, it creates a greater "pull" on the carbon flow through the pathway by increasing the rate at which pyruvate is made. Conversely, phosphofructokinase is negatively regulated by phosphoenolpyruvate. This is a form of **feedback inhibition** in which high concentrations of a product near the end of the pathway inhibit an enzyme early in the pathway (Chapter 7). Phosphofructokinase is also positively regulated by ADP levels.

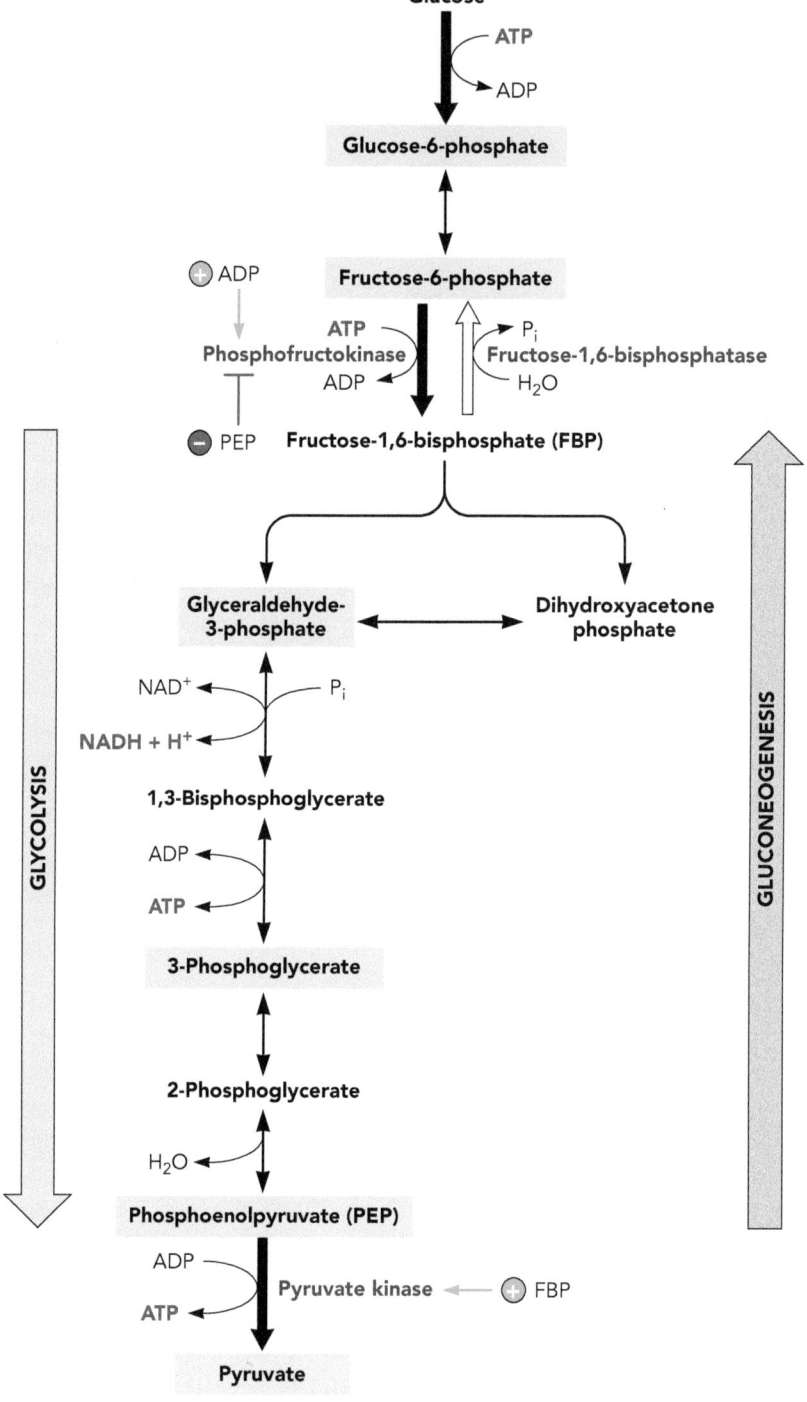

FIGURE 2.7 **Regulating the direction of carbon flow in the Embden-Meyerhof-Parnas (EMP) pathway.** All of the reactions between fructose-1,6-bisphosphate (FBP) and phosphoenolpyruvate (PEP) are physiologically reversible. The key unidirectional enzymes phosphofructokinase and pyruvate kinase are controlled by allosteric regulation to help push the carbon flow toward pyruvate during glycolysis. ADP, an indicator of low ATP levels, is a positive effector of phosphofructokinase (indicated by the green arrow), while PEP, a product formed near the end of glycolysis, is a negative (feedback) effector of phosphofructokinase (indicated by the red T bar). The precursor molecule FBP is a positive effector for pyruvate kinase. During gluconeogenesis, the direction of carbon flow is reversed. The enzyme fructose-1,6-bisphosphatase catalyzes the formation of fructose-6-phosphate during gluconeogenesis. There are also other enzymes capable of catalyzing the formation of PEP (see Fig. 2.16).

High concentrations of ADP signal low internal energy levels (i.e., low ATP). When energy levels are low, it is advantageous for the cell to increase the flow of carbon to pyruvate, thereby producing more ATP and NADH + H⁺. There is an inherent "logic" in how key unidirectional enzymes are regulated by metabolic effectors, enabling the cell to conserve essential resources for its growth and survival.

Gluconeogenesis, the formation of new glucose, is the reverse pathway to glycolysis. The capacity for gluconeogenesis in a cell ensures that in the presence of an alternative carbon source (e.g., pyruvate or a TCA cycle intermediate) all six essential precursors generated in the EMP pathway are still made. Although gluconeogenesis costs the cell energy, it is necessary to produce the essential precursors needed for growth and survival. In gluconeogenesis, alternative enzymes are used in place of the unidirectional enzymes. For example, fructose-1,6-bisphosphatase is used to reverse the carbon flow by converting fructose-1,6-bisphosphate into fructose-6-phosphate. Fructose-1,6-bisphosphatase is under negative allosteric regulation by AMP, which signals a need for the cell to generate more ATP. When AMP levels are low, the energy-consuming reactions of gluconeogenesis do not occur.

The Pentose Phosphate Pathway (PPP)

Although it does not directly produce any ATP, the PPP is another essential pathway of central metabolism (Fig. 2.8). A portion of the glucose-6-phosphate is diverted away from glycolysis into the PPP to generate the two essential precursors, ribose-5-phosphate and erythrose-4-phosphate, and the reduced coenzyme NADPH + H⁺. The carbon subsequently flows back into glycolysis at two points, as fructose-6-phosphate and glyceraldehyde-3-phosphate (Fig. 2.9). In *Escherichia coli*, about 70% of the carbon flows directly through glycolysis and about 30% enters the PPP. These reactions must occur simultaneously because both pathways produce essential precursors.

The PPP can be subdivided into three parts: the oxidation-decarboxylation reactions, the isomerization reactions, and the transketolase and transaldolase rearrangement reactions (Fig. 2.8). In the oxidation-decarboxylation reactions, glucose-6-phosphate dehydrogenase carries out a redox reaction that oxidizes three molecules of glucose-6-phosphate to three molecules of 6-phosphogluconolactone and reduces three NADP⁺ to three NADPH + H⁺. The 6-phosphogluconolactone molecules are hydrated, forming 6-phosphogluconate (6C), which then undergoes another redox reaction catalyzed by 6-phosphogluconate dehydrogenase, generating three ribulose-5-phosphate (5C) and three carbon dioxide (1C) molecules plus three additional NADPH + H⁺.

The second series of steps in the PPP are the isomerization reactions. Ribulose-5-phosphate, xylulose-5-phosphate, and ribose-5-phosphate all have the same chemical formula, but different structures. Some of the ribose-5-phosphate (an essential precursor) is used for biosynthesis as it comprises the sugar backbone in DNA and RNA. The remaining ribose-5-phosphate and xylulose-5-phosphate are consumed by the rearrangement reactions of the PPP.

In the rearrangement reactions, the backbones of the five-carbon molecules are rearranged and the essential precursor erythrose-4-phosphate is

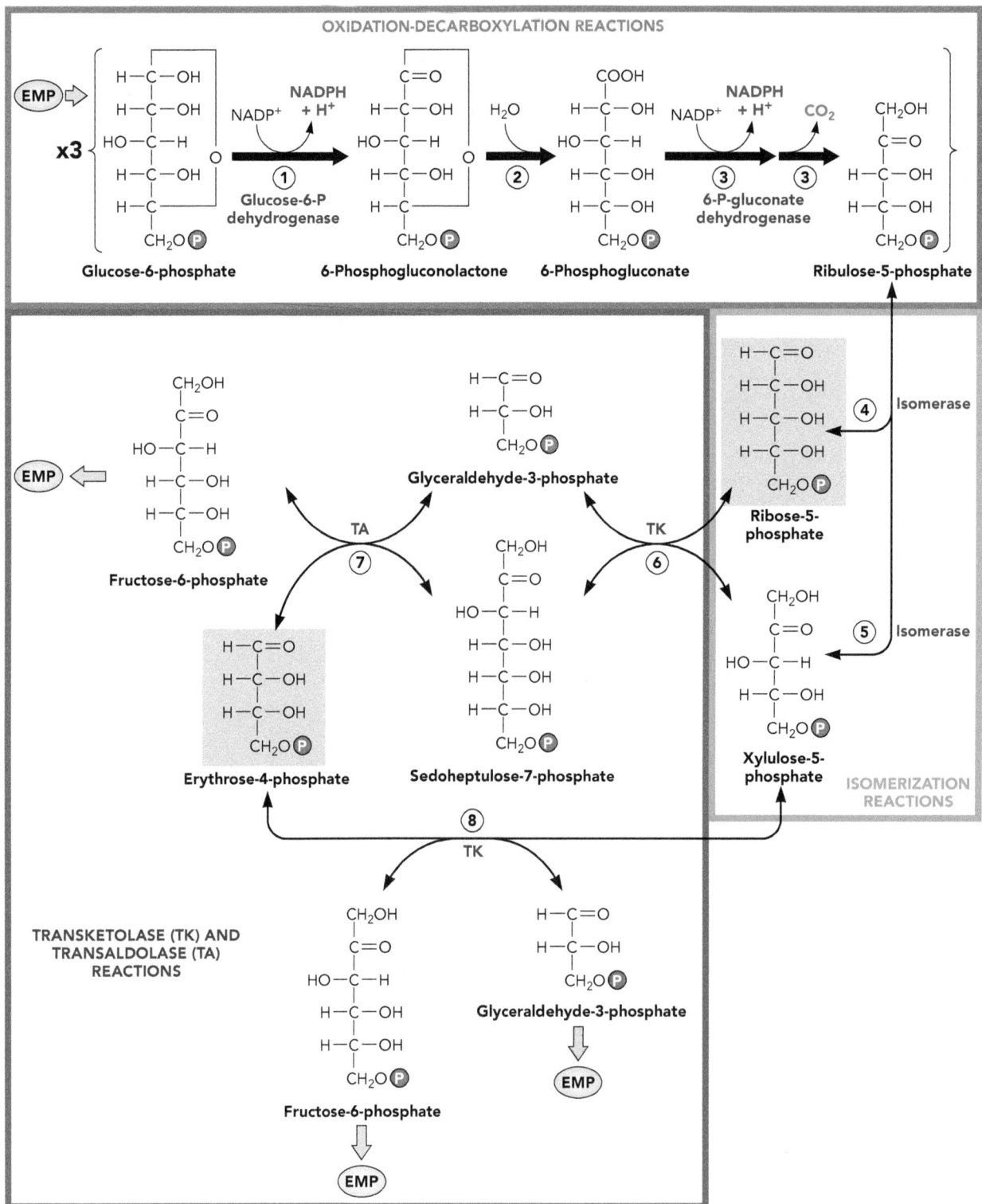

FIGURE 2.8 Reactions of the pentose phosphate pathway (PPP). Essential precursors formed in the PPP are highlighted by pink boxes. Energy outputs (e.g., NADPH + H$^+$) are highlighted in red and decarboxylation events leading to CO_2 emissions are highlighted in blue. Key unidirectional enzymes are indicated in purple and the reactions they catalyze are shown with bold arrows. Carbon flow from and to the Embden-Meyerhof-Parnas (EMP) pathway is indicated by yellow arrows. The oxidation-decarboxylation reactions in the orange box are catalyzed by (1) glucose-6-phosphate dehydrogenase, (2) gluconolactonase, and (3) 6-phosphogluconate dehydrogenase. The isomerization reactions in the green box are catalyzed by (4) ribulose-5-phosphate epimerase and (5) ribose-5-phosphate isomerase. The transketolase (TK) reactions (6 and 8) and the transaldolase (TA) reaction (7) are in the purple box.

produced. The primary enzymes that catalyze these reactions are **transketolases** and **transaldolases**. Transketolases transfer two carbon units from one molecule to another, while transaldolases move three carbon units from one molecule to another. A transketolase converts xylulose-5-phosphate (5C) and ribose-5-phosphate (5C) to glyceraldehyde-3-phosphate (3C) and sedoheptulose-7-phosphate (7C). These molecules are in turn converted into fructose-6-phosphate (6C) and erythrose-4-phosphate (4C) by a transaldolase. Erythrose-4-phosphate is an essential precursor for the biosynthesis of the aromatic amino acids tyrosine, tryptophan, and phenylalanine. The cell catalyzes these rearrangements to produce erythrose-4-phosphate even though no ATP or reduced coenzymes are generated in the process. However, the pathway allows the cell to conserve the carbon and send it back to glycolysis. Another transketolase reaction takes a molecule of xylulose-5-phosphate and some of the erythrose-4-phosphate to generate fructose-6-phosphate (6C) and glyceraldehyde-3-phosphate (3C). Starting with three molecules of glucose-6-phosphate, the two molecules of fructose-6-phosphate and one molecule of glyceraldehyde-3-phosphate produced by the final stage of the PPP can then reenter the glycolytic pathway. The fructose-6-phosphate and glyceraldehyde-3-phosphate are subsequently processed through the glycolysis reactions from which additional essential precursors and energy will be derived (Fig. 2.9).

Overall, the PPP allows cells to divert some of the carbon flow from glycolysis (i.e., glucose-6-phosphate) to gain NADPH + H$^+$ and the essential precursors ribose-5-phosphate and erythrose-4-phosphate. Most of the carbon flows back into glycolysis (i.e., as glyceraldehyde-3-phosphate and fructose-6-phosphate; Fig. 2.9), with some loss of carbon in the form of carbon dioxide.

The Entner-Doudoroff (ED) Pathway

The ED pathway (Fig. 2.10) is an alternative way to metabolize glucose that is primarily found in prokaryotes and a few simple eukaryotes; therefore, it is not typically discussed when eukaryotic biology is the emphasis. The first two steps of the ED pathway that lead to the formation of 6-phosphogluconate are shared with the PPP (Fig. 2.9). After these steps, the ED pathway diverges, with steps involving two unique enzymes. The enzyme 6-phosphogluconate dehydratase catalyzes a dehydration reaction that converts 6-phosphogluconate into 2-keto-3-deoxy-6-phosphogluconate (KDPG). In the next step, the carbon backbone of KDPG (6C) is cleaved by KDPG aldolase, producing pyruvate (3C) and glyceraldehyde-3-phosphate (3C). One way to determine if an organism has the ED pathway is to look for the presence of the genes encoding these two unique enzymes.

The pyruvate and glyceraldehyde-3-phosphate produced by the ED pathway can reenter the EMP pathway (Fig. 2.9). However, since there is only one molecule of glyceraldehyde-3-phosphate produced by the ED pathway versus two by the EMP pathway, the overall ATP generation in the ED pathway is half that of the EMP pathway. The cell does gain reduced NADH + H$^+$ and NADPH + H$^+$ coenzymes and essential precursors via the ED pathway. However, since only two ATP molecules are produced per molecule of glucose entering the ED pathway, and one ATP is consumed to generate glucose-6-phosphate, the ED pathway nets just one ATP per glucose.

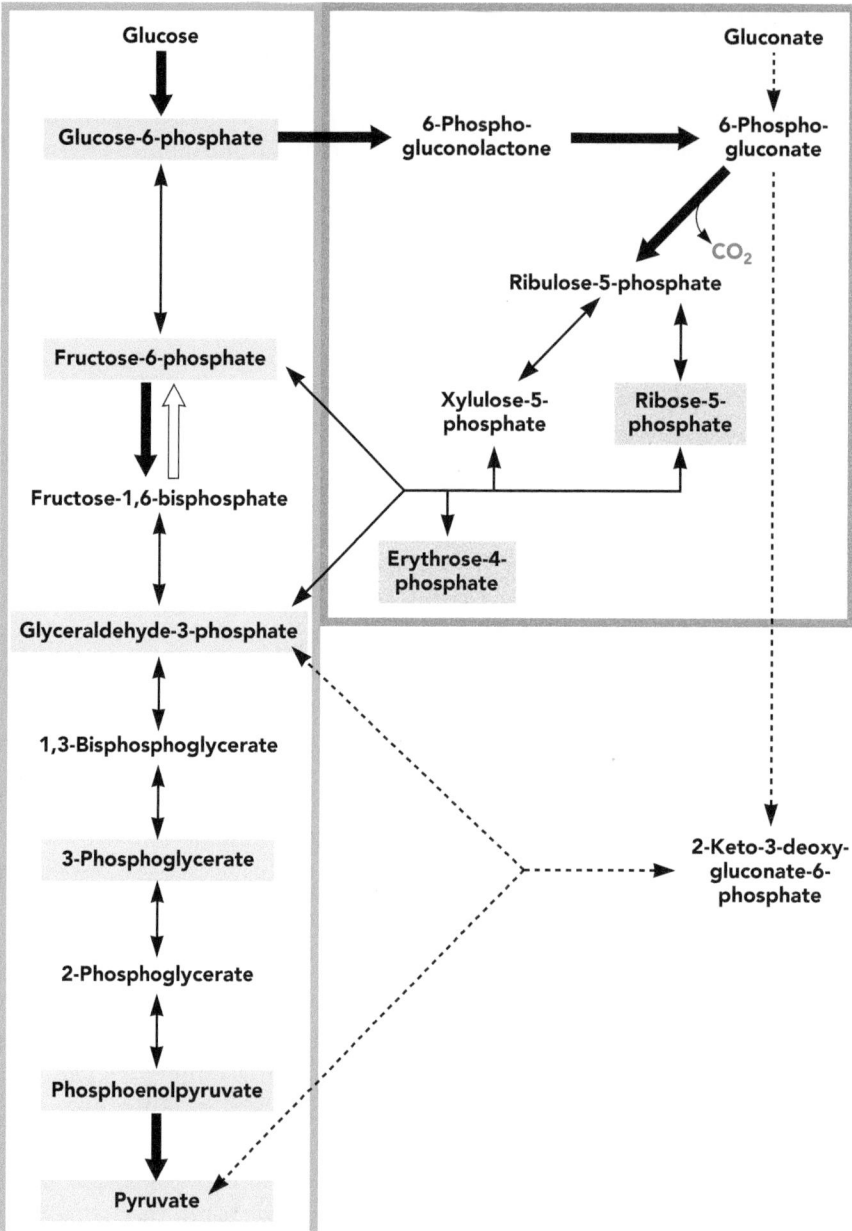

FIGURE 2.9 **Connections between the Embden-Meyerhof-Parnas (EMP) pathway, pentose phosphate pathway (PPP), and Entner-Doudoroff (ED) pathway.** Three pathways in central metabolism are linked allowing carbon flow between them, ensuring that the precursors needed from the EMP pathway (blue boxes) and PPP (pink boxes) are continually generated, even in the presence of alternative carbon sources other than glucose. The EMP and PPP pathways diverge at glucose-6-phosphate and converge at the precursors fructose-6-phosphate and glyceraldehyde-3-phosphate. The ED pathway, shown by the dotted line, diverges from the PPP at 6-phosphogluconate and converges with the EMP pathway at the precursors glyceraldehyde-3-phosphate and pyruvate. Key unidirectional enzymes are indicated by bold arrows.

Why would a cell utilize the ED pathway given its lower energy output compared to the EMP pathway? There are at least two situations in which the ED pathway would be utilized. First, some organisms have lost the ability to make phosphofructokinase or fructose bisphosphate aldolase, key enzymes required for

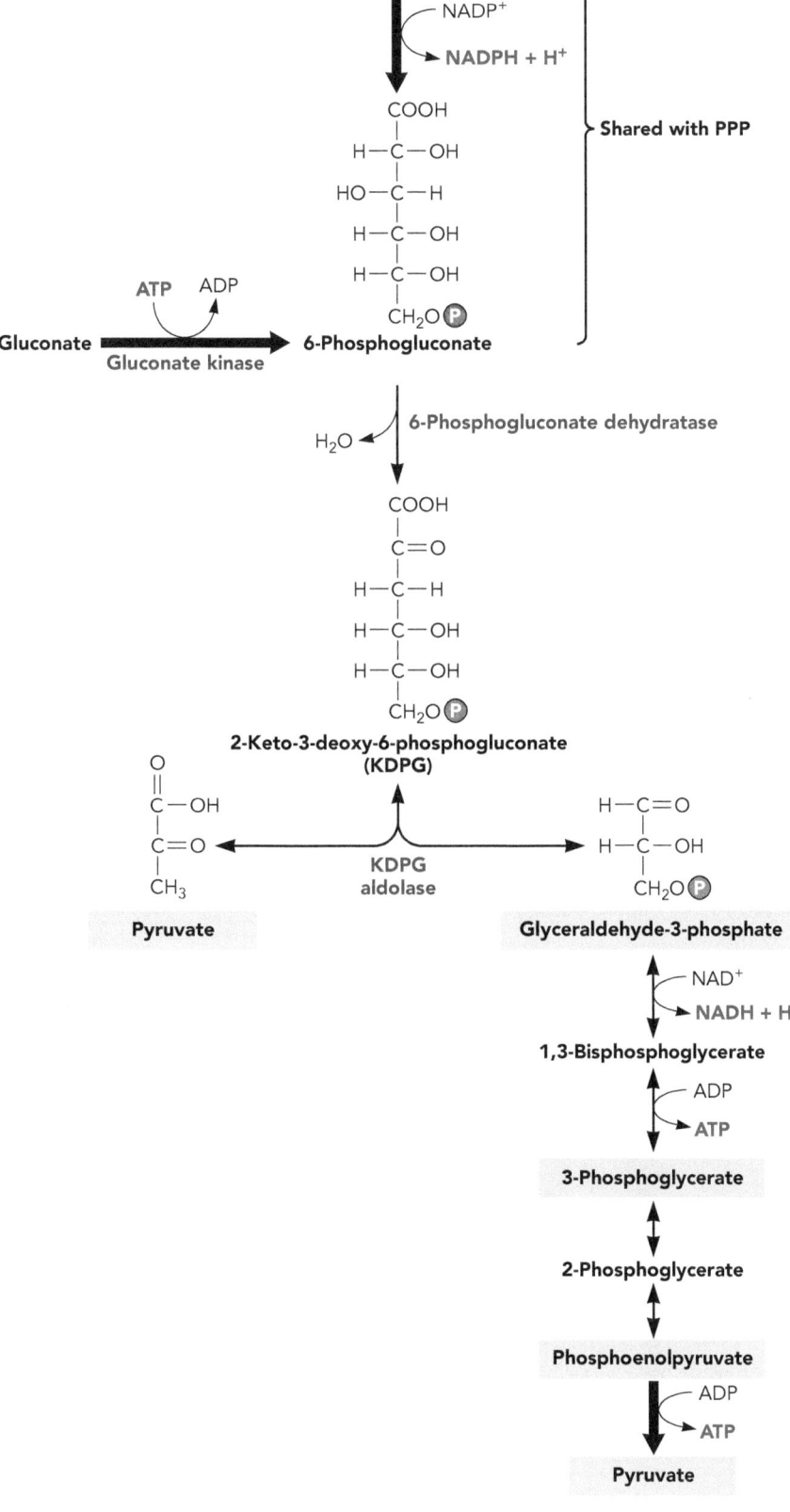

FIGURE 2.10 **Reactions of the Entner-Doudoroff (ED) pathway.** The ED pathway diverges from the pentose phosphate pathway (PPP) at 6-phosphogluconate and converges with the Embden-Meyerhof-Parnas (EMP) pathway at the precursors glyceraldehyde-3-phosphate and pyruvate. Note that only one glyceraldehyde-3-phosphate molecule is formed from the reactions of the ED pathway. The enzymes unique to this pathway, highlighted in purple, are 6-phosphogluconate dehydratase and KDPG aldolase. Gluconate kinase is used if gluconate, rather than glucose, is the carbon source. Energy inputs and outputs (e.g., NADPH + H⁺ and ATP) are highlighted in red and the essential precursors produced are highlighted with blue boxes.

the EMP pathway, which is true of organisms in the genus *Pseudomonas*. The lack of either of these enzymes creates a block in the upper portion of the EMP pathway; the ED pathway shunts the carbon flow around the defective steps. Second, some organisms use the ED pathway because it allows them to catabolize aldonic acids (e.g., gluconate), which may be more available than glucose in some environments. These bacteria use ATP and the enzyme gluconate kinase to phosphorylate gluconate, generating 6-phosphogluconate, which they then metabolize via the ED pathway. There is even some evidence that certain microorganisms will first convert glucose into gluconate to prevent neighboring bacteria incapable of utilizing the ED pathway from accessing the available glucose.

The Transition Reaction: Carbon Flow into the Tricarboxylic Acid (TCA) Cycle

The EMP, PPP, and ED pathways all ultimately lead to the production of pyruvate (Fig. 2.9). Pyruvate (3C) is converted into the essential precursor acetyl-CoA (2C) with the release of carbon dioxide by different enzymes depending on whether the organism has an aerobic or anaerobic lifestyle. Acetyl-CoA is an essential precursor needed for the synthesis of fatty acids and isoprene in the cell membranes of *Bacteria* and *Archaea*, respectively. Anaerobes may use pyruvate-ferredoxin oxidoreductase or pyruvate formate lyase to produce acetyl-CoA. Aerobes use another key unidirectional enzyme in central metabolism, pyruvate dehydrogenase (Fig. 2.11). This enzyme has a complex quaternary structure composed of multiple subunits, and the activity of the enzyme is controlled by allosteric regulation (Chapter 7). At this key unidirectional step, the cell is controlling carbon flow into the TCA cycle. Both

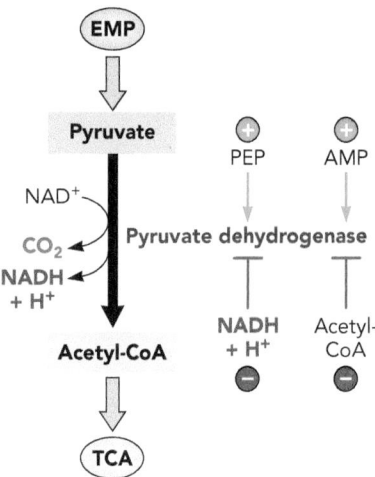

FIGURE 2.11 **The transition reaction catalyzed by pyruvate dehydrogenase.** The key unidirectional enzyme pyruvate dehydrogenase catalyzes the conversion of the essential precursor pyruvate to acetyl-CoA, another essential precursor, thereby transitioning the carbon flow from the Embden-Meyerhof-Parnas (EMP) pathway to the tricarboxylic acid (TCA) cycle and its variations. Energy outputs (i.e., NADH + H$^+$) are highlighted in red and a decarboxylation event leading to CO_2 emission is highlighted in blue. The precursor phosphoenolpyruvate (PEP) and AMP, which indicate low cellular energy levels, are positive effectors that increase pyruvate dehydrogenase activity, while the end products NADH + H$^+$ and acetyl-CoA are negative (feedback) effectors.

phosphoenolpyruvate and AMP positively regulate pyruvate dehydrogenase (Fig. 2.11). An excess of phosphoenolpyruvate signals precursor buildup, while AMP signals a low energetic state, thus driving the cell to push more carbon into the TCA cycle. On the other hand, when the concentrations of the end products NADH + H$^+$ and acetyl-CoA reach a critical threshold, they negatively regulate the pyruvate dehydrogenase activity through feedback inhibition.

The Tricarboxylic Acid (TCA) Cycle

The TCA cycle produces the essential precursors α-ketoglutarate, succinyl-CoA, and oxaloacetate. As the name suggests, under aerobic conditions this pathway functions as a cycle (Fig. 2.12), with new carbon entering in the form of the acetyl-CoA produced from the transition reaction between the EMP pathway and the TCA cycle. The TCA cycle is driven in one direction due to the presence of two key unidirectional enzymes, citrate synthase and α-ketoglutarate dehydrogenase (Fig. 2.13). The allosterically regulated enzyme citrate synthase takes the substrates acetyl-CoA (2C) and oxaloacetate (4C) and converts them into citrate (6C). Different bacteria use NADH + H$^+$, α-ketoglutarate, or ATP, all products of the TCA cycle, to negatively regulate citrate synthase. Citrate is converted into *cis*-aconitate (6C), which is subsequently converted into isocitrate (6C), and then oxalosuccinate (6C) with the production of NADPH + H$^+$ before it is decarboxylated, releasing carbon dioxide and generating α-ketoglutarate (5C). α-Ketoglutarate is an essential precursor needed to synthesize the amino acid glutamate, from which other amino acids are made (Chapter 12). The second key unidirectional enzyme in the TCA cycle, α-ketoglutarate dehydrogenase, converts α-ketoglutarate and coenzyme A into succinyl-CoA (4C) with the release of carbon dioxide and NADH + H$^+$. This reaction requires coenzyme A, which carries carbon units from reaction to reaction. Succinyl-CoA is yet another essential precursor needed for amino acid biosynthesis (lysine and methionine) as well as tetrapyrrole molecules necessary to produce cytochromes, chlorophylls, and vitamin B$_{12}$.

The next step of the pathway, the conversion of succinyl-CoA to succinate, yields ATP via substrate-level phosphorylation. Succinate is then converted into fumarate with production of the reduced coenzyme flavin adenine dinucleotide (FADH$_2$). The fumarate is converted into malate before oxaloacetate is formed with the production of the reduced coenzyme NADH + H$^+$. Oxaloacetate is the 12th precursor made by central metabolism, and it is used in the biosynthesis of aspartate and five other amino acids. Therefore, starting with one molecule of oxaloacetate and one molecule of acetyl-CoA, the aerobic TCA cycle yields one NADPH + H$^+$, two CO$_2$, two NADH + H$^+$, one ATP, and one FADH$_2$. Note that there is no net carbon incorporation from the TCA cycle; two carbons enter as acetyl-CoA and two are lost as carbon dioxide.

Anaplerotic Reactions

The essential precursors serve two roles: they are consumed via anabolic pathways to build cell macromolecules and structures, and they are also key

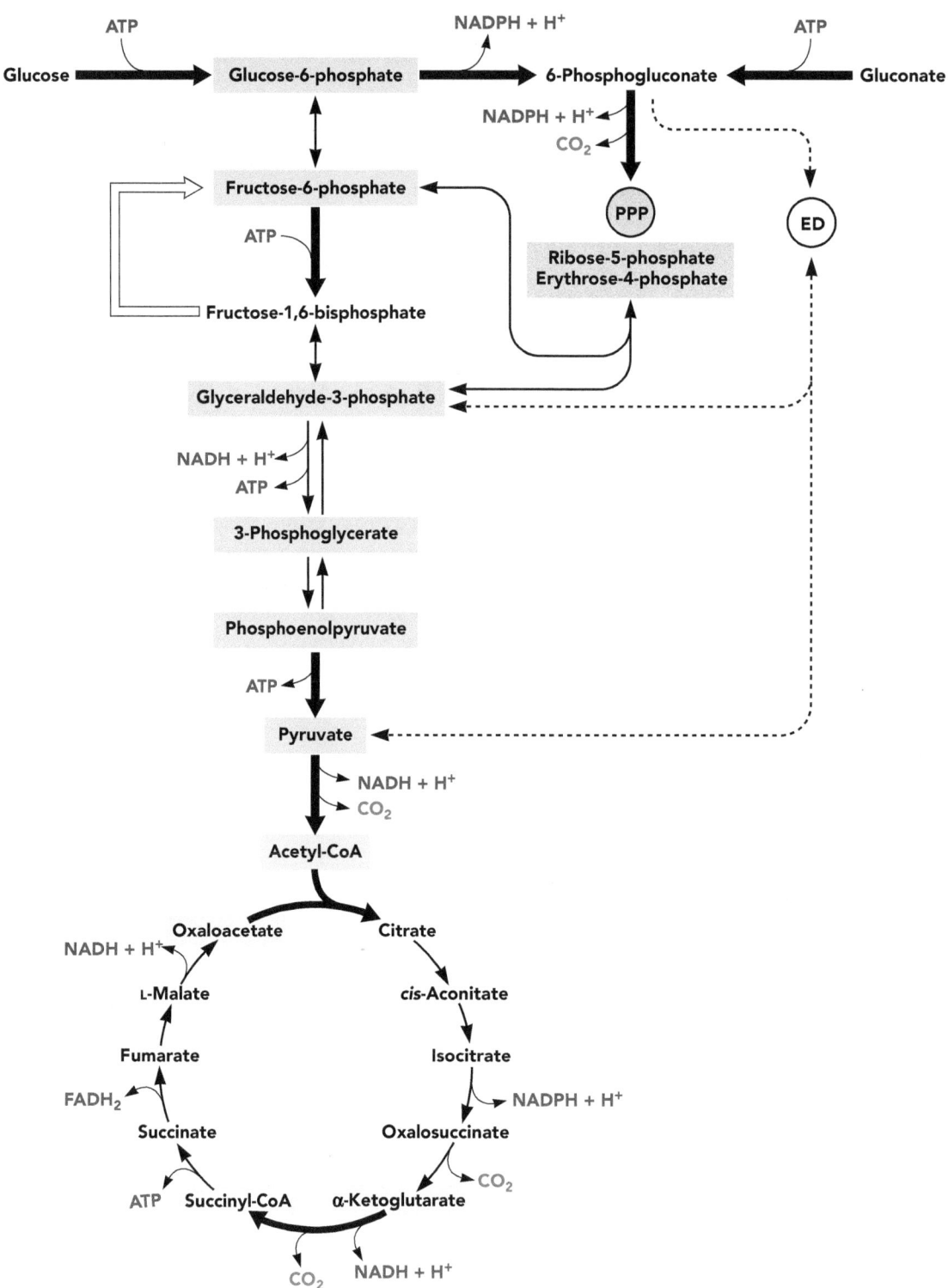

FIGURE 2.12 **Connections between the pathways associated with central metabolism.** Carbon flow between the Embden-Meyerhof-Parnas (EMP) pathway, pentose phosphate pathway (PPP), Entner-Doudoroff (ED) pathway, and tricarboxylic acid (TCA) cycle with production of essential precursors highlighted by colored boxes. Energy inputs and outputs (i.e., ATP, NADH + H[+], and NADPH + H[+]) are highlighted in red and decarboxylation events leading to CO_2 emissions are highlighted in blue. Key unidirectional reactions are highlighted with bold arrows.

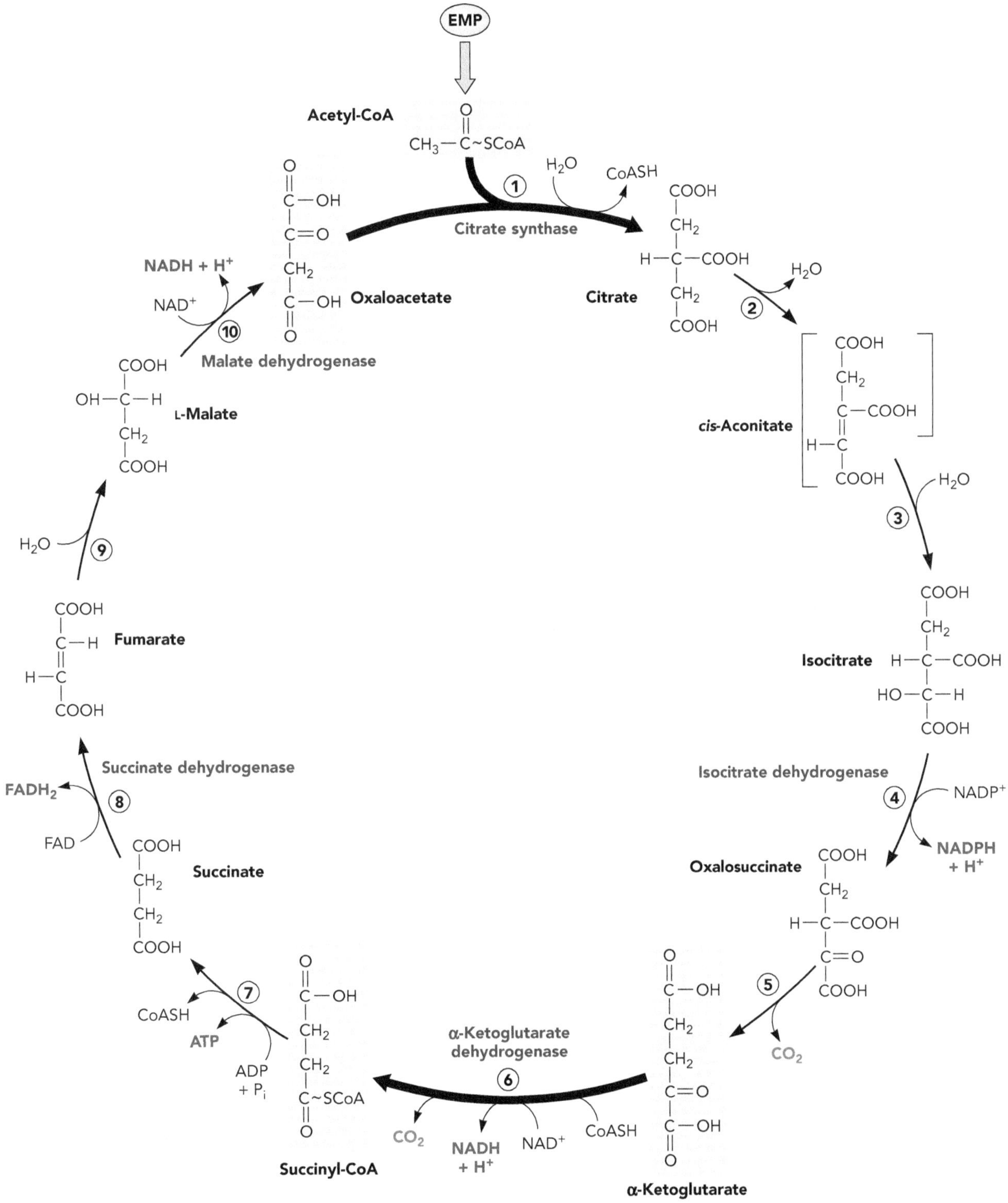

FIGURE 2.13 **The reactions of the tricarboxylic acid (TCA), Krebs, or citric acid cycle.** Essential precursors are highlighted by green and yellow boxes. Energy outputs (e.g., NADPH + H$^+$, ATP, NADH + H$^+$, and FADH$_2$) are highlighted in red and decarboxylation events leading to CO$_2$ emissions are highlighted in blue. Key unidirectional enzymes are indicated with bold arrows and highlighted in purple; enzymes involved in energy conversion are also highlighted in purple. The enzyme-bound intermediate *cis*-aconitate is bracketed. The specific enzymes catalyzing each reaction are (1) citrate synthase, (2 and 3) aconitase, (4 and 5) isocitrate dehydrogenase, (6) α-ketoglutarate dehydrogenase, (7) succinyl coenzyme A synthetase, (8) succinate dehydrogenase, (9) fumarase, and (10) malate dehydrogenase.

intermediates essential for the continuation of central metabolism. An **anaplerotic reaction** is one that replenishes pools of metabolites consumed by biosynthesis. Two examples (there are many more) of anaplerotic reactions are associated with the production of oxaloacetate by alternative enzymes since, in addition to amino acid biosynthesis (Fig. 2.14), oxaloacetate is also needed to restart another round of the aerobic TCA cycle. Phosphoenolpyruvate (3C) and bicarbonate (HCO_3^-; the soluble form of CO_2) are converted into oxaloacetate (4C) by phosphoenolpyruvate carboxylase. The high-energy bond in phosphoenolpyruvate is used to drive the reaction. Pyruvate (3C) and bicarbonate (1C) are converted into oxaloacetate (4C) by pyruvate carboxylase, and one ATP is consumed during the reaction. The cell expends energy to produce oxaloacetate because it is necessary to keep the TCA cycle functioning.

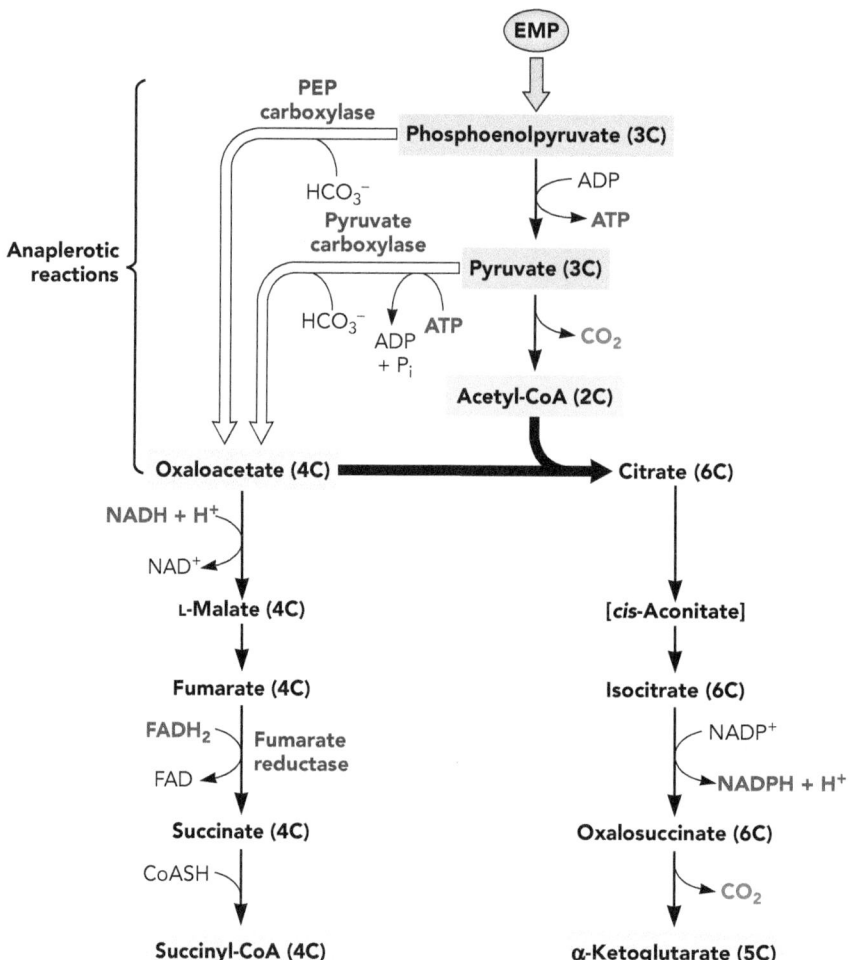

FIGURE 2.14 **The branched or incomplete tricarboxylic acid (TCA) pathway requires anaplerotic reactions.** Under anoxic or fermentative growth conditions, some bacteria use a branched TCA pathway (instead of cyclic) since the enzyme α-ketoglutarate dehydrogenase has low or no activity. The right branch of the pathway is unchanged from the aerobic TCA pathway and leads to the production of the essential precursor α-ketoglutarate. Carbon flow in the left branch is reversed due to the presence of fumarate reductase, which also results in consumption of reduced coenzymes. Energy inputs and outputs (i.e., NADPH + H^+, ATP, NADH + H^+, and $FADH_2$) are highlighted in red and decarboxylation events leading to CO_2 emissions are highlighted in blue. Levels of oxaloacetate are replenished through anaplerotic reactions catalyzed by PEP carboxylase and pyruvate carboxylase.

The Branched or Incomplete Tricarboxylic Acid (TCA) Pathway

Under anoxic conditions, the TCA cycle becomes branched or incomplete (Fig. 2.14). The gene encoding α-ketoglutarate dehydrogenase (Fig. 2.13) is not expressed in the absence of oxygen, creating a break in the cyclic pathway. Under both oxic and anoxic conditions, the "right branch" of the TCA pathway functions identically, leading to the production of α-ketoglutarate. However, the "left branch" of the pathway runs in reverse to enable production of the essential precursor succinyl-CoA. Instead of using succinate dehydrogenase to produce fumarate from succinate, succinate is converted into fumarate by fumarate reductase. The anaplerotic reactions that generate oxaloacetate (Fig. 2.14) are essential under anoxic conditions. Running the "left branch" of the TCA pathway in reverse reoxidizes reduced coenzymes, which is more difficult under anoxic conditions. However, the cell is unable to make ATP from the branched TCA pathway, which contributes to the slower growth rate of anaerobes. The cell is sacrificing the ability to make more ATP in order to recycle its coenzymes and produce the essential precursor succinyl-CoA.

The Glyoxylate Cycle

Another modification of the TCA cycle, the glyoxylate cycle (Fig. 2.15), occurs in prokaryotes, protozoa, and in plants during seed germination. Growth on acetate or fatty acids as primary carbon sources triggers multiple regulatory changes in the TCA cycle, including expression of the genes needed for the glyoxylate cycle. First, the enzyme isocitrate dehydrogenase, which converts isocitrate into oxalosuccinate, is covalently modified via phosphorylation (Chapter 7), thereby decreasing its activity. The enzyme is not completely inactivated because the cell must continue the reactions of the TCA cycle to produce the essential precursors α-ketoglutarate and succinyl-CoA. However, because the modified enzyme functions below its maximum velocity, some of the carbon flow can be shunted off into the glyoxylate cycle.

The second regulatory response of cells growing with acetate or fatty acids is to express the genes encoding isocitrate lyase and malate synthase, which are unique to the glyoxylate cycle. Isocitrate lyase, which effectively competes with isocitrate dehydrogenase for the available pool of isocitrate, cleaves isocitrate (6C) into succinate (4C) and the novel metabolic intermediate glyoxylate (2C). Glyoxylate and acetyl-CoA are then joined together by malate synthase to produce malate (4C) with the release of coenzyme A. The TCA cycle itself does not incorporate carbon into biomass since for every acetyl-CoA that enters the pathway two carbon dioxide molecules are emitted. This is not a problem when the main carbon source contains more than two carbons. However, when acetate or fatty acids serve as the primary carbon source, the carbon enters central metabolism as acetyl-CoA, and the cell would never gain biomass using only the TCA cycle. Consequently, the glyoxylate cycle permits a second molecule of acetyl-CoA to enter central metabolism through the activity of malate synthase, allowing the cell to accumulate biomass from two-carbon substrates.

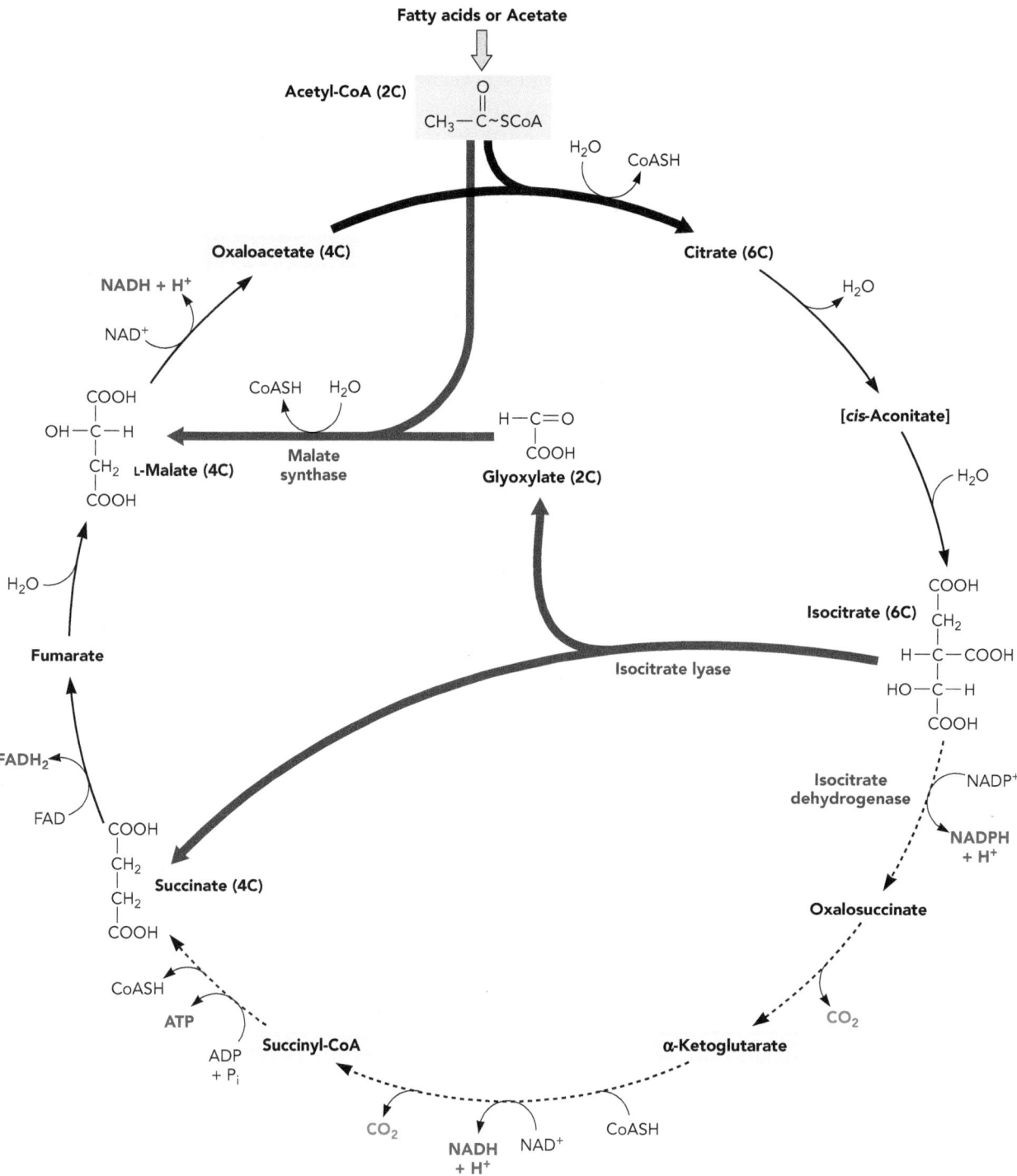

FIGURE 2.15 **Reactions of the glyoxylate cycle.** The glyoxylate cycle is used when fatty acids or acetate serve as the carbon source rather than glucose. Enzymes unique to this pathway, indicated with purple arrows, are isocitrate lyase and malate synthase. In addition, there is a downregulation of the activity of the enzyme isocitrate dehydrogenase, which decreases but does not eliminate carbon flow through the essential precursor succinyl-CoA, as indicated by the dashed arrows. Energy outputs (i.e., NADPH + H+, ATP, NADH + H+, and $FADH_2$) are highlighted in red and decarboxylation events leading to CO_2 emissions are highlighted in blue.

Reversing Carbon Flow from the Tricarboxylic Acid (TCA) Cycle to the Embden-Meyerhof-Parnas (EMP) Pathway

There are several key unidirectional enzymes that control the direction of carbon flow from glucose to pyruvate and then into and around the TCA cycle and its alternative pathways. What happens when the primary carbon source is a TCA cycle intermediate (e.g., succinate) instead of glucose? The carbon flow must be driven in reverse through the EMP pathway and the PPP so that all of the essential precursors can be produced. The key unidirectional enzyme pyruvate kinase pushes carbon flow toward pyruvate, but there are alternative enzymes employed to reverse carbon flow at this point (Fig. 2.16). Oxaloacetate and ATP can be converted into phosphoenolpyruvate, with the release of carbon

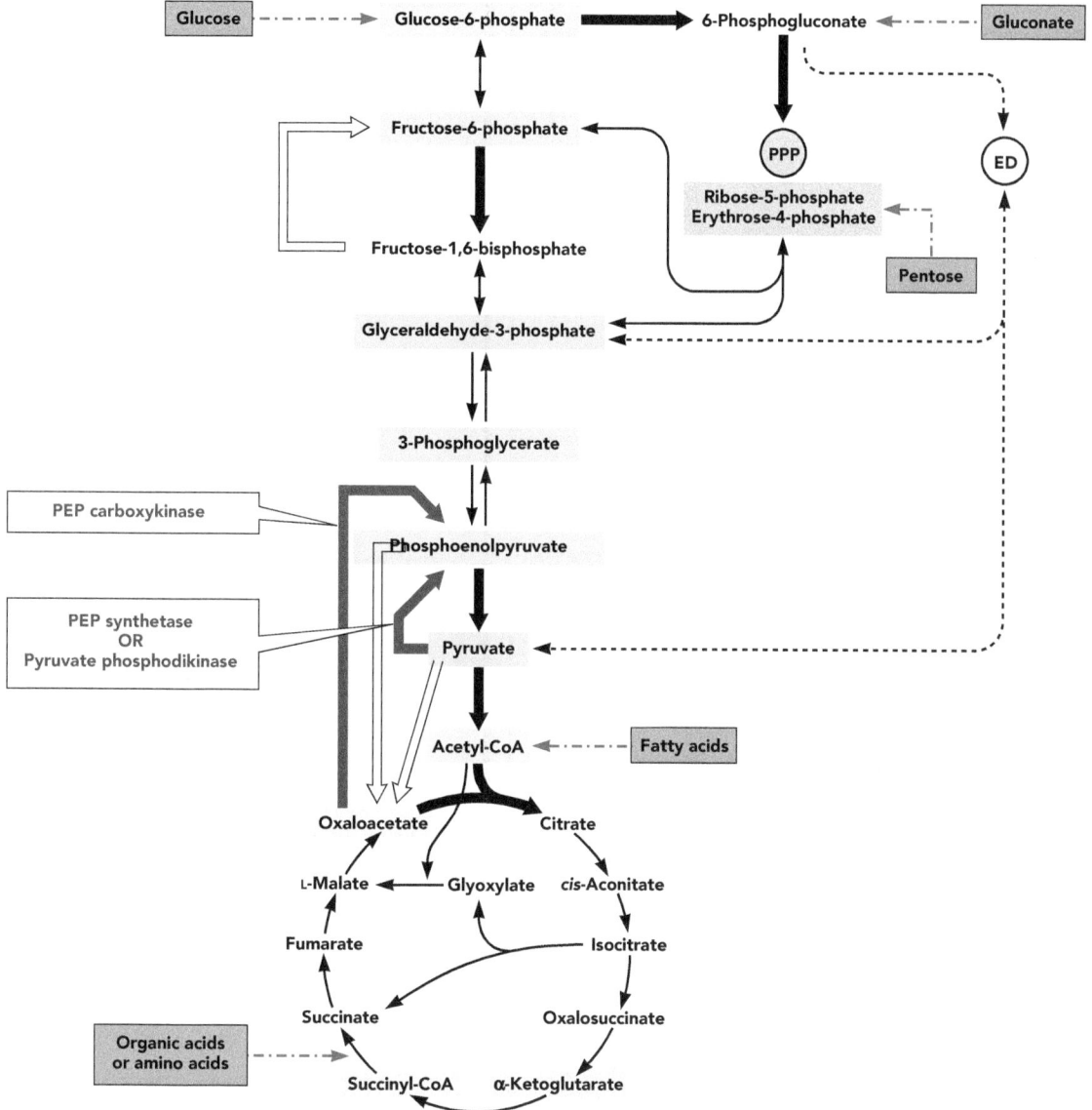

FIGURE 2.16 **Relationship between the pathways of central metabolism.** Enzymes critical for reversing the carbon flow between the tricarboxylic acid (TCA) pathway and the Embden-Meyerhof-Parnas (EMP) pathway, PEP carboxykinase, PEP synthetase, and pyruvate phosphodikinase, are highlighted with purple arrows. The entry points for alternative carbon sources are highlighted by red dashed arrows.

dioxide by the activity of phosphoenolpyruvate carboxykinase. Pyruvate can be converted into phosphoenolpyruvate by phosphoenolpyruvate synthetase or pyruvate phosphodikinase. These reactions occur at the expense of two high-energy phosphate bonds; however, this is necessary for a cell to survive because if it cannot route carbon flow from the TCA cycle into the EMP pathway and PPP through gluconeogenesis, it will not be able to make the majority of the essential precursors required for growth.

As will be discussed in subsequent chapters, prokaryotes have amazing metabolic diversity. Glucose is not always the preferred carbon source, nor is it commonly available in most environments. Molecules such as gluconate, pentose sugars, fatty acids, organic acids, and amino acids are all possible alternative carbon sources for microbes (Fig. 2.16). Consequently, distinctive pathways evolved that enable prokaryotes to grow under varying environmental conditions with a wide range of carbon compounds. Regardless of the variety of carbon compounds that bacteria and archaea can utilize, all must be metabolized through the central metabolic pathways to generate the 12 precursor metabolites to build cell material while yielding energy for the cell.

Learning Outcomes: After completing the material in this chapter, students should be able to . . .

1. identify where in the conserved pathways of central metabolism the 12 essential precursors are produced

2. explain how cells use redox reactions and substrate-level phosphorylation to generate ATP in the Embden-Meyerhof-Parnas (EMP) pathway

3. discuss the function of the coenzymes ATP, NAD(P)$^+$/NAD(P)H + H$^+$, and CoA in the cell, including the importance of recycling these compounds

4. deduce the impact of effector molecules on key unidirectional enzymes in central metabolism through allosteric regulation

5. list two possible functions of the Entner-Doudoroff (ED) pathway

6. compare and contrast the role of the tricarboxylic acid (TCA) cycle in both catabolism and anabolism

Check Your Understanding

1. Define this terminology: anabolism, catabolism, glycolysis, aldonic acids, isomerase, reduction/oxidation (redox) reactions, substrate-level phosphorylation, allosteric regulation, feedback inhibition, gluconeogenesis, transketolase, transaldolase, anaplerotic reaction

2. a) What three main outputs do the central metabolic pathways provide for the cell? (LO1)
   b) Why are these three outputs important for cellular survival?

3. a) What role do ATP and NADH + H$^+$ coenzymes play in cell metabolism? (LO3)
   b) How does ATP carry energy differently from NADH + H$^+$?
   c) How is the role of NADH + H$^+$ different from the role of NADPH + H$^+$?
   d) Why is it important for NADH + H$^+$ to be reoxidized to NAD$^+$?

4. Circle the correct words and fill in the blanks to accurately complete the following sentence. (LO3)
   In the reaction below, malate is being (oxidized / reduced) to oxaloacetate as a part of the (Embden-Meyerhof-Parnas pathway / tricarboxylic acid cycle). The enzyme _____ catalyzes the reaction and allows the transfer of electrons from (malate / oxaloacetate) to (oxaloacetate / NAD$^+$).

L-Malate → Oxaloacetate (with NAD$^+$ → NADH + H$^+$)

5. The purpose of completing the information table on the following page is to help you organize important information relative to the nomenclature, structure, enzyme names, regulation of enzymes, and cellular use of the 12 precursor metabolites. (LO1, 5, 6)
   a) Fill in the following table with the name of the pathway(s), the enzyme(s) that produced the precursor metabolite, and cellular use of each precursor.

Precursor metabolite	Stucture	Pathway(s)	Enzyme(s) that produce each precursor	Cellular use (e.g., sugar, lipid, nucleic acid, or amino acid biosynthesis)
Glucose-6-phosphate				
Fructose-6-phosphate				
Glyceraldehyde-3-phosphate				
3-Phosphoglycerate				
Phosphoenolpyruvate				
Pyruvate				
Acetyl-CoA				
α-Ketoglutarate				
Succinyl-CoA				
Oxaloacetate				
Ribose-5-phosphate				
Erythrose-4-phosphate				

b) List the unidirectional enzymes in each pathway below. (Note, there could be more than one).

- Glycolysis:
- Transition reaction:
- Tricarboxylic acid cycle:

c) Like all enzymes, the unidirectional enzymes could, theoretically, function in both directions. Using phosphofructokinase as an example, explain why these enzymes function only in one direction and why this is important for the cell.

d) Using phosphofructokinase as an example, describe how bacteria are able to run the reactions of unidirectional enzymes in the reverse direction.

6. Some of the enzymes in the pathways of central metabolism are regulated. For each regulated enzyme below, identify their effector molecule(s) from the text and indicate if the effector molecule has a positive (+) or negative (-) influence on the enzyme activity. (LO4)

- Pyruvate kinase:
- Phosphofructokinase:
- Pyruvate dehydrogenase:

7. The Entner-Douderoff (ED) and Embden-Meyerhof-Parnas (EMP) pathways both evolved in prokaryotic cells to metabolize glucose. (LO2, 5)

a) What two steps does the ED pathway share with the pentose phosphate pathway?

b) Which two enzymes are unique to the ED pathway?

c) How do these enzymes help scientists to identify bacteria that are potentially using this pathway?

d) What are two benefits for a microorganism utilizing the ED pathway instead of the EMP pathway?

e) What is one cost to a microorganism utilizing the ED pathway instead of the EMP pathway?

8. a) Describe the role of the tricarboxylic acid (TCA) cycle in both catabolism and anabolism. (LO6)

b) The glyoxylate cycle is a modification of the TCA cycle. What is the major purpose of the glyoxylate cycle?

9. Microorganisms that grow under anoxic conditions still require parts of the tricarboxylic acid cycle (TCA) to produce essential precursors for biosynthesis. However, under anoxic conditions, the cycle is broken into a branched pathway.

a) Which key enzyme is not made in the absence of oxygen? In the following figure, indicate with an asterisk where in the cycle this enzyme works.

b) What part of the TCA cycle is the same under both oxic and anoxic conditions?

c) The other part of the TCA pathway has to be run in reverse, starting with oxaloacetate. How do the anaplerotic reactions sustain this part of the TCA pathway?

d) What are the energetic consequences of this branched pathway? What products of the TCA cycle are not made?

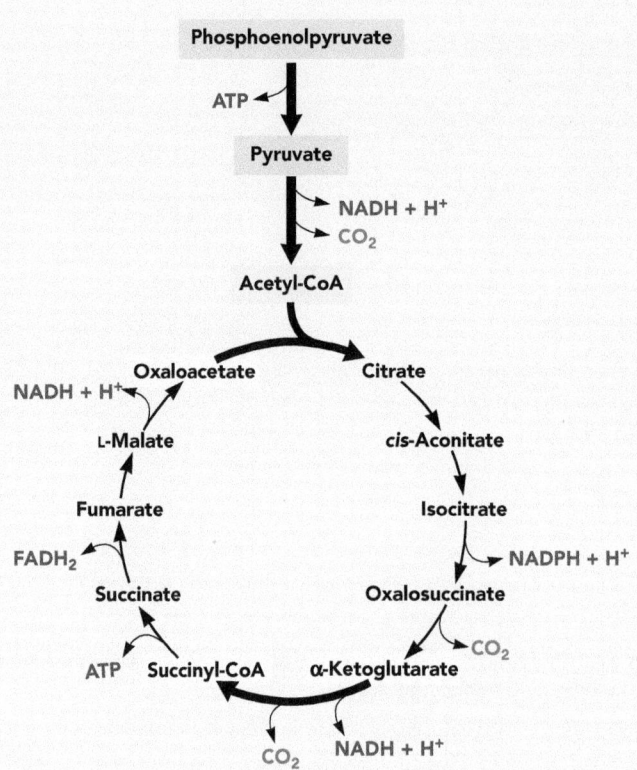

Dig Deeper

10. What parts of the central metabolic pathways would support the hypothesis that all prokaryotic cells evolved from common ancestors? (LO1)

11. Organisms that can grow in a minimal medium with a few salts and a single carbon source must be able to synthesize all 12 of the precursor metabolites to build cellular material. Microorganisms have evolved the ability to grow on many different types of macromolecules to assimilate carbon and yield energy for the cell. (LO1, 6, 7)

    a) Some bacteria can grow on each of the molecules listed below. Identify and match each molecule listed on the left with its chemical type shown in the box.

    Acetate _____

    Gluconate _____

    Ribose_____

    Aspartic acid _____

Aldonic acid
Amino acid
Carbohydrate
Organic acid

Below is a simplified diagram that shows how the central metabolic pathways connect to one another.

b) Label each colored shape with the appropriate pathway name (Embden-Meyerhof-Parnas pathway, pentose phosphate pathway, tricarboxylic acid cycle, transition reaction).

c) On the diagram, draw arrows to indicate (approximately) where acetate, gluconate, ribose, and aspartic acid would enter the central metabolic pathways for metabolism.

d) If the cells are growing on acetate, in which order would the following precursor metabolites be synthesized?

___ Erythrose-4-phosphate

___ Oxaloacetate

___ Phosphoenolpyruvate

12. a) What does the presence of a high concentration of ADP or AMP signal to the cell? (LO3)

b) Explain how the allosteric regulation of phosphofructokinase impacts the Embden-Meyerhof-Parnas pathway.

c) Explain how the allosteric regulation of pyruvate dehydrogenase impacts the transition reaction.

13. During cell growth, the biosynthesis of various polysaccharides is important for the production of bacterial capsules and biofilms. The precursor metabolite fructose-6-phosphate is an important substrate for various biosynthetic pathways that produce polysaccharide products. (LO1)

a) Based on material in the chapter about the central metabolic pathways, diagram three different reactions that would result in the synthesis of fructose-6-phosphate. In your answer, be sure to include the name of the metabolic pathway, the name of the enzyme that catalyzes the reaction, and the name of the substrate of the reaction.

b) *E. coli* can produce fructose-6-phosphate using all three of these reactions, whereas *Pseudomonas* can only generate fructose-6-phosphate using one of the reactions you diagrammed in part a. Which of these reactions does *Pseudomonas* use and why?

52 Making Connections

52 Estimating Molecular Concentrations

54 Physiologically Relevant Protein Concentrations

55 Measuring Enzyme Activity: Basic Principles of Enzyme Assays

58 Michaelis-Menten Kinetics

59 Studying the Proteome

61 The Physiological Role and Composition of Cellular RNA

63 The Physiological Role and Composition of Cellular DNA

64 Studying the Genome and the Transcriptome

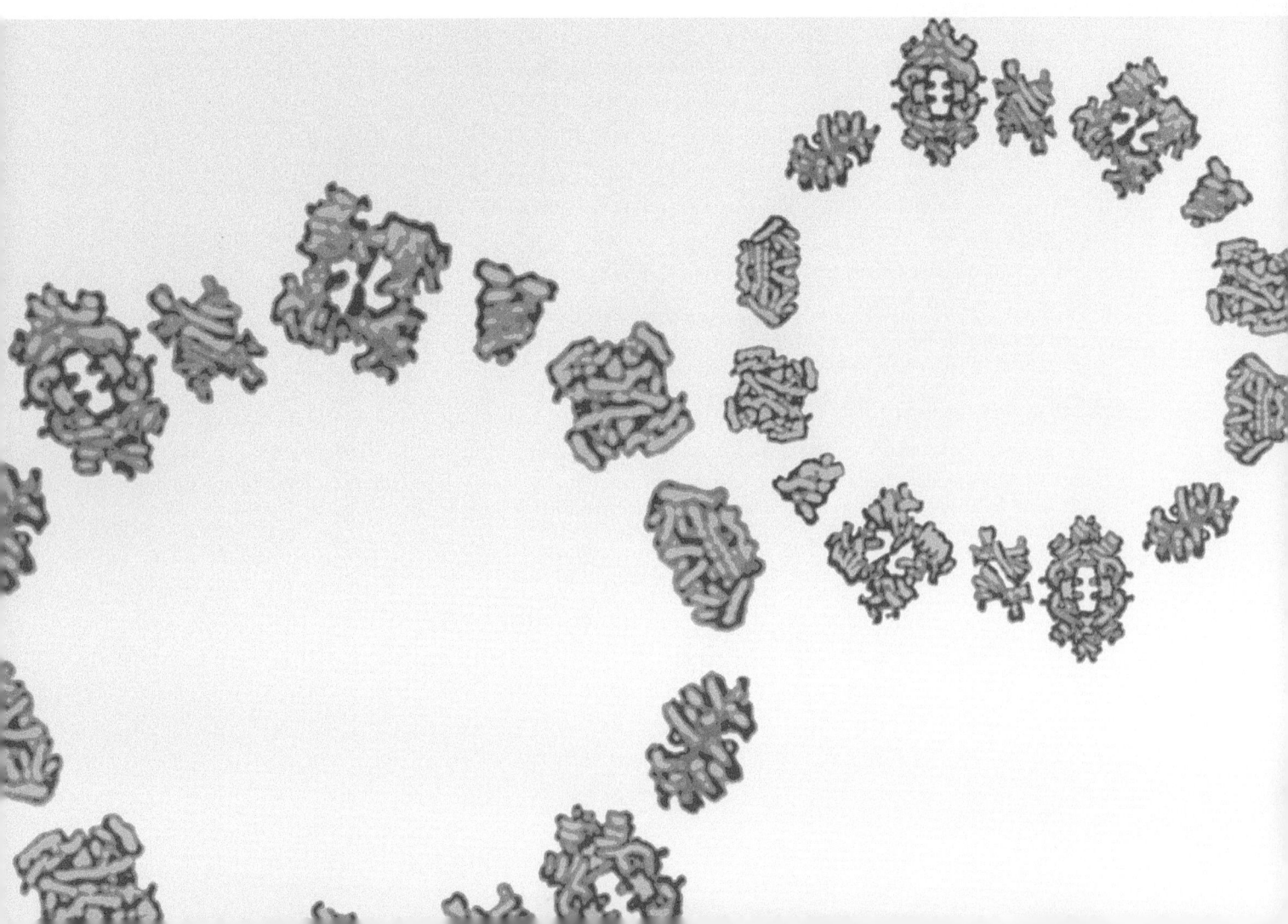

CELLULAR COMPONENTS—
WHAT'S IN A CELL

After completing the material in this chapter, students should be able to . . .

1. estimate physiologically relevant concentrations of macromolecules in the cell

2. describe how crude cell extracts and total protein concentration assays are used to determine the specific activity of an enzyme

3. list the six classes of enzymes and their functions

4. determine the V_{max} and K_m from a Michaelis-Menten curve

5. explain the methodology of 2D SDS-PAGE

6. compare and contrast the roles of DNA and RNA in a cell

Microbial Physiology: Unity and Diversity, First Edition. Ann M. Stevens, Jayna L. Ditty, Rebecca E. Parales
and Susan M. Merkel.
© 2024 American Society for Microbiology.
Companion website: www.wiley.com/go/stevens/microbialphysiology

Making Connections

The 12 essential precursors produced during the reactions of central metabolism are the building blocks for the macromolecules and higher-order structures of microbial cells (Fig. 3.1). This chapter provides a broad overview of the different cellular constituents and how to study them in the laboratory using physiologically relevant conditions. While carbohydrates and lipids are also essential for cells, in this chapter a greater emphasis will be placed on proteins and nucleic acids, as they are critical for regulating overall cellular physiology. Modern molecular-based "-omics" approaches are giving us unprecedented insights into microbial physiology. Genomic, transcriptomic, proteomic, and metabolomic methods are being used to examine the total pool of DNA, RNA, polypeptides, and metabolites, respectively, that are present in living cells under different growth and environmental conditions (Fig. 3.2).

Estimating Molecular Concentrations

It is important to have a general idea of the levels of molecules that are naturally present in living cells, as studying the individual parts of a cell can lead to a better understanding of the physiology of a microorganism as a whole. What are physiologically relevant concentrations of organic molecules, and what are appropriate "ballpark" estimates of these values? Using *Escherichia coli* as a

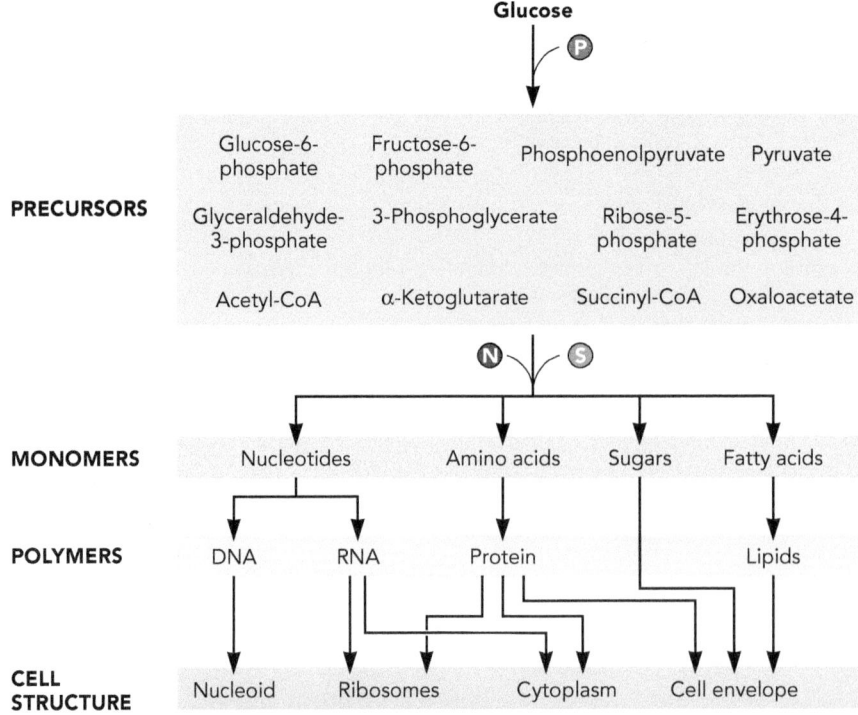

FIGURE 3.1 **Building cellular components from the 12 essential precursors.** Precursors (Chapter 2) are converted into organic monomers (i.e., nucleotides, amino acids, sugars, and fatty acids), polymers, and higher-order cell structures for the cell.

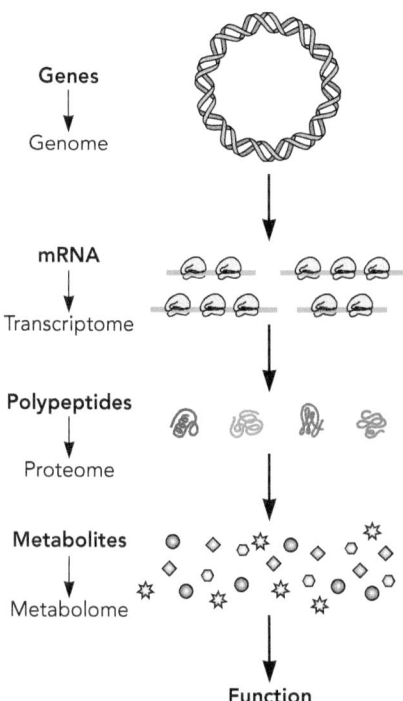

Genes
↓
Genome

mRNA
↓
Transcriptome

Polypeptides
↓
Proteome

Metabolites
↓
Metabolome

↓

Function

FIGURE 3.2 **Studying physiology using "-omics."** The total genes, mRNA, polypeptides, and metabolites in a cell are the genome, transcriptome, proteome, and metabolome, respectively. Analysis of these cellular components and how they change under different conditions and/or in mutant strains provides a detailed view of the global physiology of an organism.

model, the levels of organic molecules in an average cell have been determined experimentally with populations of cells (Table 3.1). A typical cell is 70 to 75% water by weight. A cell can be separated into its basic organic components using a Roberts fractionation. This method involves breaking open a large number of

TABLE 3.1 **Composition of an average *E. coli* cell[a]**

Components	% Total dry weight[b]	Amt (g x 10^{-15}) per cell[c]	Molecular weight	Molecules per cell	No. of different kinds of molecules
Protein	55.0	156	4.0×10^4	2,350,000	1,850
RNA	20.5	58			
23S rRNA		31.0	1.0×10^6	18,700	1
16S rRNA		15.5	5.0×10^5	18,700	1
5S rRNA		1.2	3.9×10^4	18,700	1
tRNA		8.2	2.5×10^4	198,000	60
mRNA		2.3	1.0×10^6	1,380	600
DNA	3.1	8.8	2.5×10^9	2.1	1
Lipid	9.1	25.9	705	22,000,000	
Lipopolysaccharide	3.4	9.7	4,070	1,430,000	1
Peptidoglycan	2.5	7.1	(904)n	1	1
Metabolites, cofactors, ions	3.5	9.9	–	–	>800

[a]Calculated for an average cell in a population of *E. coli* (strain B/r) in balanced growth at 37°C in aerobic glucose minimal medium with a mass doubling time of 40 minutes. The cell is defined by dividing the total biomass, or the amount of any of its measured components, by the total number of cells in the population. Adapted from Neidhardt FC et al. 1987. *Escherichia coli* and *Salmonella typhimurium* Cellular and Molecular Biology. ASM Press, Washington, DC.
[b]Relative amounts of the major components. In some cases, data from strains other than B/r, from alternative growth conditions, or both were used.
[c]Based on measurements of the total dry mass and the number of cells measured in portions of a reference culture. The total dry weight per cell is 2.8×10^{-13} g; the water content (assuming that 70% of the cell is water) is 6.7×10^{-13} g; the total weight of one cell is 9.5×10^{-13} g.

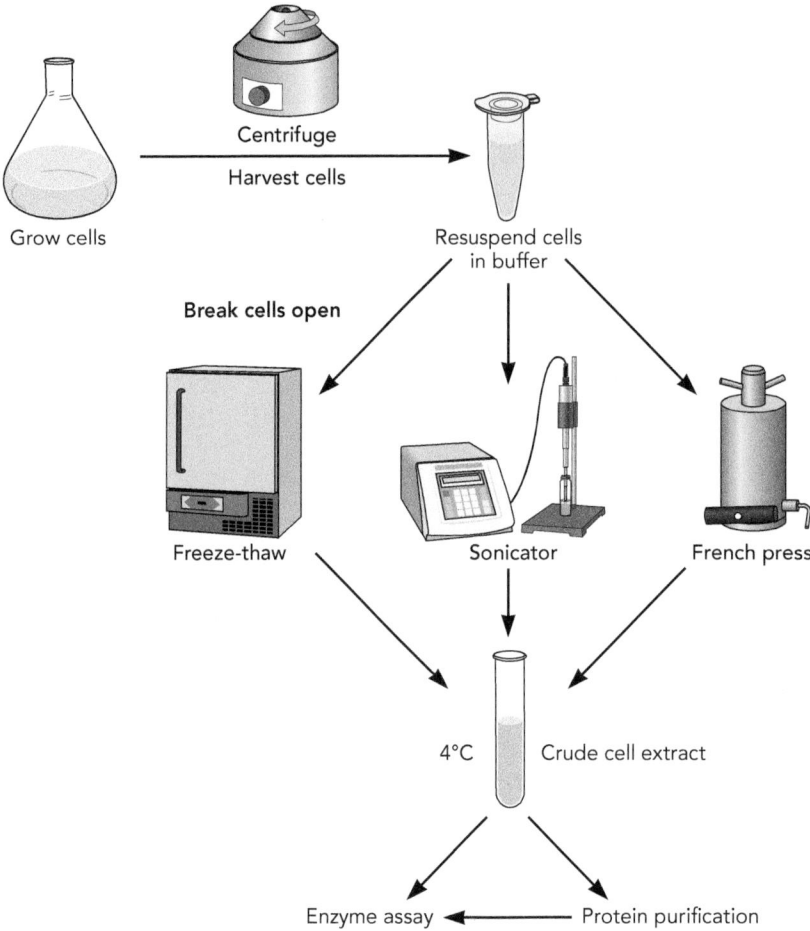

FIGURE 3.3 **Preparation of a crude cell extract from bacterial cells.** Cells grown in liquid culture or scraped off agar plates are harvested via centrifugation. The cell pellet is resuspended in an appropriate buffer, and then the bacterial cells are lysed using a lysozyme plus freeze-thaw treatment, sonicator, or French press. The resulting crude cell extract may then be directly assayed for total protein concentration and specific enzyme activity, or individual proteins may be purified and then assayed.

cells to produce a "crude cell extract" (see Fig. 3.3), and separates it into its macromolecular parts in order to quantitate the amount of protein, nucleic acid, carbohydrate, and lipid in a cell. As shown in Table 3.1, this method demonstrated that roughly 50% of the dry weight of a bacterial cell is composed of protein. Given the small size of a bacterial cell, negative exponents are often used to provide accurate values of size, mass, and volume (i.e., milli = 10^{-3}; micro = 10^{-6}, nano = 10^{-9}; and pico = 10^{-12}). A typical bacterial cell is about 1 cubic micron (1 $\times$ 10^{-6} cubic meters) in size, has a volume of approximately 1 $\times$ 10^{-12} ml, or 10^{-15} liters, and weighs around 1 $\times$ 10^{-12} g. Since 70 to 75% of the cell is water, the total dry weight of a single bacterial cell (e.g., *E. coli*) is approximately 2.5 $\times$ 10^{-13} g, and about half of this dry weight is protein.

Physiologically Relevant Protein Concentrations

How much of any particular type of protein is typically present in a bacterial cell? In an average *E. coli* cell, there are about two million protein molecules and about 2,000 different types of proteins present at any given time (Table 3.1). However, not all types of proteins are present at equivalent concentrations. The term **molarity** (M), which is moles/liter, is the most common way that concentrations are presented in scientific literature. **Avogadro's number** (6.02×10^{23} molecules/mole) enables the conversion of moles to number of molecules.

The theoretical upper limit of the concentration for a given protein within an individual bacterial cell is $\sim 10^{-3}$ M. At this concentration, the cell would be packed full of just that given protein; therefore, this value is not feasible, as the cell would not be able to survive under such conditions. The highest laboratory-determined protein concentrations achieved for a given protein are in the range of 10^{-4} M. This high concentration occurs when a protein is being "overexpressed" through a genetically engineered construct. Under these conditions, about 10% of the cell's mass is just one type of protein, which can be tolerated. However, there are negative consequences for cell growth and survival when so much biomass and energy are being applied to produce a protein that is often not of use to the cell itself. Most cellular proteins are present at the physiologically relevant concentration range of 10^{-5} to 10^{-7} M. This converts to between 60 and 6,000 molecules of the protein per cell. A practical lower concentration range is 10^{-8} M, which is on average six molecules per cell:

$$\# \, molecules/cell = \left(protein\ concentration\right)\left(Avogadro's\ number\right)\left(cell\ volume\right)$$

$$\left(1 \times 10^{-8}\ moles/liter\right)\left(6.02 \times 10^{23}\ molecules/mole\right)\left(1 \times 10^{-15}\ liters/cell\right)$$

$$= 6\ molecules/cell$$

The concentrations of regulatory proteins are often in this range to enable quick up- or downregulation through expression or degradation of the protein, depending on changes in the environment. Lower concentrations (e.g., 10^{-9} M) would mean that less than one molecule is present per cell on average; this amount is not physiologically relevant. The naturally occurring concentrations of most proteins found in living cells are between 10^{-5} and 10^{-8} M; therefore, laboratory experiments performed with protein concentrations in this range are physiologically relevant.

Measuring Enzyme Activity: Basic Principles of Enzyme Assays

In addition to understanding how much of a protein is present in a cell, it is also important to understand the specific reaction(s) a particular enzyme is catalyzing for the cell. A basic enzyme assay for a cytoplasmic protein starts with the generation of a crude cell extract (Fig. 3.3), which can be obtained by growing large numbers of cells in liquid or plate culture. The cells are then collected by centrifugation to concentrate them and remove the growth medium, followed by resuspension of the cells in an appropriate buffer. The buffer conditions should be optimized for the particular enzyme under investigation, ensuring that the cellular contents are maintained at homeostatic conditions similar to the native cytosol (e.g., pH and salt concentration). After resuspension in an appropriate buffer, the cells can be broken to release cellular contents into the controlled buffered environment.

Since both *Bacteria* and *Archaea* have rigid cell wall structures, relatively harsh methods must be used to break the cells open (Fig. 3.3). A sonicator uses high-frequency sound waves to lyse microbial cells. A French press uses a large pressure differential to break the cells; the cells are placed under ~20,000 psi of pressure (~1,360 atm) within a metal chamber, and then the pressure is changed to 1 atm of pressure when the sample is released from the chamber. The rapid change in pressure causes the cells to rupture. Alternatively, enzymes such as **lysozyme** (which breaks the β-1,4-glycosidic bonds in peptidoglycan) can be used to weaken the bacterial cell walls and then subsequent freeze-thaw cycle(s) will result in lysis of the cells. The crude cell extract generated as described above may be used directly for an enzyme assay, or a specific protein of interest may be further purified.

Although the physiological level of total protein in a cell is approximately 125 mg/ml (based on dry weight), *in vitro* enzyme assays are typically performed at a lower concentration of protein (e.g., 1 to 2 mg/ml). Experiments can be done at these lower concentrations of protein by manipulating the assay conditions (e.g., using higher levels of substrate). However, such nonphysiologically relevant protein concentrations can still be problematic if a multisubunit enzyme complex is diluted below the concentration at which the different subunits interact properly. Enzymes determine the rate, not the direction, of a reaction, and there are multiple factors besides the protein concentration that influence the rate at which an enzyme functions. The substrate and product concentrations, pH, temperature, salt, ions, and cofactors can impact the reaction rate of the enzyme and the direction of the reaction according to the laws of thermodynamics (i.e., entropy and enthalpy).

Typically, the rate of product formation or substrate disappearance is measured in the laboratory to determine enzyme activity. Substrate and product concentrations can be measured using spectroscopy if either the substrate or product absorbs light at a distinctive wavelength. This approach can also be used to monitor the reduction/oxidation of a coenzyme, like $NAD(P)H + H^+$. If a gas is consumed or produced in the enzymatic reaction (e.g., oxygen is consumed by an oxygenase), a probe may be inserted into a sealed vial to measure changes in the concentration of the gas over time. Column chromatography may also be employed to separate and quantify the substrate or product. Types of chromatography used include gas chromatography, which can be used with volatile compounds; mass spectrometry, which measures the masses of compounds; or high-performance liquid chromatography, which can separate and detect small molecules in a liquid solution. These are just a few examples of different ways that the progression of an enzymatic reaction may be monitored.

Beyond establishing the necessary parameters to measure the activity of a specific enzyme, it is also essential to quantitatively measure the total amount of protein in the crude cell extract. Two common methods to measure total protein concentration are the Lowry and Bradford assays. The Lowry assay is a classic method that involves the binding of Cu^{2+} ions to peptide bonds. This assay can be time-consuming and requires multiple chemical reagents, some of which are hazardous. Therefore, the Bradford assay is more commonly used since it is faster and involves just one reagent. In this method, the absorbance of the protein-binding dye Coomassie Brilliant Blue G-250 shifts from a wavelength of 465 nm to 595 nm when it binds to protein. In both the Lowry and Bradford assays, the protein sample is combined with the reagents to cause a visible color change that is measured using a standard spectrophotometer. The absorbance of the unknown protein sample is then compared to the absorbance of a standard curve that is produced using a known concentration of a commercially available single protein (e.g., bovine serum albumin). The concentration of the unknown sample may then be extrapolated from the linear standard curve on a plot of absorbance versus concentration, since there is a linear relationship between those parameters within a certain range (Fig. 3.4).

Knowing the total protein concentration in a sample and the rate at which an enzyme converts substrate to product(s) enables the determination of its specific activity. **Specific activity** is defined as the amount of substrate consumed or product formed per minute per milligram of protein. Therefore, enzyme activity in any particular extract is dependent on the amount and purity of the enzyme. A purified protein is useless unless it is biologically active and has a specific activity that can be measured. Vendors may use the term "unit" to define the specific activity of one of their commercial enzyme products (e.g., 1 unit of a restriction endonuclease cleaves 1 µg of a DNA substrate in 1 hour), and consumers expect that the activity of the enzyme is consistent between batches.

There are still significant numbers of proteins for which there is no known function. Establishing methods to assay the activity of a newly purified enzyme is not trivial. However, to date there are only six known classes of enzymes (Table 3.2). These six categories classified by Enzyme Commission (EC) number are class 1, oxidoreductases; class 2, transferases; class 3, hydrolases; class 4, lyases; class 5, isomerases; and class 6, ligases. There are examples of each of these six classes within the central metabolic pathways that were discussed in Chapter 2. In addition, there are a number of "common names" associated with the six enzyme classes that are useful to know (Table 3.3).

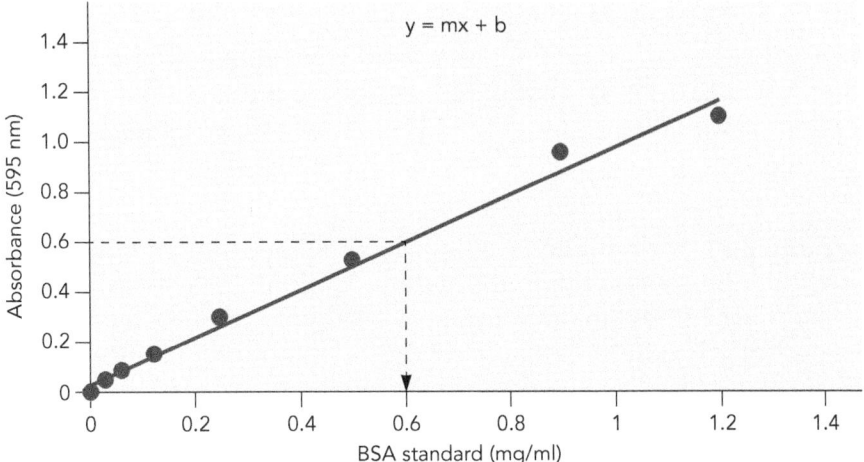

FIGURE 3.4 **Example of a standard curve for protein concentration determination.** A series of protein samples (e.g., bovine serum albumin [BSA]) of known concentration is used to establish the curve, and the equation of the resulting line may be used to calculate the protein concentration of an unknown sample based on the detected absorbance reading (see example of extrapolation shown by the dotted line).

TABLE 3.2 **Enzyme classification scheme**

Class	Reactions catalyzed	Example from central metabolism (Chapter 2)
1. Oxidoreductases	Oxidation of one molecule while the other is reduced	Glyceraldehyde-3-phosphate dehydrogenase (glyceraldehyde-3-phosphate ↔ 1,3-bisphosphoglycerate)
2. Transferases	Transfer of a functional group between molecules	Hexokinase (glucose → glucose-6-phosphate)
3. Hydrolases	Hydrolytic (involving H_2O) cleavage of C-O, C-N, or C-C bonds	Fructose-1,6-bisphosphatase (fructose-1,6-bisphosphate → fructose-6-phosphate)
4. Lyases	Cleavage of C-O, C-N, or C-C bonds by elimination, while leaving double bonds or a ring	Aldolase (fructose-1,6-bisphosphatase ↔ dihydroxyacetone phosphate + glyceraldehyde-3-phosphate)
5. Isomerases	Rearrangement of bonds within a single molecule	Triose phosphate isomerase (dihydroxyacetone phosphate ↔ glyceraldehyde-3-phosphate)
6. Ligases	Joining together of two molecules, coupled to ATP hydrolysis	Pyruvate carboxylase (pyruvate → oxaloacetate)

TABLE 3.3 **Some common enzyme names**

Enzyme	Reaction(s) catalyzed
Nucleases	Degradation of nucleic acids by hydrolyzing phosphodiester bonds between nucleotides
Proteases	Degradation of proteins by hydrolyzing peptide bonds between amino acids
Polymerases	Polymerization reactions producing molecules such as DNA and RNA
Kinases	Addition of phosphate groups to molecules
Phosphatases	Removal of phosphate groups from molecules
ATPases	Hydrolysis of ATP
Lipases	Degradation of lipids through hydrolysis of fatty acyl-ester bonds in bacterial and eukaryotic membrane lipids

Michaelis-Menten Kinetics

Michaelis-Menten kinetics define the relationship between an enzyme re-action rate and the concentration of substrate. This type of analysis requires that the particular enzyme under study be purified from all other cellular ma-terial in a crude cell extract. Protein purification is most commonly achieved through liquid chromatography, which separates the protein of interest from others based on its mass, charge, and/or a specific binding affinity. There are two key terms associated with Michaelis-Menten enzyme activity that are im-portant when discussing cellular physiology. V_{max} is the maximum velocity of a reaction. For any given enzyme, the rate at which it functions will increase as the substrate concentration increases up to a certain point, which is the V_{max}. K_m is the concentration of substrate that produces one half of the maximum veloc-ity. The K_m can be crudely considered to reflect how well a substrate binds to the enzyme, also known as **binding affinity**. The K_m and V_{max} can be obtained by plotting the velocity versus the substrate concentration (Fig. 3.5). The K_m is a constant value for an unregulated enzyme (i.e., not allosterically regulated). Interestingly, for most enzymes the K_m value is 10^{-5} to 10^{-7} M, which is similar to the naturally occurring protein concentrations in the cell.

Studying the Proteome

In addition to understanding the total amount of protein and how one specific protein functions in the cell, it is also important to appreciate the diversity of proteins that drive the physiology of the cell. Biochemical studies with proteins were feasible long before scientists were able to analyze nucleic acids. Therefore, the **proteome**, the total complement of proteins present in a cell at a given point in time, was used to analyze global physiology prior to the development of **genome**-level (DNA), **transcriptome**-level (RNA), or **metabolome**-level (metabolites) studies and is still an important tool to study bacterial physiology today (Fig. 3.2). As we will see later when gene regulation is discussed (Chapter 8), not all genes are expressed, and therefore not all proteins are produced in a cell at any given time. For example, the *E. coli* genome (~4.6×10^6 base pairs [bp]) has the capacity to produce ~4,000 different proteins, but proteome experiments have only permitted the visualization of ~2,000 proteins (Table 3.1).

One of the first experimental tools used to visualize the thousands of different proteins present in the proteome of a cell was two-dimensional sodium dodecyl sulfate-polyacrylamide gel electrophoresis (2D SDS-PAGE; Fig. 3.6). In this method, a crude cell extract is subjected to two separate electrophoresis steps. The first dimension uses a tube or strip gel that enables separation of proteins on the basis of their charge within a pH gradient. Each protein migrates in an electrical field to its **isoelectric point (pI)** in the gradient. At its pI, a protein has a neutral charge and will therefore remain in that location in the tube/strip gel. After the proteins are separated based on their charge, the tube/strip gel is laid on top of a traditional SDS-PAGE gel. In this second dimension, the proteins are coated with the negatively charged detergent SDS. The negatively charged

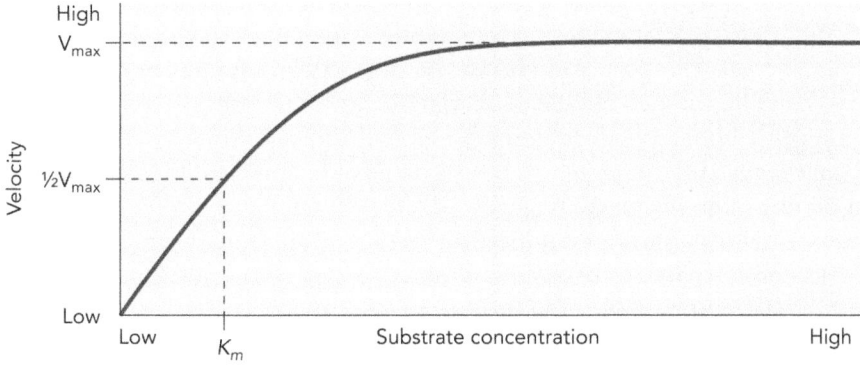

FIGURE 3.5 **Michaelis-Menten enzyme kinetics.** The velocity (rate) of a reaction will increase with the substrate concentration until the maximum velocity (V_{max}) is reached. The substrate concentration that results in $\frac{1}{2}V_{max}$ for an enzyme is a constant value called the Michaelis-Menten constant, or K_m.

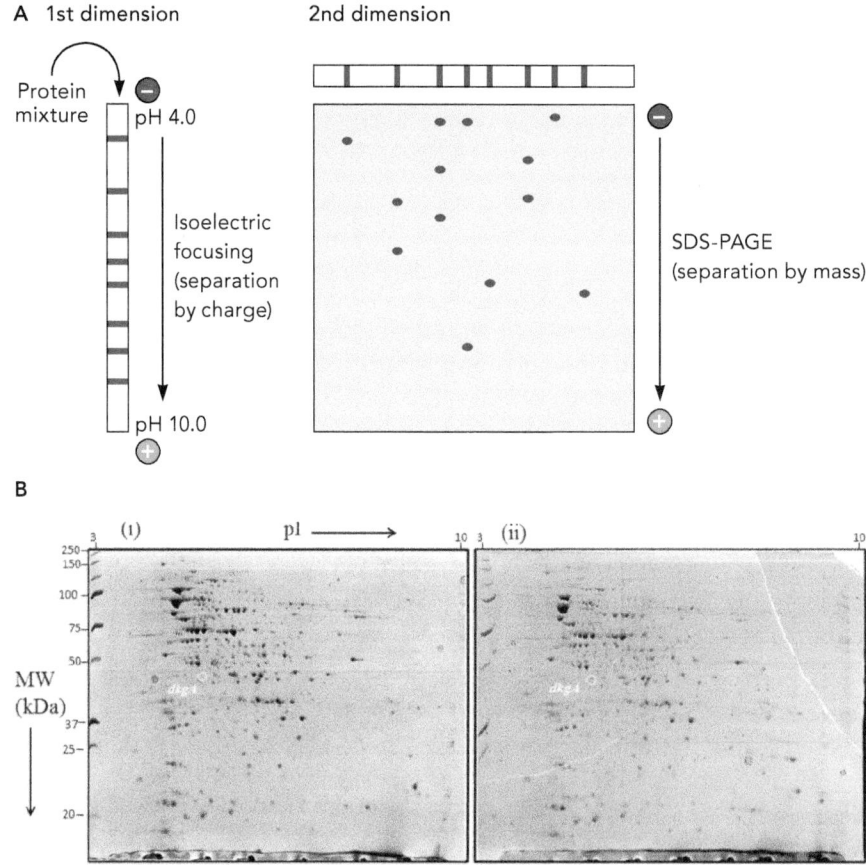

FIGURE 3.6 **Determination of the proteome via two-dimensional sodium dodecyl sulfate-polyacrylamide gel electrophoresis (2D SDS-PAGE).** (A) Proteins in a crude cell extract are separated on the basis of their isoelectric point (pI) in the first dimension and on the basis of their mass in the presence of the negatively charged detergent SDS in the second dimension. An electrical field is applied to samples moving through a polyacrylamide gel matrix. (B) Differential protein production from genes (e.g., *dkgA*, indicated by white text and circle) may be determined using cells grown under two different physiological conditions. Reprinted from Ramachandran R, Stevens AM. 2013. *Appl Environ Microbiol* 79:6244–6252.

proteins are then attracted to the positive electrode during the second round of electrophoresis, with proteins of larger mass being retained at the top of the porous gel and proteins of smaller mass migrating faster through the pores in the polyacrylamide gel. The pore size within the polyacrylamide matrix can be adjusted by changing the percentage of acrylamide and the number of cross-links within the gel. Ultimately, all proteins present in the cell are separated on the basis of their mass, and their size can be estimated based on a comparison to protein standards of known mass. Typically, 2D SDS-PAGE is used to compare the proteomes of cells grown under at least two different environmental conditions to reveal differential patterns of protein production (Fig. 3.6). Individual proteins of interest are then identified using mass spectrometry.

How many molecules of a typical protein are found in each *E. coli* cell? Based on 2D SDS-PAGE analysis, it has been estimated that the average mass of a protein in *E. coli* is ~40,000 daltons (Da). When used in biochemistry, 1 Da is defined as 1 g/mole. If the dry weight of a bacterial cell is ~2.5×10^{-13} g and 50% of that weight is protein, then there is ~1.25×10^{-13} g protein/cell. If the average mass of an individual protein is 40,000 Da (g/mole), then the number of molecules of a given protein per cell can be calculated as follows:

(amount of total protein per cell)/(mass of an average protein)

   = total moles protein per cell

$$\left(1.25\times10^{-13}\,\text{g protein/cell}\right)/\left(4\times10^{4}\,\text{g protein/mole}\right)=\left(3\times10^{-18}\,\text{moles protein/cell}\right)$$

(total moles protein per cell)(Avagodro's number)

   = total number of protein molecules per cell

$$\left(3\times10^{-18}\,\text{moles protein/cell}\right)\left(6.02\times10^{23}\,\text{molecules/mole}\right)$$

$$=\left(1.8\times10^{6}\,\text{total protein molecules/cell}\right)$$

(total number of protein molecules per cell)/(number of different kinds

    of proteins as determined from 2D SDS-PAGE gels)

    = number of molecules of a given protein per cell

$$\left(1.8\times10^{6}\,\text{total protein molecules/cell}\right)/\left(1.8\times10^{3}\,\text{different kinds of proteins}\right)$$

    =1,000 molecules of a given protein/cell

A thousand molecules of a given protein per cell matches the results from above where we calculated that 10^{-5} to 10^{-7} M is equivalent to 6,000 to 60 copies, respectively, of each type of protein per cell.

The Physiological Role and Composition of Cellular RNA

RNA comprises ~20% of the total cell dry weight and consists of ribosomal (rRNA), transfer (tRNA), messenger (mRNA), and small (sRNA) forms (Table 3.1 and Fig. 3.7). The rRNA, essential for the process of translation, makes up the vast majority (~80%) of the cellular RNA pool due to its high degree of structural stability. In *Bacteria* and *Archaea*, there are three types of rRNA (23S, 16S, and 5S), which are present at a 1:1:1 ratio since one molecule of each is required to produce a functional ribosome. There are ~20,000 ribosomes present in a bacterial cell at any given time. The precise number varies, with higher numbers of ribosomes enabling higher rates of translation and faster growth rates (Chapter 4). As was discussed in Chapter 1, rRNA has an essential physiological role in the cell. It has a high degree of secondary structure created via intramolecular hydrogen bonds. The number of rRNA gene copies varies in different microbes, and many bacteria have multiple identical copies of each rRNA gene. As examples, *E. coli* has seven copies, *Vibrio* species can have 11 copies, while some nitrifying bacteria have just one copy. Copy number is often correlated with growth rate of the organism, and bacteria with many copies of rRNA genes generally grow faster. In addition, the expression of rRNA genes is highly regulated so that cells are able to rapidly adjust their growth rates depending on the environmental conditions and the phase of growth (Chapter 4). These copy number differences

rRNA   **tRNA**

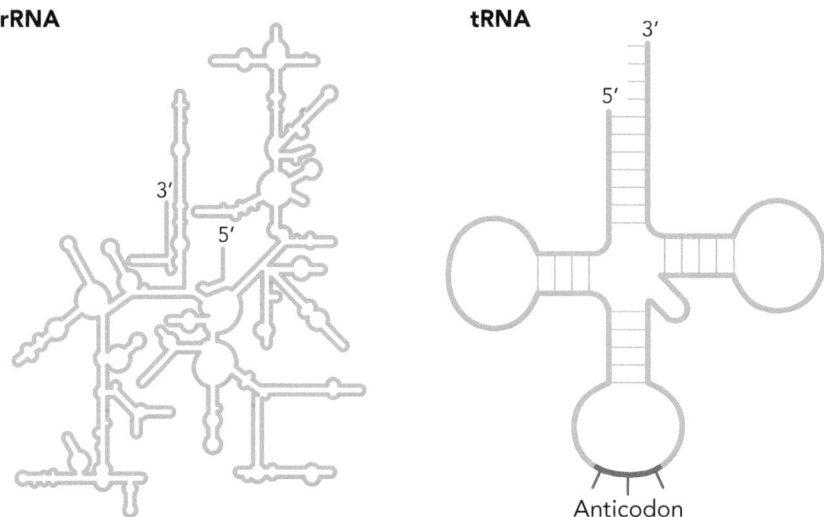

FIGURE 3.7 **Model images of 16S ribosomal (rRNA) and transfer (tRNA).** rRNA and tRNA are both important to the process of translation. These single-stranded molecules have the capacity to form secondary structures when complementary base pairing occurs due to hydrogen bond formation. The hydrogen bonds enhance the structural stability of the molecules, and thus they have longer half-lives than the template mRNA. Note that the drawings are not to scale; a molecule of 16S rRNA consists of ~1,600 nucleotides, whereas a molecule of tRNA is smaller, 76 to 90 nucleotides in length.

impact community profiles of bacteria based on 16S rRNA analysis (Chapter 1), since an organism with multiple copies of the rRNA genes will be overrepresented in comparison to another organism with lower gene copy numbers.

The tRNAs, also essential for the process of translation, have a mass of ~25,000 Da (76 to 90 nucleotides) and they also have a high degree of hydrogen bonding, producing a characteristic "cloverleaf" secondary structure that provides stability (Fig. 3.7). Thus, tRNA represents ~20% of the cellular RNA pool. There are 60 different types of tRNAs, with ~200,000 total copies per cell (Table 3.1). tRNAs couple the information in the mRNA to amino acids to facilitate the process of translation. The 3-bp codon in the mRNA aligns with the anticodon in the tRNA.

The mRNA is the most diverse portion of the RNA pool, but also the least stable, comprising just a few percent of the total RNA in a cell. The mRNA is produced from the transcription of genes, and the information stored within the mRNA molecule is ultimately converted into polypeptides through the process of translation. Unlike eukaryotic mRNA, bacterial and archaeal mRNA is typically not modified or processed prior to translation [i.e., no 5′ cap or poly(A) tail and introns are rare] and is often polycistronic, encoding multiple polypeptides. In *Bacteria* and *Archaea*, transcription and translation both occur in the cytosol and are coupled, meaning that they occur simultaneously. Consequently, bacterial mRNA has a very short half-life, on the order of just one to two minutes. This characteristic contributes to the ability of bacteria to rapidly adapt to changing environmental conditions but can also make working with bacterial mRNAs in the laboratory quite challenging. Since the majority of the RNA pool is rRNA and tRNA, these molecules are removed to enrich for the mRNAs prior to transcriptome analyses to ascertain the pool of mRNAs produced under a given set of environmental conditions.

A more recent discovery in bacteria is the presence of sRNAs, which in eukaryotes are often called microRNAs. sRNAs represent a very small percentage of the total RNA pool and, as their name suggests, are small in size, with the majority just 100 to 400 nucleotides in length. sRNAs play important roles in posttranscriptional regulation (Chapter 8).

The Physiological Role and Composition of Cellular DNA

DNA represents ~3% of the dry weight of a bacterial cell. In most *Bacteria* and *Archaea*, the genome is a single circular chromosome (here defined as a genetic unit carrying some essential "housekeeping" genes). However, some bacteria, such as *Vibrio* species, have two circular chromosomes, and some streptomycetes and spirochetes have linear chromosomes. The genomes of *Archaea* are associated with histones, like *Eukarya* but unlike *Bacteria*. In addition to their chromosome(s), *Bacteria* and *Archaea* may also have smaller extrachromosomal DNA elements known as plasmids. The vast majority of plasmids are also circular, but some bacteria have linear plasmids (e.g., *Borrelia* spp.). Plasmids can play an important role in genetic exchange through horizontal gene transfer and thereby contribute to the spread of genes amongst bacteria, including antibiotic resistance genes. Most bacterial genomes fall into the size range of 1,000 to 9,000 kilobase pairs (kbp), or 1 to 9 megabase pairs (Mbp, with 1 Mpb = 1,000 kbp) (Fig. 3.8), and archaeal genomes are similar in size. However, depending on the lifestyle and metabolic capability of the microbe, bacterial genomes are known to range from 112 kbp in obligate symbionts (e.g., *Nasuia deltocephalinicola*) to 13 Mbp in free-living species (e.g., *Sorangium cellulosum*). The model organism *E. coli* has an average genome size of ~4.6 Mbp. In contrast to eukaryotic genomes, genes are densely packed in bacterial and archaeal genomes. Typically, coding sequences represent >85% of the genome, and genes are often arranged in operons with little to no intergenic spacer DNA between the genes; noncoding DNA (introns) within genes is rare. In comparison, the human genome is much larger (~3.2 gigabase pairs [Gbp, or 10^9 bp]), but coding (exon) sequences represent only ~3% of the DNA.

If the *E. coli* genome is about 4.6 Mbp, and an average polypeptide is about 40,000 Da, what is the size of a typical gene? (Note: remember that DNA is double stranded, meaning each gene is composed of base pairs, while RNA is single stranded.)

$$(\text{mass of average polypeptide})/(\text{mass of average amino acid})$$

$$= \text{number of amino acids per polypeptide}$$

$$(40{,}000 \text{ Da per polypeptide})/(110 \text{ Da per amino acid})$$

$$= 364 \text{ amino acids/polypeptide}$$

$$(\text{number of amino acids per polypeptide})(\text{number of bases per amino acid})$$

$$= \text{number of basepairs to encode one polypeptide}$$

$$(364 \text{ amino acids/polypeptide})(3 \text{ bases/amino acid})$$

$$= 1{,}092 \text{ basepairs to encode one polypeptide}$$

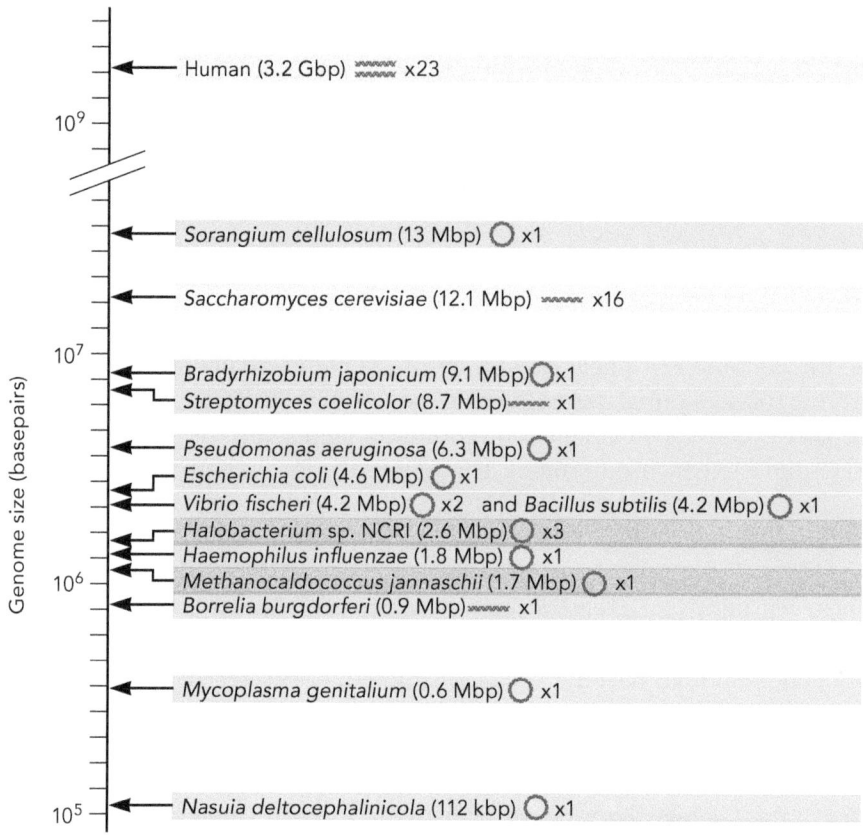

FIGURE 3.8 **Comparative genome organization and sizes.** Relative genome sizes are given in megabase pairs (Mbp), with *Bacteria* in green, *Archaea* in pink, and *Eukarya* in yellow. The bacterium *Nasuia deltocephalinicola* is an obligate endosymbiont, and the pathogen *Mycoplasma genitalium* has the smallest known nonsymbiotic bacterial genome. *Sorangium cellulosum* is a myxobacterium with one of the largest known bacterial genomes. Circular and linear chromosome arrangements are represented by circles and lines, respectively, with the actual number of chromosomes as indicated. Note that chromosome size, arrangement, or number alone cannot predict the domain of an organism.

Thus, a typical bacterial gene is ~1,000 bp in size. If the genome is ~4 Mbp then the genome encodes ~4,000 polypeptides. These are good estimates for an average bacterial genome.

Studying the Genome and the Transcriptome

The development of laboratory techniques for the analysis of nucleic acids lagged behind those developed to study proteins, although the two are intimately linked due to the principles of the central dogma (Fig. 3.2). In the 1960s, the structure of DNA was established. Maxam and Gilbert sequencing with chemical cleavage and Sanger sequencing with dideoxynucleotides enabled the first genes, and then eventually entire genomes, to be sequenced. In 1995, the pathogenic bacterium *Haemophilus influenzae* was the first organism to have its genome fully sequenced.

While having access to the entire sequence of a genome (Fig. 3.2) is important information, understanding when particular genes are being expressed (or not expressed) cannot be determined from genome sequence alone. Genomic information is also limited in that the sequence alone cannot differentiate between a gene that is functional in the cell and a gene that has accumulated mutations rendering it nonfunctional. These nonfunctional genes, known as **pseudogenes**, are very difficult to predict based on sequence analysis alone. As such, microarrays were the next technology to be developed as a tool to analyze the transcriptome. The transcriptome reveals which genes are being transcribed into RNA in cells grown under a certain condition (Fig. 3.2). One limitation of this method was that it required a genome sequence from which the open reading frames (ORFs), or potential gene sequences that encode polypeptides, could be identified. This type of analysis quickly became feasible for model organisms with complete genome sequences using commercially available "chips," but oftentimes important

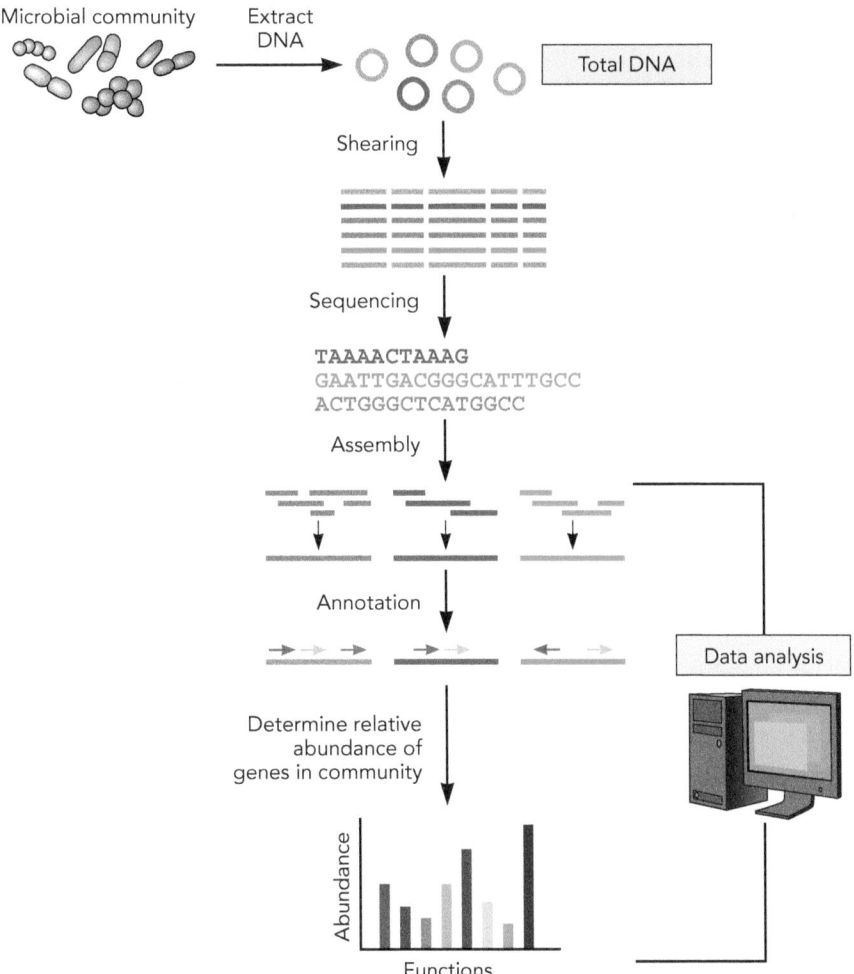

FIGURE 3.9 **Metagenomic sequencing.** Steps in the analysis of the total DNA within an environment (metagenome) are shown. After obtaining a sample from the desired environment, cells are lysed, and total DNA is extracted. The DNA is sheared into smaller fragments and sequenced. Large fragments or even complete composite genomes are computationally assembled, and open reading frames (genes) are predicted (indicated by small colored arrows). Comparison of the deduced amino acid sequences of the identified genes to sequences in databases allows the prediction of protein functions. Gene abundances are often correlated with the presence and importance of particular functions/processes in that environment.

genes in intergenic regions, such as sRNAs, were missed. Newer next-generation sequencing (e.g., 454 or Illumina sequencing) approaches permit researchers to analyze the genome and/or transcriptome of any organism. The ability to acquire a genome sequence for a typical bacterial species now takes a fraction of the time of older technologies, which took months to years to obtain, let alone annotate its content. For transcriptome (i.e., RNA-Seq) studies, genome sequences are not even required for analysis. In this technique, the total RNA in the organism is converted into complementary DNA (cDNA), which is then sequenced. The use of "barcodes" to tag the DNA or cDNA from a specific sample now allows for multiple genomes or transcriptomes to be sequenced simultaneously. In addition, microbial communities, or **microbiomes**, are now being studied using **metagenomics** (Fig. 3.9). In this method, total DNA is extracted from an environmental sample, such as a soil or water sample, instead of a pure culture. The DNA is randomly fragmented, sequenced, and assembled into larger fragments or even composite genomes. Annotation of community DNA sequences can provide information about the genetic potential of the community members. Finally, the field of metabolomics is now at the forefront of physiological studies, enabling the analysis of the small products or metabolites produced via the enzymatic reactions of the cell (made possible through the expression and activity of the proteome). The collective information obtained through these modern "-omics" approaches is providing new perspectives about the physiological capacity of microbes by linking the genome to metabolic outputs.

Learning Outcomes: After completing the material in this chapter, students should be able to . . .

1. estimate the physiologically relevant concentrations of macromolecules in the cell

2. describe how crude cell extracts and total protein concentration assays are used to determine the specific activity of an enzyme

3. list the six classes of enzymes and their functions

4. determine the V_{max} and K_m from a Michaelis-Menten curve

5. explain the methodology of 2D SDS-PAGE

6. compare and contrast the roles of DNA and RNA in a cell

Check Your Understanding

1. Define this terminology: molarity, Avogadro's number, lysozyme, specific activity, Michaelis-Menten kinetics, binding affinity, proteome, genome, transcriptome, metabolome, isoelectric point (pI), pseudogene, microbiome, metagenomics

2. Based on the topics covered in Chapter 2, what central metabolic pathways and their precursor metabolites are important for the synthesis of proteins and nucleic acids?

3. You have a newly isolated Gram-negative bacterium that produces a unique temperature sensitive amylase, which is capable of degrading starch molecules. Because amylase is a very useful industrial enzyme, you are eager to study the properties of the enzyme. Use the values in the text to calculate your answers.
   a) Based on previous research, you estimate that the concentration of amylase in your new isolate is 1 µM. Assuming the volume of each cell is comparable to that of an *Escherichia coli* cell, about how many amylase molecules are there per cell? (LO1)

   b) To study the activity of this amylase, you make a crude cell extract. Describe the steps needed to obtain a crude cell extract. (LO2)

   c) What role do French press and sonication play in this process? How does each work to accomplish this role? (LO2)

   d) Once you have your crude cell extract, you carry out a Bradford assay and measure absorbance at 595 nm. Using the standard curve shown, estimate the protein concentration in the crude cell extract if the absorbance is 0.8. (LO2)

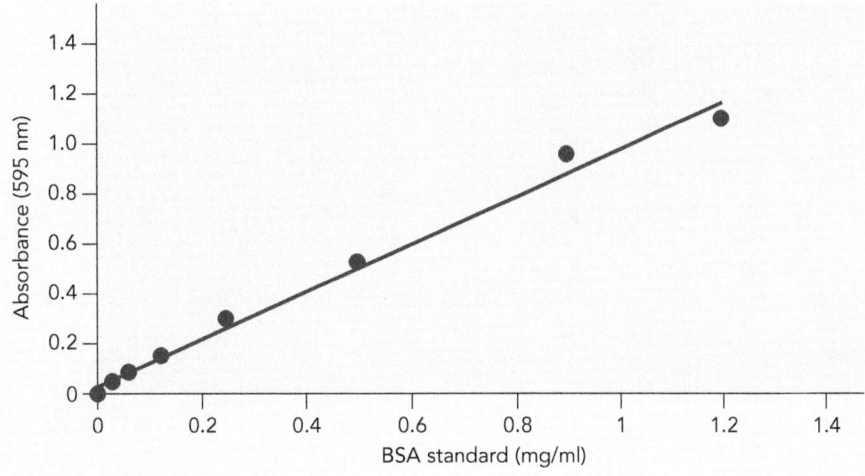

4. You can use the total protein in the crude cell extract to estimate the specific activity. (LO2)

   a) Why is it important to determine the specific activity of an enzyme?

   b) Describe two methods that would allow you to measure the rate of production of a product or consumption of a reactant, to determine specific activity.

   c) In your experiment, 4 ml of your crude cell extract converted 900 µmoles starch to product in 3 minutes. Use the value for protein concentration in the crude extract (mg/ml) from question 3 to calculate the specific activity.

5. What are the six classes of enzymes? Briefly describe the reaction carried out by each class. (LO3)

6. Circle the correct words to accurately complete the following sentences.

   a) Lactate dehydrogenase, an important enzyme for fermentations (which we will discuss in Chapter 6), is an example of (a transferase / an oxidoreductase) enzyme because it uses NADH + H⁺ to (oxidize / reduce) pyruvate to lactate.

   b) Lysozyme, which can be used to help break open cells to produce crude cell extracts, is an example of a (lyase / hydrolase) enzyme because it catalyzes the cleavage of peptidoglycan in bacterial cell walls using (water / NADH+ H⁺).

7. In the Michaelis-Menten plot of enzyme kinetics, the reaction rate is plotted as a function of substrate concentration. (LO5)

   a) What is the definition of a V_{max}?

   b) How is the V_{max} related to substrate concentration?

   c) Estimate the V_{max} using the given example plot.

   d) What is the definition of a K_m?

e) Estimate the K_m using the given example plot.

f) Why does the curve of the graph level off and reach a plateau as the substrate concentration increases?

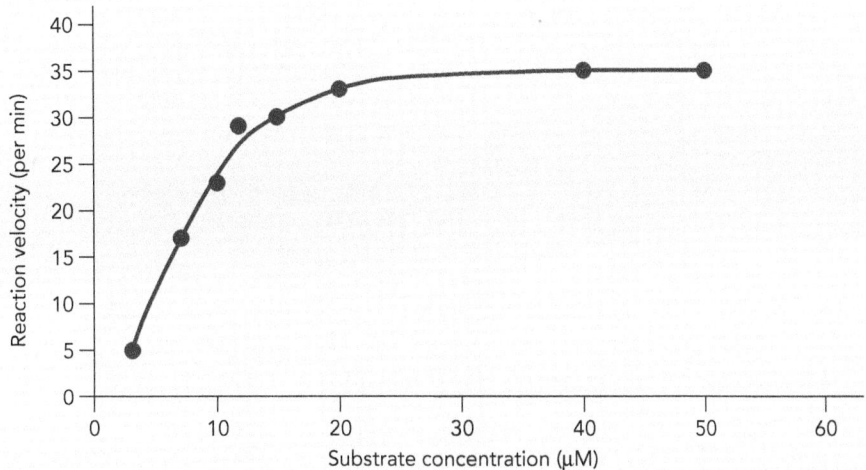

8. a) The amylase you are studying (from question 3) may be visualized as part of the total proteome using 2D SDS-PAGE. (LO5)

    This method first separates proteins on the basis of their _____

    then it separates proteins on the basis of their _____.

   b) List three general kinds of information that you can obtain from an analysis of whole bacterial cells on 2D gels. (LO5)

   c) Based on 2D gels, the average size of bacterial polypeptides is 36.3 kDa. What would the average size of a gene be in this organism? (LO1)

   d) How many molecules of a 36.3-kDa polypeptide would be present in each bacterial cell if the protein concentration was 0.2 µM? (LO3)

9. The purpose of completing the information table below is to help you organize important information relative to the role of the various nucleic acids in the cell. There can be different kinds of each type of nucleic acid in the cell; please fill in the names of each type of molecule where indicated. (LO6)

Nucleic acid	Function in cell	Molecule names and details	The molecule is relatively stable/unstable because:
DNA		DNA is organized in the cell as: 1. 2.	
rRNA		There are 3 different _____ RNAs in a prokaryotic cell: 1. 2. 3.	
tRNA		There are about 60 different _____ RNAs because:	
mRNA		There are about 400 different _____ RNAs in the cell because:	

Dig Deeper

10. Approximately 3% of a bacterial cell total dry weight is tRNA. The average molecular weight of a tRNA molecule is 2.5×10^4 g/mole. Estimate the number of tRNA molecules that would be present in a single cell. (LO1)

11. Bovine serum albumin samples containing 0, 100, 200, 400, 600, and 800 μg protein/ml were used to create a standard curve. Using the Bradford assay, absorbance values at 595 nm were 0.09, 0.21, 0.43, 0.65, and 0.87, respectively. Graph the data to create the standard curve. Use the curve to estimate the concentration of protein from a crude cell extract that, when measured in a spectrophotometer at 595 nm, gave an absorbance reading of 0.75. (LO3)

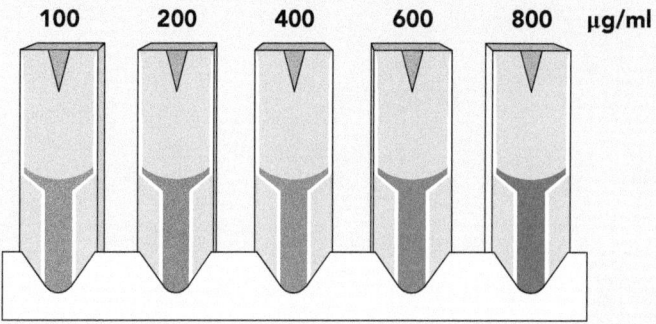

12. a) Fill in the table to compare and contrast what can be learned from different "-omics" studies. (LO6)

Method	Information obtained	Limitations
Genome sequencing		
Transcriptome sequencing		
Proteome analysis		
Metabolome analysis		
Metagenomic analysis		

b) Would there be differences in the data generated from genome, transcriptome, and proteome analyses if the cells were grown in two different environments? What information would you expect to change and why?

13. a) A newly isolated bacterial species was found to have a single chromosome with a molecular weight of 2.4 Mbp. 2D SDS-PAGE gels demonstrated that the average molecular weight of polypeptides in this species is 33 kDa. Assume that 10% of the genome encompasses noncoding sequences and 1% of the genome codes for stable RNA species. Using this information, estimate the number of coding sequences (genes) in the genome. (LO6)

b) Based on the genome size and the predicted number of encoded polypeptides relative to *E. coli*, what might you be able to infer about the lifestyle of this particular bacterial isolate?

14. a) Table 3.2 outlines an enzyme classification scheme. For each common enzyme type listed below, name the enzyme class (oxidoreductase, transferase, hydrolase, lyase, isomerase, ligase) to which it belongs.

- Nucleases
- Proteases
- Polymerase
- Kinases
- Phosphatases
- ATPases
- Lipases

b) For the reactions shown below, label each with the correct corresponding enzyme class (Table 3.2). (LO3)

Enzyme class	Reaction
	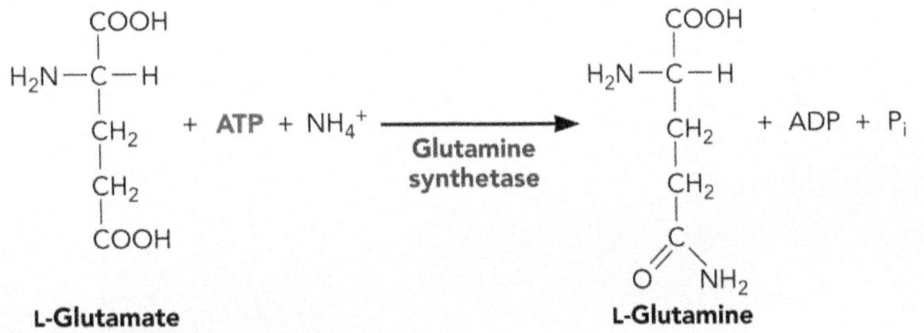 L-Glutamate + ATP + NH$_4^+$ → (Glutamine synthetase) → L-Glutamine + ADP + P$_i$
	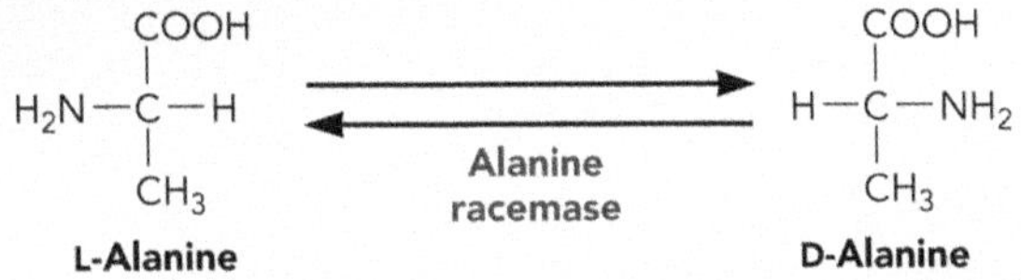 L-Alanine ⇌ (Alanine racemase) ⇌ D-Alanine
	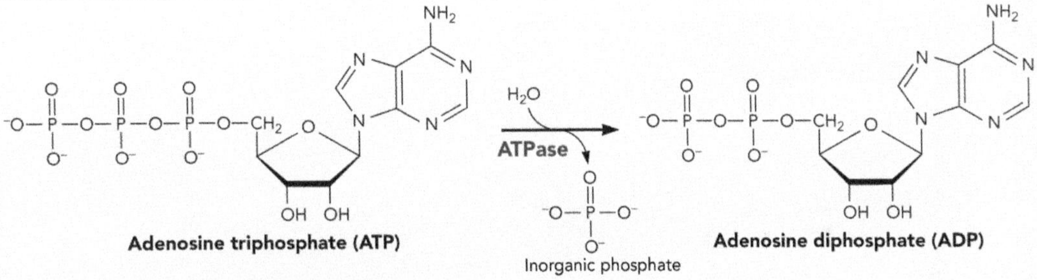 Adenosine triphosphate (ATP) + H$_2$O → (ATPase) → Adenosine diphosphate (ADP) + Inorganic phosphate
	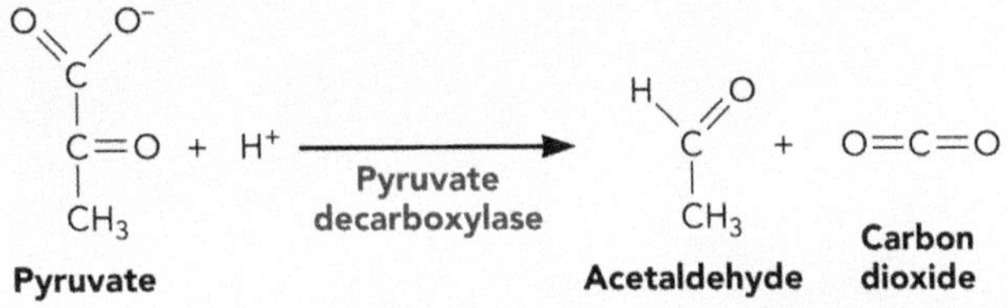 Pyruvate + H$^+$ → (Pyruvate decarboxylase) → Acetaldehyde + Carbon dioxide
	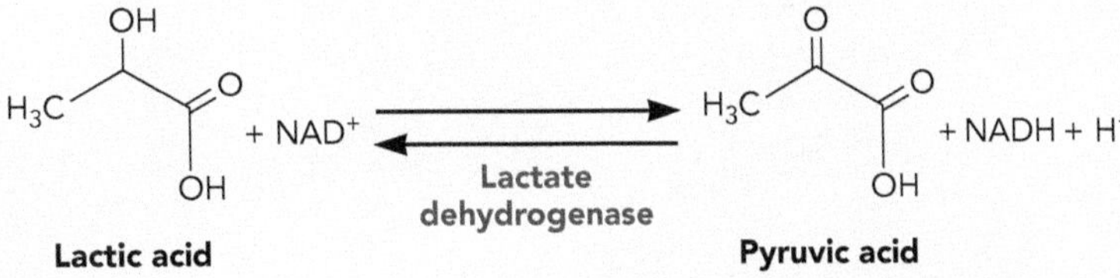 Lactic acid + NAD$^+$ ⇌ (Lactate dehydrogenase) ⇌ Pyruvic acid + NADH + H$^+$
	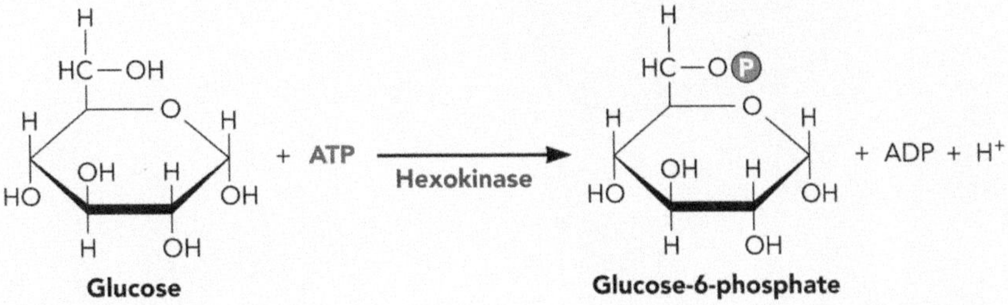 Glucose + ATP → (Hexokinase) → Glucose-6-phosphate + ADP + H$^+$

74 Making Connections

74 Methods to Monitor Bacterial Growth

78 The Phases of Bacterial Growth in Batch Culture

80 Requirements for Microbial Growth

80 Diauxic Growth

81 Exponential Growth Kinetics

83 Chemostats

84 Characteristics of Stationary-Phase Cells

85 Proteins Important for Cell Shape and Cell Division

86 Chromosome Segregation

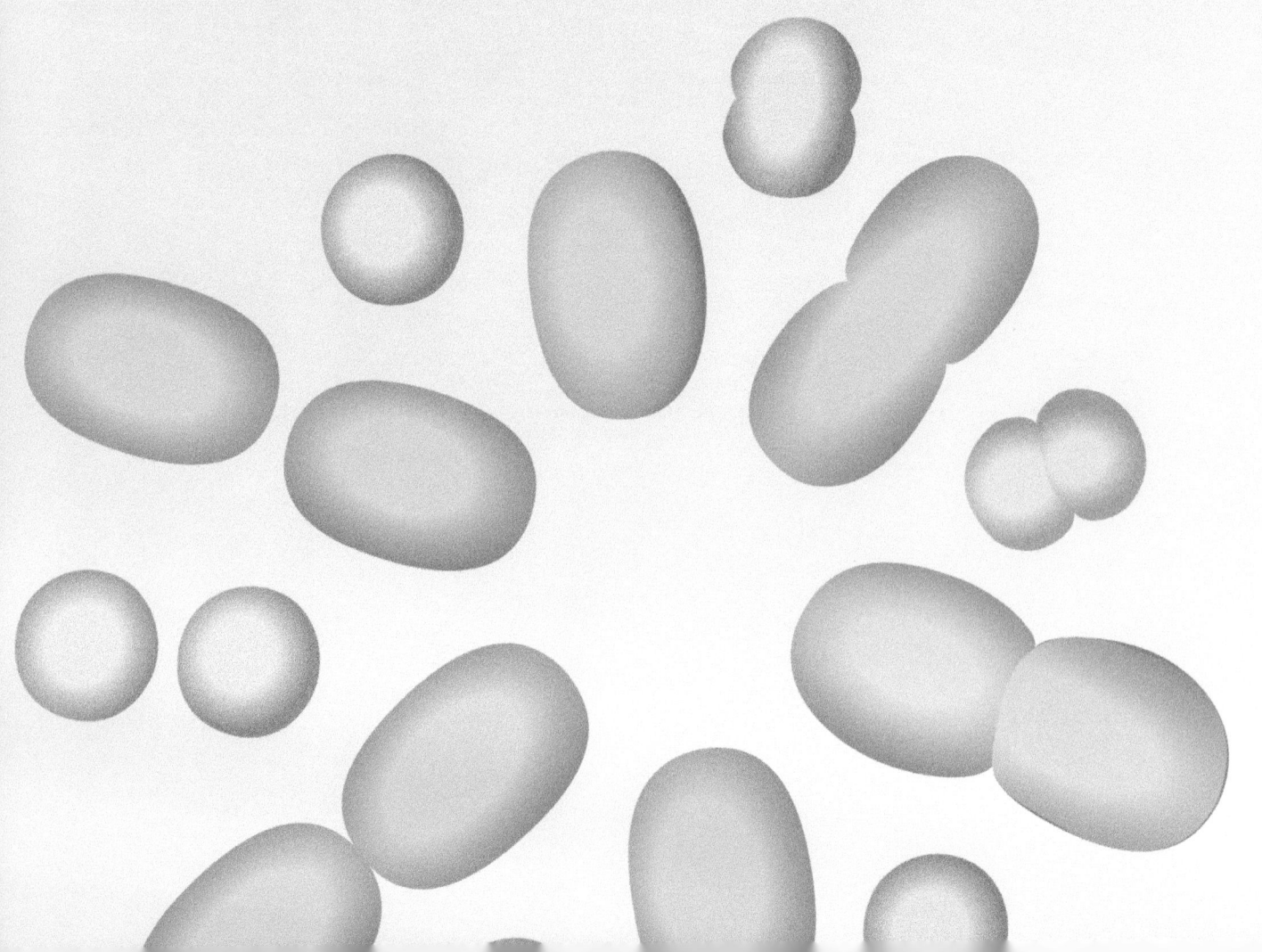

4

CELLULAR GROWTH

After completing the material in this chapter, students should be able to . . .

1. list the advantages and disadvantages of the methods used to monitor bacterial growth

2. compare the growth curves of bacteria growing with one carbon source versus two (i.e., diauxic growth)

3. calculate the bacterial growth rate in batch and chemostat cultures

4. differentiate the functions of complex versus defined media

5. discuss the characteristics of stationary-phase cells, including the stringent response

6. explain the role of proteins important for cell division, cell shape, and chromosome segregation

Microbial Physiology: Unity and Diversity, First Edition. Ann M. Stevens, Jayna L. Ditty, Rebecca E. Parales and Susan M. Merkel.
© 2024 American Society for Microbiology.
Companion website: www.wiley.com/go/stevens/microbialphysiology

Making Connections

The previous chapters focused on the metabolic pathways utilized by cells to produce the 12 essential precursors, which are used as the building blocks for the macromolecules and higher-order structures of microbial cells. In addition to building macromolecules and structures with important roles within individual cells, building cellular mass is important to support the propagation of new cells within a population. This chapter describes some of the common characteristics of cellular growth and reproduction among *Bacteria* and *Archaea*. In addition, a broad overview of the different laboratory techniques used to characterize bacterial population growth and the molecular basis for cell division are described.

Methods to Monitor Bacterial Growth

SPECTROSCOPY

Arguably the most common method to measure changes in bacterial growth involves the use of a spectrophotometer. In this instrument, a specific wavelength of light, the incident light (I_0), is passed through a bacterial culture. Some of this light will be scattered by the cells; the unscattered light (I) passes through the culture, exits on the opposite side, and is detected by a photocell (Fig. 4.1). As the number of cells in the culture increases, the bacteria deflect more light; as a result, less light passes through the sample for detection in the photocell. The units associated with this method of measuring growth can be turbidity, absorbance, or optical density (OD), which are calculated as the log I_0/I, and these measurements increase proportionally with the increase in cell density. The advantage of this method is that it is very fast, and it permits the observation of bacterial growth phases in liquid cultures in real time. However, at very high cell density the relationship between the cells and unscattered light deviates from linearity due to multiple deflections of the same photon of light as it hits more than one cell, causing an increase in unscattered light measured by the photocell (Fig. 4.2). Thus, a disadvantage of using a spectrophotometer is that it underestimates the cell density of dense bacterial cultures. Another limitation of this method is that it cannot distinguish between live and dead cells, as both will still impact the ability of the light to pass through the sample. In addition,

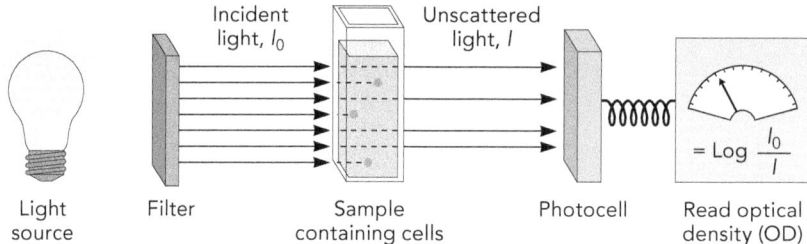

FIGURE 4.1 **Measurement of bacterial growth using a spectrophotometer.** Light from a source is filtered at a select wavelength, and this incident light (I_0) is passed through a sample containing cells. The unscattered light (I) is sensed by a photocell and then read as the optical density (OD), which is equivalent to the log I_0/I.

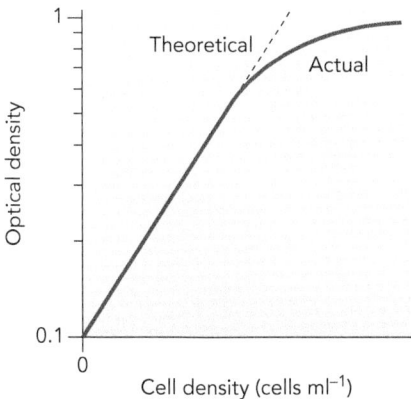

FIGURE 4.2 **Relationship between cell density and optical density.** There is a linear relationship between the cell density and optical density up to an OD of ~0.7. At higher cell densities, the actual number of cells measured by a spectrophotometer deviates from the theoretical number of cells due to photons of light being diffracted by cells more than once. This results in higher levels of unscattered light being sensed by the photocell and a lower OD reading, which leads to an underestimation of cell density.

since the spectrophotometer measures cell growth indirectly, it does not provide information about actual cell numbers unless a viable count (see below) is also performed so that cell numbers can be correlated with the OD measured by the spectrophotometer. Other limitations include the inability to count filamentous bacteria (e.g., streptomycetes) or bacteria that tend to clump in liquid cultures or when using a growth medium that is already turbid due to insoluble components (e.g., when cells are growing on polymers such as cellulose).

DIRECT COUNTS

To visualize and count individual cells, as well as determine the total population number, a **direct count method** can be employed using a phase-contrast microscope. A counting chamber (e.g., a Petroff-Hausser chamber or hemocytometer; Fig. 4.3) that holds a known volume per grid marking, etched on the slide counting chamber, is often used. A sample liquid culture is added to the chamber, and then, by counting the number of cells per grid, the total bacterial numbers in the culture per unit volume can be calculated. This method gives immediate results but it requires access to a phase-contrast microscope. Cell viability is not

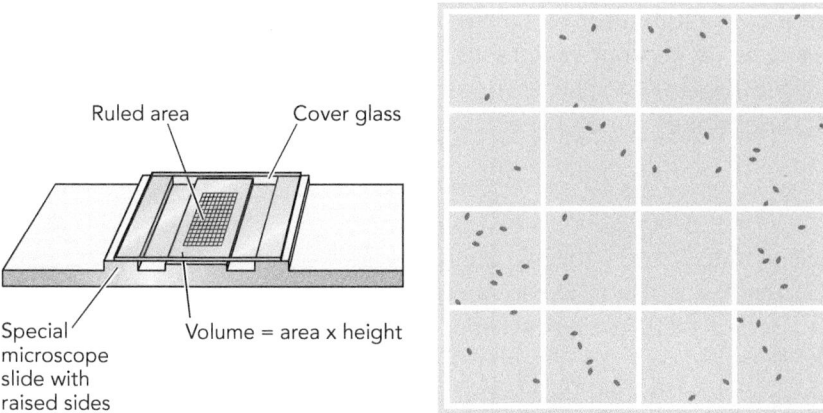

FIGURE 4.3 **Direct microbial cell counts.** A Petroff-Hausser counting chamber or hemocytometer is a specialized microscope slide with a gridded or ruled area. Each grid square holds a known volume; therefore, directly counting the cells within a certain number of squares with the aid of a microscope gives an estimate of cell numbers per unit volume.

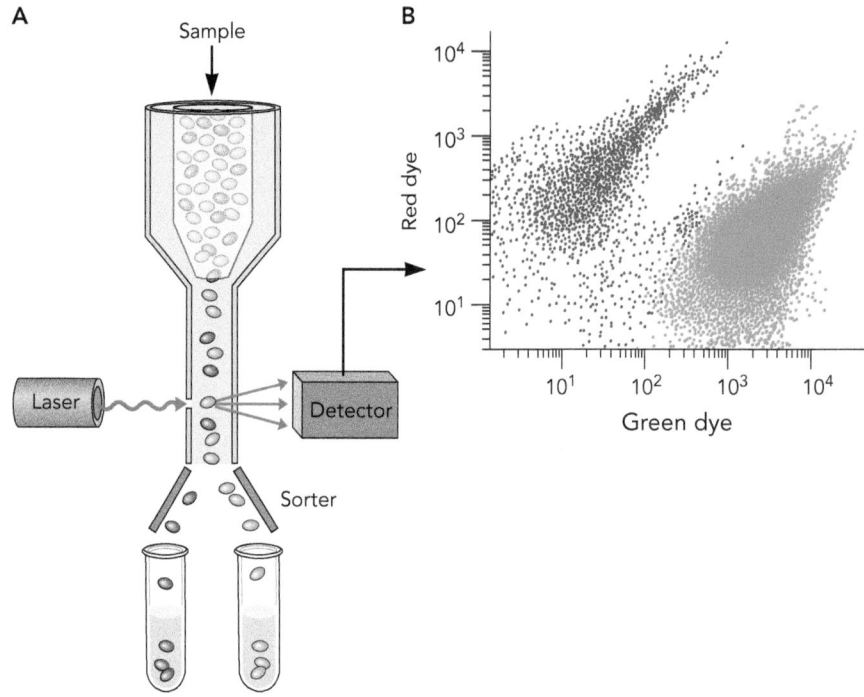

FIGURE 4.4 **Flow cytometry.** (A) In this example, microbial cells producing a fluorescent protein pass through a chamber in single file. A laser beam of a specific wavelength causes the fluorescent proteins to emit light, which is monitored by a detector. In this manner, cells with distinctive characteristics (e.g., size or fluorescence output) can be counted and sorted into separate populations. (B) These data can then be graphed to show distinct populations of cells.

apparent unless a stain is used to differentiate between live and dead cells based on integrity of the cell membrane. Alternatively, certain organisms that exhibit autofluorescence (e.g., methanogens and cyanobacteria) can be counted using a fluorescence microscope. Fluorescent *in situ* hybridization (FISH) can also be used to count specific genera/species based on fluorescent primers designed to anneal to DNA sequences within the cells. Disadvantages of direct counting methods are that they do not work well with very dilute cultures (although samples can be concentrated by filtration). If there are too few cells per grid, it is not possible to obtain statistically accurate counts. In addition, it is difficult to count highly motile cells. Electric (or automated) counting of cells using a cell sorter is also an option, but this is not as commonly used due to equipment costs (Fig. 4.4).

VIABLE COUNTS

The **viable count method** permits the enumeration of cells based on the growth of **colony-forming units (CFUs)**, each of which is assumed to have arisen from one bacterial cell. In this method, a liquid bacterial culture is serially diluted and then samples are either spread onto the surface of an agar-based medium or added to molten agar, which is then poured over an agar-based medium. The goal is, after dilution, to achieve a statistically reliable number of between ~30 and 300 colonies on a plate (Fig. 4.5). The advantage of this method is that it gives information about the number of viable bacterial cells within a population. As mentioned above, this information can be correlated

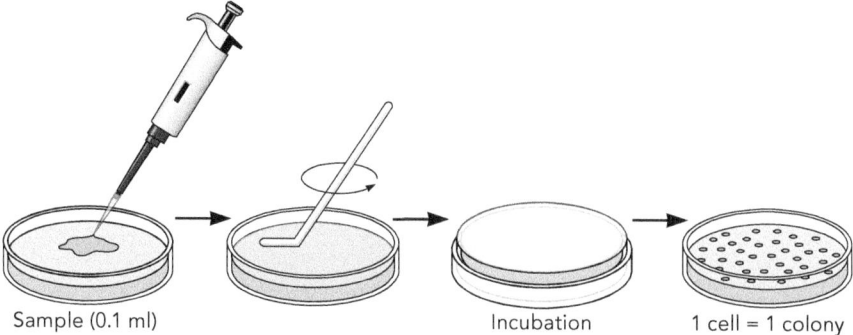

Sample (0.1 ml)   Incubation   1 cell = 1 colony

FIGURE 4.5 **Viable microbial cell counts.** A sample of cells is subjected to serial dilution and then a small aliquot (e.g., 0.1 ml) is spread onto the surface of a standard Petri dish containing an agar-based medium. Each colony that arises after incubation is assumed to have resulted from the presence of one cell that grew and divided at the location where the colony formed. Plates with between ~30 and 300 colonies on their surface are considered to provide accurate cell numbers when multiplied by the dilution factor (inverse of the dilution level) for the sample spread on the plate, resulting in a calculation of colony-forming units (CFU) per unit volume. For example, if 0.1 ml of a 10^{-6} dilution resulted in a count of 36 colonies, the concentration of cells in the sample would be 3.6×10^7 CFU/ml.

with spectrophotometer OD readings to generate a standard curve from which viable cell numbers at different OD values may be extrapolated. The disadvantage of this approach is that it is slower than the approaches listed above, as time must be allotted for the growth of colonies on the plates. In addition, sometimes the cells being plated clump together, with several cells forming a single colony, which violates the 1 cell = 1 colony assumption and leads to an underestimation of cell numbers. A second important issue to consider is plating efficiency, as environmental samples often contain organisms that are viable, but are in a nonculturable physiological state. The final issue is determining the correct growth conditions for the microbes. It is estimated that only 3 to 5% of bacteria can be grown in the laboratory. Sometimes this is due to the fact that bacterial metabolism in a community is linked between two or more different organisms, a topic discussed more in the "Diversity" section chapters in the second part of the book.

DRY WEIGHT AND PROTEIN CONCENTRATION

The next two methods of measuring bacterial growth are less commonly applied than the three described above, but they are useful under specific conditions especially with filamentous microorganisms where it is difficult to discern individual cells. Total dry weight may be used to directly determine the mass of a population of cells after all of the water is removed in a drying oven. This method is very slow and is therefore rarely used. Another, faster method to measure growth uses total protein concentration measurements (Chapter 3). However, the protein concentration per cell is dependent on the growth rate. Growth rate in turn is dependent on the number of ribosomes per cell, as more ribosomes are present in faster-growing cells. Since each ribosome contains proteins, this also impacts the total protein pool. A starving cell may even catabolize proteins, and thus the protein concentration also varies across the different stages of bacterial growth, which are described below.

The Phases of Bacterial Growth in Batch Culture

Most *Bacteria* and *Archaea* reproduce through a process called **binary fission**, whereby one cell duplicates its DNA, doubles in size, and then divides at the medial axis into two equivalent cells at each generation (Fig. 4.6). This binary growth pattern produces an exponential rate of growth on a log base 2 scale. The unit of growth from either direct or indirect methods (e.g., mass, cell number, or OD) is plotted on the *y* axis on an exponential scale versus time on a linear *x* axis. Such a semilog graph typically reveals four stages of growth when cells are grown in a batch culture: lag phase, exponential phase, stationary phase, and death phase (Fig. 4.7). On a semilog plot, the exponential stage of bacterial growth is represented as a straight line, with the slope of that line being the rate of growth or doubling time. The growth curve of a faster-growing organism will have a steeper slope, and conversely, that of a slower-growing organism will have a more gradual slope.

Batch culture occurs whenever the bacteria are placed in an environment with a limited concentration of nutrients that becomes depleted as the population grows over time. When a batch culture is started, the cells will be in lag

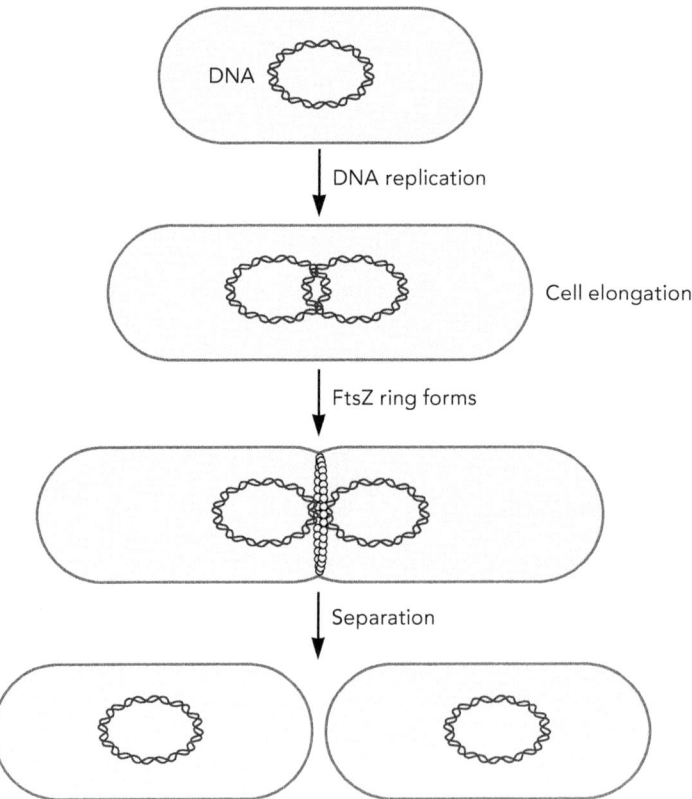

FIGURE 4.6 **Bacterial growth by binary fission.** During binary fission, one cell will divide into two equivalent cells at each generation. An individual bacterial or archaeal cell will replicate its genome while the cell grows in mass and length. When the genome has been duplicated and the cell reaches a sufficient size, an FtsZ ring will form at the medial axis (see Fig. 4.11 for more detail). As the FtsZ ring constricts, a septum will form, and this leads to the creation of two distinct cell walls and separation of the two daughter cells at each generation, resulting in logarithmic base 2 growth.

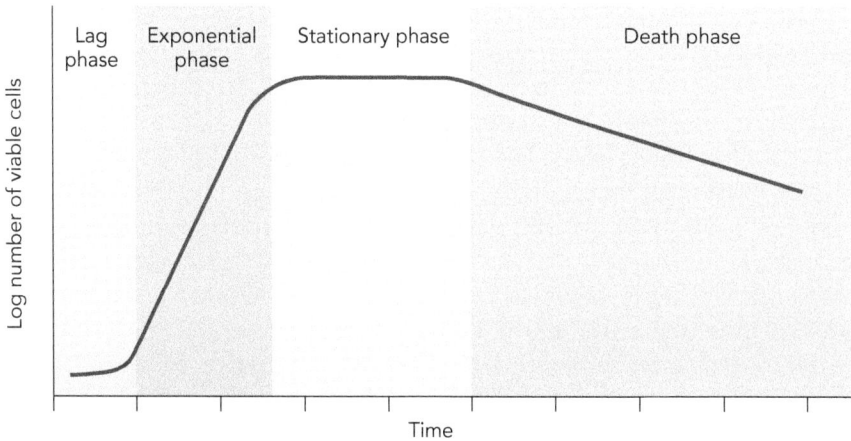

FIGURE 4.7 **A semilog plot of bacterial growth over time.** A measure of bacterial growth (e.g., OD, cell number, or cell mass) is plotted on an exponential (logarithmic) scale on the *y* axis against time using a linear scale on the *x* axis. Plotting these data generates a growth curve with a linear slope during the exponential phase from which the growth rate or doubling time can be determined. A shallow slope indicates a slower growth rate, whereas a steep slope indicates a faster growth rate. There are four distinct phases of bacterial growth that may be observed in a batch culture: the lag, exponential, stationary, and death phases. The graph of the growth curve shown in this example uses viable cell counts on the *y* axis. See the text for details.

phase if they are being moved to new environmental conditions (e.g., different carbon source or growth medium). During lag phase, the cells do not appear to be growing, but they are metabolically active and are synthesizing the appropriate metabolic enzymes and/or making other physiological adjustments appropriate for the new growth environment. For example, it is a common practice to store bacterial cultures in a −80°C freezer or in liquid nitrogen where the bacteria are in suspended animation in 20 to 25% glycerol, which prevents damage from ice crystal formation. These cultures must first acclimate to the new growth conditions before exponential growth can occur. If an older culture is taken from a Petri dish or broth culture, it too must adjust to fresh medium and a lag phase will occur. A lag phase is also observed in environmental samples being grown for the first time in laboratory conditions. Conversely, subculturing actively growing cells (i.e., cultures in exponential phase) into the same growth conditions, the cells will simply bypass the lag phase and continue with exponential growth. In contrast to growth in batch culture, a chemostat is an instrument that can be used to maintain bacterial cells in continuous active, exponential growth for long periods of time (see below).

Exponential growth occurs in batch culture when the cells are dividing via binary fission without nutrient limitation. This is considered a state of balanced growth in which a constant rate of cell division occurs (i.e., **steady state**). It is often desirable to study microbes under exponential growth conditions because the population is very homogenous, resulting in more consistent measurement of physiological parameters. The rate at which an organism grows is dependent on the physiology of the organism as well as the growth medium and environmental conditions. In the laboratory, some organisms divide every 10 minutes (e.g., vibrios), while others can take hours or days to divide (e.g., nitrifying bacteria). The amount of time it takes for the number of cells to double is called the **generation time** or **doubling time**. Again, the slope of the line on a semilog plot of growth versus time represents the growth rate or doubling time of an organism, and a greater slope correlates with a faster growth rate and shorter doubling times.

Stationary phase follows the exponential phase in batch culture. In stationary phase, there is no net growth, which is represented by a slope of zero in the semilog plot. By stationary phase the bacteria have used up one or more essential nutrients in the culture medium and can therefore no longer maintain an exponential rate of growth. For this reason, stationary phase is often considered to be a better representation of "real world" growth conditions for microbes. Microbes in the environment commonly have a feast-or-famine existence where short bursts of growth may be followed by long periods of starvation. In addition, waste products that can inhibit growth will build up in batch cultures. During stationary phase, the rates of growth and death are roughly equivalent; however, when the rate of death exceeds the rate of growth, then death phase occurs (Fig. 4.7). Not all bacteria lyse when they die, so the optical density may remain high even though the viable cell counts will drop.

Requirements for Microbial Growth

As introduced above, microbial growth is dependent on the existing environmental conditions. In laboratory conditions, this means that efforts are made to provide all of the essential components in the growth medium that are required by the cells. Unfortunately, we do not have the knowledge to grow the vast majority of the microbes on Earth in laboratory conditions. Usually, efforts to grow a heterotrophic organism, requiring a reduced carbon source, will start with a **complex** or **rich medium**, which is typically made from an "extract" of something organic (e.g., yeast cells, meat, or protein sources). This produces a rich medium with sufficient organic carbon, available nitrogen, and other nutrients that enable the growth of some of the most-studied model microbes in the laboratory. To perform experiments aimed at better understanding the physiological capacity of a microbe, specifically with regard to metabolic capabilities, a **defined** or **synthetic medium** may be used instead of a complex medium. In a defined medium, all of the components in the medium and their concentrations are precisely known. Many defined media are based on a **minimal medium,** composed of a buffer (e.g., phosphate buffer, which is also used as a phosphorus source) with carbon, nitrogen, sulfur, and trace element sources added as needed. $(NH_4)_2SO_4$ (ammonium sulfate) is commonly used as a source of both nitrogen and sulfur. However, organisms with specialized metabolic capacities may not need all of these components. For example, autotrophs are capable of CO_2 fixation (Chapter 9) and do not require the presence of an organic carbon source. Similarly, nitrogen fixers (Chapter 12) can use atmospheric N_2 as their nitrogen source. In addition, some organisms require specific growth factors, which are premade organic molecules (typically vitamins, amino acids, purines, or pyrimidines) that not all bacteria are capable of synthesizing.

Diauxic Growth

Diauxic growth is a variation of batch culture growth that occurs when a defined medium containing two different carbon sources is used. During diauxic

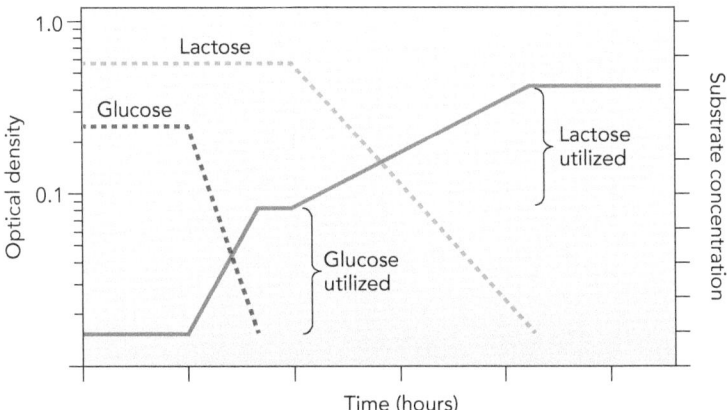

FIGURE 4.8 **Diauxic growth of** *Escherichia coli* **in medium containing two carbon sources: glucose and lactose.** The solid blue line indicates the phases of bacterial growth and the dashed lines represent the concentrations of the carbon sources as indicated. An initial lag phase is followed by an exponential phase of growth as the bacteria exclusively catabolize glucose. After the glucose is depleted in the medium, the cells undergo a second lag phase as they express genes for lactose catabolism and then a second exponential phase as they utilize the lactose. The presence of glucose in the medium results in the phenomenon of inducer exclusion and catabolite repression (Chapters 7 and 8).

growth, there are two lag phases and two exponential phases (Fig. 4.8). For *Escherichia coli*, diauxic growth occurs when the medium contains glucose (a preferred sugar) and lactose (a less preferred sugar) as the two different carbon sources. Under these growth conditions, the cells will completely consume the glucose prior to utilizing the lactose. During the first lag and exponential phases, the *E. coli* cells prepare to consume glucose and then metabolize it as the preferred carbon source. Subsequently, in the second lag and exponential phases, the enzymes necessary to degrade lactose are produced and then lactose is metabolized. The regulatory processes involved are known as inducer exclusion and catabolite repression (Chapters 7 and 8). And while glucose is the preferred carbon source for *E. coli*, this is not necessarily the case for other bacteria. For example, *Sinorhizobium* preferentially consumes succinate over glucose.

Exponential Growth Kinetics

A plot of the **growth rate (k)** for an organism versus the available substrate concentration produces a curve that resembles the kinetics of an enzymatic reaction (Fig. 4.9), indicating that the overall growth rate is limited by the kinetics of one key protein or enzyme involved in substrate utilization. Quite commonly the limitation is due to a transporter or permease protein that brings a substrate into the cell prior to metabolism. There are three formulas that can be used to define bacterial growth parameters during the exponential phase of growth, which, as mentioned above, is commonly used in the laboratory for physiological assays (Table 4.1). The first formula, $n = t/g$, indicates that the number of generations (n) is equivalent to the time that has elapsed (t) divided by the generation time (g). The second formula, $X = X_0 2^n$, highlights the growth of bacteria via binary fission. The starting value of cells or biomass (X_0) doubles every generation. The third formula demonstrates the inverse relationship between the generation time (g, unit of time per generation) and the growth

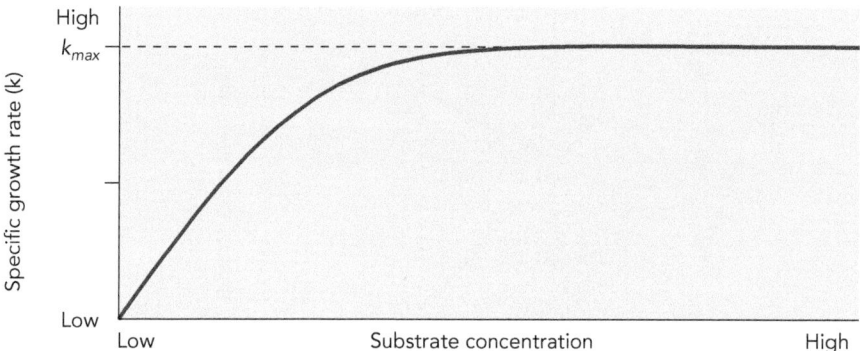

FIGURE 4.9 **The relationship between growth rate and substrate concentration.** The growth rate (represented by k) of a microbe will increase as the substrate concentration increases until a maximum growth rate (k_{max}) is reached. This curve reflects the Michaelis-Menten kinetics (Chapter 3) of a rate-limiting metabolic protein, often a transporter.

TABLE 4.1 **Exponential growth kinetics**[a]

$n = t/g$
$X = X_0 2^n$
$g(k) = 0.693$

[a] X = material that doubles each generation, X_0 = starting value of material, n = number of generations, t = time elapsed, g = generation time, and k = growth rate.

rate (k, generations per unit of time), with $g(k) = 0.693$. Thus, the growth rate k has units of inverse time (e.g., minute^{-1} or hour^{-1}). Growth rate calculations are useful for time management in the design of laboratory experiments to predict when cultures will be ready to test. An interesting exercise is to calculate whether or not just 100 *E. coli* cells growing exponentially for 48 hours and doubling every 20 minutes in rich medium at 37°C could take over the world.

$$n = t/g = 48 \text{ hours}/20 \text{ minutes per generation}$$

$$= 2,880 \text{ minutes}/20 \text{ minutes per generation}$$

$$= 144 \text{ generations}$$

$$X = X_0 2^n = 100 \text{ cells} \left(2^{144}\right) = 2.2 \times 10^{45} \text{ cells}$$

$$2.2 \times 10^{45} \text{ cells} \left(1 \times 10^{-12} \text{ ml/cell}\right) = 2.2 \times 10^{33} \text{ ml or } 2.2 \times 10^{30} \text{ liters}$$

The volume of Earth is ~3 × 10^{23} liters. Therefore, in theory the bacteria could take over the world! Fortunately, in reality this would never happen due to nutrient limitation and competition for space with other organisms.

Yield coefficient is another parameter used to measure bacterial growth efficiency. The yield coefficient (Y) equals the change in biomass divided by the change in substrate concentration. $Y = 0.5$ for *E. coli* growing aerobically, where 1.0 g glucose yields 0.5 g cells. $Y = 0.05$ for *E. coli* growing anaerobically, as the yield from 1.0 g glucose is only 0.05 g cells. Differences in ATP production under aerobic versus anaerobic conditions are critical to this difference (Chapter 5). Calculations of efficiency are vitally important to large-scale industrial growth of organisms in fermenters or chemostats.

Chemostats

Chemostats are used when steady-state microbial growth needs to be maintained and monitored over extended periods of time, in comparison to the shorter amounts of time associated with batch culture growth in liquid or on solid medium. Batch culture is considered to be a closed system with a limited amount of nutrients present. A chemostat is an open system since a source of fresh sterile medium is continuously added into the chemostat at a rate determined by the operator (Fig. 4.10). The overall system has a set volume, with the rate of fresh medium "in" being equivalent to the rate of overflow "out" of the chemostat. Hence, the flow rate of medium into the chemostat controls the growth rate of the organisms in the chemostat. There is one rate-limiting substrate in the medium that is almost instantaneously consumed by the growing organisms. When functioning properly, the system will achieve a state of balanced growth, which is an extended exponential phase. Thus, a chemostat can provide a long-term source of cells for physiological experiments. The mathematical formula $k = D = f/v$ indicates that the growth rate (k) is equivalent to the dilution rate (D) produced by the flow rate of the fresh medium (f) divided by the volume of the system (v). Since the volume is held constant, the flow rate directly controls the growth rate.

There are many practical applications of the chemostat including the following. A chemostat can be used to mimic "real world" nutrient-limited conditions and key parameters can be changed independent of the growth rate. For example, the same growth rate could be maintained at two different temperatures. A chemostat enables steady-state production of cells over long periods of time, in contrast to serial batch cultures that require repeated subculture steps. In addition, chemostats can be used to select for bacteria with certain physiological characteristics.

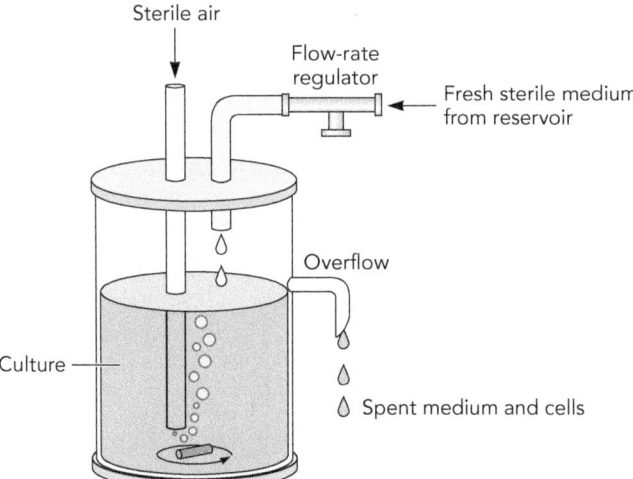

FIGURE 4.10 **Growth of microbes in a chemostat.** The growth of a culture in a chemostat is controlled by the rate at which one limiting substrate enters the growth chamber from the sterile medium reservoir. The growth chamber has a set volume; thus, as fresh medium enters the chamber, an equivalent amount of spent medium, including some cells, is removed in the overflow. The culture is held in a steady state when fresh medium is consumed by cells as it enters the chamber, leading to an extended phase of exponential growth. The rate of growth is controlled by the dilution rate (flow rate per volume) of the fresh medium entering the chamber.

As cells grow, they randomly acquire spontaneous mutations. By putting the bacteria under strongly selective growth conditions, those that acquire mutations increase their rate of growth under those growth conditions and will outcompete their neighbors. For example, to enrich for bacteria that could be used for bioremediation, cells could be grown in the presence of an herbicide or petroleum as a growth substrate, and organisms capable of degrading that compound will become dominant in the chemostat population. Organisms with naturally occurring mutations are considered safe for environmental release as opposed to genetically modified organisms that are artificially manipulated in the laboratory and whose use is tightly restricted by government policy. **Selective enrichment** is a process that is commonly used in both chemostats and batch cultures to select for the growth of organisms from environmental samples that have desired physiologies.

Characteristics of Stationary-Phase Cells

Although traditionally most physiological experiments are performed using exponential-phase cells, it is thought that most microbes in the environment exist in conditions more similar to those found in a stationary-phase batch culture. When bacteria face starvation, they make large-scale adaptive changes to their overall physiology. Some bacteria, such as *Bacillus* and *Clostridium* species, can form dormant endospores (Chapter 17). However, even nonsporulating cells like *E. coli* modify their physiology when they enter stationary phase. Cells that are normally rods become more coccoid in shape, and all types of cells decrease in size as they grow at a slower rate. This increases the surface area-to-volume ratio of the cell, giving it a better interface with the environment to obtain nutrients and eliminate waste products. Overall, the bacterial cell surface becomes more hydrophobic, which makes the cells more adhesive to surfaces (where nutrient concentrations can be higher) and to each other, causing clumping and microcolony formation important for biofilms (Chapter 18). There are also observed changes in the composition of membrane lipids that may be related to membrane fluidity. Cells strive to conserve energy by decreasing their rate of protein synthesis and anabolism, while simultaneously increasing the turnover of macromolecules such as proteins and RNA. On the other hand, there is increased production of proteins important for survival during the stress of stationary phase. Some of these are associated with nutrient acquisition (e.g., phosphate transport), while the stationary-phase sigma factor RpoS works with the core RNA polymerase (Chapter 7) to enhance transcription of >50 genes encoding stress-response proteins.

In *E. coli*, under amino acid starvation conditions, the cells will undergo a **stringent response**. If there are insufficient levels of amino acids, then there will be an insufficient number of charged tRNA molecules to continue the process of translation. Hence, the ribosomes will stall, and subsequently the ribosomal protein RelA synthesizes the **alarmones** guanosine tetraphosphate (ppGpp) and guanosine pentaphosphate (pppGpp), collectively known as (p)ppGpp. The (p)ppGpp serves as an effector for the transcription factor DskA, which in turn causes a decrease in the rate of transcription of the rRNA genes by RNA polymerase. This results in a decrease in the number of ribosomes and slowing of the growth rate. Carbon starvation will also trigger a decrease in ribosome synthesis through the activity of the bifunctional protein SpoT, which is capable of both synthesis and hydrolysis of (p)ppGpp, thus modulating (p)ppGpp levels in the cell. These mechanisms enable *E. coli* to decrease its growth rate under environmental conditions where low levels of nutrients are present, such as during stationary phase.

Proteins Important for Cell Shape and Cell Division

There are key structural proteins that control cell shape in bacteria. For example, MreB is an actin-like protein in *E. coli* that creates a helical intracellular scaffold upon which the peptidoglycan cell wall can be overlaid. This results in a bacillus or rod shape in bacteria possessing this protein. If MreB is rendered nonfunctional, *E. coli* cells will become spherical. Thus, the coccoid shape appears to be the default conformation formed by peptidoglycan in the absence of a scaffolding protein such as MreB. Curved rods such as vibrios and *Caulobacter* express a different structural protein known as crescentin. Crescentin is an intermediate filament-like protein that generates a characteristic bending of the rod. Proteins that determine cell morphology in bacteria are similar to those of eukaryotes in their fundamental functions.

Studies of cell division in eukaryotic cells preceded those in prokaryotic cells due to the generally larger size of eukaryotic cells and the idea that "simple" prokaryotic cells lacking organelles would not require a sophisticated process to divide. However, research has now demonstrated that cell division in bacteria is an intricate and tightly controlled process, and the proteins for cell division are similar to those found in eukaryotes. Since cell division is essential for growth, many of the proteins important for cell division are essential; therefore, conditional mutants are often used to study this process. A **conditional mutant** will behave as a wild-type cell at lower temperatures, but will present an observable phenotype at an elevated temperature. This is due to the fact that a mutant protein acquires a temperature-sensitive structure that becomes misfolded and nonfunctional at an elevated temperature. In *E. coli*, mutations in the gene encoding FtsZ (i.e., filamentous temperature sensitive) cause a filamentous or elongated cellular phenotype at elevated temperatures. When grown at 30°C, the mother cells will divide into two daughter cells. However, at 42°C, the cells are no longer able to divide and grow into long filamentous structures.

FtsZ is a tubulin-like protein that forms a ring around the medial axis of the bacterial cell at the point at which binary fission will occur. FtsZ recruits a number of other proteins to this location (e.g., FtsA, FtsK, FtsL, FtsN, FtsQ, FtsI, and FtsW) that collectively form a structure known as the **divisome** (Fig. 4.11). Why does the FtsZ ring form at the medial axis? In *E. coli*, the MinCDE proteins are thought to play a role. The Min proteins oscillate between the two poles of the cell, preventing the FtsZ ring from forming near the poles. Temperature-sensitive variants of the Min protein will produce two unequal-sized daughter cells at an elevated "nonpermissive" temperature. The result is that one daughter cell is much larger than the other. The smaller "mini" cells are often not viable, in some cases because they did not receive a chromosome. Min proteins are not universally found in all bacteria, and thus other mechanisms to control cell division must exist.

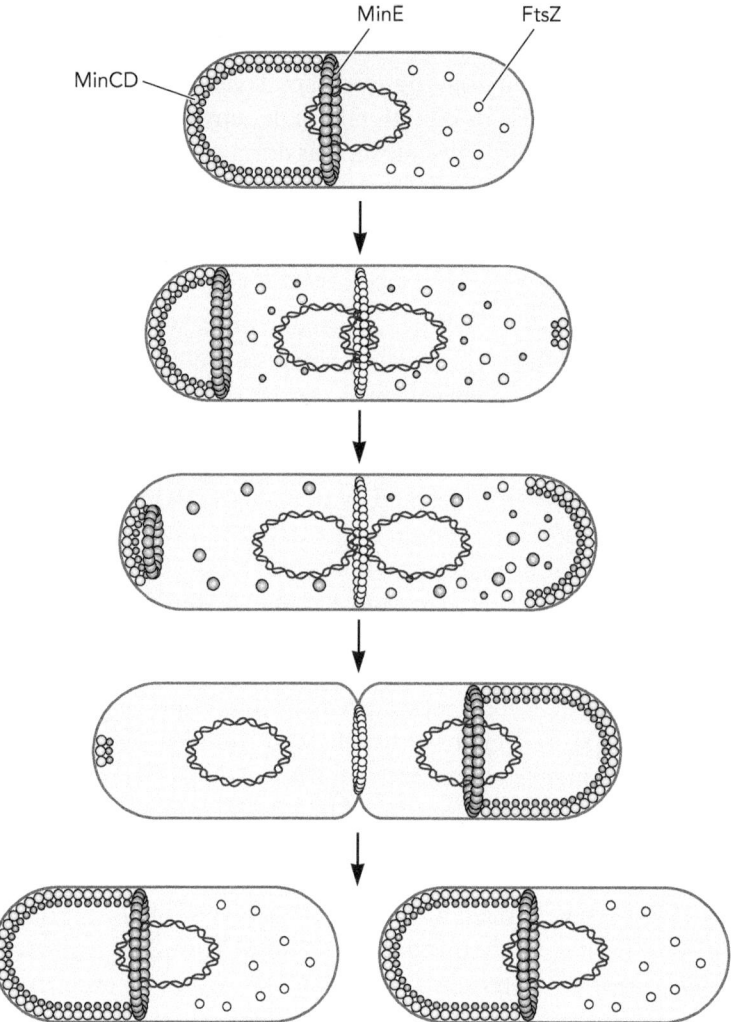

FIGURE 4.11 **Formation of the divisome during cell division in *Escherichia coli*.** In order for a cell to divide at the medial axis, a divisome protein complex must form in association with the FtsZ ring (shown as a white ring of FtsZ monomers at the cell center). The MinCDE proteins oscillate from pole to pole, preventing formation of the FtsZ ring at the poles. Only after chromosome replication is completed will the FtsZ ring form at the medial axis, leading to septum formation and the completion of cell division.

Chromosome Segregation

DNA replication must be synchronized with cell division to ensure that each daughter cell receives a complete copy of the genome, which is commonly a single circular chromosome in most *Bacteria* and *Archaea* (Chapter 3). DNA replication begins at the **origin of replication** (*ori*), and two replication forks will form and move in opposite directions around the chromosome as DNA polymerase duplicates the genetic information (Fig. 4.12). Replication will stop at termination sites (*Ter*) located on the opposite side of the chromosome from the *ori*. Some bacteria use redundant termination sites to guarantee that replication will appropriately end. During replication, the two copies of the chromosome are noncovalently linked together or catenated (i.e., like links in a chain).

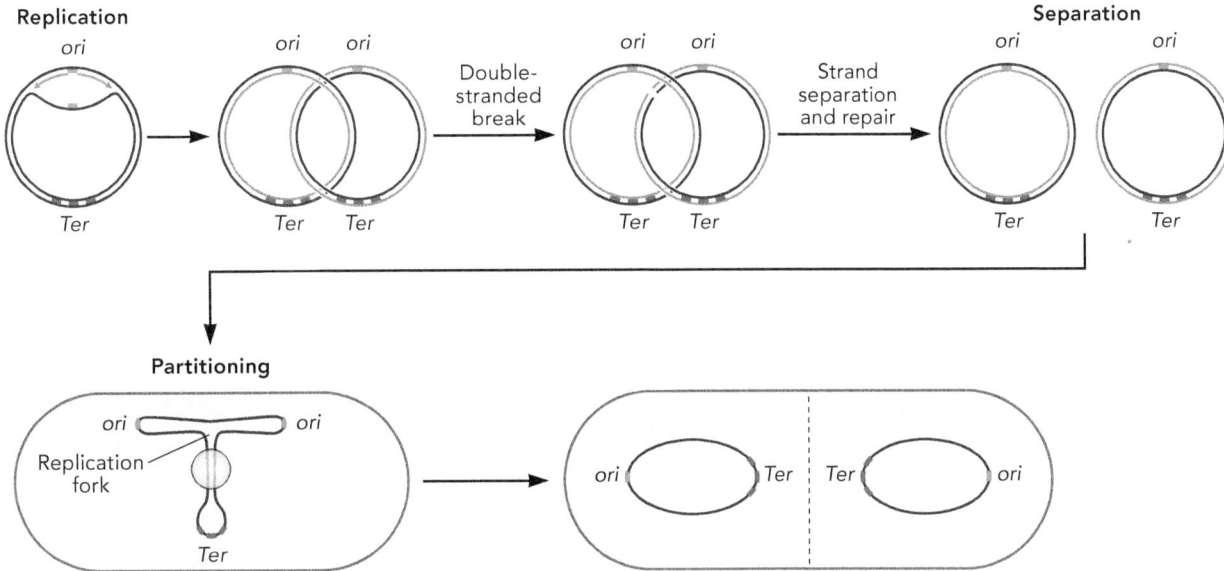

FIGURE 4.12 **Chromosome replication, separation, and partitioning.** Starting at the upper left, chromosome replication initiates at the origin of replication (*ori*) from two replication forks creating a characteristic "theta" structure until the termination (*Ter*) region is reached at the opposite side of the circular chromosome. The termination region contains multiple redundant *Ter* sequences (e.g., six in *E. coli*). The two duplicated chromosomes, now catenated, are separated from each other by the activity of type IV topoisomerase. Type IV topoisomerase catalyzes a double-stranded break in one chromosome, enabling the other chromosome to pass through. The double-stranded break is repaired to result in two fully separated chromosomes. During partitioning, the two chromosomes are replicated at the medial axis by the replisome (represented by the green circle) and become oriented such that the *ori* for each chromosome is directed toward one pole of the cell, which ensures partitioning of one chromosome to each daughter cell during cell division.

The process of chromosome separation is carried out by type IV topoisomerase, which generates a double-stranded break in the DNA of one chromosome, enabling the unbroken chromosome to pass through the gap, and then the broken chromosome is repaired. After DNA replication and chromosome separation, the two chromosomes must be partitioned to the two daughter cells. The mechanism of this process varies between organisms, but the replication machinery (i.e., replisome) often resides at the mid-cell position, first replicating the *ori* regions such that they are oriented toward the poles of the two daughter cells, enabling the correct DNA partitioning to occur and cell growth to continue.

Learning Outcomes: After completing the material in this chapter, students should be able to . . .

1. list the advantages and disadvantages of the methods used to monitor bacterial growth

2. compare the growth curves of bacteria growing with one carbon source versus two (i.e., diauxic growth)

3. calculate the bacterial growth rate in batch and chemostat cultures

4. differentiate the functions of complex versus defined media

5. discuss the characteristics of stationary-phase cells, including the stringent response

6. explain the role of proteins important for cell division, cell shape, and chromosome segregation

Check Your Understanding

1. Define this terminology: direct count method, viable count method, colony-forming units (CFUs), binary fission, batch culture, steady state, generation time (g) or doubling time, complex or rich medium, defined or synthetic medium, minimal medium, diauxic growth, growth rate (k), yield coefficient, chemostat, selective enrichment, stringent response, alarmone, conditional mutant, divisome, origin of replication

2. The purpose of completing the information table below is to help you compare the advantages and disadvantages of the methods used to estimate cell density over time. Please fill in the information as indicated. (LO1)

Method	Brief description	One advantage	One disadvantage
Spectroscopy			
Direct counts			
Viable counts			

3. Describe the defining feature of each growth phase in a batch culture: lag, exponential, stationary, death. (LO2)

4. You have a *Rhizobium* strain that prefers succinate to glucose as its carbon source. You grow it in batch culture in a medium with succinate and glucose and measure growth over time. You get the results shown. (LO2)

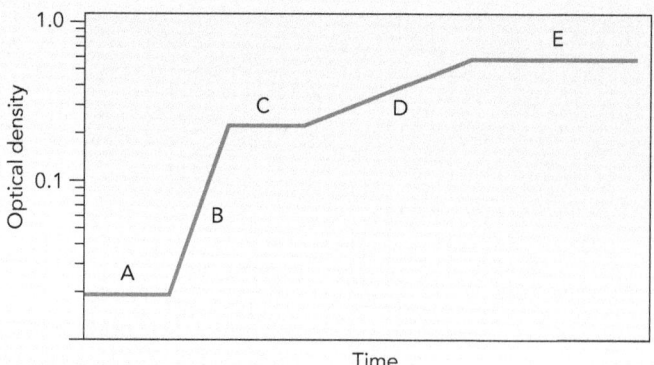

a) In which stage of growth (A to E) are the bacteria growing fastest?

b) In which stage of growth (A to E) does the medium contain the highest nutrient levels?

c) Why is there no increase in growth during A?

d) Where does the concentration of succinate approach zero? Explain how you know.

e) Why is there no increase in growth during C?

f) Where does the concentration of glucose approach zero? Explain how you know.

g) Which stage of growth (A to E) would be considered stationary phase?

5. It is often critical to know the bacterial growth rate in batch cultures. Calculate the answers to the following questions using these equations: (LO3)

$$n = t/g \left[\text{where } n = \text{number of generations}, t = \text{time that has elapsed, and } g = \text{generation time}\right]$$

$$X = X_0 2^n \left[\text{where } X_0 = \text{starting value of cell density or biomass and } X = \text{current cell density}\right]$$

$$g(k) = 0.693 \left[\text{where } k = \text{growth rate}\right]$$

a) You have a bacterial culture with a generation time of 10 hours. You let it grow for 15 hours. How many generations (doublings) will occur in this culture?

b) You are planning to run an experiment with the same culture that requires 100,000 cells/ml. If you start with a culture that has 12,500 cells/ml, how many generations will have to occur to get to 100,000 cells/ml?

c) How long will it take to get to 100,000 cells/ml?

d) What is the growth rate (k) for this culture if the generation time stays at 10 hours?

6. Instead of growing your cells in a batch culture every day, you decide to set up a chemostat. (LO3)

$$\text{growth rate } (k) = \text{dilution rate } (D) = \left[\text{flow rate of medium } (f)\right]/\left[\text{volume of the system } (v)\right]$$

a) Briefly describe the setup of a chemostat.

b) What are two advantages of growing cells in a chemostat relative to batch culture?

c) If the volume of your chemostat is 3 liters, what flow rate would you need to achieve a growth rate of 0.5 generations/hour?

7. Compare and contrast the ingredients and function of complex versus defined media. (LO4)

Medium A (per liter)	
Tryptone	5 g
Yeast extract	2.5 g
Glucose	1 g
Agar	15 g

Medium B (per liter)	
Na_2HPO_4	6 g
KH_2PO_4	3 g
NaCl	0.5 g
NH_4Cl	1 g
1 M $MgSO_4$	1 ml
0.1 M $CaCl_2$	1 ml
10% (w/v) glucose	10 ml
Agar	15 g

a) Of the media recipes that are listed, which one is complex, and which is defined? Explain your reasoning.

b) In each medium, what ingredient(s) provide carbon (C)? Nitrogen (N)?

c) Suppose you have a bacterium that can grow on both kinds of media; in which do you think the growth rate would be fastest? Explain why.

8. Answer the following questions about the stringent response in *Escherichia coli*. (LO5)
   a) What triggers the production of the alarmones, (p)ppGpp?

   b) In Gram-negative cells, (p)ppGpp binds to RNA polymerase and the regulatory protein DskA. How does this binding affect the transcription of rRNA?

   c) How can changes in the transcription of rRNA genes impact cell growth? Explain why.

   d) Explain how carbon starvation can impact ribosome production through the protein, SpoT.

9. Cell replication in prokaryotes has proven to be very complicated. (LO6)
   a) Conditional mutants have been used to study cell replication. Explain how a given protein can have different functions at different temperatures.

   b) What role does FtsZ play in cell replication? What evidence supports the hypothesis that FtsZ is involved in that role?

Dig Deeper

10. For each bacterial growth situation below, choose one method (spectroscopy, direct counts, viable counts) to determine the cell density and explain why you think your chosen method is better than the other methods. (LO1)
    a) You have a new bacterial isolate growing in pure culture in a defined medium, and you want to investigate the growth rate at different temperatures.

    b) You have several new bacterial isolates, each growing in pure culture. You want to determine whether a given antibiotic is bactericidal (kills the bacterium) or bacteriostatic (inhibits growth).

    c) You want to measure the effect of low nutrient conditions on cell size and density over time.

11. You have a Gram-negative bacterial isolate growing in a defined medium. You measure the cell density over time and notice that the cells clump together at high cell density. Consider the viable count method and the spectrophotometric method for determining cell density. If you used these cell density data to determine a growth rate, would the growth rate be accurate using these methods? If not, would it be higher or lower than the actual growth rate?

12. You have a new Gram-positive bacterial isolate with an optimal growth temperature of 32°C and a temperature range from 10 to 45°C. You grow the isolate in a defined medium with glucose at its optimal temperature. The data are shown in the graph. Under the following scenarios, predict whether each of these parameters (lag phase, slope during exponential phase, and final cell density) would likely increase, decrease, or stay the same if you changed the growth conditions as described. (LO2)

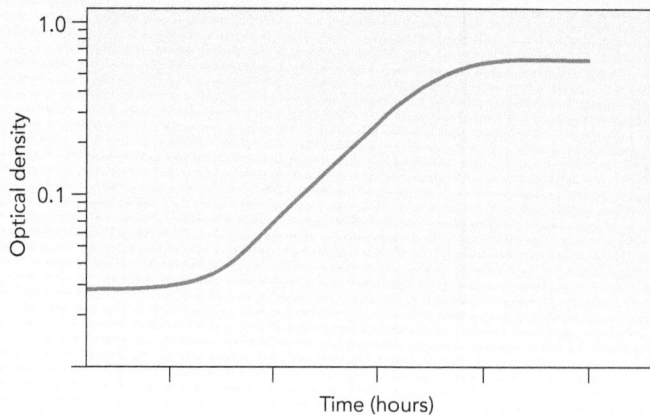

a) Predict what would change if the isolate were grown at 42°C instead of 32°C.

b) Assuming glycerol is used less efficiently, predict what would change if you switched the medium to have glycerol instead of glucose.

13. Refrigeration has dramatically improved the shelf life of perishable products, as lower temperatures slow or inhibit the growth rate of many pathogens. Unfortunately, *Listeria monocytogenes*, an important foodborne pathogen and the cause of listeriosis, is an exception. You do a series of growth experiments, growing *Listeria* at a variety of temperatures. You plot the generation (doubling) times versus temperature and get the results in the graph. (LO3)

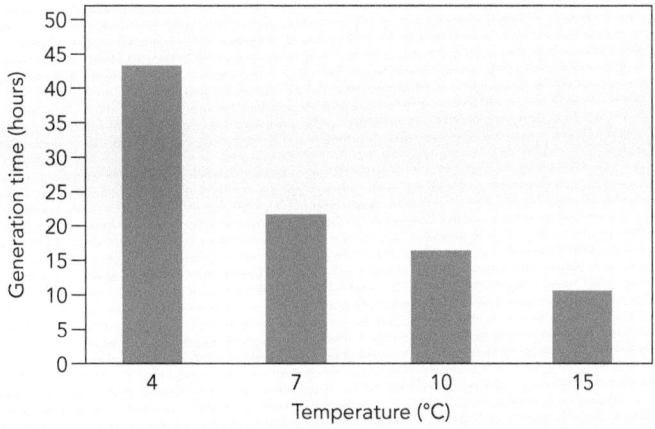

a) Which temperature allows for the fastest growth?

b) Given a generation time of 12 hours, calculate the growth rate.

c) If your yogurt were accidentally inoculated with 100 cells/ml of *Listeria* and put in the refrigerator at 4°C, how many days would it take to get to an infectious dose of 10^4 cells/ml?

14. *Lactococcus lactis* is a Gram-positive bacterium that is used worldwide as a constituent of starter cultures in the dairy industry. While *L. lactis* can grow using either glucose or cellobiose (a dimer of glucose; structure as shown), these bacteria prefer glucose, and the presence of glucose inhibits the production of enzymes required to use cellobiose. (LO2)

a) You grow a culture of *L. lactis* in a defined medium with glucose, then use those cells to inoculate a defined medium with both glucose and cellobiose. In the empty graph, draw in what you think the growth curve (cell density/time) would look like. Explain your reasoning.

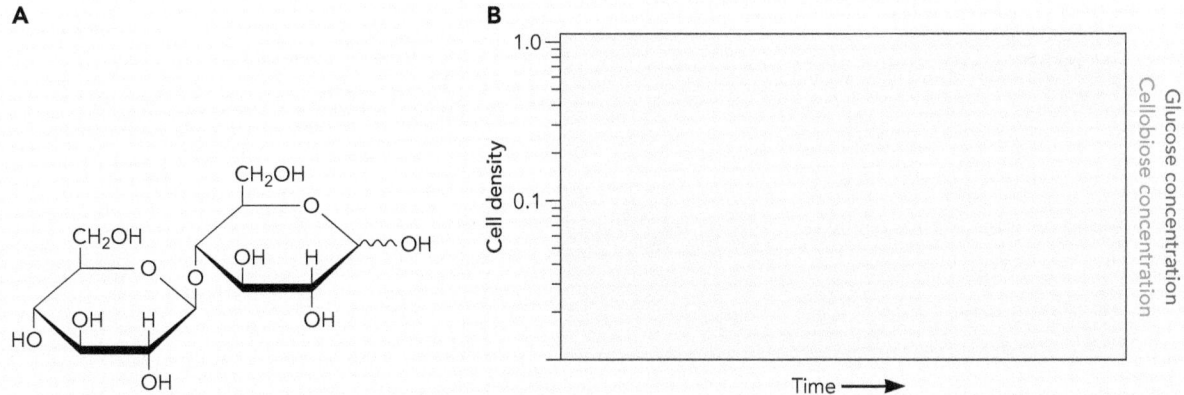

b) In the empty graph, draw in additional lines that show how the concentrations of glucose and cellobiose would change over time. Explain your drawing. (LO2)

15. Compare and contrast the ingredients and function of complex versus defined media. Consider Medium A and Medium B presented in question 7. (LO4)

    a) Which medium would you use for isolating new heterotrophs (i.e., organisms that require reduced carbon) from a sourdough sample? Explain your reasoning.

    b) Which medium would you use to examine the growth of a given isolate on different carbon sources? How would you change the medium for this experiment?

    c) Suppose you made a slurry out of soil and sterile water. You dilute and spread that mixture on agar plates of Medium A and Medium B. Predict which plate would have the most colonies after incubation. Explain your reasoning.

16. The stringent response has been found in almost all species of bacteria, with the notable exception of some obligate intracellular pathogens and obligate symbionts. Considering the evolution and lifestyle of obligate intracellular organisms, why do you think they do not have the stringent response genes? (LO5)

17. You have a conditional *relA* mutant such that the protein functions as wild type at lower temperatures but does not function at elevated temperatures. Predict how each of the following would affect the conditional mutant under different temperature conditions relative to the wild type. (LO6)

In the conditional *relA* mutant:	At low temperature	At high temperature
Is (p)ppGpp produced when amino acids are limiting?		
Are rRNAs made when there are adequate amounts of amino acids?		
Are rRNAs made when amino acids are limiting?		
Does the overall growth rate change when amino acids are limiting?		

 96 Making Connections

 96 Cellular Mechanisms for ATP Synthesis

 98 Chemiosmotic Theory

 99 ATP Synthase

 99 The Proton Motive Force (PMF)

 99 Quantifying the Proton Motive Force (PMF)

100 Cellular Proton Levels

100 Environmental Impacts on the Proton Motive Force (PMF)

101 Experimentally Measuring the Proton Motive Force (PMF)

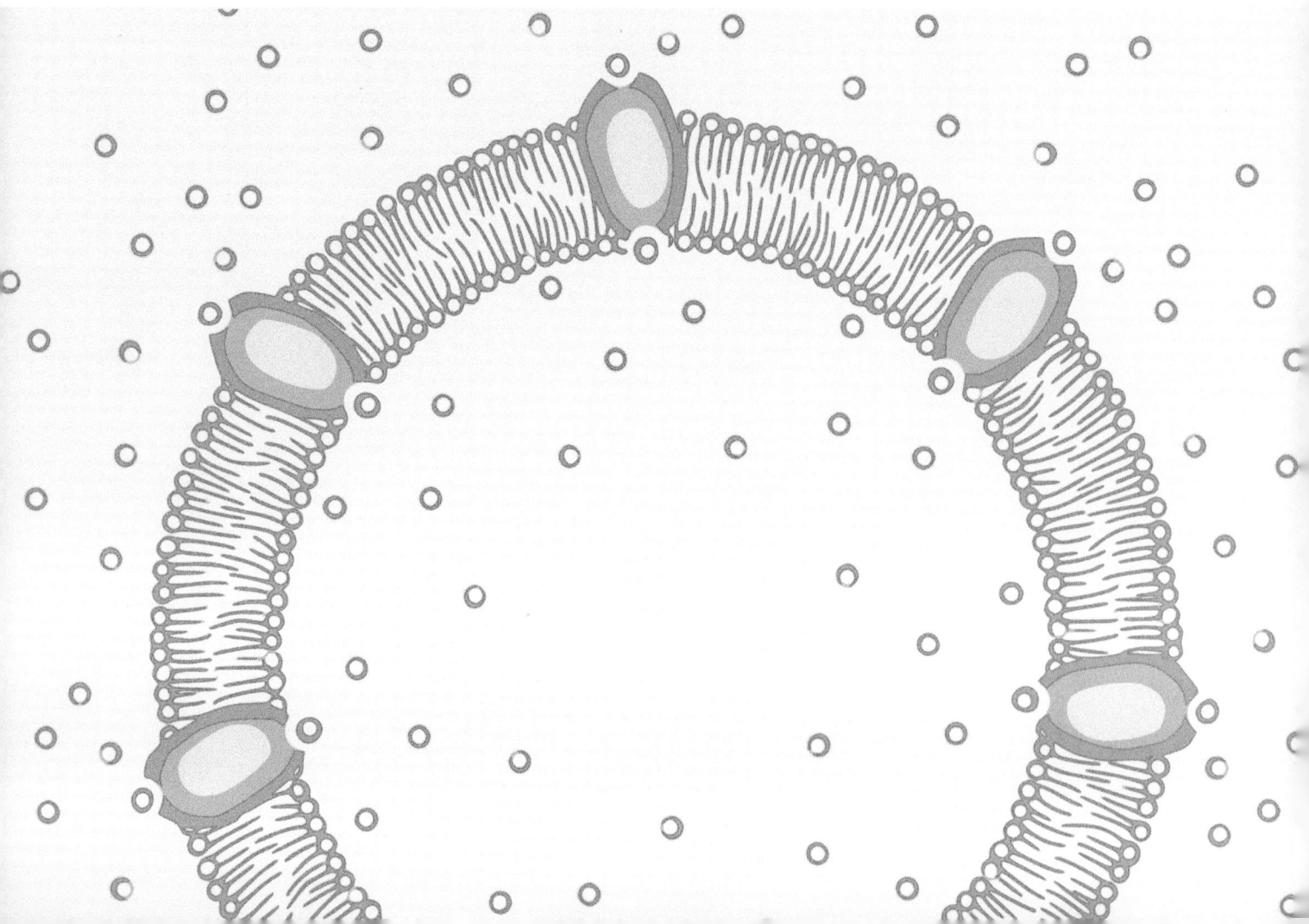

5

BIOENERGETICS AND THE PROTON MOTIVE FORCE

After completing the material in this chapter, students should be able to . . .

1. list three mechanisms by which cells synthesize ATP

2. explain the fundamental principles of the chemiosmotic theory

3. describe the cellular processes that generate and utilize the proton motive force

4. calculate the strength of the proton motive force

5. compare and contrast the proton motive force in neutrophiles, acidophiles, and alkaliphiles

Microbial Physiology: Unity and Diversity, First Edition. Ann M. Stevens, Jayna L. Ditty, Rebecca E. Parales and Susan M. Merkel.
© 2024 American Society for Microbiology.
Companion website: www.wiley.com/go/stevens/microbialphysiology

Making Connections

The biosynthetic reactions necessary for the growth of cells (Chapter 4) from precursor building blocks (Chapters 2 and 3) require energy in the form of ATP and reduced coenzymes. This chapter will explore the relationship between ATP production (i.e., bioenergetics) and the proton motive force, two forms of energy essential for the survival of every living cell.

Cellular Mechanisms for ATP Synthesis

All cells require energy in the forms of adenosine triphosphate (ATP) and the **proton motive force (PMF)**, an electrochemical gradient formed across the cytoplasmic membrane, to drive cellular functions. There are three ways that living cells can generate ATP: substrate-level phosphorylation, oxidative phosphorylation, and photophosphorylation (Fig. 5.1). These modes of energy generation are differentiated in that substrate-level phosphorylation produces energy only in the form of ATP through enzymatic reactions, whereas oxidative phosphorylation and photophosphorylation utilize electron transport chains (ETCs) (Chapter 6) to generate a PMF. The energy stored in the PMF may be converted into ATP via ATP synthase or, conversely, the energy stored in ATP may be converted into the PMF.

All living cells are capable of generating ATP via substrate-level phosphorylation through central metabolism (Fig. 2.6), whereby the energy available in a metabolic intermediate containing a high-energy phosphate bond is donated to ADP to produce ATP. For some obligate fermenters, like *Streptococcus* species, substrate-level phosphorylation is the only mechanism of ATP generation because these organisms physically lack required components of the ETC. Substrate-level phosphorylation is a conserved process in all living cells. In addition to ATP generation by substrate-level phosphorylation, organisms that possess ETCs are also capable of performing respiration via **oxidative phosphorylation** (Chapter 6). Depending on the organism, respiration may occur under oxic conditions (in the presence of oxygen) or under anoxic conditions (in the absence of oxygen). In oxidative phosphorylation, a series of reduction/oxidation (redox) reactions occurs, with a high-energy electron donor serving as the energy source that delivers electrons to an ETC. The ETC generates a PMF that can either be used itself to drive cellular work or can be harvested by the cell and converted into ATP. Finally, **photophosphorylation** is a mechanism of ATP generation that occurs in phototrophic organisms (Chapter 10). Here, photons of sunlight excite low-energy electrons that are donated to an ETC. The ETC again is capable of generating a PMF, which leads to ATP production. Figure 5.1 illustrates a comparison of these three mechanisms used by cells to make ATP.

A SUBSTRATE-LEVEL PHOSPHORYLATION

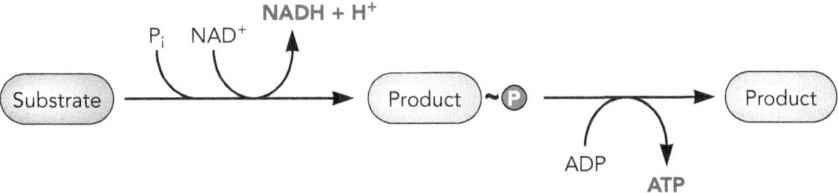

B OXIDATIVE PHOSPHORYLATION

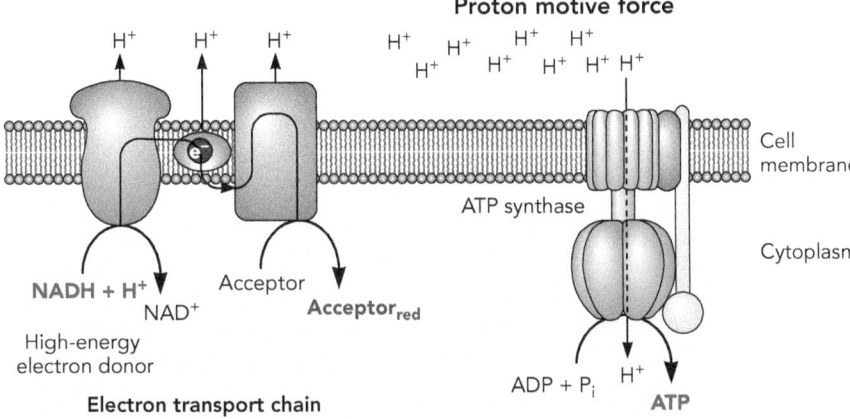

C PHOTOPHOSPHORYLATION

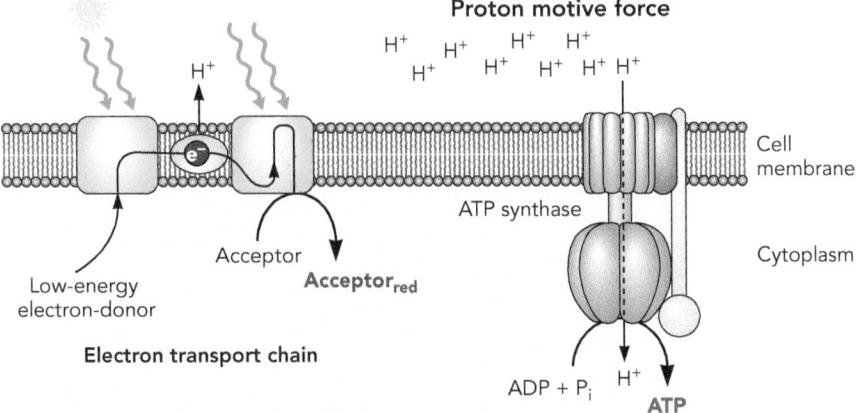

FIGURE 5.1 **Three cellular mechanisms of ATP synthesis.** (A) Substrate-level phosphorylation occurs when ATP is synthesized as part of a metabolic reaction in which an energy-rich organic substrate is cleaved and the energy that is released is conserved in the high-energy phosphate bond of ATP. While substrate-level phosphorylation is the only mechanism of ATP conservation used by fermentative organisms, it is also used by respiring organisms. (B) Oxidative phosphorylation occurs when ATP synthesis, catalyzed by the membrane-bound ATP synthase, is coupled to the entry of protons from the proton motive force (PMF) generated by an electron transport chain (ETC). Respiring organisms utilize oxidative phosphorylation under oxic or anoxic conditions. (C) Photophosphorylation occurs when ATP synthesis, catalyzed by the membrane-bound ATP synthase, is coupled to the entry of protons from the PMF generated by light-driven electron transport. Photophosphorylation is carried out by phototrophic organisms.

Chemiosmotic Theory

In the 1960s, Peter Mitchell first proposed the idea that biological energy, in the form of a PMF, could be stored across a membrane. Prior to this, substrate-level phosphorylation was the only defined mechanism of ATP synthesis. At the time, the scientific community viewed Mitchell's ideas with a great deal of skepticism. Ultimately, his work was accepted as the **chemiosmotic theory** and he was awarded the Nobel Prize. At the heart of the theory is the concept of an electro-chemical gradient across the cytoplasmic membrane of a living cell (Fig. 5.2). Protons and other ions are not evenly distributed on the two sides of a cytoplas-mic membrane. Consequently, the cell carries a positive electrical charge on the outside surface of its membrane. In most cells, protons are a major contributor not only to the charge differential but also to a chemical gradient differential across the cell membrane, which is why it is often referred to as the proton motive force. The molar concentration of protons is higher on the outside of the cytoplasmic membrane in comparison to the cytoplasmic side. Therefore, there is also a chemical or pH gradient across the cytoplasmic membrane of all living cells. Mitchell proposed that (i) an ETC permits and directs the move-ment of protons to the extracellular space across the cytoplasmic membrane; (ii) in order for the electrochemical gradient to be maintained, the membrane must be impermeable to all ions; and (iii) when ions reenter the cell through a membrane-bound **ATP synthase**, ATP is produced in the cytoplasm (Fig. 5.1). Thus, a biological membrane is like a battery, with the capacity to store poten-tial energy that can be harvested and converted into ATP or used directly to do work for the cell.

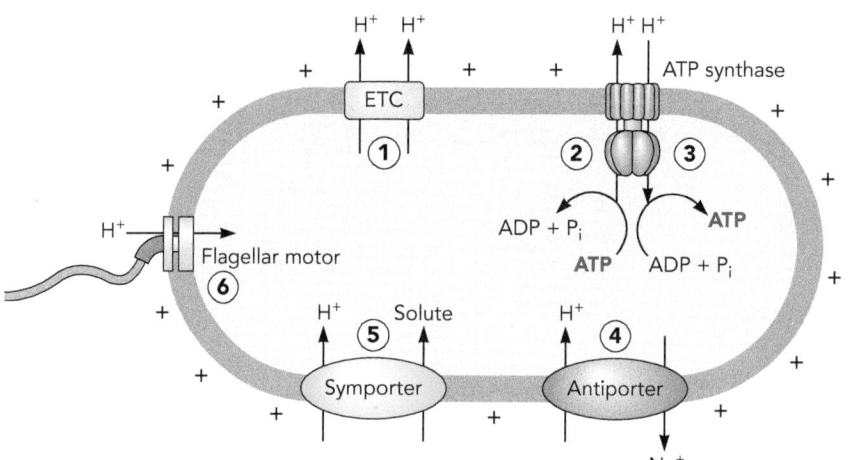

FIGURE 5.2 **Generation and utilization of the proton motive force (PMF).** (1) An electron transport chain (ETC) can transport either protons or sodium ions to the outside of the cell to generate a PMF (or sodium motive force in some organisms, not shown). The ATP synthase can run in both directions, (2) by cleaving ATP to power transport of protons, which generates a proton gradient across the membrane for fermenting organisms that have no ETC, or (3) to capture the energy released as protons enter the cell by moving down the proton gradient and generating ATP. (4) Antiporters, which are powered by the PMF (protons move down the proton gradient and enter the cell), pump charged ions out of the cell against the charge gradient. (5) Symporters, which are also powered by the PMF, allow solutes to enter the cell against the concentration gradient, moving in the same direction as the protons. (6) Rotation of bacterial flagellar motors is also powered by the PMF (or in some organisms by a sodium motive force).

ATP Synthase

ATP synthase was discovered by Efraim (Ef) Racker. Supposedly, the two multi-subunit complexes in the enzyme, F_0 and F_1, are named after him. The F_0 subunit is embedded within the cytoplasmic membrane and serves as a regulated channel for the movement of protons. The F_1 subunit is located on the cytoplasmic side of the membrane and functions as a molecular turbine, catalyzing the formation of one ATP molecule for roughly every three protons that pass through the F_0 subunit, from outside of the cytoplasmic membrane to inside the cell. Like most enzymes, ATP synthase is a reversible enzyme that is capable of not only producing ATP from the PMF but hydrolyzing ATP to generate a PMF when necessary for the cell's survival.

The Proton Motive Force (PMF)

The ETC is the primary mechanism used by respiring organisms to maintain a proton gradient across their cytoplasmic membrane and produce ATP via the ATP synthase (Fig. 5.1). However, as mentioned above, obligate fermenters produce ATP solely via substrate-level phosphorylation and lack the full repertoire of components required for a functional ETC (Chapter 6). However, fermenting organisms must still generate a PMF.

Why do all organisms need to generate a PMF? The PMF is not only used to produce ATP through ATP synthase; it also drives other cellular processes critical to survival. All living cells utilize the energy in ion gradients, including the PMF, to move ions and solutes across the membrane through ion-driven transport proteins in an energy-dependent manner (Fig. 5.2). A transporter that permits protons (or other ions such as sodium ions [Na^+]) to move inward while another molecule is moved outward is called an **antiporter**. A transporter that moves either protons or Na^+ and a solute inward simultaneously is a **symporter**. These types of ion-driven transporters are essential to life in all cells from all three domains of life (Chapter 13). In addition, motile bacteria also use the PMF to drive flagellar rotation, enabling motility (Fig. 5.2; also Chapter 15).

So, how do organisms like obligate fermenters that lack an ETC generate a PMF? The ATP that is produced via substrate-level phosphorylation is used to drive ATP synthase in "reverse," hydrolyzing ATP to ADP to power the movement of protons outward, from the cytoplasm across the cytoplasmic membrane, generating a proton gradient across the membrane (Fig. 5.2).

Quantifying the Proton Motive Force (PMF)

The PMF is also known as the ΔP. It represents the amount of potential energy stored across a membrane, which can be quantified using the formula $\Delta P = \Delta\psi - (2.3RT/F)\Delta pH$ (Fig. 5.3) and is measured in units of millivolts. $\Delta\psi$ is the electrical component of the PMF (i.e., the membrane potential) and is also in units of millivolts. The chemical component of the gradient is represented by $-(2.3RT/F)\Delta pH$, where $\Delta pH = pH_{in} - pH_{out}$. This difference in the molar

$$\Delta P = \Delta \Psi - (2.3RT/F)\Delta pH$$

proton	"ELECTRO"	"CHEMICAL"
motive	membrane	R = gas constant
force	potential	T = temperature in Kelvin
(mV)	(mV)	F = Faraday constant

FIGURE 5.3 **Mathematical equation defining the proton motive force (PMF).** The PMF (ΔP) has both an electrical component ($\Delta \Psi$), which is the difference in charge across the membrane measured in millivolts, as well as a chemical component, which is based upon the difference in proton concentration across the membrane (ΔpH). The gas constant, R, relates the energy scale to the temperature scale (R = 8.3145 J K^{-1} mol^{-1}). The Faraday constant, F, is a physical constant equal to the total electric charge carried by 1 mole of electrons (F = 96,485 coulombs mol^{-1}), which converts the chemical component of the gradient into millivolt units.

concentration of protons is converted to millivolts using the gas constant (R), the temperature in Kelvin (T), and the Faraday constant (F). Collectively, $2.3RT/F$ has a mathematical value of about 60 at 30°C or about 59 at 25°C. Thus, at 30°C, the equation for the PMF may be simplified to $\Delta P = \Delta \psi - 60\Delta pH$. A value between −140 and −200 mV is the typical range observed for ΔP in most living cells, which is also the optimal range for the production of ATP via ATP synthase. The more negative the ΔP, the more ATP can be produced.

Cellular Proton Levels

Most cells maintain their internal pH near neutrality so that their enzymes remain functional. How many protons does it take to maintain a bacterial cell's internal pH? As an example, for an *Escherichia coli* cell with an internal pH of ~8, the proton concentration is 1×10^{-8} M (M = moles/liter). Multiplying this value by Avogadro's number (6.02×10^{23} molecules/mole) and by the value 1×10^{-15} liters/cell (the approximate volume of an *E. coli* cell) results in the conclusion that there are ~6 protons/cell. This may seem like a very low value at first glance, but what is important to recognize is that this is the number of protons maintained in a cell at any given point in time. There is a large flux of protons being constantly moved across the membrane by the ETC and ATP synthase, while the internal pH remains steady.

Environmental Impacts on the Proton Motive Force (PMF)

Clearly, pH is an important variable that impacts the PMF. Thus, the environment plays an important role in the physiology of bacteria living in neutral, acidic, or basic conditions (Table 5.1). In general, neutrophiles like *E. coli* maintain an internal pH of ~8, with the chemical gradient contributing ~20 to 30% of the PMF while the majority (~70 to 80%) is contributed by the electrical gradient. Acidophiles live in low pH environments but maintain a cytoplasmic pH approaching neutrality (~6.5). In these organisms, the entire PMF is generated by the chemical gradient; the electrical gradient actually detracts from the overall strength of the PMF.

TABLE 5.1 **How the environmental pH impacts the ΔP**

				$\Delta P = \Delta \Psi - (2.3RT/F)\Delta pH$			
	pH_o	pH_i	ΔpH	$-60\Delta pH^a$	$\Delta \Psi$	ΔP
Neutrophiles	6	7.8	+1.8	−108 mV	−102 mV	−210 mV
	7	7.8	+0.8	−48 mV	−120 mV	−168 mV
	8	7.8	−0.2	+12 mV	−140 mV	−128 mV
Acidophiles	2	6.5	+4.5	−270 mV	+10 mV	−260 mV
Alkaliphiles	10	9.4	−0.6	+36 mV	−125 mV	−89 mV

^{a}At 30°C, $2.3RT/F$ equals ~ 60.

Alkaliphiles, on the other hand, have the exact opposite situation. They live in high pH environments, and although their enzymes are adapted to work at higher pH, they struggle to maintain their PMF in the optimal range for ATP synthase. In alkaliphiles, almost all of the PMF comes from the electrical component of the ion gradient, with the chemical gradient detracting from it. It is impossible for a single bacterial cell to extrude enough protons into the bulk environment around it to have any significant impact on the external pH. Thus, the ΔP is typically more positive than −140 mV. In theory, alkaliphiles should not be able to exist, but they do. How is this possible? One hypothesis postulates that the protons being moved to the outside of the cytoplasmic membrane by the ETC are not actually released to the bulk environment. They are "funneled" directly to the ATP synthase, thereby creating a higher localized concentration of protons close to the external surface of the membrane and enabling the PMF to achieve a sufficiently negative mathematical value such that ATP production can actually occur. Rather than using a PMF, some alkaliphiles (and also some halophiles, methanogens, and marine bacteria) use a sodium motive force to power various cellular processes.

Experimentally Measuring the Proton Motive Force (PMF)

Artificial membrane vesicles are commonly used to measure the parameters of the PMF (Fig. 5.4). Membrane vesicles can be generated from bacterial cells by enzymatically removing the cell wall while leaving the cytoplasmic membrane intact. Typically, bacterial cells are treated with lysozyme to weaken the structure of the peptidoglycan in the cell wall, and then an osmotic shock treatment is applied. This causes the cell to expand in a hypotonic solution and contract in a hypertonic solution to break apart the cell wall so that it may be gently removed without damaging the membrane-bound vesicle. This process results in a "right-side-out" vesicle that retains the native orientation of the cytoplasmic membrane. "Inverted" vesicles have the reverse orientation, with what was originally the cytoplasmic side of the cell now exposed to the external environment. These vesicles are generated following lysis of cells in a French press (Chapter 3). Because the vesicle is inverted, it is possible to take measurements of activity that would normally be occurring on the inside of the cell membrane since it is now exposed to the environment. The other tool used in combination

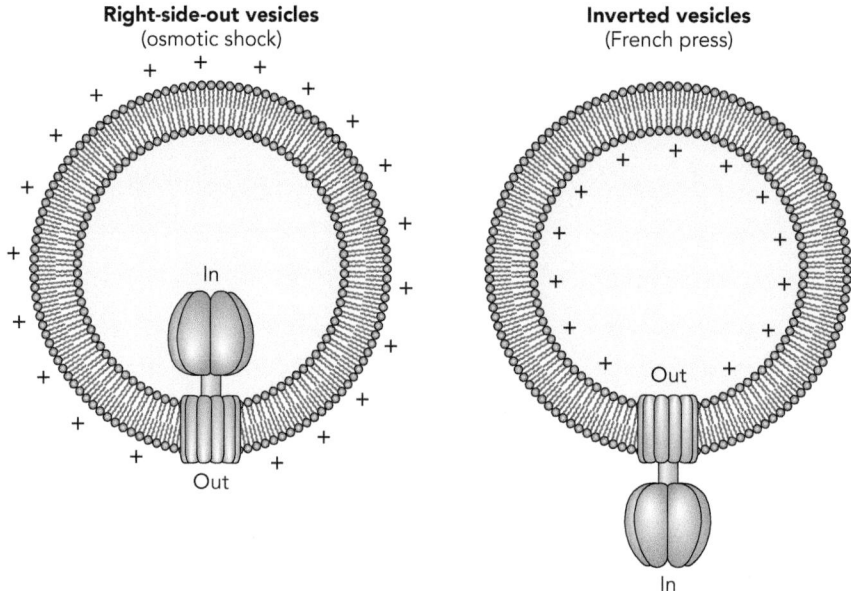

Right-side-out vesicles
(osmotic shock)

In

Out

Inverted vesicles
(French press)

Out

In

FIGURE 5.4 **Membrane vesicles can be generated in the laboratory to measure the proton motive force (PMF).** Right-side-out vesicles, which have the same orientation as the native cells (protons accumulate *outside*), can be generated by osmotic shock. Inverted vesicles can be generated by breaking cells using a French press. These vesicles have an inverted architecture in which protons accumulate *inside* of the vesicles, which is equivalent to the outside of the original cell.

with vesicles is a group of compounds known as **ionophores**. Ionophores are organic compounds that form lipid-soluble complexes with cations (e.g., H^+, K^+, and Na^+). As such, they may function as uncouplers by dissipating the ion gradients established across a membrane and producing a state of equilibrium. Using a combination of right-side-out and inverted vesicles with ionophores, it is possible to make accurate physical measurements of the values of the $\Delta\psi$ and ΔpH using the quenching of cationic or anionic fluorescent dyes or the distribution of weak acids and bases, respectively, as indicators of the physiological state of the membrane.

Learning Outcomes: After completing the material in this chapter, students should be able to. . .

1. list three mechanisms by which cells synthesize ATP

2. explain the fundamental principles of the chemiosmotic theory

3. describe the cellular processes that generate and utilize the proton motive force

4. calculate the strength of the proton motive force

5. compare and contrast the proton motive force in neutrophiles, acidophiles, and alkaliphiles

Check Your Understanding

1. Define this terminology: proton motive force (PMF), oxidative phosphorylation, photophosphorylation, chemiosmotic theory, ATP synthase, antiporter, symporter, ionophore

2. Compare and contrast substrate-level phosphorylation, oxidative phosphorylation, and photophosphorylation. List one feature they all have in common and one way they are different from one another. (LO1)

3. Electrons are required as the ultimate source of energy. What is the source of electrons in substrate-level phosphorylation, oxidative phosphorylation, and photophosphorylation? (LO1)

4. Explain how a PMF is used to generate ATP during oxidative phosphorylation. (LO2)

5. Why do all organisms need to generate a PMF? (LO3)

6. Compare and contrast antiporters and symporters. List one feature they have in common and one way they are different. (LO3)

7. Why do organisms like obligate fermenters need to be able to run their ATP synthase in reverse? What does that allow them to do? (LO3)

8. You measure the membrane potential ($\Delta\Psi$) for a bacterial cell and get -169 mV. (LO4)
   a) If the ΔpH at 30°C is +0.3 and 2.3 RT/F =60, what is the PMF (ΔP)?

   b) Given a ΔpH = +0.3, how does the pH and [H^+] inside the cell compare with those outside the cell?

   c) In this case, what contributes more to the PMF, the electrical or chemical component?

9. *Alicyclobacillus acidocaldarius* is a bacterium that can grow at a pH range of approximately 2.0 to 6.5 but maintains an internal pH of 6 regardless of the environmental pH. Assuming the volume of the cell is 1×10^{-15} liters/cell, how many protons would you expect to be present inside the cell? Explain your answer. (LO4)

10. What is a "sodium motive force"? How is it different from a PMF? (LO5)

11. Artificial membrane vesicles are commonly used to measure the parameters of the PMF. (LO4)
    a) What is the role of lysozyme in creating an artificial membrane vesicle?

    b) Suppose you are doing experiments with right-side-out vesicles to generate a PMF. Do protons accumulate inside or outside of the vesicle? Does ATP accumulate inside or outside the vesicle?

    c) Suppose you are doing experiments with inverted vesicles. Do protons accumulate inside or outside of the vesicle? Does ATP accumulate inside or outside the vesicle?

Dig Deeper

12. Suppose you have two bacterial cultures, one that grows using only substrate-level phosphorylation (Organism 1, SLP) and one that grows using only oxidative phosphorylation (Organism 2, OxP). You develop a new antibiotic ionophore allowing the free movement of protons across the membrane. In the tables below, predict how this antibiotic would affect ATP generation and nutrient transport in each of the bacteria. (LO1, 2, 3)

    a) Fill in the table below to indicate whether, in each case, ATP generation would likely increase, decrease, or remain the same. Explain your answer.

	Effect on ATP generation	Why?
Organism 1, SLP		
Organism 2, OxP		

    b) Fill in the table below to indicate whether, in each case, nutrient transport would likely increase, decrease, or remain the same. Explain your answer.

	Effect on nutrient transport	Why?
Organism 1, SLP		
Organism 2, OxP		

13. a) In general, which cell type, neutrophiles, acidophiles, or alkaliphiles, would most easily be able to maintain a PMF? Explain your reasoning. (LO5)

    b) Why might it be advantageous for alkaliphiles to use a sodium motive force? (LO5)

14. The acetic acid bacteria are obligate aerobes that grow using organic molecules such as ethanol (EtOH) as an external electron donor during aerobic respiration (Figure A). As electrons from EtOH move through an electron transport chain (ETC), protons (H^+) are pumped across the membrane with oxygen serving as the terminal electron acceptor. This process is often exploited in industrial settings because it produces large amounts of acetic acid (AcAcid) (i.e., vinegar).

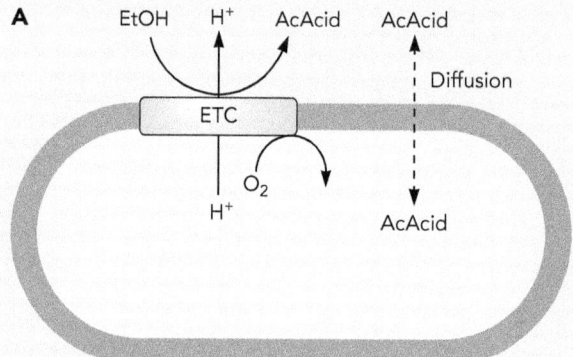

There is one problem for these bacteria—as a weak acid that is protonated at low pH, acetic acid can diffuse into cells through the cell membrane in acidic environments. Therefore, acetic acid bacteria must have a mechanism (called acetic acid resistance) to resist the toxic effect of their own waste product. (LO2,3,4)

Your hypothesis is that some acetic acid bacteria have a PMF-dependent antiporter that acts as an efflux pump, moving acetic acid out of the cell.

To study this acetic acid-resistance, you make two sets of vesicles from the acetic acid bacterium, *Acetobacter aceti*: vesicles that are right-side-out (Figure B) and vesicles that are inverted (Figure C).

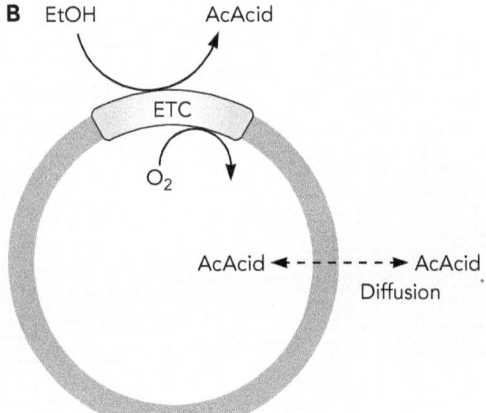

a) In your first experiment, you suspend the **right-side-out vesicles** in an isotonic solution and add ethanol (EtOH) as an external electron donor to stimulate respiration. (Figure B)

   i. In the diagram of right-side-out vesicles, draw the direction of H^+ flow that would generate the PMF.

   ii. Draw the proposed PMF-dependent antiporter efflux pump, showing the movement of acetic acid and H^+ through the pump.

   iii. When you then add carbonyl cyanide m-chlorophenyl hydrazone (**CCCP**, a proton-decoupling ionophore) to the solution, you get more acetic acid inside the right-side-out vesicle, compared to a solution with ethanol alone. Do these data support or refute the presence of PMF-dependent antiporter efflux pump? Explain your reasoning.

b) In your second experiment, you suspend the **inverted vesicles** in an isotonic solution and add ethanol as an external electron donor to stimulate respiration. (Figure C)

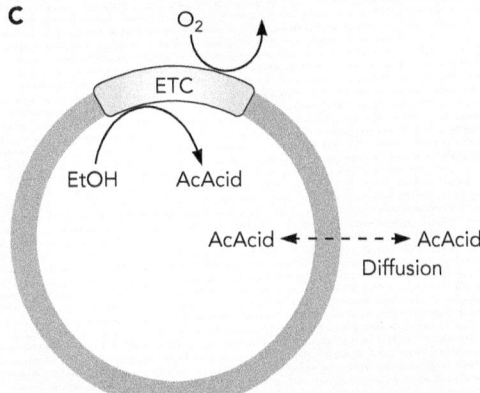

   i. In the diagram of inverted vesicles, draw the direction of H^+ flow that would generate the PMF.

   ii. Draw the proposed PMF-dependent antiporter efflux pump, showing the movement of acetic acid and H^+ through the pump.

   iii. You are able to measure ΔpH and $\Delta\Psi$ in the inverted vesicles. Calculate the PMF if the $\Delta pH = -1.3$ and $\Delta\Psi = 82$ mV in the inverted vesicles.

   iv. Assuming these inverted vesicles had a PMF-dependent antiporter efflux pump, predict how the concentration of acetic acid inside the inverted vesicles would change if you added ethanol and CCCP (a proton-decoupling ionophore) to the medium, compared to ethanol alone. Explain your reasoning.

108 Making Connections

108 The Basic Components of an Electron Transport Chain (ETC)

109 Electrode/Reduction Potential (E_0')

110 Brief Review of the Electron Transport Chain (ETC) in Mitochondria

113 Q Cycle of Mitochondria

113 Bacterial Electron Transport Chains (ETCs)

115 Q Loop of Bacteria

115 Electron Donors and Acceptors in Bacteria

117 Fermentation

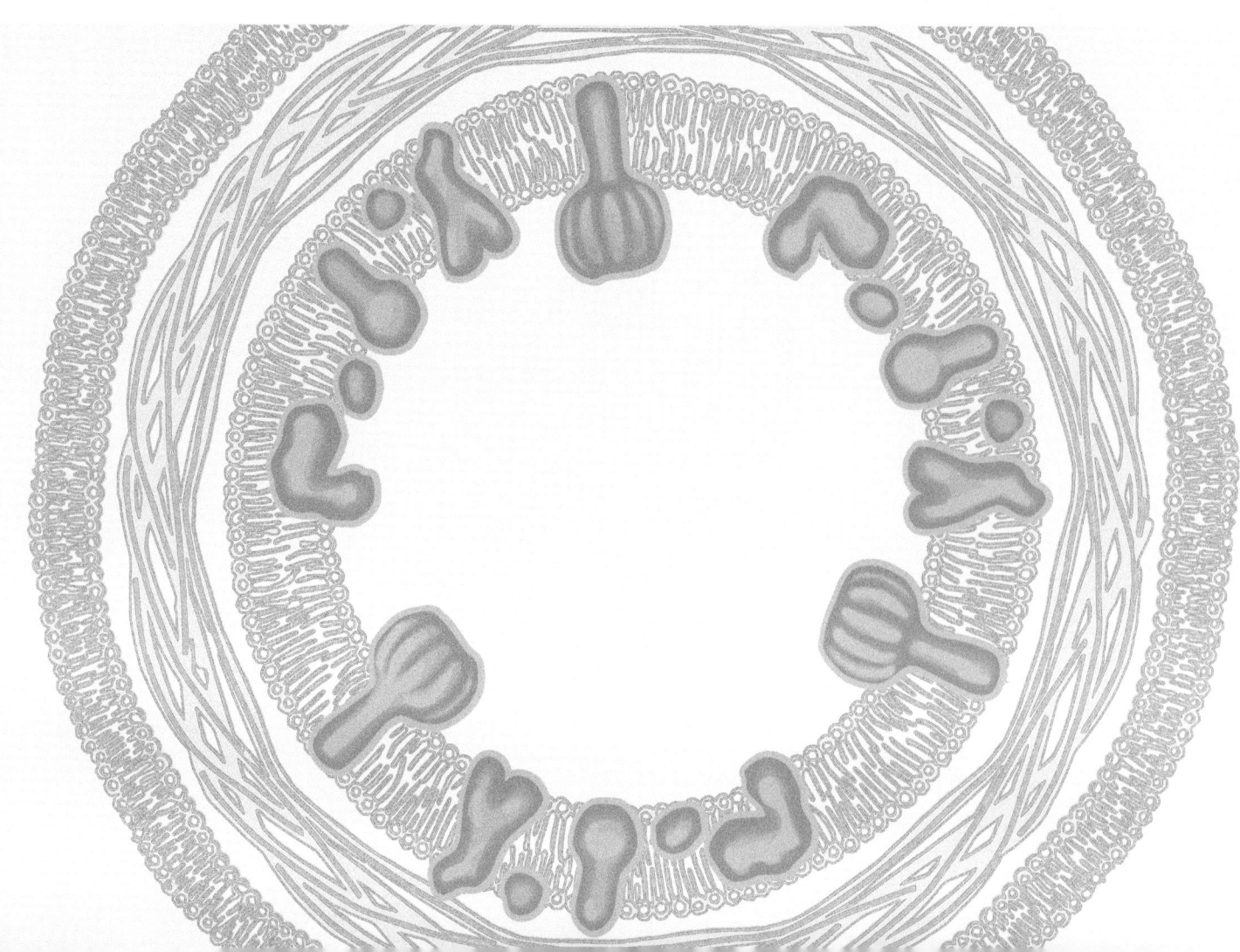

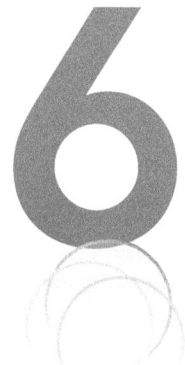

RESPIRATION AND FERMENTATION

After completing the material in this chapter, students should be able to . . .

1. describe the functions of the four basic components of an electron transport chain

2. predict differences in energy yield based on different electron donors and acceptors

3. explain how the movement of electrons through an electron transport chain is coupled to the creation of a proton motive force and ATP generation

4. compare and contrast electron and proton flow in mitochondrial and bacterial Q cycles and Q loops

5. identify key reduction/oxidation and energy-generating steps in fermentation pathways

6. differentiate the processes of aerobic respiration, anaerobic respiration, and fermentation

Microbial Physiology: Unity and Diversity, First Edition. Ann M. Stevens, Jayna L. Ditty, Rebecca E. Parales and Susan M. Merkel.
© 2024 American Society for Microbiology.
Companion website: www.wiley.com/go/stevens/microbialphysiology

Making Connections

The essential role of the proton motive force (PMF) in living cells was presented in Chapter 5. In this chapter, we will review in more detail the mechanisms whereby the PMF is formed directly through the process of respiration and indirectly by cells using fermentation to grow.

The Basic Components of an Electron Transport Chain (ETC)

As described in Chapter 5, all respiring bacteria require electron transport chains (ETCs) to generate the PMF, which is used to make adenosine triphosphate (ATP) through oxidative phosphorylation. At a fundamental level, an ETC is comprised of four components, often referred to as complexes. The components of the ETC are involved in a series of linked reduction/oxidation (redox) reactions that gradually release the potential energy carried by electrons. Three of these components, flavoproteins, iron-sulfur proteins, and cytochromes, are proteins that contain **prosthetic groups** (Table 6.1). A prosthetic group is a permanent, required, and tightly bound component of the enzyme complex that, in an ETC, functions to carry electrons or both electrons and protons. Without the prosthetic group, the protein itself is not biologically functional. A **flavoprotein** contains a flavin prosthetic group (e.g., flavin adenine dinucleotide [FAD] or flavin mononucleotide [FMN]; Fig. 6.1A); these prosthetic groups carry both protons and electrons. An **iron-sulfur (FeS) protein** utilizes an FeS cluster (Fig. 6.1B) as its prosthetic group, which is one reason that respiring organisms require sources of iron and sulfur in their growth medium for metabolism. Iron-sulfur proteins carry only electrons. **Cytochromes** are the third type of protein molecule found in ETCs. They contain heme (Fig. 6.1C) as their prosthetic group and only carry electrons. Cytochromes are named based on the kinds of heme groups they contain and the cytochrome components in the complex (e.g., Cyt c, Cyt bc_1, or Cyt aa_3). Flavoproteins, FeS proteins, and cytochromes are all protein complexes that catalyze redox reactions as electrons move through the ETC within the membrane. The fourth and final component of an ETC is a quinone. **Quinones** are lipids, and as such are hydrophobic and able to readily move within a membrane from the internal side to the external side and vice versa. They do not contain prosthetic groups but carry both protons and electrons within their own chemical structure (Fig. 6.1D).

TABLE 6.1 **Components of electron transport chains (ETCs)**

	Type of molecule	Prosthetic group	Accepts/donates
Flavoproteins	Protein	Flavin (FAD or FMN)	H^+ and e^-
Iron-sulfur proteins	Protein	FeS cluster	e^-
Cytochromes	Protein	Heme	e^-
Quinones	Lipid	None	H^+ and e^-

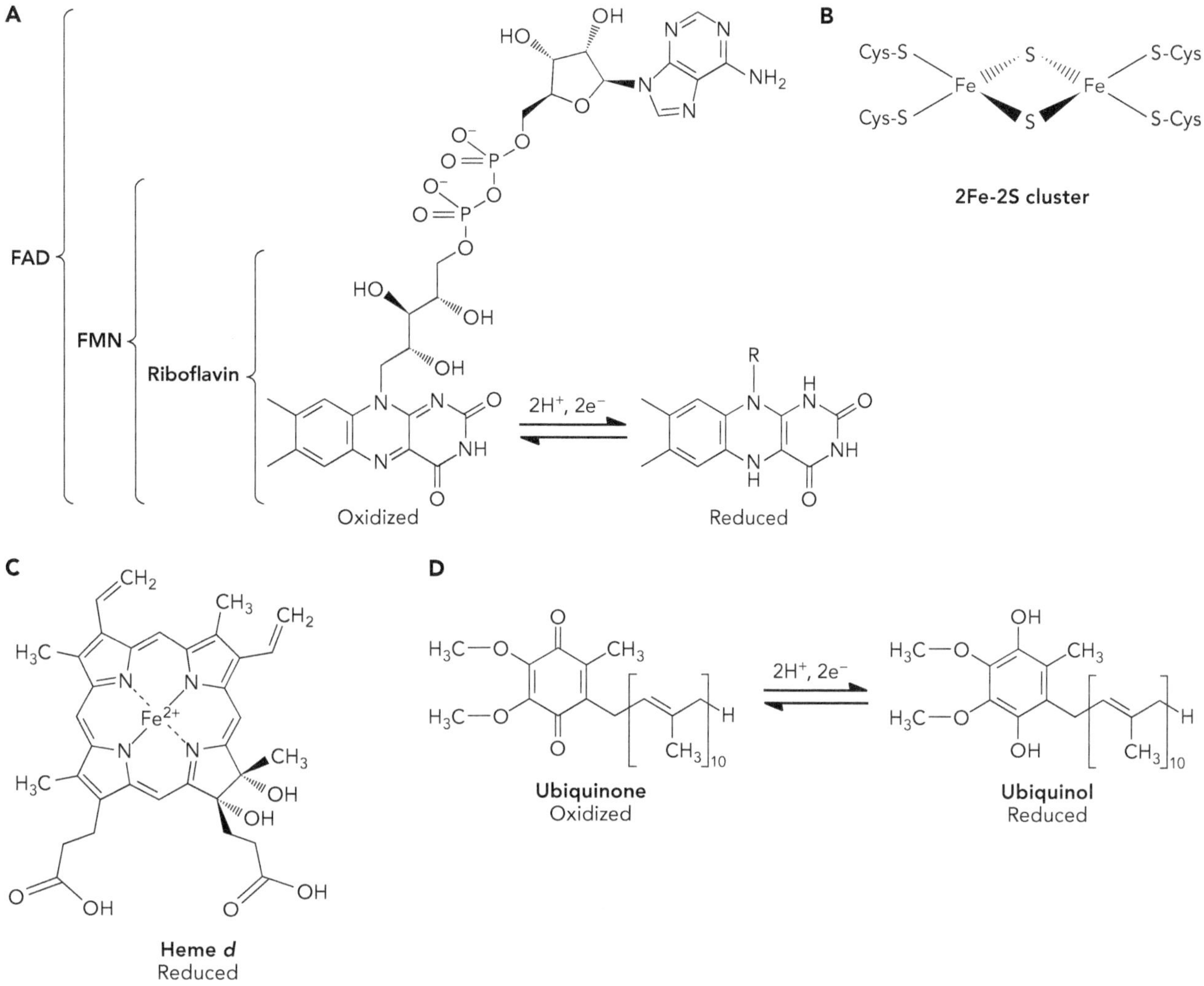

FIGURE 6.1 **Structures of molecules critical for the function of electron transport chains (ETCs).** (A) Oxidized and reduced structures of the flavin prosthetic group found in flavoproteins. Note that the molecule is capable of carrying both protons and electrons. (B) An example of an iron-sulfur prosthetic group (i.e., 2Fe-2S cluster) found in an FeS protein. The entire iron-sulfur prosthetic group functions to carry one electron. (C) Heme *d* is an example of the heme prosthetic group utilized within cytochromes to carry electrons. Hemes have a conserved ring structure with an iron molecule capable of carrying one electron; the figure shows the reduced form (Fe^{2+}, ferrous iron). The oxidized form of heme contains Fe^{3+} (ferric iron, not shown). Heme molecules can vary from one another in their side groups. (D) An example of a quinone, oxidized ubiquinone is converted into reduced ubiquinol when it accepts electrons from a donor and protons from the cytoplasm.

Electrode/Reduction Potential (E_0')

A discussion of ETCs necessitates a discussion of **electrode or reduction potential**, defined as E_0' (under standard conditions), which indicates the potential to donate or accept electrons from another molecule. For example, consider two reduced coenzymes generated during central metabolism (Chapter 2); NADH + H$^+$ ($E_0' = -320$ mV) and FADH$_2$ ($E_0' = -220$ mV). While these are both relatively good electron donors because the E_0' is relatively negative in both cases, NADH + H$^+$ acts as a stronger electron donor than FADH$_2$. Oxygen, on the other hand, has a highly positive electrode/reduction potential ($E_0' = +815$ mV) and is a good electron acceptor. The greater the absolute value of the mathematical

Table 6.2 **Standard electrode/reduction potentials**

Redox couple (oxidized/reduced)	E_0' (mV)
CO_2/formate	−430
$2 H^+/H_2$	−420
$NAD(P)^+/NAD(P)H + H^+$	−320
$FAD^+/FADH_2$	−220
Pyruvate/lactate	−190
Menaquinone (ox/red)	−74
Fumarate/succinate	+30
Ubiquinone (ox/red)	+110
Cytochrome c (ox/red)	+250
NO_3^-/NO_2^-	+430
$½ O_2/H_2O$	+820

difference between the initial donor (e.g., NADH + H$^+$ at −320 mV) and the terminal electron acceptor (e.g., oxygen at +815 mV), the more energy that can potentially be conserved by the ETC in the form of PMF and ATP. A table of standard electrode/reduction potentials (Table 6.2) lists the oxidized and reduced forms of a molecule, with the oxidized form being a potential electron acceptor and the reduced form being a potential electron donor. For example, NAD$^+$ is the oxidized form and NADH + H$^+$ is the reduced form of the coenzyme. Molecules with more negative values for E_0' serve as better electron donors. Therefore, NADH + H$^+$ ($E_0' = −320$ mV) is a "better" donor than FADH$_2$ ($E_0' = −220$ mV) because more potential energy can be harvested by the ETC (as detailed below). Molecules with highly positive E_0' values are better electron acceptors; therefore, oxygen ($E_0' = +815$ mV) is the "best" biological electron acceptor in that more energy can be yielded when it is used as a last electron acceptor in the ETC, or the **terminal electron acceptor**. Electrons will flow spontaneously from a more negative electrode/reduction potential to a more positive potential. When electrons flow in this energetically favorable direction, there is a release of potential energy that is harvested by the ETC. Therefore, rather than having one highly energetic reaction in which energy might be inefficiently lost, the ETC conserves energy by passing electrons through the various complexes, releasing the energy in smaller, more controlled increments, producing greater overall efficiency.

Brief Review of the Electron Transport Chain (ETC) in Mitochondria

The ETC of a mitochondrion is very similar to that of a bacterium carrying out aerobic respiration. This is part of the evidence for the endosymbiotic theory, whereby mitochondria evolved from what were once free-living bacteria (Chapter 1). There are four components in the ETC, with the ATP synthase as a fifth component (Fig. 6.2A). The first component is NADH dehydrogenase (also known as Complex I), a multisubunit protein complex that contains both a flavoprotein and an FeS protein. NADH + H$^+$ ($E_0' = −320$ mV) from central metabolism is the initial donor that delivers both protons and high-energy electrons to the flavoprotein. The electrons can pass from the flavin prosthetic group of the flavoprotein to the FeS cluster of the FeS protein. However, because the FeS cluster can only accept electrons from the flavin, and not protons, NADH dehydrogenase functions as a **proton pump**, which is an integral membrane protein that translocates protons across the membrane. The release of free energy is used to pump protons from the mitochondrial matrix to the outside of the inner mitochondrial membrane. When NADH dehydrogenase accepts 2 electrons, it transfers ~4 protons across the membrane, thereby contributing to the PMF.

The FeS protein next donates the electrons to a lipid ubiquinone, which then moves within the membrane to the interior (matrix) side and picks up protons. The ubiquinone transfers only the electrons to cytochrome bc_1 (also known as Complex III). The ~4 protons are carried across the inner membrane and released to the intermembrane space via a process known as a Q cycle (see below), thereby contributing to the PMF. The electrons finally move from cytochrome bc_1 (Complex III) to a **terminal oxidase complex** (also known as Complex IV) via cytochrome c (Fig. 6.2A). Here, the electrons are passed to the external terminal electron acceptor, oxygen ($E_0' = +815$ mV), and water is formed. There is sufficient energy in the system that ~2 protons are pumped across the inner membrane at the terminal oxidase complex. Thus, a total of ~10 protons are moved from the inside to the outside of the membrane every time that NADH + H^+ serves as an electron donor for the mitochondrial ETC. For every 3 protons that move back into the matrix compartment through the ATP synthase (Complex V), 1 molecule of ATP is formed. Therefore, for every NADH + H^+ that donates electrons to the mitochondrial PMF, ~3 ATP are formed.

The second component of the mitochondrial ETC is succinate dehydrogenase (Complex II), which also contains a flavoprotein and an FeS protein. $FADH_2$ produced from the tricarboxylic acid (TCA) cycle is the electron donor for succinate dehydrogenase. With an electrode potential of −220 mV, $FADH_2$ cannot donate protons and electrons to NADH dehydrogenase (Complex I). The protons and electrons carried by $FADH_2$ are donated instead to the flavoprotein component of the succinate dehydrogenase, then the electrons are passed to the FeS protein component. However, in this case it is not energetically favorable for the protons to be pumped across the membrane, so there is no contribution to the PMF. The electrons then move down the ETC from ubiquinone to cytochrome bc_1 to the terminal oxidase, in a manner identical to when NADH is the electron donor. For every molecule of $FADH_2$ that serves as an electron donor, there are only ~6 protons moved across the inner mitochondrial membrane by the ETC; ~4 protons are still moved across the membrane by the uniquinone through a Q cycle and ~2 protons are pumped across the membrane at the oxidase. Therefore, only ~2 ATP are formed when $FADH_2$ is the initial electron donor (Fig. 6.2A).

GENERATION OF THE PROTON MOTIVE FORCE (PMF)

In the mitochondrial ETC, NADH dehydrogenase, cytochrome bc_1, and the terminal cytochrome oxidase are considered to be "**coupled (or coupling) sites**". These are locations where the potential energy in the electrons moving through the ETC is coupled to the movement of protons across the membrane, which directly contributes to the PMF. NADH dehydrogenase and the terminal cytochrome oxidase are proton pumps, while proton movement by cytochrome bc_1 involves a lipid quinone. Quinones may move protons across a biological

A   MITOCHONDRIAL ELECTRON TRANSPORT CHAIN

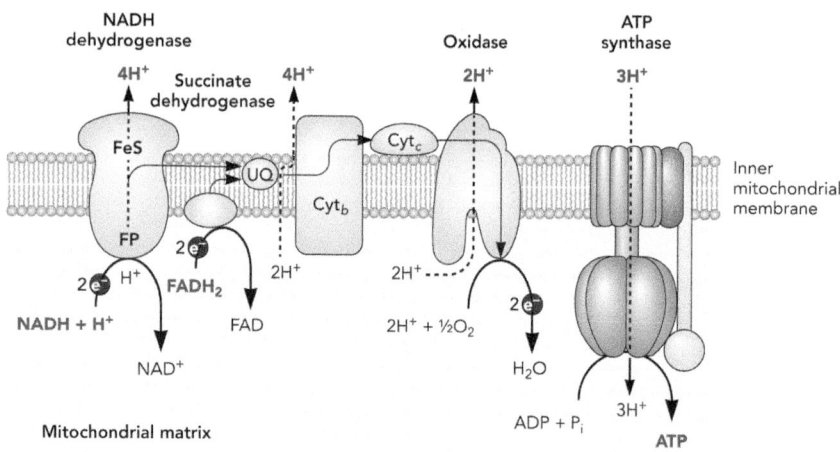

B   Q CYCLE ROUND 1

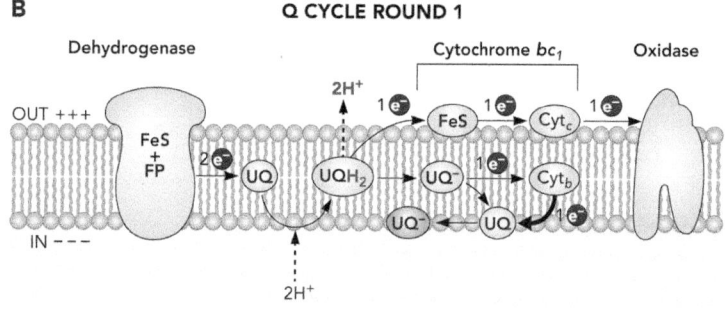

Q CYCLE ROUND 2

FIGURE 6.2 **Diagram of a mitochondrial electron transport chain (ETC).** (A) Electrons flow from the initial electron donor (e.g., NADH + H⁺ or FADH$_2$) with a negative E_0' to the terminal electron acceptor (i.e., O$_2$) with a positive E_0'. The flavoproteins (FP) within the NADH and succinate dehydrogenases (also known as Complexes I and II, respectively) accept electrons from the donor. Protons are pumped across the membrane at the NADH dehydrogenase, but not the succinate dehydrogenase. UQ represents ubiquinone, which is capable of accepting two electrons from the FeS protein within the dehydrogenase and two protons from the cytoplasm. These protons are then released outside of the membrane at the cytochrome bc_1 complex (also known as Complex III). The electrons then move to cytochrome c and finally to terminal cytochrome oxidase (e.g., cytochrome aa_3 or cytochrome c oxidase; also known as Complex IV) where the electrons are passed to the terminal electron acceptor and protons are pumped across the membrane. The potential energy in the proton motive force (ΔP) is used by ATP synthase (also known as complex V), to generate ATP as the protons move down their concentration gradient and reenter the cytoplasm. See the text for additional details. (B) A Q cycle is part of the mitochondrial ETC and is also found in chloroplasts and some bacteria. In round 1 (top) of a Q cycle two electrons are passed to UQ, which then accepts two protons from the cytoplasm as it becomes fully reduced (UQH$_2$). Just one electron is passed from UQH$_2$ to an FeS protein to cytochrome c and then to the terminal cytochrome oxidase. In this process two protons are released to the outside of the membrane. The second electron is passed from UQH$_2$ to cytochrome b, which carries the electron toward the negatively charged inner side of the membrane contributing to the membrane potential. When the electron is passed to oxidized UQ a semi-quinone (UQ⁻) is formed. During round 2 (bottom) of a Q cycle, after UQH$_2$ is formed, one electron is passed to an FeS protein to cytochrome c and then to the terminal cytochrome oxidase with the release to two protons across the membrane. The second electron passes to cytochrome b, which carries the electrons toward the inner side of the membrane contributing to the membrane potential, before passing the electron to a semi-quinone (UQ⁻). In total four protons are moved across the membrane by a Q cycle during the two rounds of electron transport. The black arrows indicate electron flow and the black dashed arrows indicate proton flow.

membrane using either a Q-cycle or Q-loop mechanism. While both Q cycles and Q loops contribute to the PMF, Q cycles result in the movement of twice as many protons per electron and therefore generate more PMF.

Q Cycle of Mitochondria

Q cycles (Fig. 6.2B) are found in mitochondria, chloroplasts, and many bacteria, including phototrophs (Chapter 10). In mitochondria, when the Q cycle is used for electron transfer events between quinone molecules and cytochrome bc_1, each cycle requires two "rounds" of electron transfer. In round 1, one of the two electrons from a reduced ubiquinone (UQH_2) is delivered to an FeS protein and then to cytochrome c (on the outer side of the membrane) prior to continued electron transfer to terminal cytochrome oxidase. The 2 protons carried by the ubiquinone are released across the membrane to the mitochondrial inner membrane space contributing to the PMF. The release of 1 electron and loss of 2 protons generates a "semi-quinone" form of ubiquinone (UQ^-) carrying just 1 electron. The cytochrome b moves the electrons through the membrane (from the outer interface to the inner interface next to the matrix); this contributes to the membrane potential (i.e., increasing the negative charge of the inside membrane surface). The electron is then passed to oxidized ubiquinone (UQ), thus generating semi-quinone form of ubiquinone (UQ^-).

In round 2, a second set of electrons from the ETC reduces the ubiquinone and similar to round 1, these electrons are delivered to cytochrome bc_1. Again, 1 electron from a reduced ubiquinone is delivered to an FeS protein and then passed to cytochrome c prior to continued electron transfer to the terminal cytochrome oxidase; in the process 2 more protons are released to the inner mitochondrial membrane space. The release of 1 electron and loss of 2 protons generates a semi-quinone (UQ^-). However, in contrast to round 1, the electron carried by cytochrome b is transferred to the semi-quinone generated in round 1. In summary, the transfer of 2 electrons through the Q cycle increases the strength of the PMF across the mitochondrial membrane by 4 protons (i.e., a 1:2 ratio). In addition, the movement of electrons within the membrane by cytochrome b also contributes to the PMF by enhancing the charge differential across the membrane.

Bacterial Electron Transport Chains (ETCs)

When *Escherichia coli* cells are growing via **aerobic respiration**, the steps of the ETC, starting with NADH + H$^+$ and ending with oxygen, are very similar in structure and function to the ETC of mitochondria (compare Fig. 6.2 and 6.3). Both the mitochondrial and bacterial ETCs rely on redox loops, where carriers for both protons and electrons alternate with carriers solely for electrons. This separation of protons and electrons results in a coupled site (i.e., proton pump, Q loop, or Q cycle) that assists in generating a proton gradient across the membrane. This PMF is then used to drive ATP production by the ATP synthase enzyme, or to drive substrate transport (Chapter 13) or flagellar rotation in some bacteria (Chapter 15).

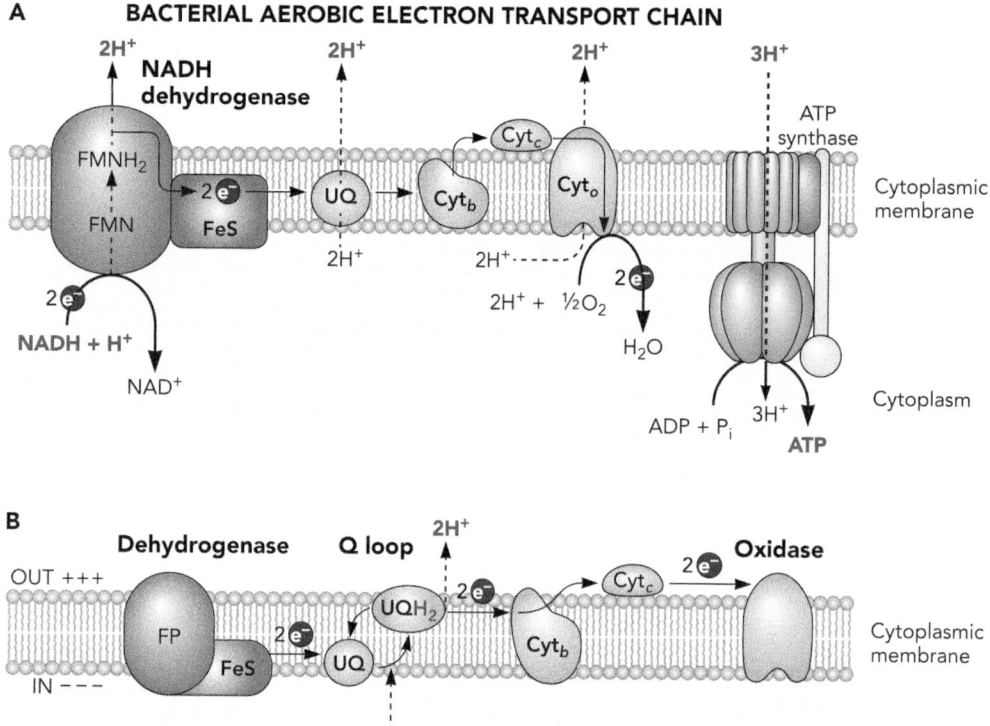

FIGURE 6.3 **An example of an** *Escherichia coli* **electron transport chain (ETC) under aerobic growth conditions.** (A) Electrons flow from the initial electron donor (e.g., NADH + H$^+$) with a negative E_0' to the terminal electron acceptor (i.e., O$_2$) with a positive E_0'. Protons and electrons are first donated to the flavoprotein within NADH dehydrogenase and are passed to the FeS protein, also within the dehydrogenase. There is sufficient energy released to pump protons across the membrane. UQ represents ubiquinone, which is capable of accepting two electrons from the FeS protein within the NADH dehydrogenase and two protons from the cytoplasm. These protons are then released outside of the membrane when the electrons are passed to cytochrome (Cyt) *b*, Cyt *c*, and then the terminal oxidase complex (Cyt *o*), where additional protons are pumped across the membrane. The potential energy in the proton motive force (ΔP) is used by the ATP synthase to generate ATP as the protons move down their concentration gradient and reenter the cytoplasm. See the text for additional details. (B) A Q loop is used in the ETC of some bacteria. When oxidized UQ accepts two electrons from a dehydrogenase and two protons from the cytoplasm it is fully reduced (UQH$_2$). When it transfers two electrons to cytochrome *b*, two protons are released to the outside of the membrane. The two electrons move from cytochrome *b* to cytochrome *c* to the terminal oxidase. In total two protons are moved across the membrane during a Q loop. The black arrows indicate electron flow and the black dashed arrows indicate proton flow.

The initial proton and electron donor to the prosthetic group of the flavoprotein of NADH dehydrogenase is NADH + H$^+$. At this coupled site, electrons are then transferred to the FeS protein component of the NADH dehydrogenase and the protons are pumped across the cell membrane. The electrons are passed from the FeS protein to ubiquinone. As part of a Q loop (see below), protons are moved to the outside of the cytoplasmic membrane. FADH$_2$ may also donate protons and electrons to succinate dehydrogenase (not shown in Fig. 6.3A). The electrons are passed to ubiquinone, but no protons are pumped across the membrane. The electrons pass from the ubiquinone to cytochrome *b*, cytochrome *c*, and then to the terminal cytochrome *o* oxidase complex. The terminal oxidase complex may serve as a proton pump in some bacteria, depending on the specific steps and molecules involved in the ETC. In aerobic respiration, oxygen is the terminal electron acceptor and water is formed as the end product.

Q Loop of Bacteria

While mitochondria, chloroplasts, and some bacteria use Q cycles, **Q loops** are found only in bacteria, including *E. coli* (Fig. 6.3B). In a Q loop, an oxidized ubiquinone (UQ) accepts 2 electrons from an FeS protein along the inner side of the cytoplasmic membrane, where 2 protons are acquired from the cytoplasm. A reduced ubiquinone (UQH_2) then donates 2 electrons to cytochrome *b* as protons are released to the periplasm, contributing to the PMF. The electrons continue down the ETC to the terminal cytochrome oxidase. In summary, the transfer of 2 electrons through the Q loop increases the strength of the PMF across the cytoplasmic membrane by 2 protons (i.e., a 1:1 ratio).

Electron Donors and Acceptors in Bacteria

Although bacterial and mitochondrial ETCs each involve flavoproteins, FeS proteins, quinones, and cytochromes, they differ in that the bacterial chains are not simply linear ETCs. Rather, bacterial ETCs may have branch points at the quinone and cytochrome components (Fig. 6.4) or dehydrogenases (Fig. 6.5). Since alternative ETC components may be utilized, bacteria also have the capacity to regulate and alter their ETC, depending on growth conditions. Therefore, based on the combination of electron carriers, the strength of the PMF will vary depending on the number of coupled sites in the complete ETC between the initial electron donor and terminal electron acceptor.

In addition to aerobic respiration, where oxygen is the terminal electron acceptor, many bacteria are capable of **anaerobic respiration**. These ETCs

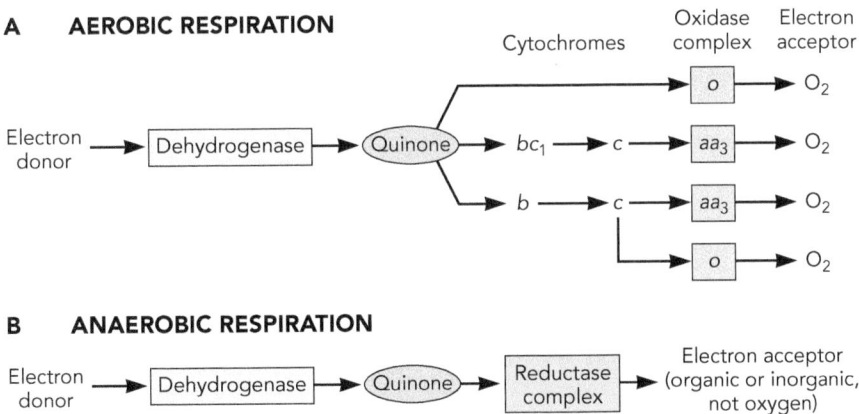

A   AEROBIC RESPIRATION

B   ANAEROBIC RESPIRATION

FIGURE 6.4 Comparison of the electron transport chains during bacterial aerobic and anaerobic respiration. (A) During aerobic respiration, the electrons are passed from the initial electron donor to a dehydrogenase, a quinone, cytochromes, and the terminal cytochrome oxidase complex, ending when the electrons reduce the terminal electron acceptor, which is always oxygen. Note that the components used in the pathway can vary depending on the organism and growth conditions and that multiple branch points for electron flow are possible. For example, some bacteria have Cyt bc_1, Cyt *c*, or oxidase aa_3 complex (similar to mitochondria), whereas other bacteria have evolved pathways utilizing different components. (B) During anaerobic respiration, the electrons are passed from the initial electron donor to a dehydrogenase, a quinone, and the terminal reductase complex, ending when the electrons reduce the terminal electron acceptor, which is never oxygen. The electron acceptor may be organic or inorganic in nature.

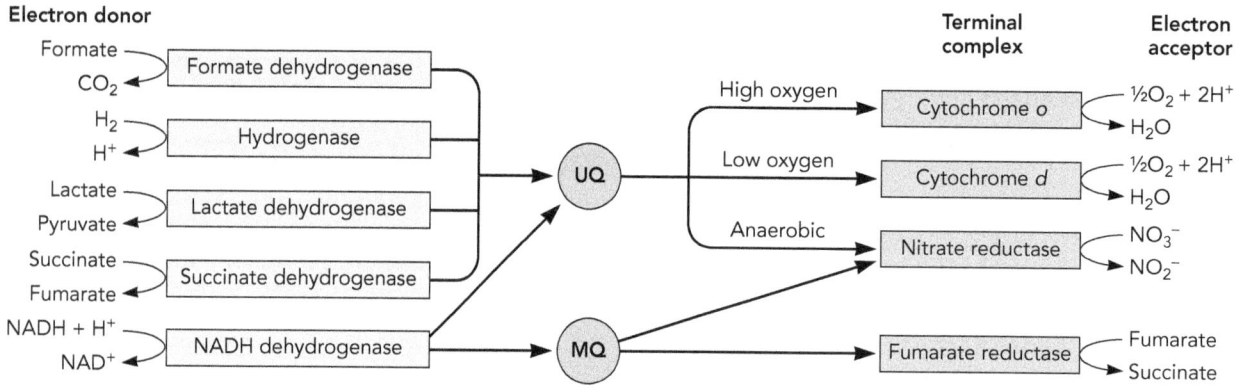

FIGURE 6.5 **The *Escherichia coli* electron transport chain (ETC) varies depending on environmental conditions.** *E. coli* has the capacity to use a number of different initial electron donors; thus, it also has the capacity to make multiple dehydrogenases and a hydrogenase. Two quinones are utilized in the *E. coli* ETC, ubiquinone (UQ) and menaquinone (MQ). Depending on the available terminal electron acceptor, *E. coli* will produce a terminal oxidase (cytochrome *o* or *d*) or reductase (nitrate or fumarate reductase) complex. The cell will produce the appropriate components to ensure that the electrons will flow in an energetically favorable manner from a negative to a positive electrode/reduction potential.

contain terminal membrane complexes known as reductases that replace the cytochromes and oxidases found in the aerobic ETC. Obviously, the terminal electron acceptor at the **terminal reductase complex** cannot be oxygen under anoxic conditions. Instead, other inorganic (e.g., nitrate) or organic (e.g., fumarate) compounds can serve as the terminal electron acceptor (Table 6.2). Chapters 11 and 12 will discuss various forms of anaerobic respiration in more detail.

Bacteria are also not restricted to using only NADH + H$^+$ and FADH$_2$ as their electron donors. There are at least seven potential electron donors and corresponding dehydrogenase or hydrogenase complexes that may be used in the ETC of *E. coli* (Fig. 6.5). *E. coli* carries the genes for not just one, but two quinones: ubiquinone and menaquinone. In addition, *E. coli* can produce multiple membrane-associated oxidases and reductases. The specific components of the ETC that are employed depend on the available electron donors and electron acceptors in the growth medium or environment; however, the electrons must always flow from a carrier with a more negative potential. For example, NADH + H$^+$ ($E_0' = -320$ mV) donates its electrons to NADH dehydrogenase, but then the electrons could flow to either ubiquinone ($E_0' = +100$ mV) or menaquinone ($E_0' = -74$ mV). The choice of which quinone is utilized is dictated by the available electron acceptors. If oxygen is available, then the electrons can flow from ubiquinone ($E_0' = +100$ mV) to either cytochrome *o* oxidase, which is produced under atmospheric oxygen conditions, or cytochrome *d* oxidase, which is produced under microoxic conditions. In both cases, oxygen ($E_0' = +815$ mV) serves as the terminal electron acceptor. If environmental conditions become anoxic, then nitrate ($E_0' = +421$ mV) is the next preferred terminal electron acceptor. Ubiquinone ($E_0' = +100$ mV) or menaquinone (-74 mV) can both donate electrons to nitrate ($E_0' = +421$ mV) at the terminal nitrate reductase complex in the ETC. However, if fumarate ($E_0' = +33$ mV) is the only available electron acceptor under anoxic conditions, then it becomes energetically unfavorable for the electrons to flow from ubiquinone ($E_0' = +100$ mV) to fumarate ($E_0' = +33$ mV). Under these circumstances, menaquinone ($E_0' = -74$ mV) will be used as the quinone component of the ETC, enabling fumarate ($E_0' = +33$ mV) to serve as the terminal electron acceptor. Bacteria that can grow under a variety of environmental conditions have evolved the ability to sense the presence of different electron donors and acceptors and then appropriately regulate the expression of genes encoding the various ETC components to optimize their energy yield (see Chapter 11 for details).

Fermentation

During respiration, the release of energy from the oxidation of organic compounds is coupled to generation of a PMF, which is subsequently used to produce ATP. The electrons ultimately land on an "external" electron acceptor that is not a product of a central metabolic pathway (e.g., oxygen under oxic conditions). In contrast, **fermentation** involves the use of balanced redox reactions where the terminal electron acceptor is an "internal" product of the catabolic pathway and ATP production occurs via substrate-level phosphorylation (see Fig. 5.1). During the fermentation of glucose, for example, glucose is oxidized to pyruvate using the reactions of the Embden-Meyerhof-Parnas pathway (Fig. 6.6). In the process, two intermediates that contain energy-rich phosphate bonds are generated: 1,3-bisphosphoglycerate and phosphoenolpyruvate. Cleavage of these high-energy phosphate bonds by the relevant kinase enzymes releases enough free energy to allow the formation of ATP from $ADP + P_i$. The oxidation of each molecule of glucose also results in the reduction of two NAD^+ molecules to $NADH + H^+$ (Fig. 6.6), which need to be reoxidized for subsequent rounds of glycolysis (Chapter 2).

Since there is no ETC functioning during fermentation, the $NADH + H^+$ must be reoxidized by an alternative mechanism. As an example, *E. coli* accomplishes this by reducing some of the pyruvate produced as the end product of glycolysis. Therefore, during fermentation, only a portion of the carbon from glucose is utilized for the production of biomass and ATP; the remainder is excreted as reduced fermentation waste products, such as organic acids and alcohols. In addition, unlike respiratory bacteria, fermenting bacteria are unable to conserve energy during the reoxidation of $NADH + H^+$.

Fermentations are generally named for the products that are formed. Figure 6.6 shows the lactic acid fermentation where pyruvate is reduced to lactate to allow the reoxidation of $NADH + H^+$. This fermentation, which is carried out by *Streptococcus* and some *Lactobacillus* species, is termed a homolactic fermentation since a single product, lactate, is formed. Members of the genus *Leuconostoc* and some *Lactobacillus* species carry out heterolactic fermentations that generate multiple fermentation products including lactate, ethanol, and CO_2. Various fermenting bacteria produce different fermentation end products, and many different compounds besides glucose can serve as fermentation substrates. It is also important to note that, because they still contain potential energy, many fermentation "waste" products can serve as substrates for other members of the microbial community. One example is the propionic acid fermentation carried out by *Propionibacterium* and *Clostridium propionicum* (Fig. 6.7). This process is termed a secondary fermentation since the substrate, lactate, is itself a fermentation end product that is then further reduced to propionate by a different group of fermenting bacteria during the reoxidation of $NADH + H^+$.

PRIMARY FERMENTATIONS:

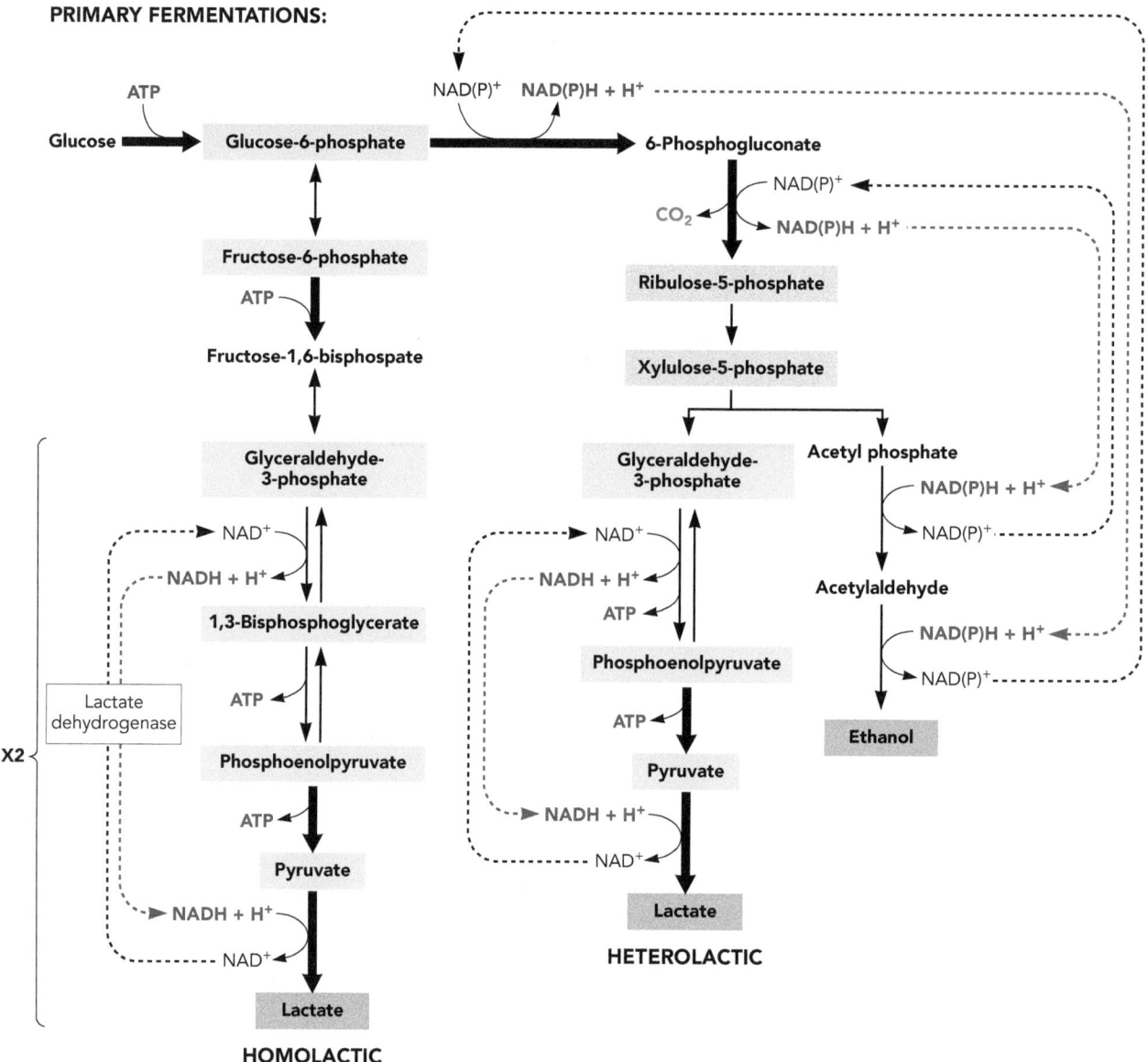

FIGURE 6.6 **Examples of primary fermentations.** Homolactic fermentations produce one end product (only lactate), whereas heterolactic fermentations produce multiple end products (CO_2, lactate, and ethanol). Note how both pathways function to reoxidize reduced coenzymes (NAD(P)H + H$^+$) produced during the reactions of glycolysis and the pentose phosphate pathway while yielding ATP via substrate-level phosphorylation.

In general, the energy yield (i.e., ATP formation) during fermentation is much lower than that of respiration. Another consequence of not having an ETC is the need to "spend" ATP by reversing the ATPase to generate the PMF, which is required to directly power essential transport systems and, in motile bacteria, the flagellar motor. It is also important to note that although oxygen is not used during fermentation and many fermentative bacteria are obligate anaerobes, some fermentations, such as the lactic acid fermentation, can take place in the presence of oxygen.

SECONDARY FERMENTATION:

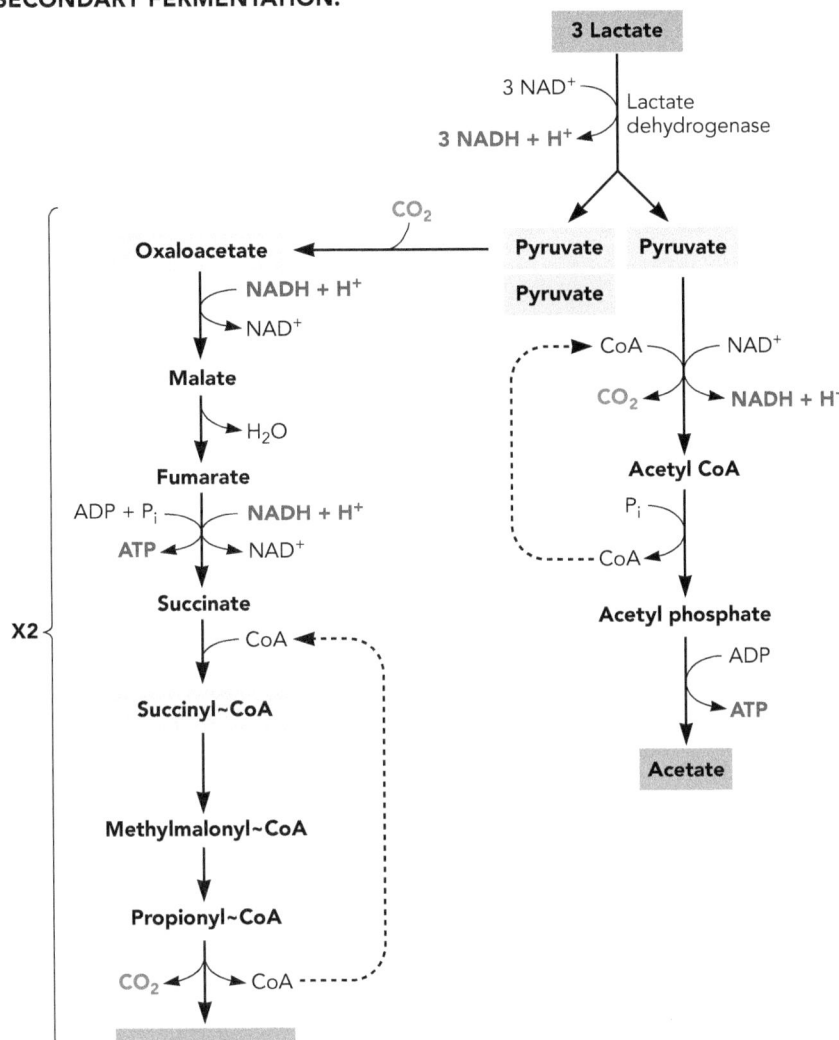

FIGURE 6.7 **An example of a secondary fermentation.** In a secondary fermentation, the end products of a primary fermentation, in this case lactate, (Fig. 6.6) are further metabolized to make new (i.e., secondary) end products (propionate, acetate, and CO_2). Note how the reactions function to reoxidize reduced coenzymes (NADH + H$^+$).

Learning Outcomes: After completing the material in this chapter, students should be able to . . .

1. describe the functions of the four basic components of an electron transport chain

2. predict differences in energy yield based on different electron donors and acceptors

3. explain how the movement of electrons through an electron transport chain is coupled to the creation of a proton motive force and ATP generation

4. compare and contrast electron and proton flow in mitochondrial and bacterial Q cycles and Q loops

5. identify key reduction/oxidation and energy-generating steps in fermentation pathways

6. differentiate the processes of aerobic respiration, anaerobic respiration, and fermentation

Check Your Understanding

1. Define this terminology: prosthetic group, flavoprotein, iron-sulfur (FeS) protein, cytochrome, quinone, electrode/re-duction potential, terminal electron acceptor, proton pump, terminal oxidase complex, coupled site, Q cycles, aerobic respiration, Q loops, anaerobic respiration, terminal reductase complex, fermentation

2. Fill in the blanks with the name(s) of each component found in an electron transport chain: cytochrome, FeS protein, flavoprotein, quinone. (LO1)
   a) Which of these components moves only electrons?
   b) Which components can move both protons and electrons?
   c) Which component contains a heme group?
   d) Which component moves freely within a membrane?

3. The purpose of completing the table below is to help you organize important information relative to components within the mitochondrial electron transport chain (NADH dehydrogenase has been done for you). (LO1)

Complex	Functional components	Electrons come from:	Electrons go to:	Does it move protons across the membrane?
NADH dehydrogenase (Complex I)	Flavoprotein and FeS protein	NADH + H⁺	Quinone	Yes
Succinate dehydrogenase (Complex II)				
Cytochrome (Complex III)				
Terminal cytochrome oxidase (Complex IV)				

4. During oxidative phosphorylation in mitochondria, the movement of electrons through the electron transport chain (ETC) results in the formation of a proton motive force. (LO3)
   a) Explain how this happens. Why is it essential that some components carry both electrons and protons?
   b) Explain how the movement of electrons through the ETC results in the production of ATP.

5. The diagram represents a mitochondrial electron transport chain. (LO3)

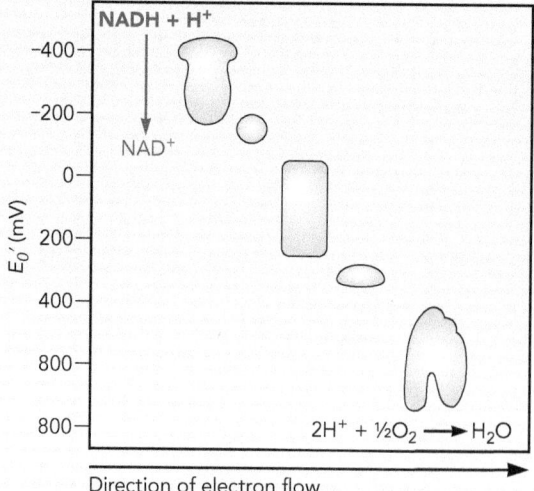

Direction of electron flow

   a) Label each complex based on its E_0' value.
   b) Draw in the pathway for electron flow with NADH + H⁺ as the electron donor.
   c) Draw in the pathway for electron flow if the electrons were being donated by $FADH_2$.
   d) Why can't $FADH_2$ donate electrons to NADH dehydrogenase?

6. The purpose of completing the table below is to help you organize important information relative to an aerobic bacterial electron transport chain. For each component, list from where the electrons come and to where they go. (LO1)

Component	Electrons come from:	Electrons go to:	Does it move protons across the membrane?
NADH dehydrogenase (flavoprotein and FeS protein)			
Cytochrome *b*			
Cytochrome oxidase			

7. Bacteria have a wide range of electron carriers, with branch points throughout the electron transport chain. (LO2)
   a) Assuming oxygen is the terminal electron acceptor, use the E_0' values to predict which energy source would release more energy. From the list below, place the energy sources in order, with the most energetic first.

	E_0' (mV)
Succinate	+30
NADH + H⁺	−320
Methane (CH₄)	−240
Hydrogen (H₂)	−410
Acetate	−290

   E_0' (mV)
   Succinate +30
   NADH + H⁺ −320
   Methane (CH₄) −240
   Hydrogen (H₂) −410
   Acetate −290

   b) Assuming glucose is the electron donor, use the E_0' values to predict which terminal electron acceptor would maximize energy conservation from the electron transport chain. From the list below, place the terminal electron acceptors in order, with the most energetically favorable first.

   E_0' (mV)
   Fe³⁺ (pH 7) +200
   Fumarate +30
   Nitrate +430
   Oxygen +820

   c) Which electron acceptors from part (b) indicate anaerobic respiration?

8. The Q loop is found in some bacteria, but not mitochondria. (LO4)
   a) In the first step of the Q loop, an oxidized ubiquinone (UQ) accepts two electrons and two protons (H⁺) to become a reduced ubiquinone (UQH₂). From where do these two electrons come? From where do the H⁺ come?
   b) In the second step of the Q loop, the reduced ubiquinone (UQH₂) then donates two electrons to regenerate oxidized ubiquinone (UQ). To where do these two electrons go? To where do the H⁺ go?
   c) How does this overall process contribute to the PMF?

9. While the Q cycle is found and probably evolved in some bacteria, it is also found in mitochondria and chloroplasts. (LO4)
   a) In round 1 of the Q cycle, two electrons and two protons reduce UQ from UQH₂. From where do the electrons and protons come in this first step of the Q cycle?
   b) After that, each electron travels a different route through different electron carriers.
      i. One electron is passed to an FeS protein in cytochrome *bc₁*, thus forming UQ⁻. Where does the electron go from there?
      ii. The other electron (now on a "semiquinone" form of ubiquinone, UQ⁻) is passed to cytochrome *b*. Where does the electron go from there?
      iii. How does this overall process contribute to the proton motive force (PMF)?
   c) In round 2 of the Q cycle, two electrons and two protons reduce another molecule of UQ to UQH₂.
      i. From where do the electrons and protons come in this step?
      ii. One electron is passed to an FeS protein in cytochrome *bc₁*, thus forming UQ. Where does the electron go from there?

iii. The other electron (now on UQ·) is passed to cytochrome *b*. Where does the electron go from there?

iv. How does this overall process contribute to the PMF?

d) Which process, the Q loop or the Q cycle, generates a stronger PMF? Explain why.

10. The pathway shows the homolactic fermentation pathway found in many lactic acid bacteria used in the dairy industry. This pathway uses the Embden-Meyerhof-Parnas (glycolytic) pathway (from glucose to pyruvate). (LO5)

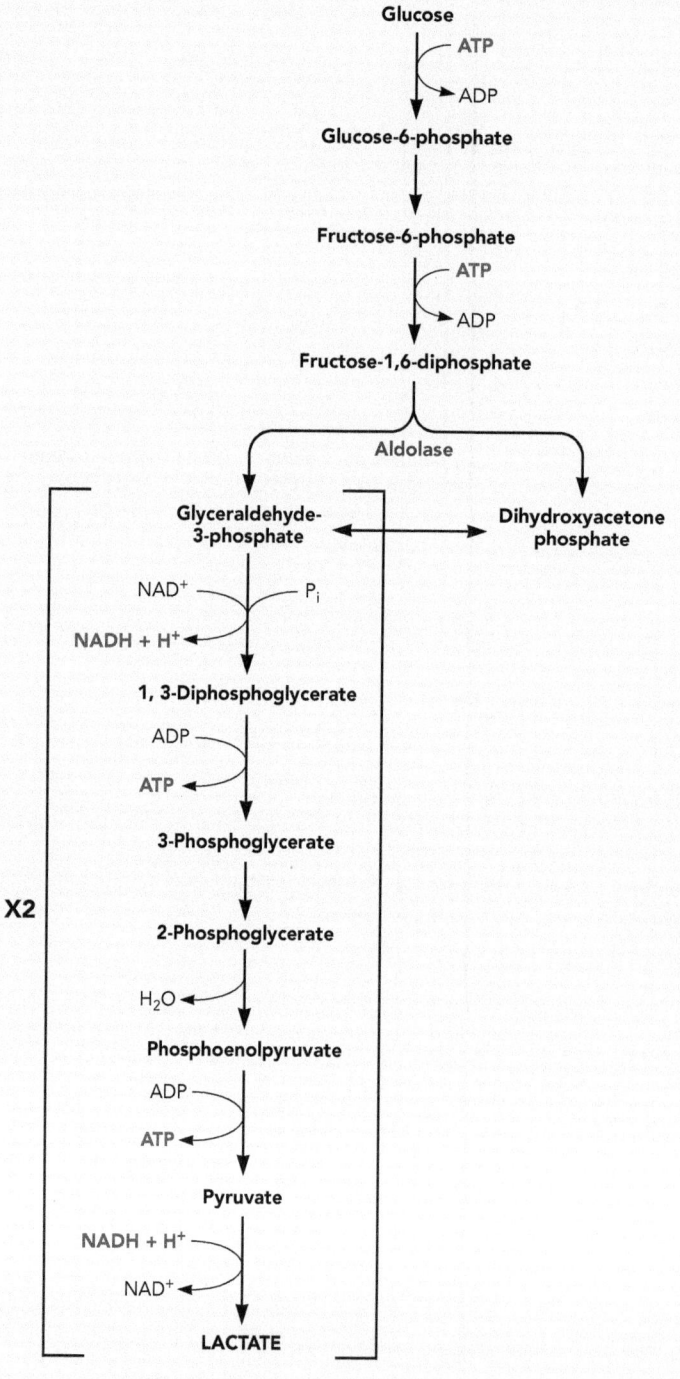

a) In which step(s) is ATP used? What is the purpose of this step? (Review from Chapter 2)

b) In which step(s) do redox reactions occur? What is the product of these reactions?

c) Where is ATP made?

d) What is the net production of ATP in this pathway?

e) Why must a bacterium using this pathway convert pyruvate to lactate? What does this part of the pathway provide for the cell?

Dig Deeper

11. Bacteria have an immense diversity of electron carriers and acceptors in their electron transport chains (ETCs). (LO6)

    a) Considering where they live and grow, why do you think bacteria have so many different electron carriers, with branch points throughout the ETCs, compared to mitochondria?

    b) *Paracoccus denitrificans* is a Gram-negative alphaproteobacterium that is commonly found in soil environments, where the oxygen concentration can vary from full atmospheric $[O_2]$ to low $[O_2]$ to no O_2. *P. denitrificans* makes at least two terminal oxidases with differing affinities for oxygen: cyt aa_3 (K_m for oxygen ~1 μM) and cyt cbb_3 (K_m for oxygen ~0.01 μM).

        i. Given the K_m values, which enzyme has a higher affinity for oxygen?

        ii. Predict which terminal oxidase it likely uses under high $[O_2]$ and which it will use under low $[O_2]$. Explain your reasoning.

        iii. Over geologic time, low levels of O_2 probably began to accumulate in our atmosphere around 2.5 billion years ago. The current levels of oxygen (~21%) did not occur until around 0.43 billion years ago. Given this history, which ancestors of cyt aa_3 or cyt cbb_3 likely evolved first?

        iv. *P. denitrificans* also makes a terminal nitrate reductase (NAR), for the reduction of NO_3^- to NO_2^-. This enzyme is only made in the cell under certain conditions. From an energetic perspective, would you predict that NAR would be made only under oxic or anoxic conditions? Explain your reasoning.

12. During your summer job investigating the microbiology of wetland soils, you were able to isolate a bacterium that uses C1 compounds (CH_4 or CO_2) and inorganic nitrogen compounds (nitrate or nitrite) to make ATP. (LO1, 2, 3) The following redox couples are involved:

$$NO_3^- / NO_2^- \left[E_0' = +430\,\text{mV} \right] \text{ and } CO_2/CH_4 \left[E_0' = -240\,\text{mV} \right]$$

    a) Your initial research shows that this bacterium respires using a membrane-based electron transport chain (ETC) coupled to an ATP synthase. If these two redox pairs were coupled together during respiration:

        i. Which molecule would be the electron donor?

        ii. Which molecule would be the electron acceptor?

    b) Below is a list of membrane components (with E_0' values) that could *potentially* be involved in an ETC for this bacterium. Your job is to decide which components are needed to build an ETC for ATP synthesis via respiration. Use as many components as possible to construct a diagram that illustrates an electron transport system that would work. Be sure to:

        • draw in the flow of electrons and include both the electron source and final electron acceptor

        • draw in the coupled sites where proton motive force is generated

	E_0' (mV)
C1: Cytochrome 1	−300
C2: Cytochrome 2	+200
FP1: Flavoprotein 1	−400
FP2: Flavoprotein 2	−100
Q: Quinone	+100
TR1: Terminal reductase 1	+500
TR2: Terminal reductase 2	+300
AS: F-type ATP synthase	N/A

c) Explain why you were not able to use all the components in the table.

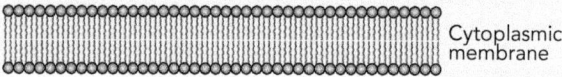

Outside cell

Cytoplasmic membrane

Inside cell

d) To further investigate this bacterium, you grow it in the presence of a new antibiotic that allows for the free movement of protons (H⁺) across the membrane. What will be the immediate effect of this antibiotic on each of these cell processes? Explain your answer.

   i. Movement of electrons through the ETC: increase / decrease / no effect

   ii. ATP synthesis via the ATP synthase: increase / decrease / no effect

13. You are working for a startup food biotech company. You have a bacterium that uses the fermentation pathway shown on the following page. Use this pathway to answer these questions. (LO5)

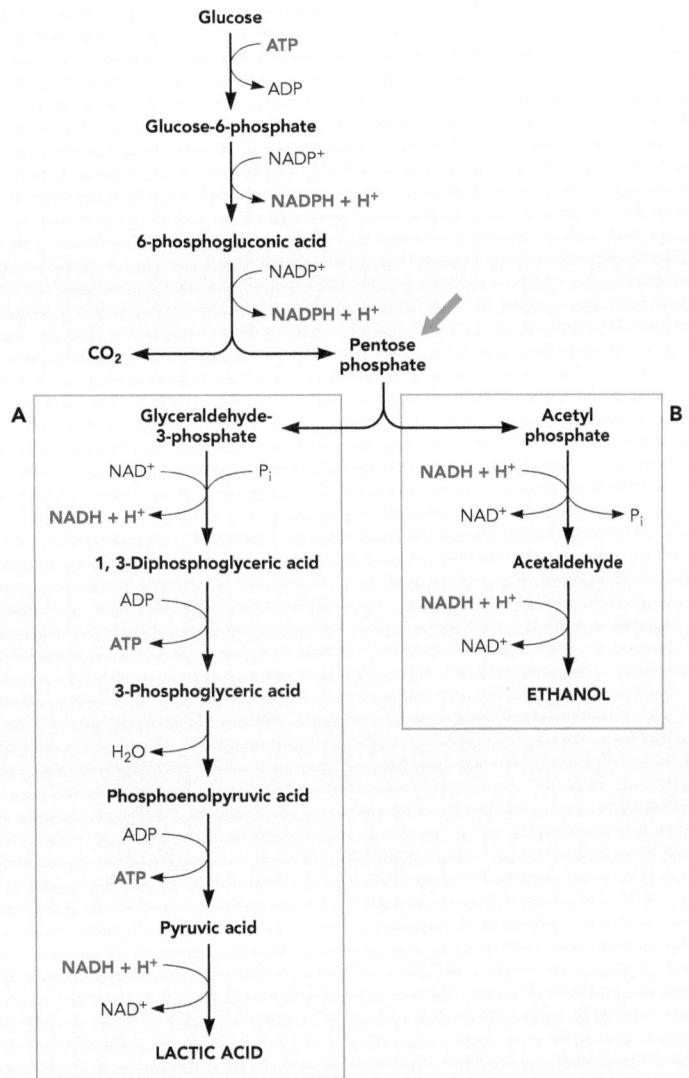

a) Where is the energy from the flow of electrons being used to make a new high-energy bond?

b) What is the net yield of ATP for this pathway (per 1 molecule of glucose)? Could a bacterium live and grow using this pathway?

c) Your boss wants the bacterium to produce more ethanol. You use genetic engineering tools to make a strain that does not have the "A" part of the pathway. Could this organism thrive making only ethanol? Explain your reasoning.

d) Now your boss wants the bacterium to produce more lactic acid. You use genetic engineering tools to make a strain that does not have the "B" part of the pathway. Could this bacterium now thrive, using glucose for its energy source and making only lactic acid, and *no* ethanol? Explain your reasoning.

e) Suppose you take that same strain that is lacking part B and grow it on ribose (a C5 pentose sugar) instead of glucose. The pathway now begins with pentose phosphate (at the blue arrow). Could this bacterium now thrive making only lactic acid, and *no* ethanol? Explain your reasoning.

128 Making Connections

128 Importance of Regulatory Processes

129 Allosteric Regulation of Enzymes

131 Allosteric Regulation of Branched Pathways

134 Covalent Modifications

138 Posttranslational Regulation in the Sugar Phosphotransferase System (PTS)

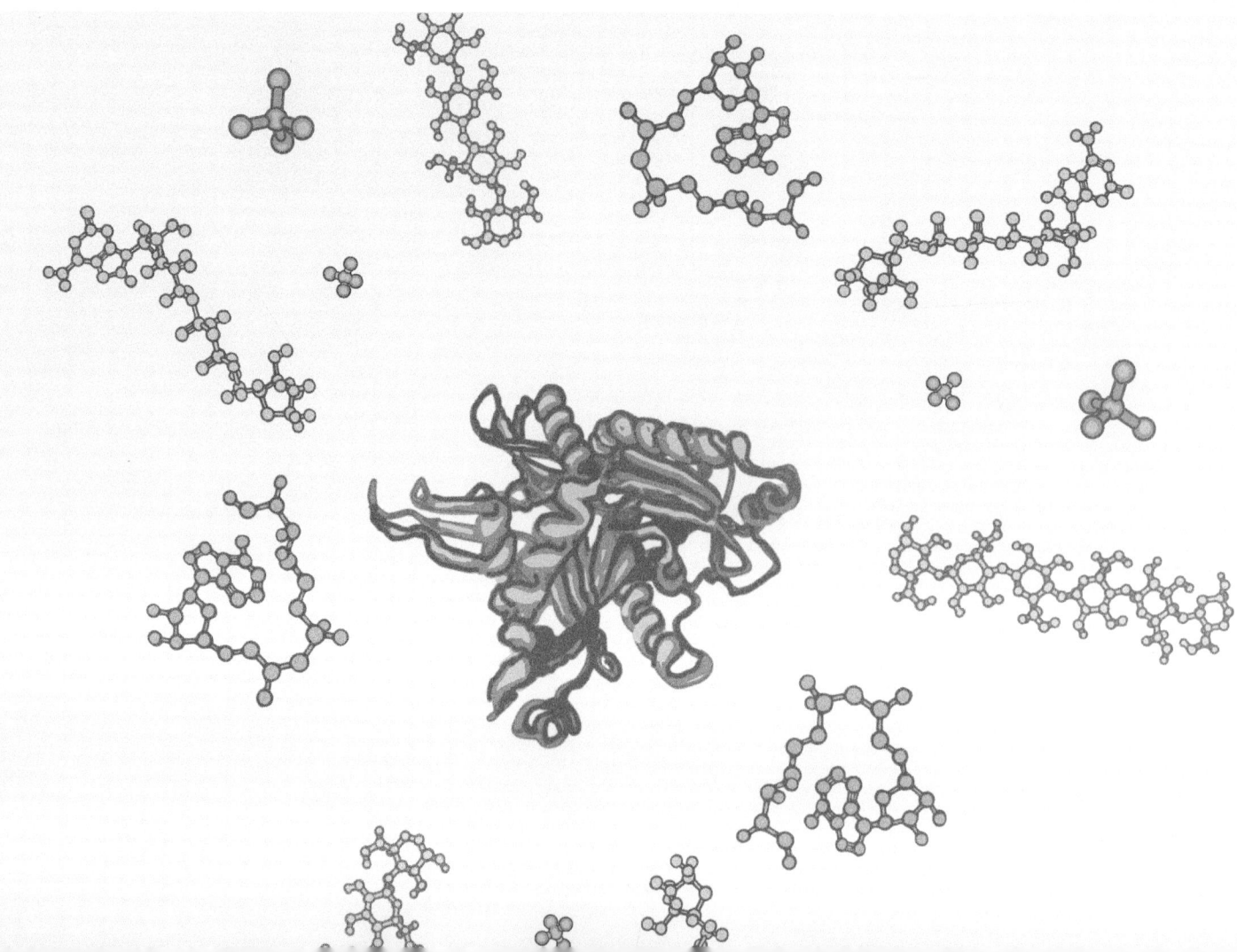

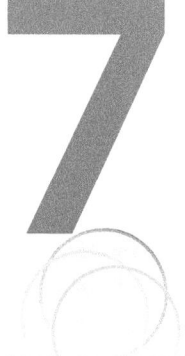

REGULATION—
POSTTRANSLATIONAL CONTROL

After completing the material in this chapter, students should be able to . . .

1. describe three key modes of regulation controlling protein levels and activity

2. explain the mechanisms of positive and negative allosteric enzyme regulation

3. give examples of feedback inhibition in which K_m or V_{max} is altered

4. discriminate between different types of feedback inhibition in branched pathways

5. describe four examples of covalent modification using different functional groups

6. outline the sugar phosphotransferase system (PTS), including how it is regulated in the presence or absence of PTS and non-PTS sugars

Microbial Physiology: Unity and Diversity, First Edition. Ann M. Stevens, Jayna L. Ditty, Rebecca E. Parales and Susan M. Merkel.
© 2024 American Society for Microbiology.
Companion website: www.wiley.com/go/stevens/microbialphysiology

Making Connections

Microbes are able to rapidly adjust to changing environmental circumstances in large part because they can regulate gene expression and enzyme activity. In Chapter 2, the regulation of key unidirectional enzymes that direct the carbon flow in central metabolism was discussed. In Chapter 6, the concept that microbes can alter their respiratory pathways based on environmental conditions was introduced. Regulation occurs at multiple points during the steps of the central dogma; this chapter focuses on posttranslational control of protein activity. Transcriptional and posttranscriptional mechanisms of control will be covered next in Chapter 8.

Importance of Regulatory Processes

If *Escherichia coli* cells are grown in a medium with ^{14}C-glycerol as the carbon source, 100% of the organic molecules in the cells will eventually be labeled with ^{14}C. However, if an unlabeled nutritional supplement such as the amino acid histidine is also added to the medium, then all of the organic molecules in the cell except histidine will be labeled. Why? Because the cells will utilize the available histidine and not expend energy to continue to synthesize it. This is due to the regulatory mechanisms in microorganisms that enable them to adapt to changing environmental conditions very quickly. There is no benefit for cells to use energy to make a product that is already readily available.

Multiple types of regulation are often simultaneously carried out by microbial cells. Regulation can be exerted at the level of gene expression, during transcription initiation, posttranscriptionally, or at the point of translation (Fig. 7.1).

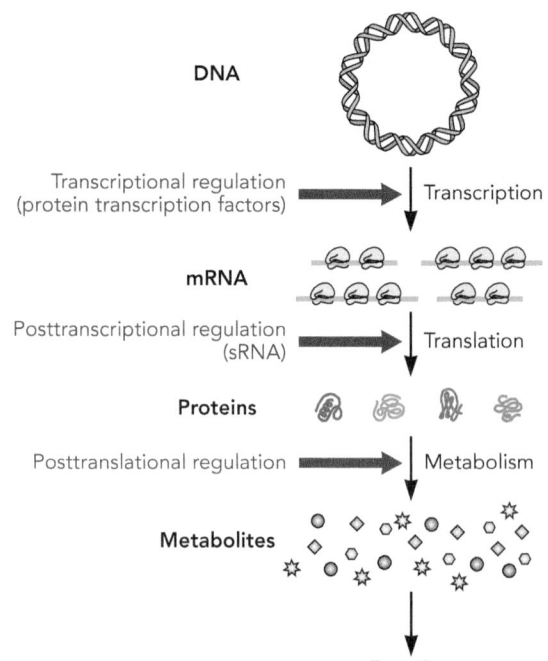

FIGURE 7.1 **Key modes of regulation that occur during the steps of the central dogma.** (Top to bottom) The process of transcription, the synthesis of mRNA from DNA, is catalyzed by RNA polymerase and regulated by protein transcription factors (e.g., activators and repressors). The process of translation, the production of polypeptides (proteins) by ribosomes using an mRNA template, is subject to posttranscriptional regulators (e.g., sRNAs and riboswitches). After a protein is translated, it may undergo posttranslational regulation (e.g., allosteric regulation or covalent modification). Enzymatic reactions catalyzed by proteins produce metabolites that influence cellular functions.

In addition, even after a polypeptide has been translated, its activity may be modulated by posttranslational regulation. In bacteria and archaea, gene expression can occur in a matter of minutes and posttranslational regulation can occur within fractions of a second. Changes in eukaryotic gene expression are slower due to the compartmentalization of transcription in the nucleus, followed by processing of the mRNA prior to transport to the cytoplasm, where translation occurs. However, posttranslational regulation, as described below, occurs at roughly the same rapid rate in all organisms.

Allosteric Regulation of Enzymes

When sufficient histidine is present in the environment, *E. coli* cells do not synthesize histidine. This is an example of **allosteric regulation**, whereby a small **allosteric effector molecule** controls (enhances or inhibits) the activity of a protein in a concentration-dependent manner. **Feedback inhibition** is one type of allosteric regulation. During simple linear feedback inhibition, the end product of an anabolic or biosynthetic pathway (e.g., histidine) functions as a negative allosteric effector to inhibit the activity of the first enzyme in its biosynthetic pathway (Fig. 7.2). This type of regulation helps the cell to conserve energy by decreasing the rate of the first reaction in the pathway, thereby preventing further energy consumption in downstream reactions. The negative effector molecule binds to a regulatory site in the enzyme, which alters the conformation of the active site and reduces the rate of the reaction (Fig. 7.3). Allosteric regulation may also be positive, where the binding of a positive effector molecule to the regulatory site in the enzyme converts the active site to a conformation that enhances the rate of the reaction (Fig. 7.3). Hence, one enzyme can have multiple conformations with both negative and positive effectors impacting its function (e.g., pyruvate dehydrogenase; Fig. 2.11).

During allosteric regulation, either the affinity of the enzyme for the substrate (K_m) or the rate of conversion of the substrate into the product (V_{max}) may be altered. The relationship between the substrate concentration and reaction velocity (i.e., Michaelis-Menten kinetics; Chapter 3) becomes a sigmoidal

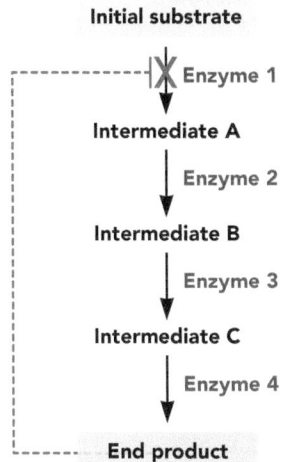

FIGURE 7.2 **Simple feedback inhibition.** During simple feedback inhibition, the end product of a pathway negatively regulates the first enzyme of the same pathway (red dashed line). The red "X" and bar indicate the enzyme is being inhibited in the pathway. This is a form of allosteric regulation.

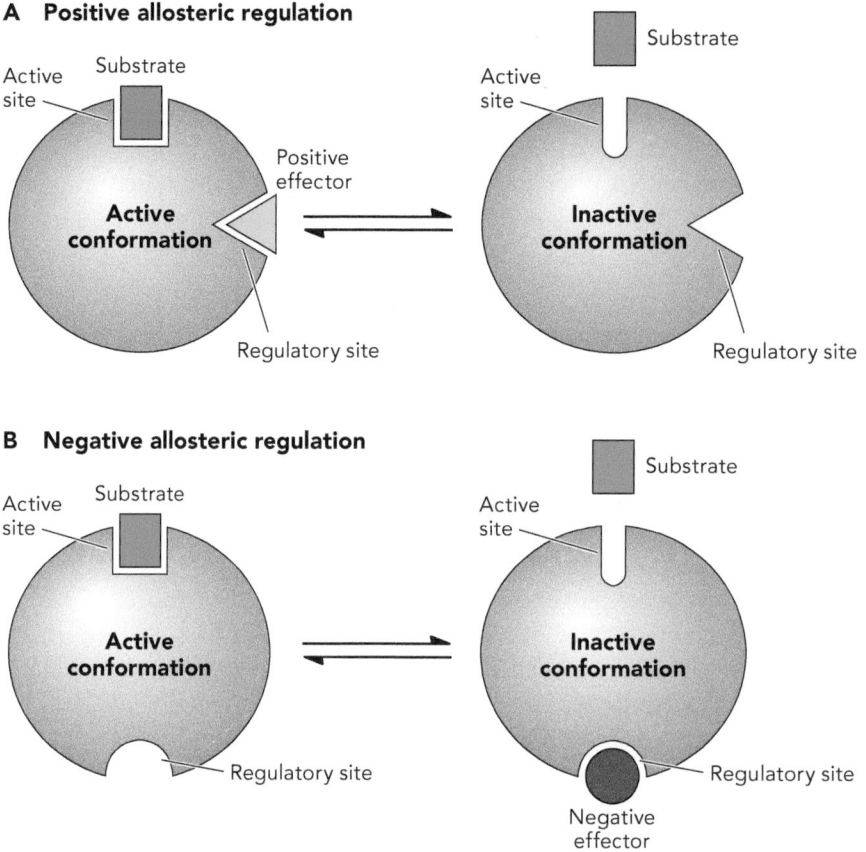

A Positive allosteric regulation

Active site — Substrate

Positive effector

Active conformation

Regulatory site

Substrate

Active site

Inactive conformation

Regulatory site

B Negative allosteric regulation

Active site — Substrate

Active conformation

Regulatory site

Substrate

Active site

Inactive conformation

Regulatory site

Negative effector

FIGURE 7.3 **Control of protein activity via allosteric regulation.** (A) During positive allosteric regulation, a protein is converted into its active conformation when a positive effector binds to a regulatory site in the protein. In the absence of the positive effector, the protein is in its inactive or less active conformation. (B) During negative allosteric regulation, a protein is in its active state until a negative effector occupies the regulatory site of the protein. In the presence of the negative effector, the protein is converted into an inactive or less active conformation.

curve for a regulated enzyme (Fig. 7.4 and 7.5). The substrate concentration at which $\frac{1}{2}V_{max}$ is achieved is still considered to be the K_m. When the K_m differs under two environmental conditions, this reflects the fact that regulation has altered the affinity of the active site for the substrate. For example, in the biosynthetic pathway for production of the nucleotide cytidine triphosphate (CTP), the first enzyme in the pathway, aspartate transcarbamoylase, is allosterically regulated by CTP via a feedback inhibition mechanism. It is counterproductive for a cell to expend energy producing CTP if it is readily available. Therefore, if exogenous CTP is added into the growth medium, the CTP serves as a negative effector molecule of aspartate transcarbamoylase, resulting in a higher K_m value for the enzyme (Fig. 7.4). The V_{max} remains unchanged, but the higher K_m reflects a decrease in the affinity of the enzyme for the substrate. Under these conditions, a higher concentration of the substrate is needed to achieve the same reaction rate. Thus, at a given substrate concentration, the rate of conversion of the substrate to the product will be lower in the presence of CTP, until the V_{max} is achieved.

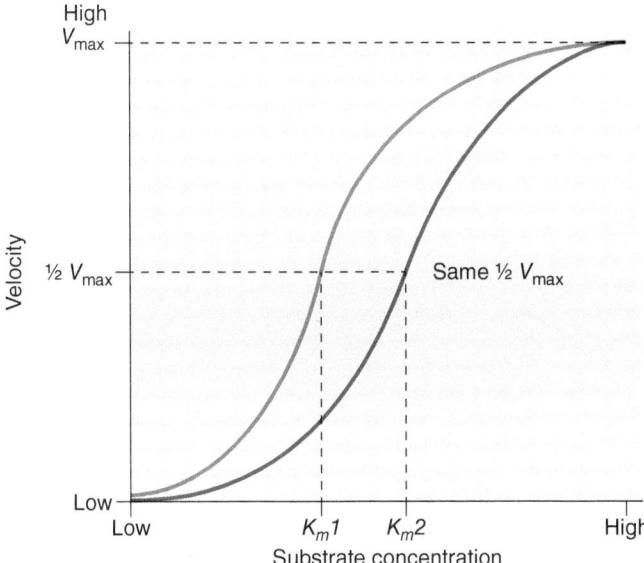

FIGURE 7.4 **Allosteric regulation may impact the K_m of an enzyme.** The activity of a regulated enzyme in the absence of an effector may be represented by a sigmoidal line when the rate of the reaction (velocity) is plotted versus the substrate concentration (blue line). The V_{max} represents the maximum rate at which the active site can convert substrate to product, and the K_m is the substrate concentration at which the reaction rate is half of the maximum ($\frac{1}{2}V_{max}$). If a negative allosteric effector binds to the regulatory site in the enzyme, then the activity plot changes (red line) such that the V_{max} remains unchanged but the K_m shifts to a higher value, indicating that a higher concentration of substrate is needed to achieve $\frac{1}{2}V_{max}$. Note that if the red line is considered to be the activity of the regulated enzyme in the absence of an effector, then the blue line would represent the enzyme activity in the presence of a positive effector. In this case, the K_m would be lower, indicating that a lower concentration of enzyme is now needed to achieve $\frac{1}{2}V_{max}$.

A less common mechanism of active-site modulation affects the V_{max}. This type of regulation occurs with the enzyme 3-deoxy-D-arabinoheptulosonic acid 7-phosphate (DAHP) synthase, the first enzyme in the biosynthetic pathway for the amino acid tryptophan. If tryptophan is provided in excess in the growth medium, it will serve as a negative effector of DAHP synthase since the cell no longer needs to expend energy to produce tryptophan. In this example, the K_m remains unchanged but the V_{max} changes (Fig. 7.5). This indicates that while the enzyme still has the same affinity for the substrate, it is unable to convert the substrate into the product with the same efficiency, and the rate of the reaction decreases. It is important to recognize that the K_m and V_{max} can also be altered due to positive allosteric regulation (e.g., PEP carboxylase; see below).

Allosteric Regulation of Branched Pathways

Biosynthetic pathways are not always organized in a simple linear sequence. Many cellular pathways are branched, with more than one end product, which complicates the process of regulation. Multiple mechanisms have evolved that enable cells to regulate these branched pathways (Fig. 7.6). One example is the use of **isozymes** (i.e., isofunctional enzymes), which are two or more

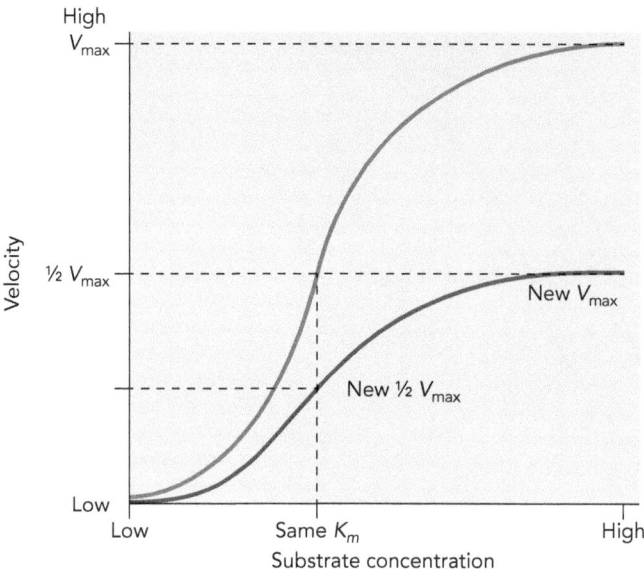

FIGURE 7.5 **Allosteric regulation may impact the V_{max} of an enzyme.** The activity of a regulated enzyme in the absence of an effector may be represented as a sigmoidal line when the rate of the reaction (velocity) is plotted versus the substrate concentration (blue line). The V_{max} represents the maximum rate at which the active site can convert substrate to product and the K_m is the substrate concentration at which the reaction rate is half of the maximum ($\frac{1}{2}V_{max}$). If a negative allosteric effector binds to the regulatory site in an enzyme, then the activity plot changes (red line) such that the K_m remains unchanged but the V_{max} and $\frac{1}{2}V_{max}$ are lower, indicating that although the substrate binds to the active site with the same affinity, the rate at which the substrate is converted into product is reduced. Note that if the red line is considered to be the activity of the enzyme in the absence of an effector, then the blue line would represent the enzyme activity in the presence of a positive effector. In this case, the K_m would remain unchanged but the V_{max} and $\frac{1}{2}V_{max}$ would be higher, indicating an increased rate in the conversion of substrate to product.

distinct enzymes that catalyze the same biochemical reaction but contain different allosteric sites that respond to different effector molecules. In pathways that generate multiple end products, the cell uses more than one enzyme to produce the first intermediate in the pathway. As an example, if there are two end products (F and G) in a pathway, there are two isozymes (enzyme 1A and enzyme 1B) that serve to generate the first pathway intermediate. Each end product then serves as a negative allosteric effector molecule to its respective isozyme that contains the matching allosteric site; this ensures that the cell will be able to make one end product (F) by keeping one isozyme active (enzyme 1A, the F-responsive isozyme) when the other end product (G) is present in excess and rendering the other isozyme (enzyme 1B, the G-responsive isozyme) inactive. The branched pathway will not be fully downregulated unless each end product is present to interact with its respective isozyme. Therefore, the isofunctional regulation process is not 100% efficient because the cell will continue to make the end products (F and G) to some extent until all end products are present above a certain threshold concentration.

Somewhat similarly, **concerted feedback** occurs when both end products of the branched pathway (F and G) must be present simultaneously to cause inhibition of the first enzyme in the pathway. In this case, one enzyme is regulated by both end products, resulting in neither end product being made. A more nuanced mechanism of regulation is **cumulative regulation**, where the two

A Negative allosteric regulation

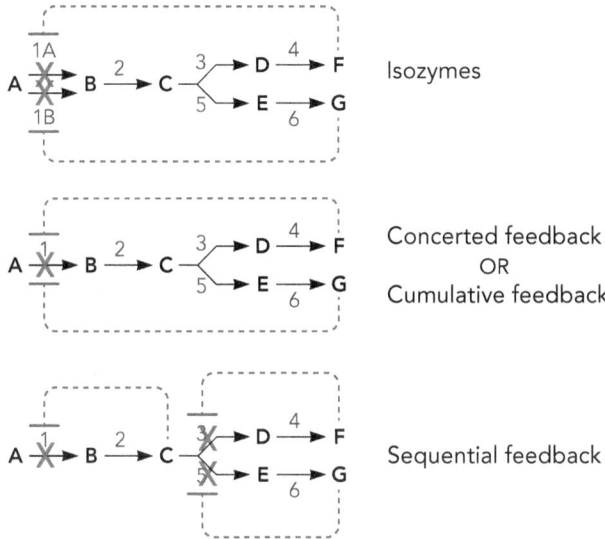

Isozymes

Concerted feedback
OR
Cumulative feedback

Sequential feedback

B Positive allosteric regulation

Precursor induction

Pathway "crosstalk"

FIGURE 7.6 **Allosteric regulation in branched pathways.** In each type of branched pathway shown, intermediates and products are labeled with letters and enzymes are labeled with numbers. (A) The red "X" and bar indicate the enzyme is being inhibited in the pathway during negative allosteric regulation. In the case of branched pathways with unique isofunctional enzymes (isozymes; e.g., 1A and 1B) that are both capable of catalyzing the first reaction of the pathway, each of the two isozymes is negatively allosterically regulated by a different end product (products F or G, respectively) of the branched pathway. Concerted feedback inhibition occurs when both end products of a branched pathway (products F and G) serve as negative effectors for the first enzyme (enzyme 1) in the pathway. Cumulative feedback inhibition results when each end product (products F and G) of the pathway functions as a negative effector to partially regulate the first enzyme (enzyme 1) in the pathway. The presence of both effectors might not be fully additive. During sequential feedback inhibition, the end product of each branch of a pathway functions as the negative effector for the first enzyme in that branch (products F and G inhibit enzymes 3 and 5, respectively, which results in a buildup of intermediate C; intermediate C inhibits enzyme 1). (B) In one example of positive allosteric regulation, precursor induction occurs when an upstream metabolic intermediate (intermediate E) builds up in the cell and serves as a positive effector for a downstream enzyme (enzyme 6; green dashed arrow). Pathway "crosstalk" can occur when the activity of an enzyme (enzyme 3; green dashed arrow) in an upstream branch of the pathway is positively allosterically regulated by an intermediate in the other branch (intermediate B) that builds up and serves as a positive effector.

end products each partially regulate the first enzyme in the branched pathway. However, this regulation is not necessarily additive. For example, one end product (F) might cause a 30% decrease in the activity of enzyme 1 and the second end product (G) might elicit a 40% decrease in the activity of enzyme 1, but together they decrease the rate of enzyme 1 to a level not necessarily equal to 70%.

The cumulative mechanism is not fully efficient, because both products are made even when one product is present above its threshold concentration. The most efficient mechanism of branched pathway regulation is when each branch is treated as a simple linear pathway, which is referred to as **sequential feedback regulation**. Here the end product of each branch of the pathway regulates the first enzyme in the relevant branch of the pathway. Therefore, if only one end product is in excess, its production is decreased without altering production of the other end product.

Positive allosteric regulation also occurs in branched pathways. One form of positive allosteric regulation is **precursor induction**. In this case, a precursor (E) increases the activity of an enzyme (enzyme 6) further down the pathway (Fig. 7.6). An example of precursor induction is the stimulation of pyruvate kinase activity by fructose-1,6-bisphosphate, which increases the rate of conversion of phosphoenolpyruvate (PEP) into pyruvate, thereby pulling more carbon through the glycolytic pathway (Chapter 2). In pathway "crosstalk," a metabolic intermediate (B) acts as a positive effector for an enzyme (enzyme 3) to increase the production of another intermediate (D) that serves as its cosubstrate in a subsequent reaction (Fig. 7.6). For example, oxaloacetate and acetyl coenzyme A (acetyl-CoA) are both required to produce citrate in the tricarboxylic acid (TCA) cycle, and acetyl-CoA is a positive allosteric effector of the enzyme PEP carboxylase, which produces oxaloacetate.

Covalent Modifications

During allosteric regulation, the effector molecule is free to move into and out of the regulatory site. However, another mechanism of polypeptide posttranslational regulation involves the addition or removal of a functional group by forming or breaking a covalent bond. Hence this mechanism is called covalent modification.

PHOSPHORYLATION

Arguably the most prevalent type of covalent modification is phosphorylation, which entails the addition of a phosphoryl group to the R group of specific amino acids (commonly histidine, aspartate, serine, threonine, and tyrosine) within a protein. Kinases are enzymes that add phosphoryl groups to polypeptides, while phosphatases are enzymes that remove the phosphoryl groups (Chapter 3). Thus, kinases and phosphatases counteract the activity of each other and maintain an appropriate balance of phosphorylated and unphosphorylated proteins in the cell. Phosphorylation is used to regulate some metabolic pathways. For example, in the glyoxylate cycle (Chapter 2), isocitrate dehydrogenase is negatively regulated by phosphorylation in the presence of fatty acids or acetate. In addition to metabolic enzymes, proteins involved in regulating gene expression are also controlled by covalent modification. This is a universal mechanism used in all living cells to control the activity of signal transduction pathways, also known as phosphorelay systems.

Bacterial **two-component regulatory systems** are commonly involved in gene regulation, and they are regulated via the process of phosphorylation.

A simple bacterial two-component regulatory system is comprised of a **sensor histidine kinase** and a **response regulator** (Fig. 7.7) that work together to modulate gene expression. Histidine kinases are often, but not always, localized to the cytoplasmic membrane. The amino (N)-terminal domain (NTD) of the protein is commonly exposed on the external surface of the cytoplasmic membrane where it is in position to bind and recognize external environmental signals. Upon binding to an environmental signal, the histidine kinase changes conformation such that information is transmitted intramolecularly across the cytoplasmic membrane to the carboxy (C)-terminal domain (CTD) of the protein. This causes the histidine kinase to autophosphorylate on a histidine aminoacyl residue, hence the name histidine kinase. After autophosphorylation, the

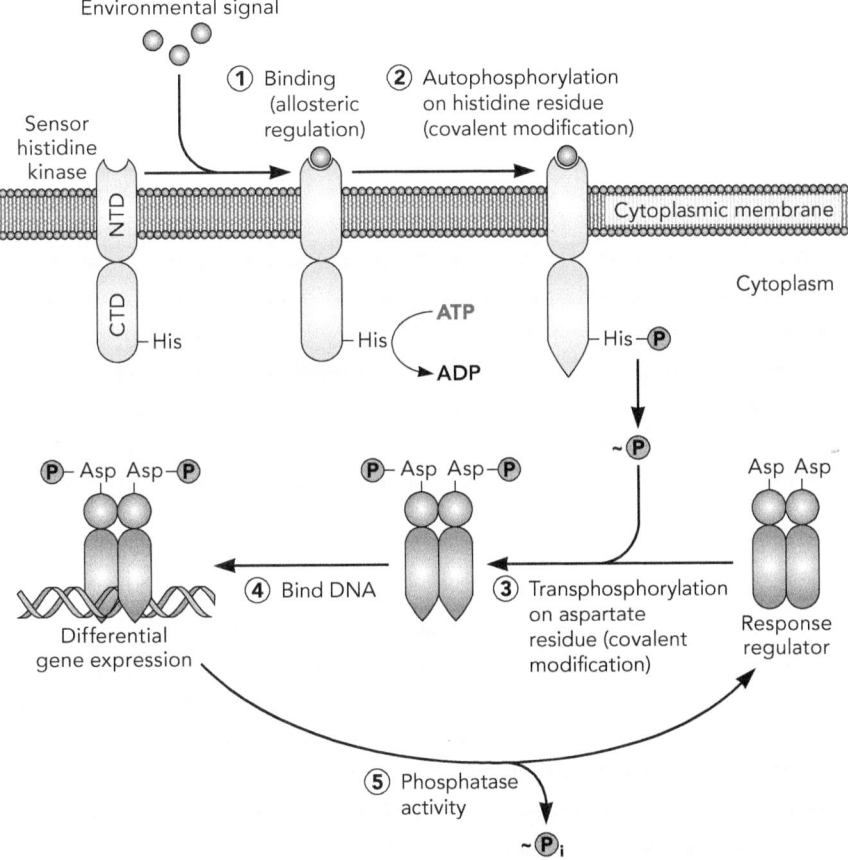

FIGURE 7.7 **Covalent modifications via phosphorylation in a two-component regulatory system.** The simplest type of two-component regulatory system, comprised of a sensor histidine kinase and a response regulator, is shown. (1) An external environmental signal (e.g., a small organic molecule) is sensed by a membrane-associated sensor histidine kinase. (2) When the signal binds to the N-terminal domain (NTD), allosteric regulation occurs within the sensor histidine kinase that results in an autophosphorylation of the kinase C-terminal domain (CTD) on a conserved histidine residue (i.e., covalent modification by the addition of a phosphate group). (3) The phosphorylated kinase protein next transphosphorylates (i.e., covalently modifies) the corresponding response regulator on a conserved aspartate residue in the NTD. (4) The phosphorylated form of the response regulator is now in a conformation capable of binding to the DNA using a helix-turn-helix motif in the CTD, altering gene expression. (5) An inherent phosphatase activity in the response regulator (or the sensor histidine kinase; not shown) removes the phosphate group, resetting the system back to its original noninduced state.

histidine kinase transphosphorylates the second component in the regulatory system, the response regulator. Response regulators generally have two domains, an NTD and a CTD. The NTD first becomes phosphorylated at a conserved aspartate residue. This phosphorylation event causes a conformational change in the response regulator that impacts the ability of the CTD to function. Most response regulators are transcription factors that contain a helix-turn-helix secondary structure in their CTD, which can bind to regulatory sequences in the DNA. Response regulators bind to the DNA as dimers and function either as activators or repressors of the process of transcription, thereby impacting the overall rate of gene expression. More details on transcriptional regulation will be discussed in Chapter 8. Specific examples of processes under two-component regulatory control will be described in the Diversity section of this textbook.

NUCLEOTIDYL MODIFICATION (E.G., ADENYLATION)

Another type of covalent modification is the addition of a nucleotidyl group to a polypeptide. Adenylation, involving the functional group adenosine monophosphate (AMP), is one example. The enzyme glutamine synthetase (GS) is required for ammonia assimilation in living cells (Fig. 7.8; Chapter 12). GS converts ammonia and glutamate into glutamine with the consumption of one ATP. GS is a complex enzyme with 12 identical subunits, and each subunit can be individually regulated by adenylation. In the presence of a low concentration of glutamine, the cell needs to synthesize more. Under these conditions, GS must be fully active and does not undergo covalent modification with AMP. However, in the presence of high levels of glutamine, between 1 and 12 molecules of AMP are added to GS to decrease its reaction rate. This is not an on/off phenomenon, but rather there are 12 activity levels increasing or decreasing the enzyme activity to the appropriate level depending on the environmental conditions.

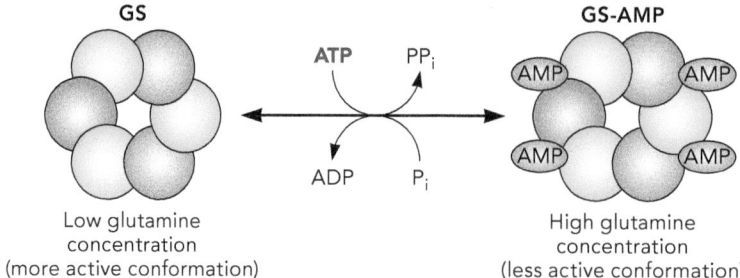

GS   ATP   PP$_i$ GS-AMP

ADP   P$_i$

Low glutamine concentration (more active conformation)   High glutamine concentration (less active conformation)

FIGURE 7.8 **Covalent modification via adenylation.** The enzyme glutamine synthetase (GS) catalyzes the formation of glutamine from ammonia, glutamate, and ATP (Chapter 12). It has 12 subunits (only 6 subunits are shown) that can be individually modified by addition of an AMP through covalent modification. In the presence of a low glutamine concentration, the enzyme has a low level of modification, resulting in a higher level of activity, enabling the cell to synthesize glutamine. As the environmental concentration of glutamine rises, the degree to which it is adenylated increases, thereby decreasing the enzyme activity and conserving energy for the cell. Note that regulation is not an on/off phenomenon, but rather there are 12 activity levels that correlate to the degree of covalent modification.

GLYCOSYLATION (E.G., ADP-RIBOSYLATION)

ADP-ribose is a large functional group comprised of a nucleotide and a sugar that can be covalently added to a polypeptide to impact its function. Two examples of the effects of ADP-ribosylation are given by the bacterial toxins diphtheria toxin and cholera toxin, which serve as virulence factors impacting target host cell activity (Fig. 7.9 and 7.10). In both of these cases, the bacterial pathogens produce a toxin that catalyzes the addition of ADP-ribose to a critical host protein. Diphtheria toxin is produced by the bacterium *Corynebacterium diphtheriae*. *C. diphtheriae* is capable of colonizing the upper respiratory tract of humans and can cause life-threatening infections. Because of this, it is recommended that the DPT vaccine (diphtheria, pertussis, and tetanus) be administered. The diphtheria toxin binds to a host epithelial cell receptor and is internalized via endocytosis. The low pH in the endosome causes the toxin to break into subunits. Subunit A moves across the endosome membrane to enter the host cytoplasm, where it modifies the essential host translation protein elongation factor-2 (EF-2) by covalent addition of ADP-ribose from NAD^+. The modified version of EF-2 is rendered nonfunctional; thus, polypeptide translation ceases and results in host cell death in the upper respiratory tract (Fig. 7.9).

Cholera toxin is produced by the bacterium *Vibrio cholerae*. Cholera is a disease manifested by profuse "rice water" diarrhea that can lead to death due to dehydration. Cholera spreads via contaminated water sources, and there is an increased risk of outbreaks following natural disasters such as hurricanes and flooding. Historically, there were no effective longterm vaccines to prevent the

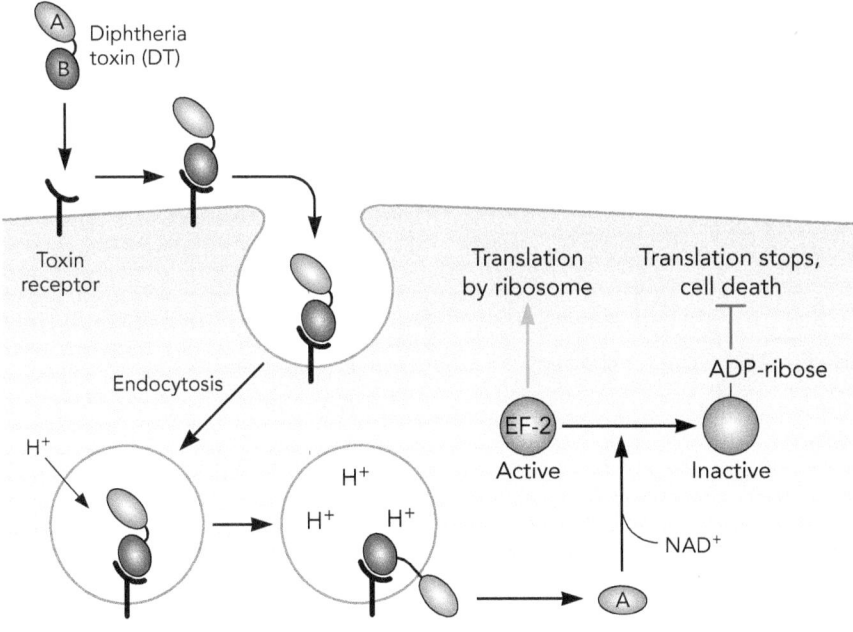

FIGURE 7.9 **Diphtheria toxin causes covalent modification by ADP-ribosylation.** The diphtheria toxin is an AB-type toxin that enters the cell through receptor-mediated endocytosis. Acidification of the endosome enables the A subunit to enter the cytoplasm, where ADP-ribosylation of the EF-2 host protein occurs (i.e., is covalently modified by the addition of an ADP-ribosyl group from NAD^+), rendering it nonfunctional so that the host cell is no longer able to perform translation and dies.

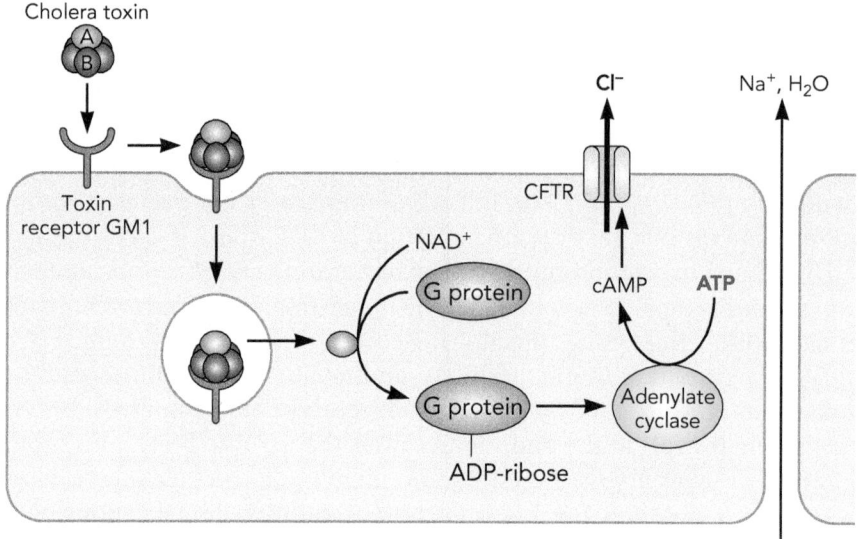

FIGURE 7.10 **Cholera toxin causes covalent modification by ADP-ribosylation.** The cholera toxin has a 1:5 A:B subunit arrangement. Following receptor-mediated endocytosis, the A subunit adds an ADP-ribosyl group from NAD^+ to a host cell G protein. This covalent modification causes the G protein to stimulate cAMP production, and this internal signaling molecule subsequently causes the CFTR (cystic fibrosis transmembrane conductance regulator) transporter to efflux chloride ions. This, in turn, causes the epithelial cell layer to also release sodium ions and water, which results in the copious amounts of rice-stool diarrhea associated with the disease cholera.

disease caused by *V. cholerae*. Although new vaccines are currently available, fluid replacement is the primary treatment in most parts of the world. Cholera toxin enters the host intestinal epithelial cell via endocytosis, then the A subunit moves into the cytoplasm where it catalyzes the covalent modification of the host G protein with an ADP-ribose. This G protein normally helps to regulate an enzyme called adenylate cyclase, which produces the internal signaling molecule cyclic AMP (cAMP). In turn, cAMP regulates an ion transporter in the host cell membrane. Modification of the G protein increases the activity of adenylate cyclase, which leads to an efflux of chloride ions into the lumen of the intestine followed by loss of water from the cell (Fig. 7.10). This results in a life-threatening diarrhea during which an individual can lose 10% or more of their body weight resulting in severe dehydration.

METHYLATION

Methylation is the addition of a methyl group to a polypeptide. Regulation through methylation is important for the process of chemotaxis by motile cells and will be discussed in Chapter 15.

Posttranslational Regulation in the Sugar Phosphotransferase System (PTS)

The phosphotransferase system (PTS) is a bacterial transport system for the uptake of carbohydrates; the process is also known as group translocation. While the PTS is not universal, it is present in a number of anaerobes and facultative bacteria such as the enteric bacteria (e.g., *E. coli*). For these organisms, the PTS sugars (most preferred) transported by the system include glucose, mannose, and mannitol. Non-PTS substrates (less preferred) include lactose, melibiose, maltose, glycerol, rhamnose, xylose, and TCA cycle intermediates. For example, the PTS is important to the process of diauxic growth (Chapter 4). During diauxic growth, *E. coli* will preferentially utilize glucose as a carbon source prior to metabolism of a less preferred carbon source such as lactose.

The PTS is regulated by multiple mechanisms of covalent modification. The energy driving the transport of PTS sugars comes from the high-energy phosphate bond in PEP. A phosphoryl group is transferred from PEP to Enzyme I (E_I) and from E_I to the phosphocarrier protein HPr (Fig. 7.11). PEP, E_I, and HPr are required for the transport of all PTS sugars. After HPr, the pathways diverge, with different Enzyme II proteins (E_{II}) involved in the transport of the different PTS sugars. These different E_{II} proteins have distinctive subunit structures (Fig. 7.11). If there are two PTS sugars present, the kinetics of the interactions of HPr with the different E_{II} proteins will determine which sugar will be utilized first. The specific E_{II} protein for glucose ($E_{II}^{glucose}$) has subunit A in the cytoplasm and subunits B and C are associated with the cytoplasmic membrane. The $E_{IIA}^{glucose}$ subunit is critical for the regulatory processes associated with PTS versus non-PTS sugar transport. The presence of any PTS sugar will trigger the same overall physiological response by the bacterial cell through the rate of phosphorylation of $E_{IIA}^{glucose}$.

In the presence of a PTS sugar, the phosphoryl group from HPr is transferred to the appropriate E_{II} and will ultimately be transferred to the PTS sugar being transported into the cell. Consequently, when a PTS sugar is present, the level of $E_{IIA}^{glucose}$ that is phosphorylated in the cytoplasm is low. $E_{IIA}^{glucose}$ in its unphosphorylated form serves as a negative regulator for the sugar permeases/transporters for non-PTS sugars. Therefore, when a PTS sugar is present, non-PTS sugars are not transported across the cytoplasmic membrane. This negative regulation is termed **inducer exclusion**. Once the PTS sugar is consumed, instead of the phosphoryl group ending up on a PTS sugar, it accumulates on $E_{IIA}^{glucose}$. When $E_{IIA}^{glucose}$ is primarily in the phosphorylated form, the negative regulation on the non-PTS sugars ceases so they may be transported into the cytoplasm.

The phosphorylated form of $E_{IIA}^{glucose}$ induces the activity of adenylate cyclase through an unknown mechanism. When active, adenylate cyclase produces the internal second messenger molecule cAMP. A **second messenger** is an internally or self-generated signal produced by cells in response to changing environmental conditions. cAMP serves as a positive allosteric effector for the transcription factor CRP (cAMP receptor protein; also known as CAP [catabolite activator protein]). When associated with cAMP, CRP undergoes a conformational change due to positive allosteric regulation that enables it to bind to sites upstream of select promoters. The promoters regulated by cAMP-CRP include those controlling transcription of the genes encoding enzymes involved in the catabolism of non-PTS sugars. When there is no PTS sugar available, cAMP-CRP forms a complex that is in a DNA-binding competent conformation and will activate the expression of genes needed to metabolize the non-PTS sugars (e.g., the *lac* operon). When any PTS sugar is present, $E_{IIA}^{glucose}$ is primarily in its unphosphorylated form, and thus CRP is in an inactive conformation and expression of the non-PTS sugar catabolism genes does not occur. This type of

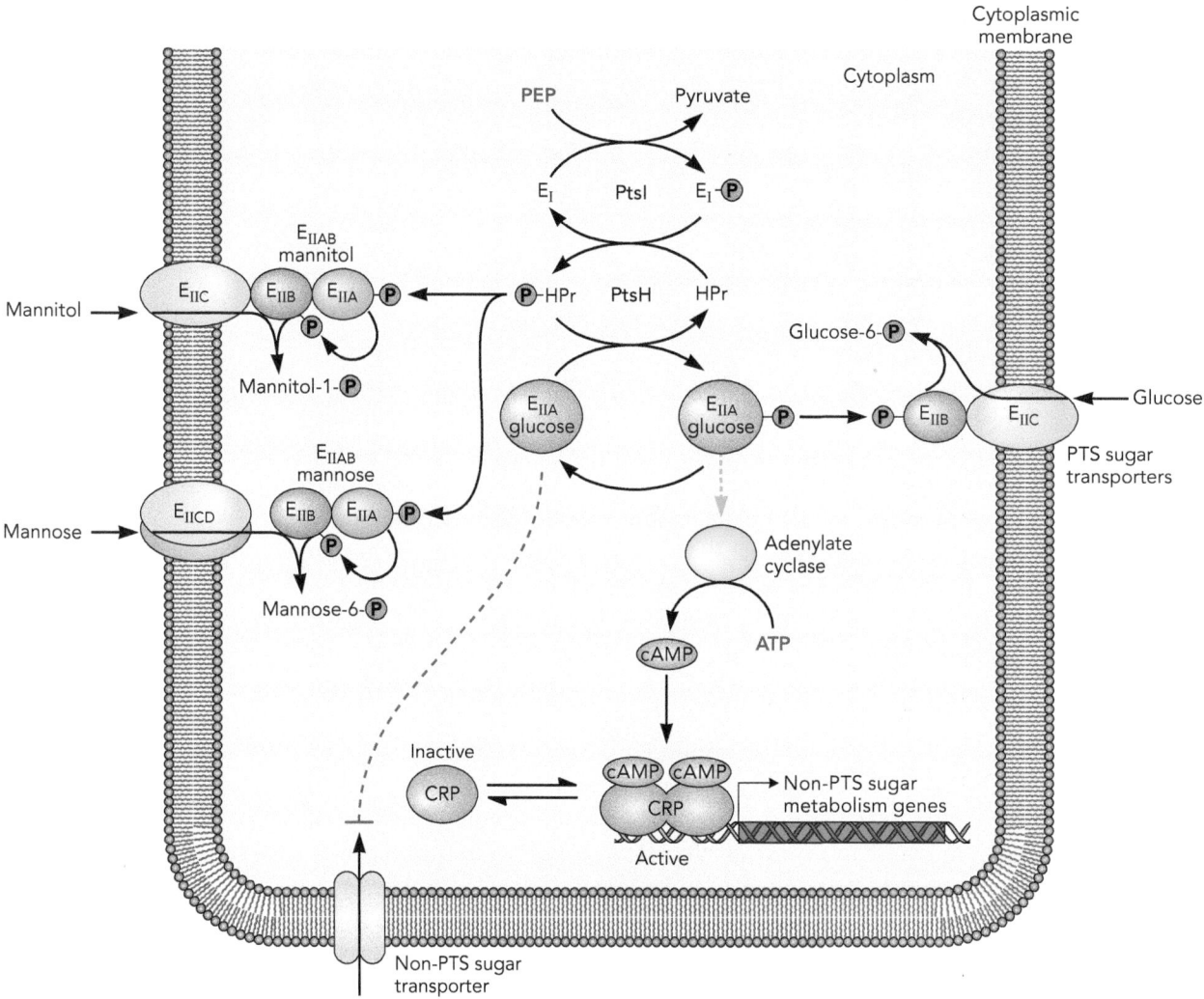

FIGURE 7.11 **Covalent modifications in the sugar phosphotransferase system (PTS).** The phosphate group from phosphoenolpyruvate (PEP) is passed to the proteins Enzyme I (E_I), HPr, and then to Enzyme IIA (E_{IIA}), used for the transport of a specific PTS sugar in a series of covalent modification reactions catalyzed by the kinases PtsI and PtsH. In the presence of a preferred PTS sugar (e.g., glucose, mannitol, or mannose), the phosphoryl group is ultimately passed to the sugar as the sugar gets translocated into the cell. Under these circumstances the $E_{IIA}^{glucose}$ is primarily in the unphosphorylated form, which inhibits non-PTS sugar transporters, leading to inducer exclusion (dashed red line). If there are no PTS sugars available for metabolism, then $E_{IIA}^{glucose}$ is primarily in the phosphorylated form, which stimulates adenylate cyclase to produce the second messenger cAMP. The regulatory protein CRP is converted into a DNA binding-proficient conformation through allosteric regulation when cAMP binds to it, which results in expression of the genes needed to metabolize a number of non-PTS sugars and ends the process of catabolite repression.

regulatory control is called **catabolite repression**. Therefore, PTS sugars are utilized prior to non-PTS sugars. When $E_{IIA}^{glucose}$ is in its unphosphorylated form, both inducer exclusion and catabolite repression will occur. Once the PTS sugar is consumed and $E_{IIA}^{glucose}$ is present in its phosphorylated form, then inducer exclusion and catabolite repression will end and the non-PTS sugar will be transported and catabolized.

In conclusion, there is a coordination of covalent modification, allosteric regulation, and transcriptional regulation that together control the use of PTS and non-PTS sugars during diauxic growth. Transcriptional and posttranscriptional regulation are discussed in more detail in Chapter 8.

Learning Outcomes: After completing the material in this chapter, students should be able to . . .

1. describe three key modes of regulation controlling protein levels and activity

2. explain the mechanisms of positive and negative allosteric enzyme regulation

3. give examples of feedback inhibition in which K_m or V_{max} is altered

4. discriminate between different types of feedback inhibition in branched pathways

5. describe four examples of covalent modification using different functional groups

6. outline how the sugar phosphotransferase system (PTS) works, including how it is regulated in the presence or absence of PTS and non-PTS sugars

Check Your Understanding

1. Define this terminology: allosteric regulation, allosteric effector molecule, feedback inhibition, isozymes, concerted feedback, cumulative regulation, sequential feedback regulation, precursor induction, two-component regulatory system, sensor histidine kinase, response regulator, inducer exclusion, second messenger, catabolite repression

2. Bacteria can regulate enzyme activity in a number of ways. (LO1)
   a) Briefly describe each control mechanism listed (Fig. 7.1):
   - transcriptional regulation:
   - posttranscriptional regulation:
   - posttranslational regulation:
   b) Which of these control mechanisms would save a cell the most energy and resources if the cell needed to make a new protein? Explain your reasoning.
   c) Compare the time scale for each of these options. Which of these control mechanisms do you think would mediate the fastest cellular response? Why are both fast and slow cellular responses a benefit for the cell?

3. Compare and contrast positive and negative allosteric regulation of an enzyme. In each case, indicate if the enzymatic reaction will occur. (LO2)

		Will the enzymatic reaction occur?
Positive allosteric regulation	Effector binds to enzyme	
	Effector does not bind to enzyme	
Negative allosteric regulation	Effector binds to enzyme	
	Effector does not bind to enzyme	

4. For each type of allosteric regulation that is listed, indicate the branched pathway that best illustrates the regulatory mechanism (as shown by the black dashed lines). (LO4)

_____ Concerted feedback

_____ Isozymes (isofunctional enzymes)

_____ Precursor induction

_____ Sequential feedback

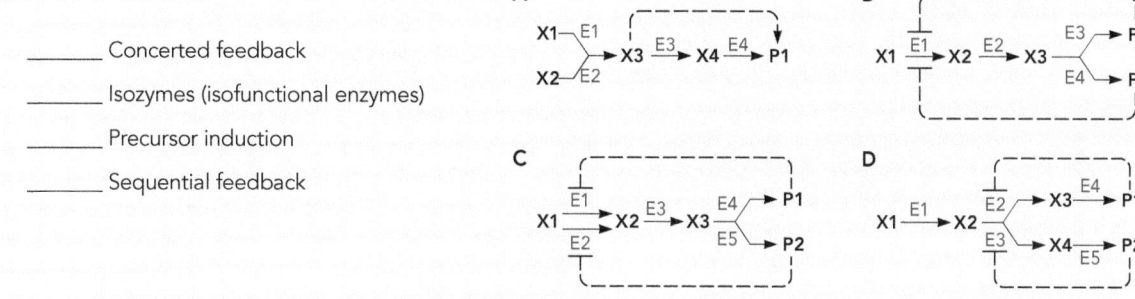

5. You are working with *E. coli* cells that have the biosynthetic pathway for production of the nucleotide cytidine triphosphate (CTP; pathway abbreviated below). (LO2, 3)

$$\text{L-aspartate + carbamoyl phosphate (CP)} \xrightarrow[\text{transcarbamoylase}]{\text{Aspartate}} \text{N-carbamoyl-L-aspartate (CAA)} \rightarrow \rightarrow \rightarrow \rightarrow \rightarrow \rightarrow \rightarrow \text{CTP}$$

a) The first enzyme in the pathway, aspartate transcarbamoylase, is allosterically regulated by CTP.

   i. Based on physiology alone, would you predict that CTP would inhibit or enhance the function of the aspartate transcarbamoylase enzyme?

   ii. Assuming your prediction above is correct, is this negative or positive allosteric regulation? Explain your reasoning.

b) You set up experiments to measure the Michaelis-Menten kinetics with and without CTP in the growth medium. You get the results shown on the graph.

   i. Comparing the experiments without CTP to those with CTP, has the K_m increased, decreased, or remained the same?

   ii. Has the V_{max} increased, decreased, or remained the same?

   iii. Explain what is happening between the effector molecule and the enzyme to give these results.

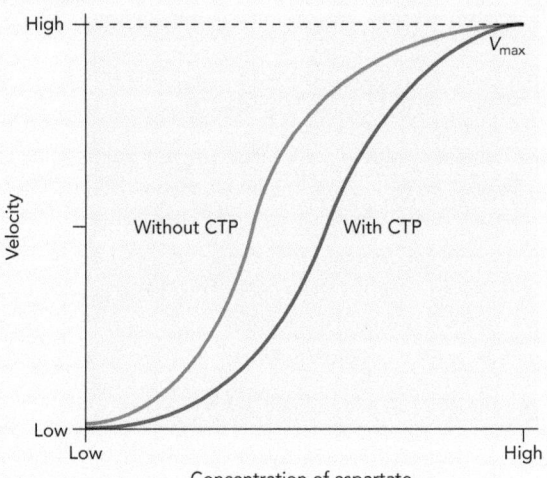

6. You are working with *E. coli* cells that have the pathway for tryptophan biosynthesis (abbreviated below).
   a) The first enzyme in the pathway, DAHP synthase, is allosterically regulated by tryptophan. (LO2, 3)

$$\text{D-erythrose-4-phosphate + PEP} \xrightarrow[\text{synthase}]{\text{(DAHP)}} \text{DAHP} \rightarrow \rightarrow \rightarrow \rightarrow \rightarrow \rightarrow \text{Chorismate} \rightarrow \rightarrow \rightarrow \rightarrow \rightarrow \rightarrow \text{Tryptophan}$$

   i. Based on physiology alone, would you predict that the presence of tryptophan would inhibit or enhance the function of the DAHP synthase enzyme?

   ii. Assuming your prediction, above, is correct, is this negative or positive allosteric regulation? Explain your reasoning.

   b) You set up experiments to measure the Michaelis-Menten kinetics with and without tryptophan in the growth medium, and get the results shown on the following graph.

   Compare the experiments without tryptophan to those with tryptophan.

   i. Has the V_{max} increased, decreased or stayed the same?

   ii. Explain what is happening between the effector molecule and the enzyme to give these results.

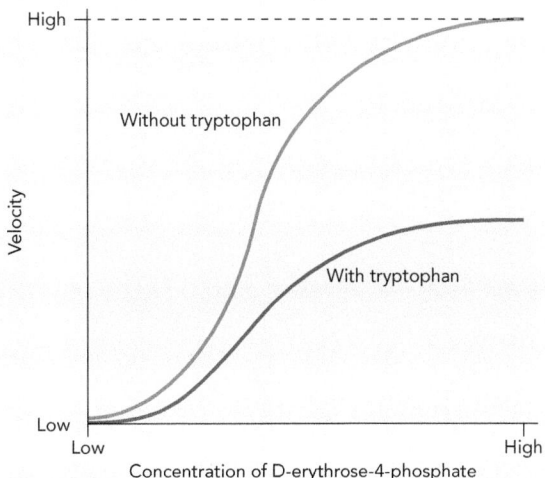

7. The following steps occur in the phosphorelay system of a typical two-component regulatory system. Put the steps in sequential order. (LO5)

_____ A phosphoryl group is transferred to a response regulator.

_____ An external environmental signal binds to a membrane-bound sensor protein.

_____ A histidine kinase domain is phosphorylated with a phosphoryl group from ATP.

_____ The carboxy-terminal domain of a response regulator binds to its target.

8. Enzyme modification is a common mechanism by which cells (and cell pathogens) regulate enzyme activity. Answer each question below related to enzyme modification. (LO5)
   a) Consider the modification of the glutamine synthetase enzyme.

      i. What is the function of glutamine synthetase?

      ii. What functional group is added to the enzyme during modification?

      iii. Under what condition (high or low concentration of glutamine) would glutamine synthetase have the highest level of modification? Does modification increase or decrease the enzyme activity?

   b) Consider the enzymatic modification of diphtheria toxin. The toxin (subunits A and B) is taken up by host epithelial cells. Once inside the host cell, the toxin breaks into subunits and a host enzyme is modified.

      i. Which toxin subunit (A or B) modifies the host enzyme?

      ii. What functional group is added to the enzyme?

      iii. What enzyme is modified, and what is its function?

      iv. Does modification increase or decrease the enzyme activity?

      v. What is the ultimate physiological consequence of modification?

   c) Consider the enzymatic modification of cholera toxin. The toxin (subunits A and B) is taken up by intestinal epithelial cells. Once inside the host cell, the toxin breaks into subunits and a host enzyme is modified.

      i. Which toxin subunit (A or B) modifies the host enzyme?

      ii. What functional group is added to the enzyme?

      iii. What enzyme is modified, and what is its function?

      iv. Does modification increase or decrease the enzyme activity?

      v. What is the ultimate physiological consequence of modification?

9. During the posttranslational regulation in the sugar phosphotransferase system (PTS), an EII enzyme complex, consisting of one or two hydrophobic integral membrane domains (domains C and D) and two hydrophilic domains (domains A and B), assists in the transport of sugars into the cell. Put the steps of the PTS listed below in sequential order. (LO6)

____ An enzyme specific to glucose catalyzes the phosphorylation and transport of glucose across the membrane.

____ Enzyme I catalyzes the phosphorylation of the small enzyme HPr.

____ Enzyme IIA is phosphorylated.

____ Glucose-6-phosphate is released into the cytoplasm.

____ PEP loses a phosphoryl group to become pyruvate.

Dig Deeper

10. You are interested in investigating the effects of uridine triphosphate (UTP) on this pathway. The last few steps of the CTP biosynthetic pathway are shown. (LO2)

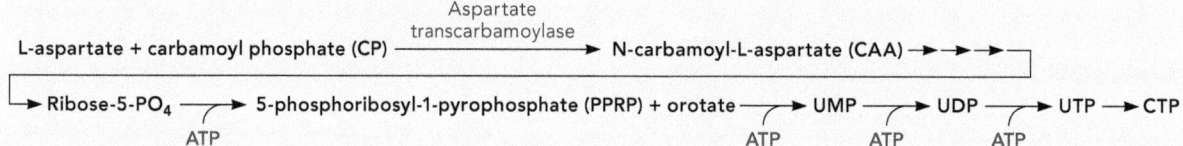

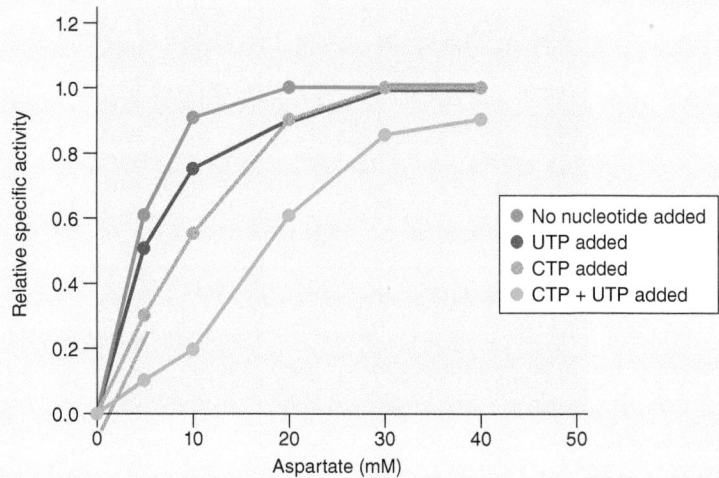

You conduct a few experiments looking at the activity of aspartate transcarbamoylase in the presence and absence of both CTP and UTP. You get the results shown in the graph.

a) Based on the data in the graph, would you consider CTP to be a positive or negative allosteric effector? Explain your thinking.

b) Is the K_m of aspartate transcarbamoylase with UTP higher, lower, or the same as the K_m with CTP alone or no added nucleotide?

c) Is the K_m of aspartate transcarbamoylase with CTP + UTP higher, lower, or the same as the K_m with CTP alone?

d) Present a logical scenario that would explain these data. Explain why these results would be useful for the cell from a physiological perspective.

11. The tryptophan biosynthetic pathway is part of a larger pathway for the synthesis of three aromatic amino acids in *E. coli* (shown below). (LO4)

a) Propose a mechanism of allosteric regulation by which the cell could shut off the production of tryptophan without stopping tyrosine or phenylalanine synthesis. When might the cell want to do this?

b) Propose a mechanism of allosteric regulation by which the cell could shut off the production of phenylalanine without stopping tryptophan or tyrosine synthesis. When might the cell want to do this?

c) While doing careful research on this pathway, you notice that *E. coli* has three different proteins (AroF, AroG, and AroH) that can catalyze the first step of this pathway. You study two of the proteins and discover that the product of the *aroG* gene catalyzes the reaction only in the absence of phenylalanine, while the product of the *aroF* gene catalyzes the reaction only in the absence of tyrosine. Based on these results, why do you think *E. coli* has three proteins that catalyze the same reaction?

12. Cholera toxin subunit A catalyzes the ADP-ribosylation of a family of proteins called G proteins found in host cells. Ribosylated G proteins can then allosterically regulate adenylate cyclase (see Fig. 7.10), which produces the signaling molecule cyclic AMP (cAMP). (LO5)

a) Would you consider this regulatory scheme to be positive or negative allosteric regulation? Explain your thinking.

b) You obtain a mutant bacterium with an altered sequence in the gene encoding the cholera toxin subunit A (CTxA). You inject some mice with the wild-type CTxA toxin and some with the mutant CTxA toxin, and measure the rates of ADP-ribosyltransferase activity and cAMP induction. You get the data in the table below. Do these data support or refute your answer to 12 a) above? Explain your reasoning.

Injected into mice	Relative ADP-ribosyltransferase activity	cAMP induction (pmol/mg)
Wild-type CTxA	4,600	793
Mutant CTxA	93	8
Saline buffer	98	9

c) Why did you include an injection of saline buffer? What do the results of saline injection tell you?

13. You are interested in investigating the sugar phosphotransferase system in *E. coli*. You grow some cells in a defined medium and monitor the states of various proteins within the cells. (LO6)

a) You notice that when you add mannitol and lactose, HPr becomes phosphorylated.

i. Would you predict that the concentration of phosphorylated $Enzyme_{II}^{glucose}$ would go up, go down, or stay the same with the addition of mannitol? Explain your reasoning.

ii. Would you predict that the concentration of phosphorylated mannitol in the cell would go up, go down, or stay the same with the addition of mannitol? Explain your reasoning.

iii. How would you predict that the concentration of phosphorylated lactose would change with the addition of mannitol? Explain your reasoning.

b) After a few hours, the mannitol is consumed and the external concentration in the growth medium drops to near zero.

i. Would you expect the concentration of phosphorylated $Enzyme_{II}^{glucose}$ to go up, go down, or stay the same with the consumption of mannitol? Explain your reasoning.

ii. Would you expect the concentration of phosphorylated lactose to go up, go down, or stay the same with the consumption of mannitol? Explain your reasoning.

iii. Under these conditions, would cAMP receptor protein (CRP) activate the expression of $Enzyme_{II}^{glucose}$? Explain your reasoning.

c) Ultimately, what is driving the transport of sugars into the cell? From where does the energy come?

148 Making Connections

148 Transcription Terminology

149 Bacterial Transcription Initiation and Elongation

151 Bacterial Transcription Termination

153 Regulatory *cis*- and *trans*-Acting Elements Impacting Transcription

154 Examples of Different Promoter Structures

156 Transcriptional Regulation of the *lac* Operon

158 Activation and Repression by the Global Regulator Cra

158 Attenuation

161 Posttranscriptional Regulation

163 Methods Used to Study Gene Regulation

164 Methods to Demonstrate Protein–DNA Interactions

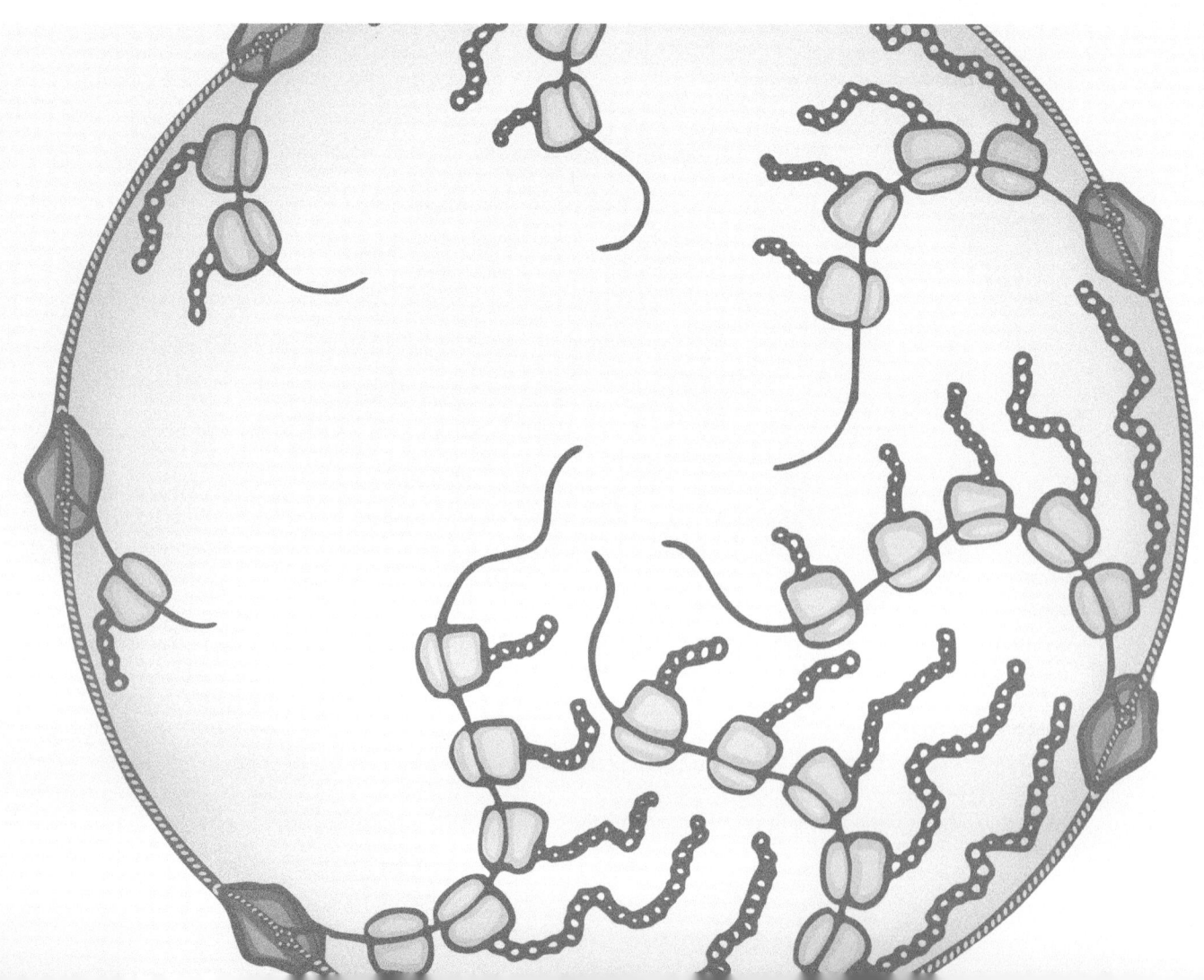

GENE REGULATION—
TRANSCRIPTION INITIATION AND POSTTRANSCRIPTIONAL CONTROL

After completing the material in this chapter, students should be able to . . .

1. explain the process of transcription initiation and elongation in *Bacteria*

2. distinguish factor-independent termination from factor-dependent termination in *Bacteria*

3. compare and contrast the structure and function of promoters controlled by activator and repressor proteins

4. explain how the regulation of the *lac* and *cra* operons is controlled under different growth conditions

5. predict how the concentration of tryptophan affects transcriptional attenuation of the *trp* operon

6. describe the three main mechanisms of posttranscriptional regulation

7. interpret results from methods for investigating levels of transcription and protein–DNA binding

Microbial Physiology: Unity and Diversity, First Edition. Ann M. Stevens, Jayna L. Ditty, Rebecca E. Parales and Susan M. Merkel.
© 2024 American Society for Microbiology.
Companion website: www.wiley.com/go/stevens/microbialphysiology

Making Connections

In Chapter 7, the regulation of translation was discussed as a shared regulatory approach used by all living cells. The regulation of gene expression at the point of transcription is another common physiological process found in all branches of the tree of life. Control at the point of transcription initiation, or even after transcription (posttranscription), but prior to translation, affords cells the ability to conserve energy. The most efficient way to regulate gene expression is by controlling transcription initiation, since it is the first step in the production of a polypeptide from a gene (Fig. 7.1). Transcriptional and posttranscriptional control mechanisms will be examined in this chapter.

Transcription Terminology

Some genes are always expressed. These are considered to have **constitutive expression** and often encode critical **"housekeeping" functions** that are essential for cell survival (e.g., DNA and RNA synthesis). However, the vast majority of genes are regulated, with varying levels of gene expression that are dependent on environmental conditions and the physiological state of the cell. Some genes are under positive regulatory control. The transcription of these genes is induced (i.e., activated) and thus these genes are transcribed by RNA polymerase (RNAP) at higher rates. Other genes are under negative control and are transcribed at lower rates (i.e., repressed).

It is common for a given pathway to be controlled at multiple levels simultaneously. As discussed in Chapter 7, feedback inhibition is a type of allosteric regulation that can be used as a mechanism to stop the activity of an entire pathway through negative posttranslational control of one or more key enzymes. A second way an entire pathway may be coordinately regulated is at the level of transcription (Fig. 8.1). In bacteria, regulation of transcription initiation occurs at the **promoter** region, which is a nonencoding, regulatory DNA sequence located upstream of one or more genes. Often, a promoter will control the expression of a group of genes encoding proteins with similar metabolic activities (e.g., encoding different enzymes within a pathway). The actual site of transcription initiation is referred to as the **+1 site**, and it is located just downstream of the promoter. Multiple genes that are cotranscribed under the control of a single promoter are known as an **operon**. Transcription of the operon produces a **polycistronic** messenger RNA (mRNA) that contains the genetic information to encode more than one polypeptide. The organization of genes into operons is common in bacteria and is also found in archaea; in contrast, eukaryotes typically control the transcription of one gene with one promoter. The mRNA for a single gene is said to be **monocistronic**, because it codes for just one polypeptide. Sometimes a single regulatory protein controls the expression of multiple promoters and thereby multiple operons. In this case, the transcription factor is called a **global regulator** and the promoters and genes it controls are referred to as its **regulon**. We will discuss the global regulators CRP (cyclic AMP [cAMP] receptor protein) and Cra later in this chapter.

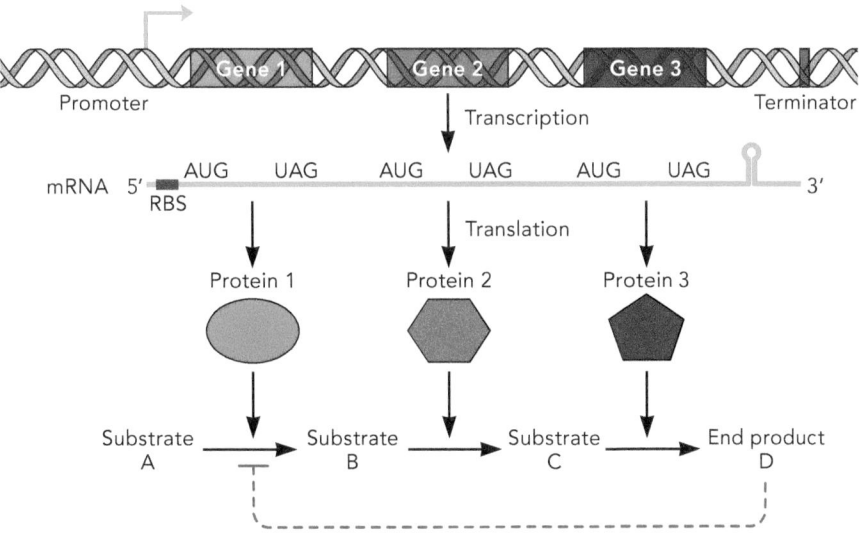

FIGURE 8.1 Different levels of regulation controlling a biological pathway. (Top to bottom) Bacterial genes are often arranged in operons with a single promoter controlling the expression of multiple genes (e.g., three genes in this operon). The rate of transcription from this promoter may be regulated, influencing expression of all of the genes in the operon. Transcription of an operon produces a polycistronic mRNA with multiple start (AUG) and stop (UAG) codons, where each open reading frame is translated into a polypeptide (e.g., three single-subunit proteins are produced from this operon). These proteins may serve as enzymes in a biosynthetic pathway, with the first enzyme in the pathway (e.g., Protein 1) being controlled posttranslationally through simple feedback inhibition by the end product of the pathway, D (dashed red line and bar). Note, the same pathway may be regulated at more than one level (e.g., transcription and posttranslation) at the same time.

Bacterial Transcription Initiation and Elongation

The mechanism of transcription in *Bacteria* is quite distinct from that used in *Archaea* and *Eukarya*. The enzyme that catalyzes the production of RNA during transcription is RNAP (Fig. 8.2). In *Bacteria*, this enzyme consists of a core catalytic enzyme with four essential subunits (beta [β], beta prime [β'], and two alpha [α] subunits) that synthesizes RNA. A fifth subunit, omega (ω), was discovered later, but it does not appear to be essential in *Escherichia coli*. The core enzyme cannot associate with the DNA unless it forms a complex with a sigma (σ) subunit, or **sigma factor**. Each bacterium has multiple sigma factors, which associate with the **core RNA polymerase** (ββ'αα) under specific physiological conditions. It is the sigma factor that recognizes and binds to the promoter region of a gene. The core RNAP enzyme plus a sigma factor is referred to as the **RNA polymerase holoenzyme**, which is essential for the initiation of transcription. The sigma factor guides the core enzyme to the correct site on the DNA by recognizing specific sequences in the promoter. The bacterial "housekeeping" promoter, which is recognized by σ^{70} in *E. coli*, consists of the −10 (Pribnow box) and −35 sites, as they are centered approximately −10 and −35 base pairs upstream from the site of transcription initiation (the +1 site), respectively. When the RNAP holoenzyme is positioned at the −10 and −35 sites

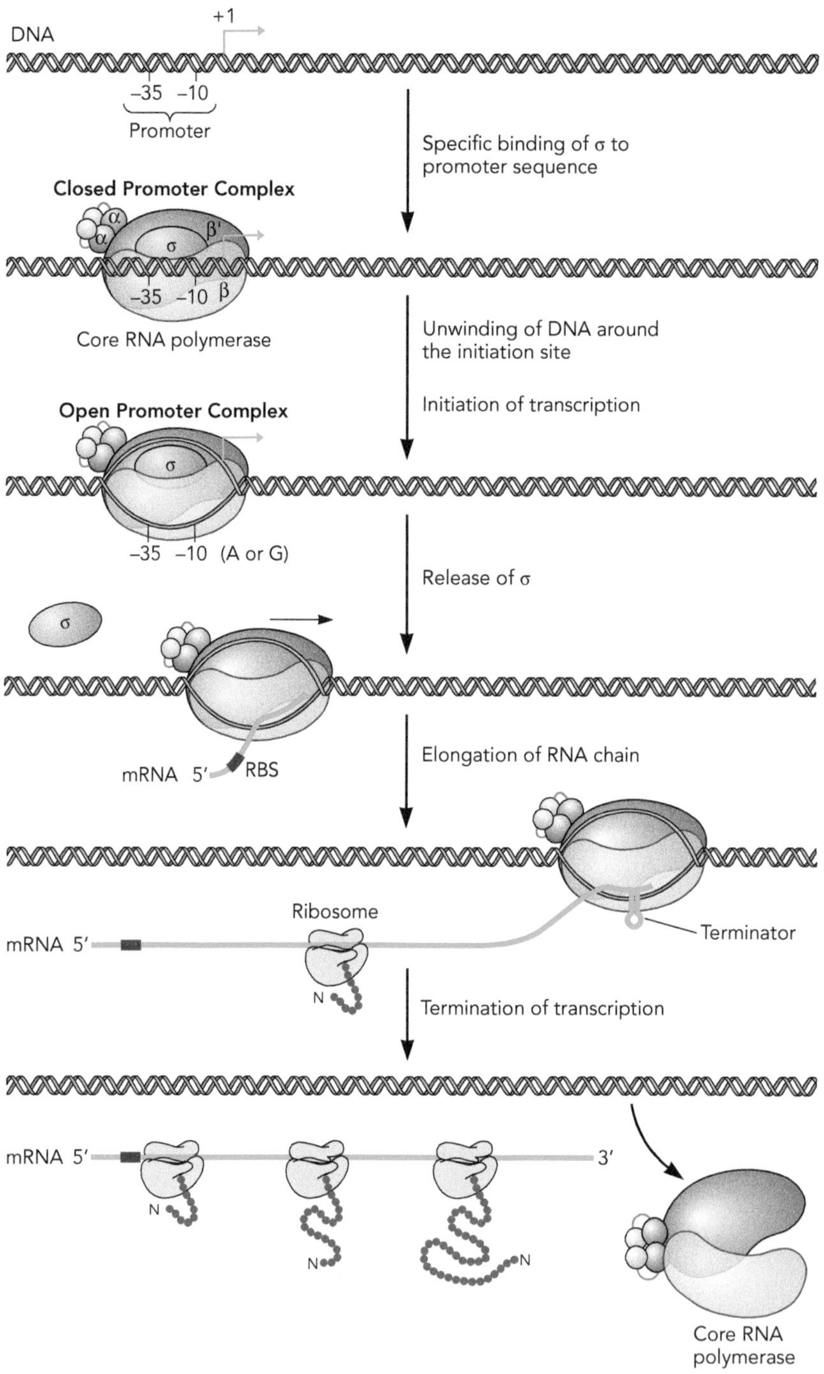

FIGURE 8.2 **Bacterial transcription.** (Top to bottom) Bacterial RNA polymerase (RNAP) will bind and initiate transcription at the +1 site of the housekeeping promoter (a region in the DNA with −35 and −10 recognition sites). The RNAP core enzyme that synthesizes RNA consists of beta (β), beta prime (β′), two alpha (α), and omega (ω) subunits, shown in brown and blue. When a sigma subunit (σ; purple) complexes with the core enzyme, the complete holoenzyme can bind to the promoter, forming a closed promoter complex. When the DNA becomes single-stranded around the +1 initiation site, the first ribonucleotide (normally A or G) will initiate transcription, enabling polymerization of the growing mRNA. During elongation of the transcript, the sigma subunit is released and polymerization of the mRNA will continue until a terminator is encountered. Translation will begin as soon as the ribosome binding site (RBS) sequence in the mRNA is transcribed, since transcription and translation are coupled in bacteria. Multiple ribosomes (a polysome) may carry out translation from a single mRNA. See the text for additional details.

in the promoter, a closed promoter complex initially forms. This closed complex is converted to an open promoter complex when the DNA template becomes single-stranded around the +1 initiation site. The first nucleotide, usually adenosine triphosphate (ATP) or guanosine triphosphate (GTP), will then be used to begin (initiate) transcription. The sigma factor is no longer needed once transcription has initiated.

Elongation of the RNA during transcription continues with the DNA template being read in the 3'-to-5' direction and the transcript being produced antiparallel in the 5'-to-3' direction. Each new ribonucleotide is added to the 3' hydroxyl group of the previously added ribonucleotide; transcription continues until a terminator is encountered in the DNA. In bacteria, transcription and translation are coupled; they occur simultaneously in the cytoplasm. As RNAP produces the mRNA, the ribosome attaches to the mRNA and performs translation. In fact, multiple ribosomes may be carrying out translation from a single mRNA in a complex called a **polysome**. Each ribosome initially attaches to the mRNA at the ribosome binding site (RBS; also called the Shine-Dalgarno sequence). The RBS, recognized by the 16S rRNA within the small subunit of the ribosome, is generally located 6 to 10 nucleotides downstream from the start of transcription. Because 16S rRNA sequences differ among bacteria, the complementary RBS sequences must also have corresponding differences. In eukaryotes, transcription and translation are compartmentalized, with transcription occurring in the nucleus and translation occurring in the cytoplasm. The capacity of bacteria to couple transcription and translation enables faster protein production, which is important due to the short half-life of bacterial mRNAs (~1 to 2 minutes). The coupling of transcription and translation also allows for unique regulatory mechanisms in bacteria, such as attenuation (see below).

Bacterial Transcription Termination

Termination is the process of stopping transcription, which occurs when a signal within the mRNA causes the RNAP to stop synthesizing the mRNA. There are two major mechanisms of transcription termination in bacteria, **factor-independent termination** and **factor-dependent termination** (Fig. 8.3). During factor-independent termination, a stem-loop structure created by intramolecular hydrogen bonds forms in the mRNA. This stem-loop is followed by a series of uracil nucleotides forming a poly(U) tail. When this secondary structure forms in the mRNA, it disrupts the association of the RNAP with the DNA template. The poly(U) tail also creates an area of weaker hydrogen-bond interactions between the RNA and DNA that leads to transcription termination. Factor-dependent termination involves an accessory protein such as Rho, Nus, or Tau that is critical to the process of transcription termination. Here, the ribosome encounters a stop codon in the mRNA, exposing an area on the mRNA between the stalled ribosome and the RNAP that is still elongating the mRNA. In this exposed area of the mRNA, there is a binding site for the accessory protein (e.g., Rho binds to the Rut site). Once bound to the mRNA, the accessory protein moves into a position behind the RNAP and dislodges it from the DNA template, resulting in the termination of transcription.

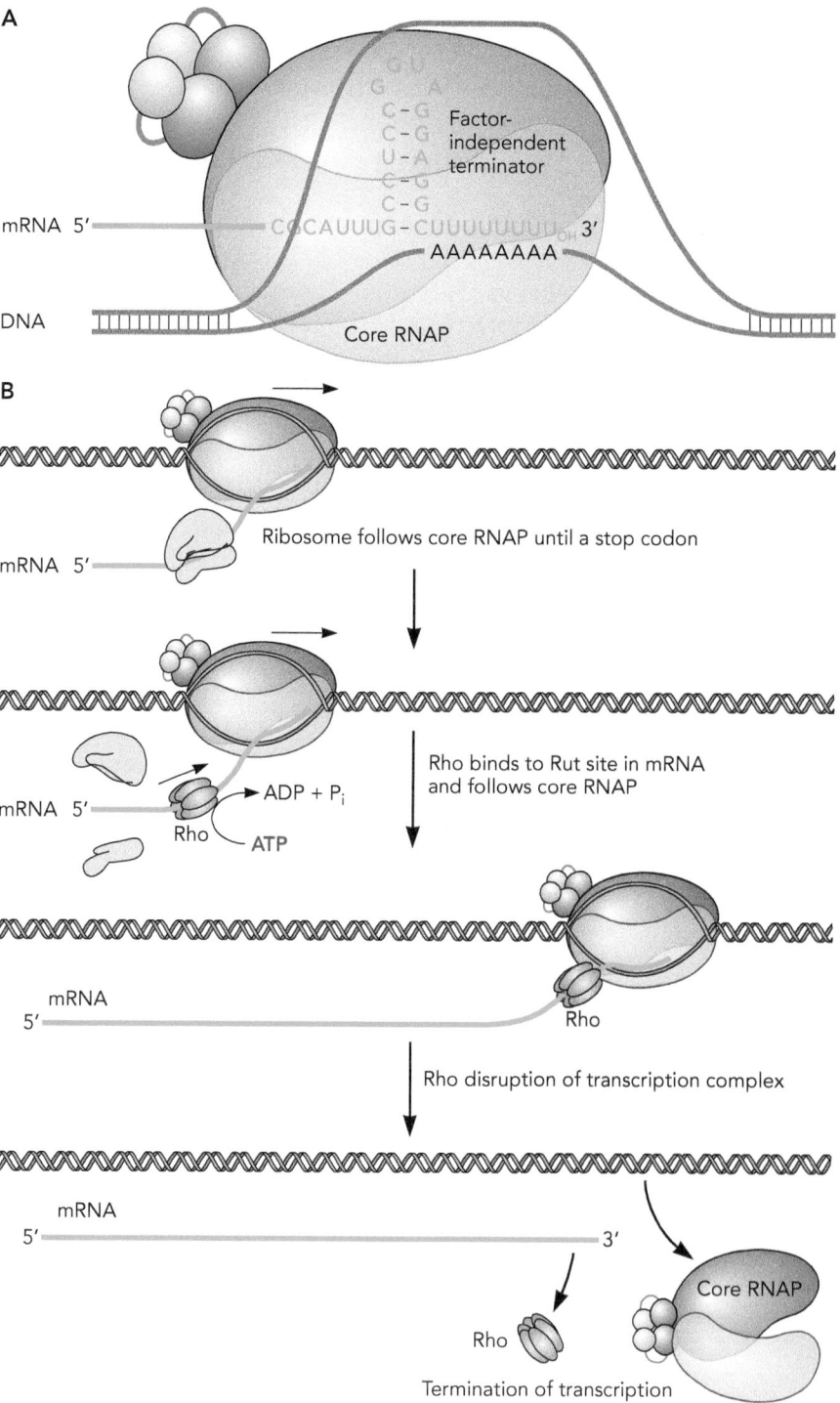

A

B

Ribosome follows core RNAP until a stop codon

Rho binds to Rut site in mRNA and follows core RNAP

Rho disruption of transcription complex

Termination of transcription

FIGURE 8.3 Termination of bacterial transcription. (A) A factor-independent terminator consists of a stem-loop region in the mRNA that forms through homologous base pairing, followed by a poly(U) tail. The poly(U) region of the mRNA is weakly associated with the DNA template, which destabilizes the mRNA–DNA interaction causing the RNA polymerase (RNAP) core enzyme (shown in brown and blue) to dislodge from the DNA template, terminating transcription. (B) Factor-dependent termination requires a protein factor (e.g., Rho, Nus, or Tau). When the ribosome carrying out translation of the mRNA encounters a stop codon, a binding site for Rho (Rho utilization site, Rut) becomes exposed and the Rho will move along the mRNA while hydrolyzing ATP until "bumping" into the core RNAP and dislodging it from the DNA template, terminating transcription.

Regulatory *cis*- and *trans*-Acting Elements Impacting Transcription

DNA sequences within or near the promoter and physically linked to the gene(s) being regulated are considered to be ***cis*-acting regulatory elements**. Proteins that diffuse throughout the cytoplasm and impact transcription at multiple sites around the genome are known as ***trans*-acting regulatory elements**.

The promoter region itself is a *cis*-acting regulatory element that is absolutely critical for determining whether or not transcription will initiate. If you take a number of promoters recognized by one sigma factor and align their sequences, a consensus binding sequence for that sigma factor can typically be identified. For the housekeeping σ^{70}, this means the –10 and –35 sites are conserved together with the spacing between them (Table 8.1). For any given promoter, the closer the sequence is to the consensus or "ideal" binding site, the greater the affinity of the sigma factor for the promoter region, which results in a higher rate of transcription. If the promoter differs from the consensus at several nucleotide positions, it will be a weaker promoter with a lower rate of transcription.

Sigma factors are *trans*-acting components essential for the initiation of transcription. All bacteria have more than one sigma factor, but the actual number varies from organism to organism. *E. coli* has a total of seven sigma factors, each with a unique consensus sequence and spacing (Table 8.1), which drive transcription of different subsets of genes in the genome. *Bacillus* species have several additional sigma factors that are used to control transcription through the different stages of sporulation in that bacterium (Chapter 17).

Regulatory proteins are additional *trans*-acting factors that also influence the rate at which the RNAP binds to a promoter (Fig. 8.4). **Activator proteins**, which increase the rate of transcription, bind to *cis*-acting regulatory elements called **activator binding sites** that are typically located upstream of promoters. **Repressor proteins**, which decrease the rate of transcription, bind to *cis*-acting regulatory elements called **operator** sequences that are typically located within or downstream of promoters. The *trans*- and *cis*-acting elements at a given promoter function together to create different types of promoter structures.

TABLE 8.1 The seven sigma factors of *Escherichia coli*[a]

Factor[b]	Gene	Number of amino acid residues	Size (kDa)	Consensus binding site[c]	Genes regulated
σ^{70} (σ^{D})	*rpoD*	613	70	TTGACA–N_{17}–TATAAT	Housekeeping
σ^{54} (σ^{N})	*rpoN* (*ntrA*)	477	54	CTGGCAC–N_{5}–TTGCA	Nitrogen metabolism
σ^{S}	*rpoS* (*katF*)	362	38	TTGACA–N_{12}–TGTGCTATACT	Stationary phase
σ^{32}(σ^{H})	*rpoH* (*htpR*)	284	32	CTTGAA–N_{14}–CCCCATNT	Heat shock
σ^{F} (σ^{28})	*fliA*	239	28	TAAA–N_{15}–GCCGATAA	Flagellar proteins
σ^{E}	*rpoE*	191	24	GAACTT–N_{16}–TCTGA	Extreme heat shock
σ^{fecI}	*fecI*	173	19	GGAAAT–N_{17}–TC	Iron transport

[a] Reprinted from Burgess RR. 2001. *In Brenner's Encyclopedia of Genetics*. Academic Press, Waltham, MA, with permission.
[b] Alternative names are given in parenthesis.
[c] N_{x} indicates any x number of nucleotides.

A Housekeeping promoter

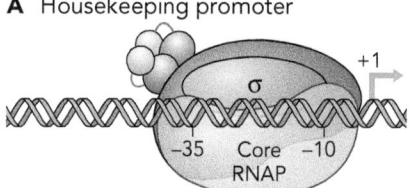

B Strong rRNA promoter

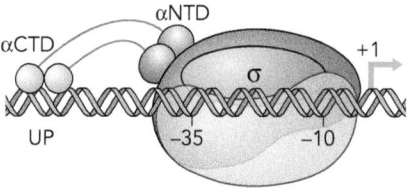

C Class I-activated promoter

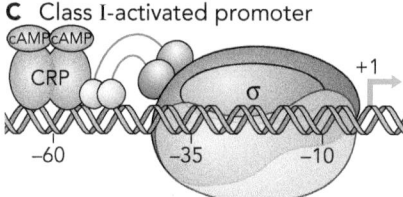

D Class II-activated promoter

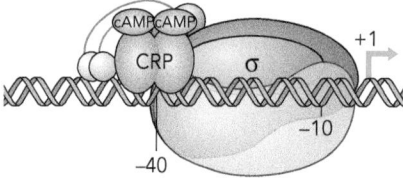

E Repressed promoter

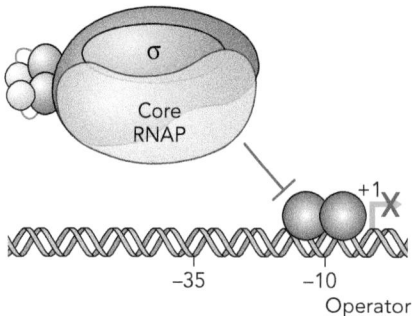

FIGURE 8.4 **Bacterial promoter structures.** (A) A housekeeping promoter contains –35 and –10 binding sites that are recognized by the sigma subunit of RNA polymerase (RNAP). (B) Strong rRNA gene promoters contain an UP element, which is bound by the C-terminal domain of the alpha subunits (αCTD) and located upstream of the –35 and –10 sites in the promoter that are recognized by sigma. (C) In a Class I type-activated promoter, there is an activator binding site located around –60 in the promoter where an activator protein (e.g., CRP) will bind. The αCTD binds to the activator protein and to the adjacent downstream (proximal) DNA. (D) In a Class II type-activated promoter, there is an activator binding site located around –40 in the promoter where the activator protein (e.g., CRP) will bind. The αCTD binds to the activator protein and to the adjacent upstream (distal) DNA. The activator will also interact with other subunits of the RNAP on its downstream (proximal) side in an "ambidextrous" manner. (E) A promoter may be repressed when a repressor protein binds to an operator sequence, most commonly located within or downstream of the promoter, either preventing the RNAP from binding or preventing its progression down the DNA. Note that this is not the only type of promoter architecture that results in activation or repression (e.g., a looping structure might form in the DNA caused by transcription factor binding at a site further away from the promoter).

Examples of Different Promoter Structures

An examination of different promoters reveals their structures and the roles of *cis*- and *trans*-acting elements in controlling transcription. In *E. coli* housekeeping promoters, the –10 and –35 sites are recognized by σ^{70} and transcription initiates at the +1 site. Some stronger promoters, such as promoters for rRNA genes, have another *cis*-acting sequence known as an UP element in addition to the −10 and −35 sites. This UP element is upstream from σ^{70} recognition sites and is bound by the C-terminal domain of the RNAP alpha subunit (αCTD; Fig. 8.4). Thus, the alpha subunit also has the capacity to recognize and bind certain DNA sequences. This additional contact point between the RNAP and promoter region leads to a greater affinity between them and results in a higher rate of transcription.

Activator proteins are also capable of binding to both the DNA and the RNAP, thereby enhancing the rate of transcription (Fig. 8.4). At a **Class I-activated promoter**, the activator protein binds to specific bases in the DNA at the activator binding site ~60 bp upstream of the main promoter sequences and enhances the rate of transcription. Class I promoters are generally weak because their sequences deviate from the consensus sequence at the −10 and −35 sites, and the presence of an activator helps to compensate for weak promoter sequences. Not only does the activator bind to the DNA, it can also interact with the RNAP αCTD. Remember, the αCTD itself also has the capacity to bind to the DNA. Therefore, the activator, the αCTD, and the sigma factor are all associated with the DNA and the activator and alpha subunit have protein–protein interactions that lead to a higher rate of transcription than would occur with just RNAP itself. At a **Class II-activated promoter**, the main promoter sequences deviate significantly from the consensus sequence; a −10 site generally exists but the −35 site may be missing. In this case, the activator binding site is near −40. In this position, the activator will have multiple protein–protein contacts with the RNAP. The αCTD associates both with the upstream (distal) side of the bound activator and to the upstream DNA. On the downstream (proximal) face, the activator associates with the alpha subunit N-terminal domain and with the sigma factor. Since the activator forms multiple protein–protein interactions with RNAP, it is considered to be an ambidextrous activator. Through the multiple interactions with the DNA and the multiple protein–protein interactions, the RNAP-activator complex compensates for the weak promoter sequence and leads to higher rates of transcription.

The activator proteins at either a Class I or a Class II promoter function to increase the rate of transcription by binding upstream of the promoter and facilitating the binding of RNAP. The activator may change the structure/conformation of the DNA (e.g., bend the DNA) in the promoter region to enhance binding by RNAP. Protein–protein interactions between the activator and RNAP lead to more efficient RNAP binding and a faster rate of formation of the closed promoter complex. Finally, some activators impact the rate at which the closed promoter complex converts to the open promoter complex so that transcription initiates faster.

At a repressed promoter, the binding of a repressor protein inhibits the ability of RNAP to carry out transcription. Often the repressor binds to an operator site that is located within the promoter near −10 or just downstream of the promoter. In this position, the repressor either blocks the ability of RNAP to recognize and bind to the promoter or blocks the RNAP from effectively progressing down the DNA template, thereby decreasing the rate of transcription. Next, we will examine the role of a variety of *cis*- and *trans*-acting regulatory components in controlling the expression of some actual *E. coli* promoters.

Transcriptional Regulation of the *lac* Operon

Previously, the concepts of diauxic growth and phosphotransferase sugar (PTS) versus non-PTS sugar transport were introduced (Chapters 4 and 7). Gene regulation is critical to these physiological processes. The *lac* operon in *E. coli*, which encodes the genes necessary for metabolism of the non-PTS sugar lactose, has been intensively studied and therefore its regulation serves as a good real-world example of both activation and repression of transcription. Joshua Lederberg performed some of the early genetic experiments on the *lac* operon in the 1940s prior to the discovery of the structure of DNA. The *lac* operon contains three genes, *lacZ*, *lacY*, and *lacA*, which encode β-galactosidase, the lactose permease/transporter, and a transacetylase, respectively (Fig. 8.5). The enzyme β-galactosidase cleaves the carbon source lactose into glucose and galactose. Inducer exclusion blocks the transport of lactose through the LacY permease when glucose is present (Chapter 7). The LacA transacetylase is not critical for the mechanism of regulation. There are two promoters, called P1 and P2, that are important for controlling the expression of the *lac* operon. When the *lac* operon is transcribed, a polycistronic mRNA is generated that contains the information for the translation of the three polypeptides, LacZ, LacY, and LacA.

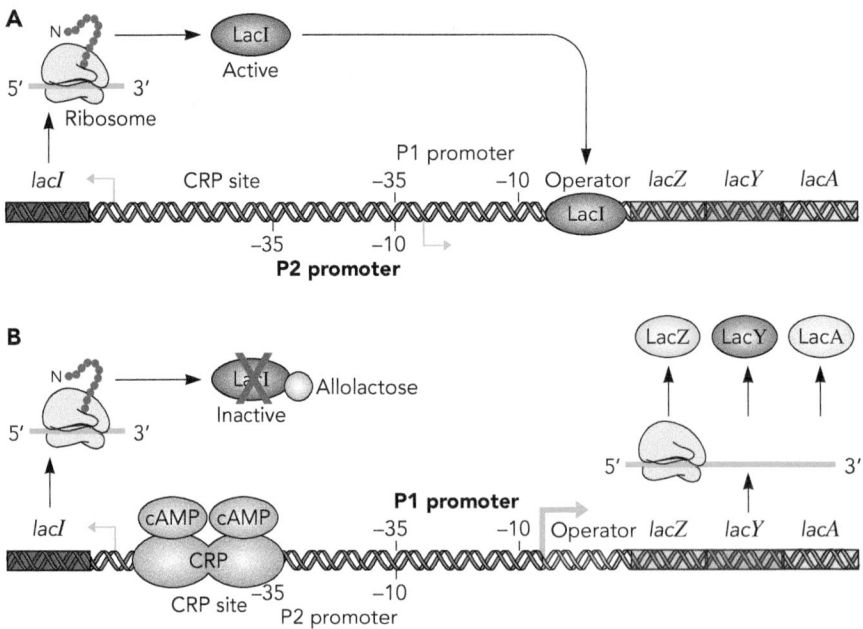

FIGURE 8.5 **Transcriptional regulation of the *lac* operon.** (A) The *lacI* gene is transcribed into a monocistronic mRNA that is translated into the LacI repressor protein. LacI is made in an active conformation capable of binding to the operator site in the *lac* operon promoter and blocking transcription of the *lac* operon genes from the P1 promoter. There is some low-level transcription (basal expression, thin green arrow) of the genes due to the weak P2 promoter. (B) When lactose is present, it is converted into allolactose, which serves as a negative allosteric effector for LacI, rendering it inactive and unable to bind to the operator site. If lactose is present and phosphotransferase system (PTS) sugars have been consumed, then inducer exclusion and catabolite repression no longer occur. The CRP–cAMP complex binds near –60 in the P1 promoter (a Class I promoter), enabling high-level transcription of the genes in the *lac* operon (thick green arrow). A polycistronic mRNA that encodes LacZ (β-galactosidase), LacY (lactose permease/transporter), and LacA (transacetylase) is synthesized. These proteins are necessary for the transport and catabolism of lactose.

The repressor protein, LacI, and the activator protein CRP (Chapter 7) regulate the *lac* operon promoters, P1 and P2 (Fig. 8.5). The *lacI* gene, which is located upstream from *lacZYA*, has its own promoter. When transcription occurs from the *lacI* promoter, a monocistronic mRNA is formed and just one polypeptide, LacI, is produced. LacI is a *trans*-acting regulatory factor; it can diffuse through the cytoplasm to act on the *lac* operon P1 promoter. When glucose is present and expression of the *lac* operon is not advantageous, catabolite repression occurs. Under these circumstances LacI is in its active form. LacI binds to the operator sequence within the *lac* operon P1 promoter near –10, functioning as a repressor and rendering the P1 promoter nonfunctional. The *lac* operon P2 promoter is functional but is weak because its sequence deviates from the consensus, and it only permits transcription at a very low rate. This enables the production of small quantities of LacY and LacZ such that the cell can begin transporting lactose and metabolizing it as soon as inducer exclusion ends.

When the glucose has been consumed and lactose needs to be catabolized, lactose is transported into the cell and the regulatory system must be reset to optimize lactose utilization. LacI is allosterically regulated by allolactose, which is produced from lactose by an alternative reaction catalyzed by LacZ. When allolactose binds to LacI, the LacI protein is rendered nonfunctional through allosteric regulation and cannot bind to its target operator site on the DNA. Thus, repression of the *lac* operon is relieved. However, this derepression of the *lac* operon is insufficient to fully activate its expression. In addition, an activator protein is needed to stimulate higher levels of transcription from the P1 promoter. Lower glucose levels result in the production of the second messenger cAMP (Chapter 7), which binds to the CRP protein and converts it to an active conformation. The CRP can now bind as a homodimer to the DNA at the CRP binding site (a position near –60 in the P1 promoter), creating a strong Class I-type promoter. Therefore, when the glucose level drops, catabolite repression ceases, and high-level expression of the *lac* operon enables the cell to begin to fully catabolize lactose.

To summarize, in the presence of high levels of glucose, inducer exclusion and catabolite repression both occur. In addition, cAMP levels are low under these conditions. Thus, LacI is active and CRP is inactive, and weak expression from the P2 promoter occurs. When the glucose concentration drops, inducer exclusion and catabolite repression cease. As the concentration of cAMP increases, CRP–cAMP complexes form, converting CRP to an active confirmation. Allolactose binds to LacI, causing the repressor to become inactive. Under these conditions, the P1 promoter becomes a strong Class I promoter activated by the CRP–cAMP complex, and the *lac* operon is highly expressed.

Expression of the *lac* operon may be artificially manipulated under laboratory conditions. The *lac* promoter region is used in many bacterial plasmids to control the expression of cloned genes for the heterologous production of proteins. As lactose is metabolized, its cytoplasmic levels as well as those of the inducer allolactose, will decrease over time and expression of the *lac* operon will decrease. To maintain the inactive state of LacI, a nonmetabolizable inducer is used instead of lactose to ensure gene expression is always turned on. The term **gratuitous inducer** is used for an allosteric effector that is nonmetabolizable. Isopropyl-β-D-thiogalactopyranoside (IPTG) is the gratuitous inducer used to regulate LacI instead of allolactose. Similarly, the effector cAMP may be added exogenously into the medium to artificially influence the activity of CRP.

CRP is a global regulator, a *trans*-acting transcription factor (i.e., an activator or a repressor) that adds another layer of regulation to control the expression of multiple genes and operons. For example, CRP controls the utilization of a variety of non-PTS sugars (Chapter 7). Each non-PTS sugar has its own catabolic operon under the control of a specific regulatory protein (e.g., as LacI controls the *lac* operon, MalT controls the maltose operon).

Activation and Repression by the Global Regulator Cra

While CRP is only known to be an activator of transcription, there are other global regulatory proteins such as Cra (catabolite repressor/activator) that can serve as either an activator or repressor, depending on the location of the DNA-binding recognition site and the available carbon source. In *E. coli*, Cra is biologically active (i.e., binds to its DNA targets) when acetate or fatty acids are available as a carbon source and there is no glucose present. In its active conformation, Cra increases transcription of the genes encoding the enzymes for the glyoxylate cycle and gluconeogenesis (Chapter 2) by binding upstream of the relevant promoters (Fig. 8.6). Simultaneously, it binds to operators within promoters of genes encoding enzymes of the Embden-Meyerhof-Parnas (EMP) and Entner-Doudoroff (ED) pathways to repress transcription. Thus, when acetate or fatty acids are present, those substrates can be metabolized via the glyoxylate cycle and the carbon flows through gluconeogenesis to generate the essential precursors.

When glucose is present, Cra mediates another type of catabolite repression that does not involve CRP and cAMP. Instead, the allosteric effectors for Cra are fructose-1-phosphate and fructose-1,6-bisphosphate, intermediates of central metabolism that are produced when glucose is the primary carbon source. When these effector molecules complex with Cra, Cra can no longer bind to its DNA targets, which results in gene deactivation and derepression (Fig. 8.6). Under these conditions, the genes encoding enzymes for the glyoxylate cycle and gluconeogenesis are no longer expressed. Simultaneously, promoters for the genes in the EMP and ED pathways become derepressed, allowing glucose to be metabolized.

Attenuation

Attenuation is a unique form of regulation in bacteria because it requires that transcription and translation are coupled. This regulatory mechanism is not widespread, but rather is used in a few specific operons to afford dual regulation of the system at the levels of both transcription initiation and transcription elongation. The tryptophan biosynthetic operon in *E. coli* is controlled by attenuation, and it serves as good model for understanding this regulatory mechanism.

A single gene, which produces a monocistronic transcript, encodes the TrpR repressor protein. TrpR remains folded in an inactive conformation until its allosteric effector is present at sufficient concentrations. The amino acid tryptophan is the natural effector for TrpR. In laboratory studies, indole acrylic acid (IAA) serves as a gratuitous inducer for TrpR. In the presence of either tryptophan or

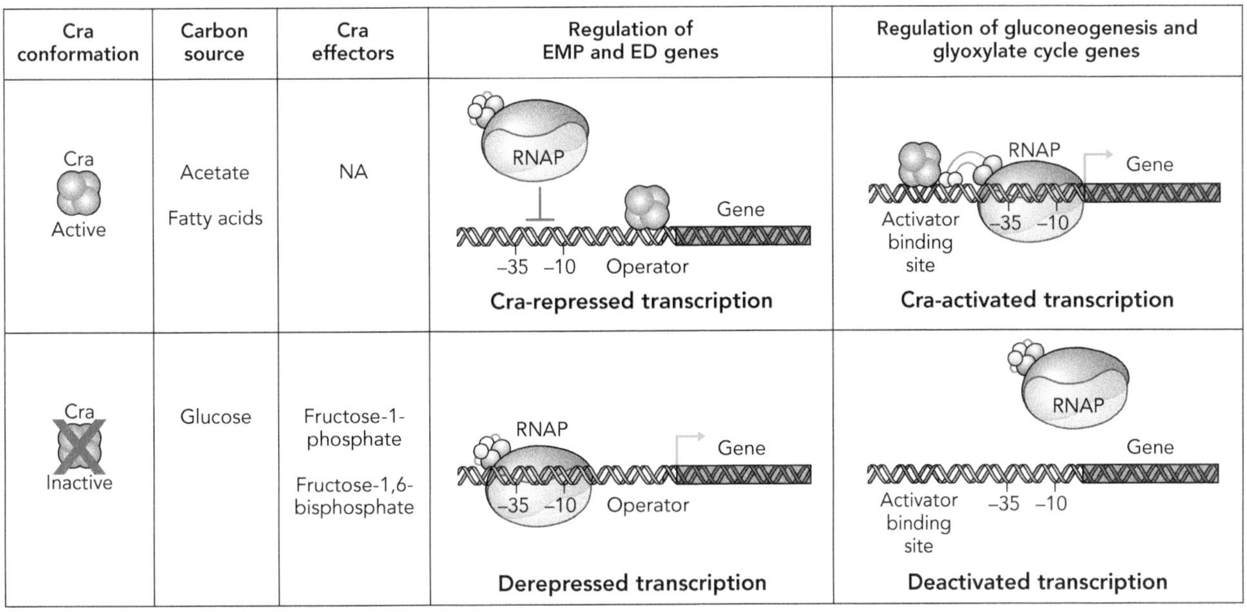

Cra conformation	Carbon source	Cra effectors	Regulation of EMP and ED genes	Regulation of gluconeogenesis and glyoxylate cycle genes
Cra Active	Acetate Fatty acids	NA	RNAP ⊥ −35 −10 Operator — Gene **Cra-repressed transcription**	RNAP — Gene Activator binding site −35 −10 **Cra-activated transcription**
Cra Inactive	Glucose	Fructose-1-phosphate Fructose-1,6-bisphosphate	RNAP −35 −10 Operator — Gene **Derepressed transcription**	RNAP — Gene Activator binding site −35 −10 **Deactivated transcription**

FIGURE 8.6 **Transcriptional regulation by the global regulator Cra in** *Escherichia coli.* Cra (catabolite repressor/activator) is a global regulator controlling carbon catabolism. (Top) When fatty acids or acetate are available instead of phosphotransferase system (PTS) sugars, Cra is in an active conformation and it serves as a repressor by binding operators downstream of the promoters for genes important in glycolysis. Simultaneously, Cra serves as an activator by binding activator binding sites upstream of the promoters of genes important for gluconeogenesis and the glyoxylate cycle. (Bottom) In the presence of a PTS sugar (e.g., glucose), Cra is in an inactive conformation due to negative allosteric regulation by fructose-1-phosphate and fructose-1,6-bisphosphate. Under these physiological conditions, genes needed for glycolysis are derepressed and transcribed, whereas genes needed for gluconeogenesis and the glyoxylate cycle are deactivated and not transcribed.

IAA, TrpR becomes biologically active and binds to the operator sequence of the *trp* operon (Fig. 8.7A). The *trp* operon consists of five genes, *trpEDCBA*, which are expressed as a polycistronic mRNA and encode the five polypeptides involved in the biosynthesis of tryptophan. Therefore, when cellular tryptophan levels are low, TrpR is inactive and the *trp* operon is highly expressed. However, if there is sufficient tryptophan present, it interacts with and activates TrpR, which represses expression of the *trp* operon, thus conserving energy for the cell.

Attenuation is a mechanism that can stop transcription after initiation but prior to complete transcription of the operon, which also helps the cell to conserve resources (Fig. 8.7A). Upstream from the first gene in the *trp* operon but downstream of the promoter is a short leader sequence that contains multiple tryptophan codons (region 1, Fig. 8.7B). The leader sequence also contains the **attenuator**, which contains three regions (regions 2, 3, and 4) that can form complementary base pairs within the mRNA transcript, generating different stem-loop structures depending on the level of available tryptophan. A stem-loop structure between regions 3 and 4, followed by a series of uracil nucleotides [poly(U) tail], serves as a factor-independent transcriptional terminator. In the presence of high levels of tryptophan (Figure 8.7B), the ribosome is closely following behind RNAP, translating the newly formed mRNA. When the leader sequence and attenuator region are quickly transcribed and translated, the stem-loop structure between regions 3 and 4 forms in the mRNA transcript, generating the factor-independent transcriptional terminator. RNAP is then

A

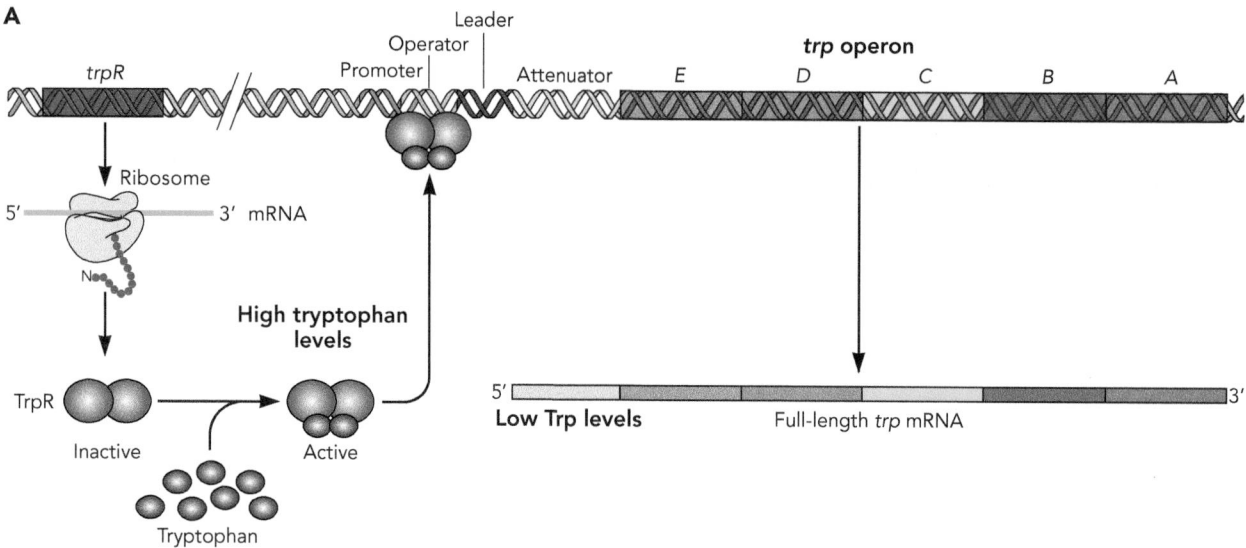

B

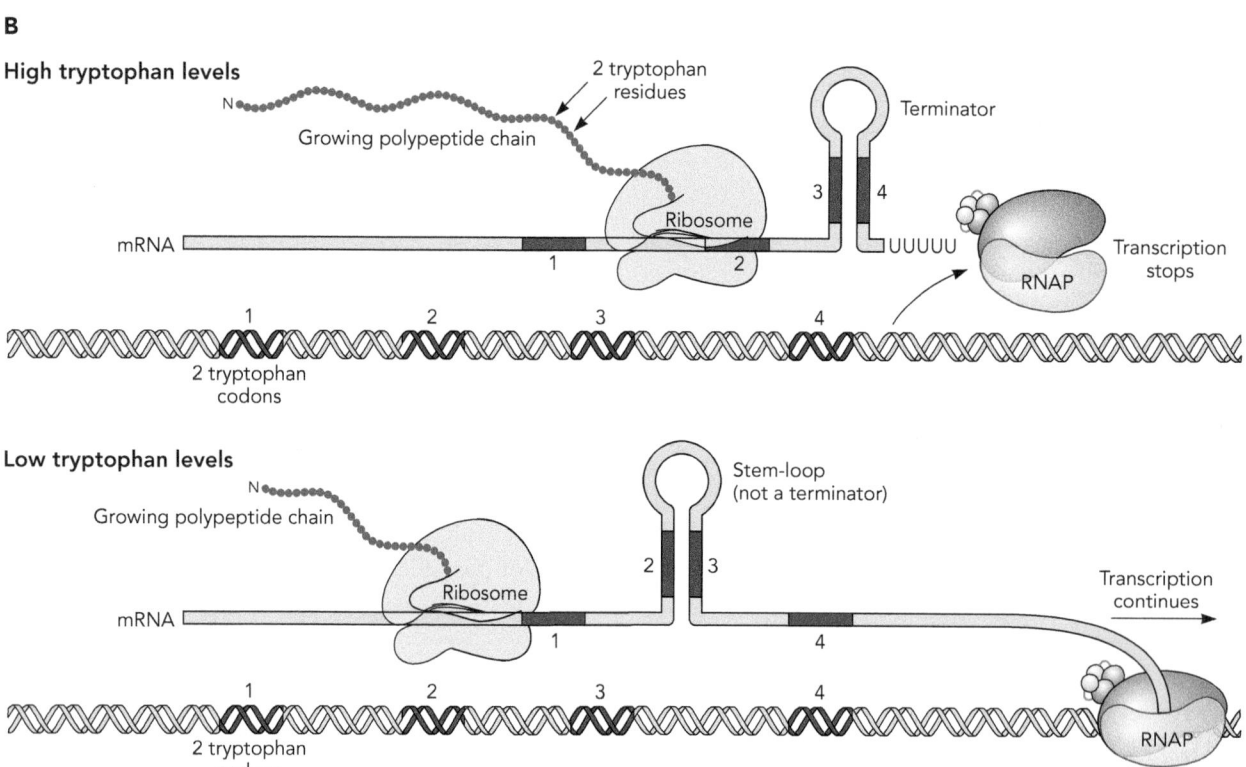

FIGURE 8.7 **Regulation of the *Escherichia coli* *trp* operon through transcription and attenuation.** (A) The *trp* operon consists of five genes (*trpEDCBA*) encoding the tryptophan biosynthetic enzymes. Transcription of these genes is controlled by the TrpR repressor. In the presence of high concentrations of the effector molecule tryptophan, TrpR is converted into an active conformation. The active form of TrpR binds to the operator upstream from the *trp* genes and represses initiation of their transcription. When levels of tryptophan are low, TrpR does not bind and the transcript is made. (B) A second regulatory element, the attenuator, enables premature termination of transcription during the elongation phase. Any mRNA transcribed from the *trp* leader and attenuator sequences is immediately translated by ribosomes. In the presence of low concentrations of tryptophan, the ribosome stalls at region 1 waiting for tryptophan-charged tRNAs. This leaves region 2 available to form a weak stem-loop structure with region 3 that does not prevent translation, and transcription of the *trp* operon genes by RNAP continues. The formation of the stem-loop structure between regions 2 and 3 (not a terminator, also known as an antiterminator) prevents the formation of the terminator stem-loop structure between regions 3 and 4. In the presence of high concentrations of tryptophan, the ribosome does not stall, thus blocking region 2. This allows the formation of the terminator stem-loop structure in regions 3 and 4 followed by a poly(U) tail (a factor-independent terminator), prematurely terminating (attenuating) transcription.

dislodged from the DNA template; transcription and expression of the *trp* operon genes downstream of the attenuator is stopped. Translation also stops, as there is no more mRNA for the ribosome to translate.

However, if tryptophan levels are low (Fig. 8.7B), then the ribosome stalls in region 1 upstream from the attenuator as it waits for transfer RNAs (tRNAs) charged with tryptophan, which are necessary for the continuation of translation. With the ribosome stalled at region 1, a stem-loop structure between regions 2 and 3 forms, preventing the factor-independent terminator from forming between regions 3 and 4. In this case, RNAP will transcribe the operon, resulting in continued production of enzymes required for the biosynthesis of tryptophan. Therefore, the process of attenuation enables the cell to synthesize tryptophan only when tryptophan levels are low.

Posttranscriptional Regulation

Posttranscriptional regulation determines whether a polypeptide will be translated from an mRNA template. This form of regulation was initially discovered in eukaryotes, where microRNAs were recognized as having key regulatory roles. Subsequently, the existence of small RNAs (sRNAs), typically between 100 and 400 nucleotides in size, was discovered in bacteria. sRNAs are single-stranded molecules, but due to hydrogen bonding they can form secondary structures that are important for their function. There are two major classes of sRNAs. One class works by base pairing with target mRNAs to modulate mRNA translation or stability. The other class functions by modifying the activity of RNA-binding proteins.

REGULATION THROUGH BASE PAIRING BY sRNAs

In *E. coli*, DsrA is a small RNA (note the same nomenclature is used for proteins and sRNAs) that controls the translation of stationary-phase sigma factor, RpoS. In turn, RpoS controls the expression of >100 genes that enhance cell survival under stress conditions (Chapters 4 and 7). During exponential growth, the *rpoS* mRNA forms a stem-loop secondary structure that prevents ribosome access to the ribosome binding site (RBS) (Fig. 8.8A). If the ribosome is unable to access the RBS, the *rpoS* mRNA will only be translated at low levels. Under conditions of stress and in the presence of Hfq (a small protein that influences mRNA expression and stability), DsrA binds to a complementary region in the *rpoS* mRNA. Because DsrA has a higher affinity for its target region on the mRNA, the binding of DsrA disrupts the stem-loop structure and changes the conformation of the mRNA so that the RBS is accessible to the ribosome. As a result, translation of the *rpoS* mRNA occurs at the higher rate needed during stationary phase or a stress event.

While the regulation described above is an example of positive regulation by an sRNA, negative regulation also occurs. In these cases, the sRNA forms complementary base pairs with the mRNA in the region near the RBS facilitated by the protein Hfq. By binding to the RBS, the sRNA prevents translation from taking place (Fig. 8.8B).

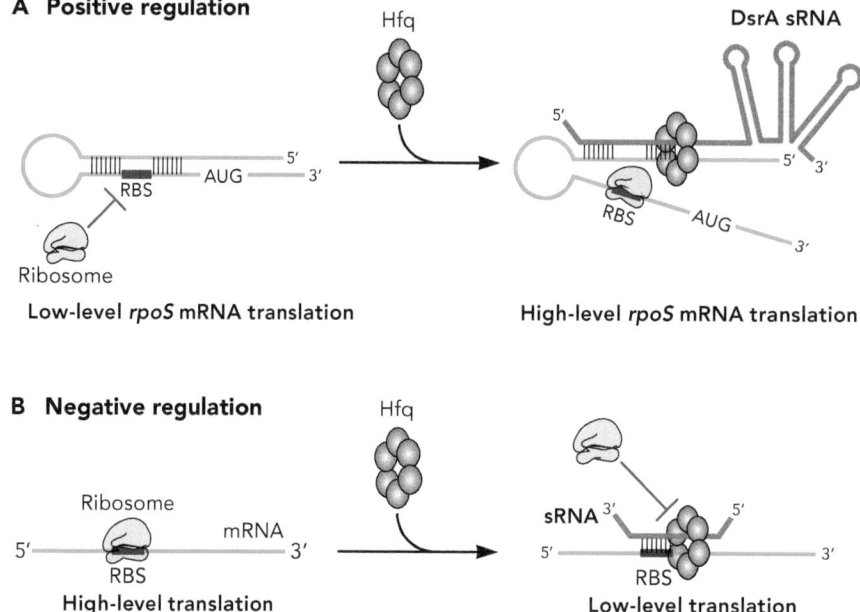

A Positive regulation

Hfq

DsrA sRNA

RBS

AUG — 3'

5'

Ribosome

Low-level *rpoS* mRNA translation

RBS

AUG

5'

3'

High-level *rpoS* mRNA translation

B Negative regulation

Hfq

Ribosome

mRNA

5' — RBS — 3'

High-level translation

sRNA 3'

5'

RBS

5'

3'

Low-level translation

FIGURE 8.8 **Posttranscriptional regulation through base pairing of sRNAs.** sRNAs are small RNA molecules that can bind to mRNA and impact translation. sRNA–mRNA interaction is facilitated by Hfq, an RNA-binding protein that stabilizes sRNA–mRNA interactions. (A) In the absence of the sRNA, the *rpoS* mRNA forms a stem loop structure that includes the ribosome binding site (RBS), blocking translation. For example, when the DsrA sRNA is present, it binds to the *rpoS* mRNA with higher affinity thus freeing the RBS and making translation possible. This is a form of positive regulation. (B) Conversely, when an sRNA binds to an mRNA in a manner that forms base pairs in the region where the RBS is located, translation will be reduced. This is a form of negative regulation.

REGULATION OF mRNA-BINDING PROTEINS BY sRNAs

The carbon storage regulators, CsrA (mRNA-binding protein) and CsrB (sRNA) together exert global regulatory changes that impact cellular physiology in relation to carbon metabolism. Free CsrA preferentially binds near the RBS of the mRNA for certain genes (e.g., *glgCAP*) that are involved in glycogen biosynthesis and blocks the ribosome from binding, resulting in lower levels of translation (Fig. 8.9). CsrB is produced under low nutrient conditions. When CsrB is present, CsrA preferentially binds to the multiple stem-loops on CsrB, which consequently lowers the internal concentration of free CsrA. As free CsrA levels drop, RBSs that were previously blocked become accessible to the ribosome and translation will initiate.

RIBOSWITCH REGULATION

Self-folding of an mRNA can also lead to posttranscriptional control of translation via a riboswitch (Fig. 8.10). In one conformation, the mRNA upstream from the coding region forms a stem-loop structure that leaves the RBS accessible to the ribosome and high rates of translation occur. However, when present at sufficiently high concentrations, select low-molecular-weight metabolites can bind to the mRNA causing an alternative stem-loop structure to form. These metabolites (such as vitamins, amino acids, or purine bases) are often the end

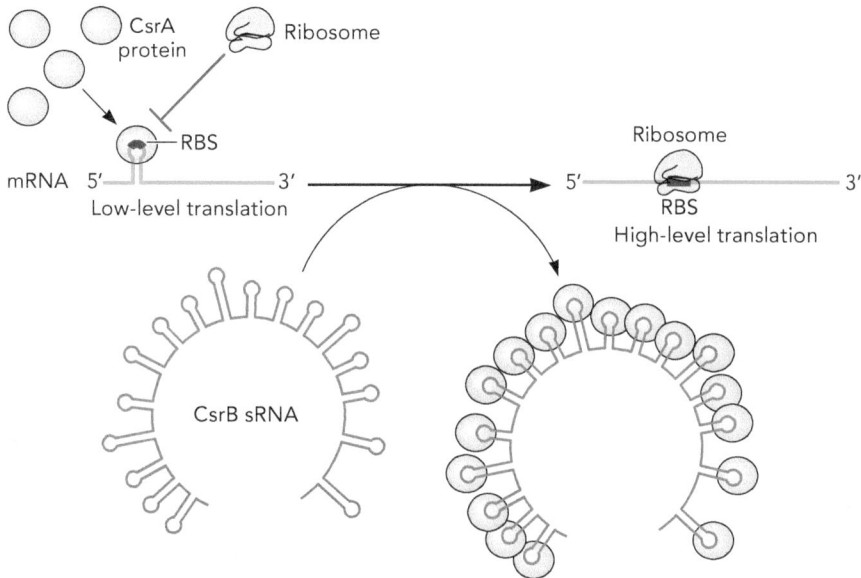

FIGURE 8.9 **Posttranscriptional regulation when sRNAs modify protein activity.** The RNA-binding protein CsrA can bind to a ribosome binding site (RBS) and block translation by the ribosome. However, if the sRNA CsrB is expressed by the cell, then CsrA associates with CsrB, leaving the RBS available for the ribosome and enabling translation to occur.

High-level translation

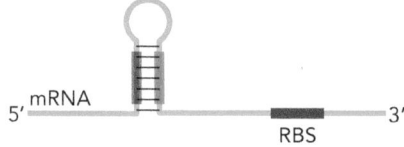

Low-level translation

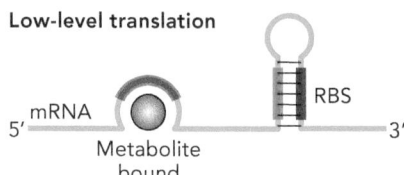

FIGURE 8.10 **Posttranscriptional regulation through riboswitches.** A riboswitch involves self-regulation of the mRNA such that in one secondary structure (top) the ribosome binding site (RBS) is accessible to the ribosome, but in the presence of a regulatory metabolite a different secondary structure forms (bottom), rendering the RBS inaccessible to the ribosome.

product of the pathway whose enzymes are encoded by the mRNA. The alternative stem-loop structure encompasses the RBS so that the ribosome is blocked from binding and a lower rate of translation occurs.

Methods Used to Study Gene Regulation

TRANSCRIPTION ASSAYS

There are several ways to measure transcription to determine when and where regulation is occurring. Some methods are based on the isolation of the mRNA produced by transcription, whereas others are based on the generation and analysis of recombinant DNA constructs known as gene fusions. Historically, a northern blot was one of the first methods used to study expression of one gene or operon at a time (Fig. 8.11A). In this approach, total RNA is isolated from cells and agarose gel electrophoresis is used to separate the RNA fragments on the basis of size. The RNA is transferred to a membrane and then the membrane is "probed" with a labeled DNA fragment of interest (using radioactivity or fluorescence). During hybridization, the labeled probe will form complementary base pairs with the target mRNA fragment of interest, enabling quantitation of the amount of that specific mRNA product.

With the publication of complete genome sequences of model organisms, microarrays have been utilized to examine the entire **transcriptome**, or the sum total of all mRNA molecules present in the cell under a particular condition (Chapter 3). However, this approach was subsequently replaced by transcriptome sequencing (RNA-Seq) (Fig. 8.11B). In this method, the total mRNA from cells is converted to complementary DNA (cDNA) and then the entire transcriptome is sequenced using next-generation sequencing methods and analyzed using bioinformatics tools. A complementary technique that permits the study of gene expression of one or several genes at a time is known as quantitative reverse transcription polymerase chain reaction (qRT-PCR) (Fig. 8.11C). Here, total RNA is converted to cDNA similarly to RNA-Seq, but then PCR is used to estimate the amounts of individual gene transcripts. This approach permits the determination of relative amounts of a specific mRNA (i.e., in sample A versus sample B) or absolute levels of the number of mRNA molecules in comparison to a standard curve with known DNA quantities.

Gene fusions, which have been extensively used in microbial genetics studies, provide another approach to quantify gene expression. This method is used to study expression from a single promoter. A **transcriptional fusion** consists of the promoter of the gene of interest being ligated to a **reporter gene** with its native RBS (Fig. 8.11D). The reporter gene encodes an enzyme or protein with activity that is simpler to assay than the product of the gene under study. Common enzyme reporters include β-galactosidase (LacZ, producing a colored product), luciferase (LuxAB, producing light), and green fluorescent protein (GFP, producing fluorescence) and its variants. A transcriptional fusion enables one to study the rate of transcription from a promoter of interest. However, *in vivo* gene expression involves not just transcription but translation as well. Therefore, a **translational fusion** can be used to measure the rate of transcription coupled to translation for a gene of interest. In this type of construct, the promoter, the RBS, and the first few codons from the gene of interest are ligated in frame to the reporter gene, which results in the production of a fusion protein (Fig. 8.11D). By comparing the rate of transcription versus translation, posttranscriptional control can be detected.

Methods to Demonstrate Protein–DNA Interactions

In order to characterize the presence and/or sequence of *cis*-acting regulatory elements (such as promoters, activator binding sites, or operators) it is necessary to determine their location on the DNA. An electrophoretic mobility shift assay (EMSA) is an *in vitro* method used to establish whether a *trans*-acting regulatory protein of interest binds within a DNA fragment (Fig. 8.12A). A DNA template

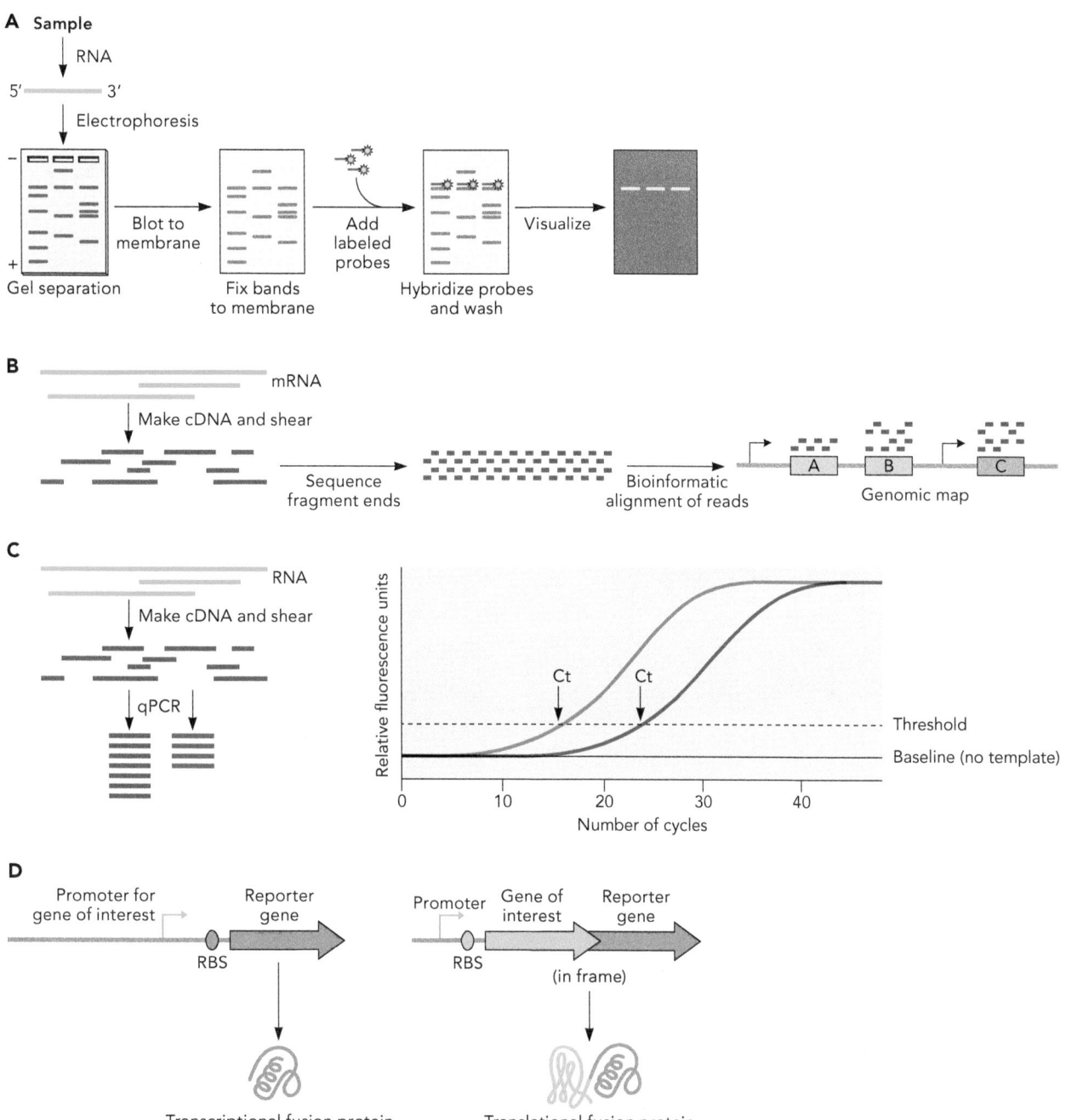

FIGURE 8.11 **Methods to measure the rate of transcription.** (A) A northern blot protocol involves extracting the total RNA from cells, using electrophoresis to separate the RNA molecules on the basis of their size, transferring the RNA to a membrane, and then hybridizing labeled DNA probes to complementary RNA sequences on the membrane to enable visualization of mRNA sequences of interest. (B) Transcriptome sequencing (RNA-Seq) requires extraction of total cellular RNA and conversion to cDNA (after depletion of rRNA and tRNA, not shown). The cDNA is then sequenced and the sequencing reads are aligned with the genome using bioinformatics, resulting in a genome map that shows where the cDNA sequences align. (C) Quantitative reverse transcription polymerase chain reaction (qRT-PCR) involves extraction of total cellular RNA and conversion to cDNA followed by quantitative/real-time PCR in an instrument that measures the Ct (threshold cycle) of product formation. qRT-PCR may be relative, comparing samples to one another as shown, or absolute, where a standard curve is used to ascertain the actual numbers of molecules of transcript in the starting sample. (D) Transcriptional fusions and translational fusions use easily assayed gene reporters to measure the rate of transcription or translation for a gene of interest.

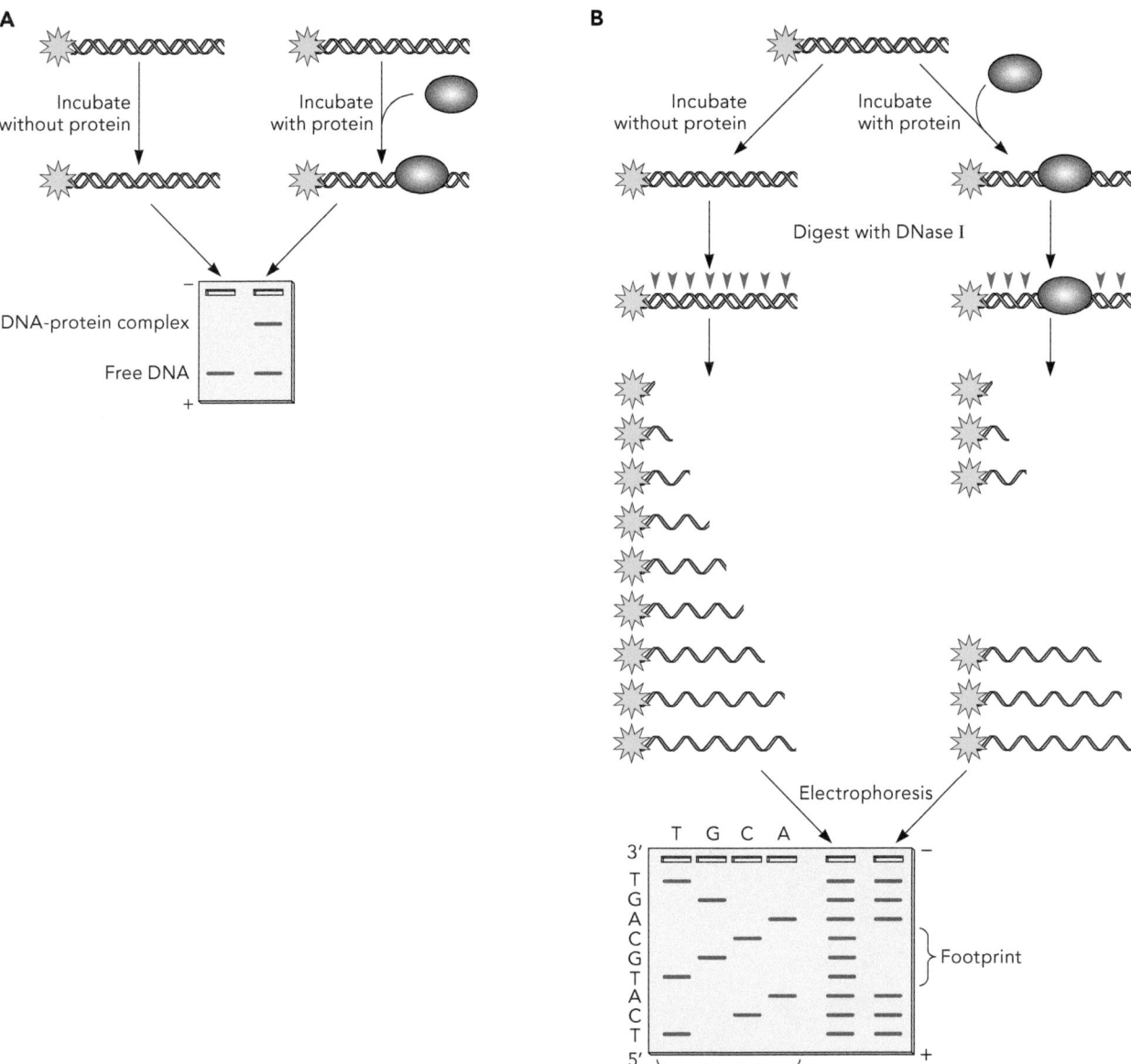

FIGURE 8.12 **Electrophoretic mobility shift assay (EMSA) and footprinting.** (A) A short fragment of DNA is labeled with a fluorescent tag and then divided such that some of the DNA is left free (naked) while some of the DNA is exposed to a putative DNA-binding protein. The two samples are analyzed via gel electrophoresis using nondenaturing polyacrylamide gels so that DNA–protein complexes remain intact. If the protein has bound the DNA, then the portion of the fragment that complexed with the protein is now larger in size, so has "shifted" to a higher position in the gel, relative to the free DNA that was not exposed to protein. (B) Footprinting, also known as a DNase I protection assay, follows a similar protocol to an EMSA, but has one additional step where the free DNA and protein-bound DNA are subjected to digestion by DNase I, which is capable of cleaving phosphodiester bonds. This creates a ladder of DNA fragments, except in the region that was protected by the DNA-binding protein. The gap in the ladder represents the precise binding site of the protein (the footprint), whose sequence may be obtained through comparison to DNA sequencing reactions.

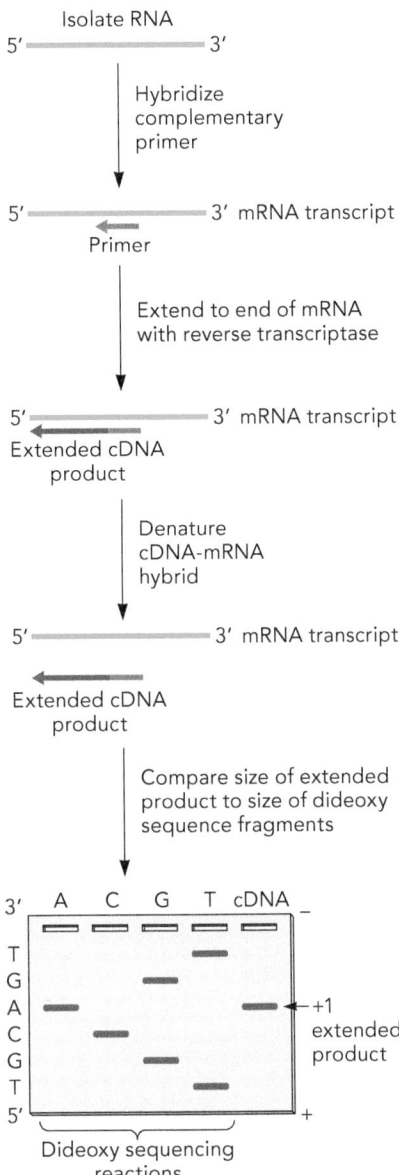

FIGURE 8.13 **Primer extension assay.** mRNA is isolated and hybridized to a primer. Reverse transcriptase is used to generate a complementary cDNA product towards the 5′-end of the mRNA template (the extended product). This extended product is denatured and analyzed by gel electrophoresis, in comparison to a dideoxy sequencing reaction.

of ~100 to 300 bp is purified and labeled with a radioactive or fluorescent tag. One DNA sample is incubated with a purified protein of interest (i.e., RNAP, activator, or repressor) and one DNA sample without protein serves as a control. The two samples are analyzed and compared via a native (nondenaturing) polyacrylamide gel and separated on the basis of their masses. The phosphate groups in the DNA carry a negative charge that is attracted to the positive electrode. If the protein is bound to the DNA fragment, the DNA–protein complex is larger than the DNA alone and will not move through the gel matrix as effectively. Thus, the mobility of the protein–DNA complex in the gel is slowed (or shifted) in comparison to the DNA-only sample. Such a result would indicate that the protein has a binding site somewhere in that DNA fragment.

If it is important to determine the precise location that a protein binds on the DNA, then another method, called a DNase I protection assay or DNA footprinting assay, must be employed. This method is similar to the EMSA in that a DNA fragment is purified, labeled, and allowed to bind to the protein of interest. However, after the DNA–protein complex is allowed to form, the sample is exposed to the enzyme DNase I. In theory, this enzyme can cleave every phosphodiester bond in the DNA fragment, although in reality there are differences in the efficiency with which cleavage occurs (Fig. 8.12B). However, wherever a protein is bound to the DNA, the protein protects the phosphodiester bonds in that DNA region from cleavage. A control DNA fragment without protein is also exposed to the DNase I enzyme. Polyacrylamide gel electrophoresis permits a comparison between the digested fragments generated with the control DNA and DNA–protein samples. A "footprint" or gap in the ladder is generated if the protein did indeed bind to the DNA and protect it from digestion in comparison to the DNA-only control. By also running a DNA sequencing reaction on the same polyacrylamide gel, the actual nucleotide bases protected by the protein can be identified. In this manner, the precise DNA sequence to which a protein binds can be determined. DNase I protection assays are now more commonly performed using automated DNA analyzer machines.

These and other methods such as primer extension (Fig. 8.13), which defines the +1 site where transcription initiates, are some of the tools that have been used to study the process of gene regulation, which is critical to understanding the complexities of microbial physiology.

Learning Outcomes: After completing the material in this chapter, students should be able to . . .

1. explain the process of transcription initiation and elongation in *Bacteria*

2. distinguish factor-independent termination from factor-dependent termination in *Bacteria*

3. compare and contrast the structure and function of promoters controlled by activator and repressor proteins

4. explain how the regulation of the *lac* and *cra* operons is controlled under different growth conditions

5. predict how the concentration of tryptophan affects transcriptional attenuation of the *trp* operon

6. describe the three main mechanisms of posttranscriptional regulation

7. interpret results from methods for investigating levels of transcription and protein-DNA binding

Check Your Understanding

1. Define this terminology: constitutive expression, "housekeeping" function, promoter region, +1 site, operon, polycistronic, monocistronic, global regulator, regulon, sigma factor, core RNA polymerase (RNAP), RNAP holoenzyme, polysome, *cis*-acting regulatory element, *trans*-acting regulatory element, activator protein, activator binding site, repressor protein, operator, gratuitous inducer, attenuation, attenuator, transcriptome, transcriptional fusion, reporter gene, translational fusion

2. Refer to Figure 8.1 to answer the following review questions about transcription and translation. (LO1)
   a) Does the figure illustrate a mono- or polycistronic operon?
   b) What is the name of the region of DNA that sigma factor recognizes?
   c) What advantage does this type of gene structure provide for a bacterium, compared to transcription and translation in eukaryotes?

3. Put the steps of transcription initiation and elongation in order from 1 to 5. (LO1)

   ___ An enzyme adds a nucleotide to the 3'-OH of another nucleotide.
   ___ The core enzyme forms a complex with a sigma factor.
   ___ The first nucleotide is added.
   ___ The holoenzyme binds to the promoter site.
   ___ RNAP separates the DNA strands.

4. Compare and contrast the two types of transcription termination in bacteria. (LO2)
   a) Briefly define each:
      - Factor-dependent termination
      - Factor-independent termination
   b) Consider factor-independent termination:
      i. What dislodges RNAP from the DNA template?
      ii. With factor-independent termination, a stem-loop structure in mRNA is followed by a series of uracil nucleotides, forming a poly(U) tail. What function does the poly(U) tail serve? Why use a series of uracil nucleotides here instead of cytosine?
      iii. Does the ribosome play a role in this process?
   c) Consider factor-dependent termination:
      i. What dislodges RNAP from the DNA template?
      ii. What is the function of the protein factor?
      iii. Does the ribosome play a role in this process?

5. Assume that the housekeeping promoter is: (LO3)

5′	...	TTGACA	...	(17 bases)	...	TATAAT	...	3′
3′	...	AACTGT	...	(17 bases)	...	ATATTA	...	5′

a) Explain why promoters that more closely match this sequence are more likely to have higher rates of transcription.

b) *E. coli* has a total of seven sigma factors, each with a unique consensus sequence and spacing. What advantage does having multiple sigma factors provide for a bacterium?

6. The purpose of the tables below are to summarize information about transcriptional regulation in bacteria, in which an effector molecule binds to a regulatory protein to influence gene expression. Fill in the tables as indicated. (LO3)

	Type of regulation	
	Negative regulation	Positive regulation
Type of regulatory protein		
Does the binding of the regulatory protein **inhibit** or **enhance** the ability of RNAP to bind to the promoter region?		

	Type of system			
	Repressed	Derepressed	Activated	Deactivated
What is the state of the regulatory protein (**active/binds** or **inactive/doesn't bind**) in each of the four possible scenarios?				
Does transcription of the gene occur?				

7. For each element listed below, indicate if it is an example of a *cis*-acting or *trans*-acting regulatory element. (LO3)

- the CRP binding site
- the housekeeping sigma factor
- the LacI repressor protein
- the *trp* promoter region

8. Fill in the table below to compare and contrast Class I- and Class II-activated promoters. (LO3)

Promoter	Strength of promoter	Where is the activator binding site relative to the +1 site?	What binds at this binding site?
Class I-activated promoter			
Class II-activated promoter			

9. The regulation of the *E. coli lac* operon serves as a good example of both activation and repression of transcription. (LO4)

a) In this regulatory system, which molecule (LacI or CRP) acts as a repressor and which as an activator?

b) How is the LacI protein allosterically controlled? What binds to it? Does that binding activate or inactivate LacI?

c) How is the CRP protein allosterically controlled? What binds to it? Does that binding activate or inactivate CRP?

d) Compare the regulation in the *lac* operon under different growth conditions. For each environmental condition below, draw in what would bind to the promoter regions, and fill in the table below.

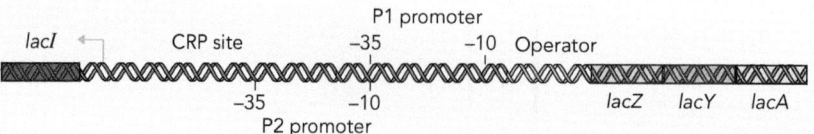

Conditions	Does LacI bind DNA?	Does CRP bind DNA?	Is *lacZYA* transcribed?
I) High glucose but low lactose			
II) High glucose and lactose present			
III) High lactose but low glucose			

e) Even when there is a high concentration of glucose in the environment, there is a small amount of LacY and LacZ in the cell. Explain how this helps the bacterium respond to changes in the environment.

10. The *trp* operon consists of five genes (*trpEDCBA*), which are expressed as a polycistronic mRNA and encode five polypeptides (TrpA, B, C, D, and E) involved in the biosynthesis of tryptophan. (LO5)

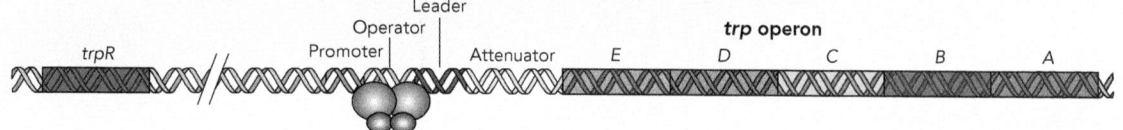

a) Fill in the table below to indicate how TrpR regulates the *trp* genes.

Conditions	Does TrpR bind DNA?	Are *trp* genes transcribed?
Cellular tryptophan levels are low		
Cellular tryptophan levels are high		

b) The *trp* genes are also controlled by attenuation, which helps the cell to conserve even more resources. The attenuator sequence lies upstream from the first gene in the *trp* operon but downstream of the promoter. This sequence contains multiple tryptophan codons and two regions that can form stem-loop structures.

    i. Explain which stem-loop structure forms when the level of tryptophan in the cell is high. Does this stop transcription or allow it to continue? How does the level of tryptophan in the cell influence this process?

    ii. Explain which stem-loop structure forms when the level of tryptophan in the cell is low. Does this stop transcription or allow it to continue? How does the level of tryptophan in the cell influence this process?

11. sRNAs are single-stranded molecules which, due to hydrogen bonding, can form secondary structures that are important for their function. (LO6)
    a) Briefly describe the two classes of sRNAs.
    b) *E. coli* RpoS is a stationary-phase sigma factor that is important for dealing with stress.

        i. Explain how the translation of the *rpoS* mRNA is inhibited during exponential growth.

        ii. Explain how the sRNA named DsrA allows the translation of the *rpoS* mRNA under stress conditions.

12. The sRNA molecule, CsrB is produced under low-nutrient conditions and influences the concentration of free CsrA (an RNA-binding regulatory protein) in the cell. Fill in the table below to indicate how CsrB and CsrA regulate certain genes involved in carbon storage (e.g., *glgCAP*, involved in glycogen biosynthesis) under high and low levels of CsrB. (LO6)

Conditions	To what does CsrA bind?	Are *glgCAP* genes transcribed?
CsrB levels are low		
CsrB levels are high		

13. Explain how riboswitches work to control the synthesis of proteins. How does the binding of the metabolite molecule influence ribosome binding? (LO6)

14. Compare and contrast different methods for transcript analysis by filling in the table below. (LO7)

Method	What is isolated?	How are transcripts detected?
Northern blot		
RNA-Seq		
qRT-PCR		

15. Consider the gene fusions shown in Figure 8.11 Panel D. (LO7)
    a) What is the advantage of using gene fusions with a reporter gene to monitor gene expression instead of monitoring the gene of interest directly?

    b) Compare and contrast transcriptional fusions with translational fusions by filling in the table below.

Type of gene fusion	Does the fusion construct contain a promoter for the gene of interest?	Does the fusion construct contain a RBS for the gene of interest?	Does the fusion construct contain any coding sequences from the gene of interest?	Does this method allow you to measure the rate of transcription, translation, or both?
Transcriptional fusion				
Translational fusion				

16. The electrophoretic mobility shift assay (EMSA) is an *in vitro* method used to establish whether a protein of interest binds somewhere within a certain sequence of DNA. In this assay, one aliquot of DNA is incubated with a purified protein of interest and another aliquot of DNA is incubated with no added protein. (LO7)
    a) Which is the control in this experiment? What information does this control provide?

    b) What does a positive result look like? That is, what results would tell you that your protein did, in fact, bind to that region of DNA? Why?

17. The DNase I protection assay, or DNA footprinting, is similar to the EMSA in that a DNA fragment is purified, labeled, and allowed to bind to the protein of interest. However, after the DNA-protein complex is allowed to form, the sample is exposed to the enzyme DNase I. (LO7)
    a) What is the control in this experiment? What information does this control provide?

    b) What does a positive result look like? That is, what results would tell you that your protein did, in fact, bind to that region of DNA? Why?

    c) What advantage does this method have over EMSA?

Dig Deeper

18. Consider the regulation of the *lac* operon when cyclic AMP is added to the cell. Draw in what would bind to the promoter regions and fill in the table below. (LO5)

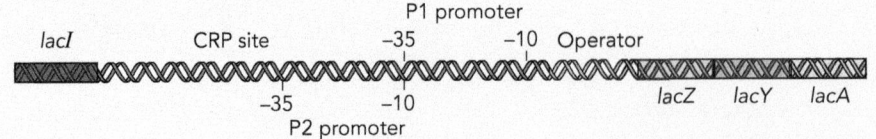

Conditions	Does LacI bind DNA?	Does CRP bind DNA?	Is *lacZYA* transcribed?
High glucose present, no lactose, but added exogenous cyclic AMP			

19. Global regulation involves the coordinated regulation of a number of different operons. Cra (catabolite repressor/activator) is an example of a global regulator in *E. coli*. (LO4)
    a) One effector molecule for Cra is fructose-1,6-bisphosphate (F-1,6-P). In which central metabolic pathway is this molecule made (see Chapter 2)?

    b) Glycolysis and gluconeogenesis complement each other in that they utilize different carbon sources. What growth substrates are metabolized by each of these pathways?

    c) Fill in the tables below to review how F-1,6-P influences Cra and the transcription of genes for glycolysis and gluconeogenesis enzymes.

Consider the control of the glycolysis genes in the presence and absence of glucose.

Conditions	Would glycoly-sis enzymes be needed?	Would F-1,6-P be made?	Would Cra be active or inactive?	Would glycolysis genes be transcribed?
When glucose is present				
When glucose is absent				

Consider the control of the gluconeogenesis genes in the presence and absence of glucose.

Conditions	Would gluconeo-genesis enzymes be needed?	Would F-1,6-P be made?	Would Cra be active or inactive?	Would gluconeo-genesis genes be transcribed?
When glucose is present				
When glucose is absent, but acetate is present				

20. There are many examples of transcriptional control in bacteria. The figure shows a partial pathway for the degradation of aromatic compounds in *Pseudomonas putida*. PcaR is an activator protein that works with the inducer, β-ketoadipate (βKA, a positive effector), to control the expression of the enzymes that constitute the *pca* branch of the β-ketoadipate pathway (*pcaK, pcaF, pcaI, pcaJ, pcaB, pcaD* and *pcaC*). When sufficient βKA is present, it binds to PcaR and the complex increases expression of the operons *pcaBDC* and *pcaIJ*, and the genes *pcaK* and *pcaF*. (LO3)

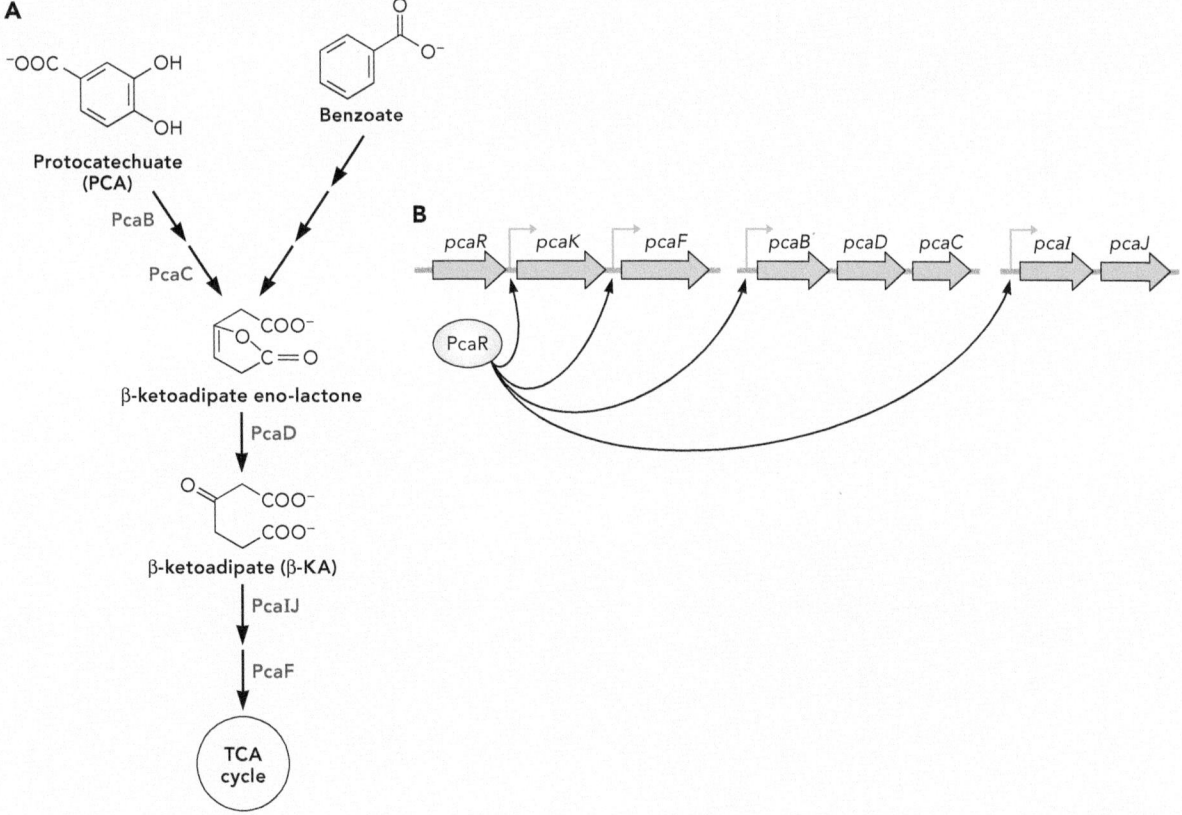

a) Fill in the table below to indicate how these genes are regulated.

Conditions	Would you expect PcaR to be in its active conformation? Why?	Would you expect *pca* genes to be transcribed? Why?
When βKA is present		
When βKA is absent		

b) Would you consider PcaR to be a *trans*-acting regulatory element or a *cis*-acting regulatory element? Explain your reasoning.

21. Your interest in the pathway for the degradation of aromatic compounds in *Pseudomonas putida* is piqued. You set out to further investigate the regulation of the pathway. (LO7)

a) To confirm that PcaR is an activator protein that works with the inducer β-ketoadipate (βKA) to control the expression of *pcaI* and *pcaJ*, you make a transcriptional fusion with the gene encoding green fluorescent protein (GFP) and the *pcaIJ* promoter. Draw out how that transcriptional fusion would be different from the *pcaIJ* promoter and genes shown below.

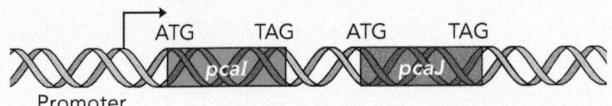

b) You have two strains of *P. putida*: one wild type and one with a nonfunctioning *pcaR* gene (mutant). You grow both in the presence and absence of βKA. Predict the relative amount of fluorescence (high or low) for your experiments by drawing bars in the graph.

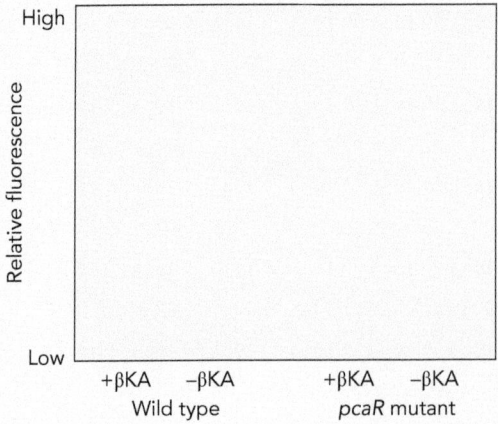

c) You know that many of the promoters in *Pseudomonas* are similar to the *E. coli* promoters. You are curious as to which sigma factor controls the *pcaIJ* promoter. You sequence the DNA upstream from the *pcaIJ* genes and find this sequence.

TTGaCc – N$_{17}$ – TATAAT

Based on the information in Table 8.1, which sigma factor likely controls this promoter?

d) You do a series of electrophoretic mobility shift assays to further explore this. Do these data support or refute your conclusions from part c? Explain your reasoning.

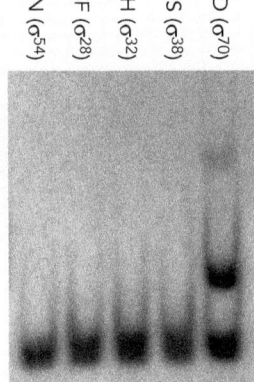

INTERLUDE: FROM UNITY TO DIVERSITY

The UNITY section of this text has introduced many fundamental concepts that apply to most living cells, including phylogenetic relationships between organisms, central metabolism, growth, energy conversion, and regulatory mechanisms. Much of the material was focused on well-studied model systems. However, while there is great unity in biology for the requirement of certain physiological processes, there is also diversity from organism to organism that permits microbes to thrive in a wide variety of environmental conditions. The DIVERSITY section of this text explores microbial diversity with regard to metabolism, cellular transport processes, motility, cell-to-cell signaling, stress response, and differentiation. We will begin by exploring the variety of metabolisms found in microbes from across the tree of life.

Microbial Physiology: Unity and Diversity, First Edition. Ann M. Stevens, Jayna L. Ditty, Rebecca E. Parales and Susan M. Merkel.
© 2024 American Society for Microbiology.
Companion website: www.wiley.com/go/stevens/microbialphysiology

Metabolic Diversity

While microbes share many common unifying physiological traits, they also have amazing diversity that permits them to exist in distinctive niches. One key contributor to microbial diversity is metabolism. In order to grow and survive, all organisms need to be able to obtain energy in their environment from either sunlight or high-energy electron donors, convert that energy into usable forms for the cell (i.e., proton motive force [PMF] and ATP), and acquire and/or generate carbon-containing precursors to build the structures of the cell and grow biomass. Metabolism can be classified based on the (i) energy sources, (ii) carbon sources, and (iii) electron source (or electron donor) (Fig. I.1).

Energy source. Any organism that is capable of using photons of sunlight as its energy source is a **phototroph** (Chapter 10). Organisms that can oxidize chemical compounds for energy are **chemotrophs** (Chapter 11).

Carbon source. All organisms need a source of carbon. Many organisms are **heterotrophs** that catabolize organic compounds and convert them to cellular material. Any organism that is capable of converting carbon dioxide into reduced carbon is an **autotroph** (Chapter 9).

Electron source. Organisms that obtain electrons from organic carbon are **organotrophs**, and if they use inorganic compounds, they are **lithotrophs**.

The energy, carbon, and electron source terms can be combined to fully describe the metabolism of any organism. For example, cyanobacteria use the energy from light to fix CO_2 as a carbon source and are therefore described as **photoautotrophs**. *Nitrosomonas europaea* is a **chemolithoautotroph** that grows using inorganic ammonia as both the energy source and electron donor,

Energy source	Carbon source	Common metabolic terminology	Electron source	Complete metabolic terminology
Light Photo–	**CO₂** –autotroph	Photoautotroph	Inorganic –litho	Photolithoautotroph
			Organic –organo	Photoorganoautotroph
	Organic –heterotroph	Photoheterotroph	Inorganic –litho	Photolithoheterotroph
			Organic –organo	Photoorganoheterotroph
Chemical compounds Chemo–	**CO₂** –autotroph	Chemoautotroph	Inorganic –litho	Chemolithoautotroph
			Organic –organo	Chemoorganoautotroph
	Organic –heterotroph	Chemoheterotroph	Inorganic –litho	Chemolithoheterotroph
			Organic –organo	Chemoorganoheterotroph

FIGURE I.1 **Definitions of the terminology used to describe metabolic lifestyles.** (Left to right) Each organism may be defined by the energy (red font) and carbon (blue font) sources it utilizes in a given environment, resulting in the assignment of a common metabolic term for its lifestyle. When also considering the specific electron source or donor (purple font) the organism is utilizing, then a more complete metabolic term is assigned.

and CO_2 as the carbon source. **Chemolithotrophs** (Chapters 11 and 12) collectively use a variety of environmental inorganic compounds as electron donors (e.g., H_2S, H_2, NH_3, and NO_2^-). *Escherichia coli* can use glucose as a source of energy, carbon, and electrons and is described as a **chemoorganohetero- troph** (often shortened to **chemoheterotroph**). On the other hand, *Rhodo- bacter sphaeroides* can grow as a **photoorganoheterotroph** (often shortened to **photoheterotroph**) using light as the energy source and succinate as both the carbon source and electron donor (Chapter 10). As can be seen in the examples above, organisms are categorized, by convention, beginning with their energy source, followed by their electron donor, and then carbon source (Fig. I.1).

Across all metabolic lifestyles, there are three mechanisms whereby the initial energy source can be converted to ATP: substrate-level phosphorylation, oxida- tive phosphorylation, and photophosphorylation (Fig. 5.1). As noted in Chap- ters 2 and 5, **substrate-level phosphorylation** occurs when a high-energy phosphate bond is transferred between metabolic intermediates. All living cells are capable of substrate-level phosphorylation, but it is the only mechanism of ATP production for obligate fermenters. **Oxidative phosphorylation** involves either aerobic or anaerobic respiration to produce a PMF via a membrane-bound electron transport chain (Chapters 5 and 6). The potential energy of the PMF can be converted into ATP through the activity of ATP synthase. **Photophospho- rylation** takes place when an electron transport chain in the reaction center of a phototrophic organism generates a PMF and ATP synthase uses that PMF to produce ATP (Chapter 10).

Global Nutrient Cycles

It is also the extensive metabolic diversity of microorganisms that allows for the carbon, nitrogen, and sulfur cycles to be maintained globally. On an ecosystem scale, carbon flows between oxidized and reduced forms due to the counterbal- anced activity of autotrophs and heterotrophs (Fig. I.2). Global levels of atmos- pheric carbon dioxide have risen due to an acceleration in the burning of fossil fuels and higher respiration rates as global populations rise, as well as loss of forests capable of autotrophic growth. The autotrophs, both photoautotrophs and chemolithoautotrophs, are capable of reducing inorganic carbon (e.g., CO_2, CO) to cellular organic carbon, converting 7×10^{16} g carbon per year to organic forms. Highly oxidized forms of carbon require input of reductant (i.e., reduced coenzymes) to "fix" them into biologically useful reduced forms. Carbon diox- ide fixation occurs via at least five different microbial pathways in addition to the well-known Calvin cycle (Chapter 9). It is also important to recognize that both autotrophy and heterotrophy can occur in both oxic and anoxic environ- ments. The anoxic world is dominated by the activity of microbes, where pro- cesses such as anoxygenic photophosphorylation (Chapter 10), sulfate reduction (Chapter 11), methanogenesis (Chapter 11), and denitrification (Chapter 12) occur. The importance of individual microbial populations working together in communities to drive the nutrient cycles on our planet will be emphasized.

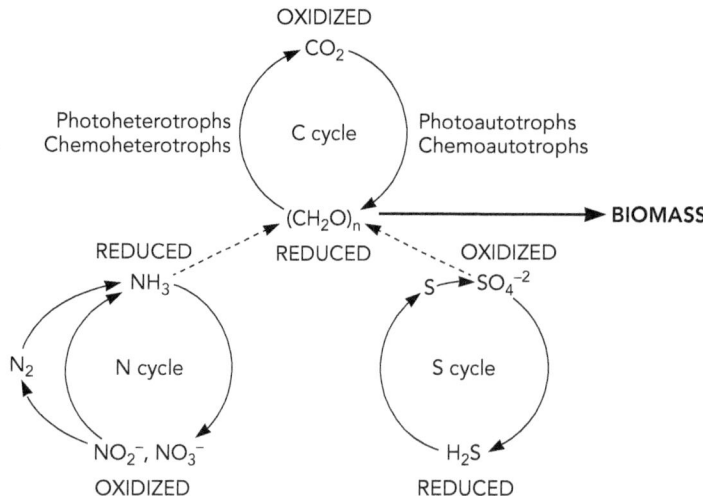

FIGURE I.2 **Broad overview of the carbon, nitrogen, and sulfur cycles.** Microbes are essential to the recycling of environmental nutrients between their reduced and oxidized states through the carbon, nitrogen, and sulfur cycles as broadly depicted. Each of these cycles will be discussed in greater detail in Chapters 9 to 12. Ultimately, the goal of the various metabolic lifestyles employed by microbes is to convert inorganic compounds into organic intermediates that will permit an increase in biomass and yield energy as the organisms grow and reproduce.

Structural and Regulatory Diversity of Microbes

Microbes also need to adapt to the wide range of environments that they inhabit using unique physical structures and distinctive regulatory pathways in order to survive. Differences in the structure of the cell envelope and mechanisms used to transport nutrients and proteins (Chapters 13 and 14) and strategies for movement in various physical environments (Chapter 15) have contributed to the structural and functional diversity of microbial species. Studying how cells sense and respond to each other via quorum sensing and to their environment through stress responses permits an understanding of the variety of regulatory mechanisms employed by microbes (Chapters 16 and 17). Finally, the topic of microbial differentiation, including biofilm formation, brings together ways that complex regulatory pathways control phenotypic outputs (Chapter 18). The physiologies described in the following chapters highlight some representative examples of the incredible diversity of the microbial world.

PART II: DIVERSITY

186 Making Connections

186 Autotrophy

187 Calvin Cycle

191 Reductive Tricarboxylic Acid (rTCA) Cycle

193 Reductive Acetyl-CoA Pathway

195 3-Hydroxypropionate (3HP) Bi-cycle

195 3-Hydroxypropionate-4-Hydroxybutyrate (3HP-4HB) and Dicarboxylate-4-Hydroxybutyrate (DC-4HB) Cycles

197 Why So Many CO_2 Fixation Pathways?

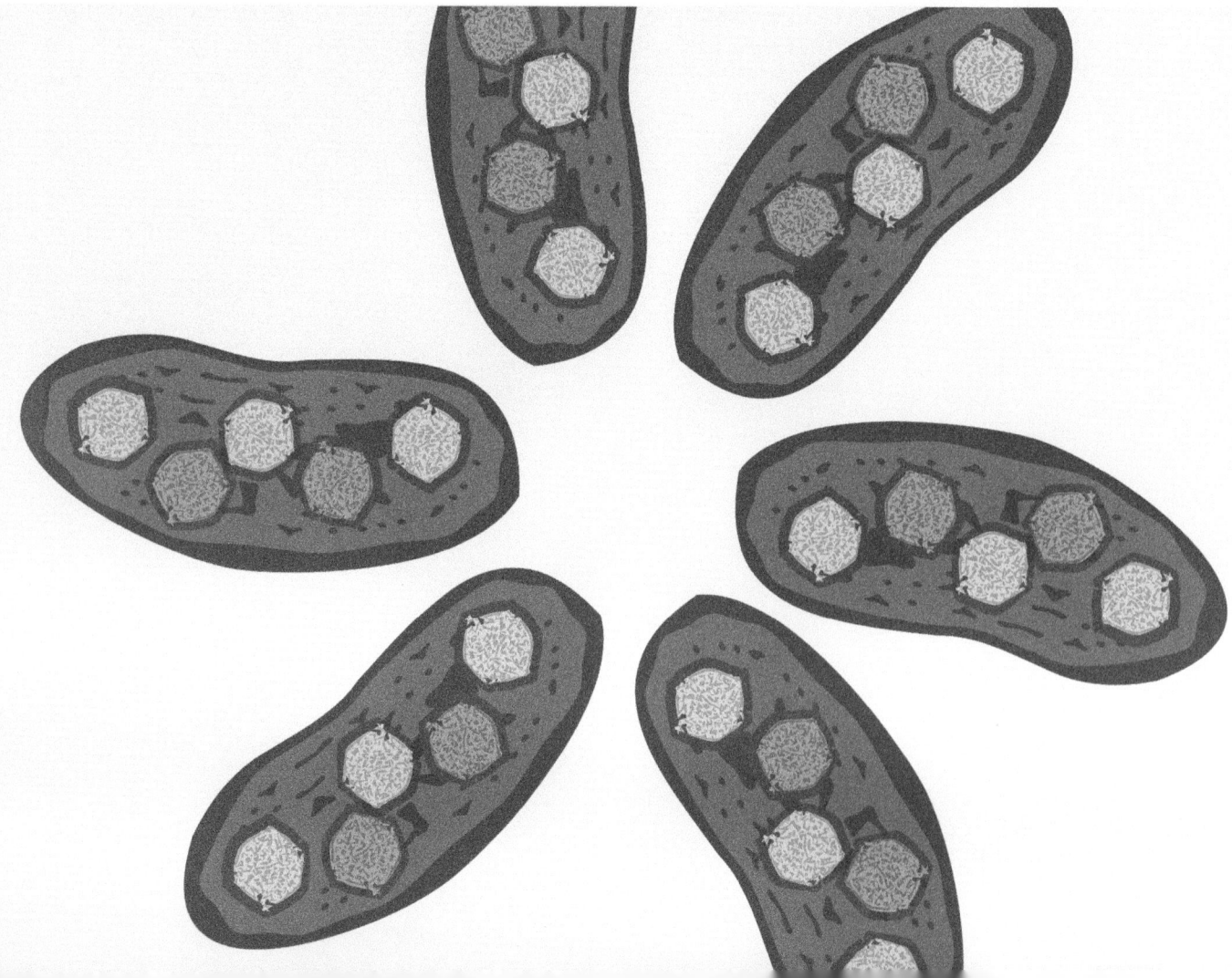

AUTOTROPHY

After completing the material in this chapter, students should be able to . . .

1. explain how the three stages of the Calvin cycle work together to fix carbon dioxide (CO_2)

2. describe how each of the five pathways of CO_2 fixation, other than the Calvin cycle, function to produce acetyl-CoA or pyruvate

3. give examples of how the limitations of critical enzymes can impact the conditions under which autotrophic organisms can grow

4. propose a theory that explains how the physiological lifestyle of a given microorganism might have influenced the evolution of its autotrophic pathway

Microbial Physiology: Unity and Diversity, First Edition. Ann M. Stevens, Jayna L. Ditty, Rebecca E. Parales and Susan M. Merkel.
© 2024 American Society for Microbiology.
Companion website: www.wiley.com/go/stevens/microbialphysiology

Making Connections

On a global ecosystem scale, carbon flows in a cycle between its oxidized (carbon dioxide) and reduced (organic carbon) forms due to the activity of autotrophic and heterotrophic organisms (Fig. I.2). As the heterotrophs produce CO_2 from respiration and fermentation, microbial photoautotrophs and chemolithoautotrophs convert CO_2 to reduced carbon in both oxic and anoxic environments. While plants and animals certainly contribute to carbon cycling, the microbial world is the predominant driver of the global carbon cycle.

In the "Unity" section of the book, we discussed how microbes build biomass using precursors generated through the central metabolic pathways (Chapters 2 to 4) and electron transport chains to conserve energy by generating a proton motive force (PMF; Chapters 5 and 6). The next three chapters present the variety of mechanisms that have evolved to use the energy and reducing power generated by phototrophy (Chapter 10) and chemolithotrophy (Chapters 11 and 12) to fix CO_2 into the reduced forms of carbon essential for building biomass.

Autotrophy

Autotrophs (photoautotrophs and chemolithoautotrophs) are the world's **primary producers**, converting atmospheric CO_2 to reduced (organic) carbon. To grow and divide, autotrophs need to generate the 12 precursors for biosynthesis (Chapter 2) from inorganic CO_2. Fixation of this highly oxidized form of carbon requires the input of reducing equivalents, typically in the form of reduced ferredoxin or $NAD(P)H + H^+$, as well as energy in the form of ATP. The sources of reducing equivalents and ATP for photoautotrophs and chemolithoautotrophs will be discussed in Chapters 10 to 12. There are currently six known pathways of CO_2 fixation (Fig. 9.1). In contrast to CO_2 fixation via the Calvin cycle, which produces 3-phosphoglycerate and other 3-carbon intermediates for use in biosynthesis, the other five CO_2 fixation pathways generate either pyruvate (3C) or acetyl coenzyme A (acetyl-CoA) (2C) from CO_2. Three of these pathways are present only in members of the *Bacteria* (Calvin cycle, reductive tricarboxylic acid [rTCA] cycle, and 3-hydroxypropionate [3HP] bi-cycle), two are unique to members of the *Archaea* (3-hydroxypropionate-4-hydroxybutyrate [3HP-4HB] cycle and dicarboxylate-4-hydroxybutyrate [DC-4HB] cycle), and only one (reductive acetyl-CoA pathway), is found in both *Bacteria* and *Archaea* (Fig. 9.1). In addition, three of the six pathways (Calvin cycle, rTCA cycle, and 3HP bi-cycle) are present in various photoautotrophs and chemolithoautotrophs, while the other three CO_2 fixation pathways are only found in specific groups of chemolithoautotrophs.

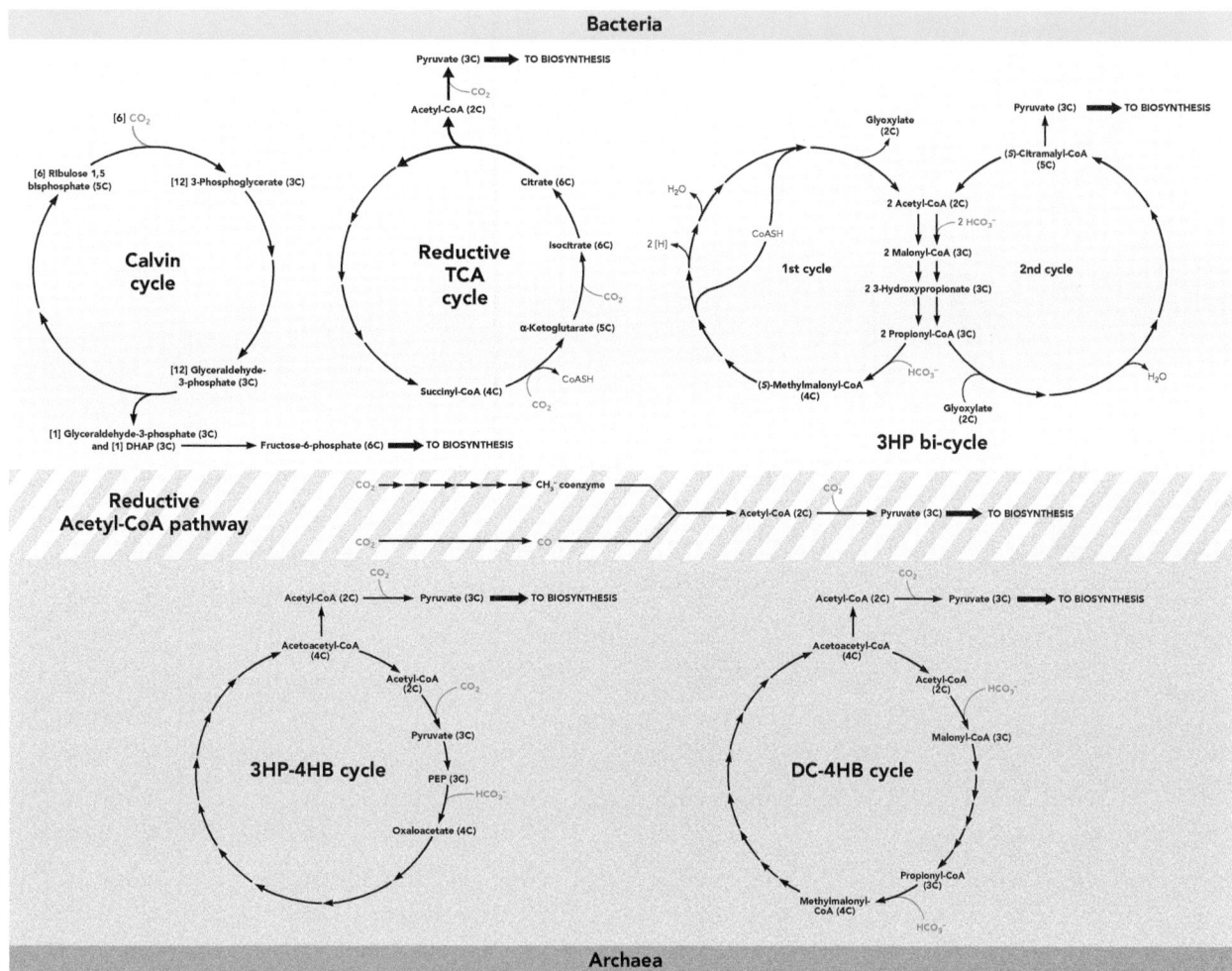

FIGURE 9.1 **Overview of the six known pathways of CO$_2$ fixation.** Pathways in the green section (top) are only found in *Bacteria*, whereas those in the pink section (bottom) have only been identified in *Archaea*. The reductive acetyl-CoA pathway is the only pathway known to be used by both *Bacteria* and *Archaea* (pink and green hashed background in center). Individual reactions are represented by arrows. Also shown in blue text are points at which CO$_2$ (or its more water-soluble bicarbonate form) are incorporated. Of the six pathways shown, only the Calvin cycle, reductive tricarboxylic acid (rTCA) cycle, and 3-hydroxypropionate (3HP) bi-cycle are used by photoautotrophs. 3HP-4HB, 3-hydroxypropionate-4-hydroxybutyrate; DC-4HB, dicarboxylate-4-hydroxybutyrate.

Calvin Cycle

The Calvin-Benson-Bassham cycle (generally known as the **Calvin cycle**) is the most common of the six known CO$_2$ fixation pathways that have evolved in living organisms to convert CO$_2$ into reduced carbon. It is found in many photoautotrophs, including plants, algae, cyanobacteria, and purple sulfur and purple nonsulfur bacteria (members of the *Alpha-*, *Beta-*, and *Gammaproteobacteria* [Fig. 9.2A]). The Calvin cycle is also present in many chemolithoautotrophic bacteria, including sulfur- and iron-oxidizing bacteria and nitrifying bacteria (Chapters 11 and 12). Because some of the intermediates are unstable at higher

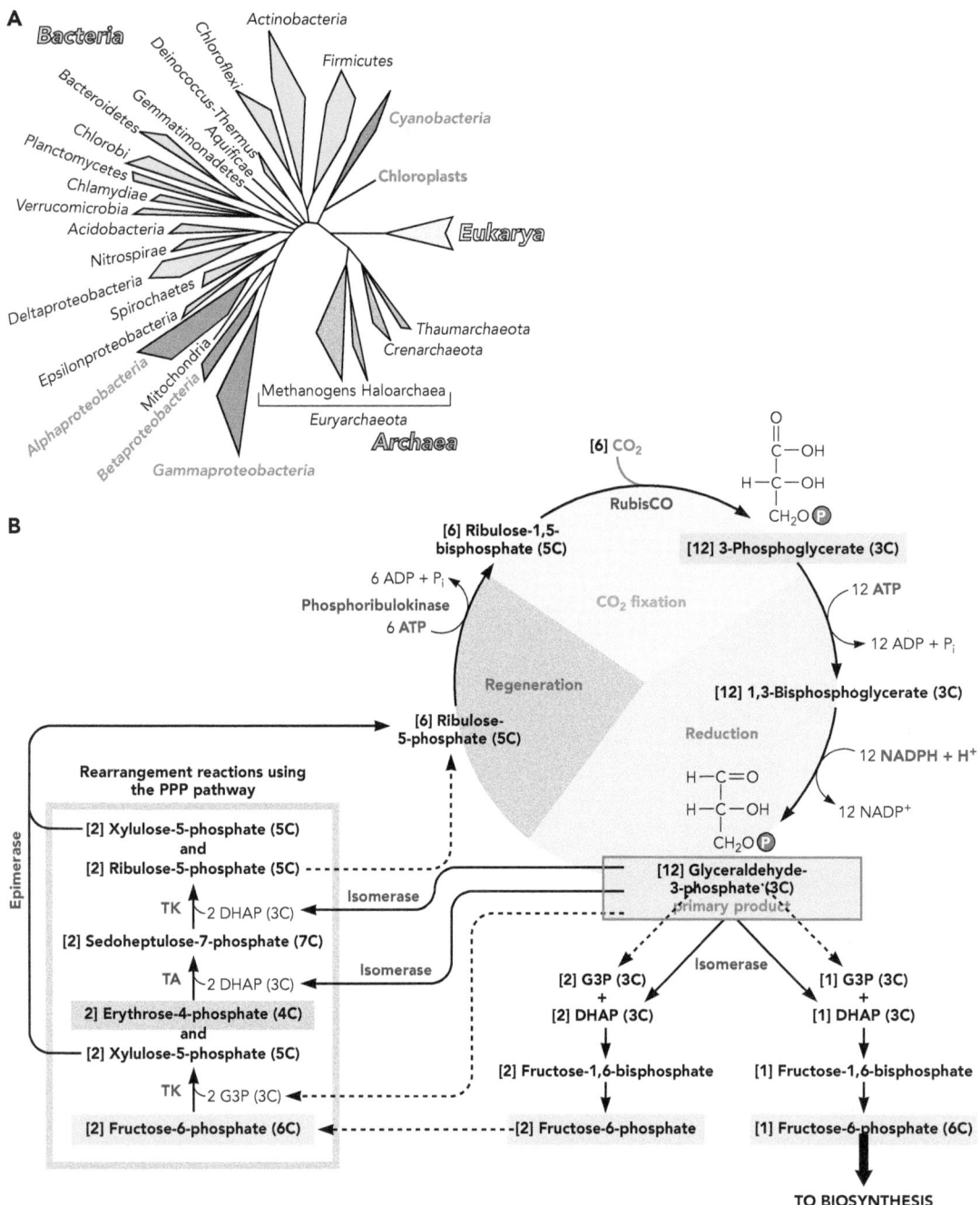

FIGURE 9.2 **Distribution and biochemistry of the Calvin cycle.** (A) Phylogenetic tree highlighting lineages with members known to use the Calvin cycle (highlighted in dark coloring and bold colored text). (B) Reactions of the Calvin cycle. The cycle can be divided into three parts. In the first part, CO$_2$ fixation (highlighted in green), a molecule of CO$_2$ is added to the 5-carbon substrate ribulose-1,5-bisphosphate by the key enzyme ribulose-1,5-bisphosphate carboxylase/oxygenase (RubisCO), to produce 3-phosphoglycerate. In the second part of the cycle, reduction (highlighted in light blue), 3-phosphoglycerate is reduced to glyceraldehyde-3-phosphate (G3P) in reactions that require both ATP and NADPH + H$^+$. In the third part of the cycle, regeneration (highlighted in pink), rearrangement reactions of the pentose phosphate pathway (PPP, boxed in pink; see Chapter 2) result in the formation of ribulose-5-phosphate. The key enzyme phosphoribulokinase catalyzes the conversion of ribulose-5-phosphate to ribulose-1,5-bisphosphate so that the cycle can continue. The enzyme triose phosphate isomerase (isomerase) catalyzes the conversion of some of the G3P to dihydroxyacetone phosphate (DHAP), which is used for both the PPP rearrangement reactions and the formation of fructose-6-phosphate. The overall reaction depicts six molecules of CO$_2$ being fixed to produce one molecule of fructose-6-phosphate for biosynthesis. Note that the primary product of the reaction is the three-carbon compound G3P. Essential precursors (Chapter 2) are highlighted in colored boxes. TA, transaldolase; TK transketolase.

temperatures, the Calvin cycle functions at maximum temperatures of 70 to 75 °C. The cycle is therefore used by some autotrophic thermophiles, but no hyperthermophiles. The enzymes of the Calvin cycle are not sensitive to oxygen, so this mechanism of CO_2 fixation is found in both aerobes (including cyanobacteria that produce O_2 as a photosynthetic byproduct) and anaerobes. Notably, the Calvin cycle has not yet been identified in any autotrophic members of the *Archaea*.

The Calvin cycle (Fig. 9.2B) has three key stages: (i) <u>CO_2 fixation</u> via the addition of CO_2 to a 5-carbon sugar; (ii) <u>reduction</u> of 3-phosphoglycerate (3C) to form other 3-carbon intermediates in glycolysis; and (iii) <u>regeneration</u> of the 5-carbon CO_2 acceptor, which allows the cycle to continue.

For CO_2 fixation to take place, two enzymes unique to the Calvin cycle are required: **ribulose-1,5-bisphosphate carboxylase/oxygenase (RubisCO)** and phosphoribulokinase. RubisCO, one of the most abundant and well-studied enzymes on the planet, catalyzes the carboxylation of ribulose-1,5-bisphosphate. This conserved enzyme likely evolved once, meaning all RubisCO enzymes have a common ancestor. RubisCO is of critical importance as it is responsible for converting massive amounts of CO_2 into organic carbon on a global scale. In the CO_2 fixation stage, carboxylation of each ribulose-1,5-bisphosphate (5C) results in the formation of an unstable 6-carbon intermediate that spontaneously hydrolyzes to form two molecules of the essential precursor 3-phosphoglycerate (3C) (Fig. 9.2B).

In the reduction stage, ATP is consumed in a phosphorylation reaction that results in the formation of 1,3-bisphosphoglycerate, which is then reduced using electrons from NADPH + H$^+$ to form glyceraldehyde-3-phosphate (3C), another essential precursor. The NADPH + H$^+$ and ATP necessary for the reactions of the Calvin cycle are produced from the light reactions of photoautotrophs (Chapter 10) or through the reactions of the electron transport chains in chemolithoautotrophs (Chapters 11 and 12).

In the regeneration stage, some of the glyceraldehyde-3-phosphate is converted to fructose-6-phosphate via gluconeogenesis (Chapter 2); this fructose-6-phosphate is incorporated into biomass. The rest of the glyceraldehyde-3-phosphate is rearranged by reactions that are part of the pentose phosphate pathway (Fig. 2.8); in fact, the Calvin cycle has also been called the reductive pentose phosphate cycle. These reactions generate ribulose-5-phosphate (Fig. 9.2B), which is converted into ribulose-1,5-bisphosphate by the second key enzyme of the Calvin cycle, **phosphoribulokinase**. This enzyme consumes ATP as it regenerates the substrate necessary to continue another round of the Calvin cycle. Since one CO_2 is fixed in each turn of the cycle, a total of six rounds of the cycle are required to generate one molecule of fructose-6-phosphate (6C).

The net overall reaction of the Calvin cycle is

$$6\,CO_2 + 18\,ATP + 12\,NADPH + 12\,H^+ \rightarrow fructose - 6 - phosphate + 18\,ADP + 12\,NADP^+ + 17\,P_i$$

RubisCO is a notoriously inefficient enzyme: it has a low affinity for CO_2 and a slow catalytic turnover rate. In addition, RubisCO can function as an oxygenase, using oxygen as an alternative substrate and catalyzing a nonproductive reaction. Because RubisCO likely evolved before oxygen concentrations built up in the Earth's atmosphere, there was no selective pressure for the active site of RubisCO to evolve to be exclusively specific for CO_2. When RubisCO uses oxygen as a substrate in place of CO_2, it converts ribulose-1,5-bisphosphate to one molecule of 3-phosphoglycerate and one molecule of the toxic side product 2-phosphoglycolate (2C). The detoxification of 2-phosphoglycolate requires the input of ATP and NADPH + H⁺. This process, termed **photorespiration**, reduces the overall efficiency of CO_2 fixation because energy is used to reoxidize the reduced carbon (2-phosphoglycolate), releasing CO_2.

To increase the efficiency of RubisCO, some organisms have evolved mechanisms for concentrating inorganic carbon and limiting access of the enzyme to oxygen. For example, to increase the local CO_2 concentration and decrease photorespiration, some aerobic autotrophs, including cyanobacteria (Chapter 10), sulfur-oxidizing bacteria (Chapter 11), and nitrifying bacteria (Chapter 12), produce internal microcompartments called **carboxysomes** (Fig. 9.3), which harbor the RubisCO enzyme. The protein shell of carboxysomes allows entry of the substrates for RubisCO, ribulose-1,5-bisphosphate and bicarbonate (HCO_3^-, the soluble form of CO_2), and the exit of the product 3-phosphoglycerate due to their negative charges.

Carboxysomes also contain **carbonic anhydrase**, an enzyme that converts bicarbonate (HCO_3^-) to CO_2, the actual substrate for RubisCO. Carboxysomes are less permeable to gases, so CO_2 is retained at a high concentration within the carboxysome, increasing the efficiency of RubisCO. In addition, the carboxysomes protect RubisCO from encountering oxygen, reducing the frequency of

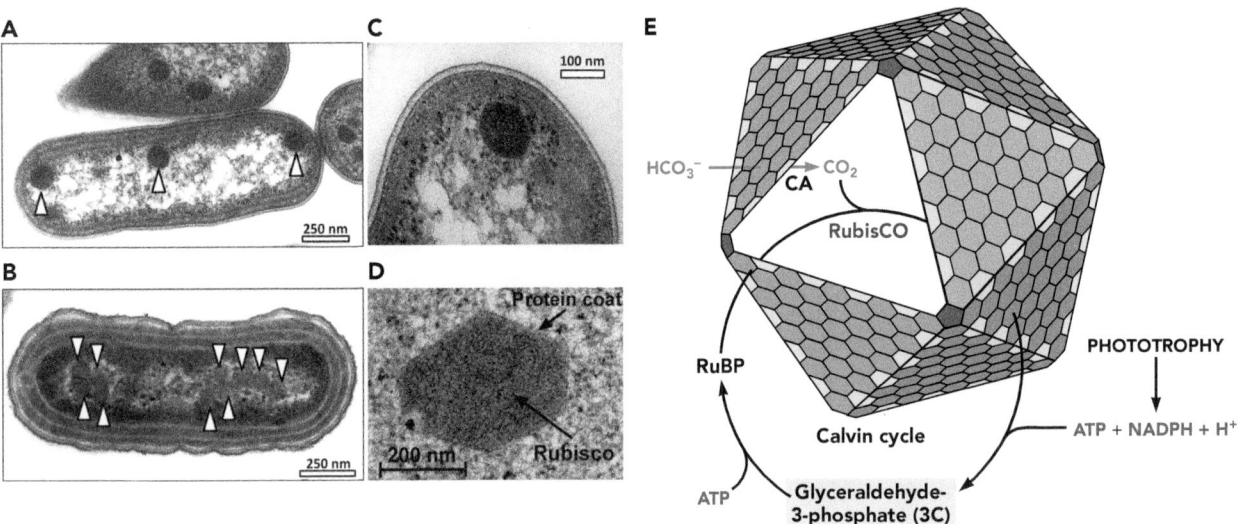

FIGURE 9.3 **Carboxysomes, the site of CO_2 fixation via the Calvin cycle in cyanobacteria and other aerobic autotrophs.** (A to D) Electron micrographs of cells containing carboxysomes (reprinted with permission from Rae BD et al. 2013. *Microbiol Mol Biol Rev* 77:357–379). Carboxysomes in (A) *Synechococcus elongatus* PCC 7942, and (B) *Cyanobium* PCC 7001 (Photo credit: Lynne Whitehead). Arrowheads in panels A and B indicate the positions of carboxysomes. (C) Close-up of a carboxysome from *S. elongatus* PCC 7942. (D) Close-up of a carboxysome from *Anabaena variabilis* M3. (E) Schematic diagram of a carboxysome. Carboxysomes have a proteinaceous polyhedral shell with pores that allow the passage of small-molecule substrates and products, and they contain carbonic anhydrase (CA), which catalyzes the conversion of soluble bicarbonate (HCO_3^-) to CO_2. Since carboxysomes are quite impermeable to gases, CO_2 is concentrated within the carboxysome and the entry of oxygen from the cytoplasm is limited. RubisCO, which is also present in the carboxysome, catalyzes CO_2 fixation. The high CO_2 concentration and low O_2 concentration in the carboxysome improves the efficiency of RubisCO. The product of the reaction, 3-phosphoglycerate, diffuses out of the carboxysome into the cytoplasm, where it is reduced to glyceraldehyde-3-phosphate using ATP and NADPH + H⁺ generated during photosynthesis. RuBP, ribulose-1,5-bisphosphate; RubisCO, ribulose-1,5-bisphosphate carboxylase/oxygenase.

photorespiration. Plants do not have carboxysomes and have no mechanism to protect RubisCO from oxygen. Increasing the efficiency of RubisCO in plants would have profound impacts on agriculture, and one strategy being explored is to genetically engineer plants to produce carboxysomes.

Reductive Tricarboxylic Acid (rTCA) Cycle

Green sulfur bacteria (*Chlorobi*, more recently named *Chlorobiota*; Fig. 9.4A) are photoautotrophs that fix CO_2 by reversing the reactions of the TCA cycle. As you may recall from Chapter 2, the forward (oxidative) TCA cycle is used to oxidize a reduced carbon substrate (acetyl-CoA) to CO_2, generating reducing power for the production of ATP. The **reductive TCA (rTCA) cycle** (Fig. 9.4B) does the opposite: it results in the formation of reduced carbon (acetyl-CoA, 2C) from two molecules of CO_2; for this reason, the pathway is also called the reverse TCA cycle. However, there are three reactions of the TCA cycle (Fig. 2.13) that are not reversible. These reactions are replaced by three enzymes unique to the rTCA cycle: **fumarate reductase** replaces succinate dehydrogenase (see the branched TCA pathway in Chapter 2 [Fig. 2.14]), **ferredoxin-dependent α-ketoglutarate synthase** replaces α-ketoglutarate dehydrogenase, and **citrate lyase** replaces citrate synthase (Fig. 9.4B).

Starting with oxaloacetate (4C), each turn of the rTCA cycle results in the incorporation of two molecules of CO_2, producing one molecule of acetyl-CoA (2C) and regenerating a molecule of oxaloacetate, which can be used in an additional cycle of the pathway. For acetyl-CoA to enter biosynthesis, **pyruvate synthase** adds another molecule of CO_2 to generate the essential precursor pyruvate (3C) (Fig. 9.4B). The ferredoxin-dependent enzyme pyruvate synthase replaces pyruvate dehydrogenase, catalyzing the carboxylation of acetyl-CoA to form pyruvate (3C). Pyruvate can be converted to phosphoenolpyruvate (PEP), which can then be used to generate other essential precursors via gluconeogenesis and the pentose phosphate pathway (Chapter 2).

Although the rTCA cycle was first found in the green sulfur bacterium *Chlorobium limicola*, it is present in several other bacterial lineages (Fig. 9.4A). For example, it is used by autotrophic members of the *Aquificae* such as the hyperthermophilic chemolithoautotroph *Aquifex*, a genus of microaerophilic bacteria that can grow at temperatures up to 95 °C, as this pathway is not as temperature-limited as the Calvin cycle. It is also found in some autotrophic members of the *Nitrospirae* (Chapter 12), as well as some autotrophic *Deltaproteobacteria* such as the sulfate-reducing bacteria in the genus *Desulfobacter* (Chapter 11). Some of the enzymes of the rTCA cycle are oxygen sensitive and the pathway is mainly present in anaerobes and microaerophiles. Although some reports have suggested the presence of the rTCA cycle in some hyperthermophilic *Archaea* that grow at temperatures >90 °C, it appears that instead these organisms use the DC-4HB cycle for autotrophic growth.

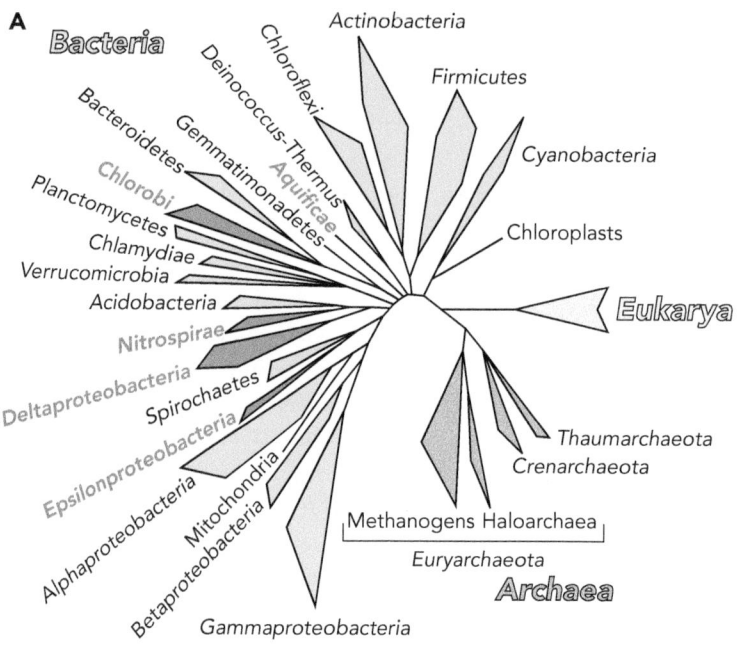

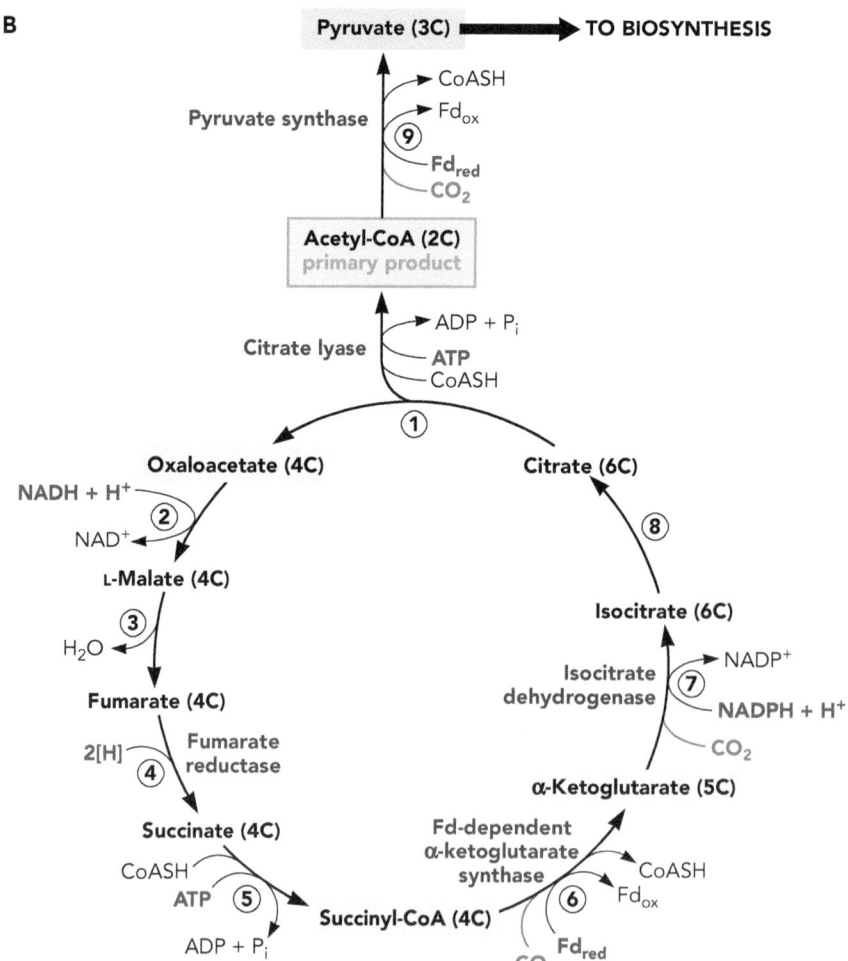

FIGURE 9.4 **Distribution and biochemistry of the reductive tricarboxylic acid (rTCA) cycle.** (A) Phylogenetic tree highlighting lineages with members known to use the rTCA cycle (highlighted in dark coloring and bold colored text). (B) Reactions of the rTCA cycle. Three key enzymes, citrate lyase (reaction 1), fumarate reductase (reaction 4), and ferredoxin (Fd)-dependent α-ketoglutarate synthase (reaction 6), are highlighted in purple because these enzymes replace irreversible enzymes of the TCA cycle. In addition, the enzymes isocitrate dehydrogenase (reaction 7), ferredoxin-dependent α-ketoglutarate synthase (reaction 6), and pyruvate synthase (reaction 9) are highlighted in purple as they catalyze CO_2 fixation reactions. Enzymes catalyzing the remaining numbered reactions are malate dehydrogenase (reaction 2), fumarate hydratase (reaction 3), succinyl-CoA synthase (reaction 5), and aconitate hydratase (reaction 8). Two molecules of CO_2 are fixed in each turn of the cycle, and result in the formation of acetyl-CoA as the primary product. Note that the enzymes catalyzing reductions (reactions 2, 4, 6, and 7) use different electron donors. There are several forms of fumarate reductase (reaction 4) that use different electron donors, so the generic electron donor 2[H] is indicated. Pyruvate synthase (reaction 9) then catalyzes the incorporation of an additional CO_2 molecule to acetyl-CoA, forming pyruvate, which is used for biosynthesis. Colored boxes indicate essential precursors.

Reductive Acetyl-CoA Pathway

The **reductive acetyl-CoA pathway**, or Wood-Ljungdahl pathway, is currently the only CO_2 fixation pathway known to function in members of both the *Bacteria* and *Archaea* (Fig. 9.5A). It is the simplest and least energy-intensive autotrophic pathway, and it is present in many anaerobes that are energy-limited. This pathway was first studied in acetogenic bacteria (Chapter 11), which are anaerobic bacteria that produce acetate from $H_2 + CO_2$. Most acetogens are members of the *Firmicutes*, but some spirochetes grow as acetogens using the reductive acetyl-CoA pathway. This pathway is also used by autotrophic sulfate-reducing bacteria (Chapter 11), the methanogenic *Archaea* (Chapter 11), and the anaerobic anammox bacteria (Chapter 12), which are members of the phylum *Planctomycetes*, recently named *Planctomycetota* (Fig. 9.5A). The reductive acetyl-CoA pathway is not known to be used by any photoautotrophs.

Unlike the other CO_2 fixation pathways, the reductive acetyl-CoA pathway is not a cycle, but rather consists of two branches (Fig. 9.5B). The product of the pathway is acetyl-CoA (2C) that is produced from two CO_2 molecules, each of which is independently reduced by a different mechanism. In one branch of the pathway, one molecule of CO_2 is reduced to the level of a methyl group. The reductions occur with CO_2 and its reduced forms bound to a coenzyme that acts as a 1-carbon molecule (C1) carrier. The other CO_2 molecule is reduced to carbon monoxide. The key enzyme **carbon monoxide dehydrogenase/acetyl-CoA synthase** catalyzes this reduction, as well as the synthesis of acetyl-CoA (Fig. 9.5B). This enzyme is very oxygen sensitive, and as a result, the pathway only functions under strictly anoxic conditions. Evidence indicates that carbon monoxide dehydrogenase/acetyl-CoA synthase evolved once and is conserved in all organisms that use the reductive acetyl-CoA pathway.

Although steps in the pathway are essentially the same in *Bacteria* and *Archaea*, there are several variations. These include the use of different C1 carriers: tetrahydrofolate is used in *Bacteria*, whereas the *Archaea* employ methanofuran and tetrahydromethanopterin, C1 carriers that are also used in methanogenesis (Chapter 11). In addition, the enzymes catalyzing the reductions to form the methyl group are different in *Bacteria* and *Archaea*; similarly, the electron donors used also vary in different organisms. Enzymes of the reductive acetyl-CoA pathway do not seem to be temperature sensitive as the pathway is found in psychrophiles, mesophiles, and hyperthermophiles. In fact, it is present in *Methanopyrus kandleri*, a hyperthermophile that can grow at temperatures up to 122°C. The pathway can be used to allow growth on other C1 compounds, including carbon monoxide, formaldehyde, methanol, and methylamine. The assimilation of acetyl-CoA into biomass requires pyruvate synthase, as described for the rTCA cycle.

In addition to using the reductive acetyl-CoA pathway for CO_2 fixation, methanogens and acetogens also use it to conserve energy by generating an electrochemical gradient, which can be used to produce ATP via ATP synthase. In the process of energy conservation, acetogens generate acetate as a waste product, whereas methanogens generate methane (Chapter 11).

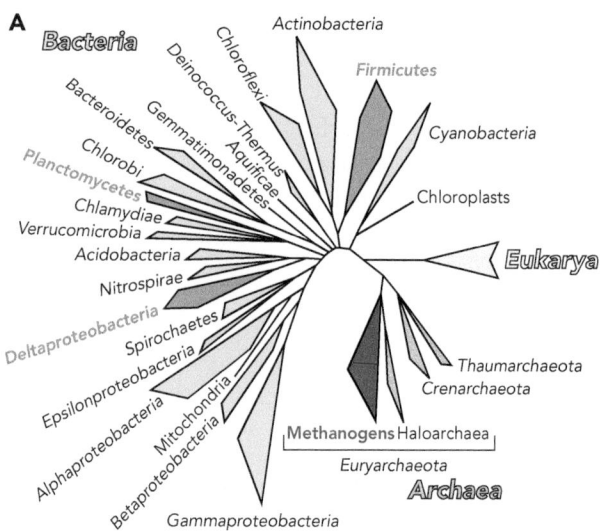

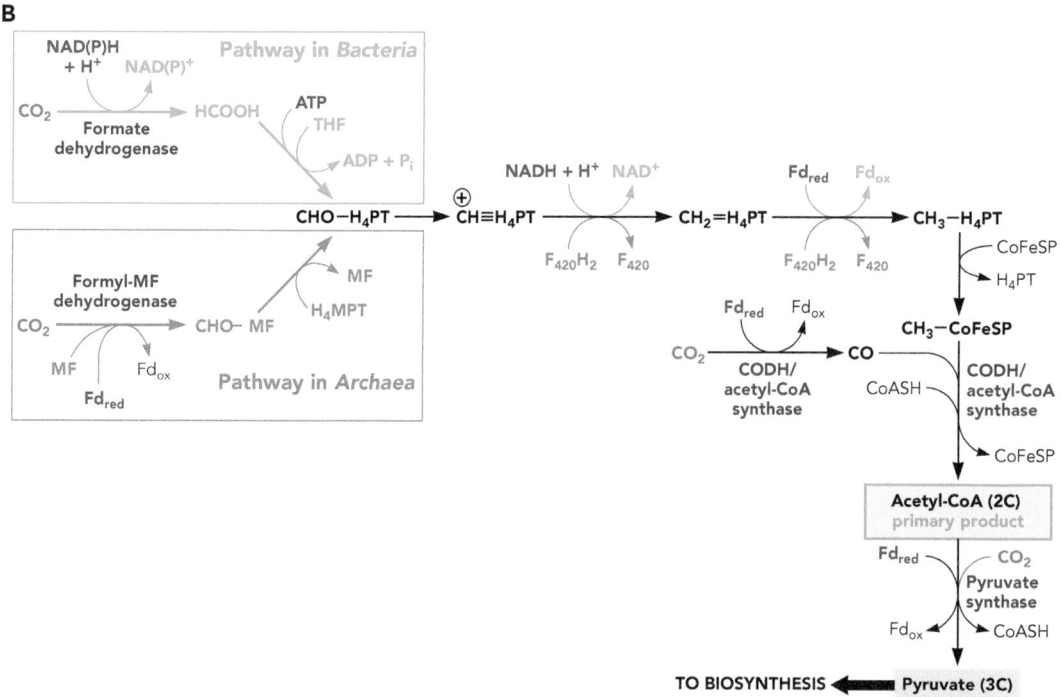

FIGURE 9.5 **Distribution and biochemistry of the reductive acetyl-CoA pathway (Wood-Ljungdahl pathway).** (A) Phylogenetic tree highlighting lineages with members known to use the reductive acetyl-CoA pathway (highlighted in dark coloring and bold colored text). (B) Reactions of the reductive acetyl-CoA pathway. In this pathway, two molecules of CO_2 are independently reduced by different mechanisms to a methyl group and carbon monoxide (CO); these are then combined to form acetyl-CoA as the primary CO_2 fixation product. Several differences occur between the versions of the pathway in *Bacteria* and *Archaea* (highlighted in green and pink, respectively). The initial reaction in the reduction of CO_2 to the level of a methyl group is catalyzed by two different enzymes in *Bacteria* and *Archaea*, formate dehydrogenase and formyl-methanofuran (formyl-MF) dehydrogenase, respectively. Note that these two enzymes utilize different electron donors. In the reactions shown, H_4PT represents a generic C1-carrying tetrahydropterin; note that THF (tetrahydrofolate) is used by *Bacteria*, whereas H_4MPT (tetrahydromethanopterin) is used by *Archaea*. Remaining enzymes that catalyze the sequential reduction reactions converting CO_2 to the level of a methyl group also differ between bacterial and archaeal reductive acetyl-CoA pathways (details not shown), and different electron donors are also used in these reactions. In contrast, in both *Bacteria* and *Archaea*, the second molecule of CO_2 is reduced to CO by the bifunctional key enzyme carbon monoxide dehydrogenase/acetyl-CoA synthase (CODH/acetyl-CoA synthase; highlighted in purple), which also catalyzes the formation of the primary CO_2 reduction product, acetyl-CoA. Pyruvate synthase then catalyzes the incorporation of an additional CO_2 molecule to acetyl-CoA, forming pyruvate, which is used for biosynthesis. $F_{420}H_2$, reduced coenzyme F_{420}; CoFeSP, corrinoid iron-sulfur protein, a protein that serves as an essential methyl-group carrier during the production of acetyl-CoA by CODH/acetyl-CoA synthase. Colored boxes indicate essential precursors.

3-Hydroxypropionate (3HP) Bi-cycle

Most members of the green nonsulfur bacteria (*Chloroflexi*; Fig. 9.6A) grow as anoxygenic photoheterotrophs under anoxic conditions or aerobic chemoheterotrophs when oxygen is present. However, some strains of the genus *Chloroflexus* can grow as anoxygenic photoautotrophs using the **3-hydroxypropionate bi-cycle (3HP bi-cycle)** for CO_2 fixation. The 3HP bi-cycle starts with acetyl-CoA, and in the first cycle, two molecules of bicarbonate are fixed, forming glyoxylate (2C) as the initial CO_2 fixation product (Fig. 9.6B). Then, in the second cycle, glyoxylate (2C) and propionyl-CoA (3C; an intermediate of the first cycle) are converted to pyruvate (3C) and acetyl-CoA (2C) via a series of reactions (Fig. 9.6B). The acetyl-CoA that is produced can then serve as the substrate for another turn of the cycle. Overall, one molecule of pyruvate is formed from three molecules of bicarbonate, using the key carboxylating enzymes **acetyl-CoA carboxylase** and **propionyl-CoA carboxylase**. The pathway has a relatively high energy cost, requiring seven ATP equivalents to produce one molecule of pyruvate. However, as a phototroph, *Chloroflexus* is not energy-limited when light is available.

Several of the enzymes used in the 3HP bi-cycle are multifunctional; only 13 enzymes are needed to catalyze the 19 reactions involved in the two intertwined cycles. As described below, several of the 3HP bi-cycle reactions are identical to those of the 3HP-4HB cycle. The 3HP bi-cycle can also be used for the assimilation of various fermentation products such as acetate, succinate, and propionate, which are metabolized via their CoA derivatives. This provides metabolic flexibility to *Chloroflexus*, which preferentially grows as a photoheterotroph when organic carbon sources are available.

3-Hydroxypropionate-4-Hydroxybutyrate (3HP-4HB) and Dicarboxylate-4-Hydroxybutyrate (DC-4HB) Cycles

The **3-hydroxypropionate-4-hydroxybutyrate (3HP-4HB)** and **dicarboxylate-4-hydroxybutyrate (DC-4HB) cycles** operate in different members of the *Archaea*. The 3HP-4HB cycle is present in autotrophic thermoacidophilic members of the *Sulfolobales*, one of the main lineages of the *Crenarchaeota* (Fig. 9.7A). The initial reactions of the 3HP-4HB cycle (Fig. 9.7B) convert acetyl-CoA to succinyl-CoA and are identical to those of the 3HP bi-cycle in *Chloroflexus* (Fig. 9.6B). However, the enzymes used for the conversion of malonyl-CoA to propionyl-CoA in the two pathways are not homologous. In addition, the mechanism for regenerating acetyl-CoA is different from that in the 3HP bi-cycle. Despite their similarities, the 3HP-4HB cycle and the 3HP bi-cycle appear to have evolved independently in the *Sulfolobales* and *Chloroflexus*. In fact, the second part of the pathway, converting succinyl-CoA to two molecules of acetyl-CoA, also occurs in the DC-4HB cycle (see below), indicating that this sequence of reactions is common to all known autotrophic *Crenarchaeota*. The enzymes of the 3HP-4HB cycle are oxygen tolerant, and this CO_2 fixation pathway seems to be used by all autotrophic members of the *Sulfolobales*, including aerobic, microaerophilic, and strictly anaerobic strains. In addition, genes encoding key enzymes of the 3HP-4HB cycle have been identified in the mesophilic marine

A

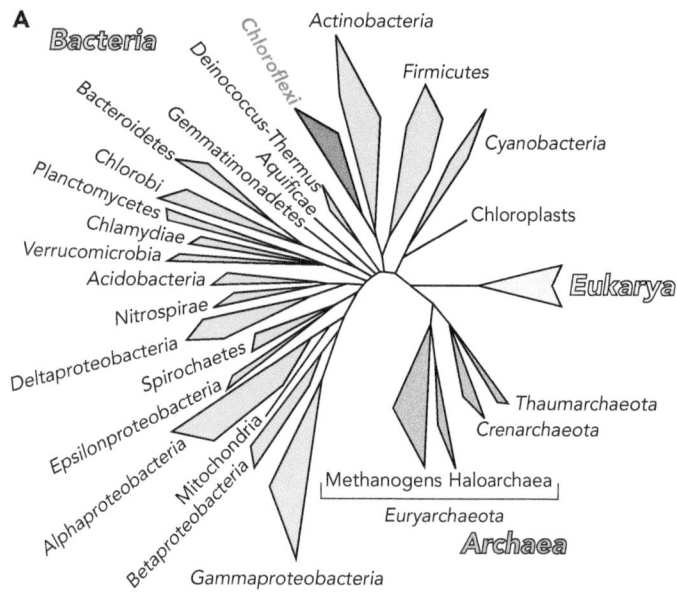

B

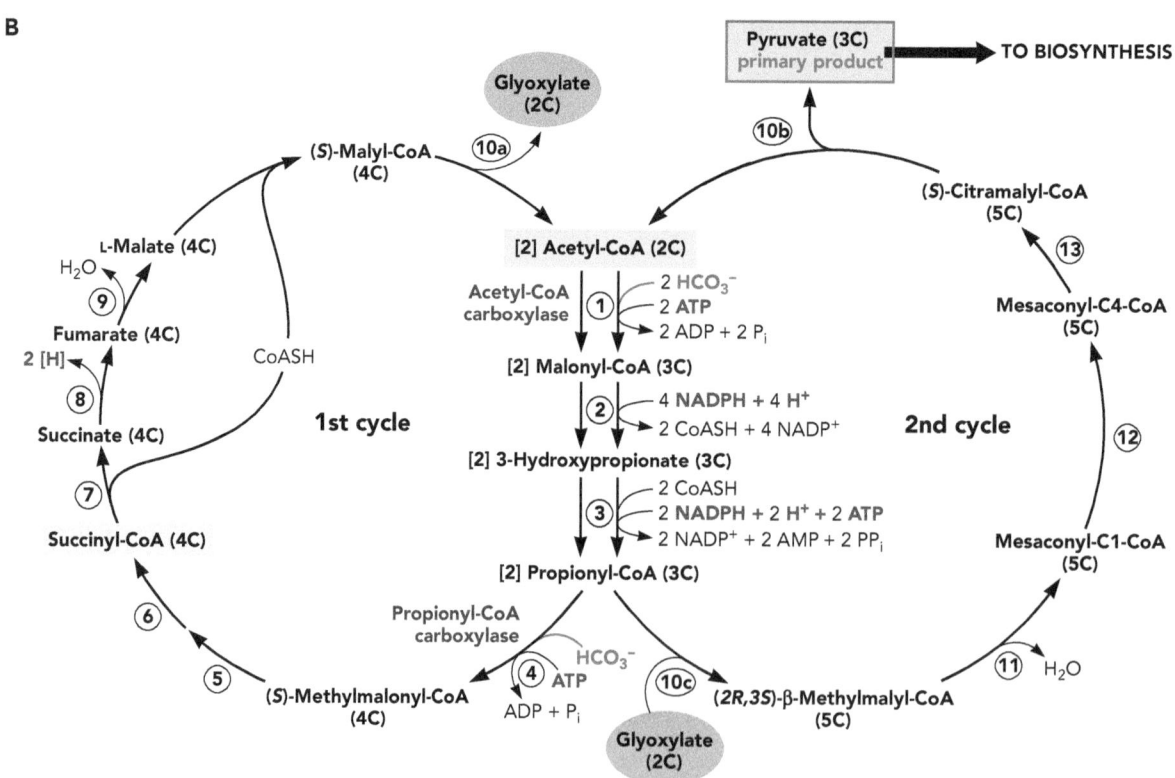

FIGURE 9.6 **Distribution and biochemistry of the 3-hydroxypropionate (3HP) bi-cycle.** (A) Phylogenetic tree highlighting lineages with members known to use the 3HP bi-cycle (highlighted in dark coloring and bold colored text). (B) Reactions of the 3HP bi-cycle. The key enzymes that incorporate CO_2 (in the form of soluble bicarbonate ions) are acetyl-CoA carboxylase (reaction 1) and propionyl-CoA carboxylase (reaction 4), both highlighted in purple. Note that the 2-carbon compound glyoxylate (in orange shaded circle), which is produced in the first cycle, is a substrate in the second cycle. The second cycle results in the regeneration of acetyl-CoA to allow the bi-cycle to continue functioning, as well as the production of pyruvate for biosynthesis. Enzymes catalyzing the remaining numbered reactions are malonyl-CoA reductase (reaction 2), propionyl-CoA synthase (reaction 3), methylmalonyl-CoA epimerase (reaction 5), methylmalonyl-CoA mutase (reaction 6), succinyl-CoA: (S)-malyl-CoA transferase (reaction 7), succinate dehydrogenase (reaction 8), fumarate hydratase (reaction 9), a trifunctional lyase (reactions 10a, 10b, and 10c), mesaconyl-C1-CoA hydratase (reaction 11), mesaconyl-CoA C1-C4 CoA transferase (reaction 12), and mesaconyl-C4-CoA hydratase (reaction 13). Colored boxes indicate essential precursors.

Thaumarchaeota (Fig. 9.7A), suggesting that these autotrophic organisms may also use the 3HP-4HB cycle to fix CO_2.

The DC-4HB cycle (Fig. 9.7C) was recently described in the thermophilic crenarchaeon *Ignicoccus hospitalis*, and it also appears to be present in other thermophilic members of the *Crenarchaeota*. The DC-4HB cycle uses some of the same enzymes as the rTCA cycle (for conversion of oxaloacetate to succinyl-CoA) as well as enzymes from the 3HP-4HB cycle that convert succinyl-CoA to acetyl-CoA. The DC-4HB cycle can be divided into two parts. In the first part, succinyl-CoA is generated from acetyl-CoA, one CO_2, and one bicarbonate molecule. Then succinyl-CoA is converted, via 4-hydroxybutyrate, to two molecules of acetyl-CoA: one that is used in the next turn of the cycle and the other as substrate for pyruvate synthase, which converts acetyl-CoA to pyruvate for use in biosynthesis. The DC-4HB cycle actually has no enzymes that are unique to the pathway, and the presence of enzymes of the rTCA cycle led to erroneous reports of the presence of the rTCA cycle in members of the *Crenarchaeota*.

Why So Many CO_2 Fixation Pathways?

The diversity in CO_2 fixation pathways reflects the ecology and physiology of the microorganisms harboring the various pathways. The concentration of oxygen is important in determining which CO_2 fixation pathways function in a given environment. Strictly anaerobic pathways likely evolved first, and adaptations to oxygen subsequently evolved. As described above, several of the pathways require oxygen-sensitive enzymes or cofactors (e.g., reduced ferredoxin), and therefore cannot be utilized under oxic conditions. Although not discussed in detail here, many of the enzymes used in the various CO_2 fixation pathways require specific metals (e.g., Fe, Co, Ni) that assist in catalysis. The reduced forms of these metals are generally more soluble in the absence of oxygen and are therefore more available in anoxic environments.

In addition, energy requirements are also important and likely played a part in the evolution of diverse CO_2 fixation pathways. In general, most energy-limited organisms, many of which are anaerobes, use the less energy-demanding pathways rather than the Calvin cycle. It is generally believed that the reductive acetyl-CoA pathway is the most ancient CO_2 fixation pathway; it is relatively simple, requires the lowest input of energy, and only functions in the absence of oxygen. In contrast to anaerobes, most aerobic chemolithoautotrophs use the oxygen-tolerant but energetically expensive Calvin cycle. Similarly, many photoautotrophs, which are not limited for energy, also use the Calvin cycle.

Environmental conditions such as temperature, salt concentration, and pH also play important roles in determining which CO_2 pathways will function. High temperatures denature and inactivate certain key enzymes, and some pathways have temperature-sensitive intermediates, thus preventing some of the CO_2 fixation pathways from functioning in (hyper)thermophiles. In addition, the solubility of CO_2 is affected by both temperature and salt concentration, and the relative concentrations of HCO_3^- and CO_2 in solution are pH dependent. Not only do the various CO_2-fixing enzymes have different affinities (K_m) for CO_2, some are specific for CO_2 and others for HCO_3^-. Thus, some CO_2 fixation pathways may be more efficient in acidic or alkaline environments depending on the preferences of the CO_2-incorporating enzymes.

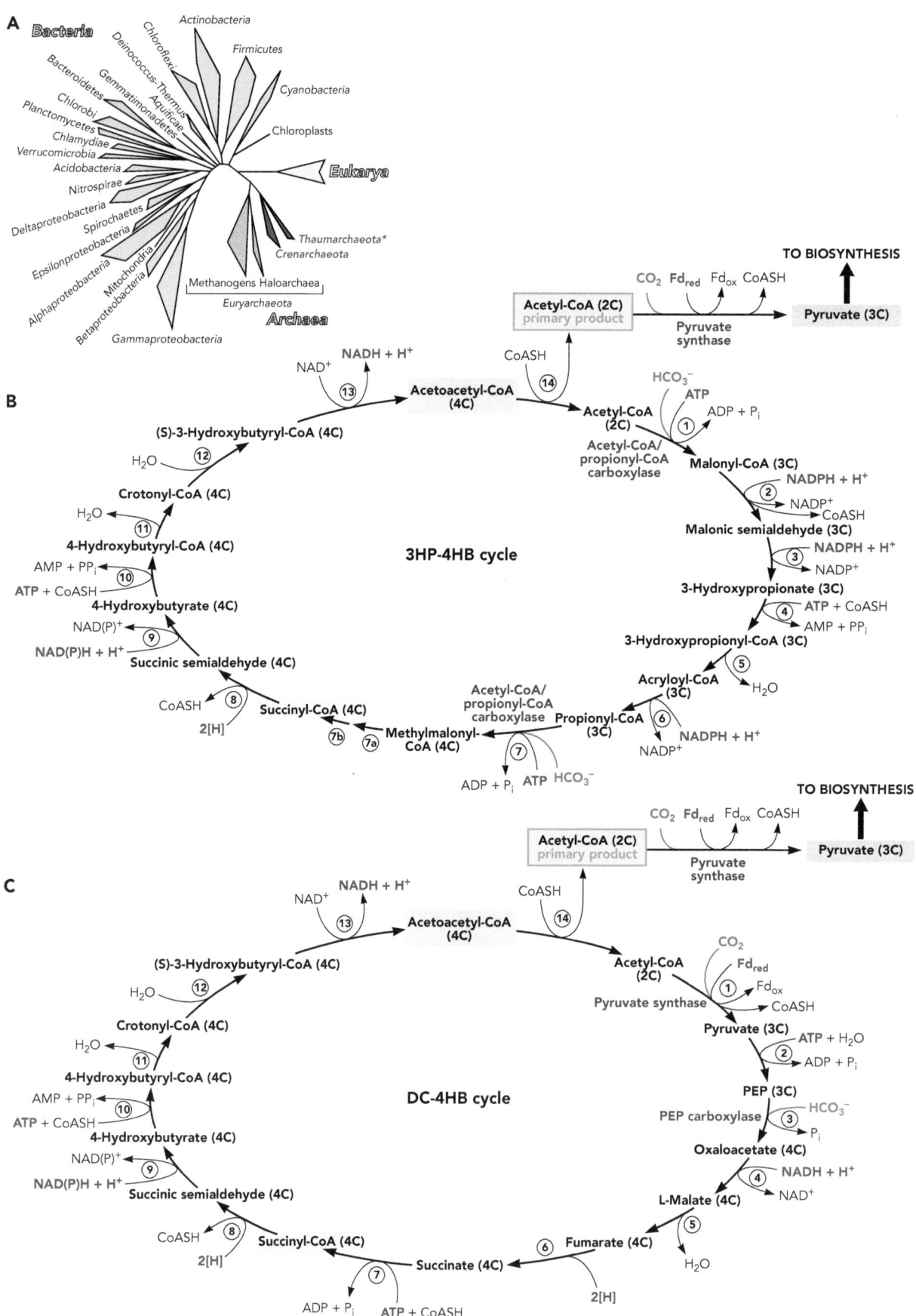

FIGURE 9.7 **Distribution and biochemistry of the hydroxybutyrate cycles.** (A) Phylogenetic tree highlighting lineages with members known to use the 3-hydroxypropionate-4-hydroxybutyrate (3HP-4HB) cycle and the dicarboxylate-4-hydroxybutyrate (DC-4HB) cycle (highlighted in dark coloring and bold colored text). Distinct lineages of the *Crenarchaeota* use one or the other of these two pathways. The asterisk indicates that, although the genes encoding enzymes of the 3HP-4HB cycle have been found in the *Thaumarchaeota*, to date, there is no experimental evidence demonstrating that the pathway functions in these organisms. (B) Reactions of the 3HP-4HB cycle. CO_2 fixation reactions are catalyzed by acetyl-CoA/propionyl-CoA carboxylase (reactions 1 and 7, highlighted in purple). Enzymes catalyzing the remaining numbered reactions are malonyl-CoA reductase (reaction 2), malonic semialdehyde reductase (reaction 3), 3-hydroxypropionate-CoA ligase (reaction 4), 3-hydroxypropionyl-CoA dehydratase (reaction 5), acryloyl-CoA reductase (reaction 6), methylmalonyl-CoA epimerase (reaction 7a), methylmalonyl-CoA mutase (reaction 7b), succinyl-CoA reductase (reaction 8), succinic semialdehyde reductase (reaction 9), 4-hydroxybutyrate-CoA ligase (reaction 10), 4-hydroxybutyryl-CoA dehydratase (reaction 11), crotonyl-CoA hydratase (reaction 12), (*S*)-3-hydroxybutyryl-CoA dehydrogenase (reaction 13), and acetoacetyl-CoA β-ketothiolase (reaction 14). (C) Reactions of the DC-4HB cycle. CO_2 fixation reactions are catalyzed by pyruvate synthase and phosphoenolpyruvate (PEP) carboxylase (reactions 1 and 3; highlighted in purple). Enzymes catalyzing the remaining numbered reactions are PEP synthase (reaction 2), malate dehydrogenase (reaction 4), fumarate hydratase (reaction 5), fumarate reductase (reaction 6), succinyl-CoA synthase (reaction 7), succinyl-CoA reductase (reaction 8), succinic semialdehyde reductase (reaction 9), 4-hydroxybutyrate-CoA ligase (reaction 10), 4-hydroxybutyryl-CoA dehydratase (reaction 11), crotonyl-CoA hydratase (reaction 12), (*S*)-3-hydroxybutyryl-CoA dehydrogenase (reaction 13), and acetoacetyl-CoA β-ketothiolase (reaction 14). In both pathways, 2[H] designates electron donors that have not been identified. Note that the two pathways have seven reactions in common (reactions 8 to 14). The primary product of CO_2 fixation in both pathways is acetyl-CoA, which is carboxylated by pyruvate synthase (highlighted in purple) to form pyruvate for biosynthesis. Colored boxes indicate essential precursors.

Finally, the type of reduced organic product generated during CO_2 fixation may be important to the overall metabolism of certain groups of organisms. In contrast to the Calvin cycle, which ultimately results in the formation of a 6-carbon sugar (fructose-6-phosphate), most of the autotrophic pathways lead to the production of acetyl-CoA and pyruvate. Although the reactions required for the production of the 12 essential precursors are generally conserved, various autotrophs may have distinct metabolic preferences due to differences in enzyme efficiencies or metabolite pools. In addition, some of the pathways allow for co-utilization of alternative substrates. For example, alternative C1 compounds can feed into the reductive acetyl-CoA pathway, and alternative organic substrates can feed into the 3HP bi-cycle, 3HP-4HB cycle, and DC-4HB cycle. These additional metabolic options may provide a selective advantage to microbes in constantly changing environments.

Learning Outcomes: After completing the material in this chapter, students should be able to . . .

1. explain how the three stages of the Calvin cycle work together to fix carbon dioxide (CO_2)

2. describe how each of the five pathways of CO_2 fixation, other than the Calvin cycle, function to produce acetyl-CoA or pyruvate

3. give examples of how the limitations of critical enzymes can impact the conditions under which autotrophic organisms can grow

4. propose a theory that explains how the physiological lifestyle of a given microorganism might have influenced the evolution of its autotrophic pathway

Check Your Understanding

1. a) Define this terminology from the Interlude: phototroph, chemotroph, heterotroph, autotroph, organotroph, lithotroph
   b) Define this terminology from Chapter 9: primary producers, photorespiration, carboxysomes, pyruvate synthase.

2. For each example below, indicate which term from the Interlude (chemoorganoheterotroph, chemolithoautotroph, photoautotroph, photoheterotroph) best describes the scenario.
   a) _____ *Aquifex* can grow exclusively on inorganic compounds, using hydrogen as an energy source and converting inorganic carbon to organic carbon.
   b) _____ *Heliobacteria* can convert sunlight energy to ATP, obtaining carbon from small organic molecules.
   c) _____ *Microcystis* harnesses light energy to generate ATP and creates cell biomass from fixing carbon dioxide.
   d) _____ *Rhodopseudomonas* can grow in the dark, using succinate as both its energy and carbon source.

3. For each CO_2 fixation pathway below, indicate if it is found in phototrophs, chemolithotrophs, or both. (LO1, 2)

   Calvin cycle

   Reductive TCA (rTCA) cycle

   Reductive acetyl-CoA pathway

   3-Hydroxypropionate (3HP) bi-cycle

   3-Hydroxypropionate-4-hydroxybutyrate (3HP-4HB) cycle

   Dicarboxylate-4-hydroxybutyrate (DC-4HB) cycle

4. The purpose of this table is to highlight some of the key enzymes in the autotrophic pathways. For each enzyme, name the pathway in which it is found and describe its function. (LO1, 2)

Enzyme	Pathway in which it is found	Function
Ribulose-1,5-bisphosphate carboxylase/oxygenase (RubisCO)		
Phosphoribulokinase		
Carbonic anhydrase		
Fumarate reductase		
Ferredoxin-dependent α-ketoglutarate synthase		
Citrate lyase		
Carbon monoxide dehydrogenase/acetyl-CoA synthase		
Acetyl-CoA carboxylase and propionyl-CoA carboxylase		

5. Briefly describe the conversions that take place in each of the three stages of the Calvin cycle. Describe where ATP and reducing equivalents are used (see Fig. 9.2B). (LO1)

   a) CO_2 fixation stage:

   b) Reduction stage:

   c) Regeneration stage:

6. In the diagram of the Calvin cycle shown, (LO1)

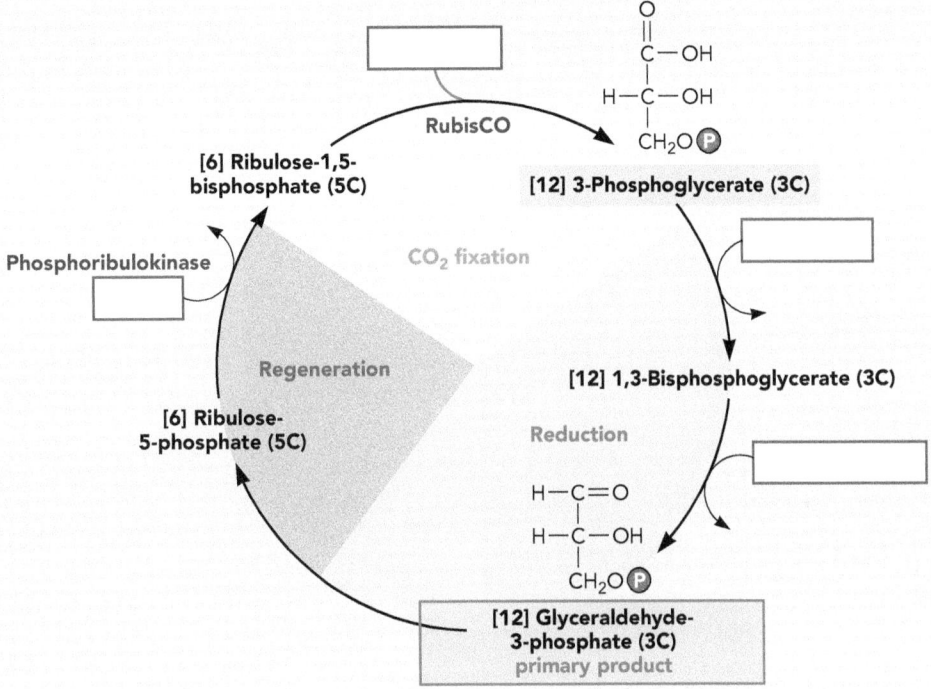

   a) Indicate where CO_2 is incorporated into the pathway.

   b) Indicate where NADPH + H$^+$ (reducing equivalent) is used

   c) Indicate which steps require ATP.

   d) What is the primary product of this cycle?

7. It has been said that RubisCO may be the most abundant single enzyme on the planet. (LO1)

   a) Does evidence support the theory that RubisCO evolved multiple times in different organisms or that it evolved from a common ancestor? What is that evidence?

   b) While extremely widespread, RubisCO is fairly inefficient. How does its function as an oxygenase contribute to the inefficiency of RubisCO?

   c) How has the evolution of carboxysomes helped to increase the efficiency of CO_2 fixation?

8. The green sulfur bacteria use a pathway called the reductive TCA (rTCA) cycle to fix carbon.

   a) Briefly review what the TCA cycle does in the process of aerobic respiration (see Chapter 2).

   b) Describe how the rTCA cycle works to assimilate CO_2. Why is the rTCA cycle sometimes called the reverse TCA cycle? (LO2)

9. In the diagram of the reductive tricarboxylic acid (rTCA) cycle shown, (LO2)

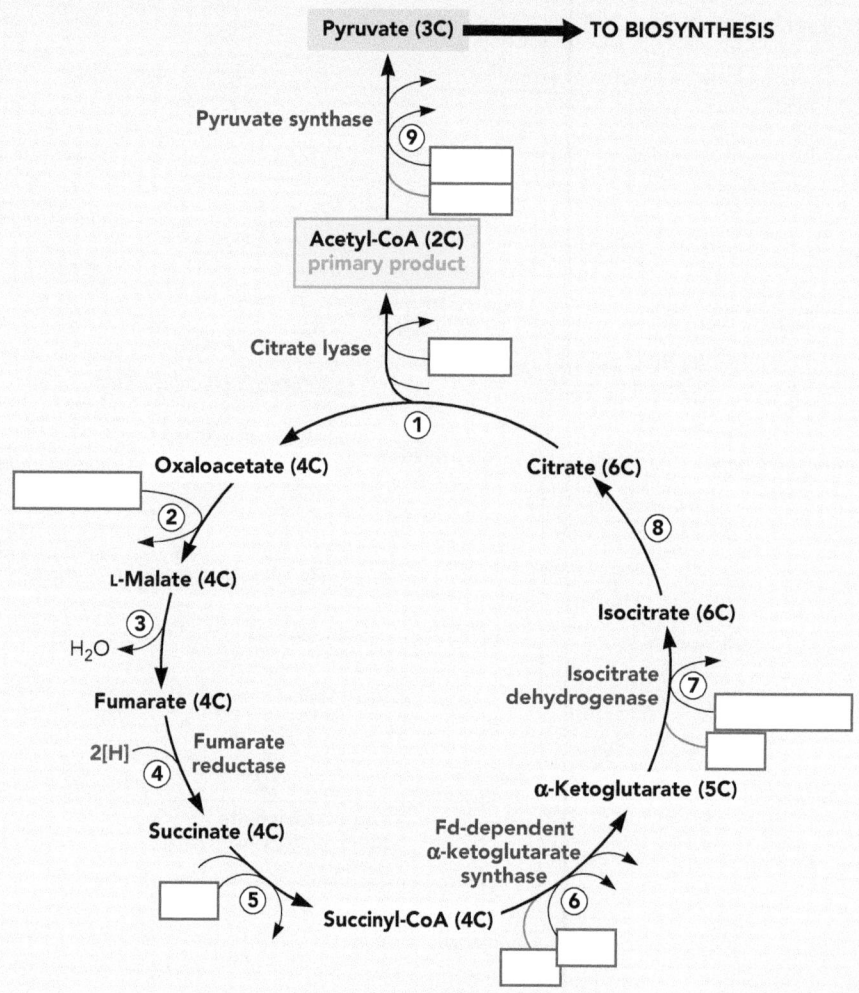

a) Indicate where CO_2 is incorporated into the pathway.

b) Indicate where NADPH + H$^+$ is used.

c) Indicate which steps require ATP.

d) What is the primary product of this cycle?

10. The reductive acetyl-CoA pathway is currently thought to be the only CO_2 fixation pathway that functions in members of both the *Bacteria* and *Archaea*. Outline the reductive acetyl-CoA pathway by choosing the correct term or filling in the blank. (LO2)

a) One molecule of carbon dioxide is (oxidized or reduced) to a methyl group.

b) A second molecule of carbon dioxide is (oxidized or reduced) to carbon monoxide.

c) These compounds are then combined by a synthase to produce the primary product, _____.

d) While both *Bacteria* and *Archaea* use this pathway, some aspects are unique. Describe three ways the reductive acetyl-CoA pathways in *Bacteria* and *Archaea* differ.

11. Certain strains of green nonsulfur photoautotrophic bacteria use the 3-hydroxypropionate (3HP) bi-cycle for CO_2 fixation. Outline the 3HP bi-cycle by choosing the correct term or filling in the blanks. [LO2]

    a) In the first shared step of the 3HP bi-cycle, the enzyme _____ uses energy in the form of _____ to combine acetyl-CoA (2C) and _____ to form malonyl-CoA (3C).

    b) In the next two steps, malonyl-CoA (3C) is (<u>oxidized or reduced</u>) to propionyl-CoA (3C) with electrons from _____.

    c) The enzyme _____ uses energy in the form of _____ to combine propionyl-CoA (3C) and _____ to form methylmalonyl-CoA (4C).

    d) The end products of the first cycle are _____ (3C), which is used for biosynthesis, and _____ (2C), which is fed back into the cycle.

12. In comparing the 3-hydroxypropionate-4-hydroxybutyrate (3HP-4HB) cycle with the 3-hydroxypropionate (3HP) bi-cycle, one can see that the initial reactions of the 3HP-4HB cycle (the conversion of acetyl-CoA to succinyl-CoA) in *Sulfolobales* are identical to those of the 3HB bi-cycle in *Chloroflexus*. Does evidence support the theory that these pathways evolved from a common ancestor? (LO4)

Dig Deeper

13. Environmental factors can impact the function of the various carbon fixation pathways.

    a) The temperature range of enzymes in a pathway can influence the growth rate of microorganisms that use that pathway. Give 3 examples of CO_2 fixation pathways that influence the temperature range of microorganisms. Include in your answer some microorganisms that use those pathways. (LO3)

    b) Describe 2 examples of groups of microorganisms that have enzymes within their CO_2 fixation pathway that are influenced by the concentration of oxygen. (LO3)

14. It is likely that the diversity of microorganisms and the environments in which they lived drove the evolution of the different pathways for CO_2 assimilation. (LO4)

    a) Of the six pathways, which pathway requires the most energy for CO_2 fixation and which requires the least?

    b) Why do you think that many of the <u>aerobic</u> chemolithoautotrophs evolved to use the Calvin cycle (as opposed to <u>anaerobes?</u>

    c) Why do you think that the reductive acetyl-CoA pathway is often found in <u>anaerobes</u> rather than <u>aerobic</u> chemolithoautotrophs?

15. The second part of the 3-hydroxypropionate-4-hydroxybutyrate (3HP-4HB) cycle, converting succinyl-CoA to two molecules of acetyl-CoA, also occurs in the dicarboxylate-4-hydroxybutyrate (DC-4HB) cycle. Suppose you sequenced the enzymes involved in both pathways. What result would support the theory that these pathways evolved from a common ancestor? (LO3)

16. Pyruvate synthase is found in many, but not all, of the CO_2 fixation pathways. What reaction does pyruvate synthase catalyze? Which of the six autotrophic pathways require pyruvate synthase? Why is this enzyme required in some, but not all, of the CO_2 fixation pathways? (LO1, 2)

17. The various CO_2 fixation pathways allow cells to generate the 12 essential precursors from CO_2. Use the following diagram of the central metabolic pathways to answer parts a through e. (LO1, 2)

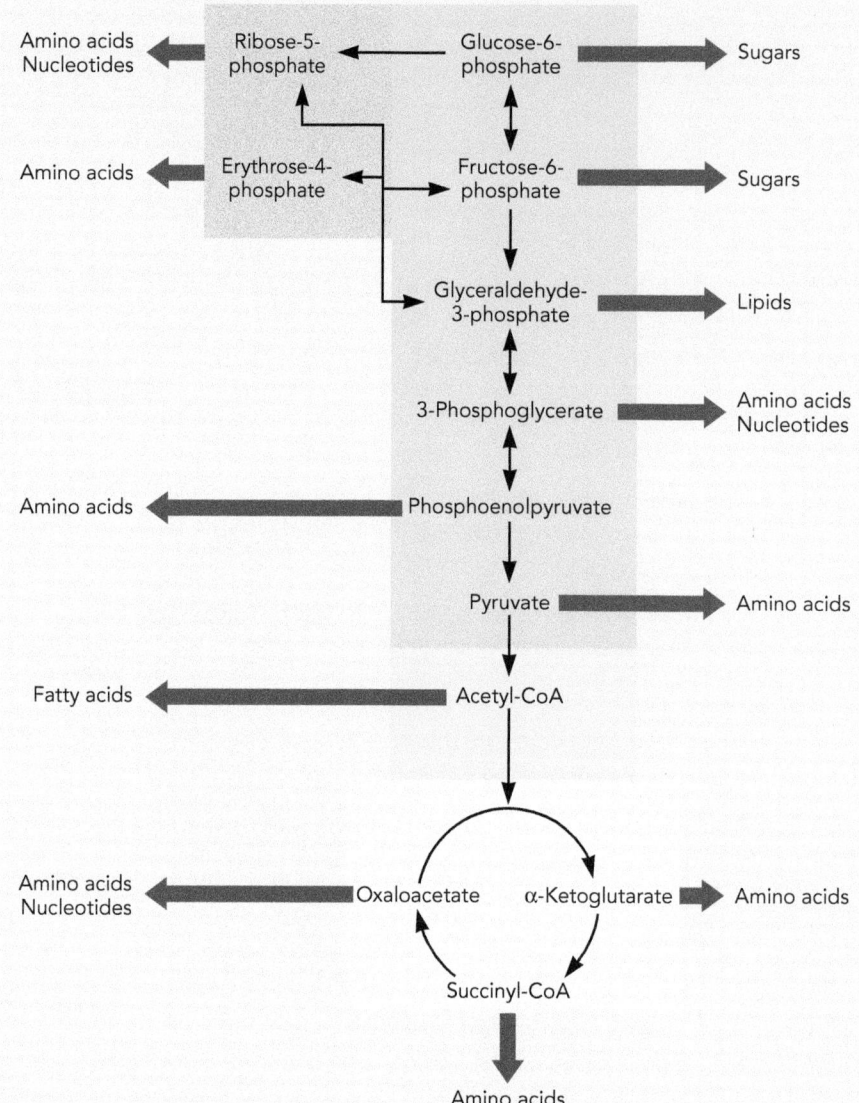

a) How would a bacterium that grows using the Calvin cycle make lipids from CO_2? Describe how the additional central metabolic pathways could connect to the Calvin cycle.

b) How would a bacterium that grows using the Calvin cycle make fatty acids from CO_2? Describe how the additional central metabolic pathways could connect to the Calvin cycle.

c) Which of the central metabolic pathways (glycolysis, pentose phosphate pathway, tricarboxylic acid (TCA) cycle, transition reaction) would anaerobic cells fixing CO_2 using the 3-hydroxypropionate (3HP) bi-cycle use to make guanine from ribose-5-phosphate?

d) How would this same microorganism use CO_2 to make oxaloacetate, then lysine?

e) How would an anaerobic microorganism that is fixing CO_2 using the reductive acetyl-CoA pathway make proline from α-ketoglutarate?

208 Making Connections

208 Phototrophy

209 Chlorophyll-Based Phototrophy

211 Cellular Structures Needed for Phototrophy: Light-Harvesting Complexes, Reaction Centers, and Unique Membrane Organizations

215 Oxygenic Photoautotrophy in the *Cyanobacteria*

218 Anaerobic Anoxygenic Phototrophy in the Phototrophic Purple Sulfur and Purple Nonsulfur Bacteria

221 Anaerobic Anoxygenic Phototrophy in the *Chlorobi* and *Chloroflexi* (Green Sulfur and Green Nonsulfur Bacteria, Respectively)

224 Anaerobic Anoxygenic Photoheterotrophy in the *Firmicutes*

224 Aerobic Anoxygenic Phototrophy

225 Retinal-Based Phototrophy

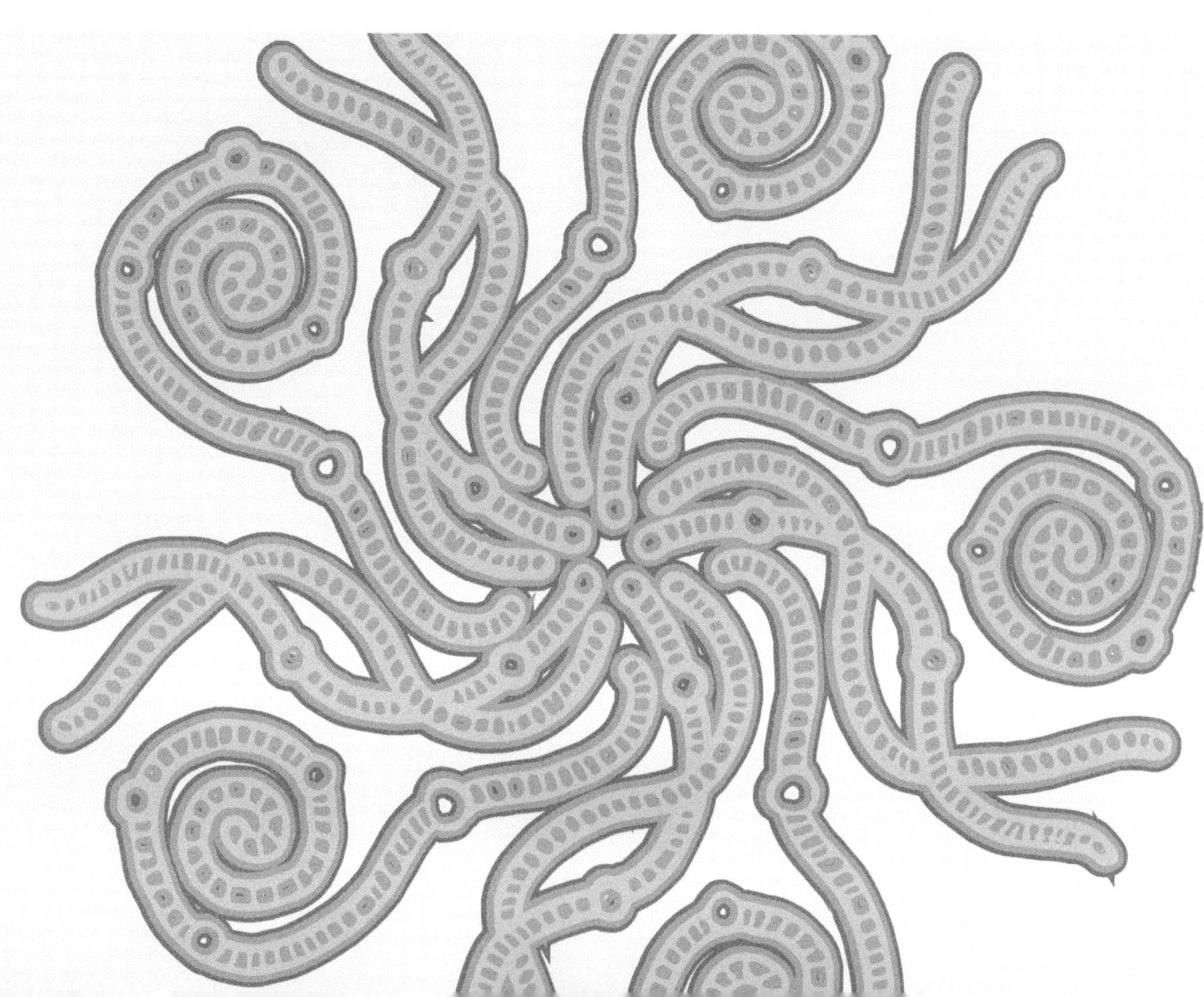

10

PHOTOTROPHY

After completing the material in this chapter, students should be able to . . .

1. differentiate the structure and function of light-harvesting and membrane assemblies among the different groups of phototrophs

2. explain how light energy is converted to ATP and NADPH + H$^+$ in oxygenic photophosphorylation

3. explain how light energy is converted to ATP and NADPH + H$^+$ in anoxygenic photophosphorylation

4. compare and contrast electron and proton flow in anoxygenic and oxygenic photophosphorylation

5. describe how cells use retinal-based photophosphorylation to supplement ATP synthesis

Microbial Physiology: Unity and Diversity, First Edition. Ann M. Stevens, Jayna L. Ditty, Rebecca E. Parales and Susan M. Merkel.
© 2024 American Society for Microbiology.
Companion website: www.wiley.com/go/stevens/microbialphysiology

Making Connections

In Chapter 9, we discussed the various autotrophic pathways of CO_2 fixation that evolved in *Bacteria* and *Archaea*, processes that require ATP and reducing power. Many microbes can utilize light as an energy source to produce a proton motive force (PMF) that drives the process of ATP synthesis through photophosphorylation (Chapter 5). Reducing power, in the form of NADPH + H$^+$, supplies the electrons necessary for converting CO_2 to reduced cell biomass (Chapter 9). Here, we will discuss the various electron donors (water, sulfur compounds, organic compounds) and the different mechanisms that bacteria use to generate reducing power via a phototrophic lifestyle. As we examine the details of the physiology, we will gain insight into the amazing diversity and utility of the phototrophic bacteria. What we will see is that two different photosystems evolved to optimize the production of either a PMF or reducing power. Organisms that use both photosystems (*Cyanobacteria* and plants) are optimized for photosynthetic growth and can balance the production of ATP and reductive power based on cellular need. Microbial species that utilize only one of these photosystems (the green and purple bacteria; see below) have adapted different mechanisms to ensure production of the energy inputs necessary for their phototrophic lifestyles.

Phototrophy

"Phototrophy" is a general term describing the metabolic ability to harness energy from sunlight by exciting electrons and using electron transport chains (ETCs) to generate PMF. The term "photosynthesis" has traditionally been used to describe plant metabolism in which light energy is converted to chemical energy, oxygen is produced when water is cleaved to provide electrons (**oxygenic phototrophy**), and CO_2 is converted to cell biomass (autotrophy; Chapter 9). Oxygenic phototrophy is carried out by many microbes, including the *Cyanobacteria*; however, bacterial and archael phototrophy is much more diverse. Most bacterial phototrophs cannot use H_2O as an electron donor, and some only use light as a supplementary energy source. In addition, not all phototrophs can fix CO_2 or only do so when organic carbon sources are not available. Rather than using the term "photosynthesis" to describe the wide range of phototrophic lifestyles that have evolved in the microbial world, we will use the terminology introduced in the Interlude (see Fig. I.1).

Photoautotrophic organisms obtain energy from sunlight; their electron donors (for the ETC) and carbon (for biomass production) come from inorganic sources. This is exemplified in the classic example of plant photosynthesis that we have all learned about since grade school, where energy from the sun is harnessed using chlorophyll to excite two electrons that then move through two photosystems. A **photosystem** is composed of the proteins, pigments, and electron carriers required to convert light energy to chemical energy. In plant photosynthesis, the electrons come from H_2O, oxygen is released as a byproduct, and organic carbon (cell biomass) is made from the fixation of CO_2. While this process of oxygenic photosynthesis occurs in the chloroplasts of plants and algae, evidence indicates that it originated in free-living *Cyanobacteria*. The evolution of oxygenic photosynthesis is believed to have been the key event that converted the anoxic atmosphere of Earth to an oxic atmosphere, thus permitting the diversity of life that now exists (Chapter 1).

Among the various lineages of photoautotrophs, three different pathways are used for CO_2 fixation: the Calvin cycle, reverse tricarboxylic acid (rTCA) cycle, and 3-hydroxypropionate (3HP) bi-cycle (Chapter 9). In addition, there are many prokaryotes, termed photoheterotrophs, that harvest energy from sunlight but, rather than fixing CO_2, utilize organic compounds as their source of carbon. Some phototrophic species even have the ability to grow as photoautotrophs or photoheterotrophs in the light, and chemoheterotrophs or chemolithoautotrophs in the dark, adjusting their metabolic lifestyle in response to changing environmental conditions. Finally, some microorganisms utilize light solely as a supplementary source of energy under adverse environmental conditions.

There are two distinct types of light-harvesting systems that are widespread among various lineages of the *Bacteria* and *Archaea*: those that, like plants, use chlorophyll (Fig. 10.1A), and a second type that uses the small molecule retinal (Fig. 10.1B) to convert light energy into chemical energy. Chlorophyll-based photosystems have been identified in seven different bacterial phyla (Fig. 10.1C) but have not yet been found in any *Archaea*. Chlorophyll-based phototrophy can be further differentiated based on the specific chlorophyll molecule(s) used in light harvesting and the number of photosystems (either one or two) utilized in energy conservation. Among the chlorophyll-based phototrophs, another major distinguishing characteristic is whether oxygen is produced (oxygenic) or not produced (anoxygenic) during the process. In fact, the only example of prokaryotic oxygenic phototrophy that utilizes two photosystems and generates oxygen is in the *Cyanobacteria*, the evolutionary precursor for the plant chloroplast. As discussed below, there are many examples of anoxygenic phototrophy that only occur in *Bacteria*. Retinal-based photosystems are widespread in both *Bacteria* and *Archaea* (Fig. 10.1D), although they were originally identified in extremely halophilic (salt-loving) members of the *Euryarchaeota*. In many cases, it appears that retinal-based phototrophy is used primarily as a survival strategy to allow microbes to maintain sufficient energy levels under less-than-optimal environmental conditions.

As we will further describe in this chapter, there is great diversity in the various modes of microbial phototrophy. Here, we categorize phototrophic species into two major groups based on the primary molecules responsible for the harvesting of light energy, chlorophyll and retinal.

Chlorophyll-Based Phototrophy

Chlorophyll is one of the two light-harvesting molecules used by phototrophic *Bacteria*. All chlorophylls are cyclic tetrapyrroles (Fig. 10.2A) similar to the heme molecules in cytochromes (Fig. 6.1). A primary difference between heme and chlorophyll is the metal ion present; chlorophyll contains Mg^{2+}, whereas heme contains Fe^{2+}. Like plants, the *Cyanobacteria* utilize chlorophyll *a* (Chl *a*) in the light-harvesting system. All other chlorophyll-based phototrophs use characteristic bacteriochlorophyll (Bchl) molecules, which are similar to Chl *a* but have slightly different molecular structures (Fig. 10.2A). These structural differences determine the wavelengths of light absorbed.

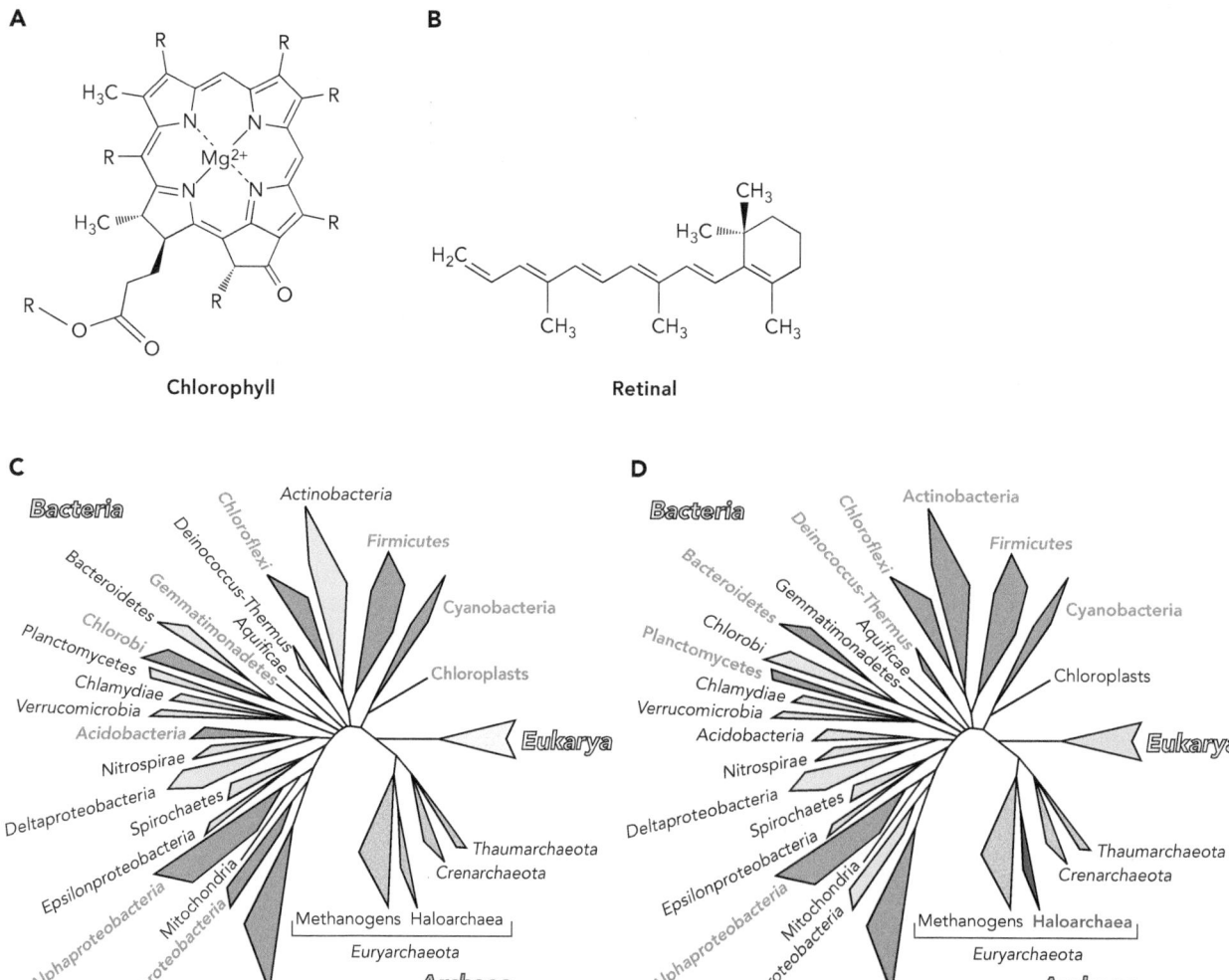

·FIGURE 10.1 **The two major types of pigments used in phototrophy and their phylogenetic distributions.** (A) Structure of a generic chlorophyll molecule. R groups indicate side chains that differ among various chlorophylls. (B) Structure of retinal. (C) Phylogenetic tree highlighting lineages known to have members that carry out chlorophyll-based phototrophy (highlighted in dark coloring and bold colored text). (D) Phylogenetic tree highlighting lineages known to have members that carry out retinal-based phototrophy (highlighted in dark coloring and bold colored text).

Among the various chlorophyll-based phototrophs, it is the suite of chlorophyll molecules and additional accessory light-harvesting pigments known as **carotenoids** and **phycobilins** (Fig. 10.2B and C) that determine the range of light wavelengths that can be absorbed by the cell. While all bacterial phototrophs utilize various carotenoids for light harvesting, the *Cyanobacteria* alone utilize phycobilin molecules (Fig. 10.2C). Not only do the various pigments absorb specific wavelengths of light, but the light that is reflected contributes to the wide range of colors that are seen when different phototrophic bacteria are grown to high density in culture (Fig. 10.3A, a to f). For example, the purple nonsulfur bacterium *Rhodobacter* (a member of the *Alphaproteobacteria*) absorbs light around 800 nm (via Bchl *a* and Bchl *b*) and 500 nm (via carotenoids) (Fig. 10.3B), and the reflected light gives them their characteristic purplish color (Fig. 10.3A, a). As a result, the macroscopic colors of the bacteria have given rise to the generic

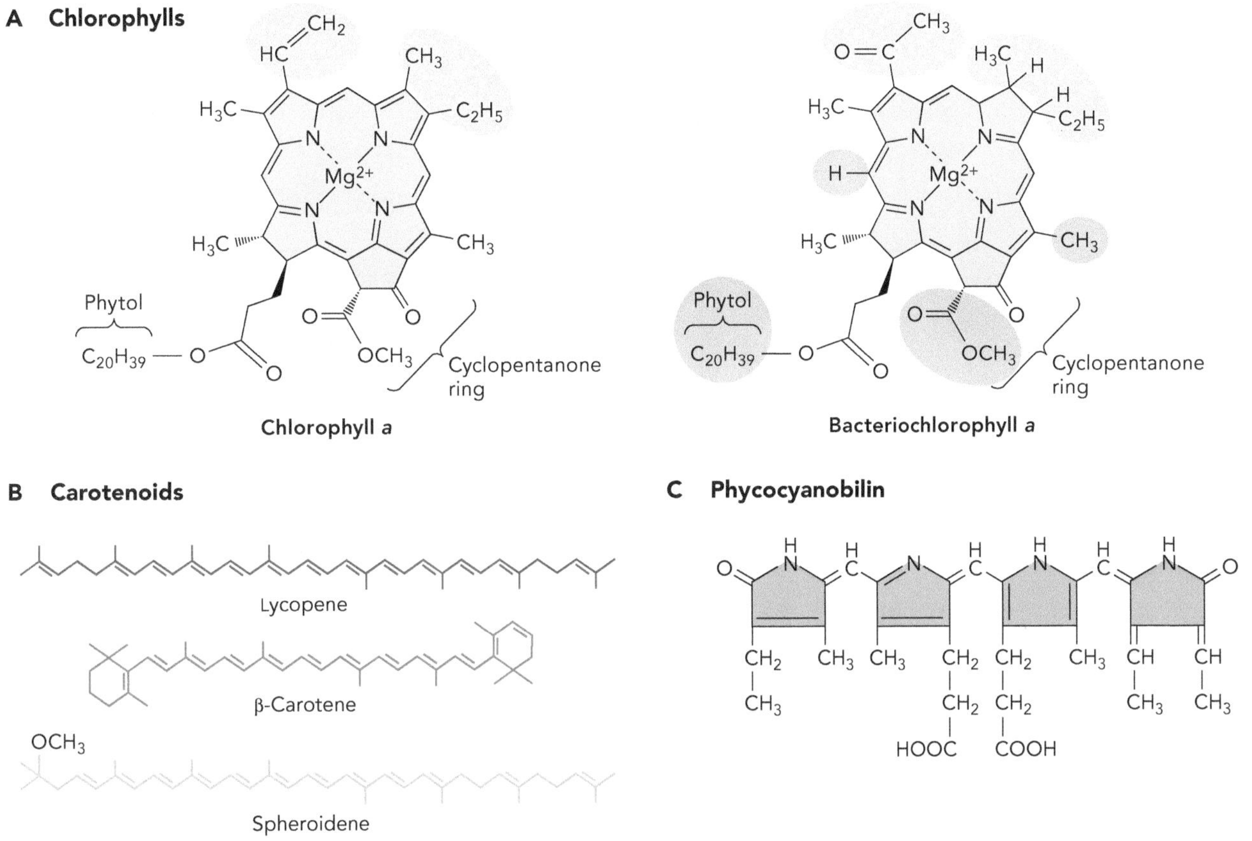

A **Chlorophylls**

Chlorophyll *a*

Bacteriochlorophyll *a*

B **Carotenoids**

Lycopene

β-Carotene

OCH₃

Spheroidene

C **Phycocyanobilin**

FIGURE 10.2 **Pigments used in chlorophyll-based phototrophy.** (A) Representative chlorophyll and bacteriochlorophyll structures (chlorophyll *a* and bacteriochlorophyll *a*). Substituents highlighted in yellow represent positions that differ among various chlorophylls and bacteriochlorophylls, and substituents highlighted in blue represent positions that differ in the various bacteriochlorophylls. (B) Representative carotenoid structures and their corresponding colors indicating absorbed wavelengths of light. (C) Linear tetrapyrrole structure of phycocyanobilin, one of several phycobilins present in cyanobacteria.

names of some of the bacterial phototrophic groups (i.e., green or purple). Since different wavelengths of light penetrate to different depths in sediments and aqueous environments, this variety of pigment molecules allows diverse phototrophs to absorb a wide range of light wavelengths.

Cellular Structures Needed for Phototrophy: Light-Harvesting Complexes, Reaction Centers, and Unique Membrane Organizations

All of the light-harvesting molecules described above are typically localized with various membrane-bound proteins that, in sum, are called **light-harvesting complexes** (Fig. 10.4). These light-harvesting complexes serve as large "satellite dishes" on the surface of the cell to gather photons of sunlight. Photon energy is collected by the various accessory pigments, which is then transferred to what is known as the "special pair" of chlorophyll molecules in the complex named the **reaction center** within a photosystem (Fig. 10.4). The reaction center is where the energy from photons is ultimately transferred from the special pair to the first electron carrier of the ETC, leading to generation of PMF

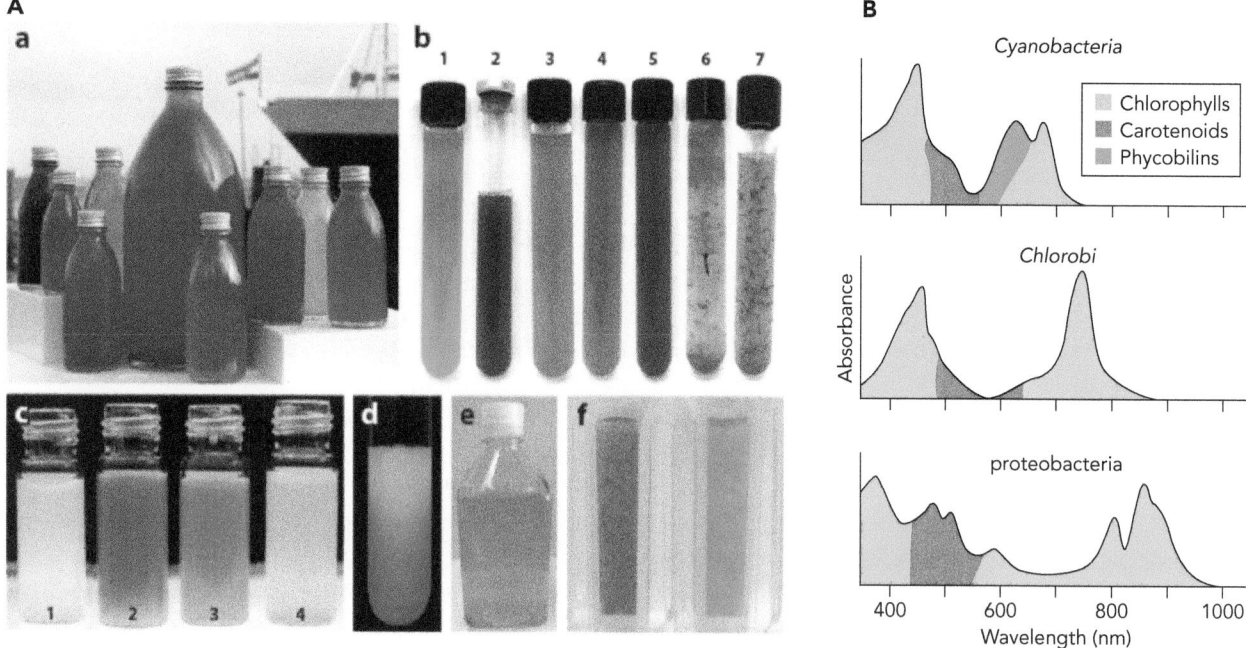

FIGURE 10.3 **Pigments give phototrophic bacteria their characteristic colors.** (A) Cultures of chlorophyll-based phototrophs (reprinted with permission from Theil V. et al. 2018. *Annu Rev Plant Biol* 69:21–49). (a) Various purple sulfur and purple nonsulfur bacteria (all proteobacteria). (b) 1. *Synechococcus* sp. PCC 7002 (*Cyanobacteria*); 2. *Heliobacterium modesticaldum* ICE-1 (*Firmicutes*); 3. *Chloracidobacterium thermophilum* (*Acidobacteria*); 4. *Chlorobaculum tepidum* (BChl *c*; *Chlorobi*); 5. *Chlorobaculum limnaeum* (BChl *e*; *Chlorobi*); 6. "green" *Chloroflexus* sp. (*Chloroflexi*); 7. "brown" *Chloroflexus* sp. (*Chloroflexi*). (c) Aerobic anoxygenic phototrophs: 1. *Sphingomonas* sp. AAP2; 2. *Erythrobacter longus*; 3. *Roseococcus thiosulfatophilus*; 4. *Roseobacter litoralis*. (d) *Gemmatimonas phototrophica*. (e) *Prochlorococcus* sp. (*Cyanobacteria*). (f) *Leptolyngbya* sp. JSC-1 (*Cyanobacteria*) grown under green light (left) and far-red light (right). (B) Absorption spectra of representative examples of the *Cyanobacteria* (top panel), *Chlorobi* (green sulfur bacteria; middle panel), and phototrophic proteobacteria (purple sulfur bacteria; lower panel). Colored regions indicate the pigment types that absorb light at the various wavelengths.

FIGURE 10.4 **A generic light-harvesting system.** Antenna pigments in a membrane-bound light-harvesting complex absorb light energy and funnel that energy to the reaction center, where electrons are excited to a higher energy state. Once excited, electrons are transferred from the reaction center to the first electron carrier (yellow tetrapyrrole) of an electron transport chain (ETC). The first electron carrier and subsequent ETC components vary in different phototrophic species.

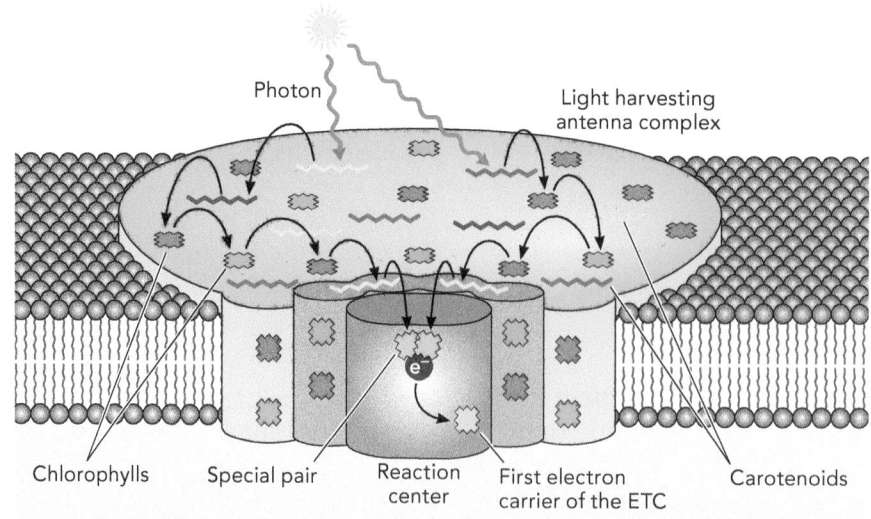

and ATP via photophosphorylation (Fig. 5.1). The components of the reaction center are oriented such that it takes fractions of a second (picoseconds to milliseconds) for the energy transfer events in the ETC to occur.

The light-harvesting components and reaction centers of photosystems are located in the membranes of phototrophic bacteria in specialized structures that are characteristic to the various phototrophic species. In phototrophic members of the proteobacteria, the accessory pigments and the reaction center are located in a highly invaginated cytoplasmic membrane, which forms vesicles, tubes, bundled tubes, lamellae, or stacks that are characteristic of specific species (Fig. 10.5). These invaginations are all continuous extensions of the cytoplasmic membrane that allow for increased surface area and increased capacity to house light-harvesting complexes and maximize the amount of light energy gathered.

As an example, the proteobacterium *Rhodobacter sphaeroides* expresses two multicomponent light-harvesting complexes in the cytoplasmic membrane, light-harvesting complex II (LH II, or P800-850) and light-harvesting complex I (LH I, or P875), which contain the structural polypeptides, bacteriochlorophylls, and carotenoid pigments that are required to harvest light energy (Fig. 10.6A). Each light-harvesting complex absorbs slightly different wavelengths of light. LH I and the reaction center represent a fixed photosynthetic unit that is set at a constant 15:1 ratio, respectively. However, LH II is differentially regulated in response to the intensity of light; the amount of LH II produced is inversely proportional to the environmental light intensity. Therefore, at lower light intensities, more LH II is produced, and vice versa. This regulation permits the cells to conserve energy by producing less LH II when sufficient photons are available. The photons of light captured by the LH complexes are funneled into the reaction center, to the special pair of Bchl *a* molecules, where the electron transfer events begin.

In the green sulfur bacteria (phylum *Chlorobi*, more recently named *Chlorobiota*), green nonsulfur bacteria (phylum *Chloroflexi*), and recently identified phototrophic members of the *Acidobacteria*, the reaction center is localized to the cytoplasmic membrane. The antenna bacteriochlorophylls (typically Bchl *c, d,* or *e*) are arranged in tube-like arrays within specialized membrane-associated structures called **chlorosomes** (Fig. 10.6B). Light energy from the antenna pigments is transferred to Bchl *a* in the reaction center via specific proteins that connect the chlorosome to the reaction center. The entire inner surface of the cytoplasmic membrane is studded with chlorosomes, each containing up to 250,000 Bchl molecules and other accessory pigments. Chlorosomes are particularly efficient at harvesting light, which allows these phototrophs to survive under low-light conditions; in fact, *Chlorobi* have been found to grow at depths of up to 100 meters below the water surface.

In the *Cyanobacteria*, the reaction center and light-harvesting phycobilosomes are localized to special multilayered membrane systems called **thylakoids** (Fig. 10.6C), which are similar to the photosynthetic membranes of chloroplasts. The thylakoids are studded with **phycobilisomes**, large light-harvesting aggregates of phycobiliproteins, which contain the pigments phycoerythrin, phycocyanin, and allophycocyanin that absorb light of different wavelengths (Fig. 10.3). Each of these accessory pigments is composed of a different open-chain tetrapyrrole molecule, or bilin, attached to the protein.

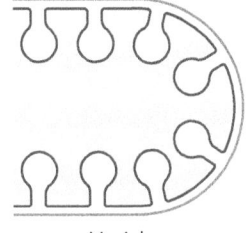
Vesicles

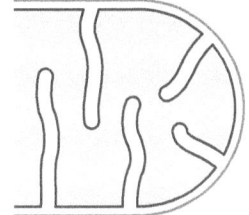
Tubules

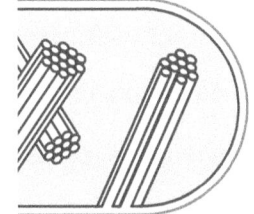
Bundled tubules

Stacks

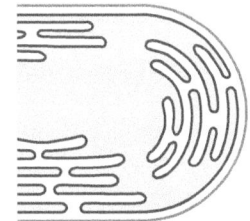
Lamellae

FIGURE 10.5 **Types of photosynthetic membranes in phototrophic purple sulfur and purple nonsulfur bacteria.** All photosynthetic membrane structures shown are invaginations of the cytoplasmic membrane (purple color).

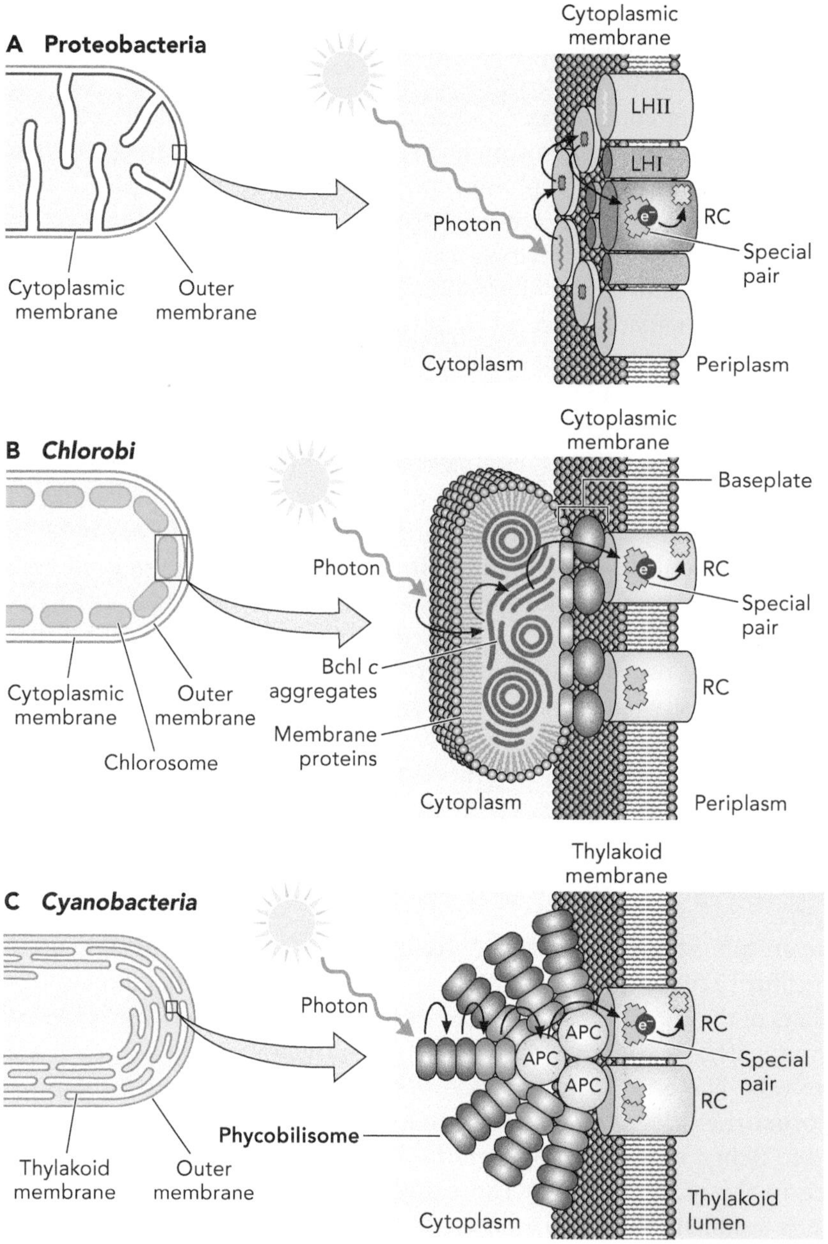

FIGURE 10.6 **Structures of light-harvesting complexes in the phototrophic proteobacteria, _Chlorobi/ Chloroflexi_, and _Cyanobacteria_.** The light-harvesting complexes are located in, or associated with, the cytoplasmic (phototrophic proteobacteria and _Chlorobi/Chloroflexi_) or thylakoid (_Cyanobacteria_) membranes of phototrophic bacteria. (A) Light-harvesting complexes in the phototrophic proteobacteria are associated with invaginations of the cytoplasmic membrane. Photons of light are absorbed via carotenoids and chlorophylls associated with light-harvesting complex II (LH II), energy is transferred (black arrows) through the accessory pigments in LH I, and ultimately transferred to the special pair (two green pyrroles) of Bchl _a_ molecules in the reaction center (RC). (B) Chlorosomes in the _Chlorobi/Chloroflexi_ have a phospholipid monolayer that is anchored to the cytoplasmic membrane by baseplate proteins. Photons of light are absorbed and energy is transferred (black arrows) between tube-like arrays of Bchl _c_, _d_, and/or _e_. Energy is then transferred through accessory pigments in the baseplate to the special pair of Bchl _a_ molecules in the RC. (C) Phycobilisomes in the _Cyanobacteria_ are located in the thylakoid membrane. Light is absorbed and energy is transferred (black arrows) via the accessory pigments phycoerythrin (red), phycocyanin (dark blue), and allophycocyanin (APC, light blue) associated with the phycobilisome proteins. The energy is then transferred to the Chl _a_ special pair in the RC. In all light-harvesting complexes, excited electrons are transferred from the RC to the first electron carrier (yellow tetrapyrrole) of the respective electron transport chain.

In all cases, the elaborate membrane structures that evolved in phototrophic bacteria increase membrane surface area and serve to optimize both light harvesting and generation of the PMF with subsequent ATP production via photophosphorylation. Although the components and cellular structures that drive photosynthesis (light-harvesting systems, reaction centers, photosystems, and ETCs) are typically drawn as a single set of components (Fig. 10.7 through 10.9), it is important to note that electrons and protons flow among and between multiple photosynthetic components that function together to generate PMF and reducing power for the cell.

Oxygenic Photoautotrophy in the *Cyanobacteria*

The *Cyanobacteria* were first considered to be "blue-green algae" (Fig. 10.3A, b1, e, and f) because their photosynthetic apparatus is similar to that of the plant chloroplast; however, the *Cyanobacteria* are a diverse group of Gram-negative aerobic *Bacteria* that form a single branch on the phylogenetic tree (Fig. 10.1B), suggesting that they arose from a common ancestor. The cyanobacteria exhibit a wide range of morphological diversity, including unicellular rods and cocci, as well as filamentous and multicellular forms. *Cyanobacteria* can be found in most environments where light is available, and they are major contributors to the overall primary productivity of the oceans. In particular, members of the genus *Prochlorococcus* are believed to be the most abundant phototrophs on Earth: it has been estimated that there are $\sim 3 \times 10^{27}$ *Prochlorococcus* cells living in the Earth's oceans.

Although a few cyanobacteria can grow slowly as chemoheterotrophs in the dark using simple sugars as carbon sources, most are obligate photoautotrophs using the energy from the sun to fix carbon dioxide via the Calvin cycle (Chapter 9). The photophosphorylation that is carried out by members of the *Cyanobacteria* is virtually identical to the process that occurs in plants (Fig. 10.7A and B). The process involves two photosystems in the thylakoid membranes, each with its own type of reaction center. Photosystems are characterized based on the type of electron acceptor within the reaction center. Photosystem II (PSII) is a quinone-type photosystem, and photosystem I (PSI) is an iron-sulfur (FeS)-type photosystem. In both types of reaction centers, electrons are excited when light of an appropriate wavelength is absorbed. The electrons are then passed through a series of electron carriers in membrane-bound ETCs, alternating between those that carry both protons and electrons, such as quinones, and those that carry only electrons (e.g., cytochromes). Note that the ETC in photophosphorylation functions in a manner very similar to the ETC utilized in oxidative phosphorylation (Chapter 5). It is the controlled release of energy from electrons in the ETC of PSII, which is optimized for generation of the PMF, that leads to ATP synthesis through photophosphorylation. In contrast, the ETC of PSI is optimized for the production of the reduced coenzyme NADPH + H$^+$, which can be used during CO$_2$ fixation (Chapter 9). However, PSI also has the capacity to contribute to the PMF under conditions when the cells need to increase ATP production.

FIGURE 10.7 **Phototrophic electron transport chain (ETC) and electron flow in the *Cyanobacteria*.** (A) ETC in the *Cyanobacteria*. Photosystem II (PSII, P680; quinone-type) and photosystem I (PSI, P700; FeS-type) reaction centers (RCs) are located in the thylakoid membrane. Photophosphorylation begins at the PSII complex when photons of light energy are absorbed via the phycobilisome and transferred to the special pair of Chl *a* molecules (two green tetrapyrroles) in the P680 RC, exciting an electron (purple circle) donated from H_2O. The electron is transferred (black arrows) in an energetically favorable direction to pheophytin (Pheo) and a series of quinones (Q_A and Q_B). Since mobile plastoquinone (PQ) electron carriers in the membrane accept two electrons from Q_B and two protons (H^+) from the cytoplasm to generate reduced PQH_2, a second electron donated from water must be transferred for this 2X-gated reaction to occur. Additional H^+ are pumped via the Q cycle as electrons are passed from PQH_2 through a series of electron carriers in the Cyt b_6f complex (details in Chapter 6). At the PSI complex, plastocyanin (PC) delivers an electron from the Cyt b_6f complex to the special pair of Chl *a* molecules in the P700 RC. After excitation, the electron is transferred in an energetically favorable direction from Chl *a* to phylloquinone (PhyQ), through a series of FeS cofactors within PSI, and then finally to the mobile electron carrier ferredoxin (Fd). In linear electron flow (black arrows), reduced ferredoxin (Fd_{red}) delivers electrons and protons to ferredoxin-NADP reductase (FR_N), which catalyzes the reduction of $NADP^+$ to NADPH + H^+, generating reducing power for carbon fixation (Chapter 9). In cyclic electron flow (dashed purple arrows), Fd_{red} delivers electrons back to the PQ pool via ferredoxin-PQ oxidoreductase (FR_Q), contributing to proton motive force (PMF) and subsequent ATP production through Q cycle reactions (details in Chapter 6). (B) Electron flow in PSII and PSI of the *Cyanobacteria* relative to electrode/reduction potential (E_0'). Solid black arrows indicate linear electron flow, resulting in the production of NADPH + H^+ in the cytoplasm. Dashed purple arrows indicate cyclic electron flow, resulting in increased PMF generation across the thylakoid membrane.

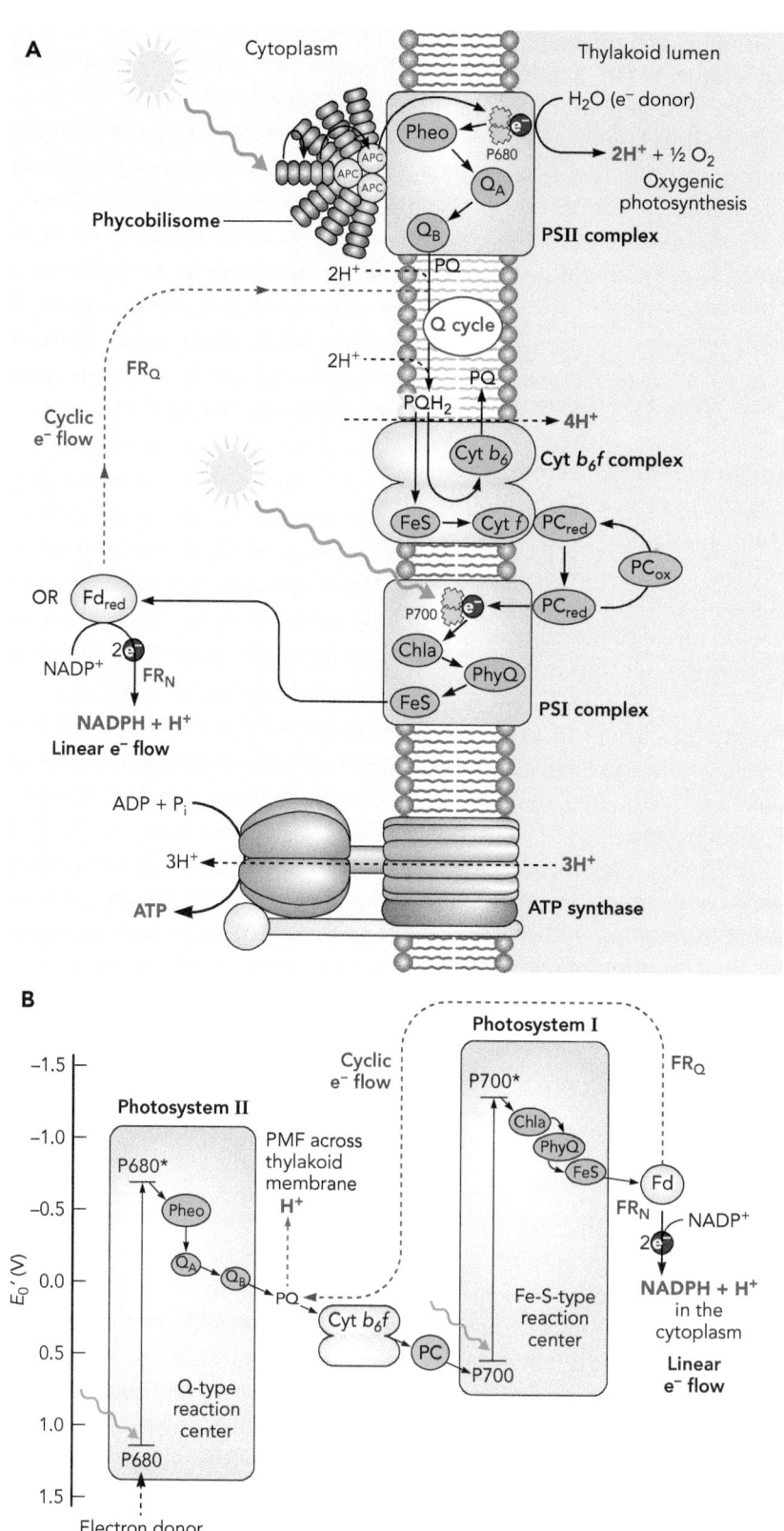

PSII (P680) REACTIONS

In the *Cyanobacteria*, photophosphorylation begins at PSII (Fig. 10.7A). PSII has a special pair of Chl *a* molecules in the P680 reaction center that become excited when light at a wavelength of 680 nm is absorbed. Excited Chl *a* is very electronegative (a good electron donor), and electrons can be transferred one at a time in an energetically favorable direction from the special pair of Chl *a* in P680 to pheophytin (a Chl *a* molecule lacking a Mg^{2+} ion) and then through two quinone molecules embedded in PSII. Loss of electrons at P680 makes it a good electron acceptor, which can then accept electrons from water to begin another series of electron transfer events. When two water molecules are split into their component electrons and protons, oxygen is released (oxygenic photosynthesis).

The next step in the ETC involves the mobile electron carrier plastoquinone (PQ). Quinones (Chapter 6) are mobile electron carriers that, when fully reduced, carry two electrons and two protons. Therefore, two electrons from the PSII reaction center must be individually transferred to the PQ before PQ can leave PSII and gain two protons from the cytoplasmic side of the membrane. Because two electrons must be individually transferred from a one-electron carrier to PQ (a two-electron carrier), this is considered to be a 2X-gated reaction. The reduced form of plastoquinone (PQH_2) delivers the electrons to electron carriers in the cytochrome b_6f (Cyt b_6f) complex. Protons are released into the thylakoid lumen, contributing to the PMF.

CYTOCHROME b_6f REACTIONS AND THE Q CYCLE

As the protons from PQH_2 are released into the thylakoid lumen, electron carriers embedded in the Cyt b_6f complex must accept the electrons. In addition to shuttling electrons to PSI, the Cyt b_6f complex is involved in electron transfer via the Q cycle (Fig. 6.2B), which is used to generate additional PMF across the thylakoid membrane. In sum, the transfer of two electrons in the Q cycle increases the strength of the PMF across the thylakoid membrane by four protons, demonstrating how the Q cycle optimizes energy conservation in the form of PMF. While some electrons are maintained in the Q cycle, some are transferred to Cyt *f*, and then mobile plastocyanin (PC) for continued electron transfer to PSI.

PSI REACTIONS

Plastocyanin molecules deliver electrons to the special pair of Chl *a* molecules in the PSI reaction center (Fig. 10.7A). The special pair absorbs additional photons of light at a wavelength of 700 nm, reenergizing the electrons to a negative electrode/reduction potential. The electrons are then transferred in an energetically favorable direction from Chl *a* to phylloquinone. In contrast to plastoquinone, phylloquinone does not directly contribute to the PMF. Instead, its electrons are passed through a series of FeS proteins within PSI, and then finally to the mobile electron carrier ferredoxin.

From ferredoxin, the electrons can take two possible paths. In **linear electron flow**, ferredoxin delivers the electrons and protons to the ferredoxin-NADP reductase, which catalyzes the production of NADPH + H⁺. It is the reduced NADPH + H⁺ that is used during autotrophy to fix CO_2 via the Calvin cycle (Chapter 9). Alternatively, ferredoxin can deliver electrons back to plastoquinone via another membrane-bound complex named ferredoxin-plastoquinone oxidoreductase (Fig. 10.7A). In this process of **cyclic electron flow**, the energy is used to generate more PMF and subsequently ATP (via cyclic photophosphorylation). Thus, PSI serves two functions: it produces NADPH + H⁺ and it contributes to the production of ATP through generation of the PMF. Again, both ATP and NADPH + H⁺ are needed to drive autotrophy via CO_2 fixation in the Calvin cycle. The option to produce more ATP is important, as approximately 1.5 ATPs are needed for every NADPH + H⁺ required to drive the reactions of the Calvin cycle (Chapter 9). Thus, the cell has the capacity to regulate whether PSI undergoes cyclic or linear electron flow.

SUMMARY

During oxygenic photophosphorylation in *Cyanobacteria*, energy from sunlight is captured by the reaction centers in the two photosystems to excite electrons to a more electronegative state. The electrons are donated by water, resulting in the release of free oxygen. The electrons move through an ETC to drive the creation of a PMF allowing the production of ATP (via ATP synthase). In addition, the electrons are used to produce reducing power (as NADPH + H⁺), which together with ATP allows the cell to carry out autotrophic growth via the Calvin cycle (Chapter 9). The utilization of both quinone-type (PSII) and FeS-type (PSI) photosystems, which are optimized for the production of PMF and reducing power, respectively, allows cyanobacteria (and plants) to grow successfully as photoautotrophs.

Anaerobic Anoxygenic Phototrophy in the Phototrophic Purple Sulfur and Purple Nonsulfur Bacteria

The process of oxygenic phototrophy described above is a more familiar concept to many than **anoxygenic phototrophy**, which does not yield oxygen as a byproduct. Anaerobic microbes called purple sulfur and purple nonsulfur bacteria are capable of anoxygenic photophosphorylation. These types of bacteria are members of the *Alpha-*, *Beta-*, and *Gammaproteobacteria*, and, when grown to high cell density, show a wide range in color including purple, orange, red, green, and brown, based on differences in their accessory carotenoid pigments (Fig. 10.3A, a). They are Gram-negative, anaerobic bacteria that have been extensively studied for their amazing metabolic diversity. While they are capable of photoautotrophy under anoxic conditions in the light, many can also grow as photoheterotrophs, chemoheterotrophs, or chemolithoautotrophs depending on the environmental conditions. In general, purple sulfur bacteria typically grow as photoautotrophs, while purple nonsulfur bacteria grow optimally as photoheterotrophs using organic acids like acetate, succinate, or benzoate as carbon sources. However, purple non-sulfur bacteria can also grow slowly as photoautotrophs, fixing CO_2 via the Calvin cycle, or as chemoheterotrophs in the dark.

Chromatium species are examples of purple sulfur bacteria that can live under high H_2S concentrations and are known to form purple "blooms" under favorable conditions, where the water column is visibly purple due to their growth. *Rhodobacter sphaeroides* is a purple nonsulfur bacterium that was initially classified as "nonsulfur" because it was believed to be unable to use reduced sulfur compounds as electron donors during phototrophic growth. However, although they are sensitive to high H_2S concentrations, most purple nonsulfur bacteria are capable of using H_2S as an electron donor during anoxygenic phototrophy. Unlike phototrophy in the *Cyanobacteria* (and plants), phototrophy in all of the purple bacteria is an anoxygenic process in which no oxygen is produced because water cannot be used as the electron donor.

P870 REACTIONS

Similar to photophosphorylation in *Cyanobacteria*, photophosphorylation in the purple bacteria involves a series of electron transfer events that result in the production of ATP and NADPH + H⁺. However, only a single quinone-type photosystem containing a P870 reaction center (similar to PSII in *Cyanobacteria*) is present in the purple bacteria. In the reaction center (Fig. 10.8A), light energy is harvested at 870 nm. Absorption at this wavelength does not provide a sufficient electrode/reduction potential to extract electrons from water, which is why these phototrophs use alternative electron sources such as H_2 or H_2S. In addition to H_2 and various reduced sulfur compounds (e.g., H_2S, elemental sulfur [S^0], and thiosulfate [$S_2O_3^{2-}$]), some purple bacteria can use ferrous iron [Fe^{2+}], nitrite [NO_2^-], or even arsenite [AsO_3^{3-}] as inorganic electron donors, or small organic electron donors such as succinate or benzoate. Within the reaction center, electrons are passed in an energetically favorable direction from the special pair of Bchl *a* molecules to bacteriopheophytin and then through ubiquinone (UQ) molecules embedded in the P870 reaction center. Similar to the reactions between PSII and plastoquinone in *Cyanobacteria*, electrons must be individually transferred to the UQ carrier in a 2X-gated manner before it can gain protons from the cytoplasmic side of the membrane and leave the P870 reaction center. The fully reduced ubiquinone, UQH$_2$, delivers the electrons to the electron carriers in the cytochrome bc_1 (Cyt bc_1) complex, which is where electron transfer continues and protons are released to the periplasm, contributing to the PMF.

CYTOCHROME bc_1 REACTIONS AND THE Q CYCLE

As the protons from UQ are released into the periplasm, electron carriers embedded in the Cyt bc_1 complex must accept the electrons. Similar to the reactions at Cyt b_6f in the *Cyanobacteria*, some electrons enter a Q cycle (Fig. 6.2B), which is used to generate additional PMF in the periplasm, while other electrons are transferred to the mobile electron carrier Cyt c_2.

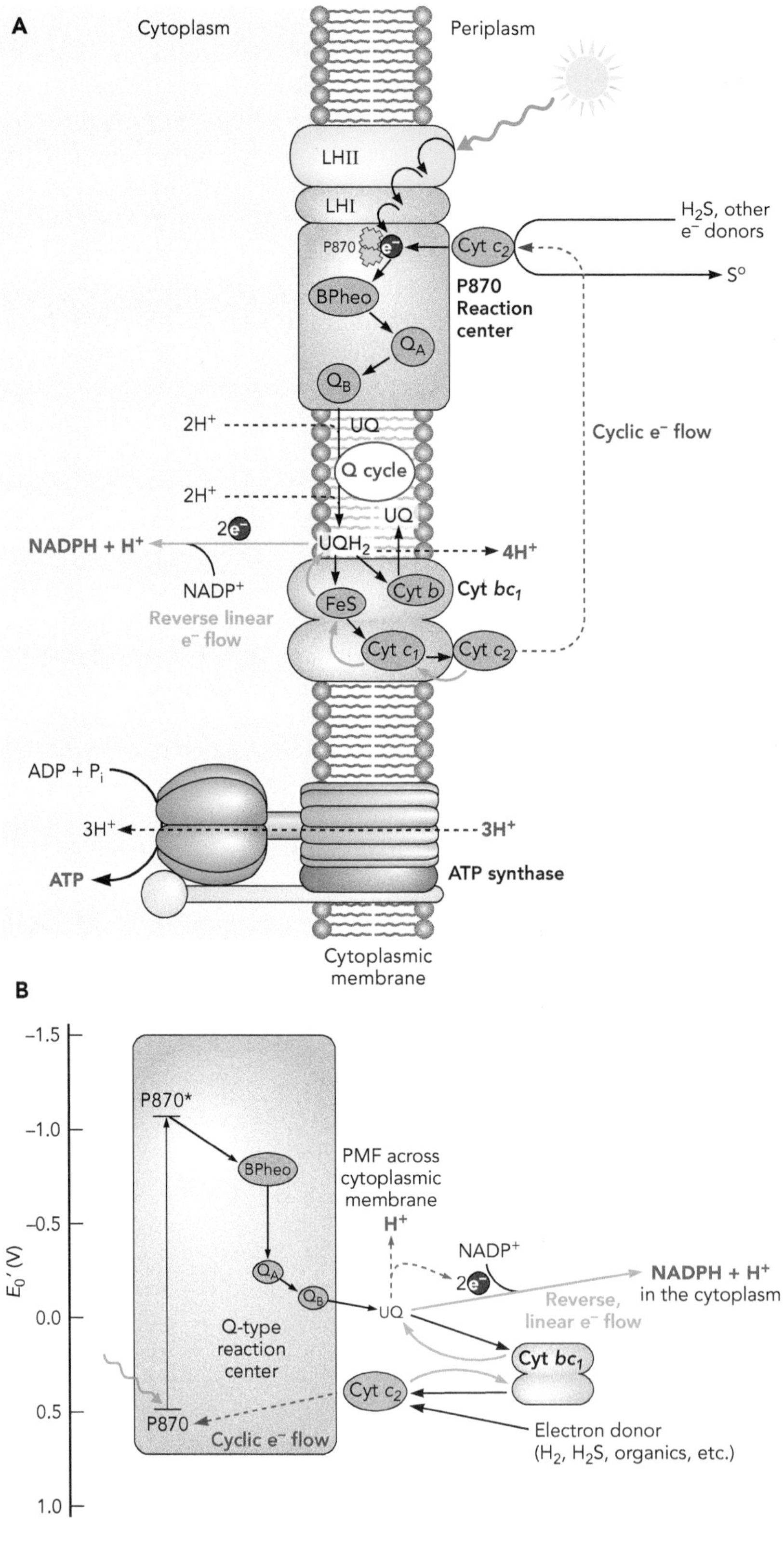

FIGURE 10.8 **Phototrophic electron transport chain (ETC) and electron flow in the phototrophic proteobacteria.** The quinone-type reaction center in the purple anoxygenic phototrophs (members of the proteobacteria) is similar to PSII in the cyanobacteria (Fig. 10.7) and is optimized to generate proton motive force (PMF) via cyclic electron flow. However, reverse linear electron flow must also occur to allow generation of NADPH + H⁺. **(A)** The ETC in the phototrophic proteobacteria. The photosystem (P870; quinone-type) reaction center (RC) is located in the cytoplasmic membrane. Photophosphorylation begins at the P870 RC when photons of light energy are absorbed via the accessory pigments of the light-harvesting complexes LH II and LH I. The energy is transferred to the special pair of BChl a molecules (two green tetrapyrroles) in the RC, exciting an electron (purple circle). Electrons are typically donated from reduced sulfur (H₂S) or small organic compounds. The electron is transferred (black arrows) in an energetically favorable direction to bacteriopheophytin (BPheo) and a series of quinones (Q$_A$ and Q$_B$). Mobile ubiquinone (UQ) electron carriers accept two electrons from Q$_B$ and two protons (H⁺) from the cytoplasm after leaving the RC to generate reduced UQH$_2$. Fully reduced UQH$_2$ delivers two electrons to the cytochrome bc_1 (Cyt bc_1) complex. One electron is transferred to an FeS cofactor, then to Cyt c_1, and on to the mobile electron carrier Cyt c_2 for delivery back to the P870 RC in cyclic electron flow. Additional H⁺ are pumped via a Q cycle (Chapter 6). Reduced UQH$_2$ is used to reduce NADP⁺ to generate NADPH + H⁺ in the cytoplasm for carbon fixation (Chapter 9). Because UQH$_2$ is not sufficiently electronegative to directly reduce NADP⁺, the input of energy in the form of PMF is used to transfer electrons in the energetically unfavorable direction. This process is called reverse linear electron flow (green arrows). **(B)** Electron flow diagram in P870 of the phototrophic proteobacteria relative to electrode/reduction potential (E_0'). Solid black arrows indicate linear electron flow. Dashed purple arrows indicate cyclic electron flow, resulting in increased PMF generation in the periplasm. Solid green arrows indicate reverse linear electron flow, which requires energy input from PMF to drive the production of NADPH + H⁺ in the cytoplasm.

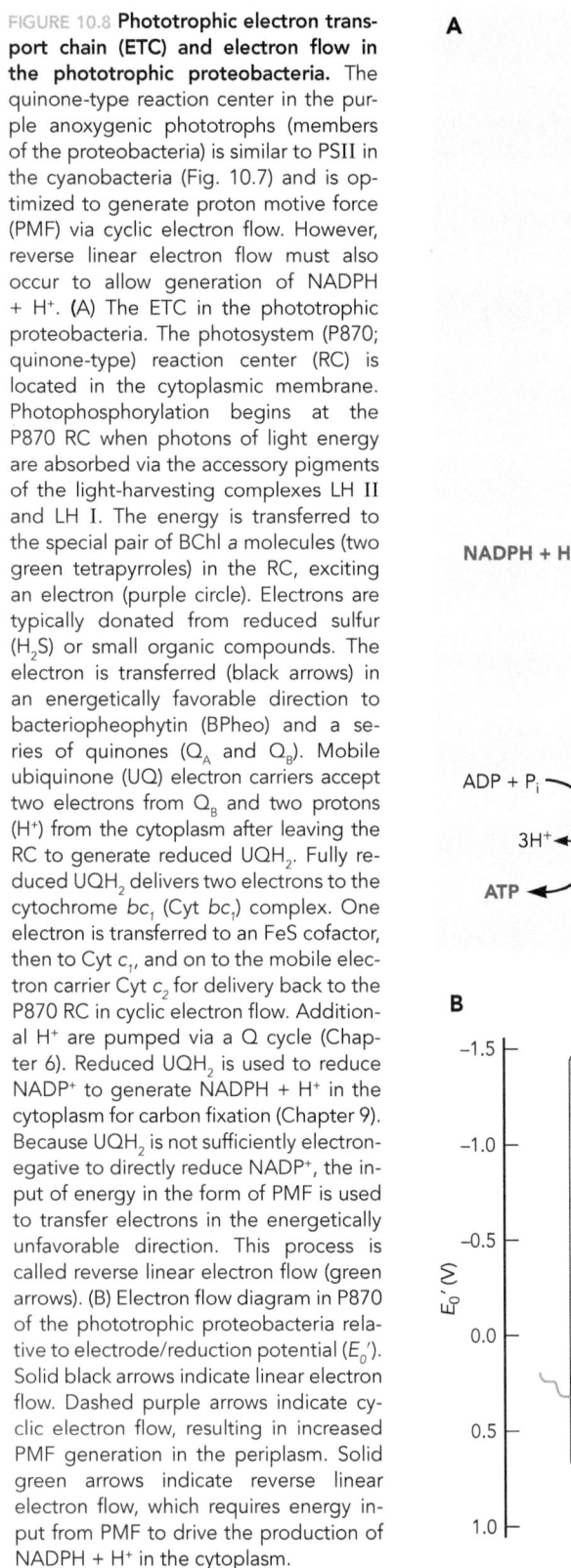

Since there is only one photosystem in the purple bacteria, there is only one possible path for the electrons carried by Cyt c_2. In **cyclic electron flow**, the mobile Cyt c_2 electron carrier ultimately returns now low-energy electrons back to the special pair of bacteriochlorophylls in the P870 reaction center. Note that during cyclic electron flow, and characteristic of a quinone-type photosystem, a PMF is generated that can be used to synthesize ATP for the cell.

The quinone-type photosystems are optimized for PMF production; however, purple bacteria must also be able to produce the reducing power necessary for CO_2 fixation. This can only occur via a **reverse electron flow** process in the purple bacteria, where the terminal electron acceptor $NADP^+$ is reduced to $NADPH + H^+$. In this process, Cyt c_2 accepts electrons from H_2S or other electron donors. These electrons are then moved, in an energetically unfavorable direction, through Cyt bc_1 to UQ to reduce $NADP^+$. This is an energetically unfavorable reaction since the electrons in the quinones are more electropositive than $NADPH + H^+$; therefore, the process is coupled to the use of a portion of the PMF to drive the electron flow in the reverse (unfavorable) direction (Fig. 10.8B). Electrons are continually replenished into the photosystem by electron donors reducing Cyt c_2. When growing as autotrophs, all purple phototrophs use the Calvin cycle to fix CO_2. Thus, they must be able to produce sufficient ATP and $NADPH + H^+$ for the process (Chapter 9).

SUMMARY

During anoxygenic photophosphorylation in purple bacteria, various inorganic (other than water) or organic compounds serve as electron donors, and oxygen is not generated as a byproduct. The energy from sunlight is conserved in the form of the PMF via a single quinone-type photosystem during cyclic electron flow. Electrons from reduced electron donors are also used to produce NADPH + H$^+$ during reverse electron flow, which costs the cell a portion of the PMF. The ATP and NADPH + H$^+$ produced provide the energy and electrons needed to drive autotrophic growth via the Calvin cycle (Chapter 9). The utilization of a single quinone-type (P870) photosystem, which is optimized to produce the PMF, suggests why the purple bacteria preferentially make their living as photoheterotrophs rather than photoautotrophs, as they are not optimized to generate reducing power using light energy.

Anaerobic Anoxygenic Phototrophy in the *Chlorobi* and *Chloroflexi* (Green Sulfur and Green Nonsulfur Bacteria, Respectively)

The green sulfur (phylum *Chlorobi*, Fig. 10.3A, b, tubes 4 and 5) and green nonsulfur (phylum *Chloroflexi*, Fig. 10.3A, b, tubes 6 and 7) bacteria are members of distinct bacterial lineages (Fig. 10.1C). Members of the *Chlorobi* and *Chloroflexi* each have a single FeS-type photosystem that resembles PSI of the *Cyanobacteria* (Fig. 10.9). Like the phototrophic proteobacteria, no oxygen is produced during photophosphorylation in the green phototrophic bacteria. The *Chlorobi* are obligate anoxygenic photolithoautotrophs that can live under

FIGURE 10.9 **Phototrophic electron transport chain (ETC) and electron flow in the *Chlorobi*.** The FeS-type reaction center in the green anoxygenic phototrophs is similar to PSI in the cyanobacteria (Fig. 10.7) and is optimized to generate reducing power (NADPH + H⁺) for use in CO_2 fixation. However, cyclic electron flow must also occur to allow generation of proton motive force (PMF). Many questions remain about the process of electron flow in these organisms, as indicated by the many question marks in the figure, and the process is likely to involve additional electron carriers that have not yet been identified. (A) Current model of the ETC in the *Chlorobi*. The FeS-type P840 reaction center (RC) is located in the cytoplasmic membrane. Photophosphorylation begins at the P840 RC when photons of light energy are absorbed via the accessory pigments of the chlorosome and transferred to the special pair of bacteriochlorophyll *a* (BChl *a*) molecules (two green tetrapyrroles) in the RC, exciting the electron (purple circle). Electrons are typically donated from a reduced sulfur compound (H_2S, thiosulfate, etc.), although the immediate electron acceptor has not been unequivocally established. The electron is transferred (black arrows) in an energetically favorable direction to Chl *a*, and through an FeS cofactor within the RC. In linear electron flow, electrons are delivered from the FeS cofactor to ferredoxin (Fd_{red}), and then to ferredoxin-NADP reductase (FR_N), which catalyzes the production of NADPH + H⁺ in the cytoplasm, generating reducing power for carbon fixation (Chapter 9). To generate PMF via cyclic electron flow, mobile menaquinone (MQ) electron carriers accept two electrons from the FeS cofactor (possibly via an additional unidentified electron carrier) and two protons (H⁺) from the cytoplasm to generate reduced MQH_2. Fully reduced MQH_2 delivers two electrons to the cytochrome bc_1 (Cyt bc_1) complex. One electron is transferred to an FeS cofactor, then to Cyt *c*, and on to the mobile electron carrier Cyt c_{555} for possible delivery back to the P840 RC in cyclic electron flow (dashed purple arrow). Additional H⁺ are pumped via the Q cycle as electrons are passed from MQH_2 through a series of electron carriers in the Cyt bc_1 complex (Chapter 6). It is also possible that electrons from Fd_{red} may be transferred back to the MQ pool as part of cyclic electron transport, but the details are not yet known. (B) Electron flow diagram in P840 of the *Chlorobi* relative to electrode/reduction potential (E_0'). Solid black arrows indicate linear electron flow, resulting in the production of NADPH + H⁺ in the cytoplasm. Dashed purple arrow indicates cyclic electron flow, resulting in increased PMF generation across the membrane. Dashed black arrows with question marks indicate incomplete understanding of electron flow to MQ and the possibility for cyclic electron flow via the MQ pool that would result in increased PMF generation similar to that in *Cyanobacteria*.

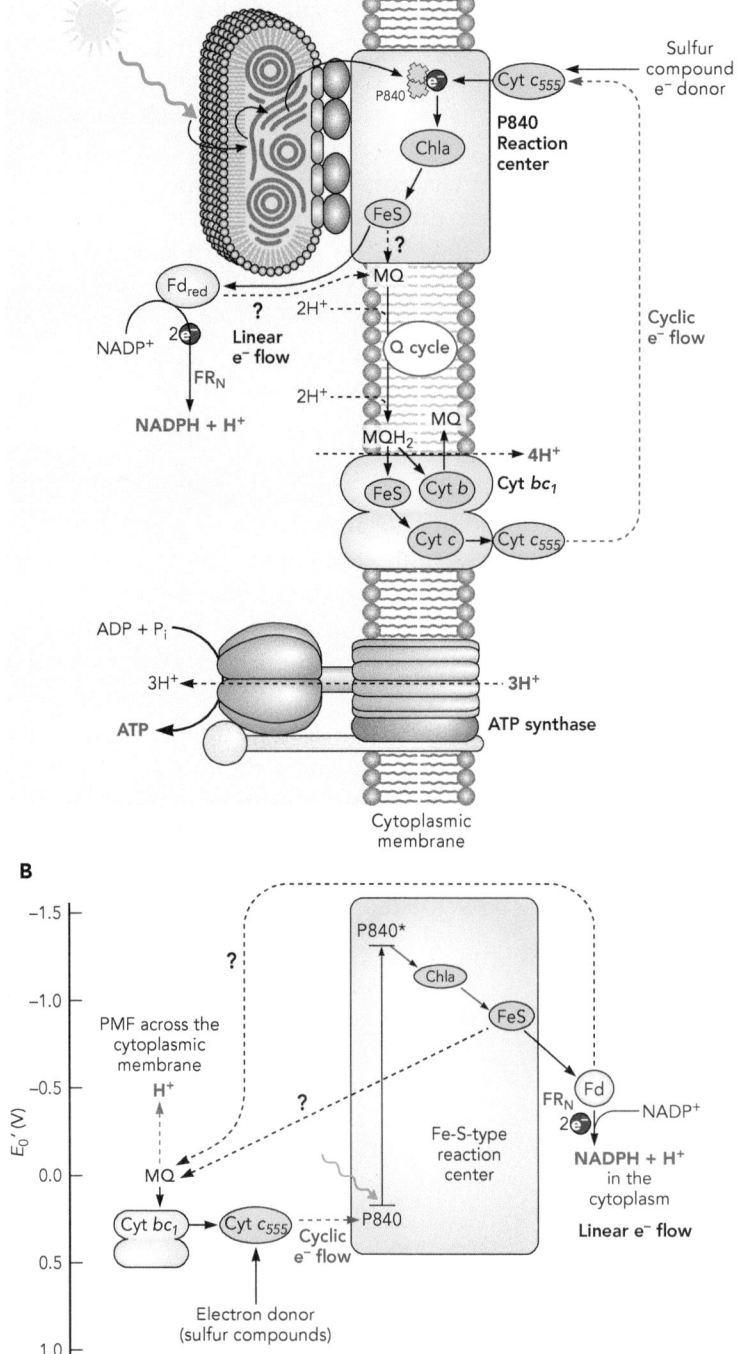

high H_2S concentrations, exceeding those tolerated by the purple sulfur bacteria. They are adapted to very low-light conditions in sediments and deep anoxic lakes. Rather than using the Calvin cycle, they use the rTCA cycle to fix CO_2 (Chapter 9). In contrast, most members of the *Chloroflexi* are facultative aerobes that can grow as anaerobic photoheterotrophs in light or as aerobic or microaerophilic chemoheterotrophs in the dark. Some can grow photoautotrophically under anoxic conditions using the 3HP bi-cycle (Chapter 9). Many members of the *Chloroflexi* are filamentous thermophiles that thrive at high temperatures.

P840 REACTIONS

More is known about the photosystem in the *Chlorobi*, so we will focus the remaining discussion on this group. Only a single FeS-type photosystem (P840, similar to PSI in *Cyanobacteria*) is present in the green bacteria (Fig. 10.9A and B). Photophosphorylation begins with the harvesting of light energy, which is transferred to the special pair of Bchl (P840) in the reaction center that is associated with the chlorosome. Similar to the phototrophic proteobacteria, and as their common name suggests, the green sulfur *Chlorobi* typically obtain electrons from H_2S, although other reduced sulfur compounds or Fe^{2+} are used as electron donors by some species. Within the reaction center, electrons are passed from the special pair in an energetically favorable direction to Chl *a* and then to an FeS center. As would be expected with an FeS-type photosystem, linear electron flow in the *Chlorobi* is similar to that in the *Cyanobacteria*. Electrons in the FeS protein are sufficiently electronegative to be transferred directly to a ferredoxin molecule, which delivers electrons to a ferredoxin-NADP reductase to produce NADPH + H^+. Again, this reducing power is used during autotrophy to build cell biomass (Chapter 9).

The next photosynthetic steps in the green bacteria are poorly understood. It is clear that electrons from P840 can be transferred to a menaquinone (MQ) carrier in a 2X-gated manner; however, it is unlikely that electrons are transferred directly to the MQ pool due to the very large difference in electrode/reduction potential (Fig. 10.9B). Regardless of the mechanism, MQH_2 delivers electrons to the electron carriers in the Cyt bc_1 complex, which is where electron transfer continues and protons are released to the periplasm, contributing to the PMF.

CYTOCHROME *bc*₁ REACTIONS AND THE Q CYCLE

As the protons from menaquinone are released into the periplasm, electron carriers embedded in the Cyt bc_1 complex must accept two electrons. Similar to the reactions at Cyt b_6f in the *Cyanobacteria* and purple bacteria, some electrons enter a Q cycle (Fig. 6.2B), which is used to generate additional PMF, while other electrons are transferred to the mobile electron carrier Cyt c_{555}.

Since there is only one photosystem (P840) in the green sulfur bacteria, there is only one possible path for the electron carried by Cyt c_{555}. In cyclic electron flow, the mobile Cyt c_{555} electron carrier ultimately returns the now low-energy electrons back to the special pair of bacteriochlorophylls in the reaction center of P840. In the *Cyanobacteria*, ferredoxin can deliver electrons back to the plastoquinone pool (Fig. 10.7A), whereby cyclic electron flow is used to generate more PMF and subsequently ATP. It is unclear if this type of cyclic electron flow is present in the green bacteria.

SUMMARY

During anoxygenic photophosphorylation in the *Chlorobi* and *Chloroflexi*, energy from sunlight is captured via a single FeS-type photosystem, exciting electrons that are typically donated by H_2S. In linear electron flow, the electrons are used to produce reducing power (as NADPH + H$^+$), which together with ATP allows the cell to carry out autotrophic growth via the rTCA cycle (*Chlorobi*) or the 3HP bi-cycle (*Chloroflexi*) (Chapter 9). The process for cyclic electron flow in the green bacteria is poorly understood. The utilization of a single quinone-type (P840) photosystem, which is optimized for the production of reducing power, suggests why the green bacteria often make their living as photoautotrophs.

Anaerobic Anoxygenic Photoheterotrophy in the *Firmicutes*

The heliobacteria (phylum *Firmicutes*; Fig. 10.1C), strict anaerobes that were first isolated from anaerobic alkaline rice paddies, are the only known Gram-positive bacteria that can utilize photophosphorylation to obtain energy for the cell. The heliobacteria are beneficial for plant growth because they also have the ability to fix nitrogen (Chapter 12). The single FeS-type photosystem in heliobacteria is similar to that of the green phototrophic bacteria described above (Fig. 10.9A). The heliobacteria possess a unique form of chlorophyll called Bchl *g* that harvests light at a wavelength of 796 nm (Fig. 10.3A, b, tube 2). This group of phototrophs is particularly interesting because they are obligate heterotrophs that use small organic molecules as carbon sources; they do not encode any of the known CO_2 fixation pathways (Chapter 9). When light is not available for photoheterotrophy, they grow as obligate fermenters like their close relatives the clostridia. Another characteristic shared with the clostridia is the ability to form heat-resistant endospores under nutrient-limiting conditions (Chapter 17).

Aerobic Anoxygenic Phototrophy

It is important to note that the ability to carry out chlorophyll-based anoxygenic phototrophy does not necessarily indicate that an organism is an anaerobe. Aerobic anoxygenic phototrophs are widespread across the *Alpha-*, *Beta-*, and *Gammaproteobacteria*, and have been found in diverse environments including oceans, lakes, hot springs, and polar regions. Members of the *Roseobacter* clade of the *Alphaproteobacteria* with this lifestyle are particularly abundant in the open ocean. Most proteobacterial aerobic anoxygenic phototrophs, like *Roseobacter*, are strictly aerobic chemoheterotrophs that harvest light as a supplementary form of energy when nutrients or oxygen becomes limiting. Unlike the anaerobic anoxygenic phototrophs, aerobic anoxygenic phototrophs produce Bchl *a* in the presence of oxygen. They have a quinone-type reaction center and produce a suite of carotenoids that give them their characteristic pink or orange colors (Fig. 10.3A, c).

Aerobic anoxygenic phototrophs have recently been identified in other bacterial lineages. For example, a strain of *Gemmatimonas phototrophica*, which is a member of the phylum *Gemmatimonadetes*, was isolated from a freshwater lake in the Gobi Desert (Fig. 10.3A, d). It is an obligate chemoheterotroph that uses light as a supplemental energy source under adverse environmental conditions; in fact, when a suitable carbon source is available, phototrophy is repressed. If energy is limiting, Bchl *a* is produced and a quinone-type photosystem is used to pump protons to augment the PMF. Analysis of its genome sequence suggests that a large gene cluster encoding all of the components necessary for phototrophy was acquired from a phototrophic proteobacterium by horizontal gene transfer. Analyses of metagenomic databases indicate that related phototrophic members of the *Gemmatimonadetes* are widespread in terrestrial and freshwater environments.

Finally, recent studies identified *Chloracidobacterium thermophilum*, which is the first phototrophic member of the phylum *Acidobacteria*. *C. thermophilum* is capable of aerobic anoxygenic photoheterotrophy. This organism, which was first identified in a metagenomic study of Yellowstone hot springs, is adapted to low-light conditions, utilizing chlorosomes for light harvesting. As its name implies, *C. thermophilum* is moderately thermophilic, growing at temperatures from 40 to 68°C. Although it requires oxygen for respiration, *C. thermophilum* is sensitive to atmospheric concentrations of oxygen and is therefore microaerophilic. It produces Bchl *a* and Bchl *c* and has an FeS-type photosystem similar to that of the *Chlorobi*. Obtaining an isolate in pure culture was challenging; in addition to being microaerophilic, *C. thermophilum* is unable to synthesize vitamin B_{12} or branched-chain amino acids (leucine, isoleucine, and valine), and is incapable of using inorganic nitrogen or sulfur sources. In fact, it appears to utilize amino acids as its primary sources of carbon, nitrogen, and sulfur.

Retinal-Based Phototrophy

An alternative type of phototrophy involves the use of retinal-binding proteins called **rhodopsins** rather than chlorophyll to convert light energy into chemical energy by generating ion gradients across the cytoplasmic membrane. Microbial ion-pumping rhodopsins were originally identified in the 1970s in the extremely halophilic microbe *Halobacterium salinarum*. At that time, the domain *Archaea* had not yet been described and, as the name implies, *Halobacterium* was considered to be a member of the domain *Bacteria*. As a result, the proton-pumping rhodopsin present in the organism was termed **bacteriorhodopsin**, while the rhodopsin that facilitates the inward movement of chloride ions was called **halorhodopsin**. Since that time, the true phylogeny of *Halobacterium* has been revealed (it is a member of the *Euryarchaeota*), and ion-pumping rhodopsins have been identified in all three domains of life. In fact, metagenomic studies have revealed the presence of "proteorhodopsins" in marine proteobacteria, and rhodopsin genes have been identified in numerous bacterial lineages, including *Actinobacteria*, *Bacteroidetes*, *Cyanobacteria*, *Chloroflexi*, *Deinococcus-Thermus*,

Firmicutes, and *Planctomyces* (Fig. 10.1D), as well as in members of the *Eukarya*, including fungi, dinoflagellates, and diatoms.

Although quite divergent in sequence and structure, all known opsins, including the opsins used for vision in vertebrates, appear to have a common phylogenetic origin. Rhodopsins are membrane-bound proteins with seven transmembrane helices (Fig. 10.10A) and the ability to bind the small molecule

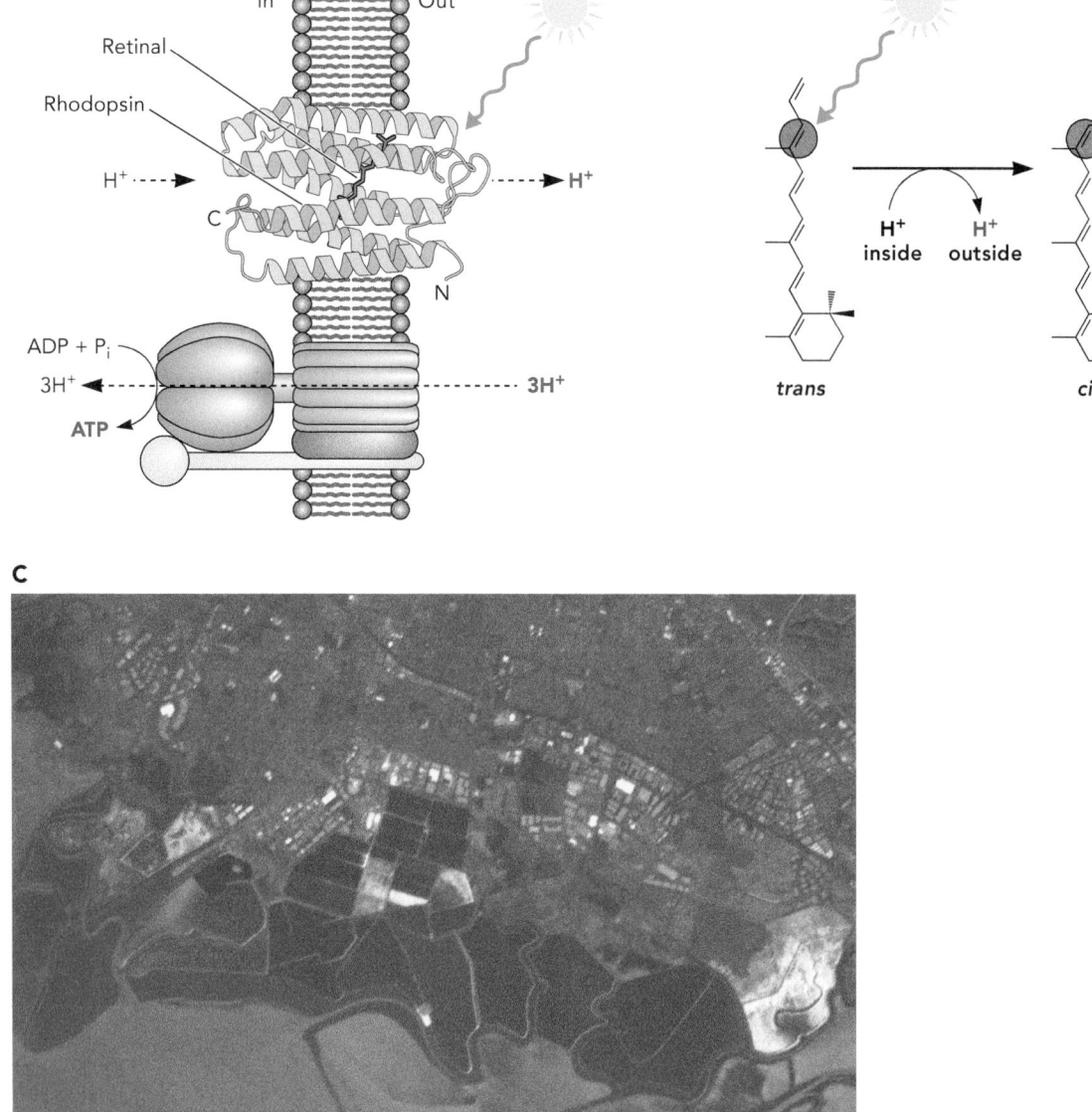

FIGURE 10.10 **Light-driven proton pump.** (A) Bacteriorhodopsin molecule with seven transmembrane helices spanning the cytoplasmic membrane is shown with retinal (orange) bound. When retinal absorbs light energy of the appropriate wavelength (~570 nm), a *trans-cis* isomerization occurs and a proton is released to the outside of the membrane. The resulting PMF can be used by ATP synthase to generate ATP for the cell. (B) Isomerization of retinal from its normal *trans* configuration to the excited *cis* form (see double bond highlighted by orange circle). The isomerization is coupled to the transfer of a proton to the outside of the cell. (C) Aerial photograph of evaporation ponds in San Francisco Bay. As the salt concentration rises, halophilic *Archaea* proliferate, turning the ponds red. Panel C courtesy of the Image and Science and Analysis Laboratory, NASA Johnson Space Center.

retinal, which can undergo *trans-cis* isomerization (Fig. 10.10B) when it absorbs a photon of light. The structural change in retinal allows for ion passage across the membrane, including outward movement of protons by bacteriorhodopsins and proteorhodopsins, outward movement of sodium ions by sodium-pumping rhodopsins, and inward movement of chloride ions all contribute to the PMF. The generation of a PMF allows the inward movement of protons through the ATP synthase (Chapter 5) to produce ATP for cellular use (Fig. 10.10A).

There are currently no known retinal-based phototrophs that are capable of coupling phototrophy to CO_2 fixation. In fact, in most cases retinal-based phototrophy appears to be used as a supplementary system of energy conservation by heterotrophic microbes to allow survival under adverse environmental conditions (e.g., limiting oxygen, carbon source). Retinal-based phototrophy is most common in aquatic microorganisms, from both freshwater and marine environments. Such environments are often oligotrophic (low in nutrients), and the ability to harvest light as a supplementary energy source can provide a survival advantage. For example, a proteorhodopsin-containing marine *Vibrio* strain was shown to survive longer under starvation conditions when placed in the light than in the dark. In addition, increases in ATP concentrations have been documented in some proteorhodopsin-containing marine bacteria when they are exposed to appropriate wavelengths of light. Proteorhodopsins that have been experimentally examined absorb light in the blue and green range, which together with carotenoids and other pigments gives rise to the characteristic purple color of high-density cultures (Fig. 10.10C).

Rhodopsin-based photosystems are quite simple relative to their chlorophyll-based counterparts. Unlike chlorophyll-based photosystems, which are composed of dozens of proteins and several accessory pigments, retinal is bound to a single membrane-bound rhodopsin protein. Chlorophyll synthesis is also complex, involving multiple enzymatic steps, whereas retinal is produced by a single reaction from the common carotenoid β-carotene. Genes coding for the production of both retinal and rhodopsin are clustered in most organisms, which might have facilitated the apparent horizontal transfer of proteorhodopsins between unrelated lineages of microorganisms.

Learning Outcomes: After completing the material in this chapter, students should be able to . . .

1. differentiate the structure and function of light-harvesting and membrane assemblies among the different groups of phototrophs
2. explain how light energy is converted to ATP and NADPH + H$^+$ in oxygenic photophosphorylation
3. explain how light energy is converted to ATP and NADPH + H$^+$ in anoxygenic photophosphorylation
4. compare and contrast electron and proton flow in anoxygenic and oxygenic photophosphorylation
5. describe how cells use retinal-based photophosphorylation to supplement ATP synthesis

Check Your Understanding

1. Define this terminology: oxygenic phototrophy, photosystems (PS), chlorophyll, carotenoids, phycobilins, light-harvesting complex, reaction center, chlorosomes, thylakoids, phycobilisomes, anoxygenic phototrophy, reverse electron flow, rhodopsin, retinal

2. For each statement, determine if it applies to oxygenic phototrophy only (OxP), anoxygenic phototrophy only (AnP), both oxygenic and anoxygenic phototrophy (B), or neither (N). (LO4)
   Cells that do this . . .

   _____ are the evolutionary precursor for the plant chloroplast

   _____ capture light energy using bacteriorhodopsin

   _____ generate a proton motive force

   _____ have two distinct photosystems

   _____ have light-harvesting molecules that are complexed with various membrane-bound proteins

   _____ use characteristic bacteriochlorophyll molecules

   _____ utilize chlorophyll a in the light-harvesting system

3. Describe at least two lines of evidence that support the hypothesis that a cyanobacterium was the evolutionary precursor to the plant chloroplast. You may want to consider information presented in both Chapters 1 and 10. (LO2)

4. In general, describe how light-harvesting complexes and reaction centers work within a photosystem to harvest light energy. (LO1)

5. What is one advantage for the elaborate membrane structures found in many phototrophs? (LO1)

6. What is the role of light energy in chlorophyll-based phototrophy? (LO1)

7. The scheme for oxygenic photophosphorylation (as carried out by *Cyanobacteria*) is shown in Fig. 10.7B. (LO2)
   a) Describe the path of an electron as it moves through photosystem II (PSII) to PSI. How do these cells use this electron movement to make ATP?
   b) In oxygenic photosynthesis, light energy excites an electron from P680 to feed into the electron transport chain. This electron is replaced by extracting an electron from H_2O.
      i. What is the electrode/reduction potential (E_0') of P680 relative to the O_2/H_2O couple? Why does this matter?
      ii. What byproduct is created when H_2O donates an electron to P680?
   c) PSI serves two functions, depending on cellular needs.
      i. Describe the path of an electron in PSI when the cell needs to make reducing power for CO_2 fixation.
      ii. Describe the path of an electron in PSI when the cell needs to make ATP.

8. In anoxygenic photophosphorylation (shown for the phototrophic proteobacteria in Fig. 10.8B), light energy is converted to cellular energy (ATP).
    a) Identify the components involved in electron transport events. How do these bacteria use this electron movement to make ATP? (LO3)
    b) Why is this process known as "cyclic" photophosphorylation? What is cycling? (LO4)
    c) Explain the process of reverse electron flow in the proteobacteria. What does it allow these cells to do? (LO3)

9. In anoxygenic photophosphorylation in the *Chlorobi* (shown in Fig. 10.9B), light energy is converted to cellular energy (ATP). (LO3)
    a) Identify the components involved in cyclic electron transport events in this photosystem. How do these bacteria use electron movement to make ATP?
    b) How do *Chlorobi* generate reducing power (NADPH + H$^+$) for biosynthesis?

10. The purpose of completing the information table below is to compare and contrast how phototrophic bacteria generate NADPH + H$^+$, the source of electrons for autotrophic growth. Fill in each blank as directed.

Type of bacteria	Number of photosystems	Type of photosystem(s): quinone-type or FeS-type	Electron donor(s)	Linear or reverse electron flow for NADPH+ H$^+$ production	Autotrophic pathway used for CO$_2$ fixation
Cyanobacteria					
Proteobacteria (purple sulfur and nonsulfur bacteria)					
Chlorobi (green sulfur bacteria)					
Chloroflexi (green non-sulfur bacteria)					

11. The *Cyanobacteria* use a quinone-type photosystem in photophosphorylation. Why don't *Cyanobacteria* need to use reverse electron flow to reduce NADP$^+$ to NADPH + H$^+$? Why does reverse electron flow occur in the proteobacteria and not in the green bacteria? (LO4)

12. a) What is a "coupled site" (Chapter 6)? Where is the coupled site in the photosynthetic electron transport chain (ETC)? (LO4)
    b) What are some of the special traits of the quinones that allow for the coupling of electron movement to the generation of the proton motive force (PMF) (Chapter 6)?
    c) The purpose of the information table below is to compare and contrast the physiology of the various groups of phototrophic bacteria. Fill in each blank as directed.

Type of bacteria	Preferred metabolic lifestyle (see Fig. I.1 in Interlude)	Oxygen requirement (aerobe/anaerobe)	Type of photosynthesis (oxygenic or anoxygenic)	Location of light-harvesting complex	Type of quinone molecule	Coupling site
Cyanobacteria						
Proteobacteria (purple sulfur and nonsulfur bacteria)						
Chlorobi (green sulfur bacteria)						
Phototrophic *Firmicutes*						

13. In retinal-based phototrophy, how is light energy converted into a PMF? (LO5)

Dig Deeper

14. Compare the diagrams of membranes involved with oxidative phosphorylation (see below) and photophosphorylation (Fig. 10.7A). Answer the questions below considering these components: ATP synthase, chlorophyll, cytochrome, FeS protein, flavoprotein, quinone, reaction center. (LO2)

BACTERIAL AEROBIC ELECTRON TRANSPORT CHAIN

a) Which four components are found in both systems? What is the function of each component?

b) How is the movement of protons similar in photophosphorylation and oxidative phosphorylation?

c) In general, how is the source and flow of electrons different in photophosphorylation and oxidative phosphorylation?

15. A number of herbicides are designed to inhibit either electron or proton flow in photophosphorylation. For each chemical described below, ATP synthesis in photophosphorylation is negatively impacted. Explain why. (LO2)
   a) Atrazine binds plastoquinone B (Q_B), thus preventing the transfer of electrons from plastoquinone A (Q_A) to Q_B.
   b) Paraquat intercepts electrons from the FeS proteins, which creates unstable, reactive products that interact with fatty acids in membranes causing membrane leakage.
   c) 2,4-Dinitrophenol, a fat-soluble compound, was once marketed as a diet pill in the 1930s, but is now used as an herbicide. It binds protons on one side of a membrane and diffuses through the membrane, where it then releases the protons.

16. Compare the photosystems and electron transport systems in PSI and PSII from *Cyanobacteria* (Fig. 10.7B) with those of the phototrophic proteobacteria (Fig. 10.8B) and the *Chlorobi* (Fig. 10.9B). These bacteria all make ATP using light energy and can fix CO_2 for cell biomass using ATP and reducing equivalents (NADPH + H$^+$). (LO4)
   a) How do the *Cyanobacteria* generate reducing equivalents (NADPH + H$^+$) for the Calvin cycle?
   b) What is one advantage of having two photosystems, rather than just one?
   c) Compare the purple proteobacteria and the green sulfur bacteria, *Chlorobi*. How do they generate reducing equivalents (NADPH + H$^+$) for fixing CO_2? What is similar and what is different?
   d) Is there evidence to support the hypothesis that oxygenic photosystems evolved from the merging of photosystems and electron transport chains from *Chlorobi* (green sulfur bacteria) and anoxygenic phototrophic proteobacteria? If not, explain why not. If yes, which do you think is PSI and which is PSII?

17. Phototrophs are found in an amazing range of habitats. Describe how each of the following strategies helps the bacterium to survive in a given environment.
   a) Figure 10.3B shows light absorbance by different pigments in different bacteria. Notice that chlorophyll molecules absorb light with multiple peaks at different wavelengths. How does having a variety of different light-absorbing pigments allow some phototrophs to live in a wider range of environments? (LO1)

b) Describe how the regulation of light-harvesting complex II (LH II) and LH I in the phototrophic proteobacterium *Rhodobacter sphaeroides* permits the cells to conserve energy under different light intensities. (LO3)

c) What allows the green sulfur bacteria (*Chlorobi*) to grow in low-level light environments at depths of up to 100 meters below sea level? (LO3)

18. Some species of *Firmicutes* grow using light energy and reduced carbon sources. Which of the terms from the Interlude best describe these bacteria? Would you predict that these *Firmicutes* use cyclic or linear electron flow to generate reducing power while growing as phototrophs? Explain your thinking.

19. You are studying a new bacterium that was isolated from seawater. You do a series of growth experiments in a minimal medium with light as the only energy source, adding either CO_2 or amino acids (results in table). (LO3)

Minimal medium in light	Growth?
With added CO_2	No
With added amino acids	Yes

a) Based on these results, what are these bacteria likely using as a carbon source for biosynthesis?

b) You conduct growth experiments to further investigate the growth patterns of this isolate under constant light (blue circles), constant dark conditions (black circles), and when dark cultures (green triangles) were exposed to light (red arrow). Based on these results, what are these bacteria likely using as an energy source?

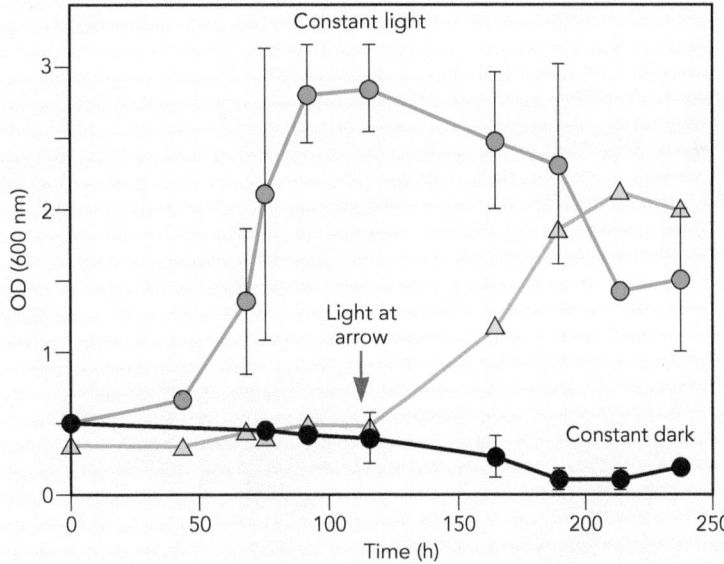

c) Given your results from 19a) and b) which of the terms from the Interlude best describe this bacterium?

d) Why would the lifestyle you described in 19c) be an advantage for an organism making a living in the open ocean, compared to an organoheterotroph?

e) Suppose that you were to grow this isolate together with a photoautotroph in the light in a minimal medium with amino acids. Which organism do you think would be more successful? Explain your reasoning.

234 Making Connections

234 The Carbon Cycle

236 The Chemoorganotrophic Degradation of Polymers

236 The Chemoorganotrophic Degradation of Aromatic Acids

241 Chemoorganotrophy in *Escherichia coli*

246 Chemolithoautotrophy

248 Chemolithoautotrophy in Methanogens

251 Methylotrophy Enables Cycling of One-Carbon (C1) Compounds

253 One-Carbon (C1) Chemolithotrophy in Acetogens

256 The Sulfur Cycle

256 Chemoheterotrophy and Chemolithoautotrophy in the Sulfur Cycle: Sulfate Reducers

259 Chemolithoautotrophy in the Sulfur Cycle: Sulfur Oxidizers

261 The Anaerobic Food Web and Syntrophy

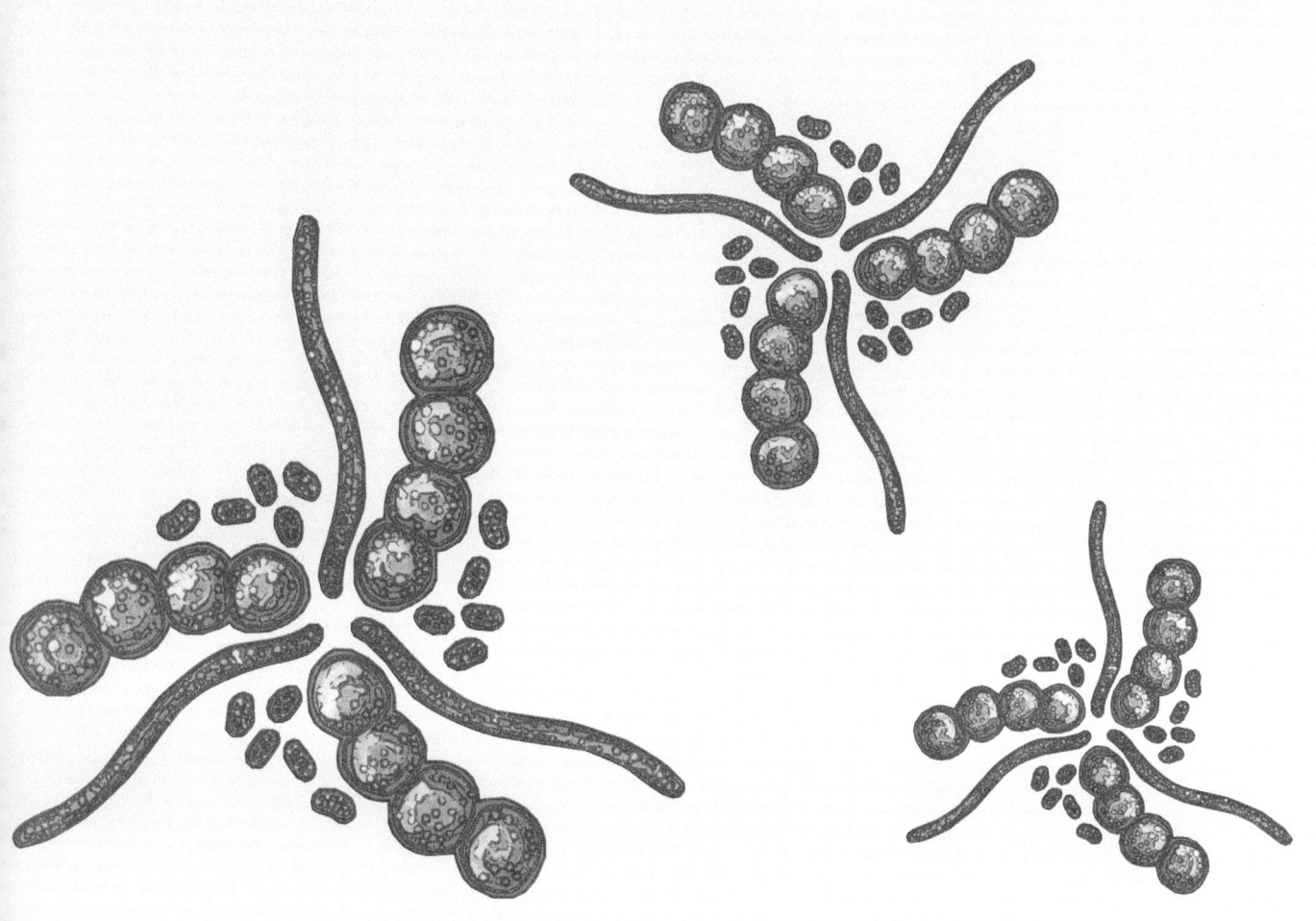

11

CHEMOTROPHY IN THE CARBON AND SULFUR CYCLES

After completing the material in this chapter, students should be able to . . .

1. identify the processes, electron donors, and electron acceptors involved in the oxidation and reduction of carbon-containing molecules during the carbon cycle

2. compare and contrast chemoorganotrophy and chemolithotrophy

3. describe the sequential degradation of aromatic compounds in oxic and anoxic environments

4. explain how a facultative organism like *Escherichia coli* regulates its choice of electron acceptors during aerobic and anaerobic growth

5. outline the major steps of the one-carbon (C1) carbon cycle (i.e., methanogenesis, methylotrophy, and acetogenesis)

6. identify the processes, electron donors, and electron acceptors involved in the oxidation and reduction of sulfur-containing molecules in the sulfur cycle

7. describe the relationships between the microbes within an anaerobic food web

Microbial Physiology: Unity and Diversity, First Edition. Ann M. Stevens, Jayna L. Ditty, Rebecca E. Parales and Susan M. Merkel.
© 2024 American Society for Microbiology.
Companion website: www.wiley.com/go/stevens/microbialphysiology

Making Connections

Over the past 3.8 billion years, microorganisms have evolved the ability to use a vast array of compounds as electron donors, sources of energy, and carbon depending on the environment in which they reside. Examples of heterotrophy and autotrophy (for acquiring carbon) and phototrophy (for energy and electron donors) have been described in Chapters 2, 9, and 10, respectively. This chapter focuses on chemotrophic microbes, which use chemicals as electron donors and a source for energy. Chemotrophy occurs via two metabolic lifestyles: chemoorganotrophs use organic compounds as both an energy source and a source of electrons to reduce electron carriers (e.g., NADH + H$^+$ and reduced flavin adenine dinucleotide [FADH$_2$]), while chemolithotrophs require inorganic molecules (e.g., H$_2$S, H$_2$, NH$_3$, and NO$_2^-$). Chemoorganotrophic and chemolithotrophic lifestyles will be introduced in the context of the carbon and sulfur cycles. Most people are familiar with the oxic portion of the carbon cycle, in which plants fix CO$_2$ and other aerobic organisms consume that fixed carbon for energy and biomass. However, the anoxic component is also critical to the carbon cycle and is dominated by the activity of microbes on a global scale. It is important to remember that microbes do not normally grow as isolated populations, as in the laboratory. Instead, they are members of complex microbial communities that depend on each other for their sources of electrons, energy, and carbon, in environments that may be oxic and/or anoxic in nature. We will use an anaerobic food web, found in environments as varied as the intestinal tract of humans and animals to freshwater and marine sediments, as an example of real-world microbial metabolic diversity and the interdependence of microbes associated with the carbon and sulfur cycles.

The Carbon Cycle

During metabolism, carbon molecules are repeatedly oxidized, reduced, and reoxidized by microorganisms that use carbon molecules as electron donors and sources of carbon and energy for biosynthesis (Fig. 11.1). Chemoorganoheterotrophs convert reduced organic carbon to CO$_2$, conserving energy and generating essential precursors in the process (Chapter 2). The autotrophs described in Chapter 9 convert CO$_2$ to reduced organic molecules via the various CO$_2$ fixation pathways for biosynthesis. Alternatively, CO$_2$ can be reduced to methane during the uniquely microbial process known as methanogenesis; the methane can then be reoxidized to CO$_2$ by methanotrophs (Fig. 11.1).

While simple sugars such as glucose can provide a source of easily metabolized carbon for biosynthesis and energy for growth, glucose is not readily available in most environments. For example, in the mammalian intestinal tract, simple sugars are rapidly consumed in the upper gastrointestinal tract (e.g., small intestine), leaving complex polysaccharides (glycans) as the primary source of carbon for organisms in the lower intestinal tract (e.g., large intestine or colon in humans). Similarly, in the bottom of a freshwater pond, plant polymers from leaves or tree branches are degraded into alcohols, organic acids, and aromatic monomers (Fig. 11.2). These molecules are then consumed by different organisms

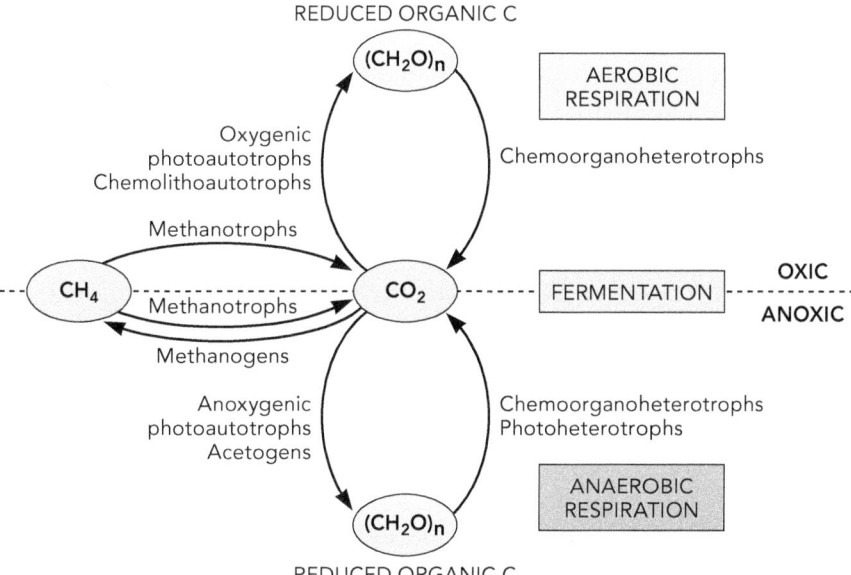

FIGURE 11.1 **Overview of the carbon cycle.** The carbon cycle occurs in both oxic and anoxic environments where oxidized and reduced forms of carbon are interconverted by different groups of microorganisms as indicated. See the text for more details.

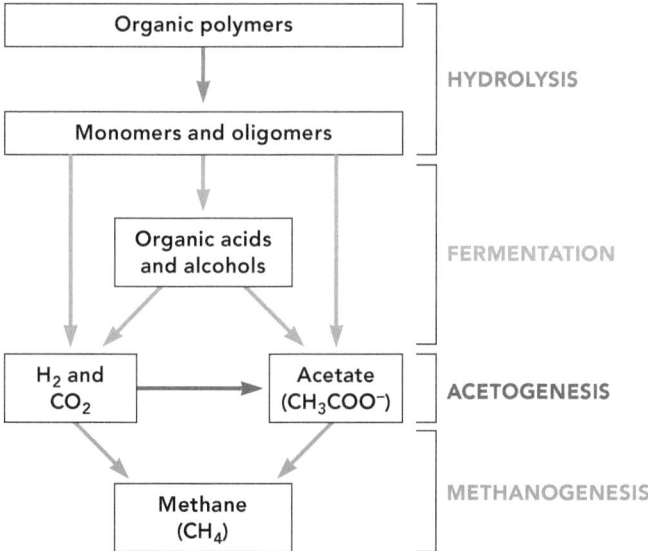

FIGURE 11.2 **Overview of the metabolic processes in an anaerobic food web.** Polymers of organic carbon are hydrolyzed into monomers and oligomers that are substrates for microbial fermentation processes. Acetogenesis and methanogenesis complete the reduction of inorganic carbon into acetate and methane, respectively.

that produce acetate, hydrogen gas, and CO_2, and those substrates are then subsequently consumed and converted by yet another group of microbes into products such as methane gas. Therefore, to fully appreciate the flow of carbon on a global scale, it is important to discuss the role of carbon-containing organic polymers. In this chapter, we will follow the stepwise degradation of these abundant organic polymers by a community of microorganisms.

The Chemoorganotrophic Degradation of Polymers

In most natural environments, much of the organic carbon exists as long polymers from plant and animal material (e.g., cellulose, lignin, starch, and proteins). These polymers must first be broken down to continue the carbon cycle (Fig. 11.3A). It takes a consortium of microorganisms to break down these carbon polymers into smaller oligomers or monomers and ultimately to CO_2. While much more abundant than glucose or other small metabolites, these polymers are too large to be taken up into cells through cellular transport systems (Chapter 13). The heterotrophs that utilize these polymers (respiring and fermentative bacteria) use **exoenzymes** (enzymes that are excreted from cells or are attached to the cell surface) to first break the polymers into smaller sugar monomers or short-chain oligomers. For example, **cellulases** degrade cellulose and **amylases** degrade starch, releasing sugar molecules that can be metabolized into molecules like organic acids, fatty acids, and alcohols. Many organisms can use other cellular polymers such as proteins and lipids as carbon sources. Proteins are converted to peptides and amino acids by **proteases**, and lipids are converted to fatty acids by **lipases**.

The Chemoorganotrophic Degradation of Aromatic Acids

Lignin makes up the woody, supportive structure of plants and is one of the most abundant carbon-containing polymers on Earth. Lignin is a complex polymer of aromatic compounds connected in a structure that resembles chicken wire (Fig. 11.3B). The presence of conjugated double bonds in the aromatic rings contributes to the strength and stability of lignin. Monomeric aromatic compounds are released from lignin by various species of fungi through the action of **ligninolytic exoenzymes** (Fig. 11.3A). In addition to lignin, there are many other sources of both natural and anthropogenic (man-made) aromatic compounds in the environment, including aromatic amino acids, plant signaling molecules (e.g., salicylate), components of petroleum and gasoline (e.g., benzene and toluene), as well as synthetic aromatic compounds that are used in industry and agriculture (e.g., pesticides and herbicides).

During the catabolism of aromatic compounds, the aromatic ring must first be destabilized to allow subsequent reactions that cleave one of the carbon-carbon bonds within the ring. Aromatic ring cleavage generates linear carbon compounds that are converted to common intermediates, which are further metabolized to form compounds that enter central metabolic pathways. This process leads to the generation of organic electron donors (e.g., NADH + H$^+$ and FADH$_2$). Pathways for the degradation of aromatic compounds differ depending on whether oxygen is available (Fig. 11.4). Under oxic conditions, oxygen itself serves as a substrate in a series of enzymatic reactions that first destabilize and subsequently cleave the aromatic ring. In the absence of oxygen, however, such reactions are not possible and the strategy to destabilize the aromatic ring involves the addition of a coenzyme A (CoA) group, followed by reactions catalyzing ring reduction (elimination of the double bonds) and ring cleavage. In the case of aerobic pathways, dihydroxylated aromatic compounds like catechol are typically formed, whereas during anaerobic aromatic compound degradation,

A

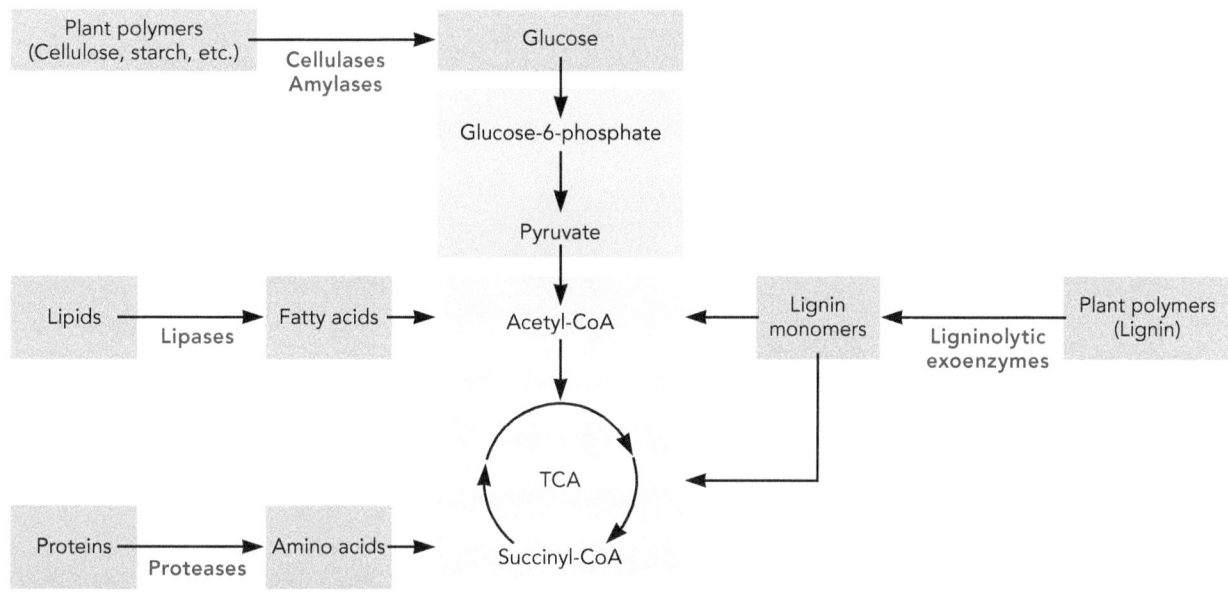

FIGURE 11.3 **Catabolism of polymers such as lignin yields essential precursors.** (A) Complex biological polymers (i.e., plant polymers, lipids, and proteins) are broken into smaller molecules by a variety of exoenzymes (highlighted in purple) prior to their catabolism through central metabolism. (B) The structure of lignin.

benzoyl-CoA is generated. In both cases, however, diverse aromatic compounds are funneled into common intermediates (Fig. 11.4).

The addition of hydroxyl groups to aromatic rings is catalyzed by oxygenase enzymes, which use oxygen as a substrate under oxic conditions (Fig. 11.5). **Monooxygenases** add one atom of molecular oxygen to the aromatic ring, whereas **dioxygenases** add two atoms of oxygen. Once there are two hydroxyl groups on adjacent carbon atoms of the ring (via the action of one dioxygenation or two

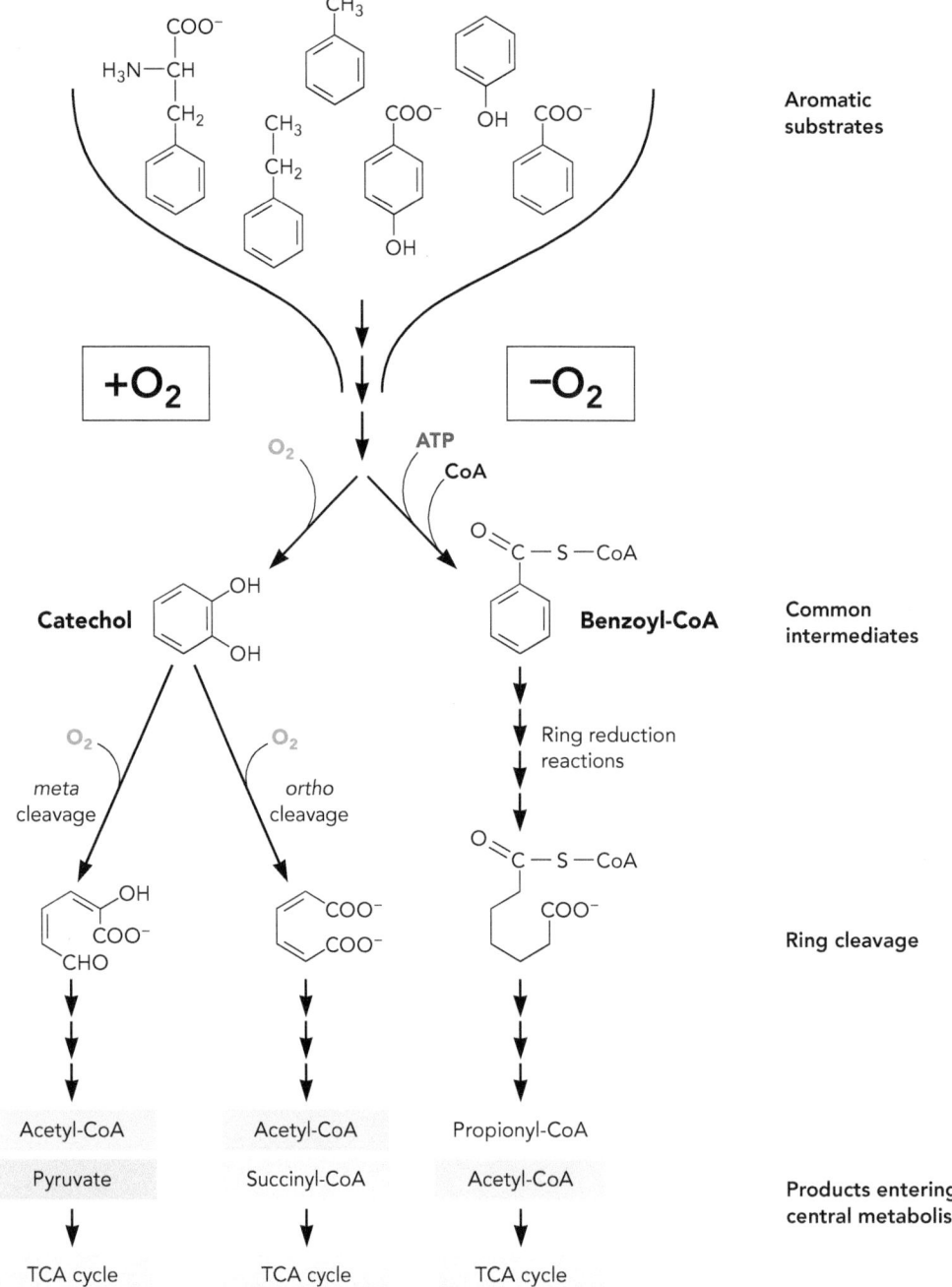

FIGURE 11.4 **Catabolism of aromatic substrates under oxic and anoxic conditions.** Under oxic conditions, aromatic compounds are converted by funneling pathways to the common intermediate catechol (or related dihydroxylated benzenes). Aerobic bacteria use oxygenase enzymes that catalyze the addition of hydroxyl groups to form catechol. Ring-cleavage dioxygenases also use oxygen as a substrate during the cleavage of the catechol aromatic ring. The position of ring cleavage can occur either adjacent to one of the hydroxylated carbons (*meta* cleavage) or between the two hydroxylated carbons (*ortho* cleavage) depending on the specificity of the dioxygenase enzyme. Additional reactions convert the linearized intermediate into compounds that enter central metabolism (acetyl-CoA, pyruvate, succinyl-CoA). In the absence of oxygen, the benzene ring is first destabilized by addition of CoA, forming the common intermediate benzoyl-CoA. Ring reduction reactions sequentially remove the double bonds in the aromatic ring, the ring is cleaved, and the linearized product is converted into acetyl-CoA, which enters central metabolism. TCA, tricarboxylic acid.

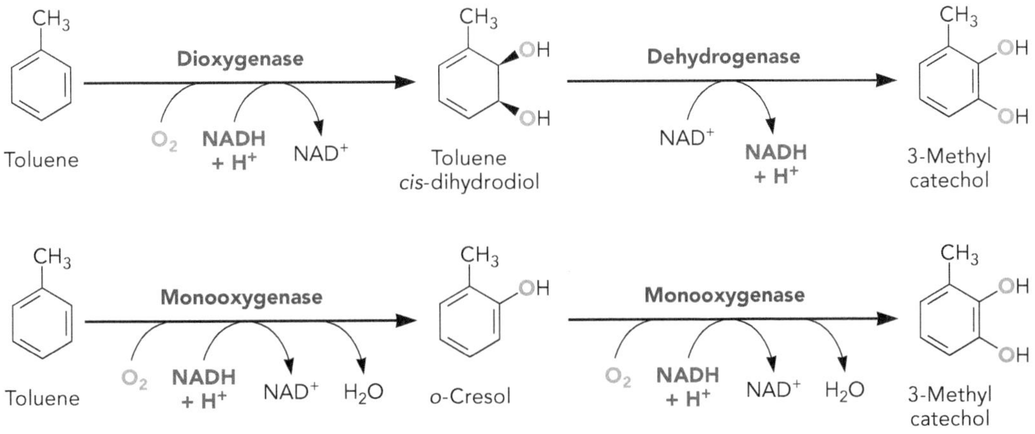

FIGURE 11.5 **Comparison of dioxygenase and monooxygenase reactions in the catabolism of toluene.** Bacterial isolates capable of growing on aromatic compounds initiate the oxidation of the substrate by either one atom or both atoms of molecular oxygen depending on the type of oxygenase enzyme present. For example, when growing on toluene, *Pseudomonas putida* F1 uses a dioxygenase enzyme (toluene 2,3-dioxygenase) to catalyze the formation of the nonaromatic product toluene *cis*-dihydrodiol, followed by a dehydrogenation reaction that forms 3-methylcatechol and regenerates NADH + H⁺ that was used by the dioxygenase in the first step (top pathway). In contrast, *Pseudomonas stutzeri* OX1 has a monooxygenase enzyme (toluene monooxygenase) that catalyzes two sequential additions of one atom of oxygen to first generate *o*-cresol and then 3-methylcatechol (bottom pathway). In this case, the formation of 3-methylcatechol requires two molecules of O_2 and the oxidation of two NADH + H⁺. Both bacterial strains have a *meta*-cleavage pathway (similar to that shown in Fig. 11.4) that converts 3-methylcatechol to compounds that enter central metabolism.

sequential monooxygenations), as in catechol, the ring can be broken by a different type of oxygen-requiring enzyme called a **ring-cleavage dioxygenase**. Ring cleavage can occur either between the two hydroxylated carbons (*ortho* cleavage) or adjacent to one of them (*meta* cleavage) depending on the specific dioxygenase enzyme used (Fig. 11.4).

One of the most well-studied aerobic pathways for aromatic acid catabolism is the β-ketoadipate pathway (Fig. 11.6A), which is found in a wide variety of soil microorganisms. In this branched pathway, benzoate and 4-hydroxybenzoate are oxidized to the dihydroxylated intermediates catechol and protocatechuate, respectively, to destabilize the aromatic ring. Both products then undergo *ortho* cleavage and the branches of the pathway converge before the ultimate products acetyl-CoA and succinyl-CoA are formed and enter central metabolism. Microorganisms that utilize the β-ketoadipate pathway are chemoorganoheterotrophs that obtain carbon from aromatic compounds and generate ATP through aerobic respiration with an organic electron donor and oxygen as the terminal electron acceptor.

In contrast, the anaerobic pathway for benzoate catabolism (Fig. 11.6B) involves the addition of a CoA group to form benzoyl-CoA to destabilize the aromatic ring, followed by reduction reactions that remove double bonds in the benzene ring. The next series of reactions leading to ring cleavage are analogous to reactions during fatty acid metabolism, which generate acetyl-CoA molecules that enter central metabolism. Since oxygen is not present, some anaerobic chemoorganoheterotrophs (like *Thauera aromatica*) obtain both carbon and energy from the catabolism of aromatic compounds by anaerobic respiration using an organic electron donor and nitrate as an alternative terminal electron acceptor. Alternatively, some anoxygenic photoheterotrophs (like *Rhodopseudomonas*

FIGURE 11.6 **Aerobic and anaerobic catabolism of aromatic acids.** (A) Aerobic degradation of benzoate and 4-hydroxybenzoate via the β-ketoadipate pathway in *Pseudomonas putida*. In this branched pathway, benzoate and 4-hydroxybenzoate are initially oxidized by oxygenase enzymes (benzoate 1,2-dioxygenase and 4-hydroxybenzoate hydroxylase; enzymes 1 and 9, respectively). Benzoate *cis*-dihydrodiol dehydrogenase (enzyme 2) catalyzes the formation of catechol. The aromatic rings of both catechol and protocatechuate are cleaved by *ortho*-cleavage dioxygenases (catechol 1,2-dioxygenase and protocatechuate 3,4-dioxygenase; enzymes 3 and 10, respectively). Muconolactone (ML) and carboxymuconolactone (CML) are formed by the action of muconate lactonizing enzyme and carboxymuconate lactonizing enzyme (enzymes 4 and 11, respectively). In the next step, the two branches of the pathway converge as both ML and CML are converted to the common intermediate β-ketoadipate enol-lactone (βKEL) by muconolactone isomerase and carboxymuconolactone hydrolase (enzymes 5 and 12, respectively). Enol-lactone hydrolase (enzyme 6) catalyzes the conversion of βKEL to β-ketoadipate, the intermediate that gives the pathway its name. Finally, in the last two steps, CoA is added to β-ketoadipate by β-ketoadipate:succinyl-CoA transferase (enzyme 7), and β-ketoadipyl-CoA thiolase (enzyme 8) generates acetyl-CoA and succinyl-CoA, which enter central metabolism. (B) Anaerobic pathway for benzoate catabolism in the anoxygenic photoheterotroph *Rhodopseudomonas palustris*. In the absence of oxygen, benzoate is activated by the addition of a molecule of CoA, forming benzoyl-CoA. This reaction, catalyzed by benzoate CoA ligase (enzyme 1), requires the input of ATP. The double bonds in the aromatic ring are then reduced and the ring is hydroxylated in a series of enzymatic reactions as follows: reaction 2 is catalyzed by benzoyl-CoA reductase, reaction 3 is catalyzed by an enzyme that has not yet been identified, reaction 4 is catalyzed by cylohex-1-ene-1-carboxyl-CoA hydratase, reaction 5 is catalyzed by 2-hydroxycyclohexane-carboxyl-CoA dehydrogenase, and reaction 6 is catalyzed by 2-ketocyclohexane-carboxyl-CoA hydrolase. The resulting pimelyl-CoA is converted to three molecules of acetyl-CoA via a series of β-oxidation-like reactions. CDC-CoA, cyclohex-1,5-diene-1-carboxyl-CoA; CC-CoA, cyclohex-1-ene-1-carboxyl-CoA; HCC-CoA, 2-hydroxy cyclohexane-1-carboxyl-CoA; KCC-CoA, 2-ketocyclohexane-1-carboxyl-CoA.

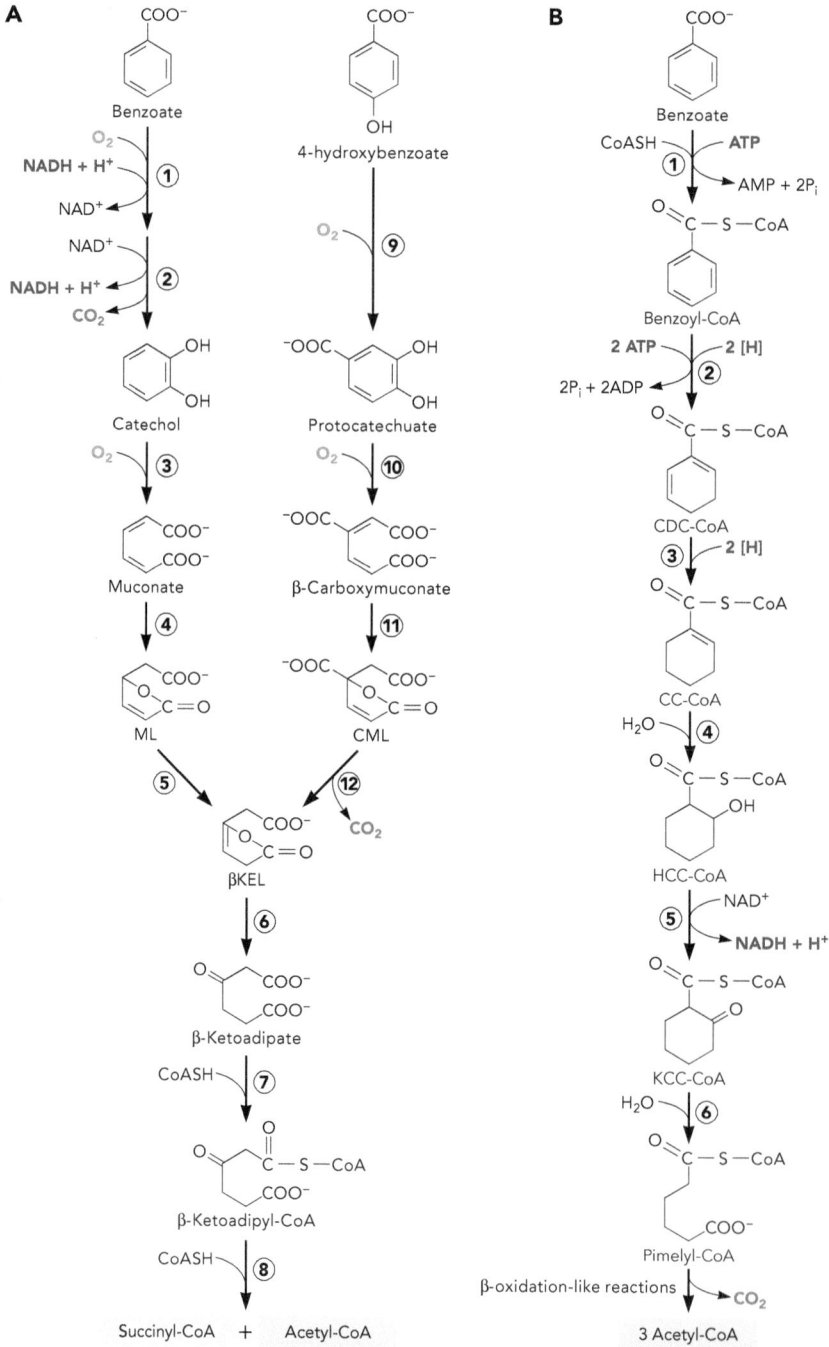

palustris) obtain carbon from aromatic molecules and generate ATP via cyclic photophosphorylation (Chapter 10).

The monomers (like sugars and organic acids) produced by microorganisms that break down polymers are quickly utilized by a variety of bacteria in both oxic and anoxic environments. In anoxic environments, secondary fermenters consume these byproducts and in turn release hydrogen gas, carbon dioxide, and acetate as byproducts that are subsequently consumed by other microbes present in the anaerobic food web, such as methanogens and sulfate reducers. Next, we will discuss growth on monomeric carbon sources by the facultative bacterium *Escherichia coli*.

Chemoorganotrophy in *Escherichia coli*

E. coli serves as a model for monomer metabolism under varying conditions (Fig. 11.7) because it can switch its metabolism to maximize growth in its current environment. As a facultative chemoorganotrophic organism, it is capable of making ATP via aerobic respiration, anaerobic respiration, and fermentation. Phosphotransferase system (PTS) sugars like glucose are a preferred carbon source for *E. coli*, and the expression of genes encoding the enzymes in these pathways is highly regulated (Chapters 7 and 8).

AEROBIC RESPIRATION

Because oxygen has a relatively positive E_0' value, aerobic respiration yields more ATP per molecule of glucose than anaerobic respiration and fermentation. Therefore, facultative bacteria will utilize aerobic pathways when oxygen is available. During aerobic respiration, glucose is sequentially oxidized in the steps of glycolysis through the aerobic tricarboxylic acid (TCA) cycle (Chapter 2), releasing CO_2 as a final product. NADH + H$^+$ ($E_0' = -320$ mV) generated through these central pathways donates electrons to NADH dehydrogenase, while FADH$_2$ (E_0' = −220 mV) donates electrons to succinate dehydrogenase (Fig. 11.7A). Under atmospheric (high) oxygen levels, the electrons are then passed to ubiquinone, to cytochromes, and ultimately to the terminal cytochrome *o* oxidase, which reduces oxygen to water. In environments that are not fully oxic (i.e., microoxic/low oxygen levels), cytochrome *d* oxidase acts as the terminal oxidase. When NADH + H$^+$ is the electron donor, ~3 ATPs are produced by oxidative phosphorylation with oxygen as the terminal electron acceptor, but when FADH$_2$ is the electron donor, only ~2 ATPs are produced (Chapter 6). The more negative the electrode/reduction potential of the electron donor, the more potential energy that can be converted into ATP. As the environment shifts from oxic to anoxic conditions, the facultative chemoorganotrophic *E. coli* transitions from aerobic to anaerobic respiration.

ANAEROBIC RESPIRATION

When oxygen is absent, *E. coli* has the capacity to use alternative terminal electron acceptors (Fig. 11.7B). Glycolysis of glucose occurs by the same process under both oxic and anoxic conditions. However, under anoxic conditions, the

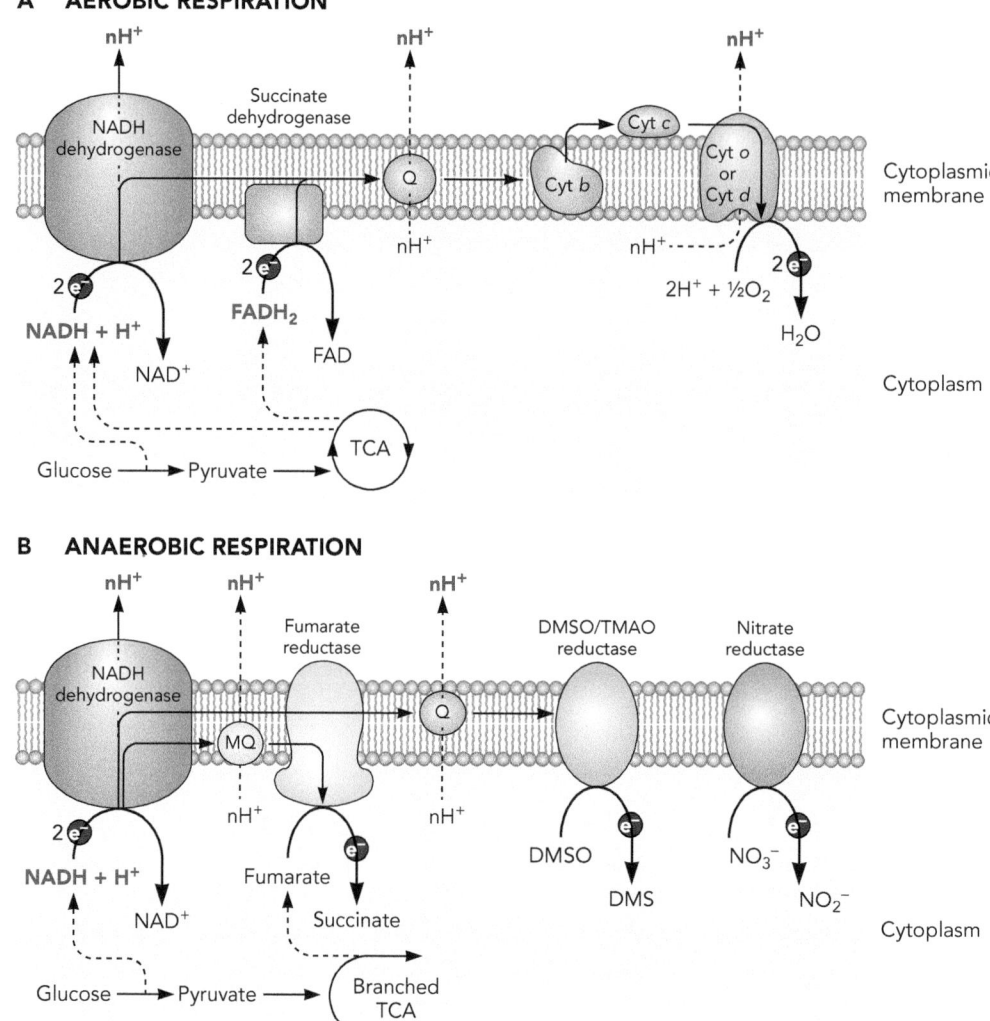

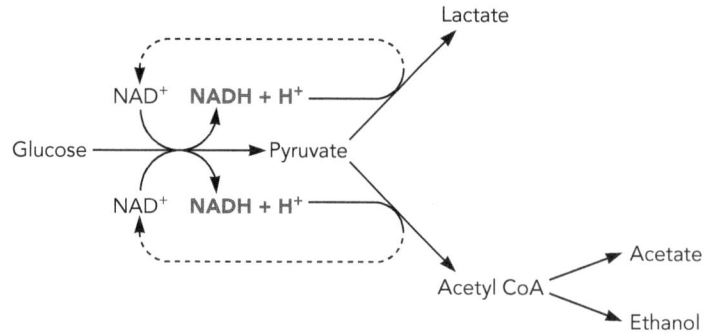

FIGURE 11.7 **Three lifestyles of the facultative chemoheterotrophic bacterium *Escherichia coli*.** (A) During aerobic respiration, reduced electron donors generated from central metabolism transfer electrons to an electron transport chain (ETC) that yields a proton motive force (PMF). An oxygenase facilitates the transfer of electrons to oxygen, the terminal electron acceptor. (B) During anaerobic respiration, electron flow through an ETC beginning with reduced coenzymes from central metabolism also yields a PMF. Two possible quinones, ubiquinone (Q) or menaquinone (MQ), might be utilized depending on which terminal electron acceptor, an inorganic or organic molecule other than oxygen, is available. The terminal complex in the ETC is a reductase that is specific for the terminal electron acceptor. (C) During fermentation, an ETC is not employed. Instead, reduced coenzymes from central metabolism are reoxidized as metabolic intermediates are reduced and then released as byproducts of the reaction. Without an ETC, no PMF is generated; therefore, no energy yield from the reoxidation of NADH + H⁺ occurs during fermentation. DMSO, dimethyl sulfoxide; DMS, dimethyl sulfide; TMAO; trimethylamine N-oxide; TCA, tricarboxylic acid.

TCA cycle becomes branched (Chapter 2) and consequently produces lower levels of reduced coenzymes. Carbon dioxide is still released by central metabolism as the cell is living a heterotrophic lifestyle. The NADH + H$^+$ (E_0' = −320 mV) serves as the electron donor to NADH dehydrogenase. However, due to the altered TCA pathway, fumarate reductase now converts fumarate to succinate and reduced coenzymes are consumed rather than produced (Fig. 11.7B). During anaerobic respiration, electrons are passed to either menaquinone or ubiquinone depending on the growth conditions, and then to the appropriate terminal electron acceptors and their associated terminal reductase complexes in the membrane (Chapter 6). There are multiple reductases encoded in the *E. coli* genome. However, they are used in a regulated manner to maximize the energy conservation for the cell and to reduce the energetic cost of making unnecessary reductase complexes. For this reason, *E. coli* has a preferred order in which it will utilize a terminal electron acceptor. First oxygen (O$_2$/H$_2$O; E_0' = +820 mV) will be used under oxic and then microoxic conditions (Fig. 11.7A). When the environment becomes anoxic, nitrate (NO$_3^-$/NO$_2^-$; E_0' = +420 mV), dimethyl sulfoxide (DMSO/DMS; E_0' = +160 mV), or trimethylamine *N*-oxide (TMAO/TMA; E_0' = +150 mV) and finally fumarate (fumarate/succinate; E_0' = +30 mV) will be utilized in that order. This hierarchy allows *E. coli* to maximize the amount of energy that it can conserve based on the availability of the various electron acceptors.

There are many alternative electron acceptors that can be used during anaerobic respiration in microorganisms other than *E. coli*. Thus, microbial anaerobic respiration not only includes the alternative electron acceptors discussed above, but also includes (and is not limited to) iron and manganese respiration, sulfate respiration (see below), respiration with chlorate and halogenated organic compounds, or with carbon dioxide or acetate as during methanogenesis (see below). These diverse pathways are carried out by microorganisms with specialized electron carriers within their electron transport chains (ETCs). As seen with *E. coli*, many microbes have multiple respiratory pathways that are utilized in a preferential order beginning with the one yielding the most energy.

FERMENTATION

If *E. coli* encounters growth conditions where there are no appropriate external electron acceptors available, then it must switch its metabolism from respiration to fermentation (Fig. 11.7C; Chapter 6). During fermentative growth, ATP is exclusively generated via substrate-level phosphorylation. However, fermentation results in less net ATP production per molecule of carbon source than respiration, because in fermentation, the carbon source is not completely oxidized. To reoxidize NADH + H$^+$ during fermentation, the electrons are transferred to fermentation products, which are excreted from the cell. In addition, there is no ETC to convert the energy in NADH + H$^+$ to proton motive force (PMF) and the cell must run ATP synthase in reverse, essentially "spending" ATP to generate the PMF needed for essential transport processes and in motile bacteria, to power flagella (Chapter 15). Thus, fermentation is only utilized when it is not possible for *E. coli* to grow via respiration because of the low energy output it yields for the cell.

REGULATION OF TERMINAL ELECTRON TRANSPORT CHAIN (ETC) COMPLEX PRODUCTION IN *ESCHERICHIA COLI*

Due to its chemoheterotrophic lifestyle, *E. coli* generates reduced organic electron donors through its metabolism. However, the external terminal electron acceptors used during respiration must be acquired from the environment. Therefore *E. coli* (and other microbes) monitors the environment and alters gene expression such that the best available terminal electron acceptor will be utilized first, maximizing energy conservation. As described above, oxygen is utilized if it is available in the environment and cytochrome *o* oxidase serves as the terminal ETC complex under atmospheric concentrations of oxygen. As the oxygen level drops and the environment becomes microoxic, cells must be able to sense this change in conditions. In *E. coli*, ArcA/ArcB make up a two-component regulatory system (Chapter 7) that senses this shift from fully oxic to microoxic conditions (Fig. 11.8A). ArcB is a sensor histidine kinase that indirectly senses reduced metabolic intermediates rather than directly sensing oxygen. Under atmospheric oxygen concentrations, ArcB is present but is in an inactive state. Upon a shift to microoxic conditions, ArcB autophosphorylates a conserved histidine residue and then transphosphorylates ArcA. ArcA is the response regulator for the system. It is also inactive under atmospheric oxygen levels. When a conserved aspartate residue of ArcA is phosphorylated by ArcB-P, ArcA-P can bind to target promoters in the genomic DNA. ArcA-P represses transcription of the *cyoABCDE* operon encoding cytochrome *o* oxidase, which cannot effectively function under microoxic conditions due to its relatively low affinity for oxygen. ArcA-P also activates transcription of the *cydAB* operon, encoding cytochrome *d* oxidase, the terminal oxidase used under microoxic conditions. Cytochrome *d* oxidase has a high affinity for oxygen and is thus capable of binding oxygen when it is present at lower concentrations. ArcA/ArcB therefore control the production of the terminal ETC complex used under normal oxygen levels (cytochrome *o* oxidase) versus lower oxygen levels (cytochrome *d* oxidase) (Fig. 11.8A).

E. coli cells switch to anaerobic respiration when they sense that all oxygen has been depleted. FNR, the fumarate nitrate reductase regulator, is a global regulator that controls multiple promoters under anoxic conditions and is the master regulator for anaerobic growth. FNR is inactive in the presence of oxygen, but becomes biologically active under anoxic conditions, as it contains an FeS prosthetic group that senses the cellular redox state. FNR controls the expression of a regulon that includes genes associated with the switch from the aerobic TCA cycle to the branched TCA pathway, and from the use of oxidases to reductases within the ETC. Structurally, FNR is similar to CRP (cAMP receptor protein; also known as CAP [catabolite activator protein]) (Chapters 7 and 8), and in the absence of oxygen FNR binds to specific promoters in the DNA via a helix-turn-helix motif. Reduced FNR represses the promoters for the genes encoding cytochrome *o* (*cyoABCDE*) and cytochrome *d* (*cydAB*) (Fig. 11.8B), resulting in no further production of the oxidases. Reduced FNR also activates all the genes encoding multiple reductase complexes (fumarate reductase, *frdABCD*; DMSO/TMAO reductase, *dmsABC*; and nitrate reductase, *narGHJI*) (Fig. 11.8B). However, it would be wasteful to produce multiple reductases simultaneously, so there are additional layers of regulatory control to specifically ensure that alternative electron acceptors are used in the order that will lead to the most efficient energy conservation for the *E. coli* cell.

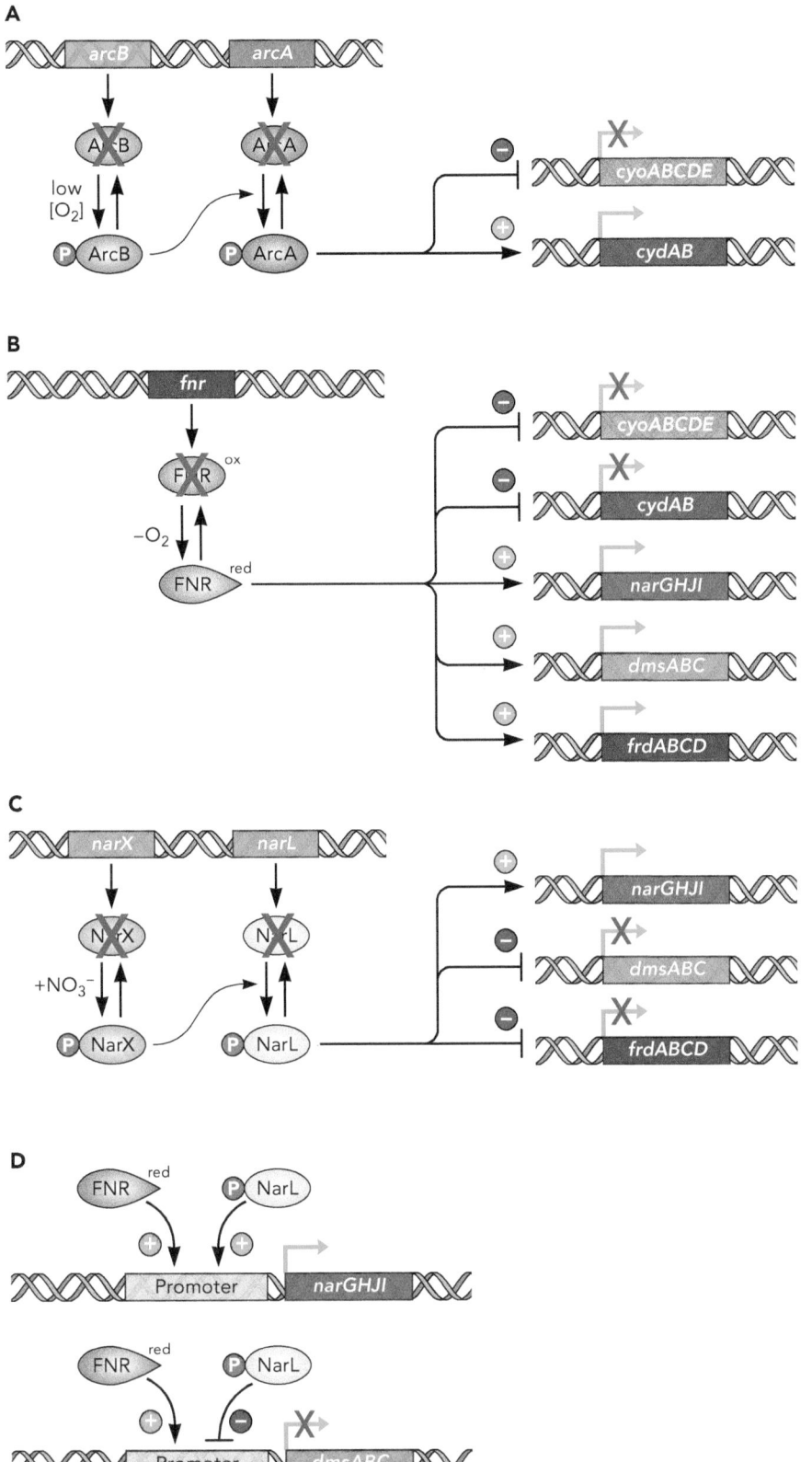

FIGURE 11.8 **Regulation of the expression of genes encoding oxidases and reductases in *Escherichia coli.*** (A) The ArcA-ArcB two-component system positively and negatively regulates the *cydAB* (encoding cytochrome *d*) and *cyoABCDE* (encoding cytochrome *o*) operons, respectively, in response to low levels of oxygen. (B) The FNR protein activates expression of multiple reductase operons (*narGHJI*, encoding nitrate reductase; *dmsABC*, encoding DMSO/TMAO reductase; and *frdABCD*, encoding fumarate reductase) while repressing the oxidase operons. (C) In the presence of nitrate, the NarX-NarL two-component system positively regulates the expression of the genes encoding nitrate reductase, while negatively regulating the operons encoding alternative reductases. (D) FNR and NarL-P coordinately regulate gene expression to activate the nitrate reductase operon under anoxic growth conditions where nitrate is available, while repressing the operon encoding the DMSO/TMAO reductase. Activation is indicated by arrows with green plus signs, while repression is indicated by T-bars with red negative signs. See the text for additional details. FNR, fumarate nitrate reductase regulator; DMSO, dimethyl sulfoxide; TMAO, trimethylamine *N*-oxide.

After oxygen (E_0' = +820 mV), nitrate (E_0' = +420 mV) is the next preferred terminal electron acceptor. The two-component regulatory system NarX/NarL contributes to the complex regulation enabling *E. coli* to sense and respond to the presence of nitrate (Fig. 11.8C). NarX is a sensor histidine kinase that directly senses nitrate. It is inactive in the absence of nitrate, but when nitrate is present, NarX autophosphorylates a conserved histidine residue and transphosphorylates the response regulator NarL. When a specific aspartate residue in NarL is phosphorylated, NarL-P can bind to target promoters in the genomic DNA. NarL-P activates the genes encoding nitrate reductase and represses the genes encoding the fumarate and DMSO/TMAO reductases. Reduced FNR and NarL-P simultaneously control promoter activity to ensure that under anoxic conditions and in the presence of nitrate, only nitrate reductase is produced (Fig. 11.8D). Additional regulatory proteins control expression of the genes encoding reductases for other alternative terminal electron acceptors to ensure that they are used in a preferential order that maximizes energy conservation.

Chemolithoautotrophy

Instead of using organic electron donors, chemolithoautotrophic microorganisms have the capacity to utilize inorganic electron donors (e.g., H_2, CO, NH_3, NO_2^-, S^0, $S_2O_3^{2-}$, and Fe^{2+}; Table 11.1) as a source of electrons for ATP synthesis via oxidative phosphorylation. Some chemolithotrophs grow aerobically using oxygen as a terminal electron acceptor to help compensate for their poor electron donors. Other chemolithoautotrophs grow anaerobically, using oxidized compounds such as CO_2 as a terminal electron acceptor (Table 11.1). Many chemolithotrophs release byproducts that serve as important intermediates in nutrient cycling and represent examples of the interdependency of different microbial populations within a community.

TABLE 11.1 **Chemolithoautotrophs**

Organism	Redox couple (oxidized/reduced)	E_0' (mV)	Electron donor	Electron acceptor	Carbon source	Reverse electron flow[a]	Byproduct(s) produced
Carbon monoxide-oxidizing bacteria	CO_2/CO	−540 mV	CO	O_2	CO_2	No	CO_2
Methanogenic archaea	$2 H^+/H_2$	−420 mV	H_2	CO_2	CO_2	NA[b]	CH_4
Acetogenic bacteria	$2 H^+/H_2$	−420 mV	H_2	CO_2	CO_2	NA[b]	CH_3COOH
Hydrogen-oxidizing bacteria	$2 H^+/H_2$	−420 mV	H_2	O_2	CO_2	No	H_2O
	$NAD(P)/NAD(P)H + H^+$	−320 mV					
Sulfur-oxidizing bacteria	S^0/H_2S SO_4^{2-}/H_2S SO_4^{2-}/S^0 $SO_4^{2-}/S_2O_3^{2-}$	−270 mV −220 mV −200 mV −50 mV	H_2S H_2S S^0 $S_2O_3^{2-}$	O_2	CO_2	Yes	SO_4^-
Ammonia-oxidizing bacteria	NO_2^-/NH_3	+340 mV	NH_3	O_2	CO_2	Yes	NO_2^-
Nitrite-oxidizing bacteria	NO_3^-/NO_2^-	+430 mV	NO_2^-	O_2	CO_2	Yes	NO_3^-

[a] Reverse electron flow to produce NADPH + H+ for Calvin cycle.
[b] NA, not applicable. Reductive acetyl-CoA pathway used to fix carbon.

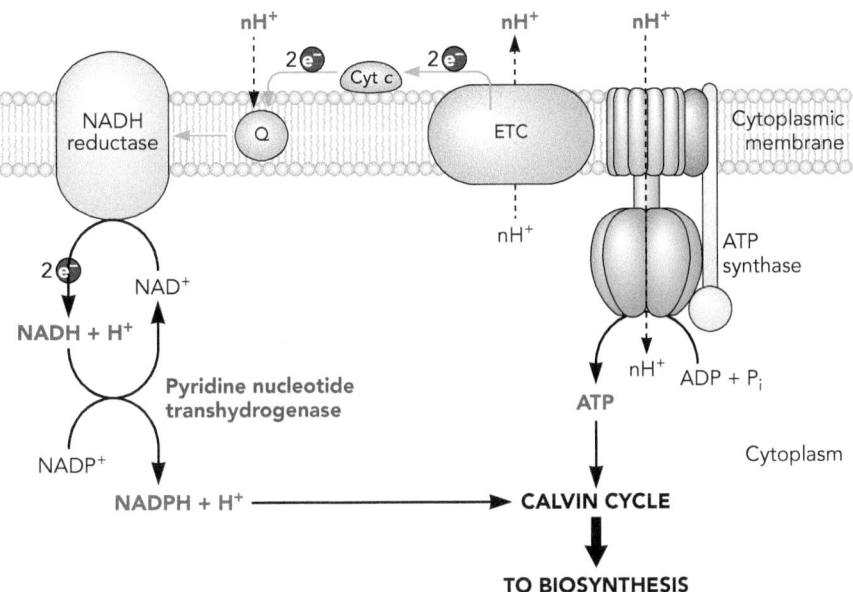

FIGURE 11.9 **Reverse electron flow.** During reverse electron flow, electrons within the electron transport chain (ETC) are moved in an energetically unfavorable direction (green arrows), from a more positive electrode/reduction potential to a more negative value, to generate NADH + H$^+$ from NADH reductase. A portion of the proton motive force is used to drive the electron flow. NADH + H$^+$ is readily converted into NADPH + H$^+$, which is needed to drive the reactions of the Calvin cycle for some autotrophs.

Many, but not all chemolithoautotrophs use the Calvin cycle, which requires ATP and NADPH + H$^+$, to reduce CO$_2$ and generate carbon for biosynthesis (Chapter 9). If the inorganic electron donor used has an electrode/reduction potential that is more electronegative than NADPH + H$^+$ ($E_0' = -320$ mV), such as CO ($E_0' = -540$ mV) or hydrogen gas ($E_0' = -410$ mV), then the cells produce NADPH + H$^+$ using forward electron flow, as it is an energetically favorable direction for the reaction (Table 11.1). However, chemolithoautotrophs that utilize electron donors with an electrode/reduction potential more positive than $E_0' = -320$ mV (e.g., NH$_3$, NO$_2^-$, S^0, S$_2$O$_3^{2-}$, and Fe^{2+}) must use reverse electron flow to generate NADPH + H$^+$ (Table 11.1). The process of reverse electron flow was described for the purple anoxygenic phototrophs where it is also used to generate reduced NADPH + H$^+$ for biosynthesis (Chapter 10). To perform reverse electron flow (Fig. 11.9), a portion of the PMF generated through respiration is used to move electrons in an energetically unfavorable direction from cytochromes and quinones to the NADH reductase (i.e., from a more positive to a more negative E_0'). This yields NADH + H$^+$ that the cell can readily convert to NADPH + H$^+$ through the activity of pyridine nucleotide transhydrogenase. Because reverse electron flow costs the cell energy, there is less energy available for oxidative phosphorylation and biosynthetic reactions. Thus, chemolithoautotrophs using reverse electron flow often have slow growth rates even during aerobic respiration.

Methanogens, sulfur-oxidizing bacteria, and nitrifying bacteria will be described (here and in Chapter 12) as examples of microbes that grow as chemolithoautotrophs and play critical roles in the global carbon, sulfur, and nitrogen cycles. Within the carbon cycle, there are specific microbes and processes associated with the oxidation and reduction of one-carbon (C1) compounds (Fig. 11.1) where methanogens play a critical role in generating the most reduced form of carbon, methane gas.

Chemolithoautotrophy in Methanogens

Methanogens serve as an example of a group of microorganisms with an anaerobic chemolithoautotrophic lifestyle. Unlike facultative organisms, methanogens are obligate chemolithoautotrophs with a single metabolic strategy for energy conservation and cell growth. Methanogens are all members of the *Euryarchaeota* (Fig. 11.10A), and all produce methane (a potent greenhouse gas) as a byproduct of their metabolism. We describe here how methane is generated, and later we will describe both aerobic and anaerobic microbes that can utilize methane as a carbon source, thus preventing much of the methane produced by methanogens from entering the atmosphere.

Much of methanogen metabolism was elucidated through the work of Ralph Wolfe's research group at the University of Illinois. Samples provided to Carl Woese helped to establish the *Archaea* as a distinctive domain on the phylogenetic tree of life (Chapter 1). Major genera of methanogens include *Methanobacterium*, *Methanobrevibacter*, *Methanococcus*, *Methanomicrobium*, *Methanospirillum*, *Methanosarcina*, and *Methanopyrus*. Methanogens are strict anaerobes that are very oxygen sensitive and are only found in anoxic habitats. Most are mesophiles, but thermophiles, hyperthermophiles, and psychrophiles are known. Morphologies also vary and include rods, cocci, irregular cocci, clusters of cocci, and filaments. Many methanogens are motile, but their flagella are unique in structure and unrelated to either bacterial or eukaryotic flagella (Chapter 15).

There are three pathways that various methanogens use to produce methane. Most methanogenic species can generate methane from C1 substrates such as CO_2 or formate, whereas relatively few species can use the methylated substrates methanol, methylamines, and methylsulfides. Even more rare is the ability to use acetate (a two-carbon compound) as a substrate for methanogenesis. Only one genus, *Methanosarcina*, is known to have all three methanogenic pathways. Here the focus will be on methanogenesis from $H_2 + CO_2$:

$$4\,H_2\,(\text{electron donor}) + CO_2\,(\text{electron acceptor}) \rightarrow CH_4\,(\text{byproduct}) + 2\,H_2O$$

As H_2 is oxidized, CO_2 is reduced to methane in a pathway involving attachment of CO_2 to a C1 carrier molecule followed by a series of reduction steps requiring several coenzymes unique to methanogens (see below). During this process, methanogens generate a PMF or sodium motive force and ATP using a highly specialized ETC. In addition to being the electron acceptor for the ETC, CO_2 is also the carbon source for these organisms. However, instead of using the Calvin cycle to fix CO_2, the reductive acetyl-CoA pathway is utilized (Chapter 9).

Methanogens carry out this unique metabolic lifestyle using highly specialized coenzymes (Fig. 11.10B). Two types of coenzymes are used: C1 carriers that carry the one-carbon molecule as it is being reduced, and electron carriers

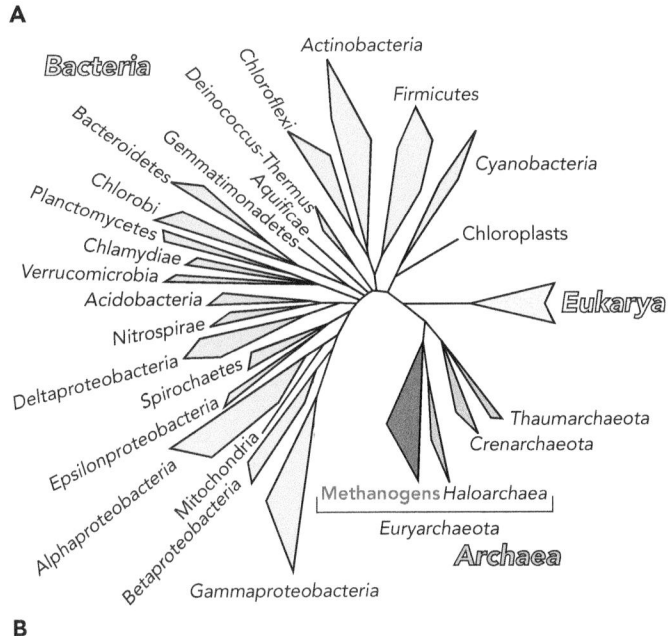

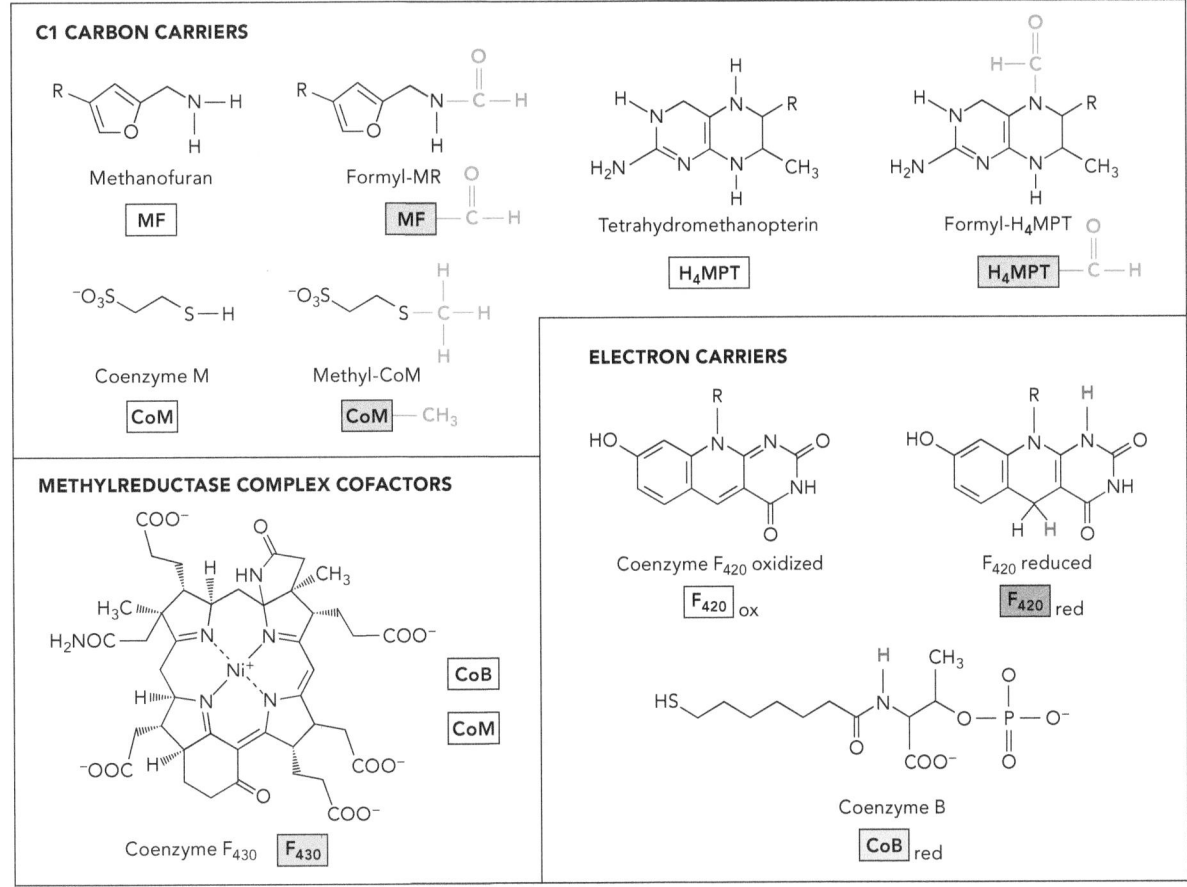

FIGURE 11.10 **Distribution of methanogenesis and specialized coenzymes used in the process.** (A) Phylogenetic tree highlighting the methanogenic *Archaea* (members of the *Euryarchaeota* highlighted in dark coloring and bold pink text). (B) Structures of specialized coenzymes that function in methanogenesis as C1 carriers and electron carriers. Also shown are cofactors associated with the methyl-reductase complex; the F_{430}, CoB, and CoM cofactors are unique to methanogens. R represents sidechains that vary.

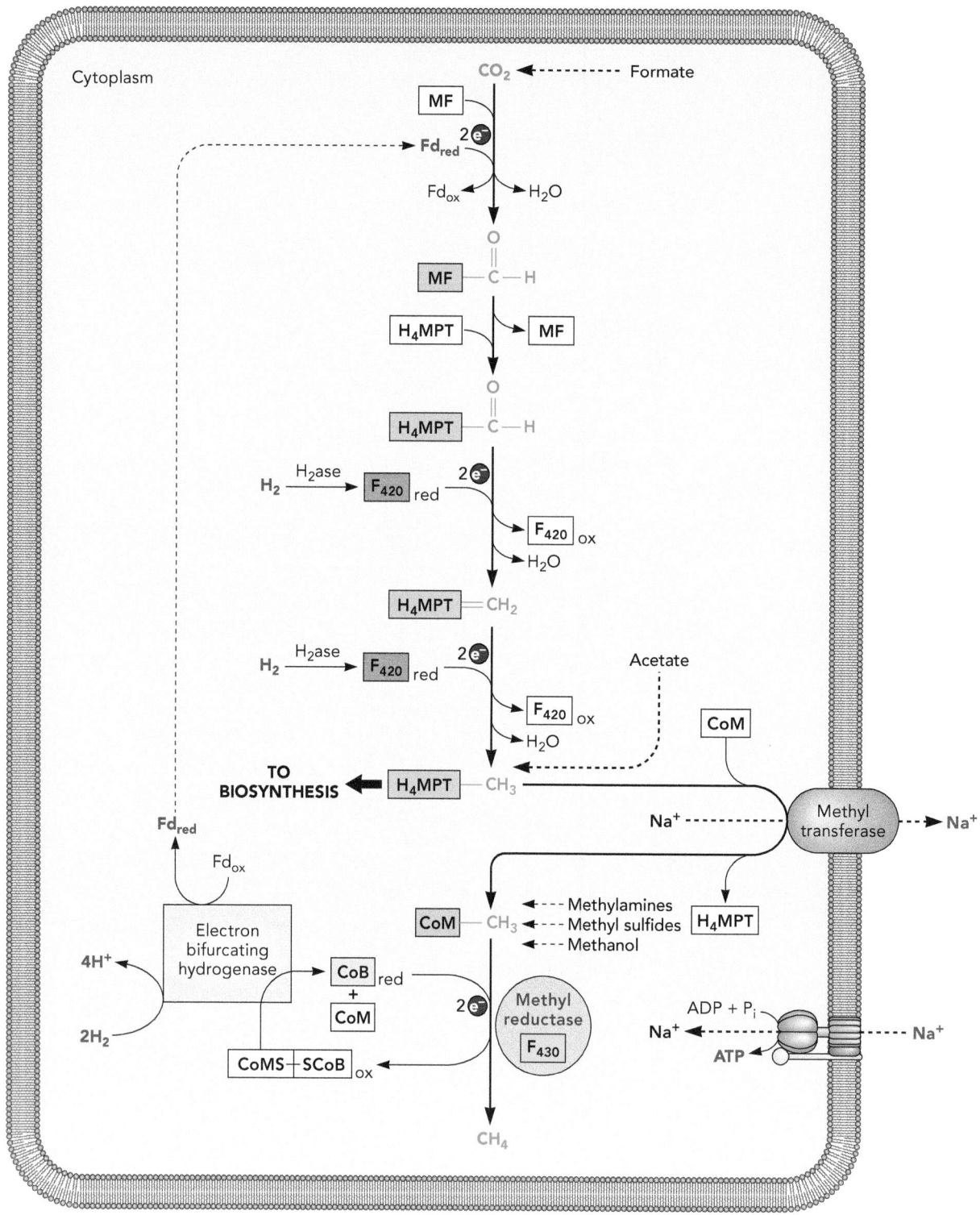

FIGURE 11.11 **Methanogenesis from carbon dioxide and hydrogen gas.** CO_2 is bound to the one carbon (C1) carrier methanofuran (MF) and is reduced to a formyl group. The formyl group is then transferred to the C1 carrier tetrahydromethanopterin (H_4MPT) and is reduced in sequential reactions to a methyl group. The methyl group is then transferred by the membrane-bound methyltransferase enzyme to the C1 carrier CoM in a process that pumps protons or Na^+ ions, contributing to the proton motive force or sodium motive force. An H^+-dependent or Na^+-dependent ATP synthase catalyzes formation of ATP. The methylreductase complex reduces CH_3-CoM to methane (CH_4) as CoM forms a disulfide complex with CoB. The CoM-S-S-CoB is reduced with electrons from H_2 gas via an electron-bifurcating hydrogenase. Dotted lines indicate entry points in the pathway for alternative carbon substrates for methanogenesis (formate, acetate, methanol, methylamines, and methylsulfides). CoM, coenzyme M; CoB, coenzyme B.

that participate in the reduction process. Methanofuran and tetrahydrometha-nopterin (H$_4$MPT) are C1 carriers found across the archaeal domain. H$_4$MTP is analogous to tetrahydrofolate, a C1 carrier that was discussed in Chapter 9 in the context of the reductive acetyl-CoA pathway for CO$_2$ fixation. Coenzyme M (CoM), coenzyme B (CoB; 7-mercaptoheptanolythreonine phosphate, also known as HS-HTP), and factor F$_{430}$, which are only found in methanogens, are part of the methylreductase complex in the ETC. Finally, coenzyme F$_{420}$, which is structurally similar to the flavin cofactor flavin mononucleotide (FMN), is an electron carrier that is also involved in methanogenesis, but it is not unique to the *Archaea*.

The methanogenesis pathway (Fig. 11.11) begins when CO$_2$ is bound to the C1 carrier methanofuran. In the first reduction step, electrons from H$_2$ are donated to CO$_2$ via a reduced ferredoxin (Fd) protein. The resulting formyl group is then transferred to the second C1 carrier, H$_4$MPT. Electrons for the next two reduction steps, converting the formyl group to a methylene and then a methyl group, are donated by reduced F$_{420}$. A cytoplasmic **hydrogenase** enzyme catalyzes the cleavage of H$_2$ gas into protons and electrons that are used to reduce F$_{420}$. The methyl group then enters the methylreductase complex, which contains CoM, CoB, and coenzyme F$_{430}$. The transfer of the methyl group from H$_4$MPT to CoM is an exergonic process that is coupled to the conservation of energy by generating a proton or sodium gradient across the cytoplasmic membrane that can in turn be utilized for ATP production via ATP synthase. The methylreductase complex catalyzes the final reduction to produce methane, the most reduced form of carbon, as a byproduct. Electrons for this final reduction are provided by reduced CoB, and the corresponding oxidation results in the formation of an oxidized CoM-S-S-CoB disulfide complex. Electrons to reduce this complex, releasing free CoM and CoB for further rounds of methanogenesis, as well as electrons to reduce the ferredoxin protein (electron donor in the first reduction step), are provided by two molecules of H$_2$ gas via a process that is known as **electron bifurcation**. In electron bifurcation, a flavin-containing multiprotein complex catalyzes the reduction of ferredoxin (an energetically unfavorable process) by coupling it to an energetically favorable reduction. In this case, the high electrode/reduction potential of CoM-S-S-CoB is used to drive the reduction of ferredoxin.

Methyl-H$_4$MPT is a critical branch point in methanogenesis, as not only is it used as a substrate by the methylreductase complex, but it also serves as the starting substrate for the reductive acetyl-CoA pathway for CO$_2$ fixation. In this process, the methyl-H$_4$MPT is converted to acetyl-CoA. Pyruvate and phosphoenolpyruvate (PEP) are subsequently produced from acetyl-CoA and CO$_2$ for biosynthesis (Chapter 9). Ultimately, gluconeogenesis and the pentose phosphate pathway (PPP) plus the branched TCA pathway are used to generate the 12 essential precursors (Chapter 2). Organisms called methanotrophs (see below) utilize the methane produced by the methanogens and reoxidize it to complete the C1 cycle on an ecosystem level.

Methylotrophy Enables Cycling of One-Carbon (C1) Compounds

Methylotrophs are organisms capable of utilizing C1 compounds other than CO$_2$ for energy and carbon sources. C1 compounds (e.g., methane, methanol, methylamines, methylsulfides, and formaldehyde) lack carbon-carbon bonds.

Like autotrophs, microbes that can utilize C1 compounds as their sole sources of carbon must synthesize organic compounds *de novo* using specialized pathways; the primary products are then used to produce the 12 essential precursors for biosynthesis (Chapter 2).

AEROBIC METHYLOTROPHS, INCLUDING METHANOTROPHS

A specialized subset of methylotrophs that can use methane as a source of carbon and energy are termed **methanotrophs**. The majority are members of the *Alpha-* and *Gammaproteobacteria*, but also include some members of the *Verrucomicrobia* (recently named *Verrucomicrobiota*; Fig. 11.12A). Methanotrophs reoxidize methane back to CO_2 to complete an important part of the carbon cycle (Fig. 11.1).

In the first step of methane oxidation, aerobic methanotrophs oxidize CH_4 to methanol (CH_3OH) using the key enzyme methane monooxygenase (MMO) (Fig. 11.12B). While some aerobic methanotrophs have a soluble form of MMO in the cytoplasm, the most common form of MMO is a membrane-bound enzyme termed particulate methane monooxygenase, or pMMO, which is typically localized to invaginations of the cytoplasmic membrane. As described above, monooxygenases are enzymes that add one atom of molecular oxygen to a substrate and H_2O is formed as a byproduct of the reaction. MMO oxidizes CH_4 by adding one atom of oxygen from O_2 to produce methanol. Electrons and protons (e.g., from NADH + H⁺) are used to produce water from the other atom of oxygen. In the second step, a methanol dehydrogenase converts methanol to formaldehyde (CH_2O). Some of the formaldehyde is further oxidized to CO_2 in a stepwise process catalyzed by additional dehydrogenases that produce reduced electron carriers (e.g., NADH + H⁺), which allows cells to generate a PMF. The remaining formaldehyde is assimilated to build biomass.

Methanotrophic members of the *Gammaproteobacteria* use the ribulose monophosphate (RuMP) cycle (Fig. 11.13) to assimilate formaldehyde into cell biomass. In the RuMP cycle, three molecules of formaldehyde (1C) are used to generate one molecule of glyceraldehyde-3-phosphate (3C), which enters central metabolism. In contrast, methanotrophic members of the *Alphaproteobacteria* assimilate formaldehyde via the serine cycle (Fig. 11.13), in which one molecule of formaldehyde and one molecule of CO_2 are used to generate one molecule of acetyl-CoA, which enters central metabolism. A third group of aerobic methanotrophs include members of the *Verrucomicrobia*. Most are thermoacidophiles that were isolated from geothermal environments such as volcanoes. These organisms are members of the genus *Methylacidiphilum*. Based on genome sequence analyses, these methanotrophs lack the RuMP cycle and appear to use the serine cycle, or they may use the Calvin cycle to fix CO_2 produced from methane metabolism (Fig. 11.13).

Aerobic methanotrophs are typically located at the oxic/anoxic interface in aquatic sediments and soil environments where they have access to oxygen and can capture the methane being generated in the anoxic sediments below. These organisms play a critical role in reducing the amount of the potent greenhouse gas methane that enters the atmosphere. Many environments also harbor methylotrophs that utilize C1 compounds other than methane. For example, aerobic methylotrophs that grow on methanol are commonly found on leaf surfaces. As plants grow, methanol derived from the breakdown of the plant cell wall

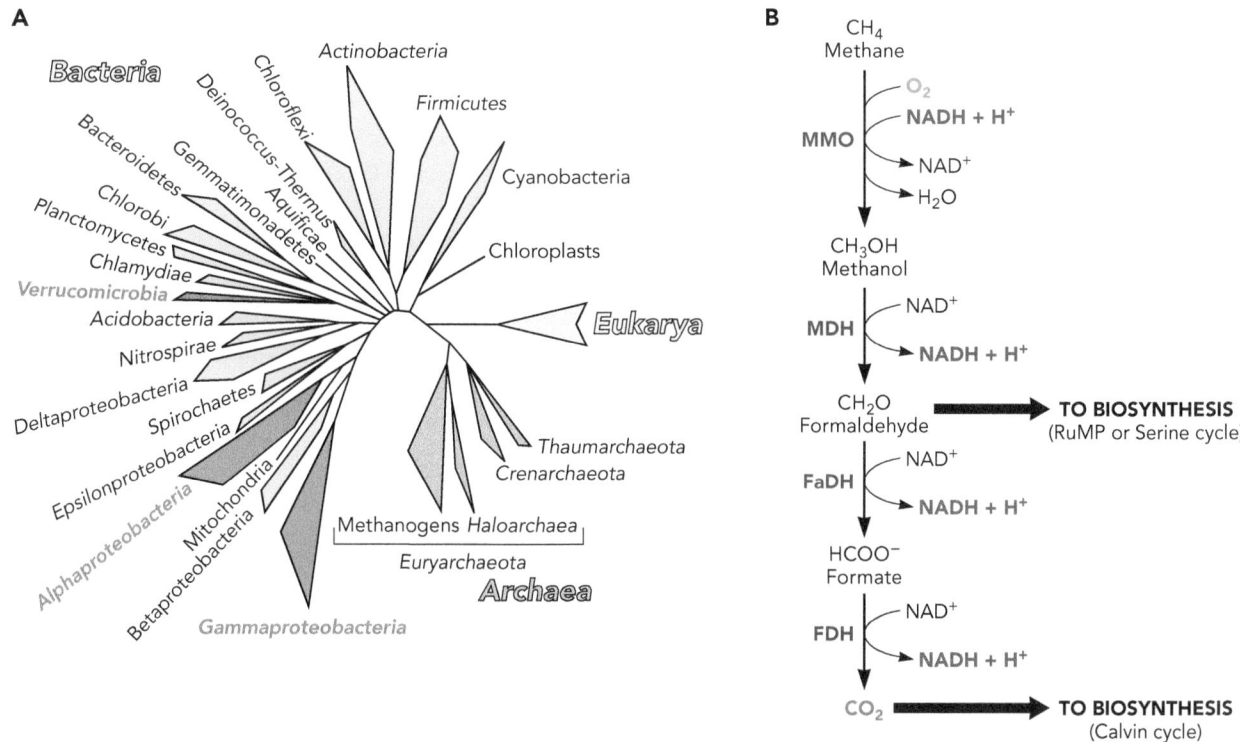

FIGURE 11.12 **Distribution and biochemistry of aerobic methanotrophy.** (A) Phylogenetic tree highlighting bacterial lineages with members known to carry out methanotrophy (highlighted in dark coloring and bold green text). (B) Oxidation of methane is initiated by the enzyme methane monooxygenase (MMO), which uses oxygen as a substrate and generates methanol. Three dehydrogenase enzymes, methanol dehydrogenase (MDH), formaldehyde dehydrogenase (FaDH), and formate dehydrogenase (FDH), complete the oxidation of methanol to CO_2. Most methanotrophs use the serine cycle or the ribulose monophosphate (RuMP) pathway to incorporate formaldehyde into biomass, but some members of the *Verrucomicrobia* are thought to use the Calvin cycle to fix CO_2 for biosynthesis.

polymer pectin is released through plant stomata. The most common methanol-utilizing bacteria on leaf surfaces are pink-pigmented members of the *Alphaproteobacteria*, which include the genera *Methylobacterium* and *Methylorubrum*. Their pink color is due to the presence of carotenoid pigments that provide protection from cellular damage caused by sunlight. These bacteria utilize the same sequence of dehydrogenation reactions described for methanotrophs to oxidize methanol to formaldehyde, formate, and ultimately CO_2 (Fig. 11.12B), and they use the serine cycle to assimilate carbon into biomass.

Until recently, methane oxidation was only known to occur under oxic conditions, but methane oxidation in anoxic environments is now known to be important for methane cycling. This process appears to require the participation of two types of microorganisms working together and will be described in the last section of this chapter in the context of the anaerobic food web.

One-Carbon (C1) Chemolithotrophy in Acetogens

Acetogenesis is another type of C1 metabolism that involves the reduction of CO_2 and formation of the reduced carbon byproduct, acetate (Fig. 11.14). Acetogenesis is an energy-conserving process that is carried out predominantly

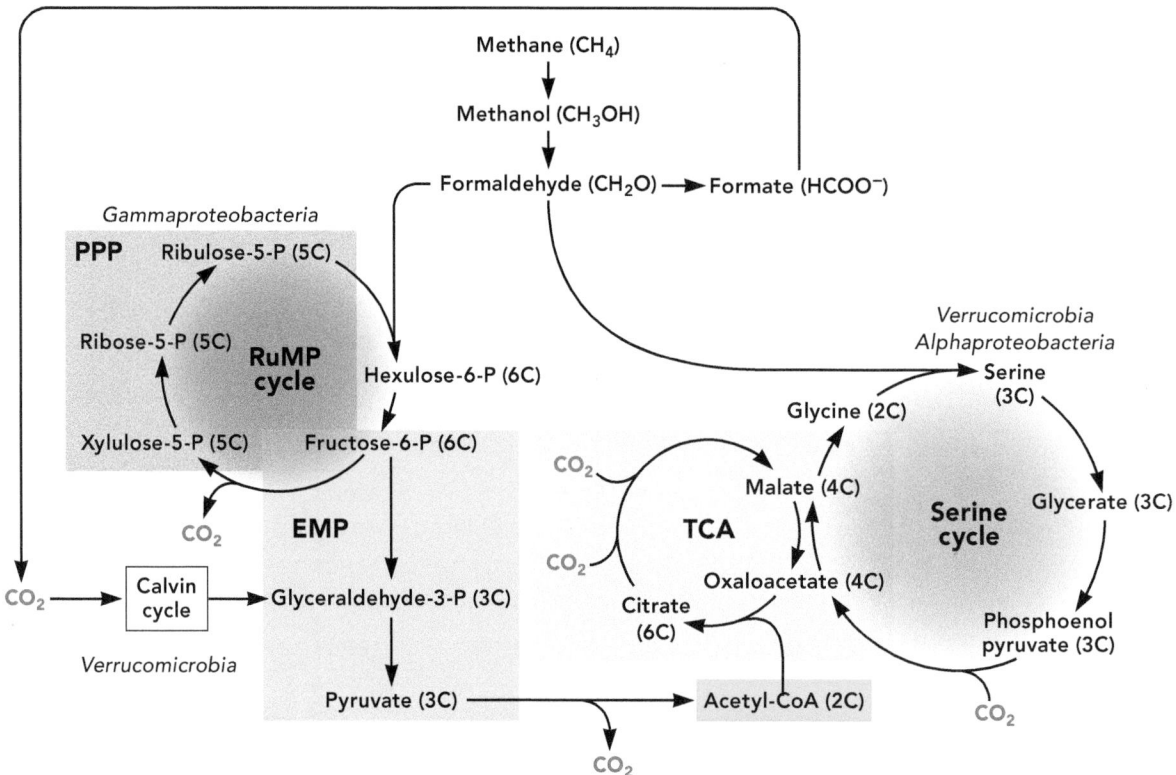

FIGURE 11.13 **Conversion of methane into essential precursors by methanotrophs.** In methanotrophs that use the ribulose monophosphate (RuMP) pathway (*Gammaproteobacteria*), formaldehyde formed from methane is added to ribulose-5-P (5C) to form a 6C sugar. Enzymes of the pentose phosphate pathway (PPP) catalyze rearrangement reactions, and after three such cycles, one molecule of glyceraldehyde-3-P (3C) is generated and enters glycolysis. Alphaproteobacterial methanotrophs and some *Verrucomicrobia* use the serine cycle for biosynthesis. In this pathway, serine (3C) is formed from one molecule each of formaldehyde (1C) and glycine (2C). A molecule of CO_2 is later added to phosphoenolpyruvate (3C) to form oxaloacetate (4C), which enters the tricarboxylic acid (TCA) cycle, ultimately producing acetyl-CoA (2C) for biosynthesis. EMP, Embden-Meyerhof-Parnas pathway; P, phosphate.

by various anaerobic Gram-positive bacteria including several *Clostridium* and *Acetobacterium* species. Acetogens use the reductive acetyl-CoA pathway (Chapter 9) to form acetyl-CoA from two independently reduced molecules of CO_2. The reduction of one CO_2 molecule to the level of a methyl group is analogous to the initial reactions of methanogenesis (Fig. 11.11). However, rather than using the C1 carrier present in methanogens, H_4MPT, acetogens use tetrahydrofolate (THF) as the C1 carrier. Like in methanogens, the other molecule of CO_2 is reduced to CO by carbon monoxide dehydrogenase/acetyl-CoA synthase; this bifunctional enzyme then catalyzes the formation of acetyl-CoA (Chapter 9). The energy present in acetyl-CoA can be conserved via substrate-level phosphorylation to produce one molecule of ATP and one molecule of acetate, a byproduct that is released from the cells. The overall reaction is:

$$4\,H_2 + H^+ + 2\,HCO_3^- \rightarrow CH_3COO^- + 4\,H_2O$$

However, because one ATP molecule was invested in an early step in acetogenesis (Fig. 11.14), there is no net ATP production. Therefore, additional energy must be conserved to power biosynthesis; this is done by generating a sodium motive force or PMF. Acetogens use an electron-bifurcating hydrogenase that splits H_2 into protons and electrons. The electrons are used to reduce NAD^+

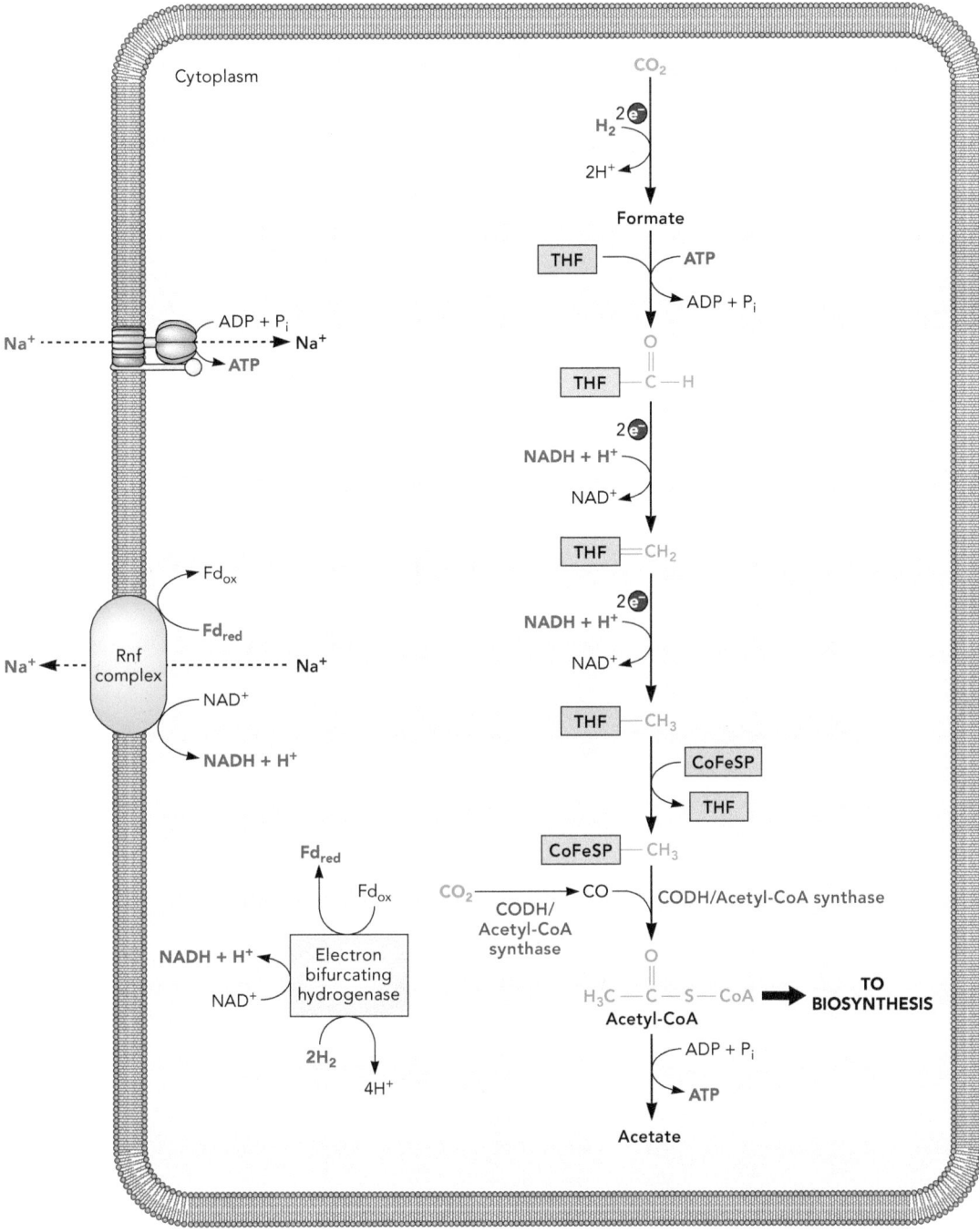

FIGURE 11.14 **Reduction of carbon dioxide into acetate during acetogenesis.** In acetogenesis, two molecules of CO_2 are independently reduced by two different mechanisms via the reductive acetyl-CoA pathway (Chapter 9). One CO_2 molecule is bound to the one-carbon (C1) carrier tetrahydrofolate (THF) as it is reduced to a methyl group. The methyl group is then transferred to a cobalt-containing iron-sulfur protein (CoFeSP). The other CO_2 molecule is reduced to CO by carbon monoxide dehydrogenase (CODH)/acetyl-CoA synthase. This enzyme also catalyzes the formation of acetyl-CoA. The ATP generated by substrate-level phosphorylation replaces the ATP invested in an earlier step of the pathway, and in this process, acetate is produced as a byproduct. Like in methanogenesis, energy is conserved by an ion-pumping membrane complex (Rnf) that pumps H^+ or Na^+ ions, contributing to a proton motive force or sodium motive force. An H^+-dependent or Na^+-dependent ATP synthase catalyzes formation of ATP. H_2 gas, the electron donor, is split into H^+ and electrons by an electron-bifurcating hydrogenase; the electrons are used to reduce NAD^+ to $NADH + H^+$ (an energetically favorable reaction) and reduce a ferredoxin (Fd; an energetically unfavorable reaction). When acetogens are growing as autotrophs, they siphon off some of the acetyl-CoA for biosynthesis.

(an energetically favorable reaction) coupled to the reduction of ferredoxin (an energetically unfavorable reaction); together, the overall process is energetically favorable. The reduced ferredoxin then interacts with an ion-translocating membrane-bound protein complex (Rnf complex), transferring electrons to NAD^+ and conserving energy by pumping either protons or sodium ions to the outside of the cell, thus contributing to the PMF or sodium motive force, respectively (Fig. 11.14).

Acetogens can grow as chemolithoautotrophs, incorporating CO_2 into biomass using the reductive acetyl-CoA pathway (Chapter 9). In this case, some of the acetyl-CoA produced is diverted to biosynthesis, and thus less ATP is made by substrate-level phosphorylation. Both methanogenesis and acetogenesis allow cells to conserve only small amounts of energy, so these organisms tend to grow slowly. Unlike the methanogens, acetogens have alternative metabolic options. Under appropriate conditions, they can grow as chemoorganoheterotrophs by fermenting sugars via glycolysis, a more energetically favorable process (Chapter 6).

The Sulfur Cycle

We've seen that acetogens and methanogens can use H_2 or organic acids produced by fermenting bacteria to generate ATP, making acetate or methane, respectively (Fig. 11.11 and 11.14). The sulfur-reducing bacteria also use H_2 or organic acids and tend to outcompete other organisms for these energy sources when there is substantial sulfate in the environment (e.g., in marine ecosystems). The metabolic activity of *Bacteria* and *Archaea* drives the sulfur cycle, in which sulfur compounds are oxidized for energy and then reduced and used as terminal electron acceptors by a variety of bacteria (Fig. 11.15). In addition to being used for energy production, sulfur is an essential element required by all living cells for some amino acids (e.g., methionine and cysteine both contain sulfur) and for those proteins in the ETCs of respiring organisms that contain FeS clusters. The activities of the sulfur-oxidizing and sulfur-reducing bacteria generate biologically available sulfur that is assimilated by other cells as the organisms grow.

Sulfur can exist in a wide range of oxidation states; therefore, many different inorganic species of sulfur are found in natural environments. Microorganisms have evolved a myriad of pathways to utilize these sulfur species as a source of energy, electron donors, and electron acceptors. The next section describes the critical processes involved in both the anaerobic reduction of sulfate and the aerobic oxidation of reduced sulfur compounds generated by the sulfate reducers during the sulfur cycle.

Chemoheterotrophy and Chemolithoautotrophy in the Sulfur Cycle: Sulfate Reducers

Many microbes can use a variety of oxidized sulfur compounds (e.g., sulfate, SO_4^{2-}; elemental sulfur, S^0; and thiosulfate, $S_2O_3^{2-}$) as terminal electron acceptors in anaerobic respiration (Fig. 11.15). In addition, there are some *Bacteria* and *Archaea* (such as hyperthermophilic members of the *Euryarchaeota*, including

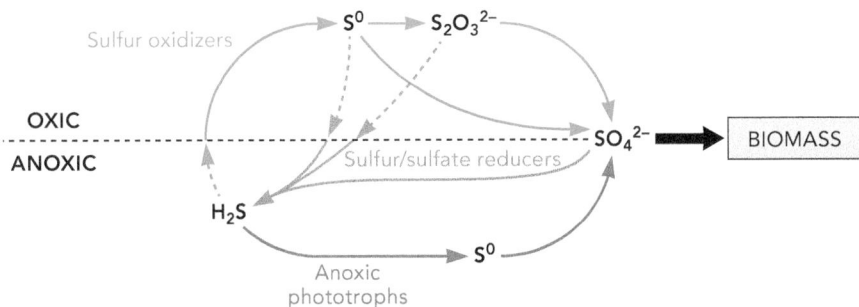

FIGURE 11.15 **Overview of the sulfur cycle.** The sulfur cycle occurs in both oxic and anoxic environments where oxidized and reduced forms of sulfur are interconverted by different groups of microorganisms as indicated. Dashed arrows indicate the diffusion of sulfur compounds to the oxic/anoxic interface. See the text for more details.

Pyrococcus and *Thermococcus*) that can reduce elemental sulfur and/or thiosulfate, but not sulfate. Since the biochemistry of sulfate reduction is understood in the most detail (although it is not yet fully understood), this section will focus on sulfate reduction.

In the process of **sulfate reduction**, SO_4^{2-} acts as a terminal electron acceptor in anaerobic respiration, producing H_2S (found as HS^- in environments above pH 7). Microbes that reduce sulfate through their respiration are widespread in anoxic aquatic and terrestrial environments where sulfate is available. Most bacteria capable of sulfate reduction are members of the *Deltaproteobacteria*; the most well-known genus is *Desulfovibrio*. However, there are a few sulfate reducers outside of this group, including endospore-forming members of the *Firmicutes* (*Desulfotomaculum* and *Desulfosporosinus*), a thermophilic member of the *Nitrospirae* (*Thermodesulfovibrio*), and a member of the *Euryarchaeota* (*Archaeoglobus*) (Fig. 11.16A).

The reduction of SO_4^{2-} to H_2S is a multistep process that requires either reduced organic compounds or H_2 as the electron donor (Fig. 11.16B). In the case of H_2, periplasmic hydrogenase enzymes cleave $H_2 \rightarrow 2\ H^+ + 2\ e^-$ in the periplasm, thus contributing to the PMF and providing electrons via cytochrome *c* for the ETC that will ultimately reduce sulfate. The first step in the reduction of sulfate, the conversion of sulfate to sulfite ($SO_4^{2-} \rightarrow SO_3^{2-}$), is not energetically favorable. This problem is solved by investing an ATP to activate sulfate to form adenosine phosphosulfate (APS) in a reaction catalyzed by ATP sulfurylase. The activated form of sulfate is a substrate for reduction in an energetically favorable reaction catalyzed by APS reductase, which forms sulfite and releases AMP. The next step, $SO_3^{2-} \rightarrow H_2S$ (catalyzed by the sulfite reductase complex and heterodisulfide reductase complex), is also thermodynamically favorable. The overall process conserves sufficient energy via the ETC to compensate for the ATP invested in the first step, and the most reduced form of sulfur, H_2S, is released as the end product of the pathway.

Most sulfate reducers are obligate anaerobes, but anaerobic respiration with sulfur compounds is not typically their only metabolic option; many sulfate-reducing bacteria can respire anaerobically with nitrate or other terminal electron acceptors, and some can ferment when external electron acceptors are unavailable. Most sulfate-reducing bacteria are heterotrophs capable of obtaining carbon for biosynthesis from a variety of organic carbon sources, such as

A

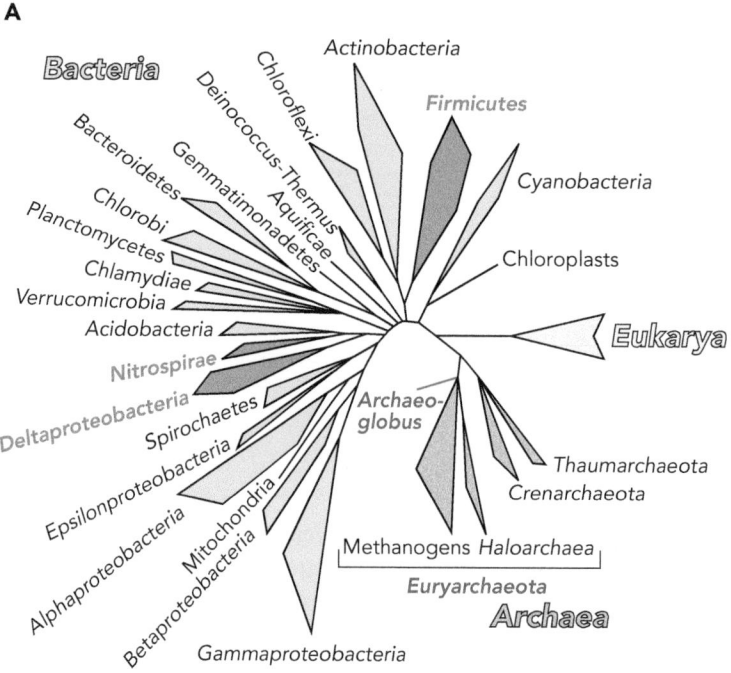

B

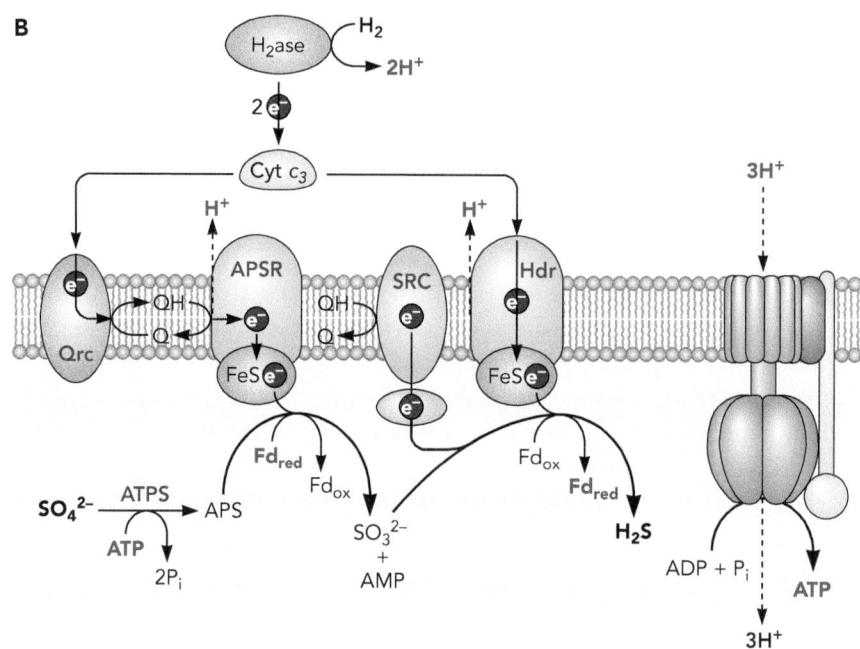

FIGURE 11.16 **Distribution and biochemistry of sulfate reduction.** (A) Phylogenetic tree highlighting lineages of *Bacteria* (dark coloring and bold green text) and *Archaea* (dark coloring and bold pink text) known to carry out sulfate reduction. (B) Energy conservation during sulfate reduction. In the cytoplasm, sulfate is first activated by the addition of AMP from ATP, which is catalyzed by ATP sulfurylase (ATPS), forming adenosine phosphosulfate (APS). A periplasmic hydrogenase (H_2ase) splits H_2 into electrons and protons. The protons are released in the periplasm and contribute to the proton motive force. The electrons are passed through an electron transport chain and used to reduce APS to sulfite (SO_3^{2-}) catalyzed by APS reductase (APSR). Sulfite is converted to H_2S by sulfite reductase complex (SRC) and heterodisulfide reductase complex (Hdr). Fd, ferredoxin.

pyruvate, lactate, and other organic acids and alcohols, which are products of fermentation. Pyruvate and lactate can also serve as organic electron donors for some sulfate-reducing bacteria. In addition, some sulfate reducers that use H_2 can grow autotrophically using the reductive acetyl-CoA pathway to fix CO_2 (Chapter 9), often competing with methanogens and acetogens for H_2 and CO_2.

Chemolithoautotrophy in the Sulfur Cycle: Sulfur Oxidizers

As chemolithoautotrophs, sulfur-oxidizing *Bacteria* and *Archaea* can use a range of reduced inorganic sulfur species for energy and electron donors for biosynthesis. *Bacteria* and *Archaea* that are capable of sulfur oxidation are physiologically diverse and include aerobes that respire with oxygen as the terminal electron acceptor, anaerobes that respire with nitrate, and anoxygenic photoautotrophs and photoheterotrophs. We have already discussed the metabolism of the anoxygenic purple and green sulfur phototrophs, which obtain energy from light and use reduced sulfur compounds as their electron source to yield ATP and NADPH + H^+ for CO_2 fixation in anoxic environments in the light (Chapter 10). In contrast, the aerobic **sulfur-oxidizing bacteria** are often termed "colorless" sulfur oxidizers to differentiate them from the pigmented sulfur-oxidizing phototrophs. Aerobic sulfur oxidizers use hydrogen sulfide (H_2S) as a source of energy, converting H_2S to elemental sulfur (S^0) or sulfate (SO_4^{2-}) with oxygen as the terminal electron acceptor. Similarly, some sulfur oxidizers can use S^0 or thiosulfate ($S_2O_3^{2-}$) as an energy source, also converting it to SO_4^{2-}. The protons and SO_4^{2-} released by many sulfur oxidizers create sulfuric acid, and therefore many sulfur oxidizers are acidophiles that thrive at low pH. Sulfur-oxidizing *Bacteria* and *Archaea* are phylogenetically diverse, but the majority are members of the *Alpha-*, *Beta-*, and *Gammaproteobacteria*, the *Nitrospirae*, and the *Crenarchaeota* (Fig. 11.17A). Two of the best-understood genera of sulfur-oxidizing bacteria are *Acidithiobacillus* and *Beggiatoa*. *Sulfolobus*, a member of the *Crenarchaeota*, is an autotrophic sulfur oxidizer that is both thermophilic and acidophilic, capable of growing at 90°C and pH 1 to 5.

This section will focus exclusively on sulfur-oxidizing bacteria that grow as chemolithoautotrophs. Because the various sulfur oxidizers are capable of using different sulfur compounds, there is no single pathway for sulfur oxidation, and much is not yet well understood. We present here a few common themes regarding sulfur oxidation that are found throughout the diversity of the microbial world.

During aerobic growth, sulfur-oxidizing microorganisms preferentially use reduced forms of inorganic sulfur as energy sources and electron donors for respiration. Since H_2S can be rapidly oxidized through abiotic chemical processes, sulfur oxidizers are often found growing in a narrow zone at oxic/anoxic interfaces where sulfide and oxygen are both present (e.g., in stratified lakes, soil layers, or the sediments of freshwater and marine environments). Here, they can access both H_2S produced in the anoxic zone by sulfate reducers and oxygen diffusing from the surface (Fig. 11.15 and 11.16).

Hydrogen sulfide is the preferred electron donor for many sulfur oxidizers (Table 11.1). In some organisms, H_2S may be oxidized to sulfate (Fig. 11.17B)

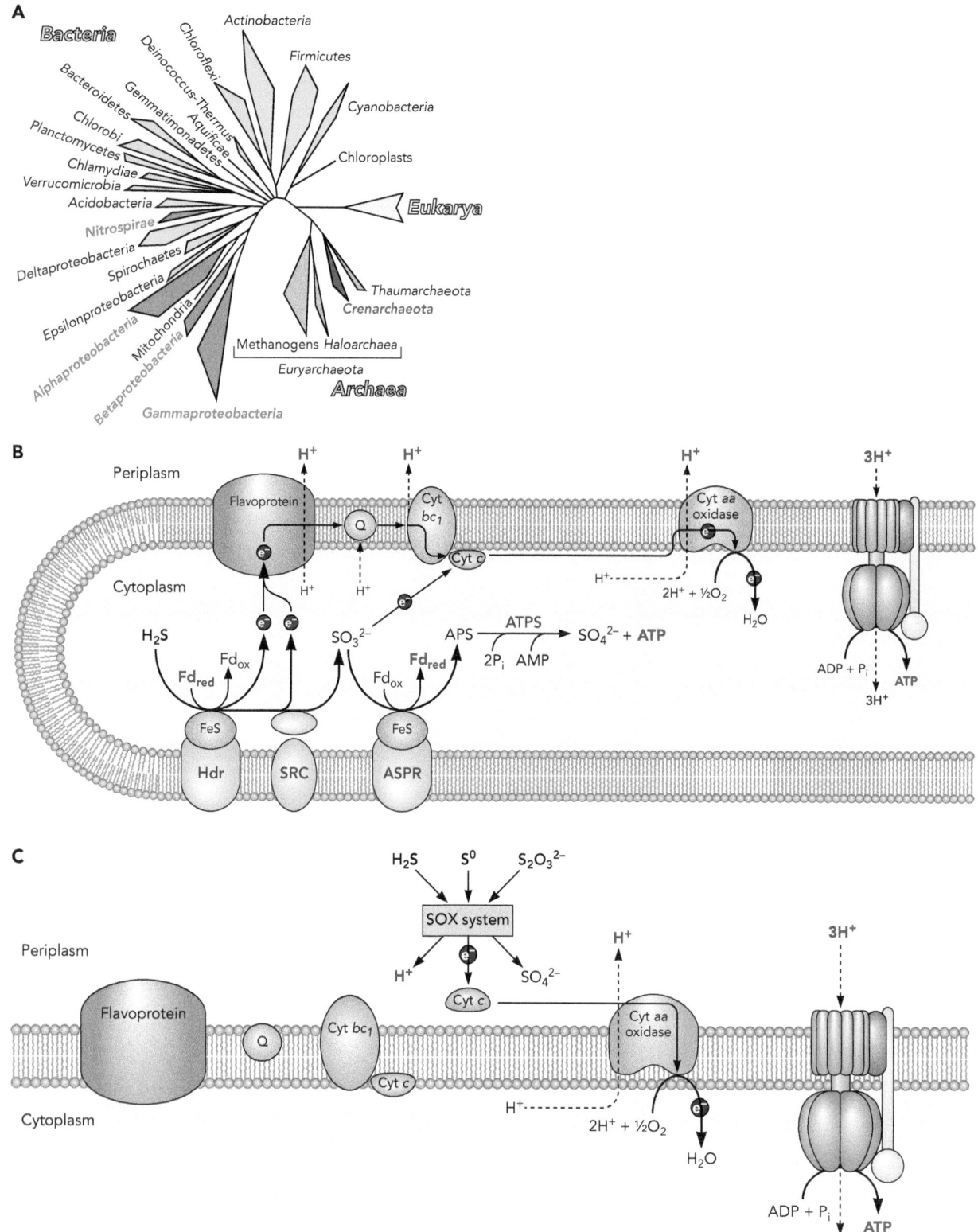

FIGURE 11.17 Distribution and biochemistry of inorganic sulfur oxidation. (A) Phylogenetic tree highlighting lineages of *Bacteria* (dark coloring and bold green text) and *Archaea* (dark coloring and bold pink text) known to carry out chemolithotrophic sulfur oxidation. (B) Energy conservation during sulfur oxidation. The oxidation of H_2S is mediated by the same enzymes utilized by sulfate reducers, but the reactions are run in the reverse direction. H_2S donates electrons to the heterodisulfide reductase complex (Hdr), the sulfite reductase complex (SRC), and APS reductase (APSR), resulting in the production of ATP. The electrons are donated to an electron transport chain (ETC) with oxygen serving as the terminal electron acceptor. Electron flow through the ETC generates a proton motive force (PMF) which is used to generate ATP via ATP synthase. (C) The SOX system allows certain sulfur oxidizers to utilize diverse forms of reduced sulfur to generate PMF. Electrons are donated directly to cytochrome *c*, bypassing the Cyt bc_1 coupling complex, contributing fewer H^+ to the PMF. Fd, ferredoxin; Q, ubiquinone; APS, adenosine phosphosulfate.

using oxygen as the terminal electron acceptor, via the following reaction, which generates sulfuric acid:

$$H_2S + 2O_2 \rightarrow SO_4^{2-} + 2H^+$$

The sulfur atom is oxidized in a stepwise manner by running the reductases discussed for sulfate reduction (Fig. 11.16B) in reverse, producing sulfite, then sulfate. Electrons from H_2S are donated to a flavoprotein, and then travel into a quinone pool generating a PMF from multiple coupling sites as they move through the ETC. The electrons are ultimately transferred to a Cyt *aa* oxidase, which reduces O_2 to H_2O and further contributes to the PMF.

Alternatively, other sulfur-oxidizing organisms utilize a complex cycle of proteins of the SOX (sulfur oxidation) system, in which the sulfur compounds are oxidized in a stepwise fashion, resulting in the production of protons and sulfate. Electrons donated from H_2S, S^0, and $S_2O_3^{2-}$ via the SOX system enter the ETC through a periplasmic cytochrome *c* complex (Fig. 11.17C), bypassing the flavoprotein and quinone pool, and contributing fewer protons to the PMF. Therefore, sulfur oxidizers conserve less energy from H_2S when the SOX system is used, or when S^0 or $S_2O_3^{2-}$ serves as the electron donor. The electrons released by the SOX system ultimately reduce oxygen, and the sulfur atom is oxidized to sulfate. The production of protons and sulfate results in sulfuric acid being generated as a byproduct, as illustrated in the following reactions:

$$2S^0 + 2H_2O + 3O_2 \rightarrow 2SO_4^{2-} + 4H^+$$

$$S_2O_3^{2-} + H_2O + 2O_2 \rightarrow 2SO_4^{2-} + 2H^+$$

In some sulfur oxidizers, excess H_2S is oxidized to S^0 ($H_2S + 2O_2 \rightarrow S^0 + H_2O$), which is then stored as intracellular inclusion bodies (sulfur granules) until it is needed. Since other bacteria present in the environment (e.g., sulfur reducers) compete for S^0 as an electron acceptor (see above), the ability to store S^0 inside the cell is beneficial for sulfur oxidizers. The sulfur oxidizers and sulfur reducers often live in close physical proximity to one another to efficiently utilize the others' end products. As was already noted, purple and green sulfur phototrophs (Chapter 10) also contribute to the sulfur cycle in environments where light is available.

The Anaerobic Food Web and Syntrophy

As discussed previously, reduced organic matter (e.g., leaves and other plant material) enters anaerobic environments mainly in the form of polymers, which are degraded by exoenzyme-producing heterotrophs to generate glucose and other small organic acids. Fermentation of these metabolites produces H_2 and CO_2. Sulfate reducers can grow as heterotrophs utilizing organic acids or as autotrophs consuming $H_2 + CO_2$, converting sulfate to S^0 and H_2S in the process. Methanogens and acetogens compete for the $H_2 + CO_2$ produced by fermenting microbes and grow as autotrophs using the reductive acetyl-CoA pathway to fix CO_2 (Fig. 11.18). Some methanogens can utilize other C1 compounds or acetate; the methane produced either diffuses to the surface, where it is oxidized by aerobic methane oxidizers, or it is oxidized anaerobically as described below. Thus, the byproducts generated by some microorganisms are used as carbon and energy sources by other microbes, resulting in the cycling of carbon and sulfur through oxidation and reduction reactions.

Instead of competing for available nutrients in the anaerobic food web, the hydrogen-producing fermenters have an interdependent syntrophic association involving **interspecies hydrogen transfer** with sulfur reducers and methanogens. **Syntrophy** is a type of metabolic dependence in which two or more organisms work together to carry out a metabolic process that they are incapable of as individuals. For example, ethanol is a common fermentation product that can be used by secondary fermenters (i.e., hydrogen producers) (Chapter 6) in a process that generates H_2, but the reaction has a positive change in free energy ($\Delta G^{0'}$), indicating that it is energetically unfavorable:

$$2\,CH_3CH_2OH + 2\,H_2O \rightarrow \mathbf{4\,H_2} + 2\,CH_3COO^- + 2\,H^+\,(\Delta G^{0'} = +19.4\,kJ)$$

$$\mathbf{4\,H_2} + CO_2 \rightarrow CH_4 + 2\,H_2O\,(\Delta G^{0'} = -130.7\,kJ)$$

In an anaerobic food web, hydrogen-producing bacteria rely on hydrogen-consuming organisms such as methanogens to use hydrogen as their electron donor, keeping H_2 levels low enough so that the overall reactions are energetically favorable. In this case, when the H_2 produced as an end product is removed from the environment, the $\Delta G^{0'}$ is negative and the overall process is energetically favorable:

$$2\,CH_3CH_2OH + CO_2 \rightarrow CH_4 + 2\,CH_3COO^- + 2\,H^+\,(\Delta G^{0'} = -111.3\,kJ)$$

Thus, both the hydrogen-producing and hydrogen-consuming organisms benefit from the transfer of hydrogen in an anaerobic food web. This example illustrates the interdependence of microbes with different physiologies and highlights the challenges of identifying appropriate culture conditions to isolate microorganisms in pure culture—remember from Chapter 1 that it is estimated that only <5% of all microorganisms can be grown in the laboratory. In addition, it demonstrates the complex relationships that exist in microbial communities.

We described the reactions involved in aerobic methane oxidation earlier in this chapter; however, recent studies have identified a syntrophic process for *anaerobic* methane oxidation. Certain members of the *Euryarchaeota* that are related to the methanogens are capable of oxidizing methane in the absence of oxygen. These organisms, commonly termed "ANME" for <u>a</u>naerobic <u>me</u>thanotrophs, are obligate anaerobes that live in anoxic marine and freshwater sediments. They use the methanogenesis pathway in reverse to oxidize methane to CO_2 (Fig. 11.18, purple arrow). However, running the methanogenesis pathway in reverse is energetically unfavorable. ANME *Archaea* live in syntrophic association with sulfate-reducing bacteria (SRB), which make a living through anaerobic respiration with H_2 as the electron donor and sulfate as a terminal electron acceptor as described above. Sulfate reduction is an energetically favorable process and, when carried out in conjunction with reverse methanogenesis, allows ANME/SRB consortia to conserve enough energy to grow slowly. In this syntrophic process, electrons in H_2 from the oxidation of methane are used by the SRB to reduce sulfate to sulfide.

$$CH_4 + 3\,H_2O \rightarrow HCO_3^- + \mathbf{4\,H_2} + H^+\,(\Delta G^{0'} = +132\,kJ)$$

$$SO_4^{2-} + \mathbf{4\,H_2} + H^+ \rightarrow HS^- + 4\,H_2O\,(\Delta G^{0'} = -150\,kJ)$$

$$\text{Overall reaction}: CH_4 + SO_4^{2-} \rightarrow HCO_3^- + HS^- + H_2O\,(\Delta G^{0'} = -18\,kJ)$$

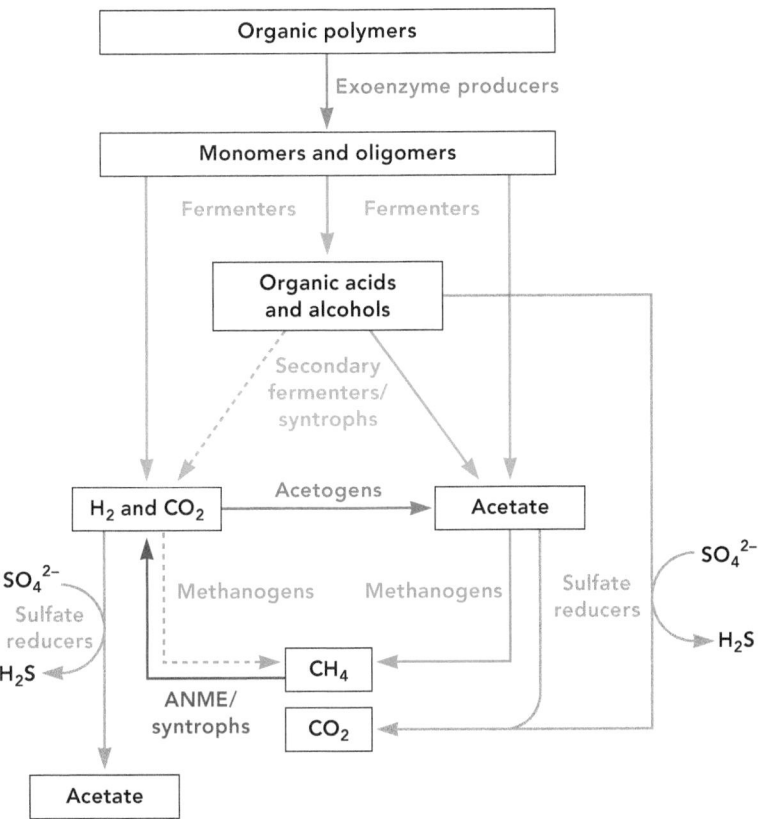

FIGURE 11.18 **The role of microbes in the cycling of carbon and sulfur during the metabolic processes in an anaerobic food web.** The steps of the anaerobic food web are carried out by a variety of microbes associated with the carbon and/or sulfur cycles. Syntrophic relationships between different groups of organisms are highlighted by the dashed arrows. See the text for details.

The absence of pure cultures of ANME combined with the very slow doubling times of ANME-containing consortia (weeks to months) have made studies of the process difficult. Evidence supporting the reverse methanogenesis process includes the identification of genes encoding the methanogenesis pathway in metagenomes of ANME consortia, as well as the demonstration that specific inhibitors of methanogenesis also inhibit reverse methanogenesis. ANME are believed to be important in preventing the release of large amounts of the potent greenhouse gas CH_4 into the atmosphere. Estimates indicate that 80 to 90% of the methane produced in ocean sediments is oxidized by ANME consortia.

Additional studies have demonstrated diversity in the anaerobic methane oxidation process. For example, ANME that couple methane oxidation to iron or manganese reduction either alone or in conjunction with a bacterial partner have been reported. In addition, one type of ANME that apparently does not need a bacterial partner couples methane oxidation by reverse methanogenesis to the reduction of nitrate to nitrite. These organisms, members of a phylogenetically distinct clade of the *Archaea* (ANME-2d), are found in nitrate-rich freshwater sediments. They live in association with denitrifying bacteria (Chapter 12) that utilize the nitrite generated during anaerobic respiration as their terminal electron acceptor. Thus, not only are the carbon and sulfur cycles intimately linked, they are also linked to the nitrogen cycle described in Chapter 12.

Learning Objectives: After completing the material in this chapter, students should be able to . . .

1. identify the processes, electron donors, and acceptors involved in the oxidation and reduction of carbon-containing molecules during the carbon cycle

2. compare and contrast chemoorganotrophy and chemolithotrophy

3. describe the sequential degradation of aromatic compounds in oxic and anoxic environments

4. explain how a facultative organism like *Escherichia coli* regulates its choice of electron acceptors during aerobic and anaerobic growth

5. outline the major steps of the one-carbon (C1) carbon cycle (i.e., methanogenesis, methylotrophy, and acetogenesis)

6. identify the processes, electron donors, and acceptors involved in the oxidation and reduction of sulfur-containing molecules in the sulfur cycle

7. describe the relationships between the microbes within an anaerobic food web

Check Your Understanding

1. Define this terminology: exoenzymes, cellulases, amylases, proteases, lipases, ligninolytic exoenzymes, monooxygenases, dioxygenases, ring-cleavage dioxygenases, methanogens, hydrogenases, electron bifurcation, methylotrophs, methanotrophs, acetogenesis, sulfate reduction, sulfur-oxidizing bacteria, syntrophy, interspecies hydrogen transfer

2. Briefly describe the difference between a chemoorganotroph and a chemolithotroph. (LO2)

3. Because life on Earth is carbon-based, carbon cycling is of critical importance to almost every aspect of life and to the survival of the biosphere. Fill in the table below by naming each process in the C cycle (indicated by the chemical reactions) using these terms: **aerobic respiration**, **anaerobic respiration**, **CO_2 fixation**, **methanogenesis**, and **methanotrophy**. Indicate whether the bolded compound in the chemical reactions below functions as an **energy source**, **electron acceptor**, or **carbon source**. (Note: Some molecules play multiple roles and some terms may be used more than once.) (LO1)

Chemical reaction	Process	Function of bold compound
$4 H_2 + \mathbf{CO_2} \rightarrow CH_4 + 2H_2O$		
Acetate [CH_3COO^-] $+ SO_4^{2-} \rightarrow HS^- + 2HCO_3^-$		
$\mathbf{CH_4} + 2O_2 \rightarrow 2H_2O + CO_2$		
$\mathbf{CO_2} + H_2O \rightarrow (CH_2O)_x + O_2$		
Succinate [$C_2H_4(COO^-)_2$] $+ O_2 \rightarrow CO_2 + H_2O$		
$\mathbf{CO_2} + 2H_2S \rightarrow (CH_2O)_x + H_2O + 2S^0$		

4. Most of the organic material in natural environments exists as organic polymers such as **cellulose**, **starch**, **lipids**, and **proteins**. (LO3)

   a) Why do bacteria need to use exoenzymes to process these organic compounds?

   b) Match the name of each polymer structure in the table below (from the polymers listed above) and name the exoenzyme that breaks it down.

Structure	Name	Enzyme that degrades it
I.		
II.		
III.		
IV.		

5. Lignin is believed to be one of the most common organic polymers on Earth. Given the structure of lignin (Fig. 11.3B), what makes it so strong and stable? (LO3)

6. The degradation of smaller aromatic compounds requires that the ring structure be destabilized to allow subsequent cleavage of the carbon-carbon bonds within the ring. (LO3)

   a) One well-studied **aerobic** pathway for aromatic acid (e.g., benzoate) catabolism is the β-ketoadipate pathway (Fig. 11.6A).

      i. How is the aromatic ring structure of benzoate destabilized in this pathway?

      ii. What end products are formed during aerobic benzoate degradation that can then feed into central metabolism for ATP synthesis and biosynthesis?

b) Benzoate can also be degraded under **anoxic** conditions (Fig. 11.6B).

    i. How is the aromatic ring of benzoate destabilized under anoxic conditions?

    ii. In this case, what end products are formed during anaerobic benzoate degradation that can then feed into central metabolism for ATP synthesis and biosynthesis?

7. As a facultative chemoorganotrophic organism, *E. coli* serves as a model for the metabolism of sugars under varying oxic/anoxic conditions. (LO4)

    a) Why is oxygen preferred as a terminal electron acceptor over other electron acceptors?

    b) Why does *E. coli* have two terminal oxidases, cytochrome *o* oxidase and cytochrome *d* oxidase? Under what conditions is each used?

8. Under certain growth conditions, *E. coli* can switch its metabolism from respiration to fermentation. (LO4)

    a) Which process (aerobic respiration, anaerobic respiration, or fermentation) generates the LEAST amount of energy from one mole of glucose? Why would *E. coli* use this mode of growth?

    b) Under fermentative conditions, how do bacteria generate a proton motive force (PMF)? Why do bacteria need to generate a PMF if they are not using it to make ATP?

9. *E. coli* monitors the environment and alters gene expression such that the best available terminal electron acceptor will be utilized first. Many of these regulatory schemes involve a two-component regulatory system. Explain what is happening at each step (1 to 5). (LO4)

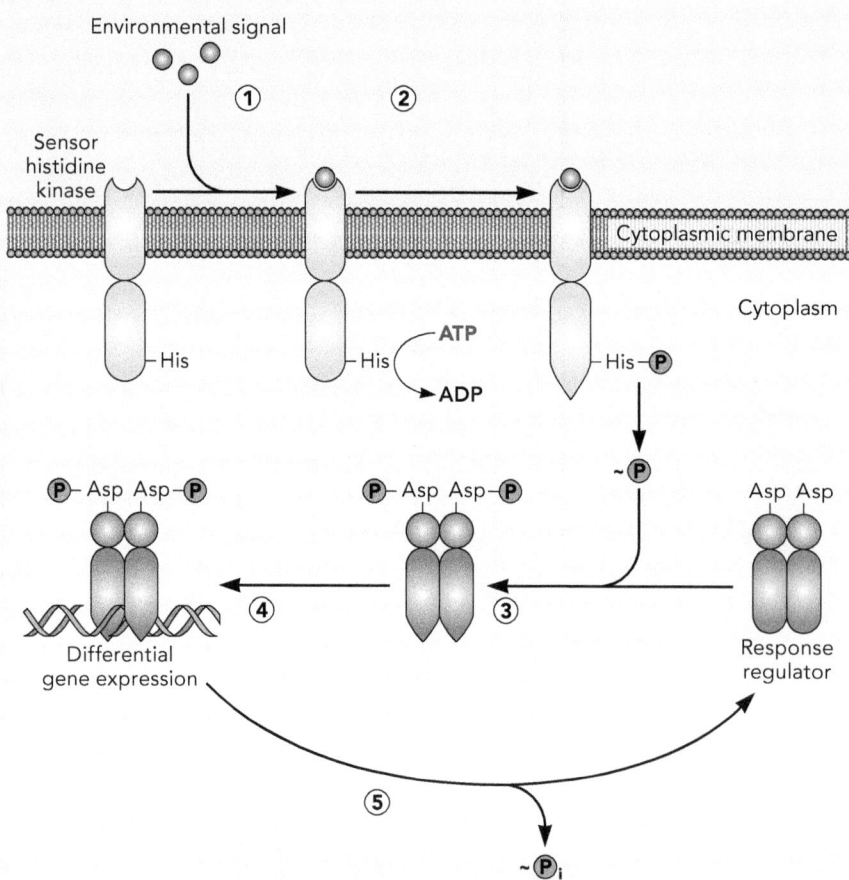

10. ArcA/ArcB make up a two-component regulatory system that senses the shift from fully oxic to microoxic conditions. (LO4)

    a) Which protein (ArcA or ArcB) acts as the sensor histidine kinase protein? What environmental signal is detected by the sensor protein?

    b) Under atmospheric oxygen concentrations, the sensor histidine kinase protein is present, but it is in an inactive state. What does this protein do when the cell experiences microoxic conditions?

    c) What function does the other protein (ArcA or ArcB) carry out? What does this protein do when it is phosphorylated?

11. FNR (fumarate nitrate reductase regulator) is a global regulator that controls multiple promoters under anoxic conditions, thereby acting as a master regulator for anaerobic growth. (LO4)

    a) Under what conditions is FNR active and inactive?

    b) FNR controls the regulation of genes associated with oxidases and reductases within the *E. coli* electron transport chain. How does FNR inhibit the production of terminal oxidases? Does this inhibition happen in the presence or absence of oxygen?

    c) How does FNR activate the production of terminal reductases? Does this activation happen in the presence or absence of oxygen?

12. After oxygen, nitrate (NO_3^-/NO_2^-; $E_0' = +420$ mV) is the next preferred terminal electron acceptor for *E. coli*. NarX and NarL are part of a complex two-component regulatory system that senses and responds to the presence of nitrate. (LO4)

    a) Which protein (NarX or NarL) acts as the sensor histidine kinase protein? What environmental signal is detected by the sensor protein?

    b) When this environmental signal is absent, the sensor histidine kinase protein is present but is in an inactive state. What does the sensor protein do when the signal molecule is present?

    c) What function does the other protein (NarX or NarL) carry out? What does this protein do when it is phosphorylated?

13. Methane is a potent greenhouse gas that is biologically produced by a select group of *Archaea*. One methanogenesis pathway can be described by the equation below. (LO5)

$$4H_2 + CO_2 \rightarrow CH_4 + 2H_2O$$

    a) In this process, which compound in the reaction above is oxidized (i.e., acts as a source of electrons)? Which compound is reduced (i.e., acts as the electron acceptor)?

    b) Outline how this process provides energy for these *Archaea*. Where is a proton/sodium motive force generated?

    c) How do these cells get carbon for biosynthesis?

14. As methane is a potent greenhouse gas, it is fortunate that a variety of *Bacteria* (methanotrophs) can use methane as a carbon and energy source, thereby decreasing the concentration of methane in the atmosphere. (LO5)

    a) The first step in aerobic methanotrophy can be described as follows.

$$CH_4 + O_2 + 2H^+ \rightarrow CH_3OH \text{ (methanol)} + H_2O$$

    i. Which compound in the above reaction is oxidized (i.e., acts as a source of electrons)? Which compound is reduced (i.e., acts as the electron acceptor)?

    ii. Which key enzyme facilitates this step?

    b) In the next step, methanol (CH_3OH) is converted to formaldehyde (CH_2O).

    i. Which enzyme facilitates this step?

    ii. Is methanol oxidized or reduced? Where do the electrons go?

    c) Formaldehyde is then converted to formate ($HCOO^-$) and then to CO_2. How does this process provide energy for these bacteria? Where is a proton motive force generated?

    d) What is the function of the ribulose monophosphate (RuMP) and serine cycles in methanotrophs?

    e) Why are aerobic methanotrophs typically found at the oxic/anoxic interface in aquatic sediments and soil environments?

15. Acetogenesis is another type of C1 metabolism that involves the reduction of CO_2 and formation of the reduced carbon byproduct, in this case acetate. The overall reaction can be described as follows. (LO5)

$$4H_2 + 2CO_2 \rightarrow CH_3COO^- + 2H_2O$$

a) In this process, which compound in the above reaction is oxidized (i.e., acts as a source of electrons)? Which compound is reduced (i.e., acts as the electron acceptor)?

b) Outline how these bacteria produce acetate from CO_2/bicarbonate.

c) How do these bacteria make ATP? How is electron bifurcation important to this process?

16. Sulfur is an essential element for all living cells, and the sulfur cycle provides this essential element. Sulfate reduction and sulfur oxidation are both important components of this cycle. (LO6)

a) Name each process in the S cycle (indicated by the chemical reactions) using these terms: **aerobic respiration**, **anaerobic respiration**, and **CO_2 fixation**. Indicate whether the bolded compound in the chemical equations below functions as an **energy source** or **terminal electron acceptor** (in anaerobic respiration). (Note: Some terms may be used more than once.)

Chemical equation	Process	Function of bold compound
H_2S + $2O_2 \rightarrow SO_4^{2-}$ + $2H^+$		
$4 H_2$ + **SO_4^{2-}** + $H^+ \rightarrow H_2S$ + $4H_2O$		
CH_3COO^- + **SO_4^{2-}** $\rightarrow HS^-$ + $2H_2CO_2$		
$2S^0$ + $2H_2O$ + $3O_2 \rightarrow 2SO_4^{2-}$ + $4H^+$		

b) Which two processes above require that the cells use reverse electron flow to generate NAD(P)H + H^+?

17. Sulfate-reducing microbes are widespread in anoxic aquatic and terrestrial environments wherever sulfate is available. (LO6)

a) In the first step of sulfate reduction, sulfate (SO_4^{2-}) is converted to sulfite (SO_3^{2-}). Why is ATP required? What is the role of ATP sulfurylase?

b) When H_2 is the energy source, electrons move from H_2 to a hydrogenase enzyme, through an ETC, and are eventually transferred to APS or SO_3^{2-}. Where in the ETC is a PMF generated?

18. Hydrogen sulfide, the product of sulfate reduction, is a colorless toxic gas that can be oxidized chemically in the presence of oxygen or biologically by aerobically respiring chemolithoautotrophic bacteria. Biological oxidation of sulfide by chemolithotrophs can be generally described by the equation below. (LO6)

$$H_2S + 2O_2 \rightarrow SO_2^{2-} + 2H^+$$

a) Like methanotrophs, sulfide-oxidizing organisms are often found at the oxic/anoxic interface in soils or sediments. Explain why.

b) Outline how these organisms use hydrogen sulfide to generate a PMF.

c) The SOX (sulfur oxidation) system allows some bacteria to use sulfide, sulfur, or thiosulfate as an energy source. However, less ATP is made when cells use the SOX system. Looking at the ETC, why do you think the SOX system generates less ATP compared to sulfide oxidizers that use a flavoprotein?

19. Hydrogen sulfide can also be oxidized under anoxic conditions by the green and purple phototrophic bacteria. What is the energy source for these organisms? How do these bacteria use the electrons from hydrogen sulfide? (LO6)

20. The recycling of organic carbon in nature is a complex process that relies on many different microbes. Fill in the blanks below with the groups of microorganisms from the anaerobic food web that produce and use the indicated byproducts of metabolism. (LO7)

a) _____ generate monomers that are used by _____.

b) _____ produce organic acids that are used by _____ and _____.

c) _____ generate H_2 and CO_2 that are used by _____, _____, and _____.

21. The hydrogen-producing fermenters in the anaerobic food web have interdependent syntrophic associations with sulfur reducers and methanogens. (LO7)

    a) For example, fermenting bacteria can convert ethanol to acetate by this chemical reaction:

    $$CH_3CH_2OH + 2H_2O \rightarrow 4H_2 + 2CH_3COO^- + 2H^+ (\Delta G^{0\prime} = +19.4kJ)$$

    i.   What does the $\Delta G^{0\prime}$ indicate about this process?

    ii. How do methanogens make this process more likely to occur?

    b) There are estimates indicating that 80 to 90% of the methane produced in ocean sediments is oxidized under anoxic conditions by ANME (<u>an</u>aerobic <u>me</u>thanotrophs). The ANME *Archaea* use the methanogenesis pathway in reverse to oxidize methane to CO_2 in a process that can be described by this chemical reaction

    $$CH_4 + 3H_2O \rightarrow HCO_3^- + 4H_2 + H^+ (\Delta G^{0\prime} = +131.8kJ)$$

    i.   What does the $\Delta G^{0\prime}$ indicate about this process?

    ii. How do sulfate-reducing bacteria make this process more likely to occur?

Dig Deeper

22. The sulfur and carbon cycles are intimately linked because all sulfur-oxidizing and sulfate-reducing *Bacteria* and *Archaea* need a carbon source for biosynthesis and some use carbon compounds as a source of energy. (LO1, 6)

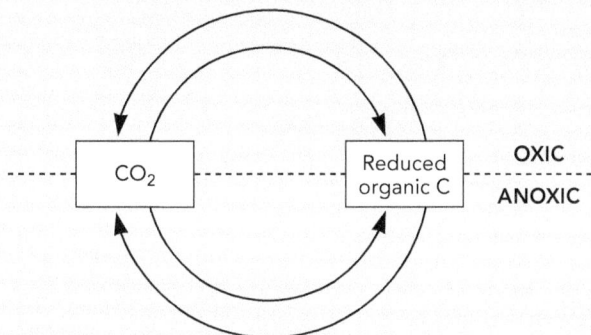

    a) Considering the carbon cycle, label the arrows in the appropriate places on the diagram with these processes: **aerobic respiration, CO$_2$ fixation, fermentation,** and **anaerobic respiration**. (Note: Some arrows may have more than one label, and some names may be used more than once.)

    b) Consider how the **sulfate-reducing bacteria** use carbon compounds. Draw three arrows on the carbon cycle diagram to indicate how sulfate-reducing bacteria use carbon compounds and explain your arrows.

    c) Consider where the **sulfur-oxidizing bacteria** get carbon for biosynthesis. Draw one arrow on the carbon cycle diagram to indicate where sulfur-oxidizing bacteria obtain carbon.

23. Aromatic molecules from petroleum and other industrial products are considered to be toxic waste, many of which are carcinogens. Not surprisingly, there is much interest in the persistence and biodegradation of these compounds in the environment. Toluene is an aromatic compound found in gasoline, petroleum, and coal and is produced for many industrial processes. There are many similarities between the degradation pathways of benzoate (Fig. 11.6A) and toluene. A toluene degradation pathway found in *Pseudomonas* is shown. (LO3)

    a) Compare and contrast the benzoate and toluene structures. How are they similar? How are they different?

    b) Do the benzoate and toluene degradation pathways occur under oxic or anoxic conditions? How do you know?

c) Compare and contrast steps 1 and 2 in the benzoate and toluene pathways. What enzymes carry out steps 1 and 2? What is the purpose of these steps?

d) Where in each pathway is the ring structure cleaved? What enzymes carry out this reaction? Is this *ortho* or *meta* cleavage in each case?

e) What products of toluene degradation (e.g., central precursors) feed into central metabolism pathways and are used for biosynthesis?

24. In the absence of oxygen, FNR (fumarate and nitrate reductase regulator) controls multiple promoters and is a master regulator for anaerobic growth. As with CRP (cAMP receptor protein), the binding of FNR to specific promoters can result in the positive or negative regulation of gene expression, depending on the position of its interaction with respect to the transcription start site, +1. The graphs indicate where different regulatory proteins bind on the DNA by showing the frequency of binding to DNA at that site. Based on what you know about how activators and repressors work, which graph (A or B) shows a regulator that is acting as a repressor? Which (A or B) is acting as an activator? Explain your reasoning. (LO4)

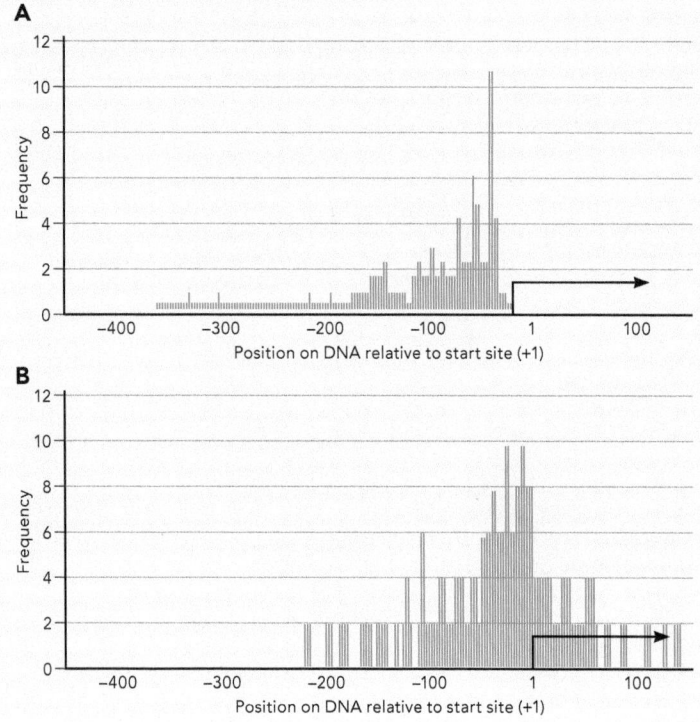

A

B

25. The graph shown represents the oxygen (red) and sulfide (yellow) concentrations **during the daytime** in an estuarine pond with light at the surface and mud at the bottom. Sulfate is present throughout the water column. (LO1, 6)

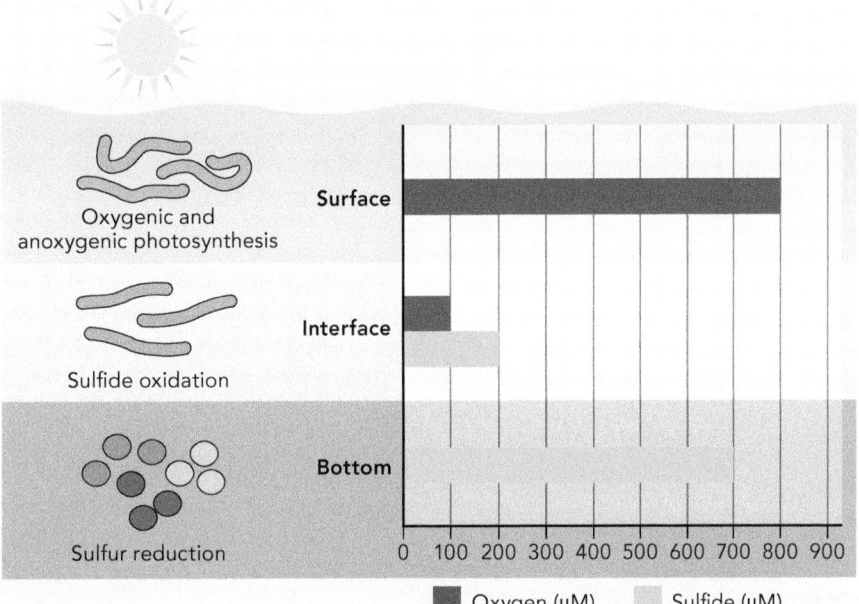

a) **During the daytime,**

    i. What group(s) of microbes contribute to the high concentration of O_2 near the surface?

    ii. Which group(s) of microbes consume O_2?

    iii. What group(s) of microbes contribute to the high concentration of sulfide at the bottom?

    iv. Which group(s) of microbes consume sulfide?

b) Predict how the concentration of O_2 and sulfide would change (increase, decrease, or stay the same) relative to the daytime concentrations in each area of the column **during the night**. Fill in the table and explain your reasoning.

Location	Change in [O_2] during night	Change in [sulfide] during night
Surface		
Interface		
Bottom		

26. Cows are ruminant animals with a four-chamber anoxic digestive system that is essentially a huge fermentation vat filled with microorganisms that break down the feed (fiber and cellulose), which the cow cannot digest on its own. The microbes provide smaller organic compounds (volatile fatty acids [VFAs]) and cell biomass that the cows can then digest and absorb, providing the ruminant with much more carbon and protein than it would otherwise obtain.

a) A schematic of the digestion process is shown. Fill in the boxes with the number indicating the types of bacteria that are likely doing each step. (Note: Numbers can be used more than once or not at all.)

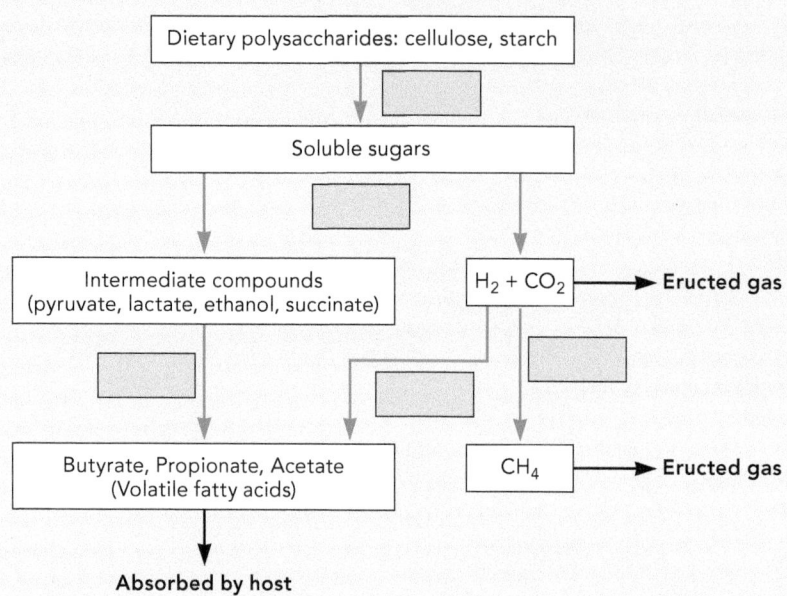

1. Fermentative aerotolerant lactic acid bacteria

2. Hydrolytic facultatively anaerobic heterotrophs like *Bacillus*

3. Methanogenic bacteria like *Methanobacterium*

4. Methylotrophic bacteria like *Methylobacterium*

5. Mixed-acid fermenting anaerobes like *Clostridium*

6. Nitrate-reducing *E. coli* growing on glucose

7. Obligately anaerobic acetogenic bacteria

b) Where in the diagram would you find a syntrophic relationship? Describe this relationship.

27. Humans have a linear digestive system, in which simple nutrients are absorbed in the small intestine while resistant polysaccharides like cellulose and starch (which humans cannot digest) move into the large intestine. Inside the large intestine are trillions of microorganisms that function as an anaerobic food web, ultimately breaking down these otherwise indigestible compounds into short-chain fatty acids (SCFAs), which are reabsorbed through the large intestine and utilized by the host as a carbon and energy source. The human colonic microbiota is mostly composed of the bacterial phyla *Firmicutes* and *Bacteroidetes*, which represent more than 90% of the total community.

a) Rather than break the polymers into simple sugars, *Bacteroides* partially degrade the polysaccharides to short-chain oligosaccharides that they then transport into the cell. How might this be an advantage for *Bacteroides*, given the competition for shared nutrients in the large intestine?

b) *Firmicutes* spp. are known to be one of the major producers of SCFAs in the colon. Recent evidence suggests that human intestinal microbiota may play a role in the development of obesity. For example, some experiments have shown that the microbiota of obese mice have significantly more *Firmicutes* relative to lean littermates. Propose a hypothesis for how an increase of *Firmicutes* could cause weight gain in the host.

c) Research has shown that the ratio of *Firmicutes* to *Bacteroidetes* can change with diet. Suppose you compared the intestinal microbiota of a vegetarian to a meat-eater. Which do you think would have a higher proportion of *Bacteroidetes*? Explain your reasoning.

276 Making Connections

276 Overview of the Nitrogen Cycle

277 Nitrogen Fixation

278 Biochemistry of Nitrogen Fixation

280 Regulation of Nitrogen Fixation

282 Symbiotic Plant-Microbe Interactions during Nitrogen Fixation

284 Assimilatory Nitrate Reduction

285 Ammonia Assimilation into Cellular Biomass

287 Nitrification: Ammonia Oxidation, Nitrite Oxidation, and Comammox

290 Anammox: Anaerobic Ammonia Oxidation

293 Denitrification

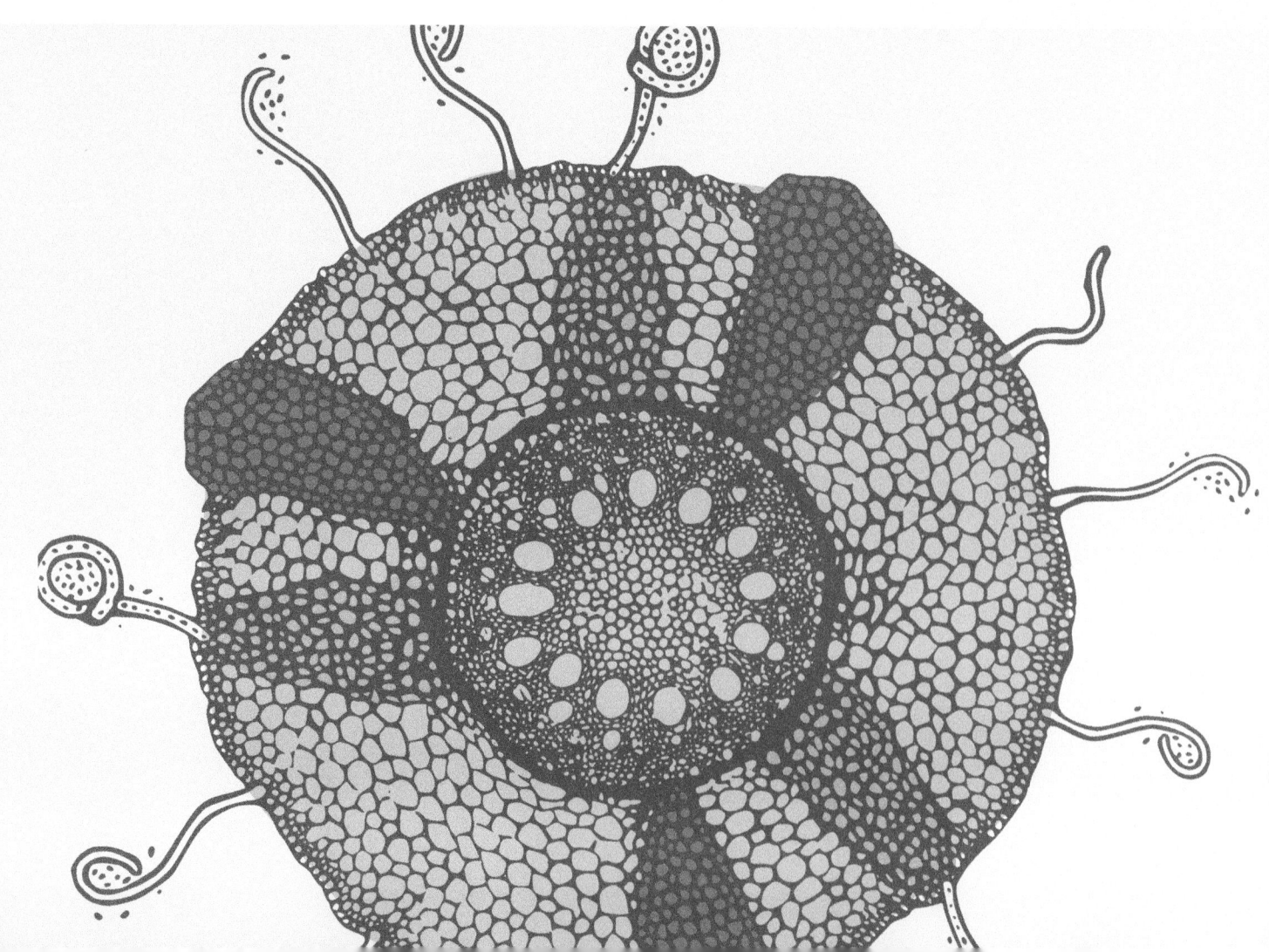

12

MICROBIAL CONTRIBUTIONS TO THE NITROGEN CYCLE

After completing the material in this chapter, students should be able to . . .

1. provide an overview of the chemical transformations in the nitrogen cycle and the lifestyles of the bacteria that carry them out

2. explain the process and regulation of nitrogen fixation

3. describe how plants and bacteria communicate and benefit from symbiotic nitrogen fixation

4. discuss the common mechanisms of nitrate and ammonia assimilation

5. compare and contrast the microbial processes that use nitrogen compounds as energy sources (oxidation) and those that use nitrogen compounds as electron acceptors (reduction)

Microbial Physiology: Unity and Diversity, First Edition. Ann M. Stevens, Jayna L. Ditty, Rebecca E. Parales and Susan M. Merkel.
© 2024 American Society for Microbiology.
Companion website: www.wiley.com/go/stevens/microbialphysiology

Making Connections

Chapters 9 to 11 introduced many different metabolic lifestyles of microbes, from autotrophy and phototrophy to chemotrophy, where the vital role microbes play during the carbon and sulfur cycles was highlighted. In this chapter, we introduce the diverse microbes associated with the nitrogen cycle. Some metabolic pathways that transform inorganic nitrogen between oxidized and reduced forms in the nitrogen cycle yield energy for the cell, including nitrate reduction during anaerobic respiration, as was described in Chapter 11. Other reactions require energy to generate reduced nitrogen compounds for cellular biomass. Nitrogen is essential for all living cells, as it is found in many cell components including amino acids, nucleic acids, and cell walls. Ammonia is used as the starting molecule for incorporating nitrogen into organic biomass. Most microorganisms rely on ammonia that is available in the environment; however, different bacterial species can convert nitrogen gas or other oxidized forms of nitrogen (e.g., NO_3^-, NO_2^-) to the ammonia required for cellular growth. In this chapter, we also describe additional mechanisms bacteria use for energy conservation using nitrogen compounds as either electron donors or electron acceptors. It is the interconversion of oxidized and reduced forms of nitrogen by microorganisms that drives the nitrogen cycle.

Overview of the Nitrogen Cycle

Like many of the metabolic processes that we have already discussed, the steps of the nitrogen cycle involve a series of oxidation and reduction reactions occurring under both oxic and anoxic conditions (Fig. 12.1). Therefore, the nitrogen cycle serves as a good opportunity to review different microbial

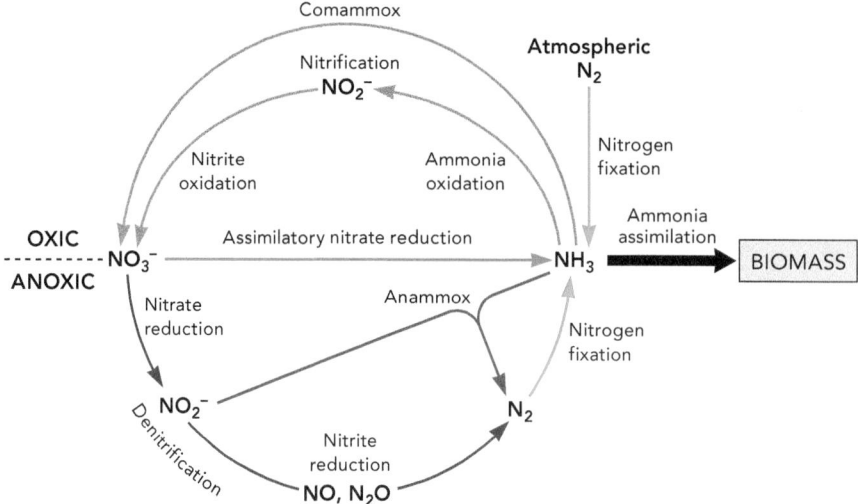

FIGURE 12.1 **Overview of the nitrogen cycle.** The nitrogen cycle occurs in both oxic and anoxic environments where oxidized and reduced forms of nitrogen are interconverted by different microbial processes as indicated. Blue, green, and black arrows indicate energy-requiring reactions, while gold, pink, and purple arrows indicate energy-yielding reactions for the cell. See the text for more details.

lifestyles in the context of the ecosystem-level cycling of nitrogen that occurs through microbial interactions. Nitrogen gas (N_2) comprises 78% of Earth's atmosphere, but it is in a form that is inaccessible to most living organisms, as it has a very strong triple covalent bond joining the two atoms of nitrogen. However, some *Bacteria* and *Archaea* have the ability to reduce N_2 to ammonia (NH_3) using a process called nitrogen fixation. Microbes that perform nitrogen fixation expend large amounts of energy converting N_2 to NH_3, which is the form of nitrogen assimilated into biomass. Ammonia (which exists primarily as ammonium ion, NH_4^+, below a pH of ~9) has the capacity to serve as an electron donor for the ammonia-oxidizing bacteria (AOB) and ammonia-oxidizing archaea (AOA). During their respiration, AOB and AOA convert NH_3 to nitrite (NO_2^-). The NO_2^- can then be used as an electron donor by the nitrite-oxidizing bacteria (NOB), which further oxidize NO_2^- to nitrate (NO_3^-). Nitrification refers to the entire process by which NH_3 is converted to NO_3^- by the collective activities of the AOB and NOB. The AOB and NOB live primarily via a chemolithoautotrophic lifestyle using oxygen as their terminal electron acceptor. In addition, bacteria capable of <u>com</u>plete <u>am</u>monia <u>ox</u>idation, or "comammox" (NH_3 to NO_3^-), have recently been identified.

The NO_2^- and NO_3^- produced during nitrification are used for two processes. Bacteria and plants can reconvert these oxidized forms of nitrogen to ammonia via energy-consuming reactions for subsequent assimilation into biomass during growth. Alternatively, there are some bacteria that can perform denitrification, a form of anaerobic respiration in which the oxidized forms of nitrogen (NO_2^- and NO_3^-) are reduced to N_2 gas. Organisms performing denitrification are removing biologically useful forms of nitrogen from the environment, and as a result, this process is detrimental for agriculture. However, since denitrification results in the formation of N_2 gas, the process is beneficial for wastewater treatment because it reduces the amount of usable nitrogen from entering water systems, thus decreasing eutrophication. More recently, the process of <u>an</u>aerobic <u>am</u>monia <u>ox</u>idation (anammox) has been discovered. In this process, nitrite and ammonia are converted to nitrogen gas ($NO_2^- + NH_3 \rightarrow N_2$). Thus, we continue to learn new ways that microbes can interconvert inorganic forms of nitrogen. Some of these processes cost the cell energy, such as nitrogen fixation and ammonia assimilation. Others conserve energy by generating a proton motive force (PMF) through an electron transport chain (ETC), such as nitrification and denitrification. The reactions of the nitrogen cycle are covered in greater detail in the following sections.

Nitrogen Fixation

There is only one known biological process that converts nitrogen gas to ammonia. Because nitrogen gas has a very strong triple bond, it is difficult to reduce it to a biologically available compound like NH_3. The reduction of nitrogen gas to ammonia is known as **nitrogen fixation**, a conserved metabolic pathway carried out by a relatively small number of *Bacteria* and *Archaea*. These organisms are phylogenetically diverse and have a variety of lifestyles, including autotrophy and heterotrophy. Some species are free-living, while others are in symbiotic associations with plants. Collectively, nitrogen-fixing bacteria are referred to as **diazotrophs**, which means "nitrogen eaters." At atmospheric temperature and pressure, they generate 65 to 75% of the global ammonia supply. However, the ability to produce ammonia comes at the cost of cellular energy levels, as biological nitrogen fixation is an energetically expensive process.

There are also two abiotic mechanisms of reducing nitrogen gas to ammonia. Lightning can thermochemically convert N_2 and O_2 to various oxidized inorganic nitrogen intermediates, which can be further converted to NH_3, which accounts for several percent of the global ammonia production annually. Secondly, an industrial method that was developed during World War I called the Haber-Bosch process uses high temperature and pressure to produce ammonia. During the war, NH_3 was needed for explosives production, but it is now used primarily as an agricultural fertilizer. This process consumes large amounts of energy in the form of fossil fuels and it can create the risk of explosion. Although it contributes only ~30% of the total available ammonia globally, the Haber-Bosch process has had a profound environmental impact on the nitrogen cycle. The practice of applying anthropogenically produced fertilizers has not only increased crop production and contributed to increased human populations, but it has also led to increased amounts of biologically accessible nitrogen entering water sources. Since nitrogen is often a limiting nutrient, an increase in available nitrogen in surface waters typically results in a massive overgrowth of microorganisms, which depletes the oxygen levels in the environment. As a result, vast anoxic regions, called "dead zones," have formed (e.g., at the mouth of the Mississippi River as it enters the Gulf of Mexico) where more complex aquatic life such as shrimp and fish can no longer exist due to the absence of oxygen. Due to these dramatic environmental impacts, considerable efforts have been made to study natural microbial nitrogen fixation processes with the goal of improving the efficiency of the biological nitrogen fixation reactions.

Biochemistry of Nitrogen Fixation

The ability to fix nitrogen is widely distributed among many lineages of *Bacteria* and *Archaea* (Fig. 12.2A), but no eukaryotes are known to carry out nitrogen fixation. All microbes that are capable of nitrogen fixation synthesize a **nitrogenase** enzyme complex that catalyzes a series of reduction reactions (Fig. 12.2B) to produce ammonia from nitrogen gas. Nitrogenase consists of two components, **dinitrogenase reductase**, which contains a four iron-four sulfur (4Fe4S) cluster, and **dinitrogenase**, which contains an FeS cluster (known as the P cluster) and an iron-molybdenum (FeMo) cofactor. Six electrons and six protons are needed to convert N_2 to NH_3. NADH + H^+, produced during the reactions of central metabolism in the cell, typically serves as an electron donor for the process of nitrogen fixation. A total of eight electrons are passed to ferredoxin and then to nitrogenase in the cytoplasm of nitrogen-fixing cells. The remaining two electrons and two protons are used to form a molecule of hydrogen gas as a byproduct of the nitrogenase reaction. A total of 16 ATPs are also consumed by nitrogenase for each molecule of ammonia produced. Therefore, the process of biological nitrogen fixation costs the cell large amounts of energy in the form of reduced coenzymes and ATP. The overall reaction catalyzed by nitrogenase is

$$N_2 + 8\,e^- + 8\,H^+ + 16\,ATP \rightarrow 2\,NH_3 + H_2 + 16\,ADP + 16\,P_i$$

A

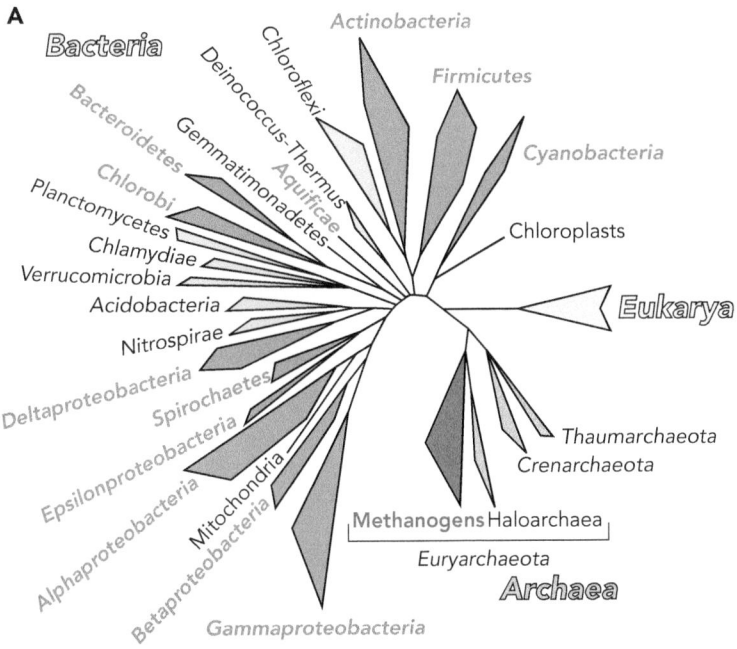

B N₂ fixation

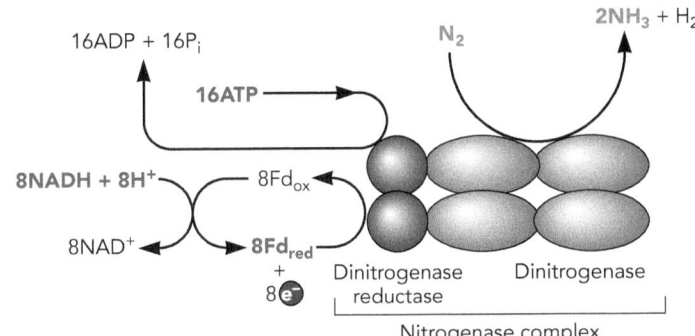

FIGURE 12.2 **Distribution of nitrogen fixation and the nitrogenase enzyme.** (A) Phylogenetic tree highlighting lineages of *Bacteria* (highlighted in dark coloring and bold green text) and *Archaea* (highlighted in dark coloring and bold pink text) that have members capable of nitrogen fixation. (B) The biochemical reaction catalyzed by the nitrogenase enzyme complex, which is composed of dinitrogenase reductase and dinitrogenase. Energy inputs are highlighted in red and inorganic forms of nitrogen in bright orange. Fd, ferredoxin.

Nitrogenase is conserved in all nitrogen-fixing bacteria and archaea; the genes encoding nitrogenase are typically colocalized in a large gene cluster together with genes that encode regulatory proteins and accessory proteins necessary for incorporation of the metal clusters. It is likely that during evolution the nitrogenase gene cluster has been lost by some lineages and acquired by other lineages via horizontal gene transfer.

Regulation of Nitrogen Fixation

The evolution of nitrogen fixation predates the evolution of oxygenic photosynthesis, and therefore nitrogenase evolved before oxygen was present in Earth's atmosphere. The FeMo cofactor is unstable in the presence of oxygen, and as a result, nitrogenase is rapidly and irreversibly inactivated by oxygen. However, both anaerobic and aerobic bacteria are capable of nitrogen fixation. Obligately anaerobic nitrogen-fixing bacteria like *Clostridium* and the anoxygenic green sulfur phototrophs (*Chlorobi*, recently renamed *Chlorobiota*) do not have a problem with oxygen inactivating their nitrogenase. In general, facultative anaerobes such as *Klebsiella* and the purple nonsulfur bacterium *Rhodopseudomonas* only fix nitrogen when they are growing as anaerobes. On the other hand, microaerophiles (e.g., *Azospirillum*) inhabit microoxic environments and require oxygen for respiration. The intracellular oxygen concentrations of such organisms remain very low because oxygen is used as the terminal electron acceptor as soon as it diffuses into the cell, so the nitrogenase in microaerophilic nitrogen fixers is unaffected. Aerobic nitrogen fixers have a more difficult problem. Under conditions in which ammonia is low, unicellular aerobes like *Azotobacter* and *Azomonas* induce the production of a thick polysaccharide capsule that slows the diffusion of oxygen into the cell. They also have a high respiration rate that consumes oxygen as fast as it enters into the cell. Unicellular rhizobia, which are also obligate aerobes, only fix nitrogen when they are in a symbiotic partnership with a plant host (see below). In contrast, unicellular cyanobacteria growing as oxygenic phototrophs have a bigger problem because they are making oxygen! One strategy used by unicellular nitrogen-fixing cyanobacteria is temporal separation of photosynthesis and N_2 fixation based on light-dark cycles. Photosynthesis (and thus oxygen production) is carried out in the light, while respiration (which consumes O_2) and nitrogen fixation are carried out in the dark. Filamentous cyanobacteria like *Nostoc* and *Anabaena* have a different strategy to protect their nitrogenase based on spatial separation of nitrogen fixation and oxygenic phototrophy involving the production of specialized nitrogen-fixing cells called heterocysts (Chapter 18).

There is also a need for tight regulatory control due to the energetic demands of nitrogen fixation. The concentrations of reduced forms of nitrogen (e.g., ammonia or amino acids) available in the environment influence whether nitrogen fixation will occur. The genes for nitrogen fixation are expressed only when ammonia production is necessary for biosynthesis. The regulatory process is well understood in the free-living bacterium *Klebsiella pneumoniae* (Fig. 12.3). A two-component regulatory system is involved (Chapter 7). NtrB is a sensor histidine kinase that senses low NH_3 levels. Under conditions of low NH_3, NtrB autophosphorylates a conserved histidine residue and then the phosphoryl group is transferred to a conserved aspartate residue on the response regulator NtrC. When NtrC is phosphorylated, it functions as an activator protein that can bind to target promoters, but it specifically requires RNA polymerase (RNAP) holoenzyme containing NtrA as its sigma factor (RNAP-NtrA). NtrA, which is also known as σ^{54}, is important for expression of genes associated with nitrogen metabolism. Transcription by the RNAP-NtrA holoenzyme requires an activator protein (in this case NtrC-P) to bind to target promoters. RNAP-NtrA and NtrC-P

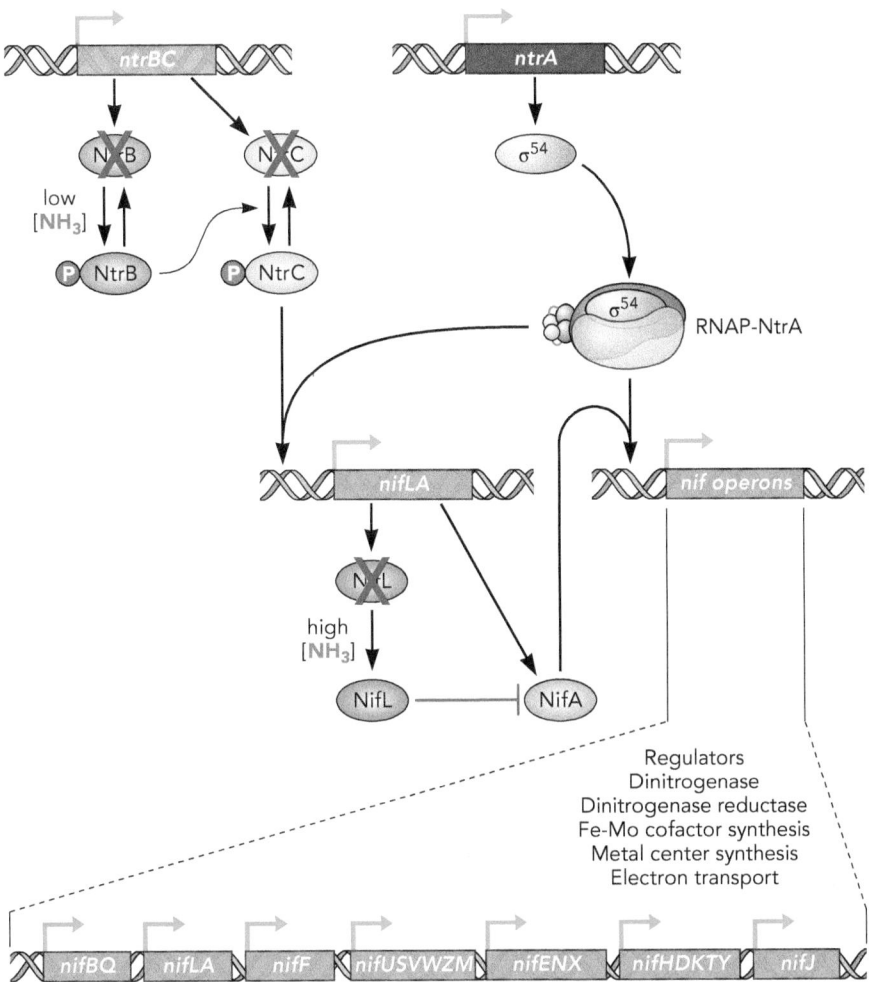

FIGURE 12.3 **Regulation of genes necessary for nitrogen fixation in *Klebsiella pneumoniae*.** The NtrB/NtrC two-component regulatory system senses environmental ammonia (NH_3) levels. When NH_3 levels are low, NtrB autophosphorylates and then transphosphorylates NtrC. NtrC-P is a transcriptional activator that works in cooperation with σ^{54} (NtrA) RNA polymerase (RNAP-NtrA) to activate transcription of *nifLA*. NifA is a transcriptional activator that works in cooperation with RNAP-NtrA to activate transcription of multiple *nif* operons encoding genes required for nitrogen fixation. When high concentrations of ammonia are present, NifL serves to repress the activity of NifA and NtrB/NtrC are in inactive conformations. Black arrows indicate activation, green arrows indicate transcription initiation, red T-bar indicates repression, and red X's indicate inactive proteins. Inorganic forms of nitrogen are bright orange.

together activate the promoter for the two-gene operon *nifLA*. NifA is a transcription factor that works with RNAP-NtrA to control the promoters of seven *nif* (*ni*trogen *f*ixation) operons encoding 20 polypeptides that are necessary to form the nitrogenase enzyme complex. Considerable energy input is required to both produce nitrogenase as well as enable it to catalyze the nitrogen fixation reaction. Therefore, when environmental conditions change such that sufficient ammonia is available, the system is turned off. NifL is an anti-activator important for regulation under conditions of high concentrations of ammonia, amino acids, or oxygen. NifL posttranslationally inactivates NifA via protein-protein interactions. When ammonia levels are high, NtrB and NtrC are unphosphorylated and therefore inactive. As a result, the cell does not expend energy expressing the genes in the *nif* operon under conditions where production of ammonia is not warranted.

Symbiotic Plant-Microbe Interactions during Nitrogen Fixation

The rhizobia are nitrogen-fixing bacteria that associate with specific leguminous plant hosts (plants that produce seeds in a pod). Members of this group are represented by both *Alphaproteobacteria* (*Rhizobium, Azorhizobium, Bradyrhizobium, Mesorhizobium,* and *Sinorhizobium [Ensifer]*) and *Betaproteobacteria* (*Cupriavidus* and *Burkholderia*) (Fig. 12.2A). Each species has the capacity to develop symbiotic associations with a specific leguminous plant (e.g., bean, pea, soybean, or alfalfa). Rhizobia can be found free-living in soil, but they only initiate nitrogen fixation after establishing a symbiotic relationship with a plant host in the rhizosphere (i.e., the vicinity of the roots). Historically in the United States, crops were rotated to enrich the soil with accessible nitrogen by alternating between growing legumes (such as soybeans) and other crops that do not form symbiotic associations with nitrogen-fixing bacteria (like corn). The use of chemically generated fertilizers now alleviates the need for this practice, but as discussed previously, farm runoff of nitrogen-based fertilizers can have profound environmental consequences.

Interestingly, the idea of crop rotation developed in other ways around the globe. In Asia, a common practice is to flood rice fields and permit the water fern *Azolla* to grow. *Azolla* is not a legume, but it can establish a symbiotic relationship with the nitrogen-fixing cyanobacterium *Anabaena*. In one farming practice, the rice fields are drained, and the *Azolla* is plowed under to serve as a "green" fertilizer to enrich the field with nitrogen. Another example of a symbiotic relationship involving nitrogen fixation is one that occurs between the nitrogen-fixing actinobacterium *Frankia* and *Alnus*, the alder tree. Nitrogen fixation by *Frankia* permits the tree to be a "pioneer" colonizer in nutrient-poor soils.

CELL-CELL SIGNALING BETWEEN RHIZOBIA AND PLANT HOSTS

During the establishment of a symbiotic relationship, both the rhizobial cells and the plant host participate in a series of cell-to-cell communications using interspecies signaling molecules (Fig. 12.4A). Each partner in the symbiosis produces signals that directly and specifically stimulate a response by the other. First, the plant produces **flavonoid compounds** and other nutrient sources such as amino acids that are sensed as chemoattractants by motile rhizobia in the rhizosphere. The free-living rhizobia move toward the source of these chemicals through motility and chemotaxis (Chapter 15) and attach to the tip of the root hair surface. Upon sensing flavonoids, the bacteria produce **nod factors**, which are oligosaccharides that signal the plant to begin developing **root nodules**. The genes encoding biosynthetic pathways for production of nod factors (*nod* genes) as well as the genes for nitrogen fixation (*fix* and *nif* genes) are often encoded on large plasmids that are required for establishment of the symbiosis and development of the nodules. In response to the nod factors, the plant root hairs curl around the bacteria and permit them to penetrate the root

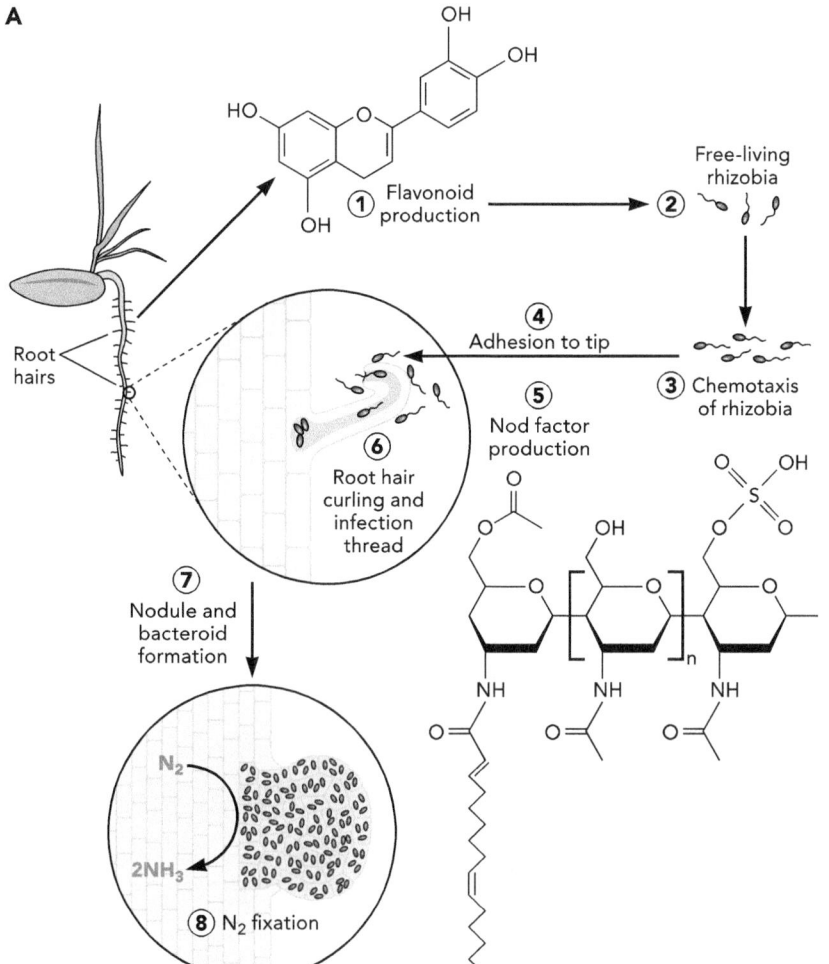

FIGURE 12.4 **Process of root nodule development in a legume.** (A) The signaling molecules and steps involved in the development of a root nodule when nitrogen-fixing rhizobia associate with the roots of leguminous plants are diagrammed and labeled in numerical order. See the text for details. (B) Photograph of root nodules that have formed on the roots of a green bean plant. Photo credit: Elena M. Oosterhuis.

hair through an **infection thread**. The bacteria multiply in the infection thread and cause the root cells to divide and ultimately form a differentiated tissue in the root known as the root nodule (Fig. 12.4B). Some bacteria housed within the root nodule terminally differentiate into nitrogen-fixing **bacteroids** that are metabolically active but no longer capable of growth and cell division. The plant is of course a photoautotroph, and as such, it fixes carbon dioxide and produces sugars and organic acids that are used as carbon sources by the bacteroids living within the cytoplasm of the plant cells in the root nodule (Fig. 12.5). Rhizobia are chemoheterotrophs that require a reduced carbon source to produce NADH + H$^+$, which serves as the electron donor for the ETC and ultimately provides energy for the bacteroid. The bacteroid conserves energy via aerobic respiration and therefore requires oxygen as the terminal electron acceptor. However, as previously mentioned, the cytoplasmic nitrogenase enzyme that fixes nitrogen is inactivated by oxygen. **Leghemoglobin** produced by plant cells within the nodule reversibly binds free oxygen, maintaining the oxygen concentration low enough to allow for continued nitrogenase activity but sufficient for the oxygen to serve as the electron acceptor during aerobic respiration in the bacteroid. The large amounts of ATP required for nitrogen fixation are generated during

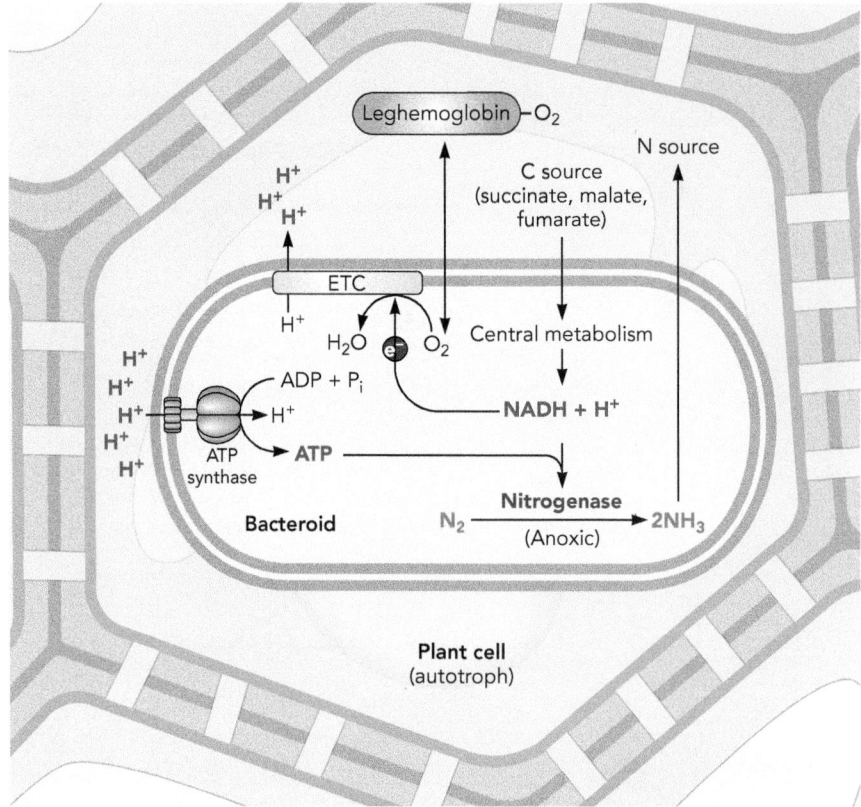

FIGURE 12.5 **Symbiosis between a bacteroid and plant cell.** Some of the rhizobia living within host plant cells in the root nodule terminally differentiate into nitrogen-fixing bacteroids. Plant cells within the root nodule are packed with multiple rhizobia and bacteroids. For simplicity, we show the interactions of a single bacteroid with a plant cell. The plant cell carries out photosynthesis and provides organic carbon sources to the bacteroid. The plant also produces leghemoglobin, which sequesters free O_2 to maintain O_2 levels compatible with nitrogenase activity. Sufficient O_2 is available for aerobic respiration within the bacteroid to generate ATP and NADH + H$^+$, which are used by nitrogenase to reduce N_2 to NH$_3$ for export to the plant cell.

aerobic respiration by the bacteroid. The ammonia produced by the bacteroid then serves as a nitrogen source for the plant. Thus, in this symbiosis, the bacterium gains its carbon source from the plant while the plant gains a nitrogen source in return. As mentioned above, not all the rhizobial cells in the nodule terminally differentiate into nitrogen-fixing bacteroids. When the plant dies, these undifferentiated rhizobial cells are released into the rhizosphere and grow as aerobic heterotrophs until they encounter and colonize another plant host.

Assimilatory Nitrate Reduction

Another biological strategy to produce ammonia for biosynthesis involves the reduction of oxidized forms of inorganic nitrogen. Many bacteria, archaea, and plants can acquire oxidized forms of nitrogen such as nitrate and nitrite from the environment. However, the nitrate and nitrite cannot be directly incorporated into biomass and must first be reduced to ammonia in a process known as

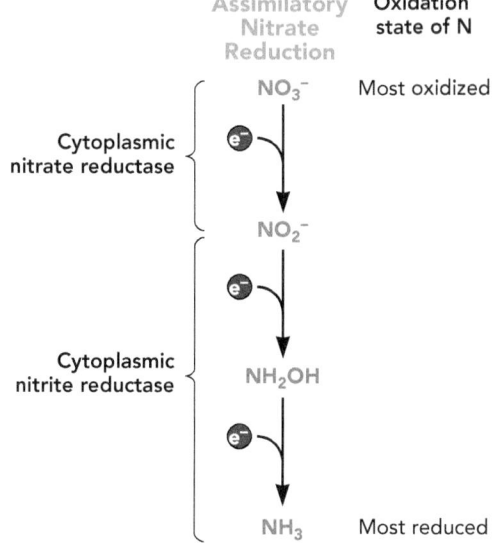

FIGURE 12.6 **Steps of assimilatory nitrate reduction. Ammonia** (NH_3) is the form of nitrogen that is incorporated into biomass. When NH_3 is not available, some organisms can convert oxidized forms of nitrogen (e.g., NO_3^-, NO_2^-) to NH_3. Cytoplasmic nitrate and nitrite reductases sequentially reduce nitrate to ammonia using electrons donated from reduced ferredoxin or NADH + H$^+$. Assimilatory nitrate reduction, which occurs in the cytoplasm and requires energy to produce the NH_3 required for biosynthesis, should not be confused with anaerobic nitrate reduction, which is a form of anaerobic respiration using a membrane-associated electron transport chain that conserves energy.

assimilatory nitrate reduction. Unlike anaerobic respiration with nitrate as a terminal electron acceptor, which conserves energy for the cell (Chapter 11), assimilatory nitrate reduction requires the input of energy and provides a source of ammonia for biosynthesis. The process occurs in the cytoplasm where organic electron donors (e.g., NADH + H$^+$ and ferredoxin) are used to sequentially reduce oxidized forms of inorganic nitrogen. Nitrate (NO_3^-) is converted to nitrite (NO_2^-) by a cytoplasmic nitrate reductase. Nitrite is converted to hydroxylamine (NH_2OH) and then to ammonia (NH_3) by a cytoplasmic nitrite reductase. The complete reduction of nitrate to ammonia consumes eight electrons (Fig. 12.6).

Ammonia Assimilation into Cellular Biomass

All living organisms require ammonia for the biosynthesis of amino acids and nucleic acids. The process by which ammonia is incorporated into biomass is known as **ammonia assimilation**. Ammonia is assimilated into glutamine by the activity of glutamine synthetase (GS) (Fig. 12.7). This process consumes ATP, and therefore the GS enzyme is highly regulated via covalent adenylylation modifications (Chapter 7). The reaction catalyzed by GS is

$$glutamate + ammonia + ATP \rightarrow glutamine + ADP + P_i$$

Ultimately, the nitrogen from glutamine is incorporated into the purine and pyrimidine bases, amino sugars, some amino acids, and NAD$^+$ (Fig. 12.7).

Glutamine is converted to glutamate by glutamate synthase also named glutamine oxoglutarate aminotransferase (GOGAT). The reaction catalyzed by GOGAT is

$$glutamine + \alpha\text{-ketoglutarate} + NADPH + H^+ \rightarrow 2 \, glutamate + NADP^+$$

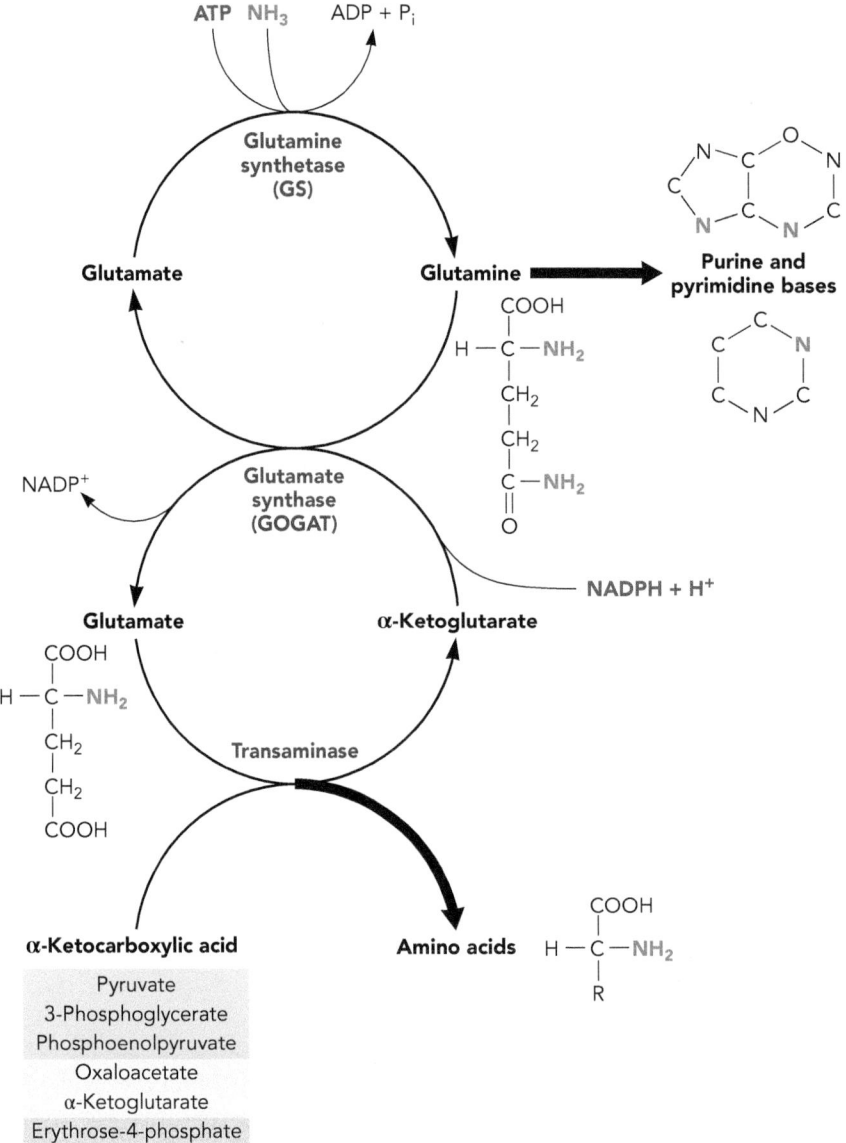

FIGURE 12.7 **Assimilation of inorganic nitrogen into biomass.** Glutamine synthetase (GS) incorporates ammonia into glutamate to generate glutamine with energy derived from ATP hydrolysis. The nitrogen in glutamine can then be incorporated into purine and pyrimidine bases, as indicated by the orange N atoms in the generic nitrogenous base structures shown. Glutamate synthase (GOGAT) catalyzes the formation of two molecules of glutamate from glutamine and α-ketoglutarate using energy from NADPH + H$^+$. Specific transaminase enzymes add the amino group from glutamate to various α-ketocarboxylic acids (e.g., essential precursors) produced during central metabolism to yield different amino acids with a variety of R groups.

Alternatively, under conditions of high ammonia, a different enzyme, glutamate dehydrogenase, can be used to synthesize glutamate. This enzyme, which has a high K_m for ammonia, catalyzes the reaction

$$\alpha\text{-ketoglutarate} + NADPH + H^+ + NH_3 \rightarrow glutamate + NADP^+ + H_2O$$

In this reaction, glutamate is produced while bypassing the ATP-consuming GS reaction.

Ultimately the nitrogen in glutamate is incorporated into amino acids by a family of enzymes known as transaminases (Fig. 12.7). Transaminases take an α-ketocarboxylic acid (an essential precursor produced from central metabolism) and add an amino group from glutamate to produce a specific amino acid:

$$\text{glutamate} + \alpha\text{-ketocarboxylic acid} \rightarrow \text{amino acid} + \alpha\text{-ketoglutarate}$$

The α-ketoglutarate produced by the above reaction can be used to generate more glutamate. Specific examples of various amino acids generated by transaminase reactions include

$$(1)\ \text{oxaloacetate} + \text{glutamate} \rightarrow \text{L-aspartate} + \alpha\text{-ketoglutarate}$$

$$(2)\ \text{pyruvate} + \text{glutamate} \rightarrow \text{L-alanine} + \alpha\text{-ketoglutarate}$$

Thus, there are several ways that ammonia can be incorporated into cellular biomass.

Nitrification: Ammonia Oxidation, Nitrite Oxidation, and Comammox

In the late 1800s, the Ukrainian microbiologist Sergei Winogradsky developed methods to enrich for microorganisms associated with the process of **nitrification** (Fig. 12.8), in which reduced nitrogen compounds serve as energy sources during chemolithotrophic growth. Nitrification is the stepwise conversion of the most reduced form of nitrogen, ammonia (NH_3), to the most oxidized form of nitrogen (NO_3^-):

$$NH_3 \rightarrow NH_2OH \rightarrow NO_2^- \rightarrow NO_3^-$$

In most environments, the nitrification process is carried out by two physiologically distinct groups of bacteria that work together to completely oxidize ammonia. The **ammonia-oxidizing bacteria (AOB)** convert ammonia to nitrite, and the **nitrite-oxidizing bacteria (NOB)** convert the nitrite produced

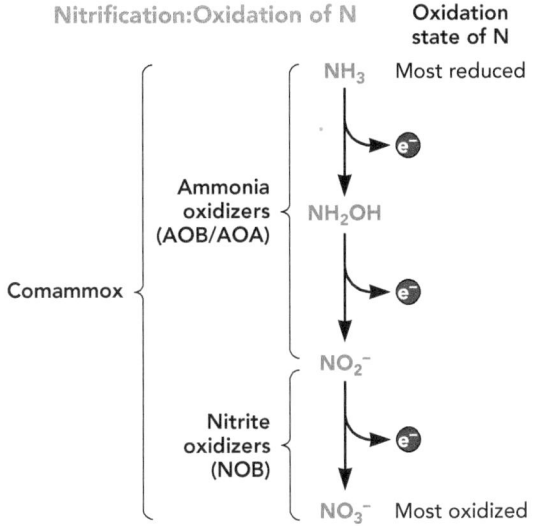

FIGURE 12.8 **Oxidation of ammonia via nitrification.** Ammonia-oxidizing bacteria (AOB) and ammonia-oxidizing archaea (AOA) oxidize ammonia (NH_3) to nitrite (NO_2^-) via hydroxylamine (NH_2OH). Nitrite-oxidizing bacteria (NOB) oxidize nitrite to nitrate (NO_3^-). Microbial ammonia oxidation and nitrite oxidation are energy-conserving reactions. Complete ammonia oxidation (comammox) is the process by which ammonia is converted to nitrate by a single organism. Ammonia and nitrite serve as electron donors for aerobic respiration in these organisms. Inorganic forms of nitrogen are highlighted in orange.

by the AOB to nitrate. Winogradsky isolated the first AOB, *Nitrosomonas*, and the first NOB, *Nitrobacter*, both Gram-negative members of the proteobacteria. While these two genera remain the easiest nitrifiers to grow in the laboratory, sequencing of 16S rRNA genes from environmental samples indicates that they are not necessarily the most abundant nitrifiers in the environment. Other genera have since been described, and AOB include genera in the *Beta-* and *Gammaproteobacteria* and the *Nitrospirae* (recently renamed *Nitrospirota;* Fig. 12.9A). The NOB include genera in the *Alpha-, Beta-, Delta-,* and *Gammaproteobacteria* and the *Nitrospirae* (Fig. 12.10A). The genus prefix *Nitroso-* is associated with the AOB and the prefix *Nitro-* is generally associated with the NOB. Most AOB and NOB are aerobic chemolithoautotrophs, although some can grow as chemoheterotrophs. When living as chemolithoautotrophs, they fix carbon dioxide using the Calvin cycle and use reverse electron flow (Fig. 11.9) to generate NADPH + H$^+$ needed for CO$_2$ fixation. Recently, specific members of the bacterial genus *Nitrospira* (members of the *Nitrospirae*) have been shown to carry all enzymes needed for the complete oxidation of ammonia to nitrate, a process termed **comammox** (for **com**plete **am**monia **ox**idation) (Fig. 12.8 and 12.10A). Not surprisingly, *Nitrospira* isolates that are capable of comammox conserve an amount of energy from the complete conversion of NH$_3$ to NO$_3^-$ that is equivalent to the sum of the energy conserved by the AOB and the NOB together:

$$\text{AOB } Nitrosomonas\text{: } NH_3 \rightarrow NO_2^- \left(\Delta G^{0'} = -275\,kJ\,/\,mol\right)$$

$$\text{NOB } Nitrobacter\text{: } NO_2^- \rightarrow NO_3^- \left(\Delta G^{0'} = -74\,kJ\,/\,mol\right)$$

$$\text{comammox: } NH_3 \rightarrow NO_3^- \left(\Delta G^{0'} = -349\,kJ\,/\,mol\right)$$

ELECTRON TRANSPORT DURING AOB METABOLISM

Ammonia is a good electron donor for the ETC in AOB, allowing the bacteria to generate a PMF (Fig. 12.9B). In the cytoplasm, ammonia is converted into hydroxylamine by the membrane-associated enzyme **ammonia monooxygenase** (AMO). Like methane monooxygenase (MMO; Chapter 11), AMO uses molecular oxygen as a substrate, adding one oxygen atom to ammonia and generating hydroxylamine (NH$_2$OH). AMO is only used in nitrification and is only found in organisms capable of ammonia oxidation. Therefore, the gene encoding AMO serves as a good biomarker or probe to identify organisms capable of ammonia oxidation. The hydroxylamine can diffuse across the membrane to the periplasm, where it is converted into nitrite by **hydroxylamine oxidoreductase** (HAO). Nitrite is an end product that is released by the AOB and consumed by the NOB, which inhabit the same environment. Electrons from the HAO reaction are passed via cytochromes and a Q cycle to the terminal oxidase complex. Oxygen is the terminal electron acceptor, and there is sufficient energy to pump protons across the membrane, creating a PMF. Both ATP and NADPH + H$^+$ are required to drive carbon dioxide fixation via the Calvin cycle (Chapter 9) for the AOB when they grow as chemolithoautotrophs. Some energy from the PMF is also used to drive reverse electron flow (Fig. 11.9), which is necessary to produce NADH + H$^+$ because ammonia has an electrode/reduction potential that is too positive to allow direct reduction of NAD$^+$. The

A

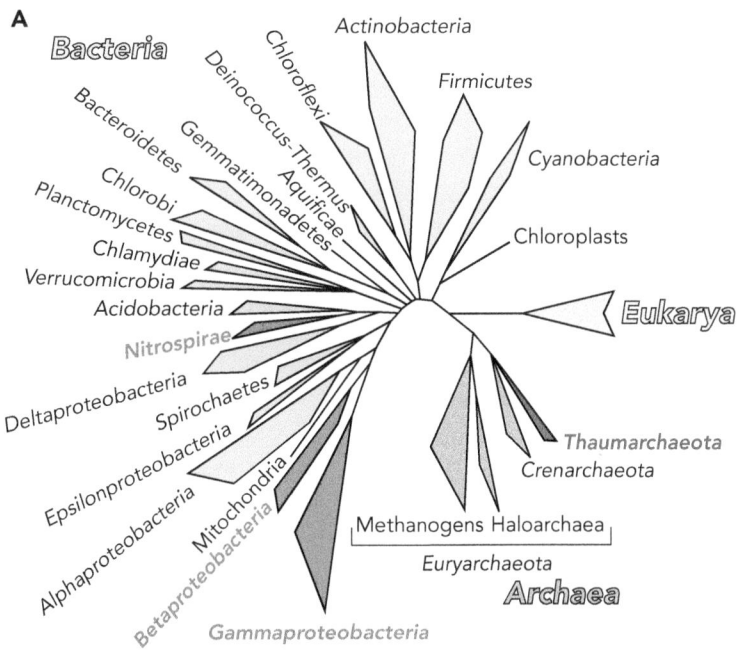

B   Nitrification:Ammonia-oxidizing bacteria

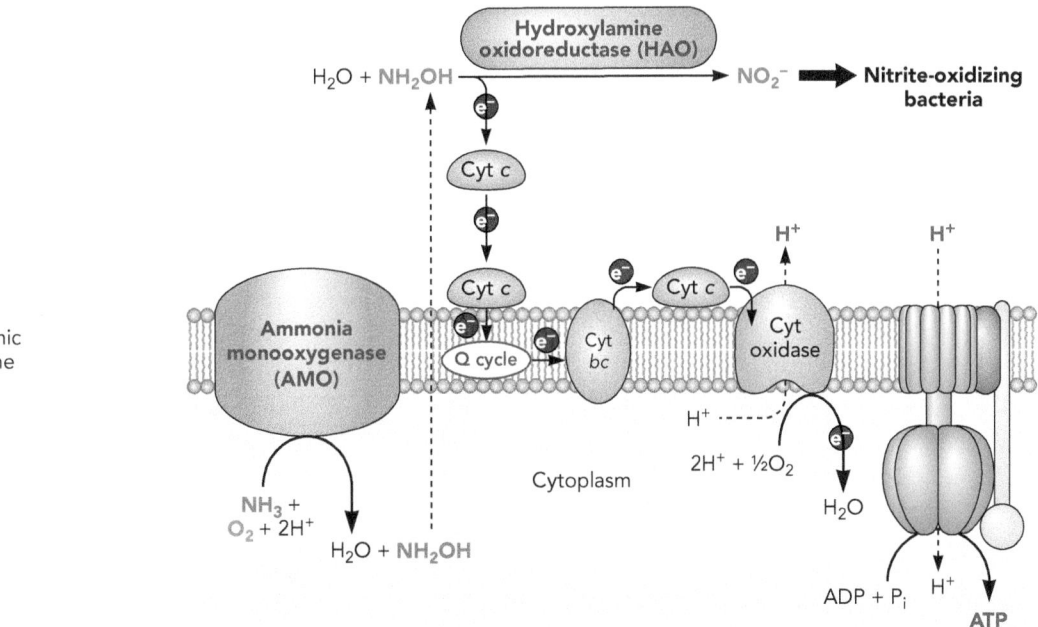

FIGURE 12.9 **Distribution of ammonia-oxidizing microbes and biochemistry of the process.** (A) Phylogenetic tree highlighting lineages of *Bacteria* (highlighted in dark coloring and bold green text) and *Archaea* (highlighted in dark coloring and bold pink text) that have members capable of ammonia oxidation. (B) Ammonia (NH_3) serves as the electron donor for a specialized electron transport system that includes ammonia monooxygenase (AMO) and hydroxylamine oxidoreductase (HAO). A proton motive force is generated at a cytochrome oxidase with oxygen serving as the terminal electron acceptor. See the text for additional details.

NADH + H$^+$ can be readily interconverted into NADPH + H$^+$. AOB are thus very slow-growing organisms: they produce little ATP since NH$_3$ is a weak electron donor and the ETC has only one coupling site. Further, they must sacrifice some of their energy to produce NADPH + H$^+$ for carbon dioxide fixation. However, the AOB have evolved to take advantage of a unique ecological niche in which they are the only organisms capable of using ammonia as an electron donor to support cellular growth and survival.

ELECTRON TRANSPORT DURING NOB METABOLISM

The NOB use nitrite (NO$_2^-$) produced by the AOB as their initial electron donor, converting it to nitrate (NO$_3^-$) and allowing the bacteria to generate a PMF (Fig. 12.10B). From the nitrite oxidoreductase complex, electrons are passed to cytochrome c in the periplasm. However, this is an energetically unfavorable event. It is hypothesized that the net positive charge on the periplasmic side of the membrane (due to the PMF) helps to overcome the energetic barrier. From cytochrome c the electrons are passed to cytochrome oxidase and then to the terminal electron acceptor, oxygen, in an energetically favorable direction. Sufficient energy is released during this "forward" electron transport so that protons are pumped across the membrane at the cytochrome oxidase. This single coupling site generates PMF for ATP production. Similar to the AOB, the NOB are slow-growing chemolithoautotrophic organisms because nitrite is a weak electron donor, their ETC has only one coupling site, and cells must sacrifice part of their PMF to drive reverse electron flow for NADPH + H$^+$ production (Fig. 11.9). The NOB are the only organisms that have evolved the ability to use nitrite as an electron donor and thus have an environmental niche with little competition. AOB are found in close proximity to the NOB, as the NOB require the nitrite produced by AOB. Other microorganisms can use the nitrite and nitrate produced during nitrification as terminal electron acceptors during the process of denitrification. Thus, aerobic nitrification and anaerobic denitrification are linked in the nitrogen cycle on an ecosystem level.

Although no nitrite-oxidizing *Archaea* have been found to date, **ammonia-oxidizing *Archaea* (AOA)**, such as *Nitrosopumilus maritimus*, are common in marine environments (Fig. 12.9A). *Nitrosopumilus* are aerobic chemolithoautotrophs with a high-affinity ammonia monooxygenase that allows them to grow at extremely low concentrations of ammonia.

Anammox: Anaerobic Ammonia Oxidation

Until recently, it was thought that ammonia oxidation only occurred in the presence of oxygen and that fixed nitrogen was only recycled to nitrogen gas by the process of denitrification (see below). However, ammonia oxidation can also occur under anoxic conditions. This process, termed **anammox** (for an̲aerobic am̲m̲onia ox̲idation), is carried out by certain members of the *Planctomycetes* (recently named *Planctomycetota*; Fig. 12.11A). Initially found in wastewater treatment plants where ammonium (NH$_4^+$) is at relatively high concentrations, and subsequently found in the Black Sea and marine sediments, these organisms

A

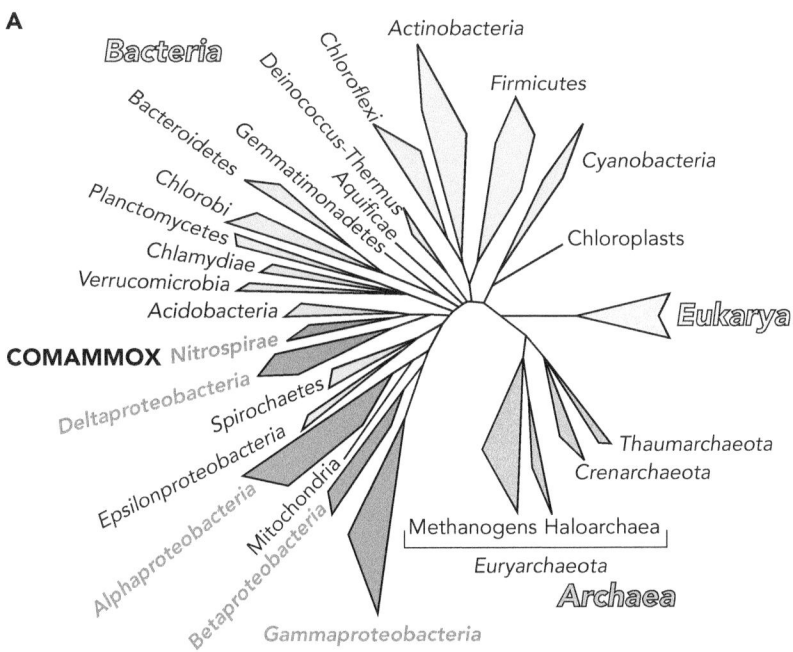

B Nitrification:Nitrite-oxidizing bacteria

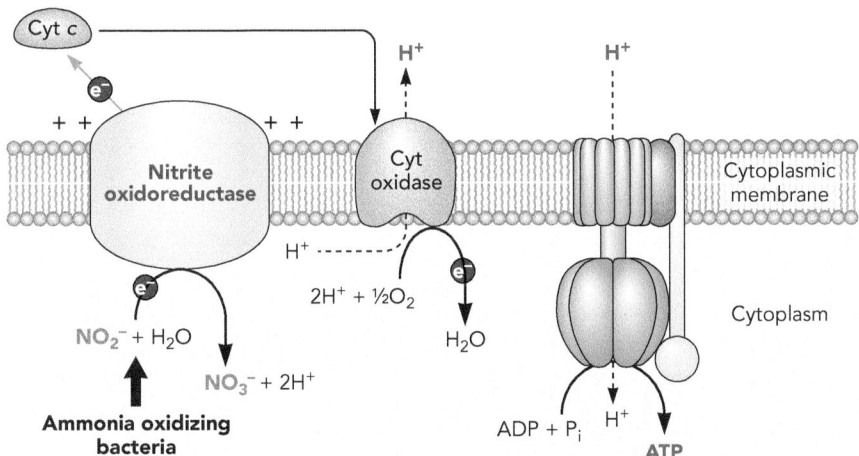

FIGURE 12.10 **Distribution of nitrite-oxidizing bacteria and biochemistry of the process.** (A) Phylo-genetic tree highlighting lineages of *Bacteria* (highlighted in dark coloring and bold green text) that are capable of nitrite oxidation. Note that known bacteria capable of complete ammonia oxidation (comammox) are members of the *Nitrospirae*. (B) Nitrite (NO$_2^-$) serves as the electron donor for an electron transport system that includes nitrite oxidoreductase. Electrons are moved in an energetically unfavorable direction to cytochrome *c* prior to the generation of a proton motive force at cytochrome oxidase with oxygen serving as the terminal electron acceptor. See the text for additional details.

catalyze the oxidation of ammonium with nitrite as the electron acceptor to yield N$_2$ gas under anoxic conditions (Fig. 12.11B):

$$NH_4^+ + NO_2^- \rightarrow N_2 + 2H_2O \left(\Delta G^{0'} = -357\,kJ\right)$$

Anammox represents an important component of the nitrogen cycle, as it is estimated to contribute up to 50% of the ammonium loss in marine sediments.

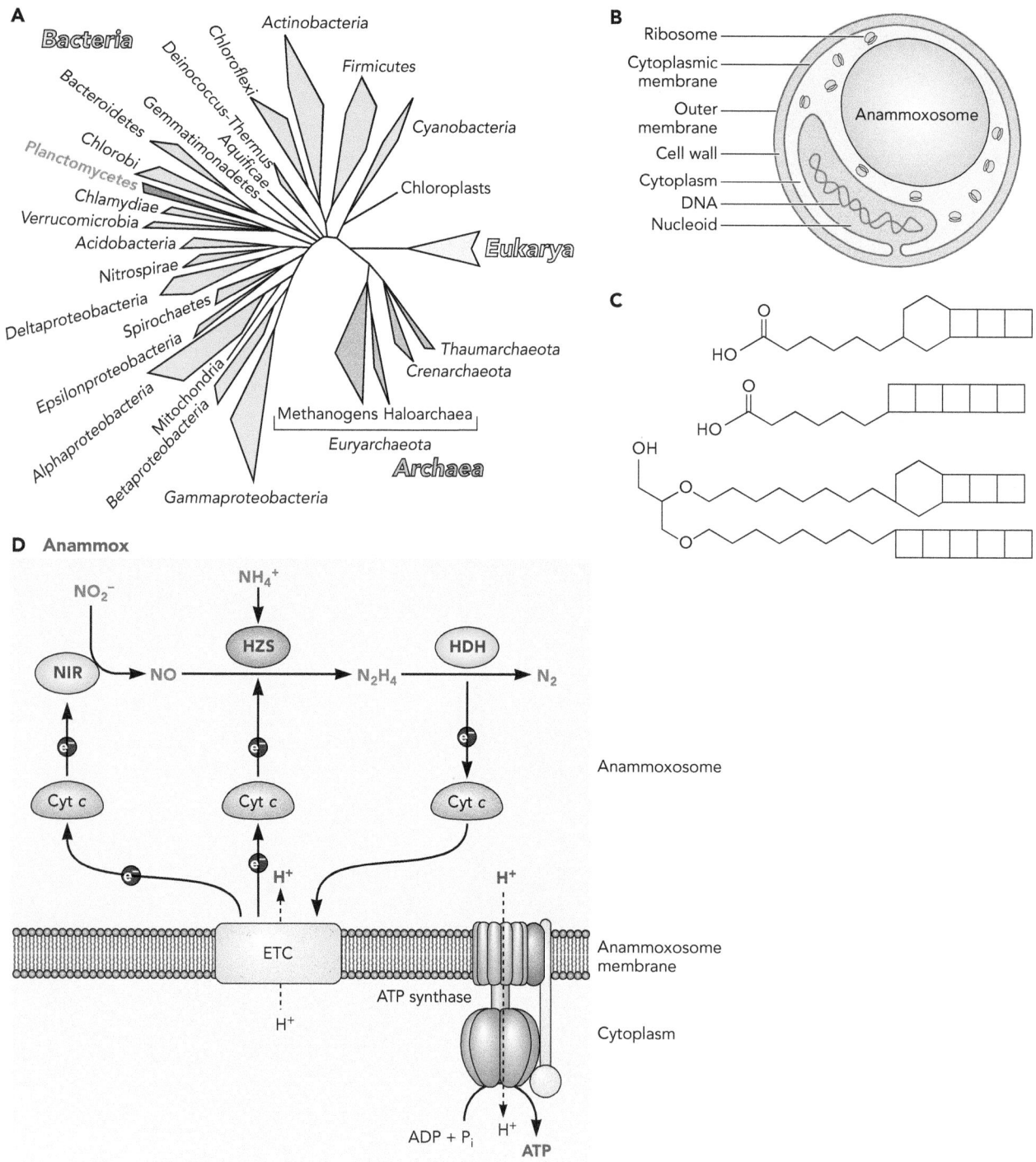

FIGURE 12.11 **Distribution of anaerobic ammonia oxidation (anammox) bacteria and biochemistry of the process.** (A) Phylogenetic tree highlighting *Planctomycetes*, the only known lineage with members capable of anammox (highlighted in dark coloring and bold green text). (B) A specialized membrane-enclosed organelle-like structure known as the anammoxosome is found in the cytoplasm of the anammox bacteria. (C) Structures of unique lipid ladderanes that provide rigidity and reduce permeability of the anammoxosome membrane. (D) During anammox, ammonium (NH_4^+) and nitrite (NO_2^-) are reduced under anoxic conditions to nitrogen gas (N_2). Nitrite is first reduced to nitric oxide (NO) by nitrite reductase (NIR), and then nitric oxide and ammonium are reduced to hydrazine (N_2H_4) by hydrazine synthase (HZS) using electrons from a poorly characterized electron transport chain (ETC) in the anammoxosome membrane. Hydrazine oxidoreductase (HDH) then oxidizes hydrazine to nitrogen gas and returns electrons to the ETC via cytochrome *c*. The ETC generates a proton motive force across the anammoxosome membrane that is used to yield ATP in the cell cytoplasm via ATP synthase.

This process also plays an important role in wastewater treatment, where it has been implemented for ammonia removal.

Most *Planctomycetes* are aerobic chemoheterotrophs, but anammox bacteria are anaerobic chemolithoautotrophs that use the reductive acetyl coenzyme A pathway (Chapter 9) to fix CO_2. Members of this phylum have some unique characteristics, including deep invaginations of the cytoplasmic membrane that surround the nucleoid. In anammox bacteria, an additional bilayer membrane surrounds an organelle-like compartment called the **anammoxosome** (Fig. 12.11B), which houses the enzymes that carry out anammox, as well as the toxic intermediates that are produced. Several unique lipids composed of multiple fused cyclobutane rings, called **ladderanes** (Fig. 12.11C), are present in the anammoxosome membrane. The ladderanes are densely packed in the anammoxosome membrane and are believed to provide additional rigidity to make the membrane impermeable to the escape of the toxic intermediate hydrazine (N_2H_4; a type of rocket fuel), which can damage cellular DNA, RNA, and proteins.

Three key enzymes are necessary to convert one molecule each of ammonium and nitrite to one molecule of nitrogen gas (Fig. 12.11D). In the first step, nitrite reductase generates nitric oxide from nitrite:

$$NO_2^- + 2H^+ \rightarrow NO + H_2O$$

Hydrazine synthase then condenses ammonium with nitric oxide to produce hydrazine:

$$NH_4^+ + NO + 2H^+ \rightarrow N_2H_4 + H_2O$$

Finally, hydrazine dehydrogenase converts hydrazine to nitrogen gas:

$$N_2H_4 \rightarrow N_2 + 4H^+ + 4e^-$$

Oxidation of hydrazine produces electrons that are passed through the ETC via ubiquinone, the cytochrome bc_1 complex, and cytochrome c in the anammoxosome membrane. A PMF is generated across the anammoxosome membrane, and ATP is produced in the cytoplasm by ATP synthase. The anammox reaction cycles 15 times to fix one CO_2 molecule, and as a result, growth of anammox bacteria is extremely slow. Although pure cultures of anammox bacteria are not yet available, enrichments of anammox bacteria that are up to 90% pure have doubling times of 10 to 14 days.

Denitrification

Denitrification is the microbial process in which nitrate and nitrite are reduced to gaseous forms of nitrogen that are released to the atmosphere (Fig. 12.12). This process has both beneficial and detrimental environmental impacts. Importantly, denitrification depletes biologically usable forms of nitrogen from the soil and is therefore detrimental for agriculture. In addition, nitrous oxide (N_2O) is a potent greenhouse gas that contributes to climate change and nitric oxide (NO) undergoes reactions that result in acid rain. In contrast, denitrification is beneficial for wastewater treatment as it eliminates usable forms of

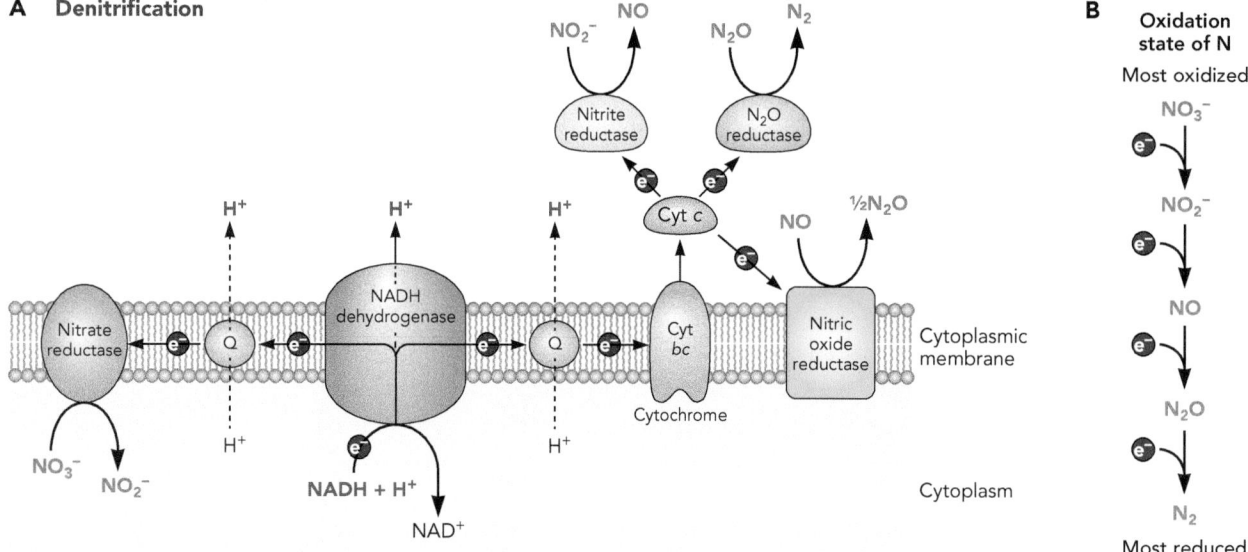

FIGURE 12.12 **Electron transport during the process of denitrification.** (A) Complete denitrification is a form of anaerobic respiration in which nitrate (NO_3^-) is reduced to dinitrogen gas (N_2). A series of reductase complexes sequentially reduce various nitrogen oxides through an electron transport chain (ETC) involving inorganic forms of nitrogen that serve as terminal electron acceptors. The initial components of the ETC, NADH dehydrogenase, ubiquinone (Q), and cytochrome *bc*, are also used in aerobic respiration. The process of denitrification is widespread and diverse; some bacteria also have a periplasmic nitrate reductase (not shown) that can be used instead of or in addition to the membrane-bound nitrate reductase. In some bacteria, the reduction of the inorganic nitrogen stops at intermediate steps depending on the reductases produced in that organism. (B) The various inorganic forms of nitrogen associated with denitrification, from the most oxidized form of nitrogen, nitrate, to more reduced forms via the addition of electrons.

nitrogen that can cause eutrophication of surface waters. The overall process of denitrification is

$$NO_3^- \rightarrow NO_2^- \rightarrow NO \rightarrow N_2O \rightarrow N_2$$

Bacteria and *Archaea* capable of denitrification are widely distributed on the phylogenetic tree, and there are also some examples of eukaryotic microbes that can carry out the process. Denitrifiers are chemoheterotrophs that use organic electron donors (e.g., NADH + H$^+$, FADH$_2$) and have membrane-associated ETCs similar to that of anaerobically respiring *Escherichia coli* (Chapter 11). Similar to *E. coli*, most denitrifiers (e.g., *Paracoccus denitrificans*, some *Pseudomonas* species, and *Alcaligenes* species) are capable of both aerobic and anaerobic respiration. When oxygen is available, organic compounds donate electrons to the ETC and oxygen serves as the terminal electron acceptor to generate a PMF. However, under anoxic conditions when nitrate is present, denitrifiers switch from aerobic to anaerobic respiration and denitrification occurs. *E. coli* has a single membrane-bound reductase (nitrate reductase) that converts nitrate to nitrite (Chapter 11). In contrast, denitrifiers have a series of reductase complexes that extend the ETC and allow the complete conversion of nitrate to nitrogen gas (Fig. 12.12). The ETC in denitrifiers is composed of dehydrogenases, quinones, cytochromes, and the appropriate reductase complexes needed to use the various forms of inorganic nitrogen as electron acceptors. The different forms of inorganic nitrogen are preferentially used from the most oxidized to the less oxidized forms (nitrate → nitrite → nitric oxide → nitrous oxide). During complete denitrification, N_2 gas is released as the end product. This brings the N cycle back to the biologically inert form of inorganic nitrogen gas.

Learning Objectives: After completing the material in this chapter, students should be able to . . .

1. provide an overview of the chemical transformations in the nitrogen cycle and the lifestyles of the bacteria that carry them out

2. explain the process and regulation of nitrogen fixation

3. describe how plants and bacteria communicate and benefit from symbiotic nitrogen fixation

4. discuss the common mechanisms of nitrate and ammonia assimilation

5. compare and contrast the microbial processes that use nitrogen compounds as energy sources (oxidation) and those that use nitrogen compounds as electron acceptors (reduction)

Check Your Understanding

1. Define this terminology: nitrogen fixation, diazotrophs, nitrogenase enzyme, flavonoid compounds, nod factors, root nodules, infection thread, bacteroids, leghemoglobin, assimilatory nitrate reduction, ammonia assimilation, nitrification, ammonia-oxidizing bacteria (AOB), nitrite-oxidizing bacteria (NOB), comammox (complete ammonia oxidation), ammonia monooxygenase, hydroxylamine reductase (HOA), ammonia-oxidizing archaea (AOA), anammox (anaerobic ammonia oxidation), anammoxosome, ladderanes, denitrification

2. The figure shows biological transformations of nitrogen commonly found in soils. Fill in each blank with the letter that best illustrates each process. (LO1)

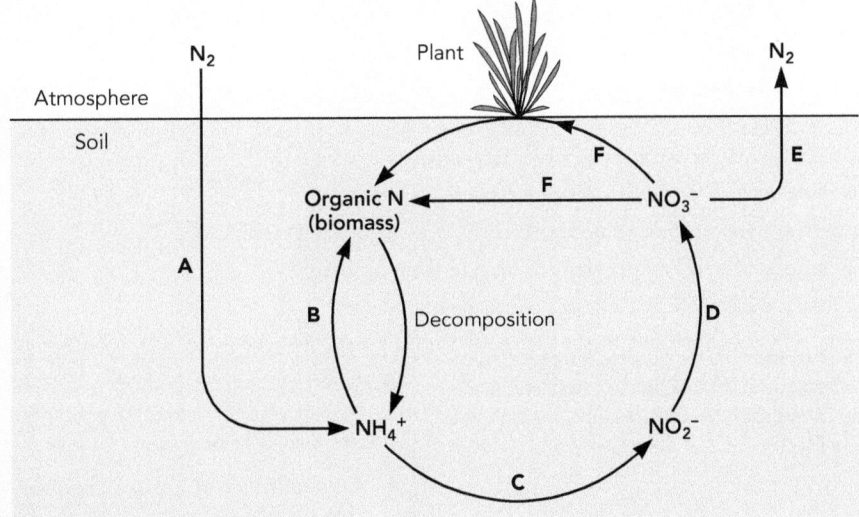

____ Nitrogen fixation

____ Ammonium oxidation

____ Nitrite oxidation

____ Complete denitrification

____ Ammonium assimilation

____ Nitrate assimilation

3. Nitrogen cycling is critically important because all cells require nitrogen for biosynthesis. In the table, name each process in the N cycle (indicated by the chemical reactions) using these terms: **ammonia oxidation**, **ammonia assimilation**, **complete denitrification (nitrate reduction)**, **nitrite oxidation**, and **nitrogen fixation**. Indicate whether the bold compound in the chemical reactions below functions as an **energy source**, **electron acceptor**, or **N source** and if the process **uses** or **conserves energy**. (LO1)

Chemical reactions (not balanced)	Process	Is the bold N compound an energy source, electron acceptor, or N source?	Does the process use or conserve energy?
$NH_3 \rightarrow$ organic N (e.g., amino acid)			
$N_2 + 8H^+ \rightarrow 2\,NH_3 + H_2$			
$NH_3 + O_2 \rightarrow NO_2^- + H_2O$			
$NO_2^- + O_2 \rightarrow NO_3^- + H_2O$			
Acetate + $NO_3^- \rightarrow N_2 + H_2O + CO_2$			

4. Nitrogen fixation is a critical process for the function of both natural ecosystems and agricultural systems. (LO2)

   a) What is the overall goal of nitrogen fixation?

   b) What is the overall reaction? (Include electrons, protons, and ATP.)

   c) The nitrogenase enzyme complex that facilitates nitrogen fixation consists of two components: dinitrogenase reductase and dinitrogenase. What is the general function of each?

   d) From where does the reducing power come during nitrogen fixation?

5. The process of nitrogen fixation is exquisitely sensitive to oxygen because of the reducing conditions required to break apart the N_2 triple bond. (LO2)

   a) Explain how each bacterium copes with the oxygen sensitivity of the nitrogenase complex.

   i. *Clostridium*, an obligate anaerobe:

   ii. *Rhodopseudomonas*, a facultative anaerobe:

   iii. *Azospirillum*, a microaerophile:

   b) Aerobic nitrogen fixers have a more difficult problem. Explain how each mechanism allows cells to fix nitrogen despite their need to respire aerobically or produce oxygen.

   i. *Azomonas* produces a thick polysaccharide capsule:

   ii. *Azotobacter* has a very high respiration rate:

   iii. Unicellular nitrogen-fixing cyanobacteria fix nitrogen only at night:

   iv. Some filamentous oxygenic phototrophs produce heterocysts:

6. Because nitrogen fixation is so energy intensive, the process is highly regulated. Match each protein involved in the regulation of nitrogen fixation in *Klebsiella* with its function from the list provided (A-D). (LO2)

   ___ NtrA

   ___ NtrB

   ___ NtrC

   ___ NifA

A. A transcription factor that works with the σ^{54} RNAP holoenzyme to control the promoters of *nif* (nitrogen fixation) operons.

B. A sigma factor (σ^{54}) that is required to work with RNAP to recognize the promoter regions controlling the operon for nitrogen fixation genes.

C. The response regulator (activator) that binds to target promoters together with the σ^{54} RNAP holoenzyme.

D. The sensor histidine kinase protein that autophosphorylates a conserved histidine residue, then phosphorylates the response regulator.

7. Assuming that the σ^{54} RNAP holoenzyme complex has formed, order the events during nitrogen fixation in *Klebsiella* from 1 (first) to 4 (last). (LO2)

   _____ A phosphoryl group is transferred to a conserved residue on NtrC.

   _____ NtrB autophosphorylates a conserved histidine residue.

   _____ NifA interacts with the σ^{54} RNAP holoenzyme to control the promoters of multiple *nif* operons that encode the nitrogenase enzyme complex.

   _____ The NtrA-RNAP holoenzyme and NtrC-P together activate the promoter controlling expression of *nifL* and *nifA*.

8. When sufficient ammonia is present, the genes encoding nitrogenase are no longer expressed. NifL is an anti-activator important in the regulation of these nitrogen-fixation genes. (LO2)

   a) With what molecule(s) does NifL interact?

   b) Which of the steps listed in Question 7 do not happen as a result of this interaction?

   c) NifL is a redox-sensitive regulatory protein which stops the transcription of some *nif* genes in the presence of oxygen. Why is this a useful strategy for nitrogen-fixing cells?

9. Some plants and bacteria have evolved a complex communication system that results in the symbiotic fixation of nitrogen that benefits both the host plant and bacteria. (LO3)

   a) For each entity related to symbiotic nitrogen fixation listed below, indicate whether it comes from the plant or bacteria, its function, and the response of the recipient organism.

Entity	Provided by plant or bacteria?	Function	Response from plant or bacterium
Flavonoid compounds			
Nod factors			
Infection thread			
Bacteroids			
Leghemoglobin			

   b) List one way that the bacteria benefit and one way the host plant benefits from this relationship.

10. Ammonia is the most common source of nitrogen for biosynthesis. (LO4)

    a) The primary reactions for the assimilation of ammonia are listed below. Using the list in the box to the right, write in the name of the enzyme that carries out each step.

    > Glutamine synthetase
    > Glutamine oxoglutarate aminotransferase (GOGAT)
    > Glutamate dehydrogenase
    > Transaminase

    i. glutamate + NH_3 + ATP → glutamine + ADP + P_i _____

    ii. glutamine + α-ketoglutarate + NADPH + H^+ → 2 glutamate + $NADP^+$ _____

    iii. α-ketoglutarate + NADPH + H^+ + NH_3 → glutamate + $NADP^+$ + H_2O _____

    iv. glutamate + α-ketocarboxylic acid → amino acid + α-ketoglutarate _____

    b) Reactions ii and iii both generate glutamate. What is one advantage for cells to use reaction iii instead of ii? Under what conditions do cells use reaction iii?

    c) Which reaction leads to the biosynthesis of purines and pyrimidines?

11. A number of bacteria can grow using inorganic nitrogen compounds as a source of energy. In the sentences below, fill in the blanks using the terms provided in the box. Terms can be used more than once or not at all. (LO5)

    > ATP
    > CO_2
    > NH_3 /NH_4^+
    > NO_2^-
    > NO_3^-
    > O_2
    > Aerobic respiration
    > Energy source
    > Proton motive force
    > Quinone
    > Reducing power
    > Reverse electron transport

    a) Ammonia-oxidizing bacteria use _____ as an energy source in _____.

    b) In nitrite-oxidizing bacteria, once the electrons are passed to cyt c, the electrons can move to _____, the terminal electron acceptor, to make _____, or to NADH reductase to make _____.

    c) Comammox bacteria use _____ and _____ as their energy sources and _____ as the terminal electron acceptor in respiration. When growing as chemolithoautotrophs, they use _____ as their carbon source and the process of _____ to generate NAD(P)H + H^+.

12. The process of ammonia oxidation under anaerobic conditions (anammox) contributes to much of the nitrogen loss in ocean ecosystems. (LO5)

   a) Anammox occurs in enclosed cellular compartments called anammoxosomes. Why do cells need to carry out the anammox reaction in enclosed compartments?

   b) The overall reaction of anammox can be described as $NH_4^+ + NO_2^- \rightarrow N_2 + 2H_2O$.

   This three-step process involves three key enzymes. Name the enzyme that catalyzes each reaction and indicate if the molecule in bold is oxidized or reduced by the reaction.

Reaction	Enzyme	Oxidation or reduction?
$\mathbf{NO_2^-} + 2H^+ \rightarrow NO + H_2O$		
$\mathbf{NH_4^+} + NO + 2H^+ \rightarrow N_2H_4 + H_2O$		
$\mathbf{N_2H_4} \rightarrow N_2 + 4H^+ + 4e^-$		

13. Bacteria that carry out nitrification and those that carry out denitrification both respire, but they use nitrogen compounds in very different ways. Fill in the table as indicated by the headings. (LO5)

Cell type	Nitrification or denitrification?	Typical electron donor	Terminal electron acceptor	Nitrogen byproduct	Are cells typically heterotrophs or autotrophs?
Ammonia oxidizers					
Nitrite oxidizers					
Nitrate reducers					

Dig Deeper

14. Carbon cycles (shown in the diagram) are intimately linked to nitrogen cycles by the processes that occur in both *Bacteria* and *Archaea*. (LO1)

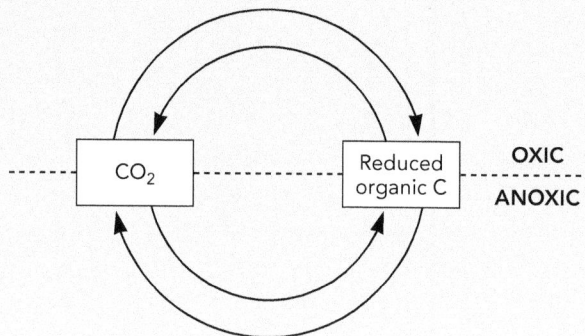

   a) Consider the ammonia- and nitrite-oxidizing (AOB, AOA, NOB, comammox, and anammox) bacteria and archaea.

   i. Do they use organic carbon compounds for biosynthesis, as an energy source, or both?

   ii. Label the appropriate arrow(s) to show from where ammonia-oxidizing (AOB, AOA) and nitrite-oxidizing (NOB, comammox) bacteria obtain carbon.

b) Consider the nitrate-reducing bacteria.

    i. Do they use organic carbon compounds for biosynthesis, as an energy source, or both?

    ii. Label the appropriate arrow(s) to show from where nitrate-reducing bacteria obtain carbon.

15. Nitrogen on the Farm

    a) Think about the consequences of the various nitrogen transformations in soils, with respect to biologically available nitrogen compounds (NH_3/NH_4^+ and NO_3^-). (See figure in Question 2.) Fill in the blanks using the terms in the box. (LO1, 5)

NH_3 and NO_2^- oxidation
NH_3 oxidation by anammox
Complete denitrification
N_2 fixation

        i. Which microbial process **adds** biologically available nitrogen to the soil ecosystem? _____

        ii. Which two microbial processes **convert** biologically available nitrogen to nitrogen gas that is utlimately lost from the soil ecosystem? _____ and _____

        iii. Which process converts a generally **inaccessible N** source to something plants and microbes can use? _____

    b) Rhea Zobium inherited a farm and decided to make a go of it. The first year, she fertilized her fields, the crops grew well, but neighbors complained that there was nitrate (which can be toxic at high concentrations) in their groundwater. She said the nitrate couldn't possibly be from her fields because she applied <u>manure</u> (organic N) as a fertilizer, not nitrate. **Ammonification** is the process by which the microbes decay organic material like manure and release ammonium into the environment. Knowing this, propose a pathway by which the manure could have been converted to nitrate.

    c) The next year, Rhea applied <u>nitrate fertilizer</u> on some fields and <u>ammonium fertilizer</u> on other fields. Unfortunately, she found that the weather was a confounding factor in trying to predict how much fertilizer to apply to her fields.

        i. If she applied <u>ammonium</u> to her fields during a wet period, when soils were waterlogged and anoxic, the NH_4^+ adsorbed to the negatively charged soil particles and remained in the field. However, if the soil was drier and well aerated, NH_4^+ was quickly lost from the soil. Explain these results for her.

        ii. Nitrate had an entirely different fate. If NO_3^- was applied to a field prior to a heavy rain, it was lost very quickly from the soils. What biological mechanism was likely at work to reduce the amount of nitrate fertilizer in the soil after a long, heavy rain? Considering the charge on nitrate, what physical mechanism likely contributed to the loss of nitrogen?

16. Because the lack of available nitrogen often limits the growth of organisms in aquatic environments, the addition of excess nitrogen can greatly alter the ecosystem. Eutrophication describes the changes that occur in a body of water due to the addition of excess nitrogen (e.g., when nitrate is washed off agricultural fields into streams and bays). (LO1, 5)

    a) The steps below explain how the addition of excess nitrate fertilizer in runoff water can cause "dead zones" (areas of water with little or no oxygen) downstream. **Number the remaining steps from 2 to 5 to indicate the chain of events.**

1	Nitrate fertilizer enters the water system in runoff.
	Aerobic heterotrophs consume oxygen as they respire organic C and N.
	As N becomes limiting, the growth of autotrophic microbes slows and cells die off.
	Autotrophic microbes like cyanobacteria enter a rapid growth phase due to high levels of nitrogen, fixing CO_2 to organic C (biomass) as they grow.
	Biomass becomes available for decomposers to use as carbon and nitrogen sources.

    b) The following graph shows how the concentrations of nitrate, oxygen, and organic C in a body of water can change with time after the addition of nitrate.

        i. Label each line. Which line on the graph represents NO_3^-? Which line represents O_2? Which line represents biomass (organic C)?

        ii. Consider each step in eutrophication described in part a). Indicate on the graph where each step is happening.

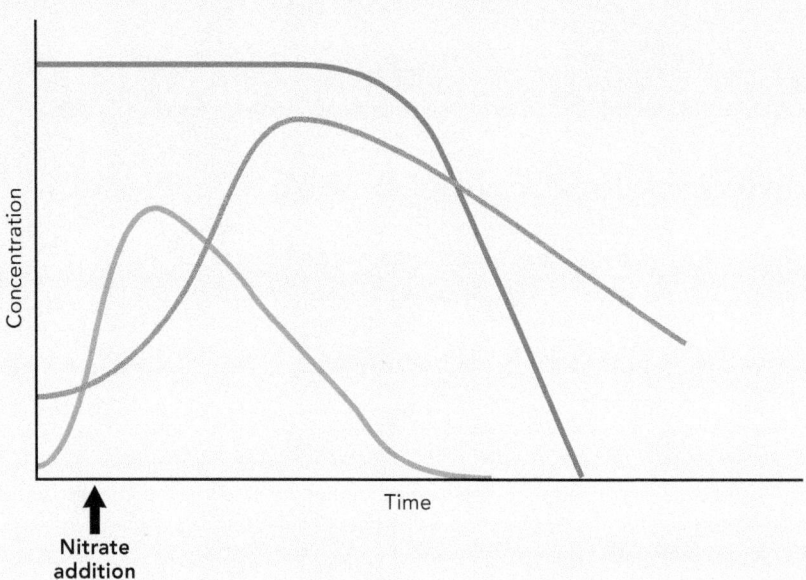

17. The nitrogenase enzyme is extremely sensitive to oxygen, and the reaction catalyzed by nitrogenase to fix nitrogen gas is very energetically expensive. As such, nitrogen fixation is highly regulated. There are a number of challenges that need to be overcome for a bacterium to be successful at fixing nitrogen, and the ways in which certain bacterial species overcome these challenges are specific to the physiology of each organism.

a) *Klebsiella* spp. are facultative anaerobes that have the ability to fix nitrogen. Anoxic conditions and low ammonia concentrations are two environmental signals that trigger nitrogen fixation in these species. Explain why *Klebsiella* only fixes nitrogen under these conditions.

   i. Anoxic conditions

   ii. Low ammonia concentrations

b) Describe the roles of NtrB, NtrC, NifL, and NifA proteins, which are responsible for detecting ammonia concentrations, in the regulation of the *nif* operon in *Klebsiella*.

c) FNR is a master regulator that is involved in controlling the expression of many genes required for growth in anoxic conditions, including reductase complexes in electron transport chains (Chapters 6 and 11). Why would the expression of the reductase complexes controlled by FNR be important in order for *Klebsiella* to fix nitrogen?

d) NifL is an anti-activator that works to detect anoxic conditions via oxidation/reduction of its FAD coenzymes. When oxidized, NifL binds to NifA, which inhibits the ability of NifA to activate the expression of the *nif* operons. When NifL is reduced by the activity of reductase complexes, it no longer binds NifA. How are the activities of FNR and NifL coordinated to regulate *nif* operon expression?

e) NifL is a complex activator that is controlled by both oxygen and ammonia levels. Summarize how *Klebsiella* coordinates nitrogen fixation under various conditions by filling in the table.

Condition	NifL active?	NifA active?	Does nitrogen fixation occur?
Oxic/low [NH$_3$]			
Anoxic/low [NH$_3$]			
Oxic/high [NH$_3$]			
Anoxic/high [NH$_3$]			

304 Making Connections

304 Fundamental Structure of the Cytoplasmic Membrane

306 Variation in Cytoplasmic Membranes

306 Transport across Cytoplasmic Membranes

311 Cell Wall Structures

315 Gram-Negative Outer Membrane

320 Periplasm

321 Additional Extracellular Layers

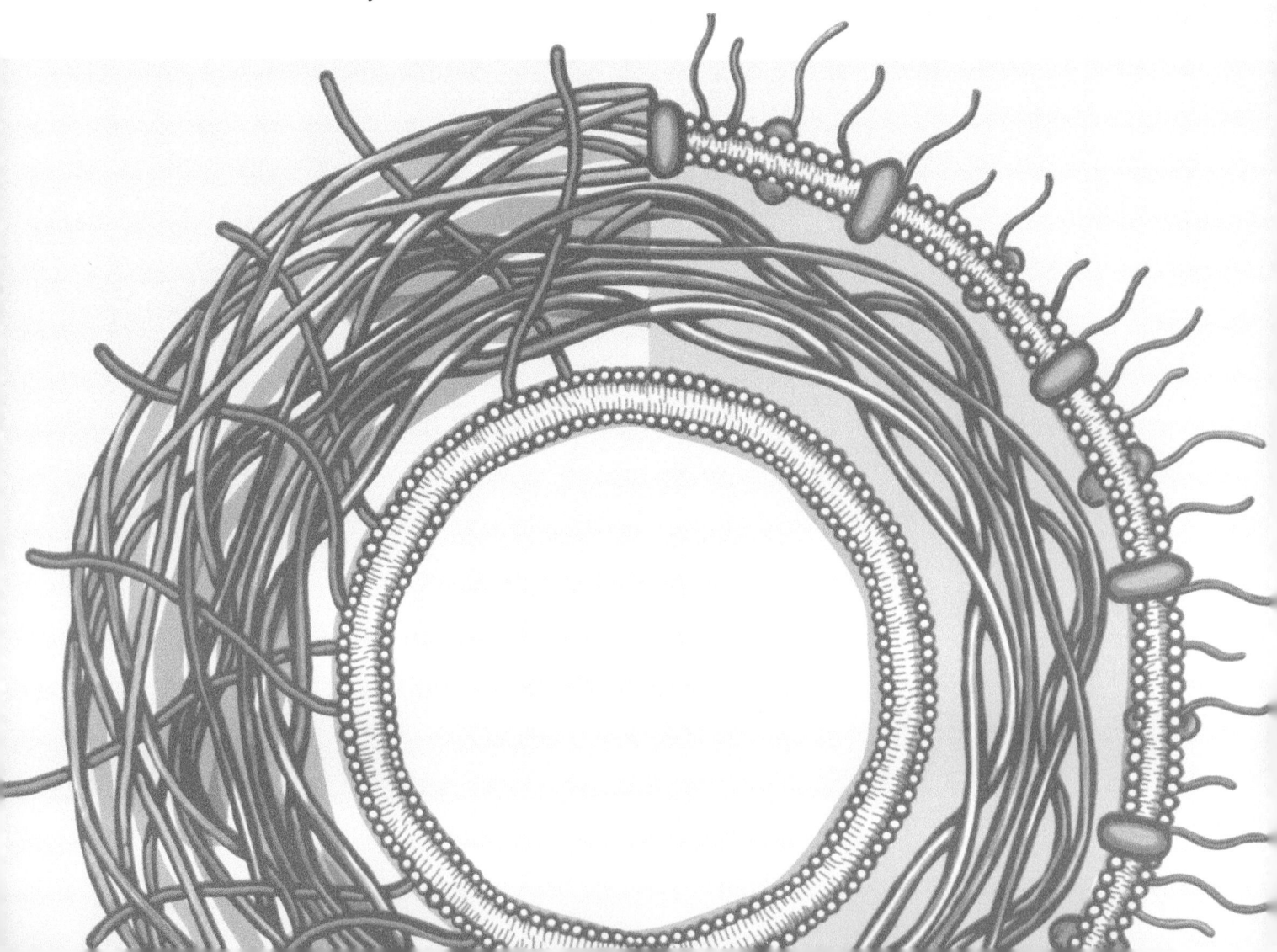

13

STRUCTURE AND FUNCTION OF THE CELL ENVELOPE

After completing the material in this chapter, students should be able to . . .

1. summarize the features of the cytoplasmic membranes of *Bacteria* and *Archaea*

2. describe the different transport processes that cells use to facilitate the movement of molecules across the cytoplasmic membrane

3. compare and contrast the structures of archaeal and Gram-positive and Gram-negative bacterial cell walls

4. detail the unique structure and function of the Gram-negative bacterial outer membrane

5. explain how porin synthesis is regulated in response to the osmolarity of the environment

6. discuss the structures and functions associated with the periplasm and capsule

Microbial Physiology: Unity and Diversity, First Edition. Ann M. Stevens, Jayna L. Ditty, Rebecca E. Parales and Susan M. Merkel.
© 2024 American Society for Microbiology.
Companion website: www.wiley.com/go/stevens/microbialphysiology

Making Connections

In Chapter 1, the molecular phylogenetic tree showing the relationship between the bacterial and archaeal domains was introduced, and some structural differences between these distinct groups of organisms were discussed. In Chapter 3, we reviewed the common macromolecules found in microbes. This chapter builds upon those topics by comparing and contrasting the cell envelopes found in prokaryotes, starting with the innermost layer, the cytoplasmic membrane, and moving outward to the cell wall, the periplasm and outer membrane of Gram-negative bacteria, and the capsule on the cell surface. The associated transport systems that function to bring nutrients across these external barriers into the cell will also be discussed. Mechanisms for protein transport and cellular localization will be described in Chapter 14.

Fundamental Structure of the Cytoplasmic Membrane

Although there is diversity in the structures of prokaryotic cell envelopes, the cytoplasmic membrane serves as a universal permeability barrier in all living cells. The cytoplasmic membrane is composed of phospholipids that are **amphipathic** in nature, meaning that they have both hydrophobic and hydrophilic characteristics. Bacterial and eukaryotic phospholipids consist of a polar head group that is hydrophilic due to the negatively charged phosphate group, and two covalently linked fatty acid hydrocarbon tails that are hydrophobic in nature (Fig. 13.1A). The cytoplasmic membrane, as described in the fluid mosaic model, is actually only ~50% phospholipids, with the remaining ~50% made up of membrane-associated proteins that typically do work for the cell as enzymes, electron carriers, receptors, or transporters (discussed below). The cytoplasmic membrane is not static, but rather has fluidity in two dimensions. Integral membrane proteins, including transmembrane proteins that span the entire membrane, are embedded in the membrane and may move laterally. The degree of membrane fluidity depends on the nature of the fatty acids that comprise the phospholipids in the membrane. Saturated fatty acids are more tightly packed due to their linear structure, and this results in decreased membrane fluidity. Unsaturated fatty acids contain one or more double bonds, and the configuration of the bonds (*cis* or *trans*) impacts the packing of the lipids in the membrane. Like saturated fatty acids, those with *trans* double bonds are linear and pack tightly, reducing membrane fluidity. In contrast, fatty acids with *cis* double bonds have a bent structure that increases fluidity (Fig. 13.1B). Transmembrane proteins are also amphipathic in nature with hydrophobic regions within the membrane and hydrophilic amino acids positioned at the inner and outer surfaces, where they are accessible to the aqueous environment of the cytoplasm or periplasm, respectively. Peripheral membrane proteins are more loosely associated with the membrane, often through protein-protein contacts with integral membrane proteins. Membrane-associated proteins are typically asymmetrical and hence give the cytoplasmic membrane two distinct sides, with each side serving a different function for the cell. As an example,

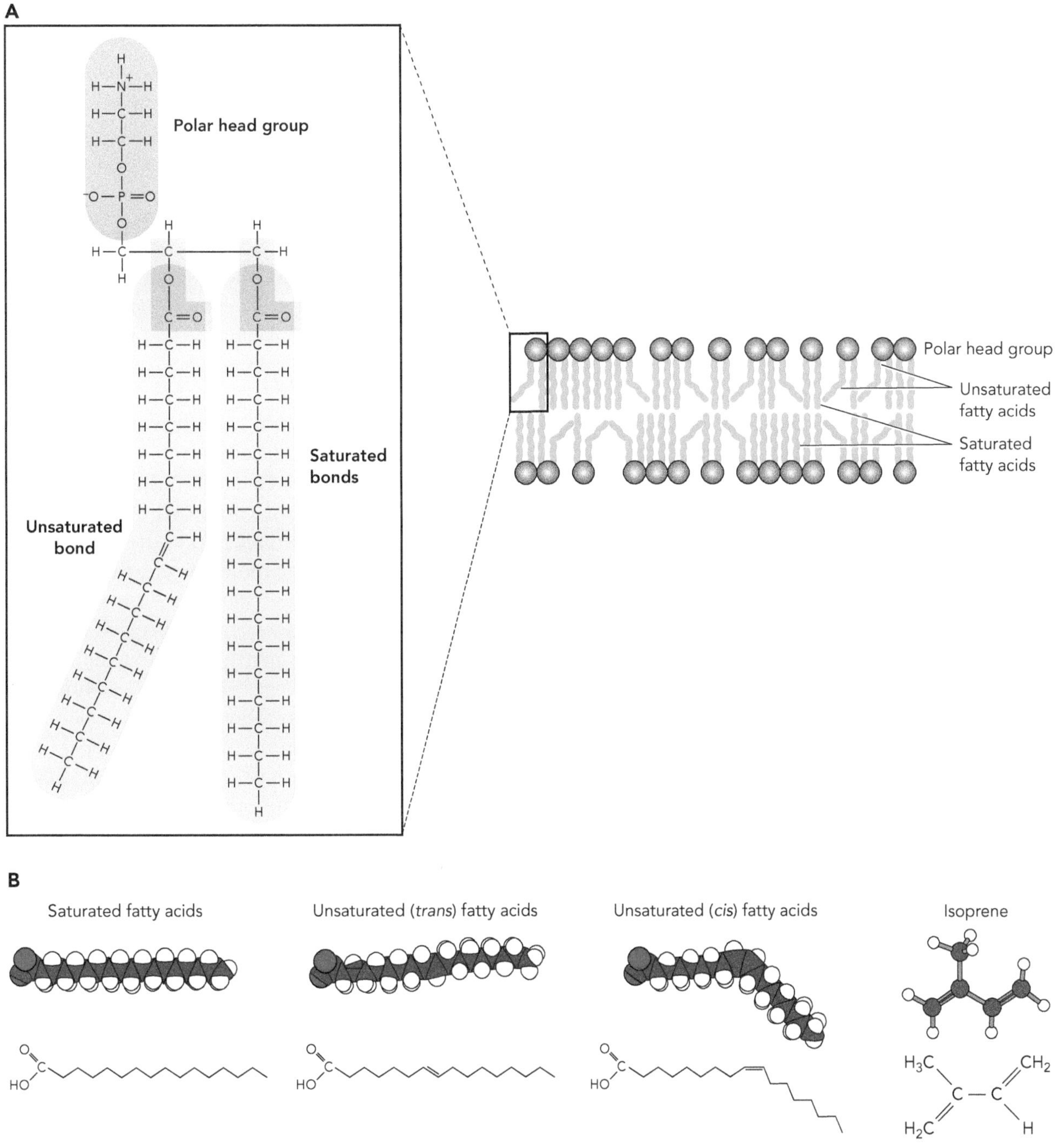

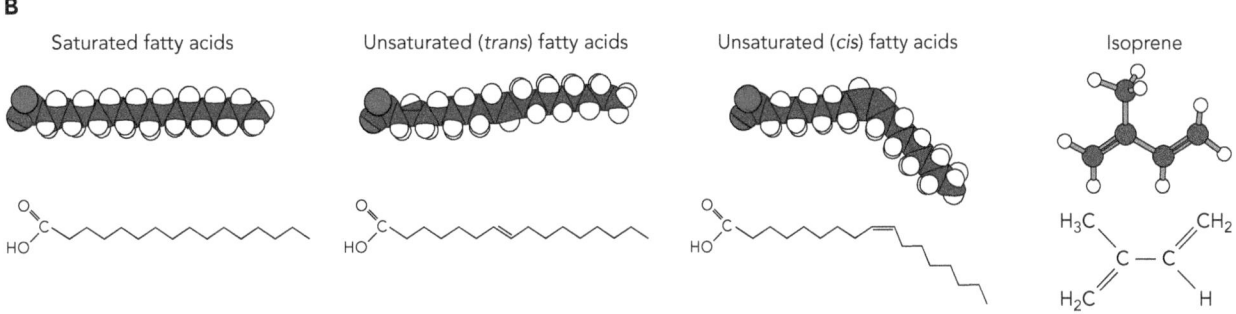

A

Polar head group

Unsaturated
bond

Saturated
bonds

Polar head group

Unsaturated
fatty acids

Saturated
fatty acids

B

Saturated fatty acids

Unsaturated (*trans*) fatty acids

Unsaturated (*cis*) fatty acids

Isoprene

FIGURE 13.1 **Structure of cytoplasmic membrane phospholipids.** (A) Bacteria and eukaryotes use phospholipids containing fatty acids to build their bilayer cytoplasmic membranes. Two fatty acids are linked to a molecule of glycerol through an ester linkage (highlighted by blue shading). The fatty acids may have either saturated or unsaturated bonds. Their hydrophobic hydrocarbon chains reside in the interior of the membrane bilayer structure. The *cis* unsaturated fatty acids have a bent structure due to the presence of a double bond and therefore they increase membrane fluidity. A charged polar head group (e.g., phosphatidylethanolamine as shown), containing a phosphate group modified by an alcohol, is also linked to the glycerol. The polar head group serves as a hydrophilic component in the phospholipid that resides on the outer face of the bilayer structure. (B) Saturated fatty acids and *trans* unsaturated fatty acids have a linear structure, while *cis* unsaturated fatty acids have a bent structure. Isoprene, another hydrophobic hydrocarbon, is the basic unit in archaeal membranes.

membrane-bound proteins that serve as receptors (Chapters 7 and 15–18) have an external domain responsible for sensing environmental signals, whereas the internal domain is responsible for transmitting that signal to various intracellular targets for an appropriate physiological response.

Variation in Cytoplasmic Membranes

The cytoplasmic membranes of organisms across the domains *Archaea*, *Bacteria*, and *Eukarya* are not identical. For example, the phospholipids in the cytoplasmic bilayer membranes of bacteria and eukaryotes have ester linkages between the fatty acids and the polar head group (Fig. 13.2A). Bacterial fatty acids are typically 16 to 20 carbons in length. In contrast, the archaeal phospholipids have ether linkages instead of ester linkages (Fig. 13.2B). Further, instead of using fatty acids as their lipids, archaeal membrane lipids have repeating isoprene units. Isoprene is a five-carbon hydrocarbon that typically forms bilayers made of phytanyl groups with diether linkages (Fig. 13.1B and 13.2B). Both fatty acids and isoprenoids are hydrophobic hydrocarbons composed of carbon and hydrogen atoms. Thus, they form the inner hydrophobic part of the membrane, away from the polar head groups and polar water molecules in the adjacent aqueous environments.

Rather than having bilayer membranes, some archaea form monolayer membranes in which two phytanyl groups are covalently linked at their ends, forming a biphytanyl molecule with tetraether linkages (Fig. 13.2C). Monolayer membranes are structurally more stable and are therefore more resilient in high-temperature environments. However, such membranes do not appear to be essential for survival at elevated temperatures since bacterial thermophiles have bilayer cytoplasmic membranes. Conversely, some mesophilic archaea also have monolayer membranes.

Transport across Cytoplasmic Membranes

Biological membranes are selectively permeable, allowing only a few molecules to move freely across the membrane to reach equilibrium (equal concentration on either side of the membrane). Small, nonpolar, uncharged molecules and gases can move across the membrane via **simple diffusion** (Fig. 13.3). Simple diffusion does not require the input of biological energy or membrane-bound proteins. However, the limitations to simple diffusion are that only select molecules of appropriate size and hydrophobicity can pass freely through the membrane, and molecules cannot be accumulated against a concentration gradient.

Although relatively uncommon in bacteria, the process of **facilitated diffusion** involves channel or carrier proteins that are selective for the molecule being transported. Both simple and facilitated diffusion are forms of **passive transport**, meaning no biological energy is expended. The movement of the molecule being transported is driven by its concentration gradient from a high concentration to a low concentration until equilibrium across the membrane is reached. For example, water can be transported via simple diffusion or via facilitated diffusion through specialized water transport proteins called aquaporins,

A Bacterial phospholipid bilayer (glycerol diesters)

B Archaeal phospholipid bilayer (glycerol diethers or phytanyl)

C Archaeal phospholipid monolayer (glycerol tetraethers ar biphytanyl)

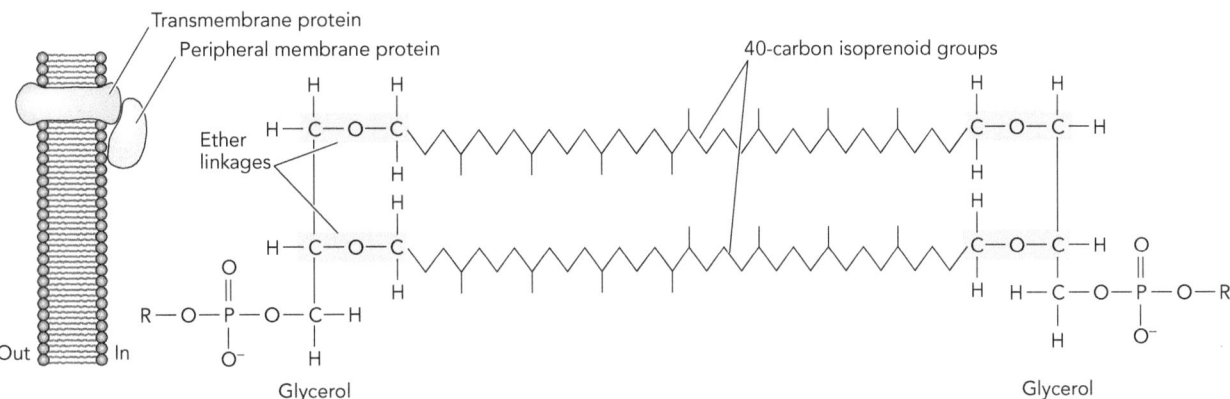

FIGURE 13.2 **The basic structures of microbial cytoplasmic membranes.** (A) *Bacteria* have bilayer cytoplasmic membranes composed of phospholipids. Each phospholipid has a hydrophilic glycerol-phosphate polar head group (R) that is covalently joined by ester linkages (highlighted by blue shading) to two hydrophobic fatty acid chains that are typically 16 to 20 carbons in length. Eukaryotic microbes (not shown) have similar cytoplasmic membrane structures. (B) Some members of the *Archaea* have bilayer membranes composed of phospholipids. Each phospholipid has a hydrophilic glycerol-phosphate polar head group (R) that is covalently joined by ether linkages (highlighted by green shading) to two hydrophobic isoprenoid chains with 20 carbons. (C) Other members of the *Archaea* have monolayer membranes composed of two hydrophilic polar head groups (R) both covalently joined by ether linkages (highlighted by green shading) to the same two hydrophobic isoprenoid chains with 40 carbons. Amphipathic transmembrane or integral membrane proteins are shown associated with each type of membrane (blue). A peripheral membrane protein is shown on the cytoplasmic (internal) face of the membrane (yellow). "R" represents a charge-modified alcohol that is part of the polar component of the phospholipid molecule.

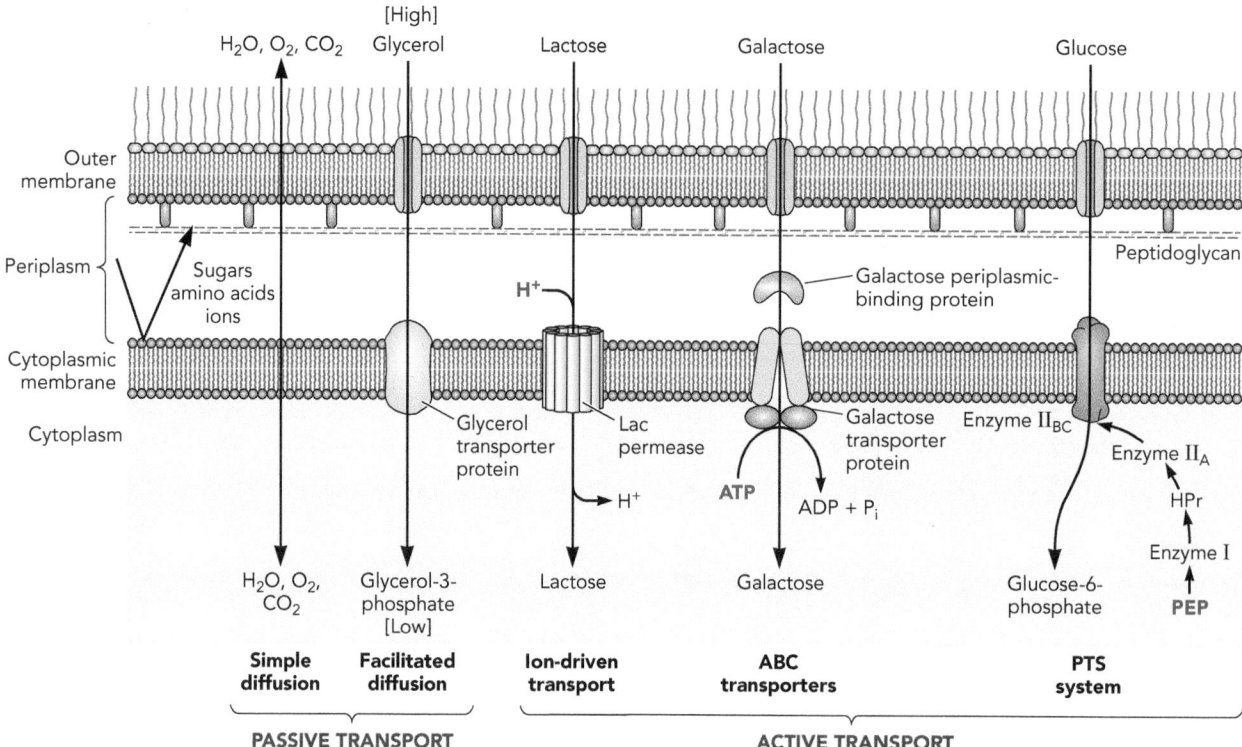

FIGURE 13.3 **Transport across the Gram-negative cell envelope.** (Left to right) Ions and other charged or polar molecules (e.g., sugars and amino acids) cannot diffuse across the cytoplasmic membrane in any living organism (a Gram-negative bacterium is shown here). However, some small nonpolar, uncharged molecules and gases do move across cytoplasmic membranes via passive transport, either by simple diffusion without the assistance of an accessory transmembrane protein or by facilitated diffusion with the assistance of an accessory transmembrane protein. Active transport systems require an input of energy in addition to one or more accessory transmembrane proteins. Examples of specific cytoplasmic membrane transporters are provided (ion-driven transport, ABC transporters, and the phosphotransferase system [PTS]); see the text for additional details. In Gram-negative bacteria, outer membrane porins (tan; see also Fig. 13.10) enable passage of small molecules, ions, and gases across the outer membrane into the periplasm by diffusion prior to transport across the cytoplasmic membrane. PEP, phosphoenolpyruvate.

which are membrane-bound proteins that form pores in the cytoplasmic membrane to allow the free movement of water molecules. In an environment of low osmolarity, water will flow into the cell, which has higher osmolarity, thus helping to maintain its turgor pressure. Carbon dioxide is another example of a molecule that can enter cells via simple or facilitated diffusion.

Glycerol transport in *Escherichia coli* is another example of facilitated diffusion (Fig. 13.3). In this case, a membrane protein facilitates the movement of the glycerol across the cytoplasmic membrane. The protein channel opens briefly to permit the solute (glycerol) to cross the membrane without loss of the proton motive force (PMF). It is the concentration of glycerol that drives the movement from a high to a low concentration. The cell modifies the glycerol immediately after it enters the cell, adding a charged phosphoryl group to prevent it from moving back across the membrane and to maintain a concentration gradient.

How do larger molecules and nutrients enter the cell? Molecules such as amino acids, simple sugars, and charged ions cannot freely diffuse through the membrane; therefore, the cell must have a transport mechanism for cell entry. In comparison to passive forms of transport, **active transport** processes require both a membrane protein and a biological energy source to move the molecules

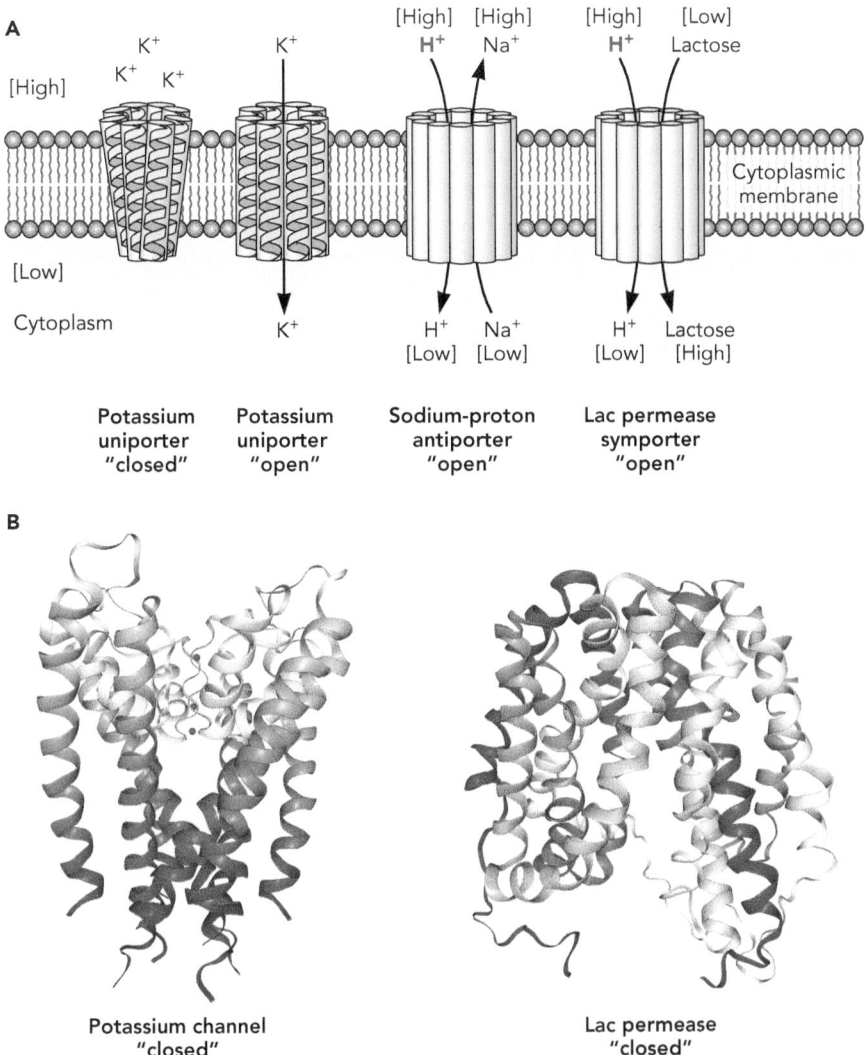

FIGURE 13.4 Examples of ion-driven transporters. (A) (Left to right) Transmembrane (integral) proteins such as uniporters, antiporters, and symporters all work by a similar mechanism involving concentration gradients. In some cases, the energetically favorable reaction of moving ions from a high to low concentration drives transport of the ion itself (e.g., potassium uniporter). Alternatively, the energetically unfavorable movement of specific ions or solutes against a concentration gradient can be coupled to the energetically favorable movement of other ions in the opposite direction (e.g., sodium-proton antiporter) or the same direction (e.g., lactose [Lac] permease symporter). See the text for additional details. (B) Channel proteins maintain the integrity of the proton motive force when in the closed conformation, but briefly open to permit passage of the molecule(s) being transported. Shown are closed conformations of the potassium channel (PDB ID 2QTO; https://doi.org/10.2210/pdb2QTO/pdb) and Lac permease (PDB ID 2V8N; https://doi.org/10.2210/pdb2V8N/pdb). Created with NGL Viewer (Rose AS et al. 2018. *Bioinformatics* 34:3755–3758), and RCSB PDB.

in an energetically unfavorable direction (Fig. 13.3). These membrane proteins typically have multiple membrane-spanning α-helices that allow specific substrates to pass through a tightly controlled channel (Fig. 13.4). The vast majority of transport systems in *E. coli* and other bacteria involve active transport. Phosphoenolpyruvate is the energy source that drives the movement of carbohydrates via the sugar phosphotransferase system (PTS) (Chapter 7) in some bacteria. Approximately 40% of transporters are ion-driven transporters that consume the PMF (Chapter 5). ATP-binding cassette (ABC) transporters, which hydrolyze ATP to provide energy for transport (see below), account for another ~40% of all bacterial transporters. Indeed, all living cells across the tree of life utilize ABC transporters and ion-driven transporters. The latter mechanism is one reason that all living cells must be able to generate a PMF across their cytoplasmic membrane and why the cytoplasmic membrane must be impermeable to ions.

Ion-driven transport, also known as chemiosmotic transport, is dependent on the energy stored in the PMF to drive the movement of molecules. A **uniporter** allows the movement of one type of ion into the cell. For example, positively charged ions like potassium ions, which contribute to the PMF, can move into the cell when concentrations outside the cell are higher. However, this movement will decrease the electrical component of the PMF. Also driven by the PMF, **antiporters** enable the movement of two molecules in opposite directions, whereas **symporters** move two molecules in the same direction. Many ion-driven transporters are members of the major facilitator superfamily (MFS). They share a similar structure where one polypeptide with 12 or 14 membrane-spanning α-helical domains (each consisting of ~15 hydrophobic amino acids) forms the membrane channel (Fig. 13.4).

Coupled reactions occur within ion-driven transporters. An unfavorable reaction with a positive ΔG can occur when it is coupled to a favorable reaction with a negative ΔG. The unfavorable reaction will not occur unless it is coupled to the favorable reaction and the sum of the two reactions has a negative ΔG. For example, in a sodium-proton antiporter, the movement of protons into the cell (PMF) drives the simultaneous transport of sodium ions from a low concentration inside the cell to a high concentration outside. Conversely, the Lac permease (LacY) functions as a symporter, with protons moving from a high concentration outside the cell to a low concentration inside while bringing lactose into the cell against the concentration gradient, even when total internal sugar concentrations are higher within the cell (Fig. 13.4). The potential energy of the PMF is used to directly drive the coupled reaction instead of using the PMF to produce ATP.

In contrast, **ABC transport** is a form of active transport that consumes ATP directly (Fig. 13.5). Both Gram-negative and Gram-positive bacteria have ABC transporters. The high-affinity binding protein that initiates the transport process is located in the periplasm, which is a large compartment in Gram-negative bacteria but is a much smaller space in Gram-positive bacteria. A well-studied ABC transporter in *E. coli* is responsible for the entry of histidine into the cell (Fig. 13.5). Amino acids are small enough to diffuse through the porins, which are protein channels in the Gram-negative outer membrane, and through the holes in the peptidoglycan matrix (see below) of Gram-positive bacteria. A periplasmic binding protein with a high affinity for the amino acid (i.e., 0.01- to 1-micromolar K_m values) binds the amino acid to ensure that it does not diffuse away from the cell. The amino acid is delivered by the binding protein to the main ABC transporter complex, which is composed of multiple subunits in the cytoplasmic membrane. ATP hydrolysis by the membrane transporter opens the channel, permitting the amino acid to enter the cytoplasm while preserving the integrity of the PMF.

Both ion-driven transporters and ABC transporters are also found in the archaea and in eukaryotes. These transporters can have a profound effect on

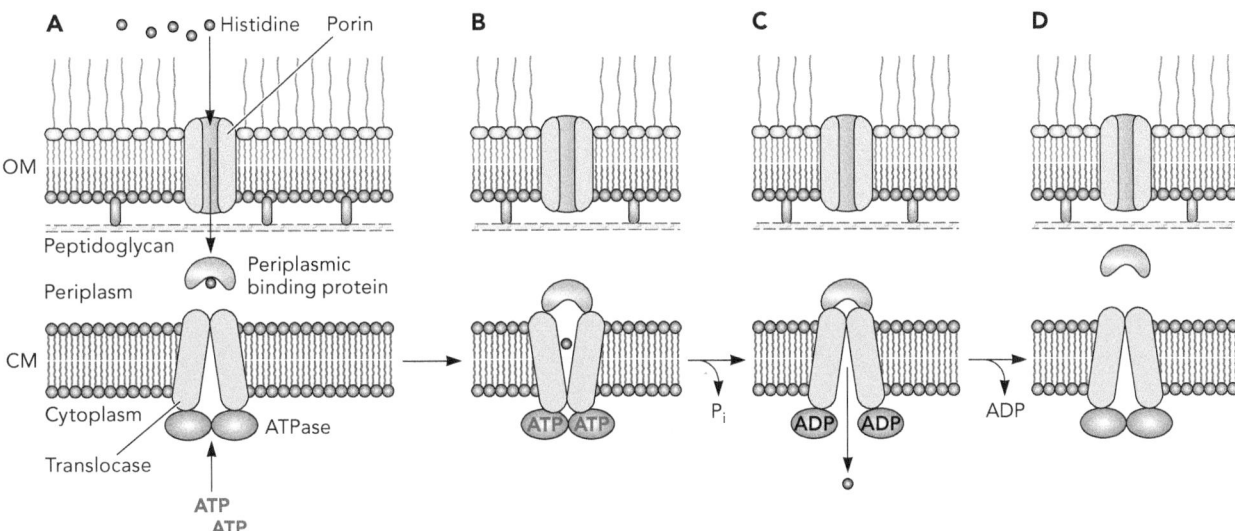

FIGURE 13.5 **Model of an ATP-binding cassette (ABC) transporter.** Movement of the amino acid histidine from the external environment across the Gram-negative cell envelope is depicted. (A) In Gram-negative bacteria, an outer membrane protein (e.g., a porin [tan]) enables transport across the outer membrane (OM). ABC transporters are composed of multiple proteins, including a periplasmic binding protein (curved gray shape) capable of binding histidine and a cytoplasmic membrane translocase (yellow) that contains an ATPase domain (orange). (B and C) Hydrolysis of ATP drives the transport of the solute across the cytoplasmic membrane (CM). See the text for additional details. (D) The transporter will return to its original unbound conformation and another molecule may then be transported across the cell envelope.

both individual cells and more broadly on human health. Examples include multiple drug resistance (MDR) transporters found in both bacterial and human cells, which have contributed greatly to the spread of antibiotic resistance and the failure of chemotherapeutic drugs, respectively. Mutations in the gene for the cystic fibrosis transmembrane conductance regulator (CFTR) transporter in human epithelial cells lead to ion imbalances associated with the genetic disease cystic fibrosis. The ubiquity of ABC transporters also demonstrates the broader implications of microbial research on different types of membrane transporters.

Cell Wall Structures

BACTERIAL CELL WALLS

Cell walls have two fundamental roles: (i) prevent lysis due to turgor pressure and (ii) maintain cell shape. Bacterial cell walls are made up of **peptidoglycan** (also called **murein**) that is a structure unique to bacteria. The carbohydrates *N*-acetylglucosamine (NAG) and *N*-acetylmuramic acid (NAM), which are joined by β-1,4-glycosidic bonds, comprise the sugar or glycan backbone of peptidoglycan (Fig. 13.6). This bond is lysozyme sensitive. **Lysozyme** is an antibacterial enzyme that is produced in tears and saliva as a first line of defense against bacterial pathogens. The glycan backbone of peptidoglycan is cross-linked by tetrapeptide side chains containing D-amino acids, which are not commonly found in biological systems. All the components necessary to build the peptidoglycan are moved across the cytoplasmic membrane via active transport using the undecaprenyl phosphate lipid carrier (Chapter 14).

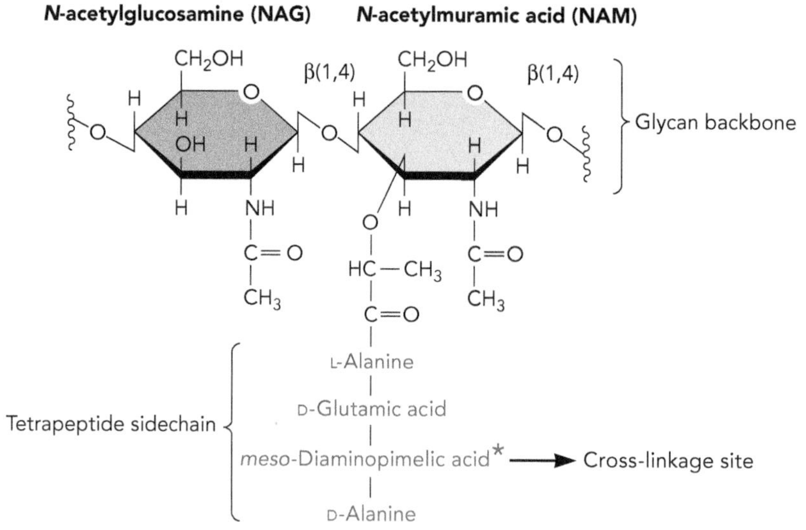

N-acetylglucosamine (NAG)   N-acetylmuramic acid (NAM)

Glycan backbone

Tetrapeptide sidechain
- L-Alanine
- D-Glutamic acid
- meso-Diaminopimelic acid* ⟶ Cross-linkage site
- D-Alanine

FIGURE 13.6 **Basic structure of bacterial peptidoglycan (murein).** In peptidoglycan, a glycan (sugar) backbone is produced when N-acetylglucosamine (NAG) and N-acetylmuramic acid (NAM) are co-valently linked through β-1,4-glycosidic bonds. The β-1,4-glycosidic bond is cleavable by lysozyme. A tetrapeptide side chain extends from NAM. An example amino acid composition is shown with a covalent cross-linking bond between the amino acid in the third position (e.g., meso-diaminopimelic acid) and the D-alanine from another tetrapeptide side chain (not shown). The amino acid at the third position in the side chain varies among different bacteria as indicated by the asterisk (*).

GRAM-POSITIVE VERSUS GRAM-NEGATIVE CELL WALLS

Although almost all bacteria have cell walls composed of peptidoglycan, it is organized very differently in Gram-positive and Gram-negative bacteria. The degree of cross-linking and the amino acids utilized in the side chains of peptidoglycan vary among different bacteria, especially the amino acid in the third position where cross-links are formed (Fig. 13.7A). Gram-negative cell walls have direct covalent cross-links between glycan strands, whereas Gram-positive cell walls have an additional glycine interbridge connecting the side chains. Gram-positive bacteria have a thick layer of peptidoglycan (20 to 100 nm thick) that lies outside of the cytoplasmic membrane, while Gram-negative bacteria have a single layer of peptidoglycan (a few nanometers thick) positioned between the cytoplasmic and outer membranes (Fig. 13.7B). The outer membrane in Gram-negative bacterial cells is anchored to the peptidoglycan via **murein lipoprotein** (also called Braun's lipoprotein). Murein lipoprotein is the most abundant protein in *E. coli* at 700,000 copies per cell, which indicates the importance of maintaining the association of the outer membrane with the rest of the cell.

The peptidoglycan layers in the Gram-positive bacterial cell envelope (Fig. 13.8A) contain molecules with negatively charged phosphate groups, such as teichuronic acid, teichoic acid, lipoteichoic acid, glycolipids, and/or neutral polysaccharides. The charged nature of these molecules prevents the passage of hydrophobic compounds, but allows small hydrophilic molecules of <40 kDa (e.g., sugars and amino acids) to readily pass through the peptidoglycan matrix. Large polymers must be degraded to subunits small enough to pass through the cell wall; this is often achieved via secreted exoenzymes (Chapters 11 and 14). In addition to influencing transport, these peptidoglycan-associated molecules are often on the surface of Gram-positive bacteria. Thus, they can play an important

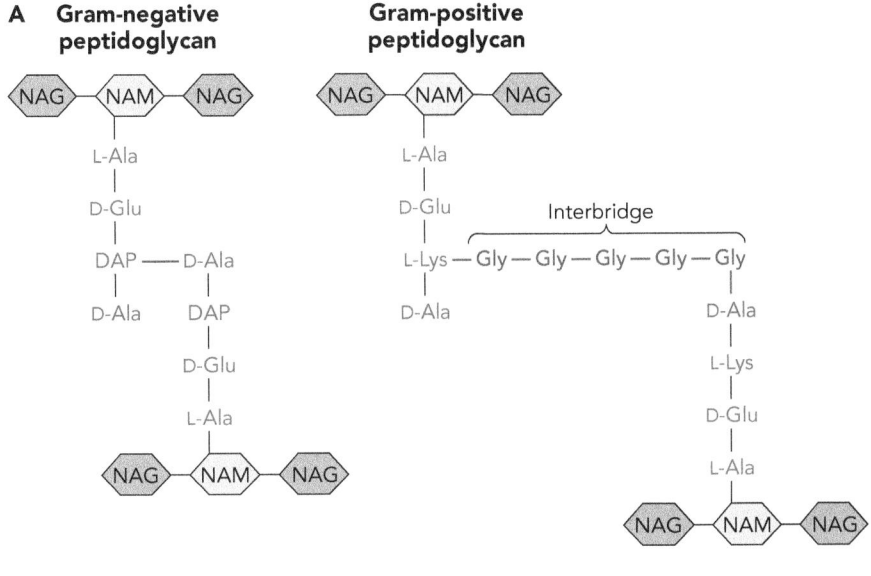

A Gram-negative peptidoglycan / Gram-positive peptidoglycan

FIGURE 13.7 Cross-links in bacterial peptidoglycan. (A) In Gram-negative bacteria, cross-links in peptidoglycan are formed directly between the third amino acid in one tetrapeptide side chain and the fourth amino acid in a second tetrapeptide side chain. In Gram-positive bacteria, cross-links are indirect between tetrapeptide side chains due to the addition of an interbridge composed of a glycine pentapeptide. DAP, diaminopimelic acid. (B) The cross-linking of the tetrapeptide side chains holds the glycan backbones together in the peptidoglycan matrix layer, which is a thin single layer in Gram-negative bacteria but a thick multilayered structure in Gram-positive bacteria.

role in adhesion and cell-cell recognition, and they have been targeted as bio-markers for clinical diagnostic tests. However, the specific roles of the various molecules associated with Gram-positive cell walls remain largely undefined.

MYCOBACTERIAL CELL WALLS

Mycobacteria, which cause diseases such as tuberculosis and leprosy, have notoriously impenetrable cell envelopes that contribute to their ability to survive in human hosts. These bacteria have a layer of branched, hydroxylated lipids called mycolic acids associated with their cell walls; this waxy layer creates a

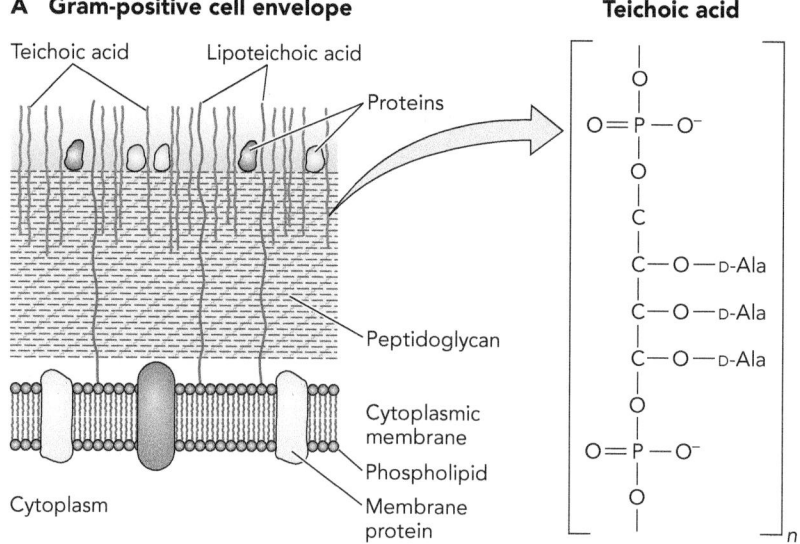

A Gram-positive cell envelope

Teichoic acid   Lipoteichoic acid

Proteins

Peptidoglycan

Cytoplasmic membrane

Phospholipid

Membrane protein

Cytoplasm

Teichoic acid

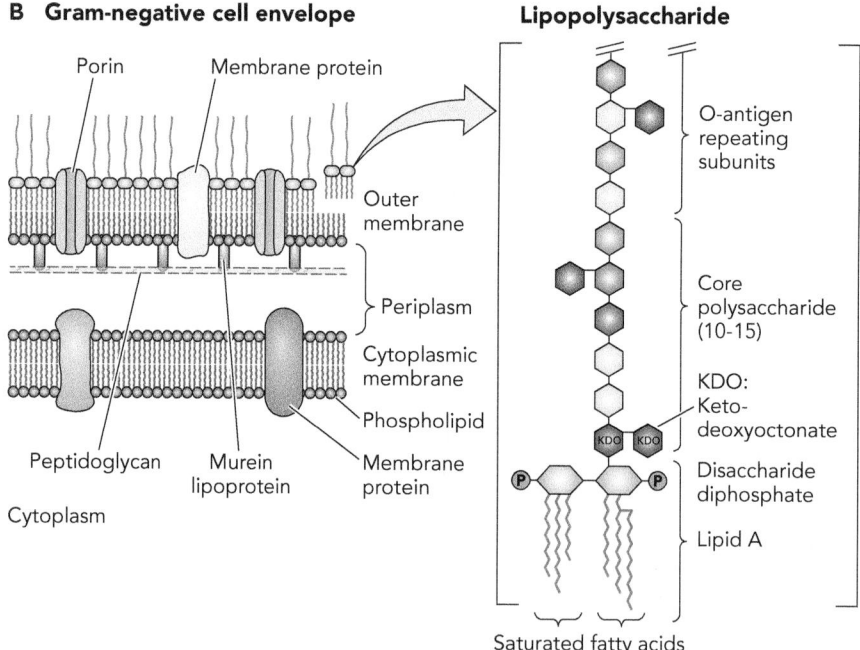

B Gram-negative cell envelope

Porin   Membrane protein

Outer membrane

Periplasm

Cytoplasmic membrane

Phospholipid

Peptidoglycan   Murein lipoprotein

Membrane protein

Cytoplasm

Lipopolysaccharide

O-antigen repeating subunits

Core polysaccharide (10-15)

KDO: Keto-deoxyoctonate

Disaccharide diphosphate

Lipid A

Saturated fatty acids

FIGURE 13.8 **Comparison of the cell envelopes of Gram-positive and Gram-negative bacteria.** (A) The cell envelope of a Gram-positive bacterium consists of a thick layer of peptidoglycan on the outside of the cytoplasmic membrane with a thin space in between that serves the same function as the Gram-negative periplasm. The peptidoglycan is associated with molecules such as teichoic acid, lipoteichoic acid, and cell surface proteins. The structure of a type of teichoic acid (ribitol) is shown in greater detail. (B) The cell envelope of a Gram-negative bacterium consists of a thin single layer of peptidoglycan located in the periplasm, outside the cytoplasmic membrane. The peptidoglycan is linked through murein lipoprotein to the outer membrane, which is found only in Gram-negative bacteria. The outer membrane has a phospholipid inner leaflet and an outer leaflet that is composed of lipopolysaccharide (LPS) with sugar groups extending out from the cell surface. LPS, shown in greater detail, is a molecule unique to Gram-negative bacteria that is located in the outer leaflet of the outer membrane. It has three parts: lipid A (a phosphorylated disaccharide with short saturated fatty acid tails), the core polysaccharides (typically including two to three molecules of ketodeoxyoctonate [KDO]), and the O-antigen repeating region, which is highly variable across different bacterial strains.

hydrophobic barrier on the cell surface that prevents penetration by Gram stain reagents (Chapter 1). An "acid-fast" staining procedure is used instead as a technique to characterize these organisms. Although the mycobacteria are members of the *Actinobacteria*, which is predominantly composed of Gram-positive bacteria, the cell envelopes of the mycobacteria are considered to be Gram-variable, neither Gram-positive nor Gram-negative.

ARCHAEAL CELL WALLS

A few lineages of *Archaea*, including some of the methanogens, have cell walls composed of a layer of **pseudopeptidoglycan** (15 to 20 nm thick; also called pseudomurein), instead of peptidoglycan. Pseudopeptidoglycan contains NAG, but instead of NAM, it contains *N*-acetyltalosaminuronic acid (NAT) linked to NAG by β-1,3-glycosidic bonds (Fig. 13.9A). Unlike the β-1,4-glycosidic bond in peptidoglycan, the β-1,3-glycosidic bond is lysozyme insensitive. Also different from peptidoglycan, the interbridges in pseudopeptidoglycan contain no D-amino acids and are typically composed of L-glutamate, L-alanine, and L-lysine. Although the overall structures of peptidoglycan and pseudopeptidoglycan are similar, the enzymes of the two biosynthetic pathways do not appear to be homologous, suggesting that the pathways evolved independently. More commonly, most archaea have a paracrystalline **surface layer**, or **S-layer**, as their cell wall. S-layers are composed of a highly organized crystalline matrix of protein or glycoprotein (Fig. 13.9B and C).

MICROBES WITHOUT CELL WALLS

Not all microbes have cell walls, and as a result, these organisms are highly pleomorphic in shape. Mycoplasmas are the only members of the phylum *Firmicutes* that lack cell wall structures. As obligate symbionts, they are found associated with animals, plants, insects, and protists, and many cause disease. Another unique characteristic of the mycoplasmas is their small genome (approximately one-fifth the size of the *E. coli* genome), which lacks genes for many metabolic pathways and cell wall synthesis. The lack of a cell wall may be an evolutionary adaptation to their obligate association with a host. There are also some *Archaea* that lack cell walls, including *Ferroplasma* and *Thermoplasma*. Both genera are acidophiles that use a monolayer membrane (Fig. 13.2C) to help stabilize the cell envelope structure.

Gram-Negative Outer Membrane

The outer membrane (Fig. 13.8B) is a distinctive characteristic of the cell envelope of Gram-negative bacteria that may have evolved to compensate for the thin peptidoglycan layer in these organisms. The asymmetric outer membrane is composed of lipid and polysaccharide and is often called a lipopolysaccharide (LPS) layer. The inner leaflet of the outer membrane is composed of ester-linked phospholipids. However, the outer leaflet of the membrane bilayer has lipid A at its base, complexed to a core polysaccharide layer, and an O-antigen (O-specific polysaccharide) projecting away from the cell surface. **Lipid A** is composed of

A Pseudopeptidoglycan

N-acetylglucosamine **N-acetyltalosaminurmic acid**
(NAG) **(NAT)**

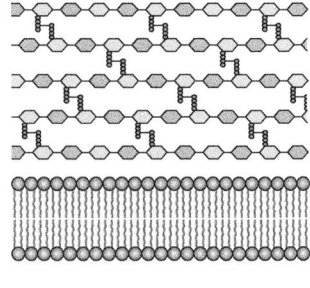

L-Glutamate
|
L-Alanine
|
L-Lysine — L-Glu
|
L-Lys — L-Glu
|
L-Ala
|
L-Glu

NAG — NAT — NAG

B S-layer cell wall

Cytoplasmic
membrane

Cytoplasm

Pseudopeptidoglycan

C

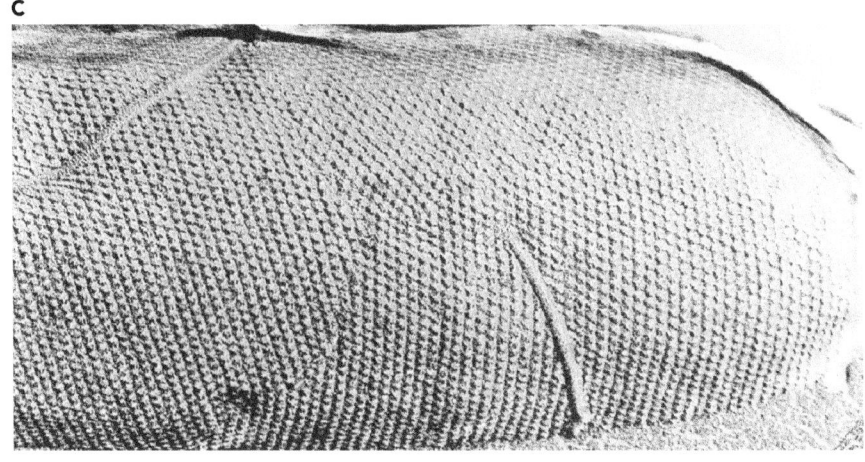

FIGURE 13.9 **Basic structures of archaeal cell walls.** (A) Some archaea have a cell wall with pseudopeptidoglycan; a glycan (sugar) backbone is produced when N-acetylglucosamine (NAG) and N-acetyltalosaminuronic acid are covalently linked through β-1,3-glycosidic bonds (which cannot be cleaved by lysozyme). A tetrapeptide side chain extends from NAT and is directly cross-linked to another tetrapeptide side chain extending from another glycan strand. (B) Other archaea have a cell wall with an S-layer (left) instead of pseudopeptidoglycan (right). S-layers are composed of glycoproteins or glycolipids arranged in a highly organized crystalline-like matrix on the cell surface outside of the cytoplasmic membrane. (C) An electron microscope image of an S-layer. Reprinted from Sára M, Sleytr UB. 2000. *J Bacteriol* 182:859–868.

a disaccharide diphosphate head group linked to six or seven saturated fatty acids. These hydroxylated, saturated fatty acids are linear molecules, shorter in length than most membrane phospholipids. Collectively, these properties make lipid A bulky and rigid in structure. Lipid A also functions as an endotoxin that triggers fever and inflammation during pathogenic infections by Gram-negative bacteria. The first sugar bonded to lipid A, ketodeoxyoctonate (KDO), is part of the **core polysaccharide** that consists of between 10 and 15 sugar groups extending outward. KDO is highly conserved in LPS. Beyond the core polysaccharide is the **O-antigen** repeat, which is exposed on the outer surface of the cell envelope and can be up to 40 sugar residues in length. The length and sequence of sugars present in the O-antigen repeat is highly variable among various bacterial species and even within a single species. In some bacteria, the composition of O-antigen is highly regulated and the sequence of sugars can be changed, enabling pathogenic bacteria to evade the immune system by altering the cell surface antigens.

Together, the lipid A, core polysaccharide, and O-antigen form the LPS. LPS can be cross-linked through divalent cations like Mg^{2+} or Ca^{2+} that help to strengthen the structure and influence the transport of molecules across the outer membrane by creating a matrix-like structure through ionic interactions. This selective permeability affords protection of Gram-negative enteric organisms against intestinal bile salts while enabling nutrients to enter the cell. In some microorganisms, the length of the polysaccharide chain extending from lipid A is shortened, and in some bacteria, the O-antigen is lacking altogether. In these cases, the term lipooligosaccharide (LOS) is often used to indicate a few (oligo) sugars in the shorter saccharide chain.

Although the proteins associated with the outer membrane are less diverse than those in the cytoplasmic membrane, they are present at high numbers of molecules per cell (e.g., murein lipoprotein). In addition, other proteins vital for transport processes are associated with the outer membrane, including general porins, specific porins, and high-affinity energy-dependent transporters.

General porins, such as the outer membrane proteins OmpC and OmpF, are each present at ~100,000 copies per *E. coli* cell. Outer membrane proteins typically form semiselective pores in the outer membrane via membrane-spanning β-barrel protein structures (Fig. 13.10). These proteins allow the nonspecific entry of any molecule below a certain size through their internal channels, which function as constantly open pores in the outer membrane. OmpC allows the diffusion of molecules <300 Da through its channel, which is ~1.1 nm in diameter. OmpF creates a larger channel of ~1.2 nm in diameter that permits the diffusion of molecules <600 Da. As an example, an average amino acid is ~110 Da in size; therefore, amino acids can cross the outer membrane via either OmpC or OmpF. Due to the porin channels, there is no PMF stored across the outer membrane and therefore no potential energy is stored to drive transport processes. The movement of solutes through the general porins is driven by facilitated diffusion from a higher to lower concentration until equilibrium is reached.

The general porins OmpC and OmpF are differentially expressed in *E. coli* in response to changing environmental osmolarity. A two-component regulatory system consisting of EnvZ and OmpR controls expression of the genes encoding

A

Porin

Outer membrane

Peptidoglycan

Channels

B

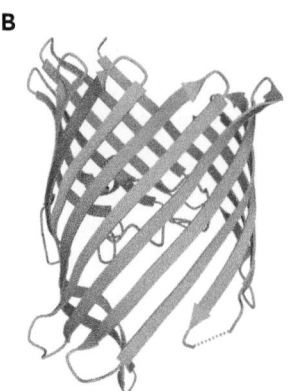

C

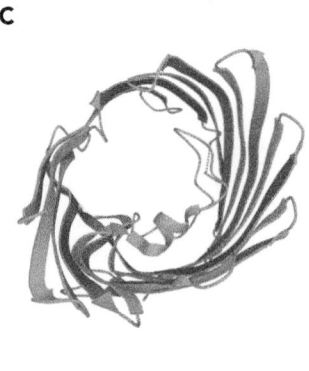

FIGURE 13.10 **Outer membrane porins.** (A) Outer membrane porins enable facilitated diffusion across the outer membrane of Gram-negative bacteria. Porins function as transport channels with a set diameter that permits molecules below a certain size to be transported across the outer membrane via diffusion using the concentration gradient. See the text for details. (B and C) Side and top views, respectively, of the general porin OmpF.

OmpC and OmpF (Fig. 13.11). The environmental osmolarity is sensed by EnvZ, a sensor histidine kinase. In environments with low osmolarity and concurrent low nutrient levels (e.g., a freshwater pond or stream), EnvZ is unphosphorylated, and unable to transphosphorylate the conserved aspartate residue on its partner response regulator, OmpR. Under these low-osmolarity conditions, only a few molecules of OmpR are phosphorylated and therefore OmpR-P only binds to high-affinity sites in target promoters. One of these high-affinity sites is upstream of the gene encoding OmpF. Increasing expression of *ompF* results in production of OmpF, optimizing the ability of the cell to take up nutrients through the porin with the larger channel. Conversely, under environmental conditions of high osmolarity where nutrient levels are high (e.g., the intestinal tracts of animals, where bile salts are present), EnvZ autophosphorylates its conserved histidine residue and transphosphorylates the conserved aspartate residue on OmpR. When the cellular levels of OmpR-P are high (i.e., when osmolarity is high), it binds to both low- and high-affinity sites upstream from its target promoters. Binding of OmpR-P to the low-affinity site upstream from *ompC* activates gene expression and results in production of the smaller porin OmpC. In addition, OmpR-P binds to a low-affinity site upstream from *ompF*, which counteracts the effect of binding at the high-affinity site and represses expression of *ompF*. Therefore, in environments with low osmolarity and low nutrients, the larger OmpF channel permits maximum nutrient entry, whereas when osmolarity and nutrients are high, the smaller OmpC channel is used, which affords protection from intestinal bile salts while allowing sufficient nutrient uptake. Interestingly, it has also been determined that the gene encoding the small RNA (sRNA) MicF is activated under conditions of high osmolarity. This gene is divergently transcribed from the *ompC* promoter. The MicF sRNA stops translation of the *ompF* transcript, adding another level of regulation to the system (Chapter 8).

Specific porins represent a second type of outer membrane channel; they permit facilitated diffusion of molecules that are specifically recognized by the porin. LamB is one example of a specific porin that permits maltose to cross the outer membrane from a high concentration to a low concentration. When *E. coli* is grown with maltose as a carbon source, LamB levels will increase. These

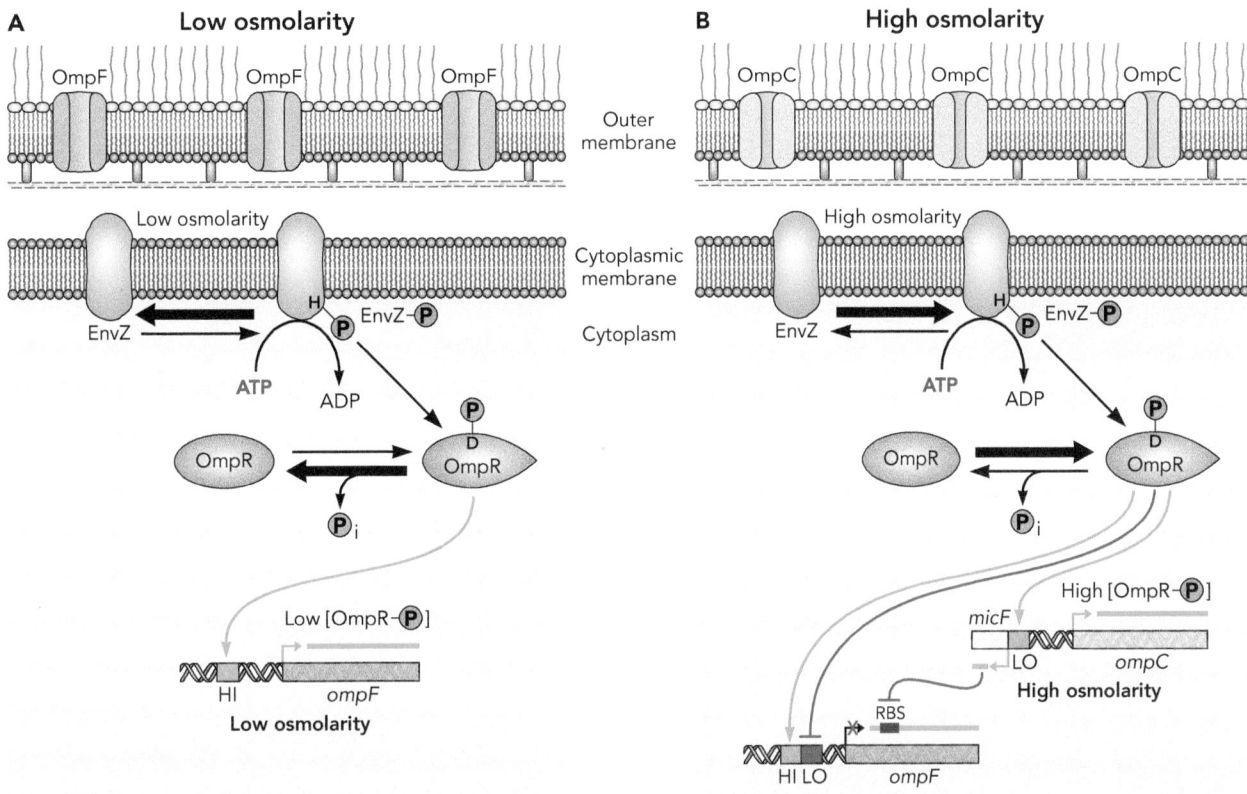

osmolarity is sensed through the sensor histidine kinase EnvZ in the cytoplasmic membrane. (A) Under conditions of low osmolarity, EnvZ autophosphorylates on a histidine residue (H) and then transphosphorylates an aspartate residue (D) on its cognate response regulator, OmpR, but at very low rates. When present at low concentrations, OmpR-P binds only at high-affinity DNA binding sites. OmpR-P binding to the high-affinity site upstream from *ompF* (green box; HI) results in activation of the gene, leading to high levels of OmpF in the outer membrane under conditions of relatively low osmolarity. The higher levels of OmpF in the outer membrane enable greater rates of facilitated diffusion across the membrane. (B) Under conditions of high osmolarity, EnvZ autophosphorylates and then transphosphorylates OmpR at much higher rates. Thus, as the osmolarity increases, the level of OmpR-P increases. Under high OmpR-P concentrations, in addition to binding at the high-affinity site in the *ompF* promoter, OmpR-P can now bind to the low-affinity binding site (red box; LO), blocking *ompF* transcription. There is also a low-affinity binding site in the *ompC* promoter (green box; LO). OmpR-P binding at this site results in the activation of the *ompC* promoter, leading to much higher levels of OmpC in the outer membrane under conditions of high osmolarity and resulting in reduced rates of facilitated diffusion. The sRNA gene *micF* is divergently transcribed from *ompC*. When the sRNA MicF is expressed, it further prevents translation of the *ompF* mRNA through posttranscriptional regulation (Chapter 8).

growth conditions are used during lambda phage studies, as LamB is not only a protein channel for maltose transport but also serves as a docking receptor for bacteriophage lambda (hence the name LamB).

A third type of outer membrane transporter is known as a **high-affinity energy-dependent transporter or TonB-dependent transporter** (Fig. 13.12). TonB ("Tone B") is the receptor for the T1 bacteriophage. As discussed above, there is no PMF across the outer membrane, and ATP is produced in the cytoplasm. Therefore, to transport molecules that cannot readily pass through porins, energy from the inner membrane is transferred to an outer membrane transporter. One way that this occurs is through the TonB protein, which spans from the cytoplasmic membrane across the periplasm to the outer membrane. Larger molecules that cannot pass through porins (e.g., siderophores such as enterochelin) are transported using a TonB-dependent transport

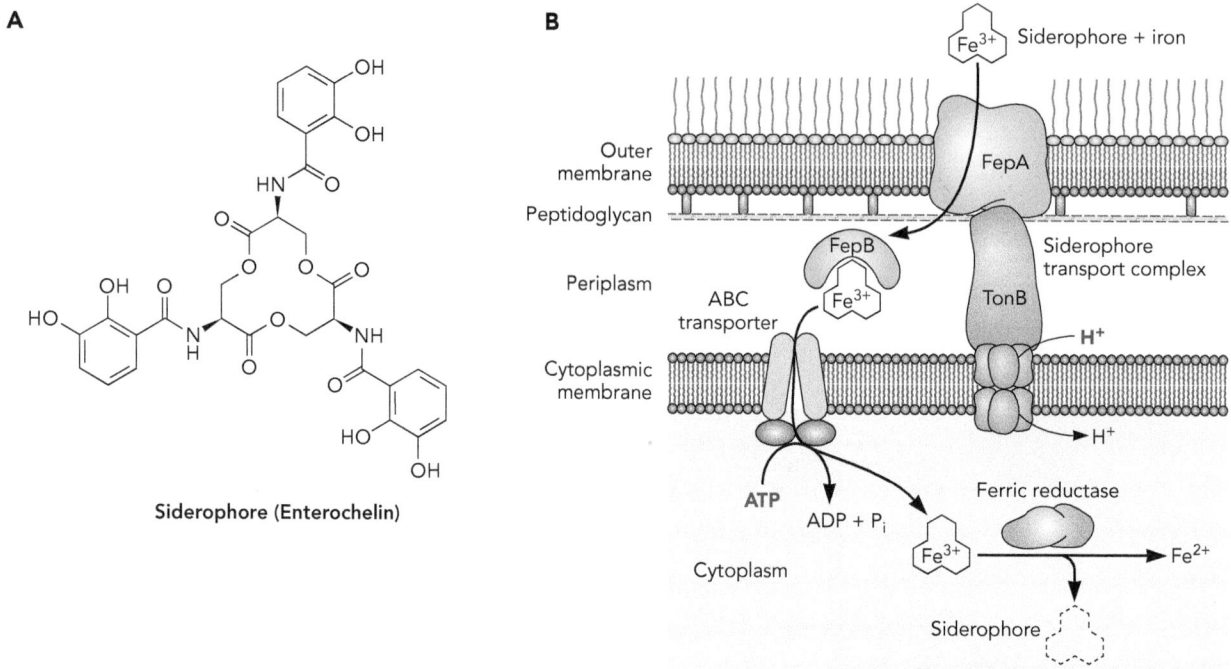

A

Siderophore (Enterochelin)

B

Siderophore + iron

Outer membrane

Peptidoglycan

Periplasm

ABC transporter

Cytoplasmic membrane

FepA

FepB

Siderophore transport complex

TonB

H^+

H^+

ATP

ADP + P_i

Ferric reductase

Fe^{3+}

Fe^{2+}

Cytoplasm

Siderophore

FIGURE 13.12 **TonB-dependent transport of a siderophore across the outer membrane.** (A) Structure of the siderophore enterochelin, which is important for iron acquisition in Gram-negative bacteria. (B) The siderophore, carrying Fe^{3+}, is transported across the outer membrane in an energy-dependent manner. TonB couples the energy from the proton motive force across the cytoplasmic membrane to a TonB-dependent transporter (FepA) in the outer membrane. After the siderophore is transported across the outer membrane into the periplasm, a periplasmic binding protein (FepB) delivers it to an ABC transporter that then moves the siderophore across the cytoplasmic membrane into the cytoplasm. Ferric reductase converts Fe^{3+} into soluble Fe^{2+} needed for biosynthesis.

mechanism (Fig. 13.12). **Siderophores** are large iron-binding molecules that are released to the outside of the cell. Because iron (critical for many enzymes such as proteins with FeS clusters) is often limiting in the environment, bacteria produce a variety of iron-scavenging siderophores. TonB takes energy from the PMF across the cytoplasmic membrane and delivers it to the siderophore transporter protein in the outer membrane to move the siderophore across the outer membrane into the periplasm. An ABC transporter then moves the siderophore from the periplasm across the cytoplasmic membrane. The ATP produced and present in the cytoplasm drives this part of the process. The trivalent iron carried by the siderophore is released into the cytoplasm and converted to divalent iron for use in biosynthesis.

Periplasm

The sizable subcompartment in a Gram-negative bacterium between the cytoplasmic and outer membranes is known as the **periplasm**. The periplasm comprises ~20 to 40% of the volume of a Gram-negative cell (Fig. 13.8). Gram-positive bacteria also have a functional periplasm, lying between the cytoplasmic membrane and the peptidoglycan, but it has a much smaller volume.

The periplasmic contents are not equivalent to the extracellular environment. In Gram-negative bacteria, the periplasm has a gel-like consistency due to the large number of molecules present, including high-affinity binding proteins (associated with ABC transporters), degradative enzymes that start breaking down nutrients, detoxifying enzymes (e.g., β-lactamases that can inactivate β-lactam antibiotics), peptidoglycan, ions/salts, oligosaccharides, protein secretion systems, and electron transport chain components (e.g., cytochrome *c*).

Additional Extracellular Layers

Many *Bacteria* and *Archaea* have extra cell envelope layers outside the cell wall or membrane that are composed of carbohydrate or protein. A **capsule** (sometimes referred to as a "slime layer" or the glycocalyx) is a polysaccharide layer tightly associated with and located beyond the cell wall in many Gram-positive bacteria and archaea or the outer membrane in Gram-negative bacteria. Traditionally, a capsule stain with India ink is used to visualize the capsule with a light microscope (Fig. 13.13A). One function of the capsule is to protect pathogens

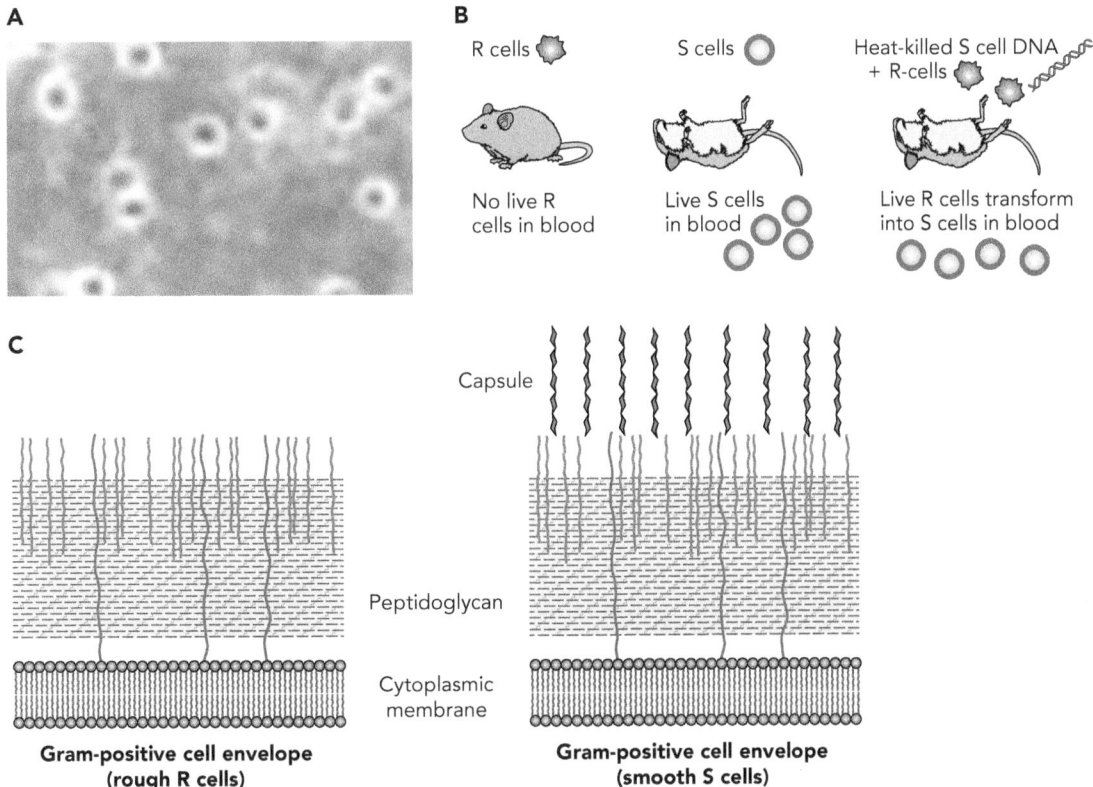

FIGURE 13.13 **A capsule is often the outermost layer of the cell envelope.** (A) Bright-field microscopy image of an India ink stain of cells with capsules showing that the capsule is impermeable to the dye molecules (gray background), creating a protective barrier around the bacterial cells (clear zone surrounding dark oval cells). Photo credit: Rebecca E. Parales. (B) Griffith's transformation experiments demonstrated the role of the capsule in bacterial virulence. In his experiments, mice survived when inoculated with cells lacking a capsule (R = rough cell morphology), whereas mice inoculated with cells that produce a capsule (S = smooth cell morphology; indicated in orange) did not survive. The addition of DNA from heat-killed S cells allowed R cells to "transform" into capsule-producing S cells that killed the mice. (C) Detailed depiction of a rough (left) versus smooth (right) Gram-positive cell envelope with the capsule layer depicted as oligosaccharides extending outward from the cell surface.

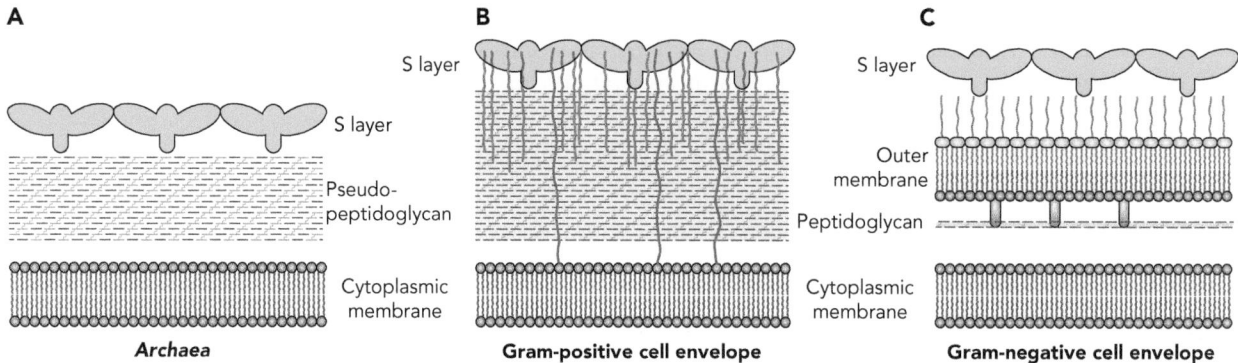

FIGURE 13.14 **S-layers on the outside of the cell envelope.** (A) Some archaea that have cell walls composed of pseudopeptidoglycan have an external S-layer composed of protein. (B) Some Gram-positive bacteria have an S-layer external to the peptidoglycan cell wall. (C) Some Gram-negative bacteria have an S-layer external to the outer membrane.

from host immune systems as it inhibits phagocytosis. This trait was studied during Griffith's famous DNA transformation (horizontal gene transfer) experiments with smooth (with capsule) and rough (without capsule) cells of *Streptococcus* (Fig. 13.13B and C). Cells that carried the genes for capsule production were able to evade the immune response, resulting in the death of the mouse host. The capsule also provides protection against dehydration, enabling bacteria to survive until more favorable conditions are encountered. In addition, the capsule can protect cells from physical damage (e.g., shear force) as well as chemical stressors by inhibiting the diffusion of potentially toxic molecules into the cell (e.g., antibiotics).

In the case of microbial biofilms in which communities of microbes are attached to surfaces (Chapter 18), the cells are embedded in an organic matrix composed of polysaccharide, extracellular DNA, proteins, and lipids. This organic matrix is commonly referred to as **extracellular polymeric substance**, or EPS (Chapter 18).

Finally, as mentioned previously, some *Archaea* have a protein or glycoprotein S-layer that functions as a cell wall. In contrast, some *Bacteria* and *Archaea* have an S-layer composed of protein or glycoprotein in addition to a cell wall. Examples of *Archaea* with S-layers outside of their pseudopeptidoglycan cell walls have been documented (Fig. 13.14A). In addition, some Gram-negative and Gram-positive bacteria have protein S-layers that are located external to the outer membrane or peptidoglycan layer, respectively (Figure 13.14B and C). These S-layers function as an additional layer of protection for the cell.

Learning Outcomes: After completing the material in this chapter, students should be able to . . .

1. summarize the features of the cytoplasmic membranes of *Bacteria* and *Archaea*

2. describe the different transport processes that cells use to facilitate the movement of molecules across the cytoplasmic membrane

3. compare and contrast the structures of archaeal and Gram-positive and Gram-negative bacterial cell walls

4. detail the unique structure and function of the Gram-negative bacterial outer membrane

5. explain how porin synthesis is regulated in response to the osmolarity of the environment

6. discuss the structures and functions associated with the periplasm and capsule

Check Your Understanding

1. Define this terminology: amphipathic molecule, passive transport, active transport, peptidoglycan, lysozyme, murein lipoprotein, pseudopeptidoglycan, surface or S-layer, lipid A, core polysaccharide, O-antigen repeat, general porins, specific porins, high-affinity energy-dependent transporter (TonB-dependent transporter), siderophores, periplasm, capsule, extracellular polymeric substance (EPS)

2. While all cells have cytoplasmic membranes, their structures may differ. (LO1)
   a) What is the function of the cytoplasmic membrane?

   b) These diagrams show two different cytoplasmic membrane phospholipids. Fill in the boxes on the diagrams using the following terms: **ether linkages, ester linkages, saturated fatty acids, unsaturated fatty acids, glycerol, phosphate head.**

c) Which diagram shows a bacterial phospholipid and which shows an archaeal phospholipid? Explain how you know.

d) Label the parts of the molecules as being hydrophobic or hydrophilic. Why does this matter with respect to the function of the membrane?

3. Define each of the following types of transport and indicate if it is an example of passive or active transport. (LO2)

Transport type	Definition	Active or passive transport?
Simple diffusion		
Facilitated diffusion		
ABC transport		
Ion-driven transport		

4. Transport proteins can depend on coupled reactions to move molecules into or out of the cell. (LO2)

a) What is a coupled reaction?

b) Describe each of the following and explain how each transporter works. Indicate whether or not the transporter depends on a coupled reaction.

   i.   Uniporter

   ii. Antiporter

   iii. Symporter

5. Cells must be able to move molecules across the cytoplasmic membrane without allowing the uncontrolled movement of protons in and out of the cell. (LO2)

a) Why is the controlled movement of protons so important?

b) How do cells utilize ATP-binding cassette (ABC) transport proteins to get amino acids into and out of the cell while prohibiting the free movement of protons?

6. While many organisms have cell walls, the structures differ between archaeal cell walls and those found in Gram-positive and Gram-negative bacteria. (LO3)

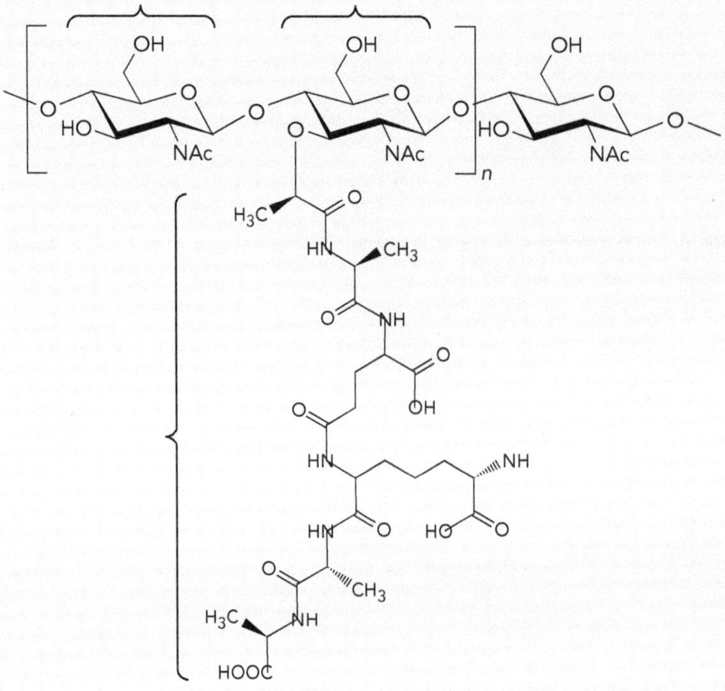

a) What are two functions of the cell wall?

b) What molecule is shown?

c) Label the parts of the molecule shown by the brackets.

d) Indicate and name the bond that the enzyme lysozyme targets.

e) List two ways this molecule is different from pseudopeptidoglycan found in some archaeal cell walls.

7. The purpose of this table is to compare and contrast the structures of Gram-negative and Gram-positive cell envelopes. Fill in the blanks below. (LO3)

Cell envelope feature	Gram-negative cell	Gram-positive cell
Thick or thin cell wall?		
Connection between the sugar backbones?		
Has outer membrane?		
Has LPS?		
Has porins?		

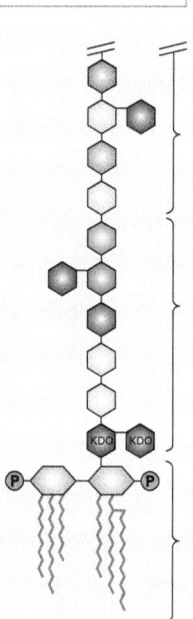

8. The Gram-negative outer membrane has a number of unique molecules, such as the one shown. (LO4)

a) What molecule is shown?

b) Label the parts of the molecule shown with these terms: **lipid A, O-antigen, core polysaccharide**.

c) The fatty acids in this molecule tend to be fully saturated. Would this produce a more rigid, more tightly packed membrane or one with increased fluidity, relative to fatty acids with unsaturated bonds?

d) The outer layer of this molecule attracts divalent cations like Mg^{2+} or Ca^{2+}. Explain how these cations help to support the survival of bacteria in harsh environments (including against bile salts within the intestines of human hosts).

9. Gram-negative bacteria also need mechanisms to get molecules across the outer membrane. (LO3)

a) What determines which molecules can move through general porins? Is this movement active or passive?

b) What is the advantage to cells having *specific* transporters in their cytoplasmic membranes and *general* porins in their outer membranes?

c) Can cells maintain a PMF across the outer membrane? Why or why not?

10. One type of outer membrane transporter is known as a high-affinity energy-dependent transporter or TonB-dependent transporter. Siderophore transporters represent one example of this type of transporter. (LO4)

a) What energy source is used to move the siderophore across the outer membrane, from outside the cell to the periplasm? Why is it critical for TonB to span the cytoplasmic membrane for this to happen?

b) Why does the siderophore transport protein require TonB to move the siderophore across the outer membrane?

c) What energy source is used to move the siderophore across the cytoplasmic membrane, from the periplasm to the inside of the cell?

11. *E. coli* can adjust the types of general porins in its outer membrane in response to changing osmolarity in the environment, as illustrated by the regulation of OmpF and OmpC. (LO5)

a) Expression of the genes encoding OmpF and OmpC is controlled by a two-component regulatory system. In the case of the regulation of *ompF* and *ompC*:

i. What is the environmental signal?

ii. What molecule acts as the sensor histidine kinase?

iii. What molecule acts as the response regulator?

b) The purpose of this table is to help you understand the two-component regulatory system that controls porin gene expression (Fig. 13.11). Fill in the table below by indicating the state of the protein at relatively high or low osmolarity.

Osmolarity level	Is EnvZ more or less phosphorylated?	Is OmpR more or less phosphorylated?	Does OmpR bind to high-affinity binding sites, low-affinity binding sites, or both?	Is the transcription of *ompF* activated?	Is the transcription of *ompC* activated?
Low osmolarity					
High osmolarity					

c) Describe the role of EnvZ in the phosphorylation of OmpR.

d) How do the functions of OmpF and OmpC differ? Why would *E. coli* make OmpF in a low-nutrient pond environment and OmpC in a high-nutrient animal intestine?

12. In the table below, indicate if each trait is true for the periplasm and the capsule. (LO6)

Trait	Periplasm	Capsule
Found in Gram-negative and Gram-positive bacteria		
Composed of polysaccharides and/or protein		
Contains high-affinity binding proteins and degradative enzymes		
Includes components of the electron transport chain		
Provides protection against dehydration		

13. Griffith's famous experiments demonstrated how the presence of a capsule can inhibit an immune response in mammals. (LO6)

a) In Griffith's experiment (Fig. 13.13), which cells (smooth or rough) had a capsule?

b) Fill in the table below to indicate the results when mice were injected with each component.

Cell type	Did infected mice live or die?
Rough bacteria	
Smooth bacteria	
Rough bacteria + heat-killed smooth bacteria	

c) Explain why the conditions of some of the above experiments caused mice to die while others lived.

14. Using words from the list provided, match each function with the structure that carries out that function.

Couples the potential energy in the PMF to move solutes across a membrane: _____

Hydrolyzes ATP to move molecules across a membrane: _____

Allows the nonspecific entry of any molecule below a certain size: _____

Anchors the outer membrane to the peptidoglycan layer: _____

Prevents cell lysis in the presence of osmotic pressure: _____

Provides a permeability barrier between the inside and outside of the cell: _____

Secures LPS in the outer membrane: _____

Links NAG-NAM backbone strands: _____

Breaks β-1,4-glycosidic bonds: _____

ABC transporter
Amino acid cross-bridges
Murein lipoprotein
Cell membrane
Cell wall
Ion-driven transporter
Lipid A
Lysozyme
Porin

Dig Deeper

15. Given the two membranes with differing proportions of saturated lipids, which membrane structure would you most likely find in a bacterial psychrophile? Explain your reasoning. (LO1)

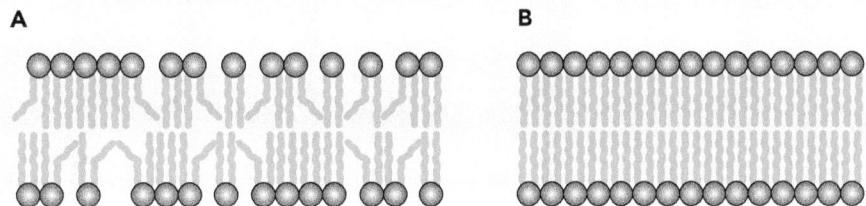

16. The different components of cell membranes and cell walls can inhibit the movement of various molecules. Considering each layer independently, indicate with **yes** or **no** which of the following compounds can diffuse freely (passive transport) through each layer? For each compound that <u>cannot</u> diffuse through a given layer, explain why it cannot get through. (LO2, 3)

Compound	Cytoplasmic membrane	Cell wall	Gram-negative outer membrane
Protons			
Oxygen			
Water			
Glucose			
110-Da amino acid			
1,450-Da antibiotic			
200,000-Da protein			

17. You find four different unlabeled antibiotics in your lab. You set up four experiments growing three different microorganisms with (red bars) and without (blue bars) each antibiotic. You then measure the percent survival of each organism in each case (results shown).

   Given the information about each microorganism and each antibiotic, determine which cell types were killed in each experiment, and predict which experiment was conducted with each antibiotic. Explain your reasoning. *(Hint: Think about the different barriers associated with each cell envelope type and how they would be affected by each antibiotic.)* (LO3, 4).

Microbe types:

- *Staphylococcus aureus*, a Gram-positive bacterium
- *Vibrio cholerae*, a Gram-negative bacterium
- *Methanosarcina*, a member of the *Archaea*

Information about potential antibiotics:

- Ampicillin: Targets peptidoglycan biosynthesis; approximate size 500 Da
- Cationic antimicrobial peptide (or CAMP): a small, positively charged, hydrophobic antibiotic that inserts into membranes and creates pores
- Polymyxin B: Binds to and destabilizes LPS
- Vancomycin: Targets peptidoglycan biosynthesis; approximate size 1,500 Da

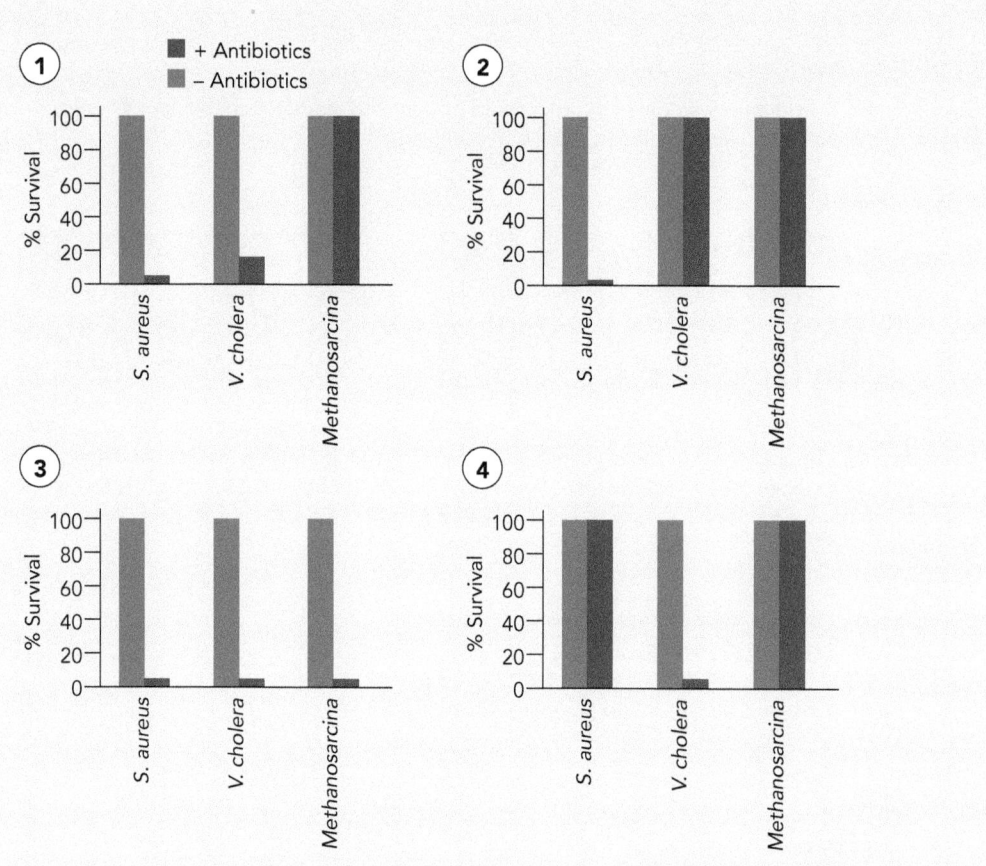

Experiment	Which antibiotic was used?	Your reasoning
Experiment 1		
Experiment 2		
Experiment 3		
Experiment 4		

18. Lysozyme is an enzyme that breaks the β-1,4-glycosidic bonds in peptidoglycan. (LO3)

    a) What do you think would happen if you treated a bacterial cell with lysozyme under hypotonic (more salt inside the cell) conditions? Explain why.

    b) What do you think would happen if you treated a bacterial cell with lysozyme under isotonic (same osmotic pressure inside and outside the cell) conditions? Explain why.

    c) What do you think would happen if you treated an archaeal cell with lysozyme under isotonic conditions? Explain why.

19. You grow *E. coli* under high- and low-osmolarity conditions, with varying amounts of the antibiotic tetracycline (~440 Da), and measure the percentage of bacteria that survive. You get the results shown. (LO5)

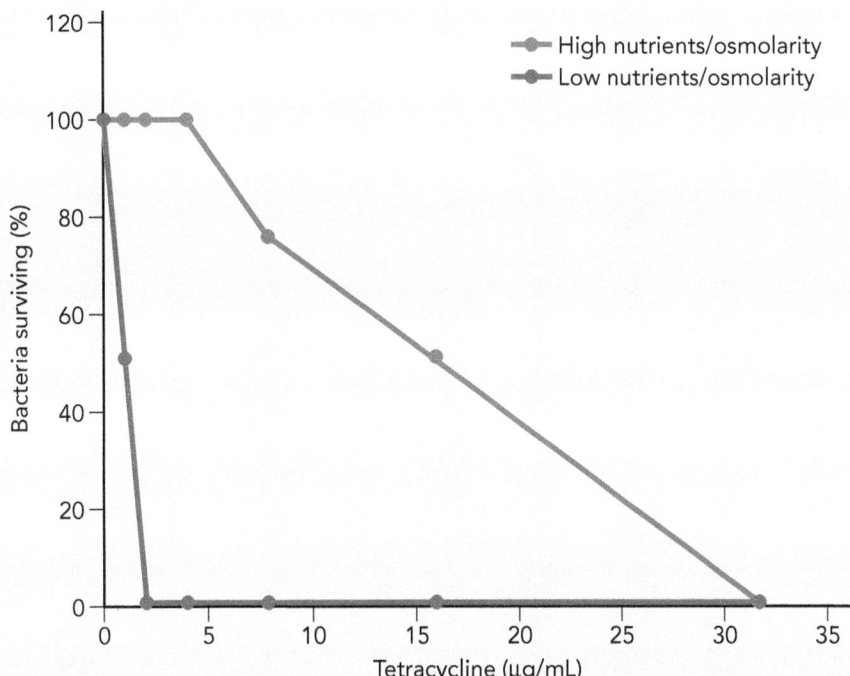

a) Under which conditions (low- or high-nutrient/osmolarity) was the tetracycline more effective at killing cells?

b) Under which conditions (low- or high-nutrient/osmolarity) would you expect the cells to make more OmpF and less OmpC? How might this help explain the results?

c) Predict the results for a strain of *E. coli* in which the *ompF* gene was deleted (such that it cannot make the OmpF protein) when the culture is grown under low-nutrient conditions.

20. Griffith's famous experiments demonstrated how having a capsule can help cells avoid an immune response in mammals. Predict the results if Griffith had added either DNase or proteinase to the heat-killed smooth cells before mixing them with rough cells and then injecting them into mice. Explain your prediction. (LO6)
    a) DNase
    b) Proteinase

21. The diagram illustrates how solute X is accumulated within a bacterial cell. Use the diagram to answer the following questions. (LO2)

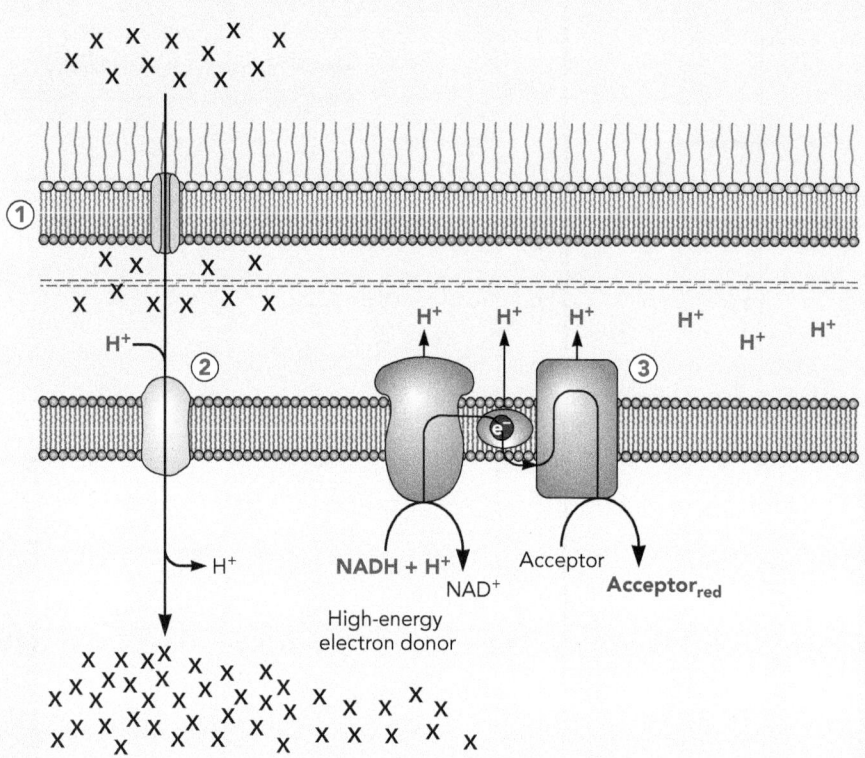

a) Is the transport of solute X occurring in a Gram-positive or Gram-negative cell? Explain how you know.

b) Which of the transport processes labeled in the diagram (1, 2, or 3) best illustrates a facilitated solute transporter? What drives the transport process in facilitated transport? Explain your answer.

c) Which of the transport processes labeled in the diagram (1, 2, or 3) best illustrates an active solute transporter? What is the source of energy for this transport? Explain your answer.

d) Which of the transport processes labeled in the diagram (1, 2, or 3) are coupled? What is the reason for these two transport processes to be coupled?

22. Your research team has identified a gene that you think encodes a transport protein (PcaK) for 4-hydroxybenzoate (4-HBA; an aromatic acid) in *Pseudomonas*. To study this putative transporter, you make a mutant strain in which the *pcaK* gene is deleted. You add 4-HBA to the media, then measure the uptake of 4-HBA in the wild-type strain and in the *pcaK* mutant strain (see graph). (LO2)

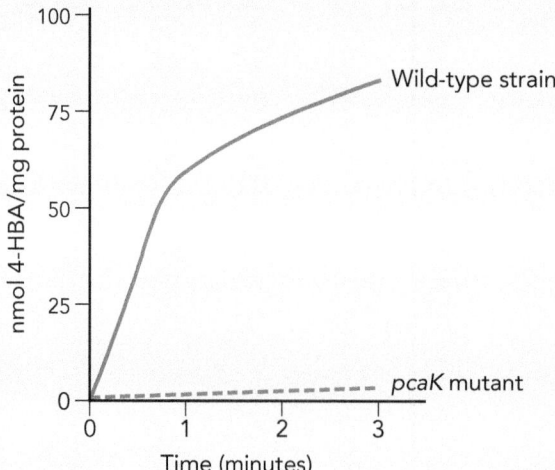

a) Do these results support your hypothesis that PcaK is a transport protein for 4-HBA? Can you determine if uptake involves simple diffusion, facilitated diffusion, or active transport?

b) To investigate whether energy is required for 4-HBA transport, you conduct a similar experiment measuring the uptake of 4-HBA in the wild-type strain (solid line). At the arrow you take a subsample of cells and add dinitrophenol (DNP, which acts as an ionophore and allows the free movement of H^+ across biological membranes). The dotted line shows the activity of the subsample treated with DNP over time. Based on these data, can you now determine if 4-HBA uptake involves simple diffusion, facilitated diffusion, or active transport?

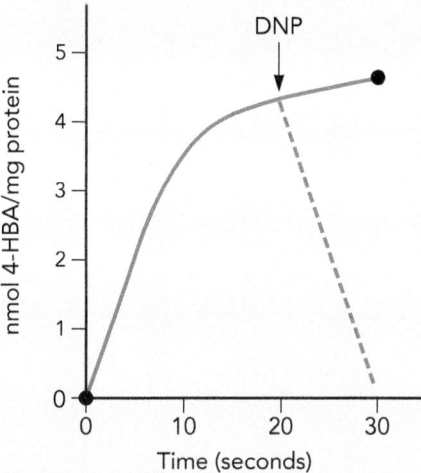

334   Making Connections

334   Introduction to Cytoplasmic Membrane Protein Transport Systems

334   Secretory (Sec)-Dependent Protein Transport System

337   The Secretory (Sec)-Dependent Protein Transport Process

338   Signal Recognition Particle (SRP)-Dependent Protein Transport Process

339   Twin-Arginine Translocation (Tat) Protein Transport Process

340   Integration of Cytoplasmic Membrane Proteins

341   Gram-Negative Bacterial Outer Membrane Protein Secretion Systems

341   Secretory (Sec)- and Twin-Arginine Translocation (Tat)-Dependent Protein Secretion Systems

343   Secretory (Sec)-Independent and Mixed-Mechanism Protein Secretion Systems

347   Importance of Disulfide Bonds

348   Transport and Localization of Other Cell Envelope Components

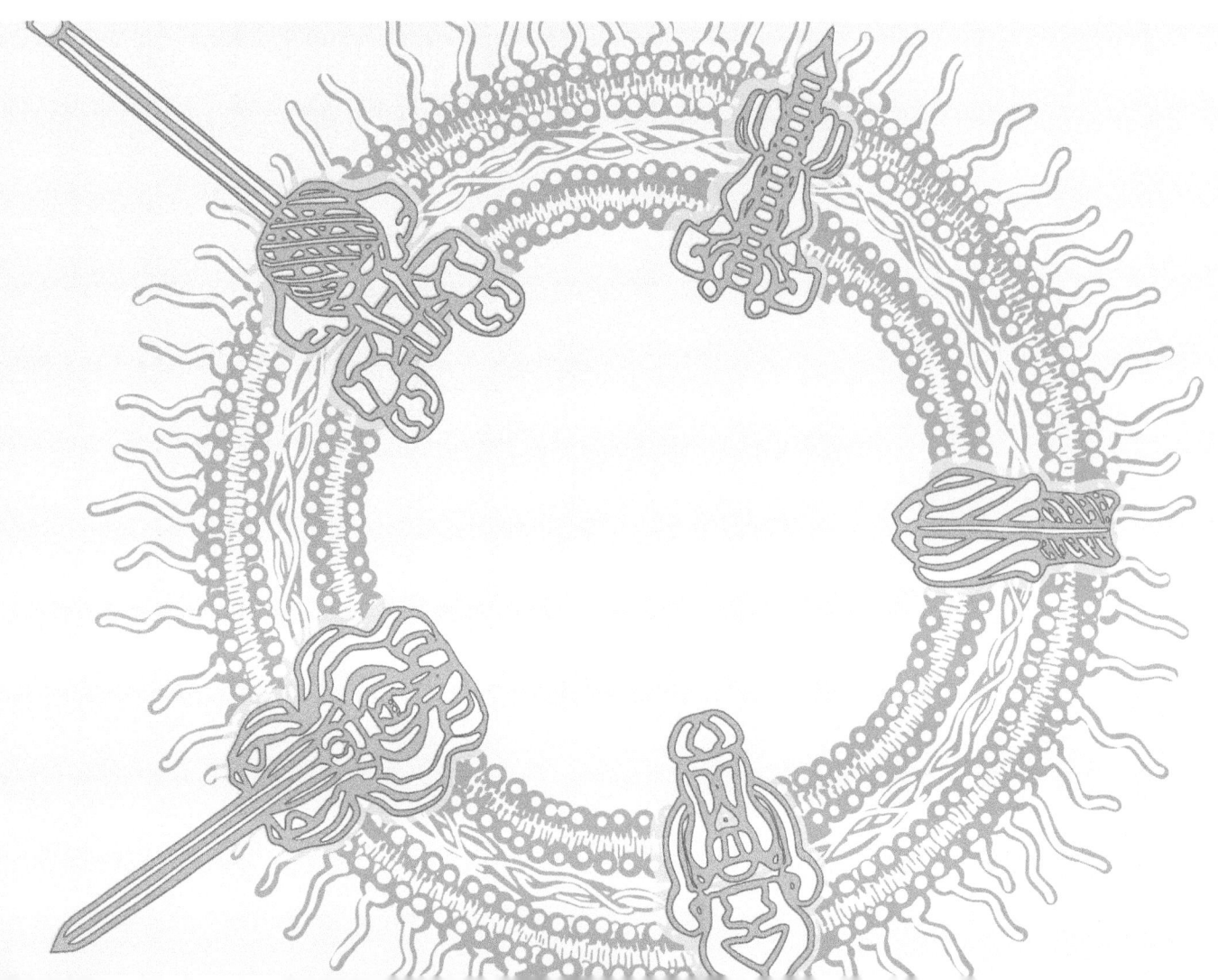

14

TRANSPORT AND LOCALIZATION OF PROTEINS AND CELL ENVELOPE MACROMOLECULES

After completing the material in this chapter, students should be able to . . .

1. compare and contrast the signal, accessory proteins, and energy sources utilized by the three cytoplasmic membrane protein transport systems: secretory (Sec)-dependent, signal recognition particle (SRP)-dependent, and twin-arginine translocation (Tat) pathways

2. outline the process of protein secretion or localization of a protein via the Sec-dependent, SRP-dependent, and Tat protein transport systems

3. predict the orientation of a cytoplasmic membrane protein from the membrane-spanning domains in its sequence

4. state key characteristics of the type I to VI secretion systems that are associated with transport through the Gram-negative outer membrane

5. describe how the oxidation state of the cellular environment and the presence of disulfide bonds impact protein folding

6. explain how components of the cell envelope move from the cytoplasm to their final locations in the cell

Microbial Physiology: Unity and Diversity, First Edition. Ann M. Stevens, Jayna L. Ditty, Rebecca E. Parales and Susan M. Merkel.
© 2024 American Society for Microbiology.
Companion website: www.wiley.com/go/stevens/microbialphysiology

Making Connections

In Chapter 13, the structures found in bacterial and archaeal cell envelopes were introduced and the transport of solutes across the cell envelope was examined. Building from knowledge of the cell envelope structure, protein transport from the cytoplasm through and into the different compartments of the cell envelope is a next step in further understanding cellular transport processes. Delivery and localization of membrane and cell wall components from the cytoplasm into the cell envelope will also be discussed.

Introduction to Cytoplasmic Membrane Protein Transport Systems

The process of polypeptide synthesis (translation) occurs in the cytoplasm. In bacteria, cells must have a way to transport and localize required proteins across the cytoplasmic membrane to the periplasm, into membrane(s), and outside the cell. There are three primary mechanisms by which bacterial polypeptides are transported from the cytoplasm across or into the cytoplasmic membrane. The transport systems that will be discussed are (i) secretory (Sec)-dependent, (ii) signal recognition particle (SRP)-dependent, and (iii) twin-arginine translocation (Tat). Although these are distinct systems, they share certain commonalities that will be discussed. Each system has (i) information (signals) controlling transport that is within the protein being transported; (ii) required "accessory" proteins that facilitate the transport process; and (iii) an energy source to drive the active transport of the protein, as it is not energetically favorable to transport a protein across a membrane.

Secretory (Sec)-Dependent Protein Transport System

The **Sec-dependent transport** of proteins, also known as the general secretory protein pathway, is the dominant form of protein transport across the cytoplasmic membrane in bacteria. This transport system is involved as a key first step for moving proteins to the periplasm, inserting integral membrane

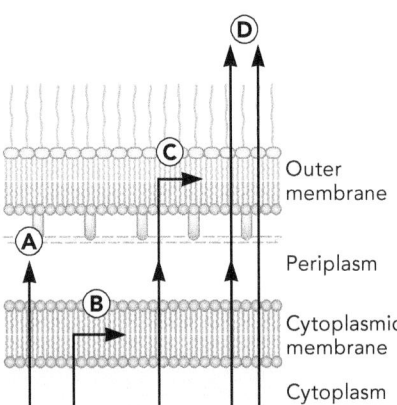

FIGURE 14.1 **Overview of protein transport and localization pathways in bacteria.** There are multiple mechanisms of protein transport for (A) transporting periplasmic proteins, (B) embedding integral membrane proteins into the cytoplasmic membrane, (C) moving proteins to the outer membrane (in Gram-negative bacteria), and (D) secreting proteins across the entire cell envelope in either one or two steps and releasing them into the external environment (or into a target cell, not shown). See text for details.

proteins into the cytoplasmic membrane, delivering proteins to the outer membrane in Gram-negative bacteria, and exporting proteins across the entire cell envelope to the external environment (Fig. 14.1). The process initiates when the Sec secretion system recognizes a polypeptide with a signal sequence that designates it for transport out of the cytoplasm. In addition, there are several key accessory proteins that facilitate the transport process, including a chaperone or chaperonin, a translocase, and, in some cases, a signal peptidase. The process also requires both ATP and the proton motive force (PMF) to power protein transport. The steps involved in Sec-dependent transport are covered in greater depth in the following sections.

SIGNAL SEQUENCE

The **signal sequence**, which is also known as the **leader sequence** or **leader peptide**, is an ~20-amino-acid sequence located at the N-terminus of a polypeptide that is targeted for transport by the Sec-dependent pathway (Fig. 14.2). The amino acids within this region of the protein contain conserved features that are important for transport to occur. Thus, if a signal sequence is added to a protein that is not normally secreted, it will be targeted for Sec-dependent protein transport. Closest to the N-terminus is a basic region that contains 3 to 8 positively charged amino acids (e.g., lysine and arginine). The net positive charge of this region helps to attract the protein being transported to the negative charge that exists on the internal face of the cytoplasmic membrane due to the PMF. Adjacent to these basic residues are 15 to 18 amino acids that are hydrophobic in nature. This hydrophobic core forms a membrane-spanning domain in the protein that inserts into the membrane once transport has been initiated. Finally, some, but not all, polypeptides contain a **signal peptidase cleavage site** that is recognized by a signal peptidase enzyme, which cleaves a peptide bond, and in the process forms a new N-terminus for the polypeptide. The preprotein containing the signal sequence is processed by the peptidase to generate the mature active form of the protein. Whether or not the signal sequence is removed is important for localization of the polypeptide to the appropriate subcompartment in the cell and/or to convert it to a biologically active form.

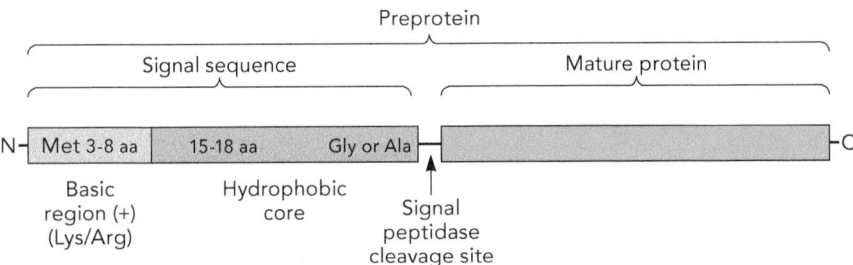

FIGURE 14.2 **Key features of a signal sequence in a polypeptide transported by the secretory (Sec)-dependent system.** A Sec-dependent signal sequence in a polypeptide has a basic region of 3 to 8 amino acids (aa) carrying a net positive charge (often Lys and Arg) located near the N-terminus, followed by a stretch of 15 to 18 amino acids that form a hydrophobic core, which spans across the cytoplasmic membrane. Some, but not all, signal sequences contain a signal peptidase cleavage site (shown by the arrow) at the end of the hydrophobic core after a Gly or Ala residue. Preproteins are processed into mature proteins upon cleavage of the signal sequence by the signal peptidase.

ACCESSORY PROTEINS

There are several accessory proteins required to facilitate the Sec-dependent transport of a polypeptide across the cytoplasmic membrane. SecB is an example of a **chaperone**, a monomeric protein that facilitates the folding or unfolding of another protein. Interestingly, SecB does not appear to recognize the signal sequence, but the signal sequence may help to keep the protein in a conformation recognizable by SecB. SecB recognizes the unfolded preprotein in the cytoplasm and associates with portions of the preprotein that will be retained in the mature form of the protein. The association of SecB with the polypeptide to be transported helps to keep the target polypeptide in an unfolded, translocation-competent conformation. It is easier to move the polypeptide through the membrane in an extended, linear shape. Another type of chaperone, called a **chaperonin**, functions as an oligomer to enhance conditions for the proper folding of denatured proteins, preventing aggregation. The universal cellular chaperones and chaperonins DnaK (heat shock protein 70 [Hsp 70]), GroEL (Hsp 60), and GroES (Hsp 10) may also facilitate transport processes.

SecA is an ATPase that binds to SecB, delivers the SecB-preprotein complex to the membrane channel, and provides energy for transport via ATP hydrolysis (Fig. 14.3). The membrane channel is known as the **translocase**, which is composed of SecY, SecE, and SecG. These integral membrane proteins span the cytoplasmic membrane and create a regulated channel through which other polypeptides may be transported. Opening of the channel is tightly controlled

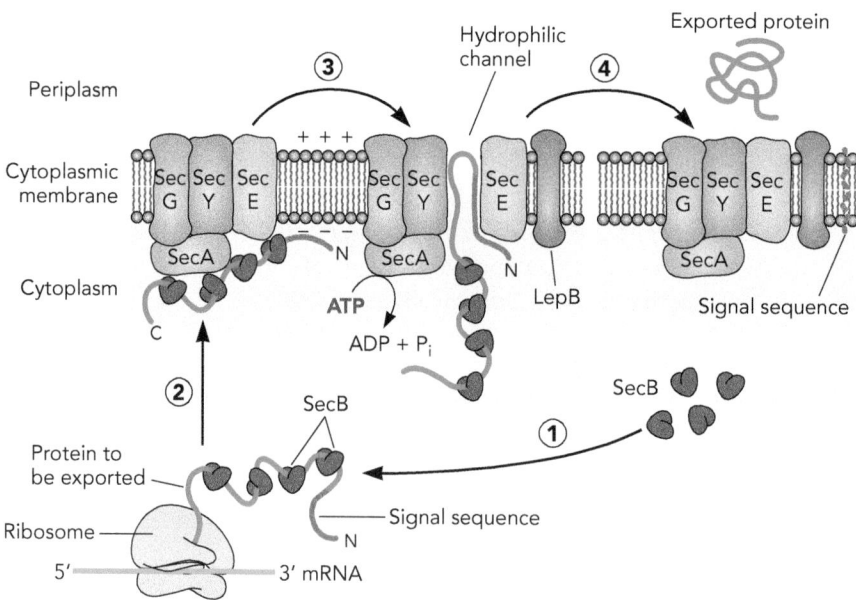

FIGURE 14.3 **The process of secretory (Sec)-dependent protein transport.** (1) A polypeptide with a signal sequence is recognized by the SecB chaperone and kept in an unfolded conformation to enable transport. (2) After translation is complete, SecB delivers the protein to the ATPase SecA. (3) Upon hydrolysis of ATP, the hydrophobic core of the signal sequence becomes embedded in the cytoplasmic membrane and the SecYEG translocase channel creates a hydrophilic environment through which the protein passes. In this example, (4) the LepB signal peptidase cleaves a peptide bond at the end of the hydrophobic core within the signal sequence, which releases the mature form of the protein into the periplasm.

to prevent dissipation of the PMF during polypeptide transport. There are ~500 translocase complexes distributed around the cytoplasmic membrane of an *Escherichia coli* cell. Also present at ~500 molecules per *E. coli* cell and associated with the translocase is the signal peptidase LepB. LepB is an integral membrane protein with its catalytic domain on the external face of the cytoplasmic membrane positioned to cleave off the signal sequence. However, cleavage only occurs with polypeptides that have the appropriate signal peptidase cleavage site, a glycine or alanine at the end of the hydrophobic core, in the signal sequence (Fig. 14.3). Therefore, some, but not all, signal sequences are removed by the signal peptidase during Sec-dependent transport.

ENERGY SOURCE

Both ATP and the PMF are required for Sec-dependent transport, as described in more detail below.

The Secretory (Sec)-Dependent Protein Transport Process

Although the process of Sec-dependent protein transport is best understood in *Gammaproteobacteria* like *E. coli*, it is believed to be a widespread secretion mechanism in bacteria. Sec-dependent transport of a polypeptide occurs posttranslationally (Fig. 14.3). The protein is translated in the cytoplasm and is recognized by the SecB protein, which keeps it in a translocation-competent conformation. SecA recognizes the SecB-polypeptide complex and delivers it to the SecYEG translocase. ATP binding by SecA initiates transport and the hydrophobic core starts to insert into the membrane. Then, through a poorly understood mechanism, the signal sequence is laterally transferred from the channel into the adjacent phospholipid bilayer. There is *in vivo* evidence suggesting that ATP is required to initiate Sec-dependent protein transport, but after the ATP is hydrolyzed and SecA is released from the translocase, the process is completed with energy derived from the PMF. However, *in vitro* experiments have demonstrated that transport can occur in the absence of PMF (i.e., in the presence of uncouplers). Under these conditions it is hypothesized that multiple rounds of ATP binding and hydrolysis exclusively provide the energy required for transport.

Once the signal sequence has been transported, the signal sequence will be removed if it is recognized by a signal peptidase, and the mature processed protein will be released into the periplasm. At this point, the signal sequence is no longer needed and its amino acids may be recycled by the cell. However, if the signal sequence is not cleaved, the polypeptide will remain anchored to the membrane by the signal sequence and most of its structure will be in the periplasm. Thus, proteins transported by a Sec-dependent mechanism may be released into the periplasm or become cytoplasmic membrane proteins, depending on whether the signal sequence is removed during transport.

In *E. coli*, there are ~10^6 periplasmic and outer membrane proteins required for growth at each generation. With ~500 translocase complexes per cell, it can be calculated that ~80 polypeptides are secreted per minute by each translocase complex, more than 1 per second!

Signal Recognition Particle (SRP)-Dependent Protein Transport Process

The **SRP-dependent protein transport system** (Fig. 14.4) is used by bacteria to move proteins destined for the cytoplasmic membrane. The process is similar to the mechanism used by eukaryotic cells to transport proteins across the membrane of the rough endoplasmic reticulum. It is a cotranslational mechanism in which translation and transport occur simultaneously.

SIGNAL

Polypeptides to be secreted by this system have an N-terminal hydrophobic signal that is recognized by SRP.

ACCESSORY PROTEINS

SRP is an accessory molecule made of RNA and protein. When SRP binds to its recognition signal in the N-terminal amino acid sequence, it stops translation of the protein that will be transported until the entire translation complex docks onto the cytoplasmic membrane. FtsY serves as the accessory protein receptor on the *E. coli* bacterial cytoplasmic membrane. FtsY recognizes SRP bound to the protein and delivers the protein-ribosome-mRNA complex to the SecYEG translocase (described above). Translation resumes after SRP is released

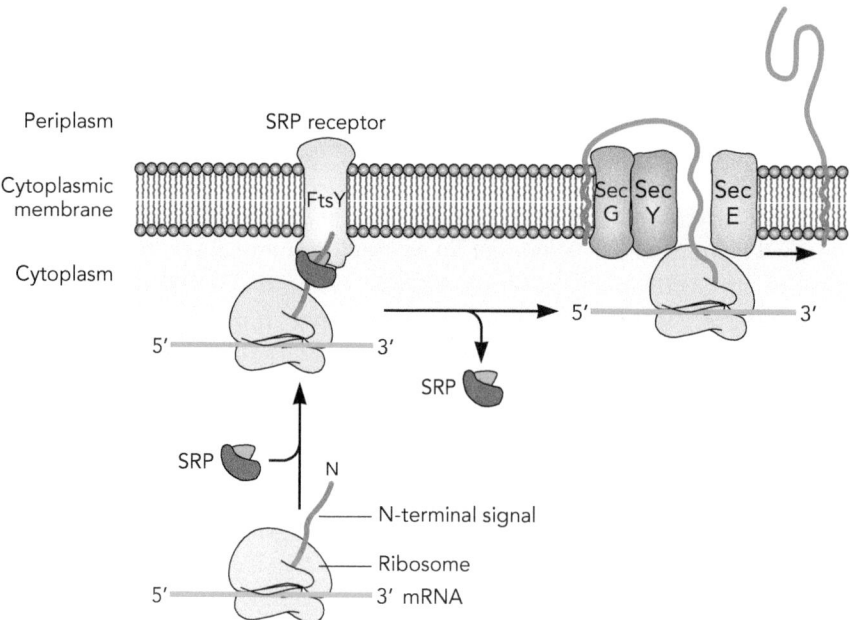

FIGURE 14.4 **The process of signal recognition particle (SRP)-dependent protein transport.** A polypeptide with the appropriate N-terminal signal is recognized by the signal recognition particle (SRP), a hybrid molecule composed of protein (purple) and RNA (green). SRP stops the process of translation and brings the translation machinery to the SRP receptor protein, FtsY, in *Escherichia coli*. SRP is then released, and translation by the ribosome resumes. Transport of the nascent polypeptide occurs in a cotranslational manner through the SecYEG translocase. The N-terminal signal moves laterally and becomes the transmembrane-spanning domain for the integral cytoplasmic membrane protein.

from the complex, and the N-terminal signal of the protein is inserted laterally into the membrane via the translocase channel through a poorly understood mechanism.

ENERGY

SRP-dependent transport occurs simultaneously with translation; hence it is considered a cotranslational process. The protein is pushed through the transport channel as it is elongated during translation (which is powered by ATP/GTP hydrolysis).

Twin-Arginine Translocation (Tat) Protein Transport Process

The **Tat protein secretion system** (Fig. 14.5) is a Sec-independent system that exports folded proteins. Examples of proteins that are folded prior to transport include those with a prosthetic group or a cofactor that must be added in the cytoplasm. Two such examples include cytochrome *c*, which contains a heme prosthetic group and is destined for the periplasm, and redox-sensitive cytoplasmic membrane proteins that carry iron-sulfur clusters.

SIGNAL

The Tat recognition signal on the protein is an N-terminal sequence with two (twin) conserved arginine residues.

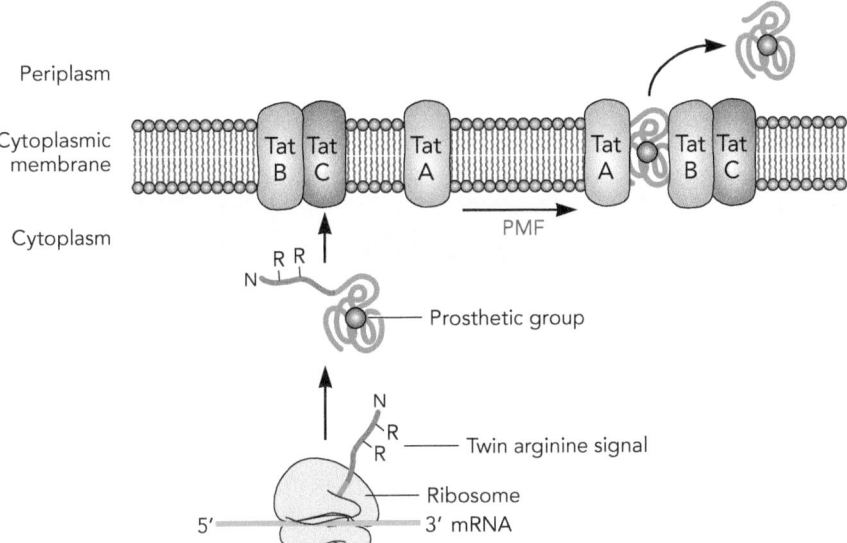

FIGURE 14.5 **The process of twin-arginine translocase (Tat)-dependent protein transport.** A folded polypeptide containing a covalently bound prosthetic group is recognized by the TatBC translocase at a twin-arginine (R-R) signal near the N-terminus. In *Escherichia coli*, TatA associates with the translocase and harnesses the proton motive force (PMF) to enable transport of the fully active form of the protein from the cytoplasm to the periplasm.

ACCESSORY PROTEINS

The N-terminal sequence is recognized by the accessory proteins TatB and TatC in *E. coli*. They recruit another accessory protein, TatA, to form the Tat translocase channel through which larger-diameter folded proteins may pass without dissipating the PMF. In Gram-positive bacteria, TatA and TatB are combined into one multifunctional protein.

ENERGY

TatA utilizes the PMF as the energy source for transport through the Tat translocase.

Integration of Cytoplasmic Membrane Proteins

The SRP-dependent and Sec-dependent systems are both used to facilitate the insertion of transmembrane proteins into the cytoplasmic membrane. For the SRP-dependent system, the signal at the N-terminus recognized by SRP also serves to anchor the protein in the cytoplasmic membrane during transport (Fig. 14.4). For the Sec system, the polypeptide being transported must have either a cleavable or a noncleavable signal sequence at the N-terminus to initiate transport. The process whereby hydrophobic sequences in proteins move laterally from the translocase channel into the membrane is not fully understood. Nevertheless, a polypeptide with a noncleavable signal sequence becomes inserted into the membrane such that the N-terminus is in the cytoplasm and the C-terminus is in the periplasm. In this case, the hydrophobic core of the signal sequence anchors the polypeptide in the cytoplasmic membrane (Fig. 14.6A). The opposite orientation may also be achieved, resulting in

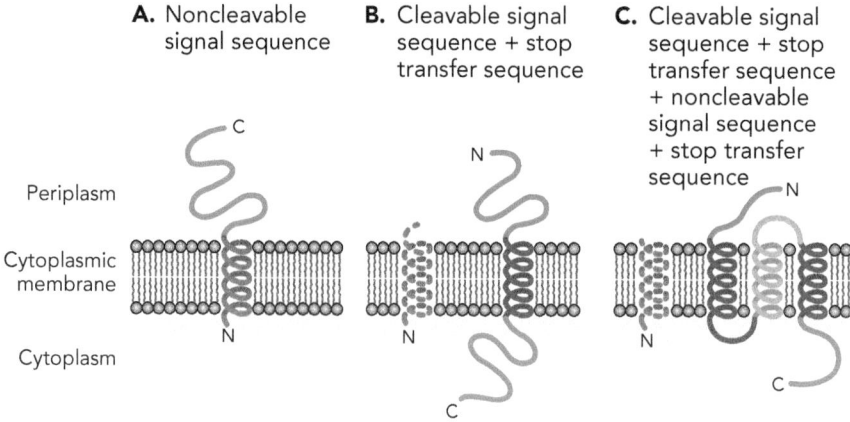

FIGURE 14.6 **Signals involved in the secretory (Sec)-dependent localization of integral cytoplasmic membrane proteins.** (A) A polypeptide (blue) with a noncleavable signal sequence (orange) will have one transmembrane-spanning domain and be oriented with the N-terminus in the cytoplasm. (B) A polypeptide with a cleavable signal sequence and a stop transfer sequence (red) will have one transmembrane-spanning domain and be oriented with the N-terminus in the periplasm. The cleaved signal sequence is degraded by the cell (dashed orange). (C) A polypeptide with three transmembrane-spanning domains may have a cleavable signal sequence, as shown, followed by a stop transfer sequence, a noncleavable signal sequence (gold), and a second stop transfer sequence.

a polypeptide with the N-terminus in the periplasm and the C-terminus in the cytoplasm (Fig. 14.6B). Polypeptides with this membrane orientation contain a cleavable signal sequence and a **stop transfer sequence**. The stop transfer sequence follows the initial signal sequence (which is removed by a signal peptidase) and is composed of a segment of ~15 to 18 hydrophobic amino acids that is not cleaved off and forms a transmembrane-spanning domain in the polypeptide. Just as the name implies, the stop transfer sequence stops the transport of the remainder of the protein, which remains in the cytoplasm unless there is another signal that reinitiates the transport process. This is the case for polypeptides that span the membrane more than once. Polypeptides with multiple transmembrane domains contain alternating signals within their amino acid sequence. The first signal is always either a noncleavable or a cleavable signal sequence. The second signal is a stop transfer sequence that is then followed by a noncleavable signal sequence and then another stop transfer sequence (Fig. 14.6C). Larger membrane proteins have extended numbers of alternating signals; for example, LacY has 12 membrane-spanning domains (Fig. 13.4).

Gram-Negative Bacterial Outer Membrane Protein Secretion Systems

For Gram-negative bacteria, the outer membrane creates an additional barrier in the cell envelope across which polypeptides must be transported. At least nine (I to IX) outer membrane secretion systems have been discovered. Key features about several of them are provided below. Some use the Sec system or Tat system described above to first transport the protein to the periplasm, and then the protein is moved across the outer membrane (e.g., type II, type V, and sometimes type IV). The other outer membrane secretion systems are said to be Sec-independent (e.g., type I, type III, type VI, and sometimes type IV). In the following section, the Sec- and Tat-dependent secretory systems are described first, followed by the Sec-independent and mixed-mechanism secretion systems.

Secretory (Sec)- and Twin-Arginine Translocation (Tat)-Dependent Protein Secretion Systems

TYPE II

Type II secretion systems (T2SSs) are conserved across most Gram-negative bacteria and are an important route for the secretion of folded polypeptides across the outer membrane (Fig. 14.7). Polypeptides are first translocated across the cytoplasmic membrane into the periplasm through either the Sec or Tat system described above. Tat system proteins are already folded in the cytoplasm, while proteins secreted via the Sec pathway must fold after they are transferred into the periplasm. From the periplasm, the proteins are then transported across the outer membrane by the T2SS. The T2SS has an inner membrane complex including an ATPase, an ATP-dependent pseudopilus, and an outer membrane channel. The outer membrane channel is primarily composed of a β-barrel protein known as a **secretin**. As the pseudopilus is polymerized from individual subunits, using energy derived from the hydrolysis of ATP by the ATPase, it pushes the secreted protein through the outer membrane secretin channel, releasing it to the outside of the cell (Fig. 14.7). Components of T2SSs are very similar to those of type IV pili, which are responsible for gliding motility in myxobacteria and twitching motility in *Pseudomonas* (Chapters 15 and 18). Bacterial pathogens commonly use T2SSs to deliver virulence factors, such as the cholera toxin of *Vibrio cholerae* (Chapter 7). Some Gram-positive bacteria have T2SS protein components, but how proteins are transported across the thick peptidoglycan layer is poorly understood.

TYPE V

In type V secretion systems (T5SSs; Fig. 14.7), the polypeptides are first transported across the inner membrane by the Sec-dependent pathway, and their N-terminal signal sequence is cleaved off as they enter the periplasm. Part of the polypeptide being secreted folds into a β-barrel structure in the outer membrane; this is called the **helper domain** (also known as the translocation domain), as it facilitates transport of the rest of the polypeptide. The portion of the polypeptide that is to be secreted to the cell surface is called the **passenger domain**. In many cases, the passenger domain is cleaved and released outside the cell, while in other cases, it remains attached to the cell surface. The T5SS was known as an autotransporter system since initially no additional outer mem-

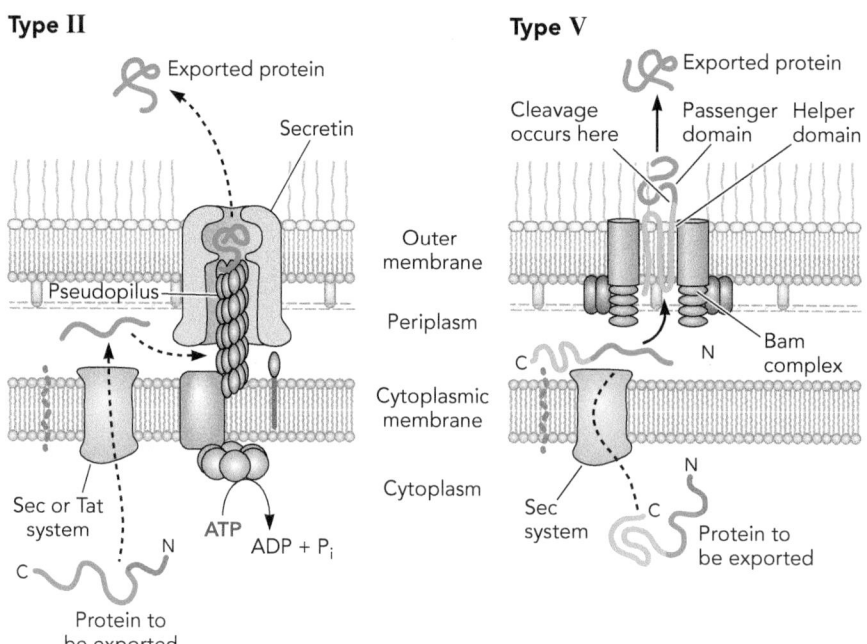

FIGURE 14.7 **Gram-negative outer membrane type II and type V protein secretion systems.** (Left) A polypeptide that will be secreted across the outer membrane by a type II secretion system is first transported across the inner membrane by a secretory (Sec)-dependent or twin-arginine translocation (Tat) pathway and released into the periplasm. The polypeptide then moves through an outer membrane secretin complex by polymerization of an ATP-dependent pseudopilus and is released to the external environment. (Right) A type V secretion system relies on Sec-dependent transport across the cytoplasmic membrane. Once in the periplasm, the polypeptide is autotransported across the outer membrane and cleaved, assisted by the helper domain (green). The Bam outer membrane complex helps to facilitate secretion of the passenger domain (blue) to the external environment.

brane accessory protein complex was identified. It was believed that the polypeptide being transported contained within its sequences all the components necessary for outer membrane transport. However, the Bam (β-barrel assembly machinery) complex is now known to be required for insertion of the β-barrel structure of the type V helper domain, and periplasmic chaperones are believed to also facilitate this process. In the case of immunoglobulin A (IgA) protease, a bacterial virulence factor that acts against immunoglobulins, the transported protein contains a cleavable N-terminal signal sequence. The C-terminal helper domain of the polypeptide assists with transport of the passenger domain to the exterior of the cell, where it serves as a virulence factor for pathogenic bacteria encoding it.

Secretory (Sec)-Independent and Mixed-Mechanism Protein Secretion Systems

TYPE I

The type I secretion system (T1SS) is a Sec-independent secretion system that transports polypeptides from the cytoplasm directly to the outside of the cell without releasing them into the periplasm. The recognition signal for transport is in the C-terminal domain of the protein to be transported, and three accessory proteins that form a regulated transport channel are required: (i) an ABC transporter (Chapter 13), which also provides the energy for transport via ATP hydrolysis; (ii) a membrane fusion protein that transverses the periplasmic space; and (iii) an outer membrane channel protein that facilitates transport across the outer membrane (Fig. 14.8). The genes encoding these components of the secretory apparatus are often, although not always, linked to the gene encoding the protein to be transported (located near each other on the chromosome), and there is specificity in the capacity of the transport machinery to recognize the protein being transported. One exception to this rule is the TolC protein, which facilitates transport across the outer membrane. A gene encoding TolC is not always linked to other genes encoding a particular T1SS, and a single TolC can be used for multiple T1SSs. Hemolysin, an enzyme from *E. coli* that lyses red blood cells, is an example of a protein that is transported by a T1SS.

TYPE III

Type III secretion systems (T3SSs) in bacteria are important to the process of infection by both animal and plant pathogens. T3SSs function in a Sec-independent manner to transfer polypeptides from the cytoplasm of a Gram-negative bacterium into the cytoplasm of a target eukaryotic cell. The transported proteins must traverse three membranes in the process, two in the bacterium and one in the eukaryotic host (Fig. 14.8). The polypeptide being transported has a special N-terminal signal and cytoplasmic chaperone. The accessory proteins forming the transport channel are arranged into a complex multipolypeptide structure (~20 polypeptides) known as the **injectisome**, which contains components that resemble a syringe and needle. The syringe components attached to the cytoplasmic membrane form a structure like the basal body of a bacterial flagellum (Chapter 15). A needle-like structure projects outward from the syringe across the entire cell envelope and into the eukaryotic cell. The syringe

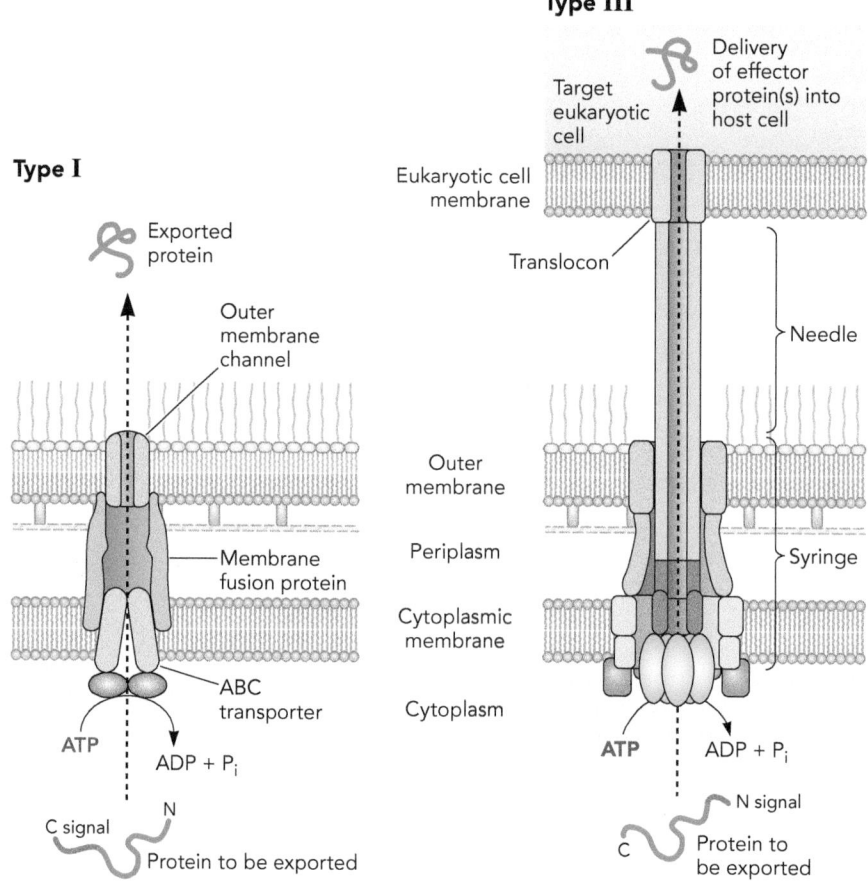

Type III

Delivery of effector protein(s) into host cell

Target eukaryotic cell

Eukaryotic cell membrane

Translocon

Needle

Outer membrane

Periplasm

Syringe

Cytoplasmic membrane

Cytoplasm

ATP   ADP + P$_i$

N signal

Protein to be exported

Type I

Exported protein

Outer membrane channel

Membrane fusion protein

ABC transporter

ATP   ADP + P$_i$

C signal   N

Protein to be exported

FIGURE 14.8 **Gram-negative outer membrane type I and type III secretion systems.** Both the type I and type III secretion systems function in a secretory (Sec)-independent manner with the polypeptide being transported across both bacterial membranes without being released into the periplasm. (Left) The type I secretion system has three accessory proteins: an ABC transporter, a membrane fusion protein spanning across the periplasm, and an outer membrane protein channel. Polypeptides with the appropriate C-terminal signal are secreted directly from the cytoplasm to the external environment. (Right) A type III secretion system is composed of a syringe, needle, and translocon channel, which together form an injectisome. The injectisome spans the entire bacterial cell envelope and extends outward into the target eukaryotic cell membrane. The translocon is inserted into the eukaryotic membrane, creating a channel through which effector proteins are delivered. In this manner, pathogenic effector proteins are injected directly from the bacterial cell cytoplasm into the target eukaryotic cell cytoplasm. The effector proteins disrupt the function of the eukaryotic cell and can lead to cell death.

is normally capped off to prevent loss of polypeptides and the PMF until the appropriate environmental signal is detected. This may be direct contact with a eukaryotic cell and/or changes in the environment (e.g., calcium ion levels or temperature) that trigger complete synthesis of the needle structure. Effector polypeptides (not to be confused with allosteric effectors) are synthesized in the bacterial cytoplasm and are transported in an unfolded state through the syringe. They are passed from the needle through a translocon channel in the eukaryotic cell membrane and into the cytoplasm of the eukaryotic cell.

The bacterial effectors then induce changes in the eukaryotic cytoskeleton (e.g., stop actin-based engulfment by phagocytic white blood cells) or alter internal regulatory processes (e.g., interfere with the eukaryotic cell phosphorelay signal transduction systems). Thus, the effectors may inhibit an appropriate host immune response and/or cause damage to the target eukaryotic cell. *Yersinia* outer proteins (Yops) are T3SS effectors produced by the pathogens *Y. pestis* and *Y. enterocolitica*, which cause the bubonic plague and gastroenteritis, respectively. Interestingly, the bacterial flagellum is evolutionarily related to T3SSs and the two structures share several homologous proteins (Chapter 15).

TYPE IV

Type IV secretion systems (T4SSs) are found in both Gram-negative and Gram-positive bacteria and are evolutionarily related to bacterial DNA conjugation systems transporting DNA, but some T4SSs transport proteins rather than DNA. Thus, T4SSs are quite functionally diverse. There are documented examples of their roles in conjugation to transmit single-stranded DNA to other eukaryotic or bacterial cells and DNA release to the extracellular environment, which contributes to DNA exchange and biofilm formation. However, other T4SSs deliver effector proteins or toxins outside the cell or directly into eukaryotic hosts and competing bacteria. Common accessory components of T4SSs include inner and outer membrane proteins that form the cell envelope-spanning channel and cytoplasmic ATPases that may process substrates and/or cause structural changes in the channel to allow substrate passage (Fig. 14.9). Many T4SSs have an extracellular pilus, and they may also have additional protein components that are specific to each system. Some T4SSs are Sec-dependent, such as the mechanism used to transport pertussis toxin produced by *Bordetella pertussis*, the causative agent of whooping cough, to the outside of the cell. Other T4SSs use Sec-independent mechanisms to transport DNA rather than proteins across the cell envelope. For example, T-DNA from the Ti plasmid of *Agrobacterium tumefaciens* is transferred into a plant cell via conjugation involving a T4SS. The bacterial T-DNA is associated with a nuclear localization protein that mediates T-DNA integration into the plant genome, where it causes the plant cell to proliferate in an accelerated manner and produce nutrients for the bacteria, resulting in crown gall tumor disease. This natural mechanism of T-DNA transfer has been applied through biotechnology to permit the production of genetically modified plants.

TYPE VI

Type VI secretion systems (T6SSs) are thought to have evolved from phages and are widespread in Gram-negative bacteria. T6SSs function to deliver effector proteins and toxins into the cytoplasm of other bacterial and eukaryotic cells. The injected proteins are used to manipulate host cell metabolism, cause disease, or control the growth of competing bacteria. For example, *V. cholerae* uses a T6SS to resist phagocytosis by injecting a protein that cross-links the host cytoskeletal protein actin, which kills the host cell. Some *Vibrio* species also make toxins that can kill other susceptible *Vibrio* species. In fact, all *Bacteria* with active T6SSs must produce an immunity protein to prevent self-intoxication.

Type IV

A Sec-dependent

B Sec-independent

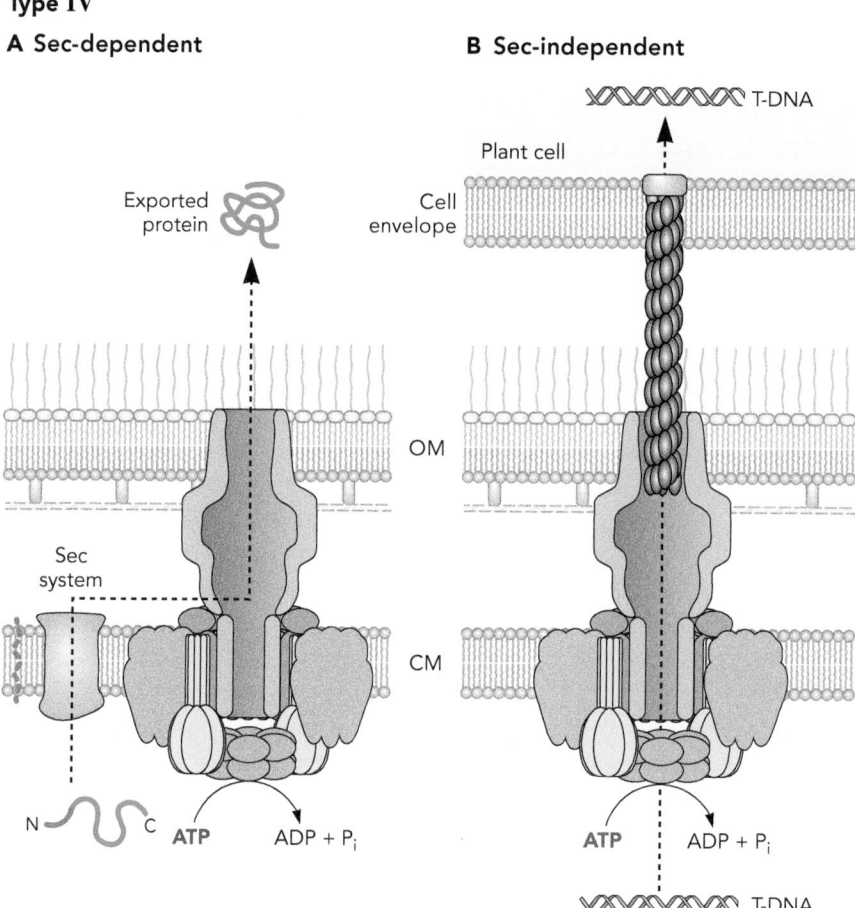

FIGURE 14.9 **Gram-negative outer membrane type IV secretion systems.** (A) In some type IV secretion systems, the polypeptide is transported into the periplasm in a secretory (Sec)-dependent manner. The periplasmic protein then moves across the outer membrane via the type IV machinery in an ATP-dependent manner. (B) In some microorganisms, like *Agrobacterium tumefaciens*, the type IV secretion system works independently of Sec. Here, the system evolved to transport DNA and associated proteins from the cytoplasm of a bacterial cell directly into the cytoplasm of a eukaryotic plant cell in a process similar to conjugation.

The T6SS complex functionally resembles a T4 bacteriophage tail and utilizes a syringe-like mechanism to inject proteins into the target cell (Fig. 14.10). This molecular machine is made up of at least 13 proteins that form a sheath surrounding the hollow needle structure and a baseplate that anchors the complex to the cytoplasmic membrane. The needle and sheath are built in an extended state. Upon contraction of the sheath, the needle structure exits the bacterial cell, penetrates the target cell, and "fires" to inject the effector proteins into the target cell. The T6SS is then disassembled by the ATP-dependent protease ClpV.

Gram-negative bacteria are known to use at least nine distinctive secretion systems (I to IX) to move proteins across the outer membrane; here we have examined six of them. All these secretion systems share the following characteristics: they require an internal signal in the protein being transported, accessory proteins, and an energy source.

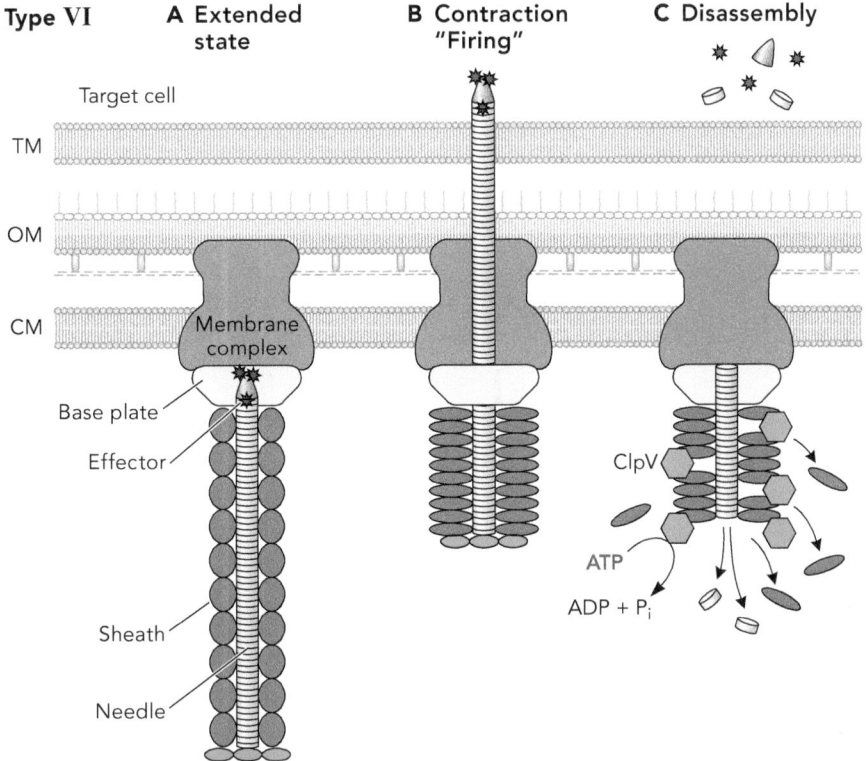

FIGURE 14.10 **Gram-negative outer membrane type VI secretion systems.** Type VI secretion systems use a phage-like injector to transport polypeptides, often toxins, from the cytoplasm of the toxin-producing bacterial cell into the cytoplasm of a target cell (bacterial or eukaryotic). (A) Initially, the sheath and needle structure is in an extended state; (B) then the sheath contracts, "firing" the needle across the target cell membrane (TM) and injecting bacterial toxins (red stars) into the target cell cytoplasm. (C) The sheath is then disassembled by an ATP-dependent protease, ClpV. The toxin-producing bacterial cell must also express an immunity protein for self-protection against the toxin.

Importance of Disulfide Bonds

A very important and fundamental issue to consider when discussing polypeptide transport is the formation of disulfide bonds (S-S) between cysteine residues. The cytoplasm is a highly reduced environment, and cysteine residues will have reduced sulfhydryl groups (–SH) under these conditions. Therefore, when working with cytoplasmic proteins in the laboratory, it is vital to provide reduced buffer conditions (e.g., addition of a reducing agent like dithiothreitol) to maintain the protein in its active conformation. Compared to the cytoplasm, the environment outside of the cell membrane, including the periplasm, is an oxidized environment. Under oxidized conditions, disulfide bonds will form and contribute to the higher-ordered structure of the polypeptide (i.e., tertiary and quaternary structure). Dsb (disulfide bond) proteins assist with proper protein folding in the periplasm (Fig. 14.11). DsbA catalyzes disulfide bond formation by oxidizing the target protein; DsbA is reduced in the reaction. DsbB reoxidizes DsbA and then reduced DsbB donates the electrons to the electron transport chain (ETC), reoxidizing DsbB and keeping the system functioning.

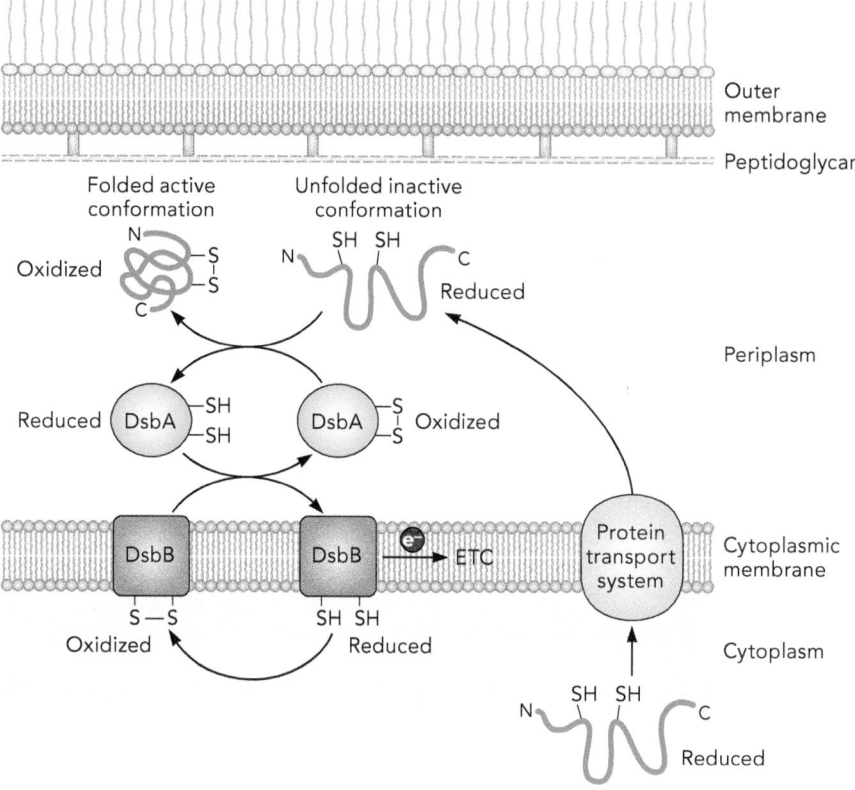

FIGURE 14.11 **Facilitation of disulfide bond formation in polypeptides transported out of the cytoplasm.** Proteins produced in the reduced environment of the cytoplasm have sulfhydryl groups (–SH) that are converted into disulfide bonds (S-S) in the oxidized environment that exists outside of the cytoplasmic membrane. DsbA facilitates formation of disulfide bonds, DsbB reoxidizes DsbA, and DsbB donates the electrons to the electron transport chain (ETC).

Transport and Localization of Other Cell Envelope Components

Components of the cell envelope containing lipids and sugars must also be transported from the cytoplasm to their final destinations in the cell envelope. Below are brief descriptions about how this occurs.

CYTOPLASMIC MEMBRANE COMPONENTS

Phospholipids are synthesized in the cytoplasm. The newly formed phospholipids are added to the inner leaflet of the cytoplasmic membrane and then transferred to the outer leaflet by membrane proteins such as flippases (Fig. 14.12). Details of the mechanism are poorly understood, including the energy requirements of the process.

CELL WALL COMPONENTS (PEPTIDOGLYCAN)

The two sugar groups used in peptidoglycan, *N*-acetylglucosamine (NAG) and *N*-acetylmuramic acid (NAM), are generated in the cytoplasm as UDP nucleotide-linked precursors (Fig. 14.13). UDP-NAM with a pentapeptide side chain associates with a lipid carrier molecule known as undecaprenyl phosphate.

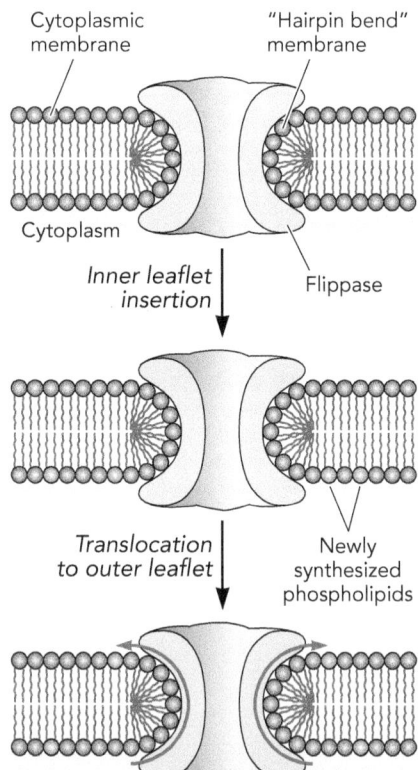

Cytoplasmic membrane

"Hairpin bend" membrane

Cytoplasm

Inner leaflet insertion

Flippase

Translocation to outer leaflet

Newly synthesized phospholipids

FIGURE 14.12 **Production of the cytoplasmic membrane in bacteria.** Cytoplasmic membranes composed of phospholipid bilayers grow and expand as new phospholipids (red) are added to the cytoplasmic side of the membrane. The new phospholipids are subsequently translocated to the outer leaflet of the membrane by flippase enzymes, through a poorly defined mechanism.

NAG is then added to UDP-NAM-pentapeptide. The disaccharide-pentapeptide molecules are translocated across the cytoplasmic membrane via the undecaprenyl phosphate cycle, likely with the assistance of other membrane proteins. Penicillin-binding proteins (PBPs) then covalently attach the new disaccharide-pentapeptide into the growing peptidoglycan. Some PBPs act as transglycosylases that form glycosidic bonds between NAG and NAM in the sugar backbone. Other PBPs act as transpeptidases that form cross-links between the peptide side chains. As the name implies, penicillin (and other β-lactam antibiotics) binds and inactivates PBPs (Fig. 14.13C). In Gram-negative bacteria, the thin layer of peptidoglycan is anchored to the inner half of the outer membrane through murein lipoproteins. In Gram-positive bacteria with a multilayer peptidoglycan, the growth occurs from the side closest to the cytoplasmic membrane and older peptidoglycan is pushed toward the cell periphery, where it is hydrolyzed and sloughed off.

OUTER MEMBRANE COMPONENTS

Lipopolysaccharide is synthesized in the cytoplasm via two separate pathways that converge (Fig. 14.14). The O-antigen is polymerized using an undecaprenyl phosphate lipid carrier molecule (also used to synthesize peptidoglycan, as described above). The undecaprenyl phosphate facilitates transport across the cytoplasmic membrane. The core saccharides are built on lipid A, which serves as both a base and a carrier molecule for transport. Exactly how these molecules and the phospholipid inner leaflet of the outer membrane are integrated into

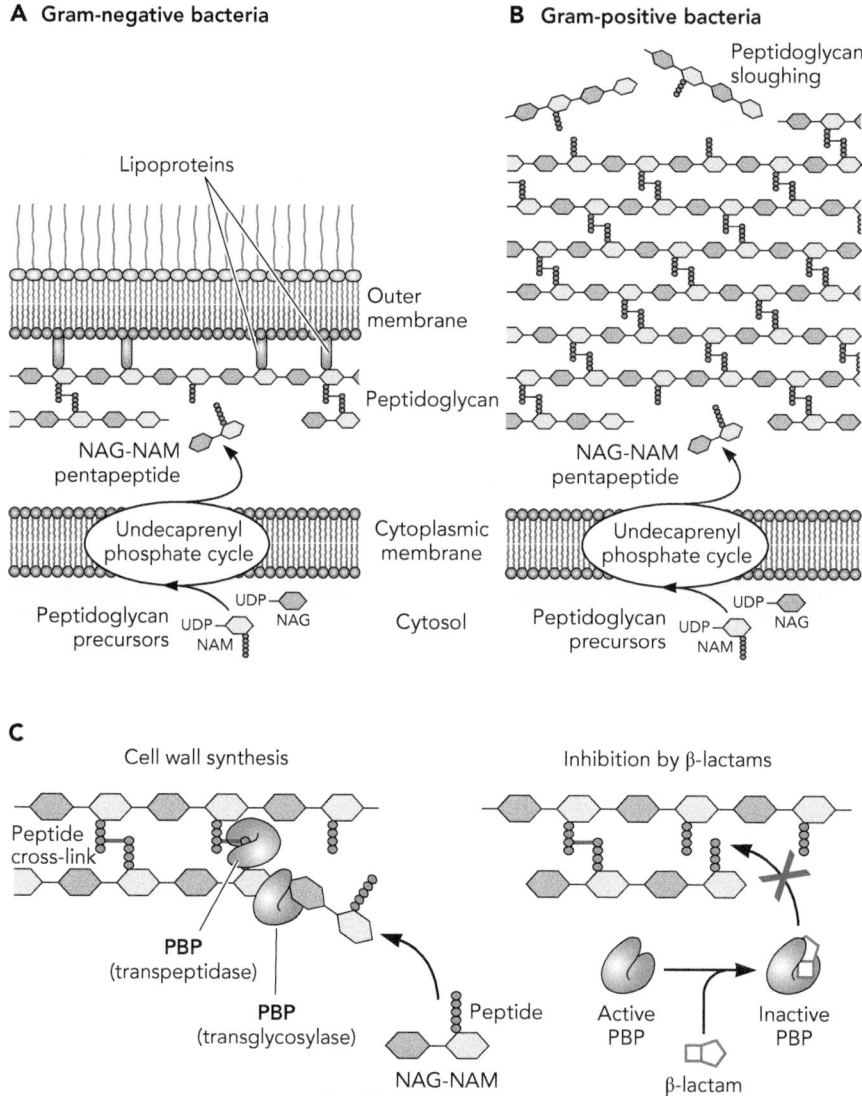

A Gram-negative bacteria

B Gram-positive bacteria

Peptidoglycan sloughing

Lipoproteins

Outer membrane

Peptidoglycan

NAG-NAM pentapeptide

Cytoplasmic membrane

Undecaprenyl phosphate cycle

Cytosol

Peptidoglycan precursors

UDP–NAG

UDP–NAM

C

Cell wall synthesis

Inhibition by β-lactams

Peptide cross-link

PBP (transpeptidase)

PBP (transglycosylase)

Peptide

NAG-NAM

Peptidoglycan precursor

Active PBP

Inactive PBP

β-lactam antibiotic

FIGURE 14.13 **Production of peptidoglycan in bacteria.** (A) In Gram-negative bacteria, peptidoglycan precursors are transported as *N*-acetylglucosamine-*N*-acetylmuramic acid (NAG-NAM) dimers with attached pentapeptide side chains into the periplasm by the lipid carrier molecule undecaprenyl phosphate. The NAG-NAM dimers are then assembled into the peptidoglycan layer. (B) Peptidoglycan addition in Gram-positive bacteria is similar to that in Gram-negative bacteria except that the peptidoglycan is continually moved toward the cell periphery as components of the oldest layers of the cell wall are continually sloughed off into the environment. (C) Peptidoglycan precursors are added into the growing cell wall through the activity of penicillin-binding proteins (PBPs). PBPs include transglycosylases that create glycosidic bonds in the glycan chain, while other PBPs function as transpeptidase enzymes, generating peptide bond cross-links between the NAG-NAM chains. β-lactam antibiotics interfere with the activity of PBPs, thereby weakening the cell wall.

the outer membrane is not fully understood. The presence of membrane adhesion sites between the inner and outer membranes has been proposed as one mechanism.

How are proteins inserted into the outer membrane of Gram-negative bacteria? Proteins destined for the outer membrane are typically moved by the Sec pathway across the cytoplasmic membrane. In the periplasm, chaperones (Skp)

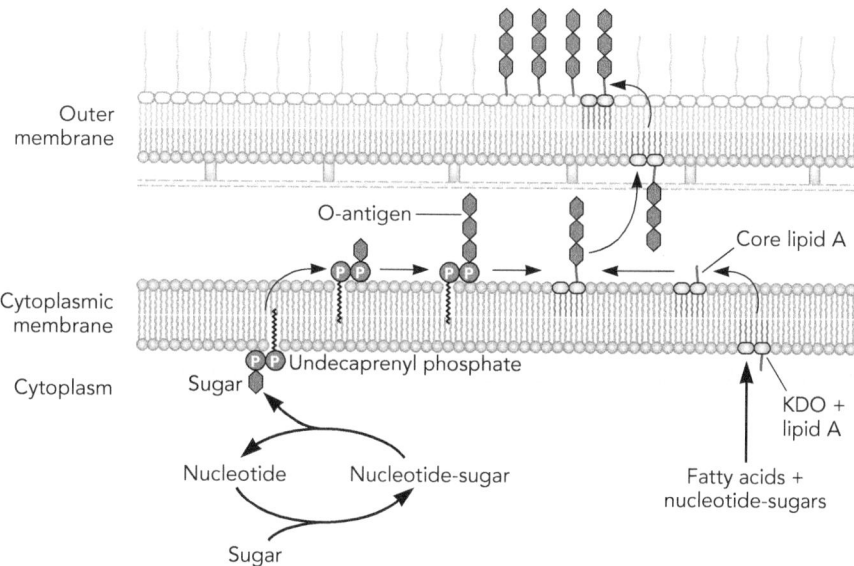

FIGURE 14.14 **Production of lipopolysaccharide (LPS) in Gram-negative bacteria.** LPS is produced from two parallel pathways, one that generates the repeating O-antigen polysaccharide components using undecaprenyl phosphate as a carrier and a second that generates the core polysaccharide using lipid A as the base. The two pathways merge at the outer membrane where the LPS components are inserted in the outer leaflet.

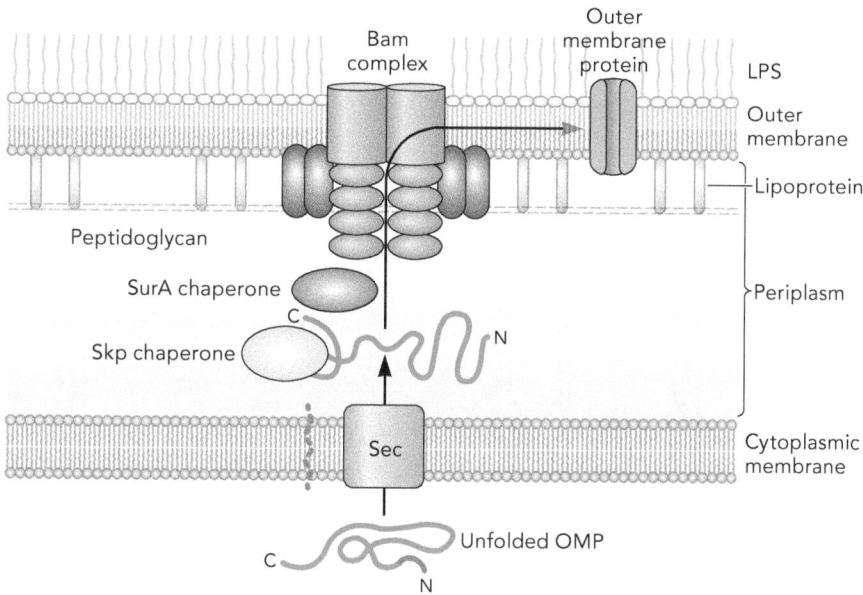

FIGURE 14.15 **Insertion of outer membrane proteins in Gram-negative bacteria.** Outer membrane proteins (OMPs) are transported to the periplasm in a secretory (Sec)-dependent manner and peri-plasmic chaperones (Skp and SurA) facilitate their translocation through the Bam complex (see type V secretion system). Integral outer membrane proteins (e.g., porins) then move laterally from the Bam complex into the outer membrane.

bind to the protein and escort it to the Bam complex in the outer membrane (Fig. 14.15) (see also T5SS above). Another chaperone (SurA) then helps the Bam complex fold the protein and insert it into the outer membrane.

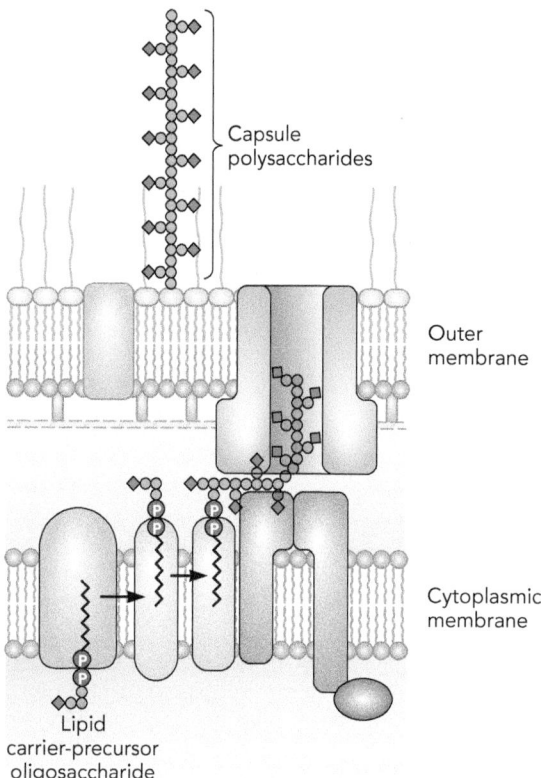

FIGURE 14.16 **Production of the capsule in bacteria.** In the one mechanism shown, precursor oligosaccharides destined for the capsule layer are attached to lipid carriers to facilitate transport across the cytoplasmic membrane into the periplasm. The oligosaccharides are then polymerized in the periplasm and secreted to the cell surface.

CAPSULE

Two different strategies are used by microbes to synthesize and transport capsule polysaccharides. In some microorganisms, the oligosaccharide precursors are synthesized in the cytoplasm and secreted as preformed polymers, possibly through the activity of ABC transporters. In other microbes, the precursors are secreted and then polymerized by transport proteins or by extracellular enzymes. This second mechanism requires a specialized secretory complex and lipid carrier molecules to facilitate their transport across the cytoplasmic membrane (Fig. 14.16). Similar mechanisms are used to export the polysaccharide component of extracellular polymeric substance (EPS) from cells in biofilms.

Learning Outcomes: After completing the material in this chapter, students should be able to . . .

1. compare and contrast the signal, accessory proteins, and energy sources utilized by the three cytoplasmic membrane protein transport systems: secretory (Sec)-dependent, signal recognition particle (SRP)-dependent, and twin-arginine translocation (Tat) pathways

2. outline the process of protein secretion or localization of a protein via the Sec-dependent, SRP-dependent, and Tat protein transport systems

3. predict the orientation of a cytoplasmic membrane protein from the membrane-spanning domains in its sequence

4. state key characteristics of the type I to VI secretion systems that are associated with transport through the Gram-negative outer membrane

5. describe how the oxidation state of the cellular environment and the presence of disulfide bonds impact protein folding

6. explain how components of the cell envelope move from the cytoplasm to their final locations in the cell

Check Your Understanding

1. Define this terminology: Sec-dependent protein transport, signal sequence (also known as leader sequence or leader peptide), signal peptidase cleavage site, chaperone, chaperonin, translocase, SRP-dependent protein transport, Tat protein transport system, stop transfer sequence, secretin, helper (translocation) domain, passenger domain, injectisome

2. The purpose of this table is to compare and contrast the most common protein transport systems in bacteria. Answer each question below with respect to the secretory (Sec)-dependent, signal recognition particle (SRP)-dependent, and twin-arginine translocation (Tat) protein transport systems. (LO1)

System	What is the overall general function?	Is transport posttranslational or cotranslational?	Which sequence controls the initiation of transport?	Which required accessory proteins facilitate transport?	What energy source drives transport?
Sec					
SRP					
Tat					

3. The secretory (Sec)-dependent transport system is complex, with many components. (LO2)

   a) Polypeptides being transported through the Sec-dependent transport system require a signal sequence. What is the function of each part of the signal sequence listed below?
      i. Basic region of the signal sequence
      ii. Hydrophobic core
      iii. Peptidase cleavage site

   b) What is the function of each accessory component of the Sec-dependent transport system?
      i. Chaperone (SecB)
      ii. ATPase (SecA)
      iii. Translocase (SecYEG)
      iv. Signal peptidase (LepB)

4. Put the steps of the process of secretory (Sec)-dependent transport in order from 1 (first) to 6 (last). (LO2)

   _____ ATP binds SecA and the hydrophobic core enters into the membrane.
   _____ SecA binds to the polypeptide complex.
   _____ The polypeptide complex is delivered to the SecYEG translocase.
   _____ SecB recognizes and binds to a translated polypeptide.
   _____ The polypeptide is released into the periplasm.
   _____ The signal peptidase binds and cleaves the signal sequence.

5. What is the function of each component of the signal recognition particle (SRP)-dependent protein transport system listed below? (LO2)

   a) SRP
   b) N-terminal hydrophobic signal
   c) Cytoplasmic membrane receptor (FtsY)

6. Put the steps of the process of signal recognition particle (SRP)-dependent protein transport in order from 1 (first) to 5 (last). (LO2)

   _____ The FtsY complex associates with translocase.
   _____ The polypeptide is inserted into the transport channel.
   _____ The polypeptide-ribosome/mRNA complex binds to the receptor on the cytoplasmic membrane.
   _____ The SRP binds the recognition signal in the N-terminal recognition sequence.
   _____ Translation of the polypeptide to be transported begins.

7. Describe the function of each component of the twin-arginine translocation (Tat) protein transport system listed below. (LO2)

   a) N-terminal amino acid sequence with two Arg residues
   b) TatBC
   c) TatA

8. The positioning of proteins in a membrane is dependent on stop transfer sequences and noncleavable and cleavable signal sequences. On the diagram shown, indicate the position of the **stop transfer sequences** and the **noncleavable signal sequences**. Which polypeptide must have had a **cleavable signal sequence**? (LO3)

9. The purpose of the table below is to compare and contrast the various systems for protein transport through the Gram-negative outer membrane. Answer each question with respect to the six protein transport systems discussed in the chapter. (LO4)

System	Is it secretory (Sec)-dependent or -independent?	Is the polypeptide moved to the periplasm first before moving across the outer membrane or across both membranes in one step?	Is the protein folded in cytoplasm, periplasm, or outside the cell?	What energy source drives this process?
Type I				
Type II				
Type III				
Type IV				
Type V				
Type VI				

10. Match each type of outer membrane transport system (I to VI) to the statement that best describes it. (LO4)

   Helper domain facilitates autotransportation of the polypeptide passenger domain: _____

   Is related to bacterial conjugation systems: _____

   Resembles a bacteriophage: _____

   Uses a pseudopilus to push the protein through the transport channel: _____

   Utilizes an injectisome to transport effector polypeptides: _____

   Consists of an ABC transporter, membrane fusion protein, and outer membrane protein: _____

11. In polypeptides with cysteine residues, the polypeptide folding is influenced by the relative reducing conditions of the environment. (LO5)

    a) Under which conditions (oxidized or reduced) do disulfide bonds form?

    b) Where (the cytoplasm or outside the cell) are disulfide bonds more likely to form?

    c) How does the formation of disulfide bonds impact protein folding?

12. Bacterial cell wall components are made in the cytoplasm and must be transported through the cytoplasmic membrane. (LO6)

    a) What is the function of the undecaprenyl phosphate cycle in the process of peptidoglycan biosynthesis? What cell wall precursor is transported across the cytoplasmic membrane?

    b) What does PBP stand for? Why are they called PBPs? What two functions do PBPs carry out for the cell?

13. Put the following steps relating to peptidoglycan biosynthesis in order from 1 (first) to 6 (last). (LO6)

    ___ N-Acetylglucosamine (NAG) and N-acetylmuramic acid (NAM) are made in the cytoplasm.
    ___ NAG and NAM are linked to uridine diphosphate (UDP) nucleotide-linked precursors.
    ___ NAG is added to UDP-NAM-pentapeptide.
    ___ The disaccharide-pentapeptide molecules are translocated across the cytoplasmic membrane.
    ___ Transpeptidase enzymes cross-link the peptide side chains of the peptidoglycan strands.
    ___ UDP-NAM-pentapeptide associates with a lipid carrier molecule (undecaprenyl phosphate).

Dig Deeper

14. The positioning of proteins in a membrane is dependent on stop transfer sequences and noncleavable and cleavable signal sequences. (LO3)

    a) On the diagram shown, label the **stop transfer sequences** and the **noncleavable signal sequences**. Draw in any **cleavable signal sequences**.

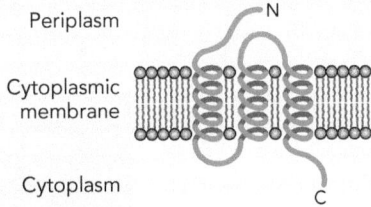

    b) For each sequence, indicate how the polypeptide would be positioned in the membrane, including the N- and C-termini.

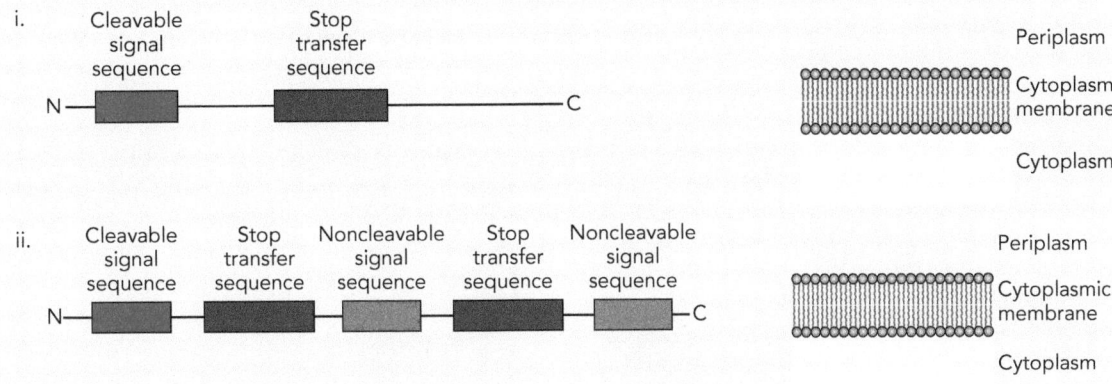

iii. Predict if and how the orientation would change if the cleavable signal sequence in part i were replaced with a noncleavable signal sequence.

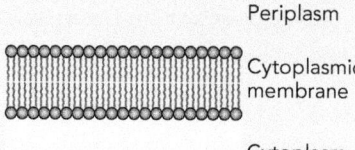

Periplasm

Cytoplasmic membrane

Cytoplasm

15. Because *Escherichia coli* grows well and can be genetically manipulated, it is often used as a "bacterial factory" to produce proteins for industrial and medical applications. A major challenge in this process is getting the protein out of the bacterial cell into the growth medium. The secretion of target proteins into the medium simplifies product recovery, making the purification and downstream processing steps significantly easier. (LO1, 2)

You are working for a biotech company that wants to produce a protein of interest. Your project is to get *E. coli* to make the protein of interest and transport it out of the cell. To get *E. coli* to secrete the protein of interest, you construct a recombinant fusion gene with the sequence to a specific signal peptide attached to the amino-terminal end of the gene for the protein of interest. You decide to work with a type II secretion system that uses Sec-dependent transport.

a) What are the three critical parts of a cleavable signal sequence for a type II secretion system? Describe the traits of each.

b) You engineer the gene encoding the protein of interest to have a signal sequence called SS1 (amino acids listed below). You introduce the engineered gene into your strain of *E. coli* (brackets indicate the three regions).

SS1: Met-[Lys-Asp-Glu]-[Thr-Ala-Ala-Ile-Ile-Ala-Val-Ala-Val-Ala-Gly-Phe-Ala-Thr-Val]-[Ala-Gln-Gln]

Your early experiments indicate a transport efficiency of 15% (that is, 15% of the protein made by *E. coli* is transported outside the cell). Not satisfied, you try again with the sequence SS2 (amino acid changes compared to SS1 are in blue).

SS2: Met-[Lys-Asp-Glu]-[Thr-Ala-Ala-Ile-Ile-Ala-Val-Asp-Gln-Asp-Gly-Phe-Ala-Thr-Val]-[Ala-Gln-Gln]

Would you expect the protein export efficiency of SS2 to increase, decrease, or stay the same, relative to SS1? Explain your reasoning. *(Hint: Consider the properties of different amino acid residues.)*

c) Still not satisfied, you try again with the sequence SS3 (amino acid changes compared to SS1 are in blue).

SS3: Met-[Lys-Lys-Arg]-[Thr-Ala-Ala-Ile-Ile-Ala-Val-Ala-Val-Ala-Gly-Phe-Ala-Thr-Val]-[Ala-Gln-Gln]

Would you expect the protein transport efficiency of SS3 to increase, decrease, or stay the same, relative to SS1? Explain your reasoning.

16. Penicillin kills susceptible bacteria by binding to and inhibiting transpeptidase (PBP) enzymes. (LO6)

a) Indicate on the diagram where the transpeptidase enzymes make new bonds.

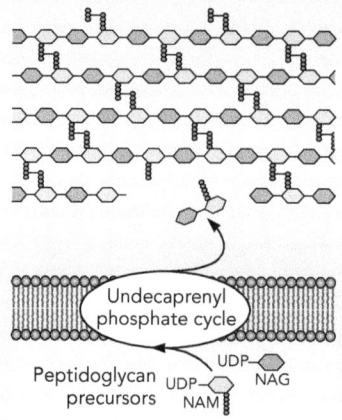

Undecaprenyl phosphate cycle

UDP–NAG

Peptidoglycan precursors

UDP–NAM

b) Why would inhibiting transpeptidase (PBP) enzymes kill these cells?

c) You grow bacterial cells in a constant amount of penicillin but with varying growth rates and measure the amount of cell lysis (see graph). Which cells are more affected, rapidly growing or nongrowing cells? Explain these results.

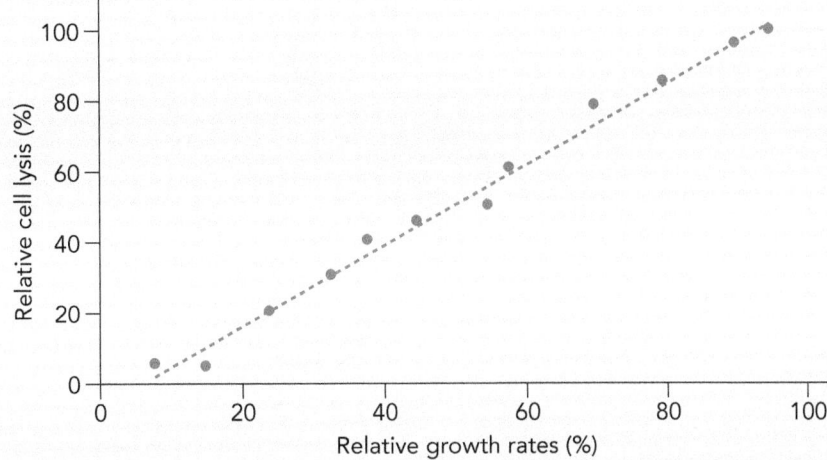

d) The study of mutant bacteria that cannot make critical amino acids (auxotrophs) has provided much of our insight into fundamental biological pathways. However, it is very difficult to select for these mutants because they don't grow in medium without the addition of the amino acid in question. You are interested in isolating a proline auxotroph of *Escherichia coli*.

   i. You first grow cells in proline-free medium. Which type of cells, wild type or proline auxotroph, would grow in this medium? Explain why.

   ii. You then add penicillin to your medium and incubate for a day. Which type of cells, wild type or proline auxotroph, would be killed in this medium and which would survive? Explain your reasoning.

   iii. You then plate the cells that survived part ii on agar plates to search for auxotrophic mutants. Which type of medium should you use: growth medium with proline or without proline? Should you add penicillin or not? Explain your reasoning.

   iv. In this experiment, how could you differentiate the auxotrophs from any wild-type cells that might be in the mix? Explain your reasoning.

e) The central structure of penicillin consists of a β-lactam ring (in red). Penicillin-resistant bacteria are fairly common, and there are a number of mechanisms that can confer resistance.

   i. Suppose a bacterium acquired a gene encoding an enzyme that breaks the β-lactam ring structure. Would this likely confer resistance to penicillin? Explain why or why not.

   ii. Would a mutation that altered the permeability of the cytoplasmic membrane be likely to confer resistance to penicillin? Explain why or why not.

   iii. Describe a mutation in the transpeptidase (PBP) enzyme itself that could confer penicillin resistance.

17. Vancomycin is an alternative antibiotic used when bacterial pathogens are resistant to β-lactam antibiotics like penicillin. Vancomycin works by **binding** with high affinity **to the D-Ala-D-Ala** termini of the NAG-NAM-pentapeptide precursor units. (LO6)

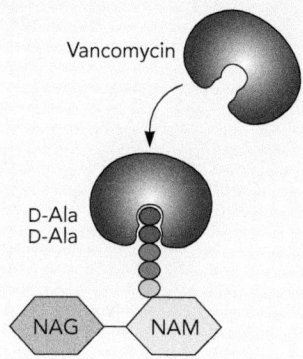

a) Why would binding to the D-Ala-D-Ala terminus kill these cells?

b) You have access to the same kind of smooth and rough cells Griffith used in his transformation experiments (Chapter 13). Predict which cells, rough or smooth, would be more resistant to vancomycin. Explain your reasoning.

c) Describe a mutation that could confer resistance to vancomycin in a bacterial cell.

18. Answer the following questions with regard to Gram-negative protein secretion systems (types I to VI) (LO1, 4)

a) Bacterial wilt in corn is a serious agricultural problem caused by the bacterium *Pantoea stewartii*. This Gram-negative bacterium has a type III secretion system that exports an effector protein called WtsE. WtsE allows the bacteria to colonize plant xylem, where the bacteria grow and form biofilms that block sap flow, causing leaf wilting. Predict whether each of the following mutations in *P. stewartii* would decrease pathogen virulence compared to wild type. Explain your reasoning.

   i.   A mutation in the N-terminal region of the effector protein, WtsE

   ii. Deletion of the gene encoding SecA

   iii. A mutation in the gene for the translocon protein

   iv. Deletion of a gene encoding a protein required for extracellular polymeric substance (EPS) biosynthesis

b) *Aeromonas hydrophila* is a Gram-negative opportunistic pathogen of fish that secretes an array of virulence proteins into the surrounding environment. One key virulence factor is a hemolysin (an enzyme that cleaves red blood cells [RBCs]) that has contributed to *A. hydrophila* outbreaks in channel catfish ponds, causing huge annual losses in catfish production.

   i.   You set up a series of experiments to test for hemolytic activity (lysis of RBCs) under these conditions:

        • Exp. 1: grow wild-type *A. hydrophila* cells in medium with added RBCs

        • Exp. 2: grow wild-type *A. hydrophila* cells in medium, spin down the cells, and mix just the supernatant with RBCs

        • Exp. 3: grow wild-type *A. hydrophila* cells in medium, lyse the bacterial cells, then add RBCs

The bars in the graph indicate the level of hemolytic activity observed in each experiment. Explain the results for each experiment.

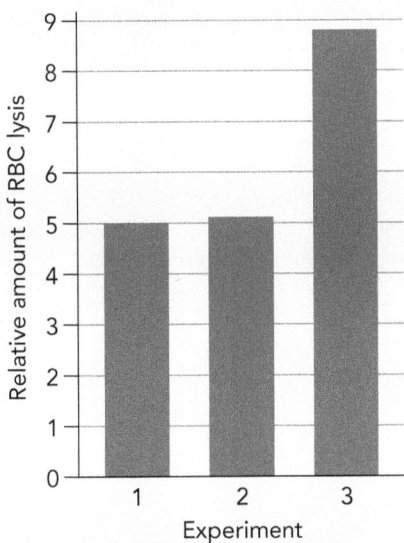

ii. You hypothesize that the transport system involved is a type II protein export system. To test this hypothesis, you create a strain in which the *exeD* gene has been deleted; *exeD* codes for secretin. Predict the results for each experimental condition assuming your hypothesis is true.

19. A type IX secretion system has recently been found in bacteria from the Gram-negative *Fibrobacteres-Chlorobi-Bacteroides* superphylum. In some pathogens, the type IX system plays a critical role in pathogenicity as it transports multidomain virulence proteins (100 to 650 kDa in size) that fold in the periplasm prior to their transport across the outer membrane. (LO4)

   a) Would you predict that the periplasmic proteins are transported in a secretory (Sec)-dependent or Sec-independent manner? Explain your reasoning.

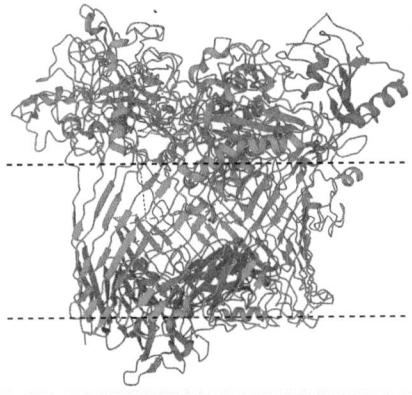

   b) This type IX secretion system has an outer membrane translocon channel (SprA), a 267-kDa protein with 36 membrane-spanning domains that create a β-barrel structure (shown in green; PDB ID 6H3J, https://doi.org/10.2210/pdb6H3J/pdb). Compare SprA to the lactose permease (LacY), which spans the membrane 12 times. Would you predict that the channel created by SprA is larger or smaller? Explain your reasoning.

   c) The SprA protein has a "plug" on the inner side of the outer membrane called the Plug protein (shown in purple). Use this information to interpret the following data.

Strain	Relative levels of secreted proteins	Resistance to vancomycin
Wild type	High	High
sprA deletion	Low	High
Plug gene deletion	High	Low

i. Are SprA and Plug required for protein transport? Explain your reasoning.

ii. How does vancomycin work to kill cells? Propose reasons for why the wild-type and *sprA* deletion strains are resistant to vancomycin, while the strain lacking Plug is sensitive.

364 Making Connections

364 Motility in Microorganisms

364 Bacterial Flagella and Swimming Motility

367 Regulation of Flagellar Synthesis in *Escherichia coli*

369 Mechanism of Swimming Motility

370 Archaeal Flagella

371 Bacterial Surface Motility

372 Chemotaxis

380 Conservation and Variation in Chemotaxis Systems among Bacteria and Archaea

381 Methods to Study Bacterial Motility and Chemotaxis

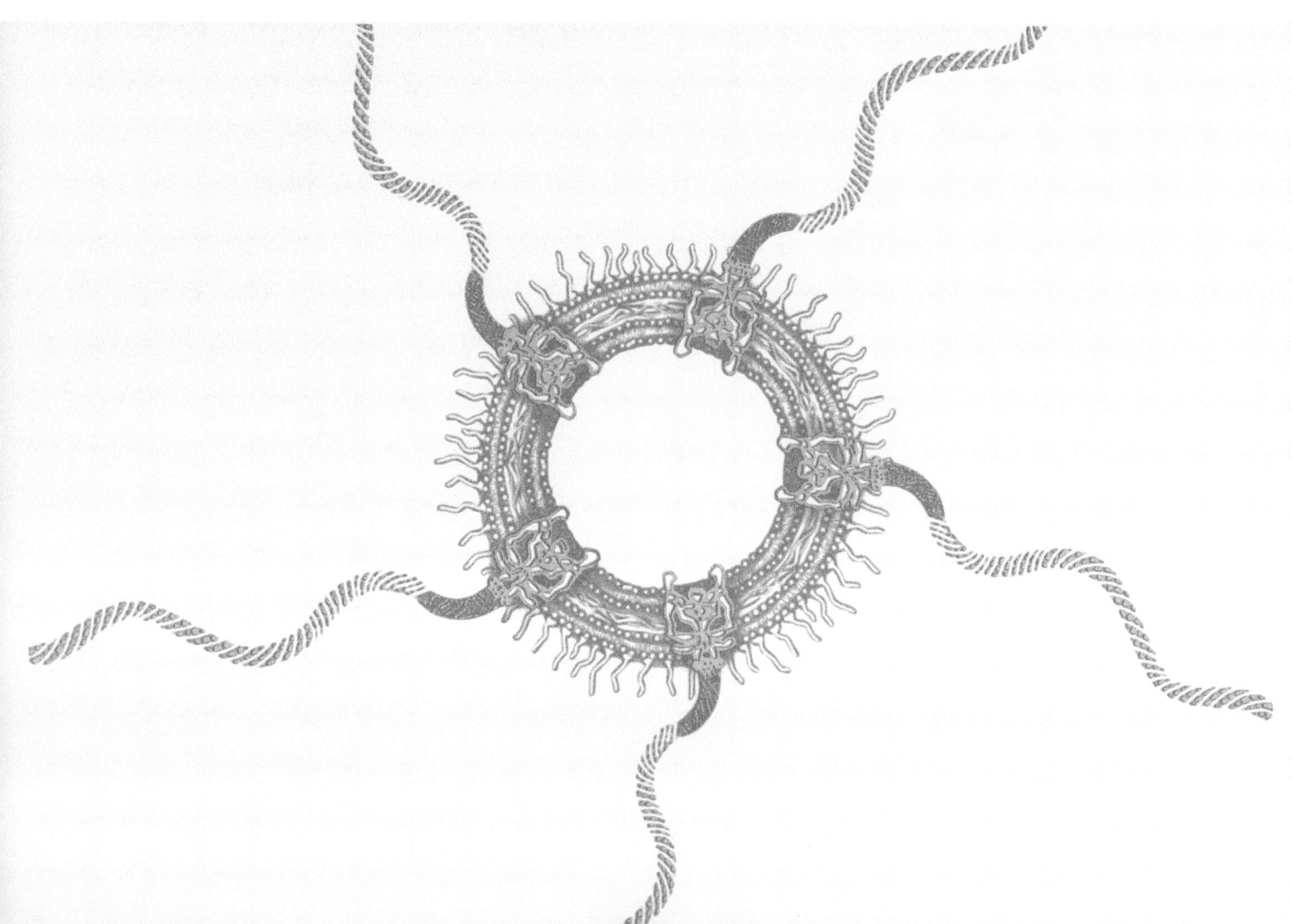

15

MICROBIAL MOTILITY AND CHEMOTAXIS

After completing the material in this chapter, students should be able to . . .

1. compare and contrast the different types of motility in *Bacteria* and *Archaea*

2. describe how the various parts of bacterial flagella function to produce swimming motility

3. explain how the synthesis of a bacterial flagellum is temporally regulated

4. discuss how chemotaxis allows bacterial and archaeal cells to respond to chemical concentrations in their environment

5. interpret experiments exploring bacterial motility and chemotaxis

Microbial Physiology: Unity and Diversity, First Edition. Ann M. Stevens, Jayna L. Ditty, Rebecca E. Parales and Susan M. Merkel.
© 2024 American Society for Microbiology.
Companion website: www.wiley.com/go/stevens/microbialphysiology

Making Connections

Motility and chemotaxis play a key role in the establishment of symbiotic relationships with plants (Chapter 12) and bacterial pathogenesis, including biofilm formation (Chapter 18). The mechanisms of protein transport and localization covered in Chapter 14 are intimately linked to the proteinaceous appendages used for some forms of motility that are discussed in this chapter. Bacterial flagella are structurally similar to type III secretion systems, while archaeal flagella and bacterial type IV pili are more similar to type II secretion systems. In *Escherichia coli*, the synthesis of flagella is tightly controlled through a master regulator and the temporal regulation of a sigma factor (Chapter 8). The direction of rotation of the flagella is under the control of a complex two-component regulatory system and posttranslational regulation of proteins (Chapter 7). Flagellar rotation is driven by energy provided by the proton motive force (PMF) in bacteria or by ATP in archaea (Chapter 5). Thus, coordination of physiological processes associated with bioenergetics, protein and gene regulation, and protein secretion are closely linked to enable the motility and chemotaxis of microbes in response to their environmental conditions.

Motility in Microorganisms

While motility is not ubiquitous among microorganisms, it most certainly provides a survival advantage, as some microbes invest a great deal of biological energy to synthesize and power the machinery that drives motility. Motile cells can respond to their environment through a highly regulated process. The best-understood forms of motility through liquid involve one flagellum or multiple flagella. Eukaryotic flagella are flexible, whiplike structures, whereas bacterial flagella are rigid helical structures that rotate in a manner similar to a boat propeller. Archael flagella are structurally and functionally distinct from both bacterial and eukaryotic flagella.

Some bacteria can move across a solid surface using gliding and/or twitching motility. Many bacteria capable of gliding motility have no obvious cell surface appendages. Exceptions are the myxobacteria and filamentous cyanobacteria (Chapter 18), which have type IV pili-based gliding motility systems. Type IV pili are also involved in twitching motility, which generates a jerky motion, hence the term "twitching." This type of movement is common in the pseudomonads (discussed later in this chapter).

Bacterial Flagella and Swimming Motility

Bacterial flagella are similar in structure to type III secretion systems (Chapter 14), with a basal body fixed in the cell envelope and a helical filament that extends well beyond the cell surface (Fig. 15.1). Bacterial species capable of swimming in liquid each have a characteristic number of flagella (generally between 1 and 10), which are located at certain positions on the cell surface. **Polar flagella** (Fig. 15.2A) are located at one end of a cell (as with *Vibrio* species), while bipolar flagella are found on both poles of the cell (as with some spirilla). Bacteria with **monotrichous** flagella have a single flagellum at one pole. Bacteria

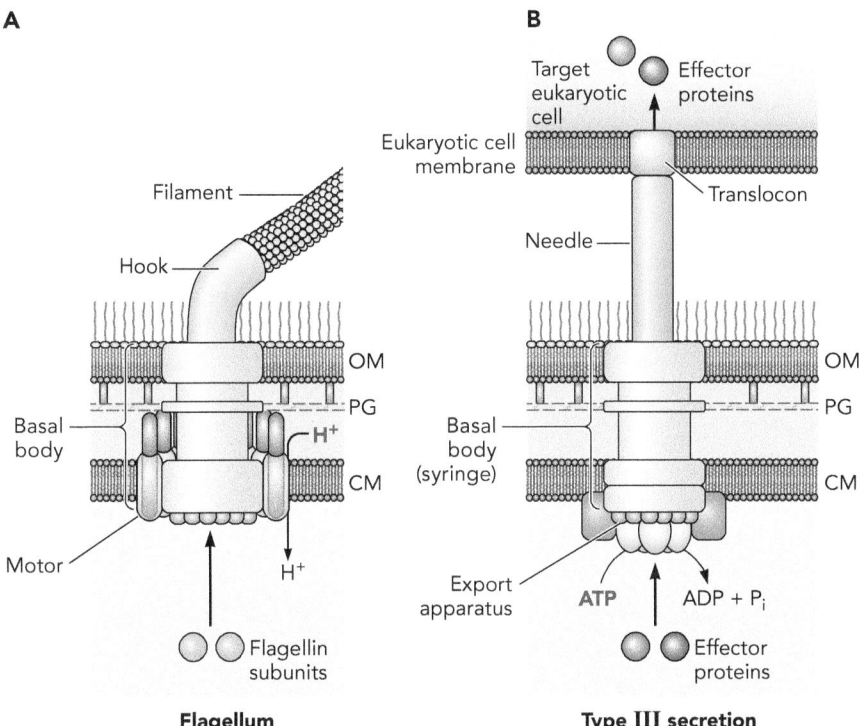

FIGURE 15.1 **The structures of the bacterial flagellum and type III secretion systems (T3SSs) are made of homologous components.** Homologous protein components form a basal body that anchors the bacterial flagellum and the T3SS to the bacterial cell envelope. (A) The hollow basal body, hook, and filament of the flagellum provides a route for flagellin monomers to traverse the cell envelope for polymerization at the tip. (B) In T3SSs, a needle extends from the basal body and penetrates the cytoplasmic membrane (CM) of the target eukaryotic cell. Effector proteins move through the hollow needle and are released through the translocon into the target cell (Chapter 14). The proton motive force provides the energy needed to rotate the flagellum, whereas ATP hydrolysis powers secretion by the T3SS. OM, outer membrane; PG, peptidoglycan.

with **lophotrichous** flagella (Fig. 15.2B) have a tuft of flagella at one pole of the cell (e.g., *Pseudomonas putida*), whereas amphitrichous flagella are located at both poles. Many bacteria (including *E. coli*, *Salmonella* spp., and *Bacillus subtilis*) have **peritrichous** flagella that are randomly distributed over the cell surface (Fig. 15.2C). Some bacteria, like *Vibrio* species and *Helicobacter pylori*, have sheathed flagella. The sheath is an extension of the outer membrane of these Gram-negative bacteria. A unique characteristic of the spirochetes is the presence of **endoflagella** (Fig. 15.2D), which are located within the periplasm. Endoflagella allow spirochetes to easily move in a drill-like manner through viscous environments such as sediments or between host tissue cells. With both endoflagella and sheathed flagella, the flagellar proteins are not exposed on the cell surface.

Bacterial flagella, which have a diameter of ~20 nm, cannot be seen with the resolution of a light microscope unless they are stained with a mordant that thickens the flagella. Alternatively, live cells can be visualized under dark-field microscopy with a high-intensity light source that makes the cell and flagella appear white against a dark background.

The bacterial flagellum (Fig. 15.3) is a complex proteinaceous structure with many genes and gene products required for its formation. The basal body and hook create a protein transport channel through the cell envelope that permits

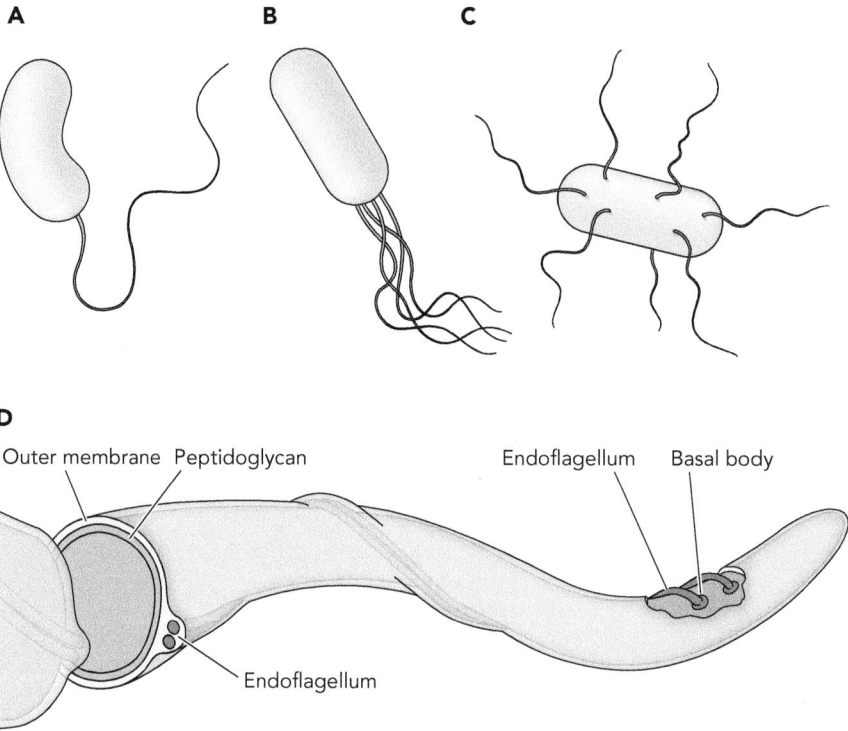

A

B

C

D

Outer membrane Peptidoglycan   Endoflagellum   Basal body

Endoflagellum

FIGURE 15.2 **Arrangement of bacterial flagella.** (A) Polar; (B) lophotrichous (tuft of polar flagella); (C) peritrichous; and (D) periplasmic endoflagella (red) in spirochetes.

synthesis of the filament on the cell surface. The filament is composed of flagellin subunits assembled in a rigid helical structure that grows from the tip. The flagellin monomers move through the central channel of the basal body, hook, and growing filament, and are polymerized at the tip of the filament (Fig. 15.3A). The basal body contains multiple proteinaceous "rings." The M/S ring is closely associated with and traverses the cytoplasmic membrane. The C ring in the cytoplasm serves as the switch controlling the direction of rotation of the flagellum (i.e., clockwise [CW] or counterclockwise [CCW]). Gram-negative bacteria have a P ring, which is associated with the peptidoglycan layer, and an L ring, which is associated with the lipopolysaccharide (LPS). A rod extends through the center of the rings, functioning like an axle in a car, and is connected to the hook. The flagellar filament extends beyond the hook and it is made of one to three types of flagellin proteins, depending on the bacterial species. The tip of the filament is capped off by a specific protein that prevents loss of flagellin (and the PMF) from the end of the hollow filament and helps to ensure that the flagellin is properly assembled at the growing tip.

The M/S ring is associated with the motor proteins (MotA and MotB) that serve as the "force-generating units" producing rotation of the flagellum and driving movement. MotA is an ion channel located adjacent to the M/S ring. MotB acts as a stationary (stator) complex that serves as an anchor to the peptidoglycan cell wall and enables thrust to be generated. Between 12 and 16 MotA/B complexes are arranged around the M/S ring to form the motor (Fig. 15.3B). The movement of ions (H^+ or in some cases Na^+) through the MotA

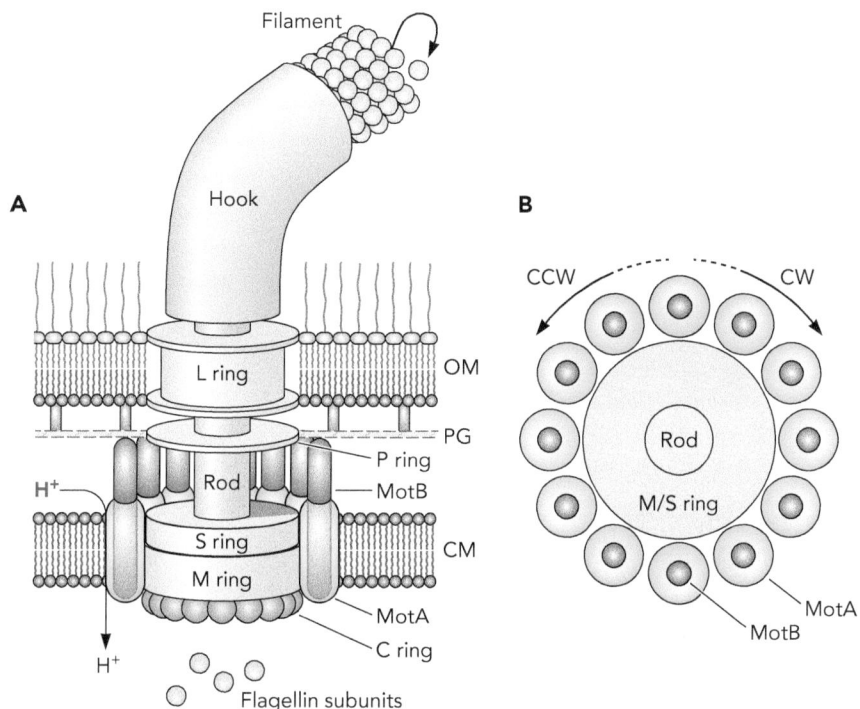

FIGURE 15.3 **Structure of the bacterial flagellum in a Gram-negative bacterium.** (A) Side view of the flagellum. The basal body is composed of several different proteins and is anchored to the cytoplasmic membrane (CM), peptidoglycan (PG) layer, and outer membrane (OM) by the M/S ring, P ring, and L ring, respectively. The hook attaches the flagellar filament to the basal body. Flagellin monomers produced in the cytoplasm pass through the hollow basal body and the flagellar filament is polymerized at the tip, driven by the proton motive force (PMF). Flagellar rotation is powered by the PMF as protons enter the cell through the flagellar motor proteins, MotA and MotB. The C ring functions as the "switch" that changes the direction of flagellar rotation. (B) Top view showing the arrangement of the motor proteins MotA and MotB and the directions of flagellar rotation. CW, clockwise; CCW, counterclockwise.

channel causes the flagellum to rotate (like "ticks" in a clock). The direction of rotation (CW or CCW) is controlled by the C ring, which functions as the switch.

Regulation of Flagellar Synthesis in *Escherichia coli*

Proper assembly of the flagellum requires ~40 genes in *E. coli*. Approximately half of the genes encode structural proteins and the remainder encode regulatory proteins. The flagellum is built in a precise order from the inside out, and thus the requisite proteins must be synthesized and added at the appropriate time (Fig. 15.4). The cell produces the relevant flagellar components as they are needed, since large amounts of energy and precursors are expended in the process. Thus, flagellar synthesis is under tight temporal regulation, with the process divided into three stages that require the expression of early, middle, and late genes. The process begins with the expression of the early genes encoding the master regulator FlhDC, which is required for activation of the entire flagellar regulon (Fig. 15.4A). Once produced, the master regulator functions along with the housekeeping σ^{70} RNA polymerase (σ^{70} RNAP) holoenzyme to express genes from 10 different operons, known as the middle genes, producing flagellar proteins for the M ring and S ring (known as the M/S ring) and the C ring, in

A. Early gene operon

Master regulator

σ^{70}

FhIDC

D C

B. Middle gene regulon

Flagellar assembly

D C σ^{70}

M/S/C ring (4 operons)

1 M ring inserts in membrane **2** S ring added **3** C ring added

D C σ^{70}

σ^{28} and FlgM (2 operons)

D C σ^{70}

rod, P/L ring, hook (4 operons)

4 Rod added and capped **5** P ring added **6** L ring added

7 Hook added

σ^{28} FlgM Inactive

8 Hook finished

FlgM

σ^{28} Active

FlgM

C. Late gene regulon

Late flagellar assembly

D C σ^{28}

Flagellin/cap (2 operons)

9 Flagellar filament, flagellin subunits added

D C σ^{28}

Motor (1 operon)

D C σ^{28}

Chemotaxis genes (5 operons)

Flagellin

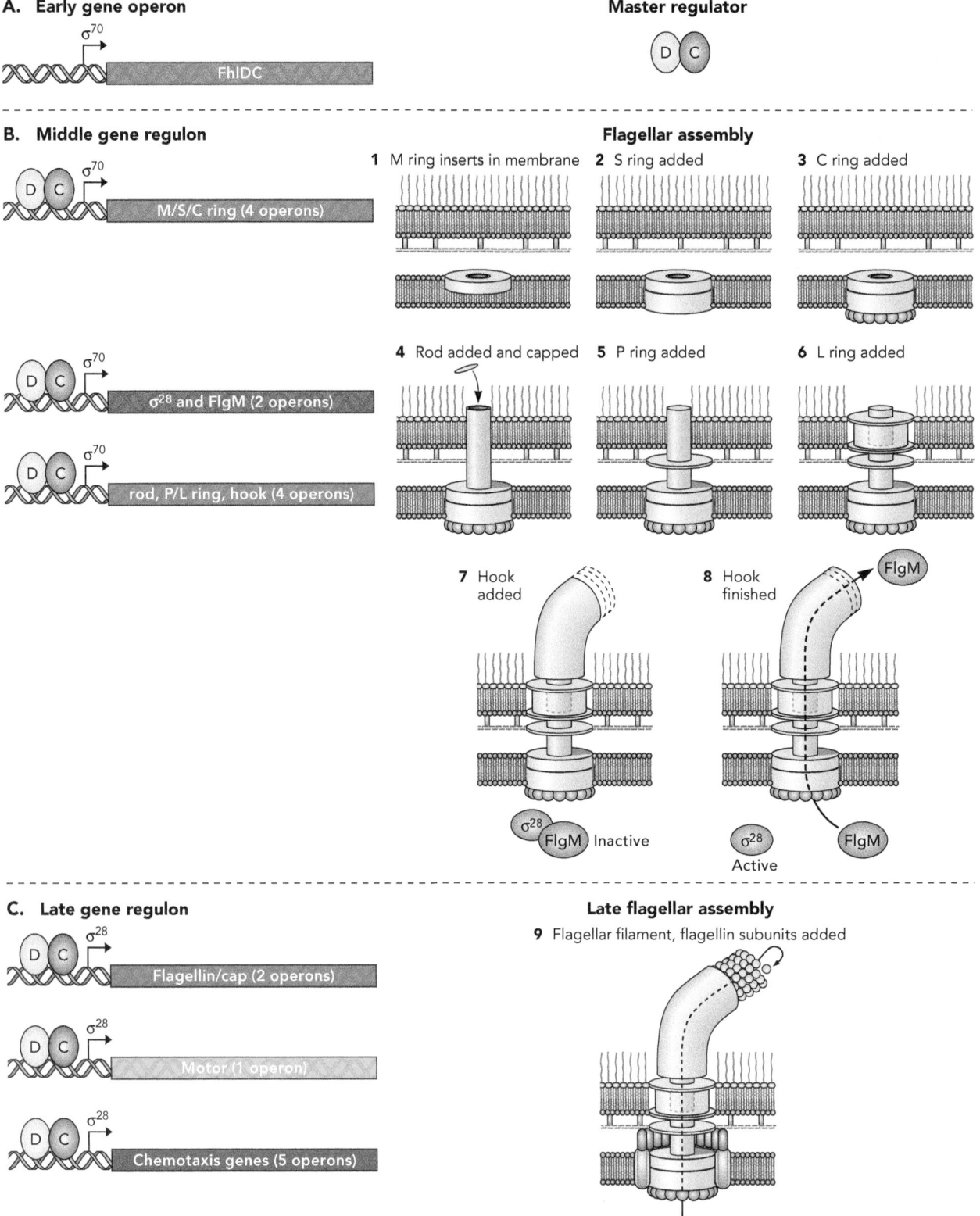

that order (Fig. 15.4B). These components of the flagellum have homology to type III secretion systems, and this hollow structure must be formed first so that the other polypeptides can be transported across the cytoplasmic membrane during further flagellar synthesis steps. Once the first components of the flagellum are assembled, FlhDC and σ^{70} RNAP then activate transcription of the gene regulon encoding the rod, P and L rings, hook, and two important regulatory proteins, FliA and FlgM. FliA is σ^{28}, which, when combined with RNAP, is responsible for the expression of the late flagellar genes. FlgM is an **anti-sigma factor** that inactivates σ^{28} until the middle flagellar assembly is complete (Fig. 15.4B). This regulatory mechanism ensures that the flagellar filament is not produced until after the basal body and hook are in place. The late genes are only expressed under the direction of FlhDC and σ^{28} RNAP. Once hook assembly is complete, FlgM is secreted outside of the cell, releasing σ^{28} in its active form. In the presence of FlhDC, σ^{28} RNAP transcribes the late operons to complete the flagellar structure (Fig. 15.4C). The late genes also include genes in five different operons that encode the machinery for chemotaxis, which will be described below.

Mechanism of Swimming Motility

The mechanism of *E. coli* swimming motility is not universal, but it serves as a good model for understanding how bacteria move using flagella. *E. coli* cells have 5 to 10 helical flagella per cell in a peritrichous arrangement. The biological energy for rotation comes from the PMF; a transmembrane gradient of H^+ is converted into physical movement and force. The bacterial flagellar motors are the smallest biological motors known to exist. When all the flagella rotate CCW, they bundle together, allowing the cell to swim forward at speeds in the range of ~10 to 20 cell lengths per second (Fig. 15.5A). When the direction of flagellar rotation switches to CW, this causes the flagellar bundle to fly apart and the cell will exhibit **tumbling behavior**. During the tumble, Brownian motion serves to randomly reorient the cell so that it swims in a random direction once the flagella return to CCW rotation.

In contrast, some bacteria (e.g., *Pseudomonas aeruginosa*) with a single polar flagellum typically move forward during CCW rotation and reverse direction during CW rotation (Fig. 15.5B). Other monotrichous bacteria (e.g., *Rhodobacter sphaeroides*) move by rotating their flagellum in one direction with intermittent

FIGURE 15.4 **Biosynthesis of a flagellum in *Escherichia coli*.** (A) Early genes encoding the FhlDC master regulator are expressed in response to environmental signals. (B) The middle genes, under control of FhlDC and σ^{70} RNAP, are expressed next. First, the M, S, and C ring proteins are produced. The M and S rings are inserted into the cytoplasmic membrane and the C ring is then added. Then the genes encoding proteins that make up the rod, P and L rings, and hook are expressed and the proteins are assembled into the flagellar structure in that order. The middle genes also encode σ^{28} and its anti-sigma factor, FlgM. FlgM remains bound to σ^{28} until the rod, P and L rings, and hook assembly are completed. Once complete, FlgM is released to the outside of the cell through the hollow basal body structure. (C) σ^{28}, now in its active form, associates with RNAP and works with FhlDC to transcribe the late genes, which encode flagellin, the motor proteins, and the chemotaxis machinery. Flagellin monomers pass through the hollow basal body and are polymerized at the tip of the flagellum, which has a cap protein at the end. Once the flagellar filament is complete, the motor proteins MotA and MotB are added, completing the structure.

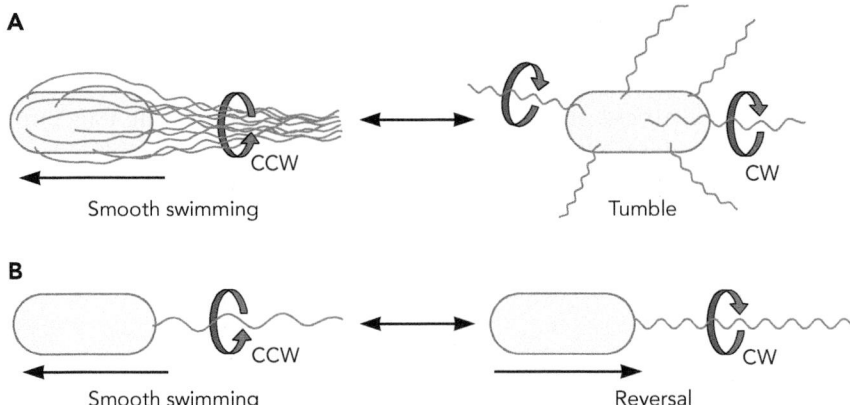

FIGURE 15.5 **Movement of peritrichously and polarly flagellated bacterial cells**. (A) When peritrichous bacteria like *Escherichia coli* are swimming smoothly, the flagella rotate together as a bundle with counterclockwise (CCW) rotation. When the direction of flagellar rotation changes to clockwise (CW), the structure of each helical filament changes, disrupting the tight flagellar bundle. As a result, the filaments in the bundle fly apart, causing the cell to tumble. (B) A cell with a single polar flagellum (like *Pseudomonas aeruginosa*) exhibits smooth swimming behavior when its flagellum is rotating CCW. When the direction of rotation changes to CW, the cell reverses direction for a short time.

pauses in flagellar rotation, which permits reorientation via Brownian motion before they resume swimming.

Archaeal Flagella

Although many archaea are capable of swimming motility, their flagella are completely different from bacterial flagella. Rather, these external cell structures are evolutionarily related to both type II secretion systems and type IV pili (Fig. 15.6; Chapter 14 and below). To emphasize the difference between the bacterial flagellum and archaeal flagellum, an alternative name— **"archaellum"**—has been proposed for the latter. Archaella (Fig. 15.6A) are ~10 nm in diameter and are made up of glycosylated protein subunits that are homologous to the pilin proteins that make up type IV pili (Fig. 15.6B). These "archaellins" are synthesized as precursor proteins with an N-terminal signal peptide and inserted into the cytoplasmic membrane. Addition of the archaellins to the base of the archaellum requires ATP hydrolysis and removal of the signal peptide by a specific peptidase that is homologous to type IV prepilin peptidases. Archaella, like bacterial flagella, rotate as a bundle to push the cell through liquid environments. Reversing the direction of rotation "pulls" the cell in the opposite direction. This is in contrast to type IV pili, which do not rotate, but rather extend and retract during twitching or gliding motility (see following section). In addition, unlike bacterial flagella but similar to type IV pili, motility mediated by archaella is powered by ATP hydrolysis. Interestingly, although their motility systems are completely different, archaea and bacteria use a conserved chemotaxis system to sense and respond to environmental signals (discussed later in this chapter).

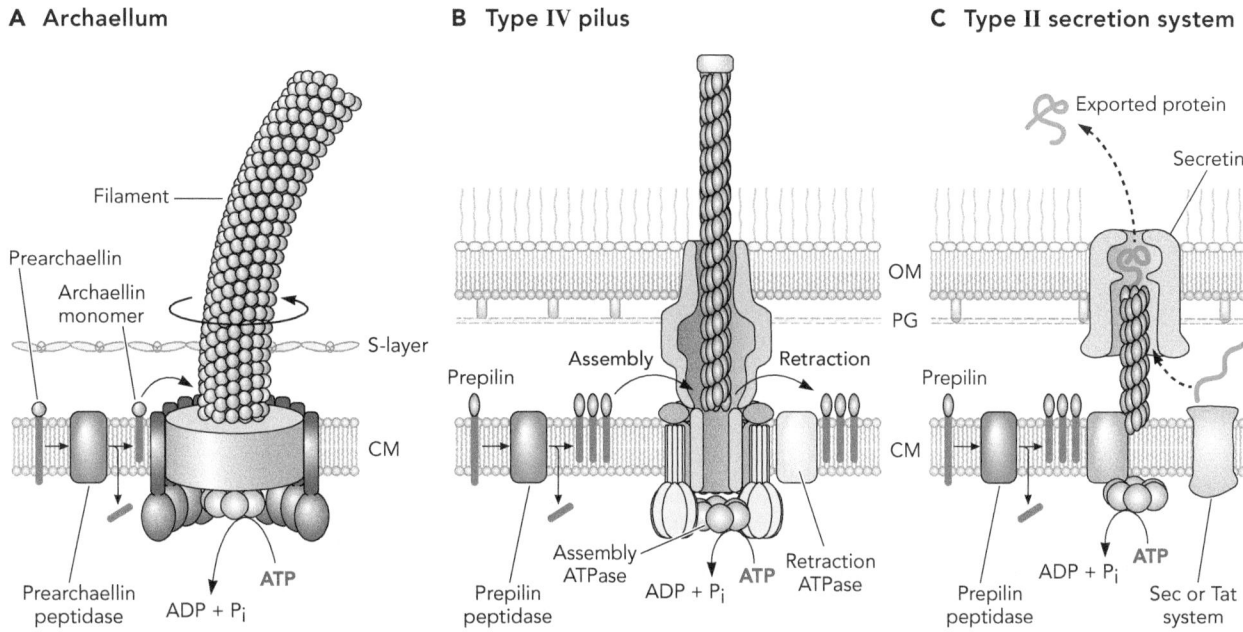

A Archaellum

Filament

Prearchaellin

Archaellin monomer

S-layer

CM

Prearchaellin peptidase

ADP + P$_i$

ATP

B Type IV pilus

OM

PG

Prepilin

Assembly

Retraction

CM

Assembly ATPase

Prepilin peptidase

ADP + P$_i$

ATP

Retraction ATPase

C Type II secretion system

Exported protein

Secretin

OM

PG

Prepilin

CM

Prepilin peptidase

ADP + P$_i$

ATP

Sec or Tat system

FIGURE 15.6 **The structures of the archaellum, type IV pilus, and type II secretion system are made of homologous components.** (A) The archaellum is a flagellar structure unique to the *Archaea* that has homology to (B) type IV pili and (C) type II secretion systems. In all structures, the filament/pilus/pseudopilus are assembled from membrane-embedded prearchaellin/prepilin monomers that are cleaved to mature monomers for assembly at the base of the structure by respective peptidases. ATP is the energy source used to drive the function of these homologous systems that underlie swimming motility, attachment and twitching motility, and secretion for various cells, respectively.

Bacterial Surface Motility

In addition to swimming, there are multiple alternative mechanisms of bacterial motility on solid surfaces. The best-studied mechanism of surface motility involves the extension and retraction of type IV pili (Fig. 15.6B). Type IV pili are widespread in Gram-negative and Gram-positive bacteria as well as in *Archaea*. Type IV pili promote attachment to surfaces and can also mediate a type of surface motility called **twitching motility**. Using a type II-like secretion system (Fig. 15.6C; Chapter 14), the pili are extended outward, where they attach to the surface of a substrate or another cell and then are retracted in a manner that causes a jerky forward movement (Fig. 15.7). Extension and retraction of pili are mediated by ATP-dependent polymerization and depolymerization of pilin subunits. Pilins are synthesized as prepilin proteins with signal peptides targeting them to the cytoplasmic membrane. As the pilin subunits are added at the base of the pilus, the signal peptide is cleaved by a specific prepilin peptidase enzyme (Fig. 15.6B). During retraction, the depolymerized pilin monomers remain in the membrane and are recycled in the next round of extension. This mechanism of surface movement is important for the pathogenesis of some organisms such as *P. aeruginosa*.

There are several other poorly understood mechanisms of surface motility that are termed "gliding motility." In some bacteria (e.g., *Flavobacterium*),

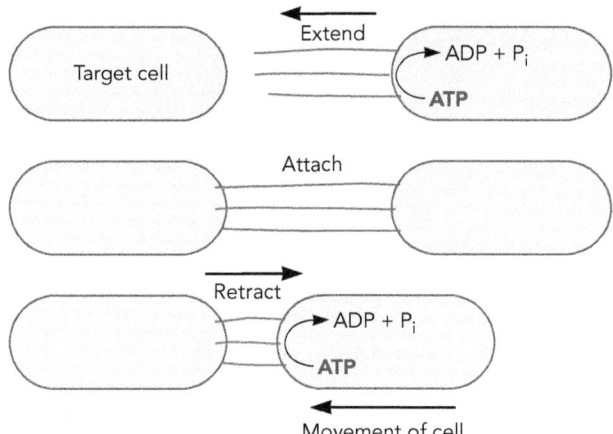

FIGURE 15.7 **Movement on surfaces by twitching motility in *Pseudomonas aeruginosa*.** The type IV pili extend and attach to a target cell or surface. Pili then retract, which results in the movement of the cell toward the target. Both extension and retraction are powered by ATP hydrolysis.

it is proposed that adhesion proteins in the outer membrane rotate around a helical intracellular track that interacts with gliding motors creating a continuous propulsion system, causing the cell to rotate as it moves forward. Gliding motility in filamentous cyanobacteria (Chapter 18) involves both pili and the extrusion of a polysaccharide slime from pores at the poles of each cell within the filament. Myxobacteria exhibit two other types of gliding motility, social and adventurous, involving type IV pili and slime extrusion, respectively (Chapter 18).

Chemotaxis

Chemotaxis is the movement of cells toward or away from a chemical signal. The mechanism of bacterial chemotaxis during swimming motility is best understood in *E. coli*. Rather than sensing a chemical gradient between the two poles of the cell, *E. coli* monitors the chemical concentration difference between two locations approximately once every second as it swims. Since *E. coli* moves at a rate of ~10 to 20 cell lengths per second (proportionally much faster than humans), this enables sensing of a chemical gradient over distance. This rate of swimming is impressive because at their small size, bacteria experience water as a much more viscous environment compared to humans.

In an environment where there is no net chemical concentration gradient, *E. coli* moves via a **random walk** (Fig. 15.8A) in which it alternates between smooth swimming and tumbling. Smooth swimming, when the flagella are rotating CCW and form a bundle, is interspersed with tumbles, which occur when the direction of flagellar rotation switches to CW rotation and the flagellar bundle flies apart (Fig. 15.5). Upon the next switch to CCW rotation, the cell swims

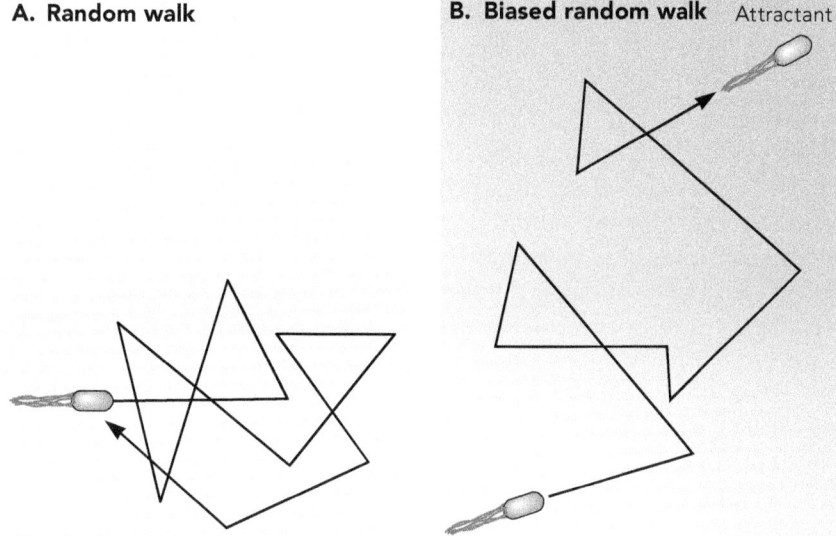

A. Random walk

B. Biased random walk Attractant

FIGURE 15.8 **Bacterial random walk in a uniform environment and biased random walk in a chemical attractant gradient.** (A) The track of the movement of one cell in a uniform liquid environment by a "random walk", which consists of short bouts of smooth swimming (runs, indicated by line segments) punctuated by tumbles (points between line segments). After each tumble, Brownian motion reorients the cell, and the cell swims off in a random direction. (B) In the presence of a chemical gradient (depicted by the increase in purple shading toward the top right), the cell biases the random walk so that the cell swims smoothly for longer periods of time toward the increasing concentration of attractant. Note that runs are shorter if the cell reorients in a direction that is away from the source of attractant.

off in a random direction. This pattern of swimming and tumbling therefore continues without generating any net movement in a particular direction. In comparison, in the presence of a gradient of attractant (i.e., a carbon source or other useful chemical), the bacteria are capable of net movement up the gradient. Similarly, they will move away from the source of a repellent (i.e., harmful chemical). Directed movement in the presence of an attractant or repellent is called a **biased random walk** (Fig. 15.8B). In the case of an attractant, the amount of time spent swimming toward the attractant increases (the frequency of tumbles decreases), while movement toward a repellent will instead increase the frequency of tumble events. Although the biased random walk results in net movement toward an attractant or away from a repellent, cells do not swim in a straight line. Because of the small size of bacterial cells, Brownian motion in the liquid can cause them to move "off course" and therefore cells need to correct their trajectory by tumbling occasionally. The direction after a tumble event is always random, and if the cell reorients and moves in an unfavorable direction (i.e., away from an attractant), the cell will tumble more frequently until it randomly reorients such that it is moving toward the attractant. Other bacteria exhibit different types of taxis, including movement toward light (phototaxis) or away from darkness (scotophobia), movement toward oxygen (aerotaxis), movement in response to osmotic strength (osmotaxis), or movement toward environments that allow optimal energy conservation (energy taxis).

CHEMOTAXIS IN *ESCHERICHIA COLI*

How do *E. coli* cells sense their environment and control their motility in response? The amount of a particular attractant or repellent does not drive the bacterial response during chemotaxis. Rather, bacteria detect a change in the concentration of the chemotactic signal over time that triggers changes in bacterial behavior. As described above, in a uniform environment with no concentration gradient of attractant or repellent, cells alternate at a steady rate between swimming and tumbling (CCW and CW flagellar rotation, respectively). Because they move in random directions after tumble events with no net overall direction to their movement, there is no chemotaxis occurring (Fig. 15.8A).

However, after the addition of an attractant, the bacteria exhibit an **excitation** response (Fig. 15.9). Within ~200 milliseconds of sensing the attractant, the bacteria will suppress switching of the direction of flagellar rotation and maintain CCW rotation (resulting in smooth swimming) for a longer period of time (Fig. 15.8B). The excitation phase is followed by a period of **adaptation** (Fig. 15.9), which occurs within seconds or minutes depending on the attractant. Bacteria exposed to a repellent also exhibit an excitation response; however, the repellent response results in an increased frequency of tumbling due to CW flagellar rotation, followed by a period of adaptation. The initial excitation response is due to the posttranslational regulation of protein activity via a phosphorylation cascade, whereas adaptation is due to the regulation of protein activity via methylation.

Molecules Involved in the Chemotactic Response

Attractants and repellents are known as **chemoeffectors**, which are the environmental signals that trigger an excitation response. In *E. coli*, a variety of sugars and amino acids serve as chemoattractants. The presence of these chemicals

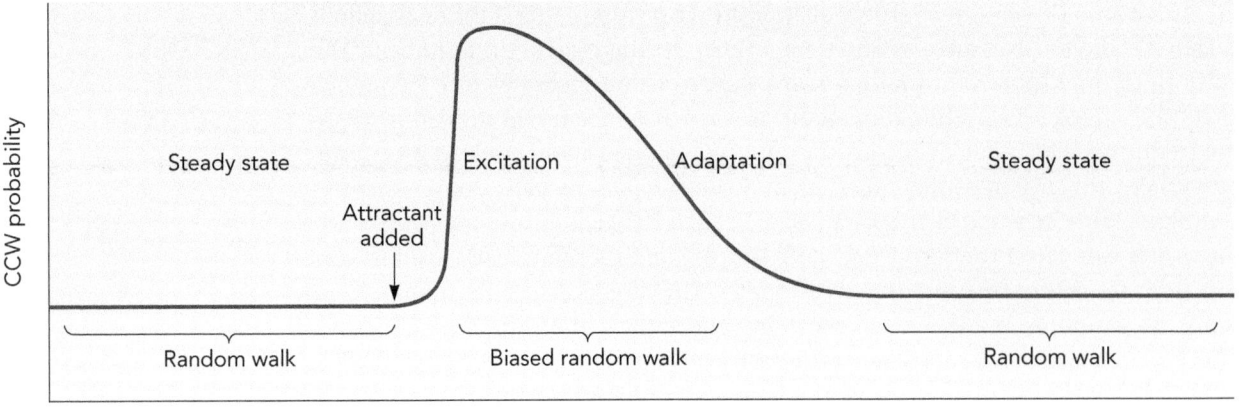

FIGURE 15.9 **Timeline of a chemotactic response.** In the absence of attractant, cells move in a random walk, changing direction frequently by alternating clockwise (CW) and counterclockwise (CCW) flagellar rotation, and are physiologically in a steady state. When attractant is added, an excitation response occurs, and cells bias the random walk by increasing the length of time that the flagella rotate in the CCW direction. After a few seconds, the cells begin an adaptation phase whereby they acclimate to the presence of the attractant and begin to return, over time, to the steady-state prestimulus random walk behavior.

is detected by membrane-bound receptor proteins called **methyl-accepting chemotaxis proteins** (MCPs). As the name suggests, MCPs may be methylated at conserved glutamate residues during the chemotaxis response. Four different MCPs have been identified in *E. coli*: Tar, Tsr, Trg, and Tap; each MCP is responsible for mediating the response to a specific set of attractants and/or repellents (Table 15.1). MCPs are composed of a periplasmic ligand-binding domain, which detects the extracellular signal, and a cytoplasmic signaling domain (Fig. 15.10A). Trimers of MCP homodimers (Fig. 15.10B) are clustered in chemosensory arrays at one pole of the cell (Fig. 15.10C), where they either directly bind the chemoeffectors or bind to periplasmic binding proteins that are bound to specific chemoeffectors. Conformational changes in the cytoplasmic signaling domain initiate the excitation response via a phosphorylation cascade involving a set of cytoplasmic chemotaxis (Che) proteins that function to mediate changes in swimming behavior.

The underlying mechanism of chemotaxis is based on a complex two-component regulatory system that modulates the excitation and adaptation outputs (Fig. 15.11). As discussed in Chapter 7, the simplest form of a two-component regulatory system is a single sensor histidine kinase-response regulator pair that regulates gene expression in response to extracellular signals (Fig. 15.11A). In chemotaxis, the regulatory system is more complex. The MCP

TABLE 15.1 **Components of the *Escherichia coli* chemotaxis system**

Component	Definition
Chemoeffector	Chemical being sensed (e.g., amino acids or sugars)
Methyl-accepting chemotaxis protein (MCP)	Cell surface chemoreceptor. Typically has two membrane-spanning domains and a periplasmic ligand-binding domain that directly binds to specific chemoeffectors or periplasmic binding proteins bound to chemoeffectors; modified on conserved glutamate residues in the cytoplasmic C-terminal domain by methylation
Tar	MCP that binds aspartate, glutamate, and repellents
Trg	MCP that binds ribose (via periplasmic ribose-binding protein) and galactose (via periplasmic galactose-binding protein)
Tsr	MCP that binds serine, alanine, glycine, and repellents
Tap	MCP that binds dipeptides and pyrimidines
Aer	Energy taxis/aerotaxis receptor with a domain that binds flavin adenine dinucleotide (FAD)
Periplasmic binding protein	Binds to some chemoeffectors (e.g., sugars in *E. coli*) and interacts with a specific MCP when bound to chemoeffector
CheW[a]	Relays information between MCP and CheA
CheA[a]	Cytoplasmic histidine kinase; transphosphorylates CheY and CheB
CheY[a]	Response regulator that controls the direction of flagellar rotation; when phosphorylated it causes a switch to clockwise (CW) rotation (tumble); when unphosphorylated it does not interact with flagellar motors and the flagella continue to rotate counterclockwise (CCW, smooth swimming); possesses some intrinsic phosphatase activity
CheZ[b]	Phosphatase that removes phosphoryl groups from CheY-P
CheR	Methyltransferase; catalyzes the addition of methyl groups to MCPs at a steady rate
CheB[b]	Methylesterase and response regulator; catalyzes the removal of methyl groups from MCPs; only active when phosphorylated

[a] Deletion of the genes encoding CheW, CheA, or CheY results in cells that constitutively exhibit a smooth swimming phenotype.
[b] Deletion of the genes encoding CheZ or CheB results in cells that constitutively exhibit a tumbly phenotype.

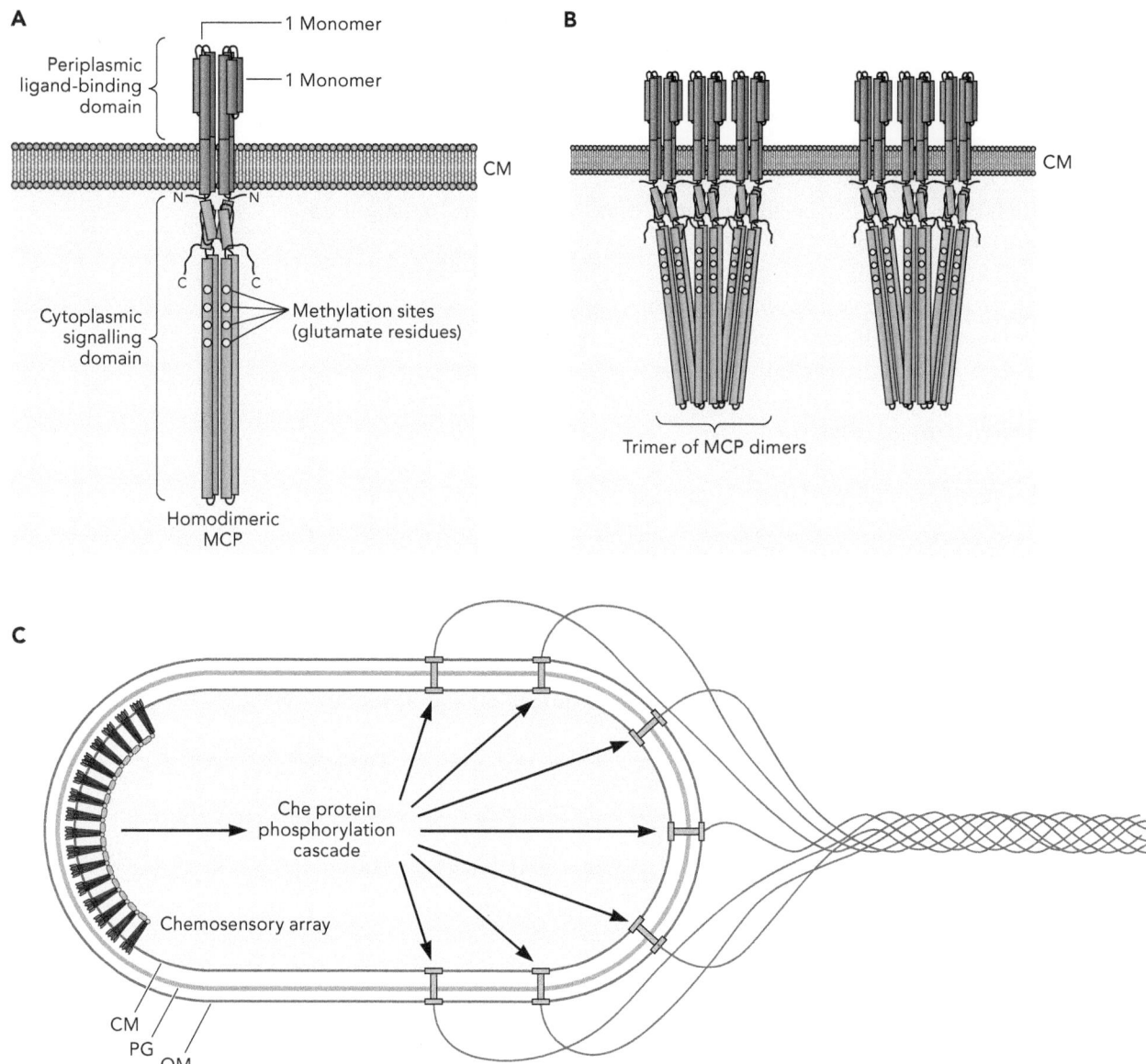

A

1 Monomer

1 Monomer

Periplasmic ligand-binding domain

CM

N N

C C

Cytoplasmic signalling domain

Methylation sites (glutamate residues)

Homodimeric MCP

B

CM

Trimer of MCP dimers

C

Che protein phosphorylation cascade

Chemosensory array

CM
PG
OM

FIGURE 15.10 **Structure and localization of methyl-accepting chemotaxis proteins (MCPs) and the process of signal transduction to the flagellar motors in *Escherichia coli*.** (A) Structure of an MCP homodimer in the cytoplasmic membrane (CM). The ligand-binding domain (red) is located in the periplasm and the signaling domain (green) is located in the cytoplasm. Each MCP monomer is anchored in the CM by two membrane-spanning domains (purple). MCPs form homodimers in the membrane. (B) MCPs function as trimers of MCP dimers; two MCP monomers make up each homodimer. (C) Trimers of MCP dimers are located in a large array within the CM at the cell pole. Incoming chemical signals detected by the ligand-binding domains are transmitted to the flagellar motors via a Che protein phosphorylation cascade that coordinates the response. OM, outer membrane; PG, peptidoglycan.

acts as the sensor that indirectly interacts with a histidine kinase (CheA) via a coupling protein (CheW). When CheA autophosphorylates on a histidine residue, it can then transfer phosphoryl groups to conserved aspartate residues on two different response regulators (CheY and CheB; Fig. 15.11B). CheY-P and CheB-P do not change gene expression; rather, they work via protein-protein interactions to rapidly change motile behavior in response to changing environmental conditions. Another Che protein, CheZ, has phosphatase activity that removes phosphate groups from CheY (Table 15.1).

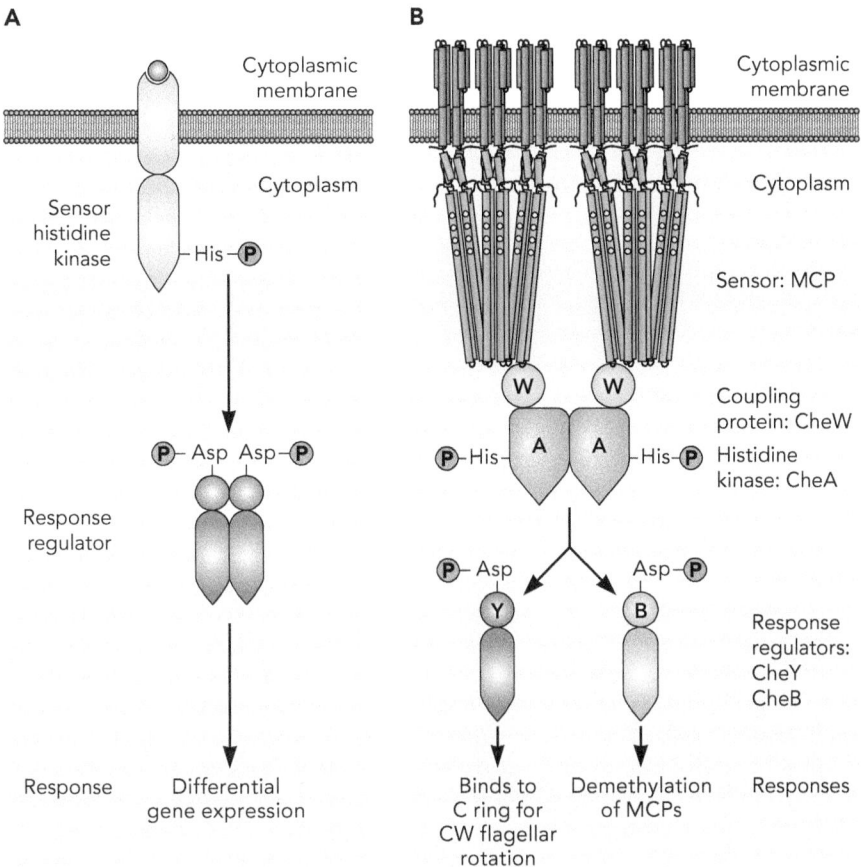

A

Cytoplasmic membrane

Cytoplasm

Sensor histidine kinase

His–(P)

(P)–Asp Asp–(P)

Response regulator

Response

Differential gene expression

B

Cytoplasmic membrane

Cytoplasm

Sensor: MCP

(W) (W) Coupling protein: CheW

(P)–His–(A)(A)–His–(P) Histidine kinase: CheA

(P)–Asp Asp–(P)

(Y) (B) Response regulators: CheY CheB

Binds to C ring for CW flagellar rotation

Demethylation of MCPs

Responses

FIGURE 15.11 **Two-component signal transduction systems control gene transcription and drive the chemotactic response.** (A) Typical two-component regulatory system with a membrane-bound sensor histidine kinase that autophosphorylates on a conserved histidine (His) amino acid residue when it binds an extracellular chemical signal (blue). The phosphoryl group is then transferred to a conserved aspartate (Asp) residue on the cytoplasmic response regulator and the phosphorylated response regulator binds to appropriate promoters to control gene expression (Chapter 7). (B) In bacterial chemotaxis, methyl-accepting chemotaxis proteins (MCPs) generally serve as sensors of extracellular signals. MCPs are associated with the CheA kinase (labeled A) via the coupling protein CheW (labeled W). In the absence of attractant, CheA is active and autophosphorylates a conserved His residue. The phosphoryl group can then be transferred to a conserved Asp residue on one of two different cytoplasmic response regulators, CheY (labeled Y) and CheB (labeled B). When phosphorylated, CheY-P functions as a tumble generator by binding to the C ring of the flagellar motor to drive clockwise (CW) flagellar rotation. Under these conditions, CheB-P functions in the adaptation process by demethylating MCPs and the cell continues to exhibit random walk behavior (see the text and Fig. 15.12 for details).

The Excitation (Phosphorelay) Response during Chemotaxis

Instead of controlling gene expression like other two-component systems, the chemotaxis response mediated by the MCP and Che proteins modulates cellular swimming behavior. As will be described in more detail in the following sections, changes in chemoeffector concentration shift the distribution of phosphoryl groups from CheA to the response regulators, which determines the direction of flagellar rotation (via CheY), and adaptation in response to chemoeffector concentrations (via CheB). In its unphosphorylated form, CheY cannot bind to the flagellar C ring, resulting in continuous CCW rotation of the flagella and smooth swimming behavior. In its phosphorylated form, CheY-P binds to the flagellar C ring, resulting in CW rotation of the flagella and tumbling behavior (Fig. 15.11B). The roles of the Che proteins were determined in part through examining the phenotypic impacts of mutations in the *che* genes. For example, deletion of *cheA* or *cheY* results in a continuous smooth swimming phenotype since there would never be any CheY-P present (Table 15.1).

The Adaptation (Methylation) Response during Chemotaxis

To sense changes in chemoeffector concentrations over time, *E. coli* cells have an adaptation mechanism that "resets" the phosphorylation cascade system. Under all environmental conditions, the methyltransferase CheR (Table 15.1) continually methylates conserved glutamate residues in the cytoplasmic signaling domain of MCPs. When CheA is phosphorylated, CheA-P transfers phosphoryl groups to another response regulator, CheB (Fig. 15.11B). CheB-P, the active form of the protein, is a methylesterase that counteracts the activity of CheR by demethylating MCPs. A change in the level of MCP methylation resets the phosphorylation cascade system by sending a signal through CheW to CheA, impacting the rate of CheA autophosphorylation.

CHEMOTACTIC RESPONSE TO AN ATTRACTANT OVER TIME

Steady State

In order to understand what triggers the excitation and adaptation responses in *E. coli* (Fig. 15.9), fundamental experiments were performed in defined *in vitro* experimental growth conditions (e.g., buffer with or without specific chemoeffectors). In a uniform environment with no chemoeffector present, cells move via a random walk. Under these conditions, the MCP/CheW/CheA complex is in a conformation that favors high CheA kinase activity, which results in CheA autophosphorylation. CheA-P rapidly transfers its phosphoryl group to the response regulator CheY (Fig. 15.12). CheY-P binds to the flagellar motor switch (C ring), causing CW rotation (tumble). Thus, CheY-P is considered a "tumble generator" for the cell. Over time, the intrinsic phosphatase activity within CheY-P plus the phosphatase activity of CheZ leads to the dephosphorylation of CheY-P. When CheY is not phosphorylated, it can no longer bind to the flagellar switch, and this results in CCW rotation and the initiation of smooth swimming. Thus, the relative concentrations of CheY and CheY-P are constantly changing, resulting in alternating CW and CCW flagellar rotation, which produces the random walk behavior.

Excitation Response

In the presence of an attractant gradient, cells modify their swimming behavior to a biased random walk by decreasing the frequency of flagellar rotation in the CW direction, resulting in fewer tumbles. Fewer tumbles leads to longer periods of smooth swimming toward higher concentrations of attractant. The excitation response is accomplished when attractant molecules bind to the sensory domain of MCPs, causing a conformational change in the MCP/CheW/CheA complex that results in a decrease in CheA kinase activity (Fig. 15.12). With less CheA-P to donate a phosphoryl group to CheY, the level of CheY-P decreases. Since unphosphorylated CheY cannot interact with the flagellar motors, the rotation of the flagellar motors remains CCW for longer periods of time, which results in

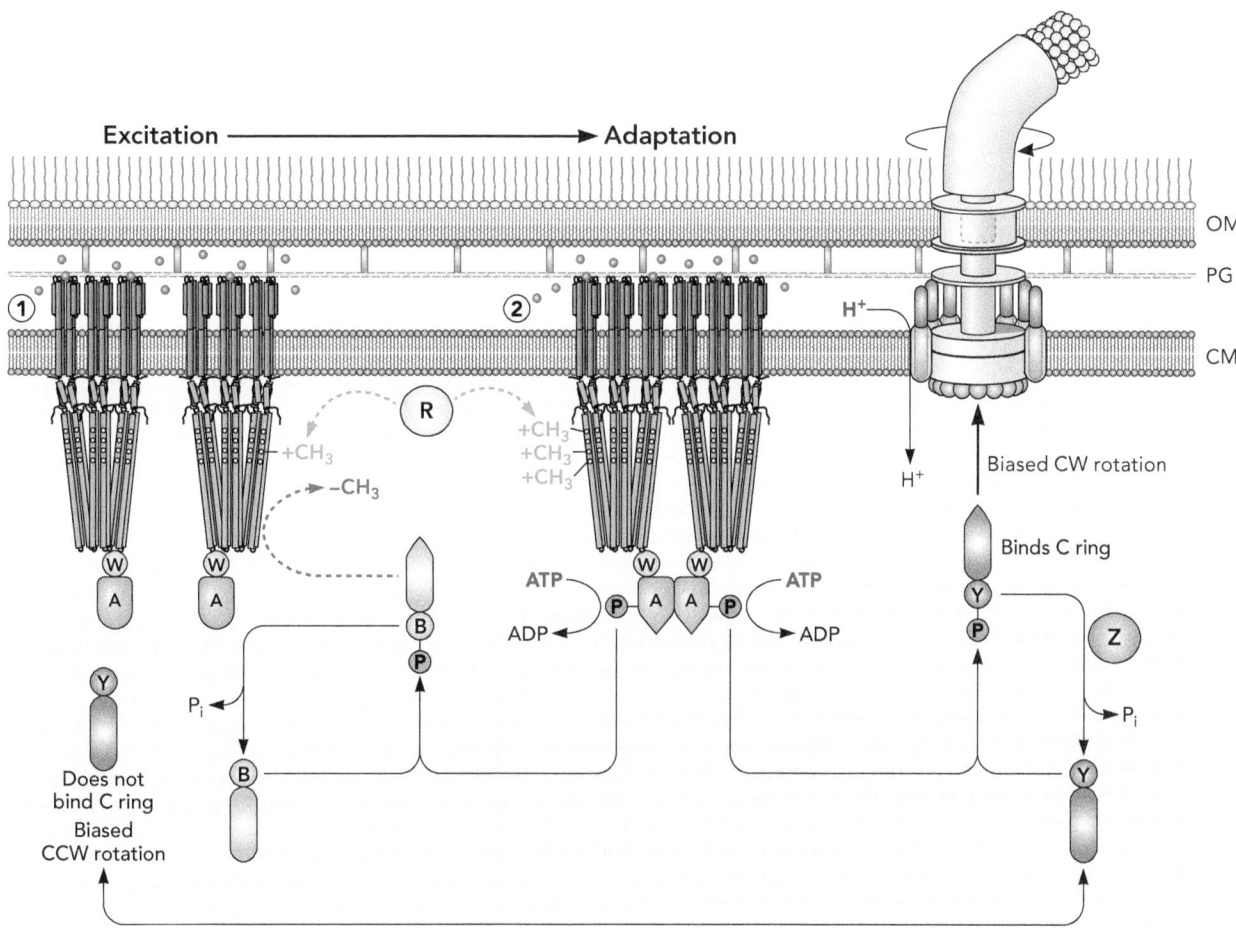

FIGURE 15.12 **Signal transduction in response to an attractant during bacterial chemotaxis.** Before attractant is sensed, the cells are in a physiological steady state (random walk) where they alternate between counterclockwise (CCW, swimming behavior) and clockwise (CW, tumbling behavior) flagellar rotation (see Fig. 15.11). The excitation phase (1) occurs in the presence of an attractant gradient, and biased random walk is induced. Attractant molecules bind to the periplasmic binding domain of MCP trimers. The signal is transmitted across the cytoplasmic membrane (CM), causing a conformational change in the MCP/CheW/CheA complex that results in a decrease in CheA kinase activity. As levels of CheA-P decrease, the majority of CheY remains unphosphorylated and no longer interacts with the flagellar C ring. This leads to biased CCW flagellar rotation and longer runs of smooth swimming, causing net movement of the cell up the attractant concentration gradient. CheB remains unphosphorylated and is inactive. After a few seconds, cells enter the adaptation phase (2) to reset the chemotaxis system. Under these conditions, CheR continuously adds methyl groups to MCPs, resulting in highly methylated MCPs. When MCPs are highly methylated, the conformation of the MCP/CheW/CheA complex once again favors high CheA kinase activity, which leads to higher concentrations of CheY-P and CheB-P. CheY-P binds the flagellar C ring, which leads to biased CW flagellar rotation and more frequent tumbling behavior. Higher CheB-P levels result in demethylation of MCPs. CheZ dephosphorylates CheY-P. At this point, the entire chemotaxis system has been reset to pre-excitation status and the cells resume random walk behavior and are poised to detect further increases in attractant concentration. OM, outer membrane; PG, peptidoglycan.

extended periods of smooth swimming and concurrently reduced tumbling. The levels of CheY-P remain low because phosphoryl groups are continuously being removed from CheY-P by the phosphatase activity of CheZ and CheY itself. Therefore, upon excitation, the phosphorylation cascade system is turned "off," resulting in the majority of CheA and CheY molecules being unphosphorylated, which causes cells to swim smoothly up the attractant concentration gradient as part of the biased random walk behavior.

Adaptation Response

To be able to sense increasing attractant concentrations within a chemical gradient, cells reset the phosphorylation cascade system through the CheB/CheR adaptation mechanism. As discussed previously, during the excitation response, attractant molecules bind to the sensory domain of MCPs, causing a decrease in CheA kinase activity. As the level of CheA-P decreases, the level of CheB-P decreases. Under these conditions, CheR continues to add methyl groups to MCPs, and unphosphorylated (inactive) CheB is unable to remove the methyl groups; this results in highly methylated MCPs. When MCPs are highly methylated, the conformation of the MCP/CheW/CheA complex once again favors high CheA kinase activity, which leads to higher concentrations of CheY-P and CheB-P. Higher CheY-P levels increase the frequency of cellular tumbles, whereas higher CheB-P levels result in demethylation of MCPs. At this point the entire chemotaxis system has been reset to pre-excitation status and the cells resume random walk behavior and are poised to detect further increases in attractant concentration.

Comparison to a Repellent Response

The presence of some repellents is also sensed through the MCPs. The mechanism for sensing a repellent is considered an excitation event; however, the excitation response results in increased frequency of CW flagellar rotation. The repellent causes a conformational change in the MCP/CheW/CheA complex that results in an increase in CheA kinase activity and CheY-P levels, which increases the frequency of tumbling. In this way, when cells are swimming toward higher concentrations of repellents, the rate of tumbling increases and cells reorient by Brownian motion before swimming again in a random direction. This behavior results in the cells ultimately swimming away from the source of the repellent. Adaptation via methylation events through CheR and CheB allows cells to sense changing repellent concentrations.

Summary

Chemotaxis in *E. coli* is mediated by a complex series of protein phosphorylation and methylation events that drive changes in swimming behavior. The resulting shifts in the phosphorylation levels of CheY and CheB protein pools stimulate excitation and adaptation responses, respectively. It is important to recognize that under real-world physiological conditions MCP arrays function to simultaneously detect many different chemoeffectors. At present, we do not have a good understanding of how bacteria integrate multiple environmental signals and prioritize their response in complex environments with multiple attractants and repellents present.

Conservation and Variation in Chemotaxis Systems among Bacteria and Archaea

Chemotaxis is widespread in motile bacteria and archaea; the overall mechanism and most of the components are generally conserved. However, some bacterial genera, like *Bacillus*, do not have CheZ proteins and rely on other dephosphorylation mechanisms to affect the appropriate cellular response. In addition, some bacteria have alternative mechanisms of adaptation that use a variety of different Che proteins. Many bacteria have a large repertoire of MCPs that detect a wide range of chemoeffectors. For example, some bacterial species with broad metabolic capabilities (e.g., soil bacteria that experience widely diverse and changing environmental conditions) produce between 20 and 60 different MCPs that respond to aromatic acids, hydrocarbons, metals, plant exudate chemicals, and/or host signaling molecules, to name a few.

Methods to Study Bacterial Motility and Chemotaxis

There are several methods that can be used to study bacterial motility and chemotaxis. The most common ones are discussed here. The **capillary assay** (Fig. 15.13A) involves the use of small (1μl) capillary tubes that are filled with different potential chemoeffectors that are then placed into a suspension of bacterial cells in buffer. If a tube filled with the same buffer is placed into the culture, the bacteria will move via random walk behavior and over time the concentration of the bacteria in the capillary tube will become equal to that in the original culture. However, if a capillary tube containing an attractant is placed into the culture suspension, motile chemotactic bacteria will sense the gradient of attractant diffusing from the capillary and move toward and into the capillary via a biased random walk. In this case, the response can be qualitatively assessed by visualizing the cloud of cells that accumulates within or at the mouth of the capillary by microscopy. The number of bacteria that accumulate within the capillary after a certain period of time can be quantitatively enumerated using viable counts. In the presence of an attractant, the capillary tube will ultimately contain a larger number of bacterial cells than the initial suspension. The opposite is true for a repellent. The number of bacterial cells in the capillary will be lower than in a control capillary containing buffer only because the bacteria move away from the mouth of the capillary tube from which the repellent is diffusing via a biased random walk.

A **swim plate assay** (Fig. 15.13B) involves examining flagellar-based motility in a semisolid agar plate (typically 0.25 to 0.3% agar) containing a test attractant chemical, which also serves as the sole source of carbon and energy. The bacteria are inoculated at a single point in the plate, where they will catabolize the attractant chemical and generate a concentration gradient at the point of inoculation. Motile chemotactic bacteria will sense the gradient of attractant and swim away from the inoculation point, forming a ring of growth that gradually moves outward from the point of inoculation.

Tethered cell assays (Fig. 15.13C) utilize anti-flagellin antibodies to tether cells by a single flagellum, and the rotation of the cell body is monitored under phase-contrast microscopy in response to the addition of a chemoeffector. Bacteria with multiple flagella must be pretreated, usually by mixing the culture for a set amount of time in a blender, so that on average, each cell has only one flagellum. Addition of an attractant will decrease the frequency of changing the direction of rotation of the flagellum. **Computer-based tracking and modeling** of bacterial movement is a quantitative approach that can be used to accurately measure the behavior of cells using a computer program to analyze digital video. Within a population of swimming or tethered cells, the program quantifies the frequency of individual cells changing direction in response to the addition of a chemoeffector. Swimming cells in a wet mount are observed

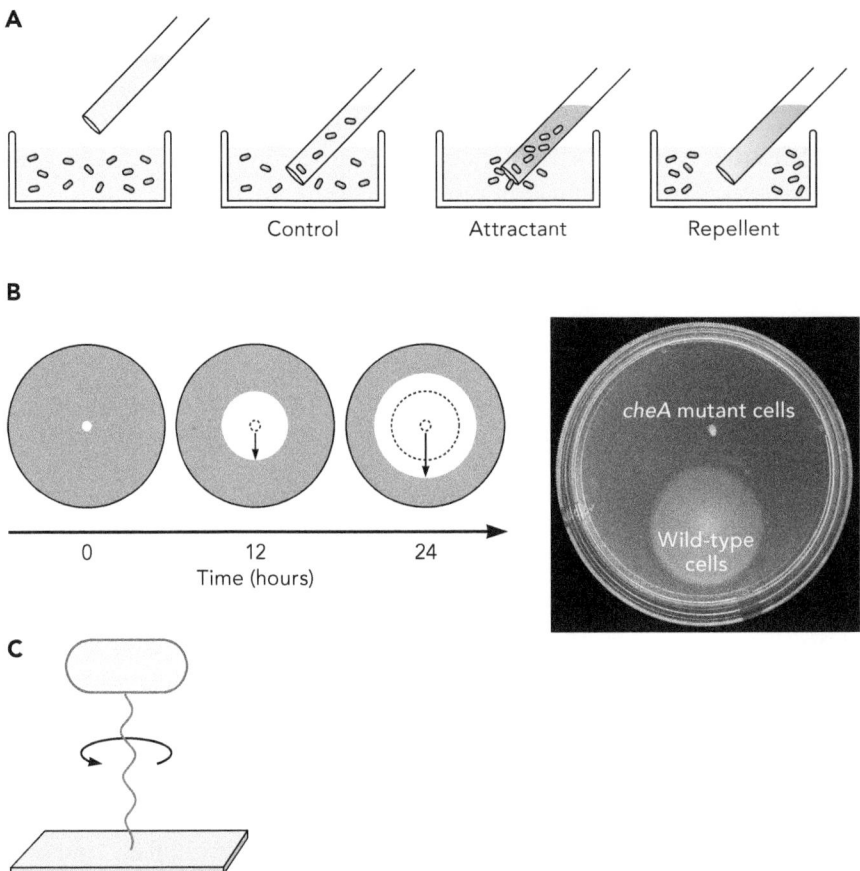

FIGURE 15.13 **Methods for monitoring and assessing bacterial chemotaxis behavior.** (A) Capillary assay. A small capillary tube containing an attractant or repellent chemical or buffer only (control) is inserted into a suspension of motile cells. In the case of the buffer control, approximately equal numbers of cells will be present in the suspension and the capillary. However, if the cells sense the chemical as an attractant, they will accumulate in the capillary, whereas if the chemical is sensed as a repellent, cells will avoid entering the capillary. (B) Swim plate assay. Cells are inoculated into soft agar medium (~0.25% agar instead of the typical 1.5% agar), which cells can swim through. The medium also contains a chemical that serves as both a carbon source and an attractant. On the left is an illustration of wild-type cells responding to attractant over 24 hours. Wild-type cells consume the attractant, generating a gradient over time that they sense and follow, allowing for cellular growth in a concentric ring from the original point of inoculation. On the right is a representative swim plate showing wild-type and *cheA* mutant cells. Wild-type cells show the typical concentric ring growth. The *cheA* mutant cells remain near the point of inoculation as they can grow and are motile, but cannot detect the concentration gradient. Photo credit: Rebecca E. Parales. (C) Tethered cell assay. In this assay, cells are tethered by their flagellum to the surface of a microscope slide that has been treated with an anti-flagellin antibody. Before and after an attractant is added, the direction of rotation of a population of cells is monitored over time and quantified via phase-contrast microscopy and computer-assisted motion analysis software. The frequency of counterclockwise flagellar rotation will increase in response to the addition of attractant.

under the microscope. The computer then tracks individual bacteria moving in three dimensions and quantifies the various aspects of cellular movement (swimming speed, amount of time spent smooth swimming, frequency of changes of direction, etc.). Both the tethered cell assay and the computer-based tracking assay are examples of temporal assays in which the cells experience an immediate upshift in attractant concentration rather than a gradient of attractant. Cell behavior is monitored before and after the addition of the attractant.

Learning Outcomes: After completing the material in this chapter, students should be able to . . .

1. compare and contrast the different types of motility in *Bacteria* and *Archaea*
2. describe how the various parts of bacterial flagella function to produce swimming motility
3. explain how the synthesis of a bacterial flagellum is temporally regulated
4. discuss how chemotaxis allows bacterial and archaeal cells to respond to chemical concentrations in their environment
5. interpret experiments exploring bacterial motility and chemotaxis

Check Your Understanding

1. Define this terminology: polar flagella, monotrichous flagellum, lophotrichous flagella, peritrichous flagella, endoflagella, anti-sigma factor, tumbling behavior, archaellum, twitching motility, chemotaxis, random walk, biased random walk, excitation, adaptation, chemoeffectors, methyl-accepting chemotaxis proteins (MCPs)

2. Describe the location and function of each component of an *Escherichia coli* flagellum listed below. (LO2)

Component	Location	Function
M/S ring		
C ring		
P ring		
L ring		
Rod		
Hook		
Flagellar filament		
Cap		

3. Briefly explain how the motor proteins (MotA, MotB) work together to cause the rotation of the flagellum in *Escherichia coli*. (LO2)

4. How does the direction of flagellar rotation (clockwise or counterclockwise) impact motility in *Escherichia coli*? (LO2)

5. *Escherichia coli* has a complicated regulatory scheme for flagellum biosynthesis. (LO3)

   a) Describe the functions of the proteins involved in flagellum biosynthesis listed below.

Protein	Function
FhlDC	
FliA	
FlgM	

   b) Order the events that take place temporally during flagellum biosynthesis.

   ___ FhlDC together with σ^{70} RNAP activate transcription of the genes encoding the rod, P/L rings, hook, and FliA (σ^{28}) and FlgM.

   ___ FhlDC together with σ^{70} RNAP holoenzyme activate expression of the genes encoding the M ring, S ring, and C ring, in that order.

   ___ FlgM binds to and inactivates σ^{28}.

   ___ Once hook assembly is complete, FlgM is secreted outside of the cell.

   ___ The genes that encode the machinery for chemotaxis are expressed under the direction of FhlDC and σ^{28} RNAP.

   c) Why does *E. coli* have such a complicated regulatory scheme for flagellum biosynthesis?

6. Bacteria and archaea both exhibit swimming motility. Archaella are in some ways similar to bacterial flagella and in some ways similar to type IV pili. Fill in the table below with "Yes" or "No" and include your reasoning. (LO1)

Archaella characteristic	Similar to bacterial flagella?	Similar to type IV pili?
Rotate to push the cell through liquid environments		
Made up of glycosylated protein subunits		
Powered by ATP hydrolysis		

7. Some bacteria exhibit twitching motility on solid surfaces. Briefly describe how type IV pili work to generate twitching motility and the energy source used. (LO1)

8. Chemotaxis is the movement of cells toward or away from a chemical gradient. (LO4)

   a) Describe the difference between smooth swimming and tumbling in *Escherichia coli*. In each case, what does the behavior look like? How does the rotation of the flagellum impact swimming and tumbling behavior?

   b) Bacterial movement can be described as a random walk or a biased random walk. Both are the result of the cell swimming and tumbling, swimming and tumbling, in succession. Describe the difference between a random walk and biased random walk.

   c) Describe how a biased random walk by *E. coli* cells in the presence of an attractant results in the cells moving up the concentration gradient. How does the balance between swimming and tumbling over time result in a biased random walk?

9. *Escherichia coli* cells are able to sense their environment and control their motility response. In the case of an attractant, *E. coli* goes through phases of **random walk** (no new concentration gradient sensed), **excitation** upon sensing the attractant, and **adaptation** to the attractant concentration. Which statement below best describes each phase? (LO4)

   The bacterium suppresses switching the direction of flagellar rotation and maintains counterclockwise rotation for a longer period of time. _____

   The frequency of cellular tumbles begins to increase. _____

   The bacterium moves with a steady rate of flagellar motor switching between counterclockwise and clockwise rotation. _____

10. The presence of attractants and repellents is detected by membrane-bound receptor proteins called methyl-accepting chemotaxis proteins (MCPs), which relay signals to the cell through a variation of a two-component regulatory system. The response to attractants and repellents involves cycles of both phosphorylation/dephosphorylation and methylation/demethylation. (LO4)

   a) Considering the phosphorylation cycles in chemotaxis in *Escherichia coli*:

      i. Which molecule acts as the sensor?

      ii. Which molecule acts as the histidine kinase?

      iii. Which molecule relays information between the sensor and the histidine kinase?

      iv. Which two molecules does the histidine kinase phosphorylate?

      v. Which molecule removes phosphate groups?

   b) In an environment where there is no net chemical concentration gradient, *E. coli* moves via a random walk in which it alternates between smooth swimming and tumbling. In the absence of chemoeffectors, the MCP/CheA/CheW complex is in a conformation that favors high kinase activity. Describe the chain of events through which high kinase activity results in this alternating tumbling-smooth swimming behavior.

   c) In the presence of attractant, a conformational change in the MCP/CheA/CheW complex results in a decrease in kinase activity. Describe the chain of events through which lower kinase activity results in a biased random walk behavior.

d) Fill in the table below to summarize the effects of changes in CheA-P and CheY-P concentrations on swimming motility in the presence of an attractant.

Condition	Does CheY-P increase or decrease?	Impact on tumbling and swimming
Increase in CheA-P		
Decrease in CheA-P		

11. The response to chemoeffectors involves cycles of phosphorylation/dephosphorylation and methylation/demethylation. Consider the methylation/demethylation cycles that drive adaptation in chemotaxis in *Escherichia coli*. (LO4)

    a) In the methylation/demethylation system:

       i. Which Che protein methylates MCPs?

       ii. Which Che protein demethylates MCPs?

    b) In the absence of attractant, the MCP/CheA/CheW complex is in a conformation that favors high kinase activity. Describe the chain of events through which high kinase activity results in demethylation of the MCPs.

    c) In the presence of an attractant, attractant molecules bind to the MCP, causing a conformational change in the MCP/CheA/CheW complex that results in a decrease in CheA kinase activity. Describe the chain of events through which lower kinase activity results in higher methylation of the MCPs.

    d) How do cells adapt to the presence of an attractant, allowing the chemotaxis system to reset to pre-excitation status?

12. Briefly describe each method to study bacterial motility and chemotaxis listed below. Include in your description what a positive response to an attractant looks like. (LO5)

    a) Capillary assay

       i. Describe method:

       ii. Positive response:

    b) Swim plate assay

       i. Describe method:

       ii. Positive response:

    c) Tethered cell assay

       i. Describe method:

       ii. Positive response:

    d) Computer-based tracking and modeling of bacterial movement

       i. Describe method:

       ii. Positive response:

Dig Deeper

13. A unique characteristic of spirochetes is the presence of endoflagella. (LO1)

    a) Compare and contrast the endoflagella of spirochetes to *Escherichia coli* flagella by filling in the table below.

Characteristic	*E. coli* flagella	Endoflagella
Where are flagella located in the cell?		
Name the flagellar protein subunits.		
Do flagella move by rotation?		

    b) Vaccines often consist of proteins from pathogens that antibodies (immune proteins made by the host) recognize and bind to, initiating an immune response. You are working at a biotech company, and your supervisor gives you a choice of developing new vaccines against the flagella of *Treponema pallidum* (the spirochete that causes syphilis) or *Salmonella* (a Gram-negative relative of *E. coli*). Which project do you think is more likely to succeed? Explain your reasoning.

14. You are investigating flagella synthesis and function in *Escherichia coli*. (LO2)

   a) You generate a series of random *E. coli* mutants and use a swim plate assay to screen for nonmotile mutants. Results for wild-type cells in a soft agar plate containing fructose are shown.

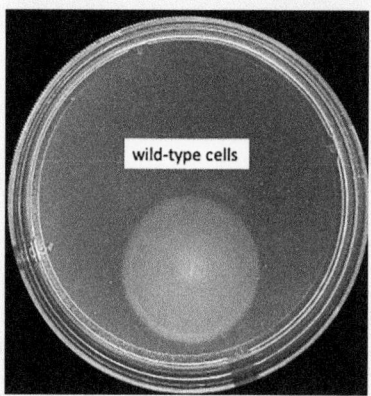

wild-type cells

   i. Why does this assay require soft agar?

   ii. Why do wild-type cells move away from the inoculation site?

   iii. How would results in this assay for nonmotile mutants look different from wild-type cells?

   b) You go to the literature to research the phenotypes of previously isolated flagella mutants. Predict the phenotypes for each nonfunctional class of mutant listed below. One answer has been provided.

Mutated gene (mutant protein)	Would genes for M/S ring be transcribed?	Would genes for flagellin be transcribed?	Would cells be motile?
flhC (regulatory protein)			
fliA (σ^{28})			
fliM (anti-sigma factor)			No

   c) In the case of FliM, why do you think these cells would not be motile?

   d) You find and isolate three nonmotile strains that you believe have mutations in the genes for *flhC*, *fliA*, and *fliM*. To study the expression of flagellar genes in your mutant strains, you create transcriptional fusions of the *lacZ* (β-galactosidase) gene to two flagellar genes: *fliF* (M ring protein) and *fliD* (cap protein). You measure the β-galactosidase activity of these transcriptional fusions (*fliF-lacZ* and *fliD-lacZ*). Your results are reported as the percent β-galactosidase activity relative to wild-type cells.

   From your data, indicate which strain has a mutation in *flhC*, *fliA*, or *fliM*. Explain your reasoning.

Cell type	Relative β–galactosidase activity		Are cells motile?
	fliF-lacZ	*fliD-lacZ*	
Wild-type strain	100%	100%	Yes
Mutant strain 1	0	0	No
Mutant strain 2	100%	100%	No
Mutant strain 3	100%	5%	No

   i. Mutant strain 1 has a mutation in the gene _____ because:

   ii. Mutant strain 2 has a mutation in the gene _____ because:

   iii. Mutant strain 3 has a mutation in the gene _____ because:

15. The underlying mechanism of chemotaxis is based on a complex two-component regulatory system that modulates swimming and tumbling behavior. (LO4)

    a) The deletion of the genes encoding CheW, CheA, or CheY results in cells that constitutively exhibit a smooth swimming phenotype. For each protein, list its function and explain why a deletion of the gene would give a smooth swimming phenotype.

       i. CheY:

       ii. CheA:

       iii. CheW:

    b) The deletion of the genes encoding CheZ or CheB results in cells that constitutively exhibit a tumbly phenotype. For each protein, list its function and explain why a deletion of the gene would give a tumbly phenotype.

       i. CheB:

       ii. CheZ:

16. Typically, a two-component regulatory system uses phosphorylation to activate gene expression. In chemotaxis, the two-component regulatory system does result in the phosphorylation of CheY and CheB. However, CheY-P and CheB-P do not alter gene expression, but instead act directly on other cell components. (LO4)

    a) Briefly review the proteins with which CheY-P and CheB-P interact and explain how they impact smooth swimming behavior.

       i. CheY-P:

       ii. CheB-P:

    b) Why is this system of posttranslational regulation better suited to controlling the chemotaxis response, compared to controlling gene expression?

17. Describe the expected results for each chemotaxis assay control listed below. (LO5)

    a) Capillary assay: A tube filled with buffer but no attractant is placed into a suspension of bacterial cells in buffer.

    b) Tethered cell assay: A tethered cell with no attractant is monitored under phase-contrast microscopy.

    c) Computer-based tracking and modeling of cell movement: Cells in a wet mount with no added attractant are monitored.

18. Many rhizobia species (common inhabitants of soil) and legume host plants have evolved a symbiotic relationship in which bacteria in plant root nodules convert N_2 to bioavailable ammonium in exchange for fixed carbon from the plant (Chapter 12). Motility and chemotaxis likely provide a competitive advantage to bacteria in this rhizobia-host symbiosis, which depends on the rhizobia finding the plant root hairs for infection. You have a strain of *Rhizobium meliloti* that forms root nodules on alfalfa. Because plant flavonoids are involved in the initial signaling events that lead to nodule formation, you hypothesize that plant flavonoids will be strong chemoattractants for *R. meliloti*. (LO5)

    You set up a series of capillary assay experiments to study chemotaxis in *R. meliloti* to chemicals commonly associated with alfalfa root (see results). The concentration of bacteria in your assay (outside the tube) started at 1×10^3 cells/ml.

Chemical in tube	Concentration of bacteria in tube ($\times 10^3$ cells/ml)
Chemotaxis buffer	?
Glutamate (positive control)	100
Alfalfa root exudate	250
Alfalfa-derived flavonoids	10

    a) Given the setup of the capillary assay, why did you do an experiment with only chemotaxis buffer? What approximate result should you get for the chemotaxis buffer?

    b) According to these data, which chemical is the strongest chemoattractant? Do these data support your hypothesis that plant flavonoids are strong chemoattractants for *R. meliloti*?

    c) What would your next experiment be to further investigate chemotaxis in *R. meliloti*?

390   Making Connections

390   Fundamentals of Quorum Sensing

391   Quorum Sensing and Bioluminescence in the *Vibrio fischeri*-Squid Symbiosis

395   Basic Model of Quorum Sensing in Gram-Negative Proteobacteria

398   Basic Model of Quorum Sensing in Gram-Positive Bacteria

400   Interspecies Communication: the LuxS System

400   Regulatory Cascade Controlling Quorum Sensing in *Vibrio cholerae*

402   Quorum Quenching

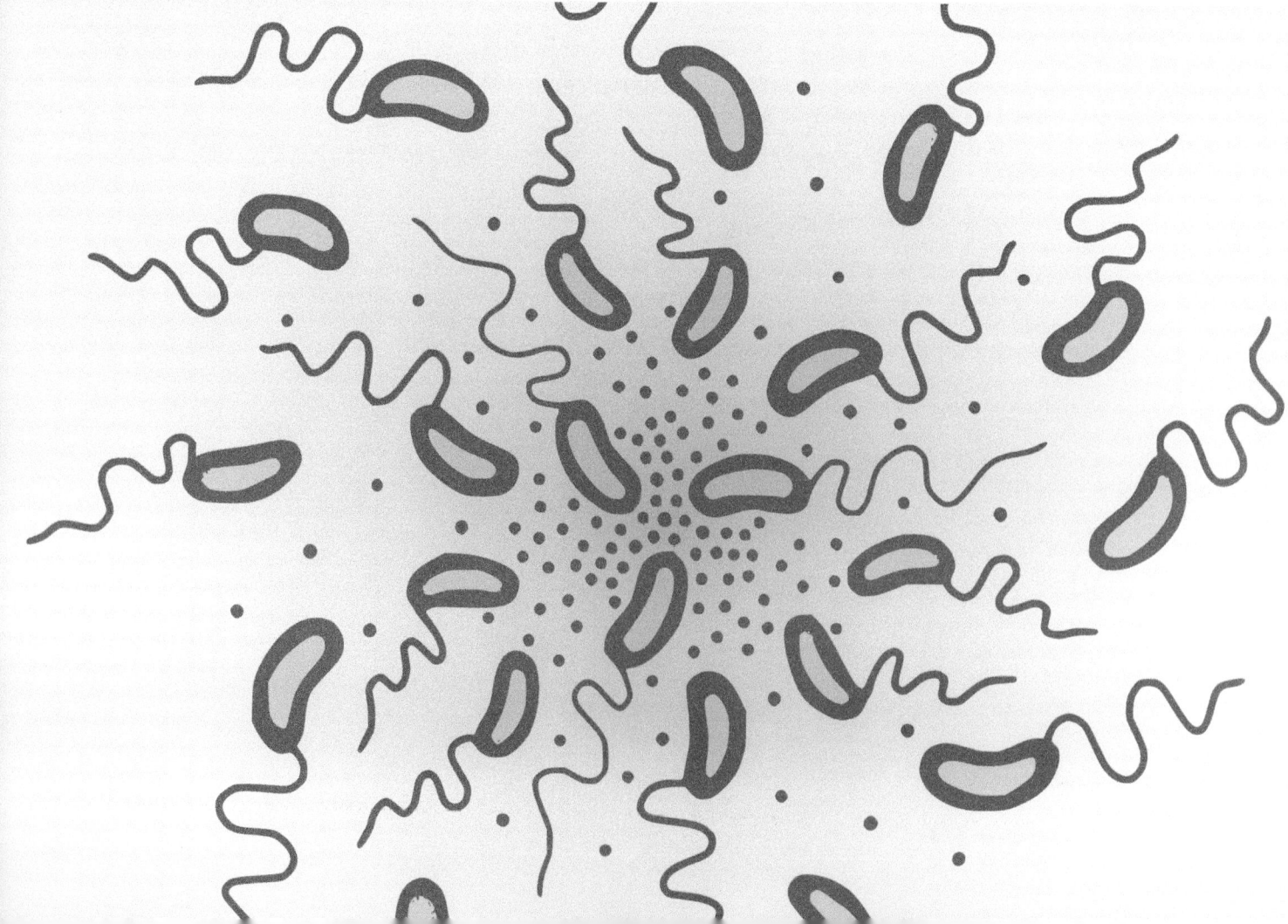

16

QUORUM SENSING

After completing the material in this chapter, students should be able to . . .

1. define the concept of quorum sensing

2. describe the symbiotic relationship between *Vibrio fischeri* and squid

3. explain the biochemical mechanism and regulation of bacterial bioluminescence

4. compare and contrast the fundamental components for the quorum-sensing systems of Gram-negative proteobacteria and Gram-positive bacteria

5. apply quorum quenching to manipulate bacterial phenotypes

Microbial Physiology: Unity and Diversity, First Edition. Ann M. Stevens, Jayna L. Ditty, Rebecca E. Parales and Susan M. Merkel.
© 2024 American Society for Microbiology.
Companion website: www.wiley.com/go/stevens/microbialphysiology

Making Connections

Motile microbes exhibit specific behaviors in response to external signals, such as chemotactic responses to attractants or repellents (Chapter 15). Interestingly, many bacteria produce signaling molecules that elicit physiological self-responses and/or responses from neighboring cells. This chapter focuses on the phenomenon of **quorum sensing**, whereby microbial cell-to-cell signaling molecules are used to facilitate coordinated behaviors across an entire population of cells. Quorum-sensing communication systems enable the microbes to function as a coordinated multicellular unit when cell densities reach a threshold level. Quorum sensing is involved in the regulation of bacterial bioluminescence, virulence, and genetic competence (this chapter), sporulation (Chapter 17), and biofilm formation (Chapter 18), as well as many other processes.

Fundamentals of Quorum Sensing

Quorum sensing is a form of bacterial cell-to-cell communication across a population that was originally known as autoinduction. As a bacterial population grows, cells produce, release, and sense chemical signals called **autoinducers**, which are classified based on their chemical composition. At low cell density, the concentration of autoinducer present in and around bacterial cells remains low. However, at high cell density, a high concentration of autoinducer may accumulate (Fig. 16.1). Rather than directly sensing the number of cells present in the local environment, bacterial cells indirectly measure population density via the concentration of the autoinducer molecules present. When the autoinducer concentration reaches a critical **threshold level**, the cells sense the autoinducer signals and they coordinate behaviors that benefit the bacterial population as a whole. Quorum sensing was first discovered in the 1960s and 1970s in Gram-positive bacteria performing conjugation and Gram-negative

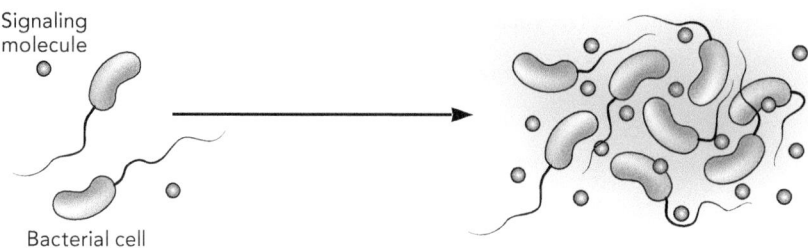

Signaling molecule

Bacterial cell

FIGURE 16.1 **Process of bacterial quorum sensing.** Individual bacteria cells, such as *Vibrio fischeri*, produce and release signaling molecules commonly termed autoinducers. At low cell density, there is a low autoinducer concentration, but at high cell density, the autoinducer concentration rises and triggers a quorum-sensing physiological output across the population, such as bioluminescence, as shown by blue-green shading.

marine bioluminescent bacteria, respectively. However, it is now appreciated that quorum-sensing regulatory mechanisms are widespread, although not universal, across the domain *Bacteria*. Some of the bacterial phenotypic outputs controlled by quorum sensing include bioluminescence, virulence factor production, exoenzyme production, biofilm formation, antibiotic synthesis, and genetic competence. These physiological outputs are all beneficial to the cells at high cell density; production by individual cells would have little impact and would consequently be a waste of energy. Therefore, quorum sensing is like a quorum in politics, whereby a certain number of individuals is required to pass a law or policy. In the case of bacteria, a quorum is necessary to generate a shared phenotypic output that permits the bacteria to function in essence as a multicellular unit.

Quorum Sensing and Bioluminescence in the *Vibrio fischeri*-Squid Symbiosis

The phenomenon of quorum sensing has been studied intensively in bioluminescent bacteria in relation to their symbiotic association with animal hosts (e.g., fish and squid). In particular, the relationship between *Vibrio fischeri* (recently named *Aliivibrio fischeri*) and the Hawaiian near-shore bobtailed squid, *Euprymna scolopes*, has been examined in detail. These squid have a specialized **light organ** that houses a pure culture of *V. fischeri*. In this environment, the bacteria can grow to high cell densities using nutrients provided by the squid. Under conditions of high cell density, the *V. fischeri* cells produce blue-green light, which is used by the squid for counterillumination. Counterillumination is a defense mechanism whereby the squid projects light downward at an intensity equal to that of moonlight shining from above (Fig. 16.2A), allowing the squid to hide its shadow from predators when it comes out to forage for food at night. Thus, there is a strong selective pressure for the squid to form a symbiotic association with bioluminescent *V. fischeri*.

Juvenile squid (Fig. 16.2B and C) lack bacterial symbionts when they hatch and must acquire *V. fischeri* from the environment. The association is highly specific; of all the microbes in the open ocean, only *V. fischeri* is permitted to enter the squid tissues that will develop into the light organ. Colonization requires bacterial motility and chemotaxis (Chapter 15). During the colonization process, *V. fischeri* senses specific chemoattractants produced by the squid. The squid also produces antimicrobials that inhibit bacteria other than *V. fischeri*. Interestingly, the symbiotic association is essential to the life cycle of the squid. *V. fischeri* colonization and light production are required for normal squid gene expression and development. In response to the presence of the bacteria, the light organ will mature (Fig. 16.2C and D); it will not develop in the absence of *V. fischeri*. In addition, only bioluminescent-proficient *V. fischeri* are maintained, as research has shown that dark mutant strains are outcompeted by bright wild-type strains within the light organ. This symbiotic relationship has been likened to that of rhizobia and legumes, which leads to root nodule development (Chapter 12). In this case, the *V. fischeri*-squid system is considered a model for studying interactions between a bacterial monoculture (versus a complex microbiome) and an animal host. If egg clutches are moved to an environment away from adult animals, then the juvenile animals that hatch can be artificially inoculated in the laboratory with genetically manipulated *V. fischeri* cells to study the functions essential for the symbiosis.

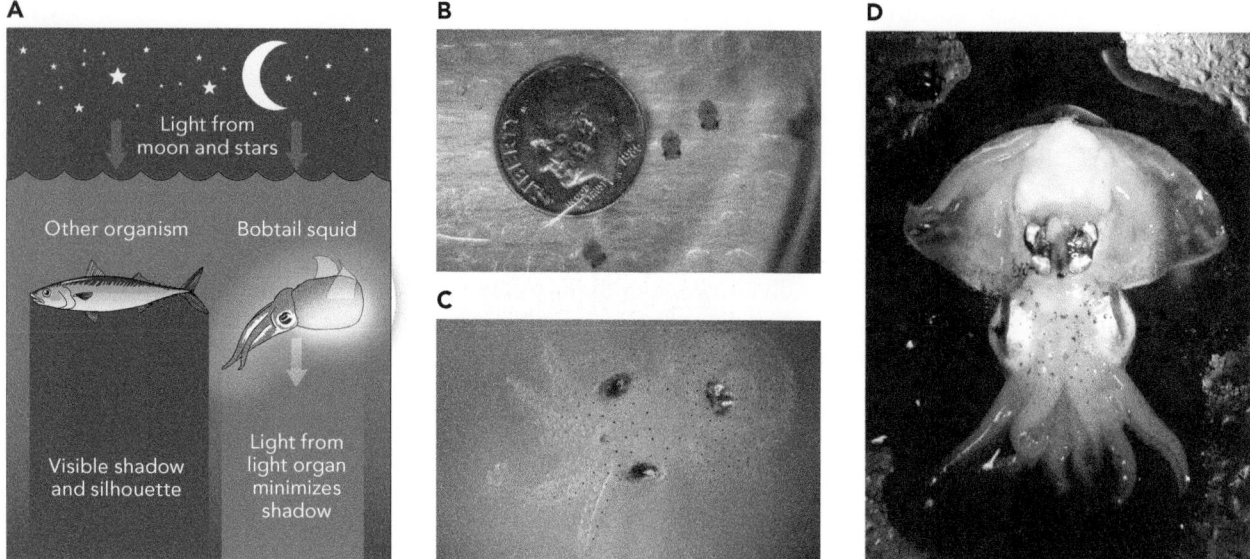

FIGURE 16.2 **Symbiotic relationship between *Vibrio fischeri* and squid.** (A) When a symbiotic relationship forms between *V. fischeri* and squid, the bacteria are provided with a nutrient source and in turn they provide light that the squid uses for counterillumination. Counterillumination permits the squid to mask its presence by the downward projection of light that is equivalent to the intensity of moon and star light. (B) Juvenile squid hatched from egg clutches, shown in proportion to a dime. (C) A magnified photo of a juvenile squid with the developing light organ (silver bilobed structure) in the center of the mantle cavity. (D) The bilobed light organ (silver in color) that houses *V. fischeri* is shown in the mantle cavity of a dissected adult animal (~3 to 4 cm in size) adjacent to the black ink sac. Panels B to D courtesy of Eric V. Stabb, University of Illinois-Chicago.

Within the light organ, *V. fischeri* reaches concentrations of ~10^{10} to 10^{11} cells/ml utilizing nutrients provided by the squid host. In turn, the host gains the capacity for defensive counterillumination (Fig. 16.2A). The production of light costs *V. fischeri* significant biological energy, but when living in the squid, *V. fischeri* cells are provided with sufficient resources from the host not only for growth but also for the energy needed to make light. Since the squid are nocturnal, they purge most of the bacteria from their light organ every morning when they bury themselves in sand and go to sleep for the day as they do not need bacterial light production. By evening, the bacterial population has regrown to a high cell density and is thus capable of producing light for the animal. When *V. fischeri* is released into the ocean from the daily release of bacteria at dawn, the cell density drops by many orders of magnitude, to ~100 to 1,000 cells per milliliter in squid habitats (with concentrations much lower in the open ocean). In the open ocean, the bacteria encounter low nutrient levels and face starvation conditions and can no longer afford to expend energy to make light. The regulation of light production by quorum sensing permits the bacteria to determine whether they are at high cell density within a light organ or at low cell density in the open ocean.

BIOCHEMICAL BASIS OF BACTERIAL BIOLUMINESCENCE

Bioluminescence is exhibited by both bacteria and eukaryotes, but the two processes are biochemically different. Bacterial bioluminescence utilizes a distinctive **luciferase** enzymatic reaction that produces blue-green bioluminescent light (Fig. 16.3):

$$\text{FMNH}_2 + \text{RCHO}\left(\text{long-chain aldehyde}\right) + \text{O}_2 \rightarrow$$
$$\text{FMN} + \text{RCOOH}\left(\text{long-chain acid}\right) + \text{H}_2\text{O} + \text{light}$$

A

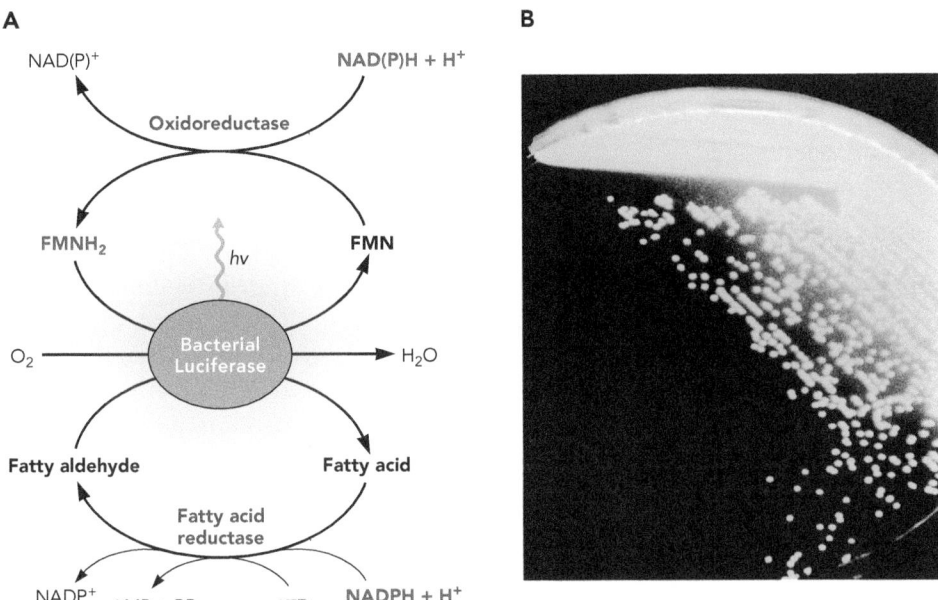

B

FIGURE 16.3 **Enzymatic reaction catalyzed by bacterial luciferase.** (A) Bacterial luciferase is a mixed-function oxidase that requires the reduced coenzyme $FMNH_2$, a fatty aldehyde, and oxygen as substrates to catalyze blue-green light production. The fatty aldehyde is regenerated by a fatty acid reductase complex using a fatty acid, $NADPH + H^+$, and ATP as substrates. $FMNH_2$ is reduced by a flavin oxidoreductase that requires $NAD(P)H + H^+$. Thus, large amounts of biological energy are required indirectly to catalyze the luciferase reaction. (B) Streak plate image of bioluminescent *V. fischeri* colonies photographed in the dark. Image reprinted from the cover of the *Journal of Bacteriology*, Volume 184, Issue 16.

Bacterial luciferase is a mixed-function oxidase that uses oxygen as the electron acceptor in a redox reaction under aerobic conditions. To complete another cycle of this reaction, both the oxidized long-chain acid (RCOOH) and the flavin mononucleotide (FMN) must be recycled back to the reduced forms.

$$FMN + NAD(P)H + H^+ \rightarrow FMNH_2 + NAD(P)^+$$
$$RCOOH + NADPH + H^+ + ATP \rightarrow RCHO + NADP^+ + AMP + PPi$$

These reduction reactions are energetically expensive since reducing power is used to produce $FMNH_2$ via a flavin oxidoreductase, and ATP plus reducing power is used to regenerate the aldehyde substrate (i.e., tetradecanal in the case of *V. fischeri*) via a fatty acid reductase. The overall process requires a huge amount of energy; ~50 ATPs are required to produce just 1 photon of light. A *V. fischeri* cell commits the equivalent of 6,000 to 60,000 ATP per second to light production. Regulation via quorum sensing ensures that the energy-consuming reactions necessary to produce light occur only under conditions where the bacteria have sufficient energy levels to drive the production of light. In comparison, eukaryotic luciferase, such as that found in fireflies, also generates light, but the process is not regulated by quorum sensing, and the eukaryotic enzyme directly consumes ATP.

Why did bacteria evolve bioluminescence? There are several ideas that have been proposed. It may have evolved as an inefficient way to remove oxygen when the atmosphere on Earth became aerobic and the bacteria had to deal with oxidative stress (Chapter 17). In addition, the blue-green wavelength of light generated by bioluminescent vibrios has the capacity to cause photoreactivation, repairing DNA damage by reversing pyrimidine dimer formation caused by UV light. However, at some point, the bacterial light production evolved into an essential component of symbiotic relationships with marine animals, and those relationships are still maintained.

HISTORY OF THE DISCOVERY OF QUORUM SENSING AND THE *LUX* GENES

Some of the initial studies of quorum sensing in relation to bioluminescence were performed by Nealson, Platt, and Hastings in the 1970s. These researchers did a simple but elegant experiment (Fig. 16.4A) in which they first grew the *V. fischeri* cells to high cell density where the cells were bioluminescent. They harvested the cell-free supernatant, also known as spent medium, and then added it to a low-cell-density *V. fischeri* culture that would normally not be bioluminescent. The supernatant contained a chemical they termed "autoinducer" that had been released by the bacteria at high cell density. The addition of the autoinducer tricked the *V. fischeri* cells into producing light at low cell density. This process, which is known today as quorum sensing, was initially called autoinduction.

In the 1980s, Engebrecht and Silverman cloned an ~9-kb fragment of *V. fischeri* DNA (Fig. 16.4B) into *Escherichia coli*. The *E. coli* cells not only produced light, but they regulated the light via quorum sensing. Fortuitously, all the genes necessary for the density-dependent production of light were organized into one contiguous DNA fragment instead of being scattered around the chromosome (Fig. 16.4B).

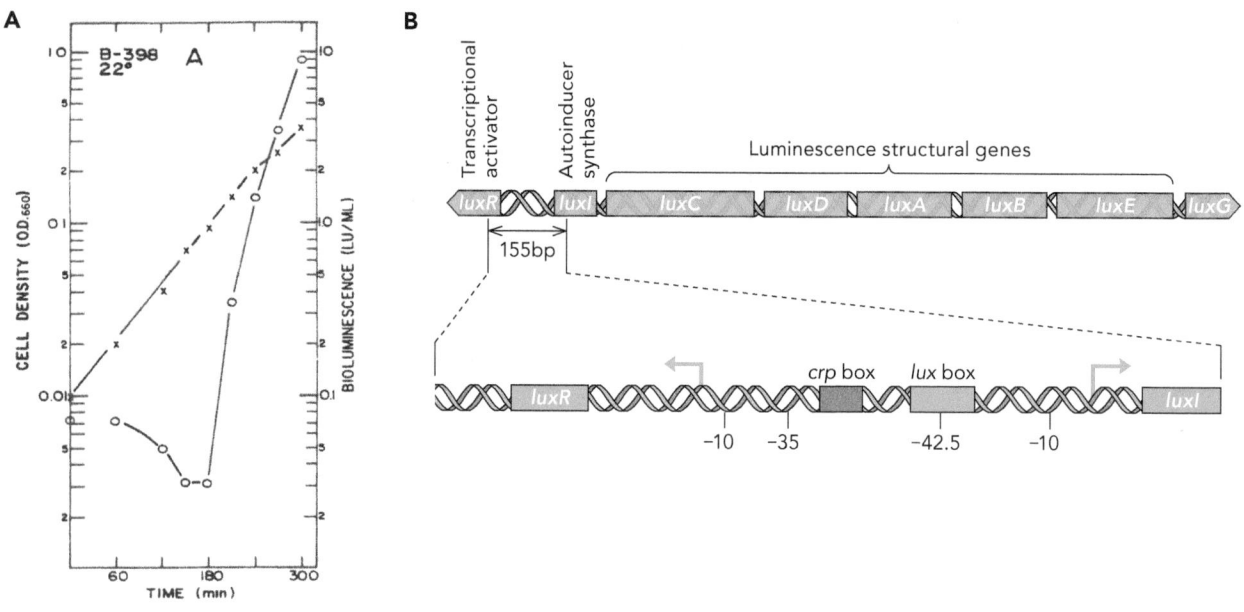

FIGURE 16.4 **Discovery of quorum sensing and the required genes in *Vibrio fischeri*.** (A) Nealson, Platt, and Hastings found that *V. fischeri* produced high levels of bioluminescence beginning in mid-exponential phase once the autoinducer molecules hit a critical threshold concentration. Reprinted from Nealson K. 1977. *Arch Microbiol* 112:73-79, with permission. X, cell density; O, bioluminescence. (B) Engebrecht and Silverman determined that the *V. fischeri lux* genes encoding bioluminescence and quorum-sensing regulation are located in two divergent transcriptional units; one encodes the transcriptional activator LuxR and the other, the *lux* operon, encodes the autoinducer synthase LuxI and the proteins needed for light production, including the luciferase enzyme composed of LuxA and LuxB. The promoter region for the two transcriptional units and its key features are highlighted, including the *lux* box where LuxR binds to activate transcription of the *lux* operon. See the text for additional details.

The genes are arranged in an operon, *luxICDABEG*; the gene designation *lux* is short for Luxor, the Egyptian city of light. The genes *luxAB* encode the heterodimeric luciferase enzyme, *luxCDE* encode the fatty acid reductase complex needed to regenerate the aldehyde substrate, *luxG* is not essential to the process in *E. coli* (its product likely helps to regenerate reduced coenzymes), and *luxI* encodes the **autoinducer synthase**, an enzyme that produces autoinducer. The *luxI* gene is the first gene in the *lux* operon. Transcription of the *lux* operon is regulated by LuxR, a transcription factor that binds at position –42.5 bp upstream of the Class II *lux* operon promoter (Chapter 8) to activate transcription. The *luxR* gene is transcribed from its own Class I promoter, which is activated by CRP-cAMP (Chapter 8), and is divergently transcribed from the *lux* operon.

Basic Model of Quorum Sensing in Gram-Negative Proteobacteria

V. fischeri has served as a model system for understanding the basic regulatory components required for the quorum-sensing response in Gram-negative proteobacteria (Fig. 16.5). At low cell density, there is a low level of transcription of the *luxI* gene (and other genes in the *lux* operon). The LuxI enzyme (autoinducer synthase) produces the autoinducer signal, 3-oxo-hexanoyl-homoserine lactone, an **acyl-homoserine lactone** (AHL) molecule also known as 3-oxo-C6-HSL. LuxI produces AHL using intermediates from amino acid and fatty acid biosynthesis (Fig. 16.5A). The structure of the AHL permits it to freely diffuse through the cell envelope of *V. fischeri*. At low cell density, AHL diffuses out and away from the cell, driven by the AHL chemical gradient; therefore, the concentration of AHL remains below the threshold concentration required to activate transcription of the *lux* genes.

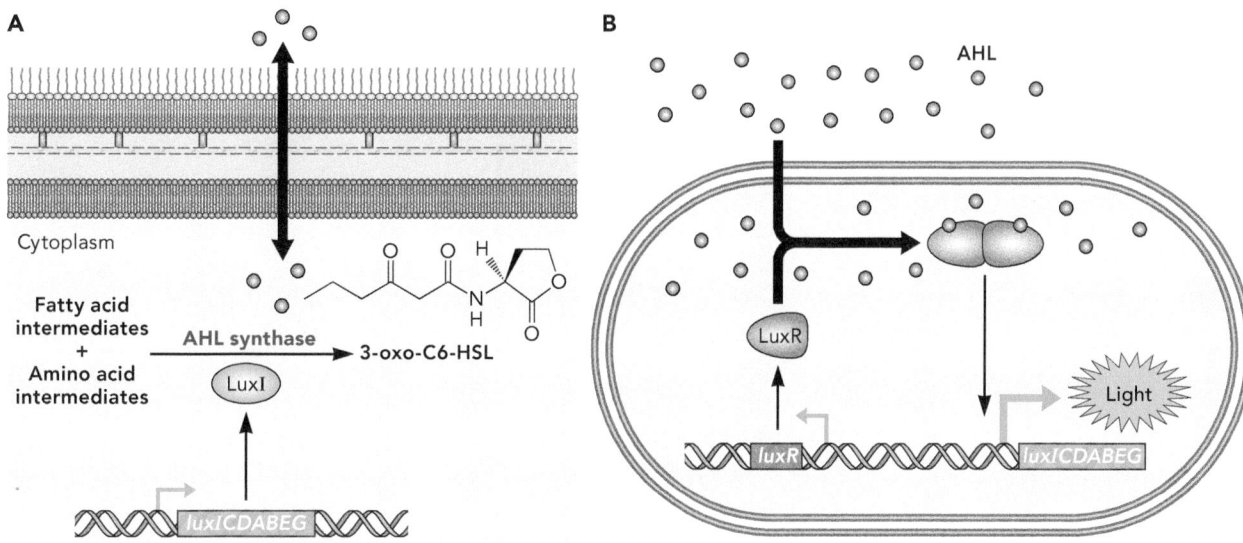

FIGURE 16.5 *Vibrio fischeri* **as a model for quorum sensing in Gram-negative proteobacteria.** (A) The *luxI* gene encodes the autoinducer synthase enzyme LuxI, which produces the 3-oxo-hexanoyl-homoserine lactone (3-oxo-C6-HSL) signal molecule. This molecule is able to diffuse across the cell envelope. (B) At high cell density, there is a high concentration of 3-oxo-C6-HSL, which then form complexes with LuxR. Homodimers of LuxR associated with 3-oxo-C6-HSL bind to target sites in the DNA including upstream of the *lux* operon promoter. The *lux* operon is activated through a positive feedback loop, which increases the transcription of the *lux* operon, resulting in an increase in light production.

At high cell density, because there are many more cells producing AHL, AHL accumulates and diffuses back into the cells. When AHL reaches a threshold level in the cytoplasm, it binds to the regulatory protein LuxR (Fig. 16.5B). LuxR is a transcription factor that is inactive in the absence of AHL (at low cell density), but in the presence of the AHL (at high cell density) forms an active homodimer. In its active AHL-associated conformation, LuxR binds target promoters including the *lux* operon promoter. This positive feedback loop produces more LuxI and thereby more AHL, and results in very rapid induction of light production once the threshold concentration of AHL accumulates (Fig. 16.4A).

It is now known that many other proteobacteria use similar quorum-sensing systems. Most of these systems have one, phylogenetically related, LuxR **homolog** serving as a regulator of transcription and one LuxI homolog that functions as an autoinducer synthase, but some organisms, like *Pseudomonas aeruginosa*, have more than one LuxR/LuxI homolog pair (see below). Most, but not all, AHL molecules are capable of diffusing across the cell envelope. They differ from one another in length (between 4 and 18 carbons) and functional groups on the acyl side chain, which may have keto or hydroxyl side groups. Some side chains also contain double bonds. The lactone ring, on the other hand, is highly conserved.

Most LuxR homologs function as activators, but some may also function as repressors. An example of one such LuxR homolog is EsaR from the plant pathogen *Pantoea stewartii* subsp. *stewartii*. EsaR is biologically active and is capable of binding DNA targets in the absence rather than the presence of its cognate AHL. EsaR both activates and represses genes at low cell density; as the AHL concentration rises, EsaR is inactivated, leading to gene deactivation and derepression, respectively. Thus, quorum sensing has evolved in two parallel ways, with some LuxR homologs being stimulated by AHL while others are inhibited by it. Despite these variations, bacterial quorum-sensing systems play important roles in regulating a variety of bacterial physiological outputs including virulence and biofilm formation. There is great interest in understanding the role of quorum sensing in causing bacterial infections.

QUORUM SENSING AND VIRULENCE IN *PSEUDOMONAS AERUGINOSA*

From the 1970s until the early 1990s, it was thought that AHL-based quorum sensing was limited to bioluminescent marine bacteria. Then in 1992, it was discovered that the opportunistic human pathogen *Pseudomonas aeruginosa* used quorum sensing to control expression of its virulence factors. Since then, many other human, animal, and plant pathogens have been demonstrated to use quorum sensing (Table 16.1). The *P. aeruginosa* system also demonstrated that quorum-sensing systems can be more complex and have more than just one set of LuxR- and LuxI-like proteins. Two regulatory protein sets were initially described in *P. aeruginosa*: LasR/LasI and RhlR/RhlI. Collectively, these regulatory systems control hundreds of genes in a coordinated manner, including genes essential for virulence and biofilm formation (Chapter 18). The two quorum-sensing systems in *P. aeruginosa* function in a hierarchical fashion, with the Las system controlling the Rhl system (Fig. 16.6). LasR and its associated AHL (3-oxo-C12-HSL) increase expression of genes for virulence, biofilm production, and RhlR. RhlR is activated upon binding to a four-carbon AHL (C4-HSL) produced by RhlI. Once activated, RhlR increases the expression of *rhlI* as well as

TABLE 16.1 **Examples of the diversity of quorum-sensing signaling**

Genus	Autoinducer signal molecule	Physiological output
Gram-negative		
Vibrio	Acyl-homoserine lactones, AI-2	Bioluminescence, pathogenesis
Pseudomonas	Acyl-homoserine lactones, quinones	Pathogenesis, biofilms
Pantoea	Acyl-homoserine lactones	Biofilms, motility
Agrobacterium	Acyl-homoserine lactones	Conjugation
Rhodopseudomonas	Aryl (*p*-coumaroyl)-homoserine lactones	Chemotaxis
Xanthomonas	Diffusible signal factor (α-, β-unsaturated fatty acid)	Pathogenesis (plant)
Proteus	Putrescine	Swarming
Anabaena	Peptides	Cell differentiation
Myxococcus	Amino acids	Fruiting body development
Gram-positive		
Bacillus	Peptides	Competence, sporulation
Enterococcus	Peptides	Conjugation, plasmid maintenance, pathogenesis
Streptococcus	Peptides	Competence, pathogenesis
Streptomyces	γ-butyrolactones	Secondary metabolism, antibiotic production

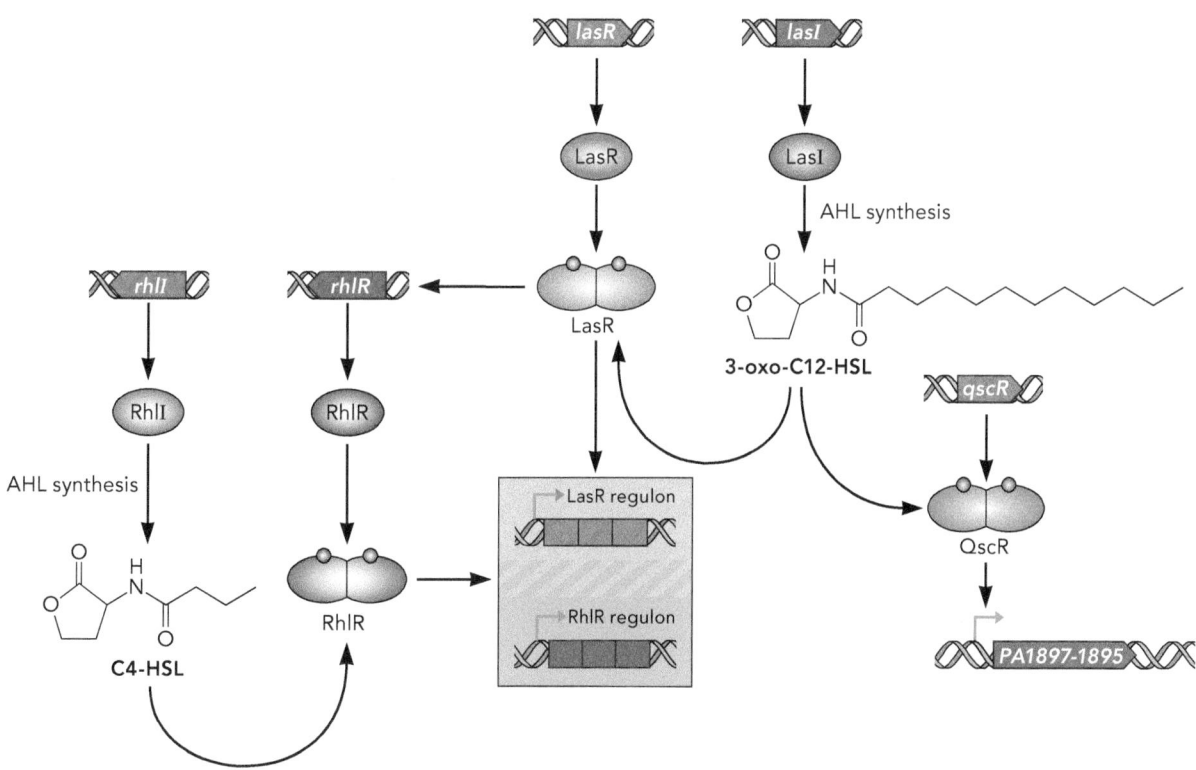

FIGURE 16.6 **Hierarchical quorum-sensing system in *Pseudomonas aeruginosa*.** *P. aeruginosa* utilizes more than one quorum-sensing regulatory system. The 3-oxo-C12-HSL signal produced by the LasI autoinducer synthase binds to the LasR transcriptional regulator. LasR activates expression of its regulon (indicated by purple), including *rhlR*. The RhlR transcriptional regulator binds to C4-HSL, the product of the RhlI autoinducer synthase, and then activates expression of its regulon (indicated by teal). The hatched area indicates genes regulated by both LasR and RhlR. An orphan LuxR homolog, QscR, also recognizes 3-oxo-C12-HSL to activate a three gene operon.

additional virulence genes, including *rpoS* (encodes the stationary-phase sigma factor; Chapter 4). Strains with *lasR/I* mutations have defective biofilm formation, and mutants lacking LasR/I and/or RhlR/I are significantly less pathogenic in animal models. By inducing virulence factors and biofilm production once cells reach high cell density, the population can avoid the immune response until enough bacteria have accumulated to overwhelm the immune system and result in a successful infection. Mice inoculated intranasally with wild-type *P. aeruginosa* get pneumonia, bacteremia, and die, while quorum-sensing mutants colonize, but do not cause pneumonia or bacteremia, and the mice recover.

In addition to having more than one set of LuxR/I homologs, *P. aeruginosa* also has an "orphan" LuxR homolog. The 3-oxo-C12-HSL signal produced by LasI also binds to and activates a LuxR homolog called QscR (Fig. 16.6), which serves as an additional quorum-sensing regulator controlling a promoter for a three-gene operon (*PA1897-1895*). QscR is considered to be an orphan LuxR because it has no QscI-associated autoinducer synthase.

While AHL autoinducer signals were among the first to be studied, many non-AHL signals also play a role in quorum sensing in a variety of microbes (Table 16.1). For example, subsequent to the discovery of LasR/LasI and RhlR/RhlI, it was determined that *P. aeruginosa* has the ability to produce a non-AHL signal, *Pseudomonas* quinone signal (PQS; 2-heptyl-3-hydroxy-4-quinolone). PQS is sensed via PqsR, a transcriptional regulator in the LysR protein family that is unrelated to LuxR. Interestingly, PQS is thought to play a role in inducing membrane curvature that leads to the formation of outer membrane vesicles (OMVs). OMV formation mediates PQS packaging and transport. OMVs are ubiquitous in Gram-negative bacteria; they serve to mobilize periplasmic components and have been implicated in virulence-associated behaviors. Thus, PQS is an example of the many non-AHL signals that have been discovered (Table 16.1).

Basic Model of Quorum Sensing in Gram-Positive Bacteria

The mechanism of quorum sensing in Gram-positive bacteria (Fig. 16.7) evolved in a completely different manner than in the Gram-negative proteobacteria (Fig.16.5). Autoinducers in Gram-positive bacteria are typically small peptides rather than AHLs. These autoinducers are produced as peptide precursors that are processed and modified to the mature form before being actively transported outside of the cell (Fig. 16.7A) by an ABC transporter (Chapter 13). These **peptide signals** (also called pheromones) are sensed by a membrane-associated sensor histidine kinase that autophosphorylates and then transphosphorylates a response regulator (Chapter 7). The phosphorylated response regulator can then bind to target promoters through its helix-turn-helix motif (Fig. 16.7B). Often one of these promoters controls the gene encoding the peptide signal, thereby creating a positive feedback loop. Thus, in comparison to Gram-negative cells, the Gram-positive quorum-sensing systems use a peptide signal that must be transported out of the cell (rather than an AHL signal that diffuses) and a two-component signal transduction cascade (rather than a LuxR homolog). Gram-positive bacteria control a variety of genes using quorum sensing, including genes encoding

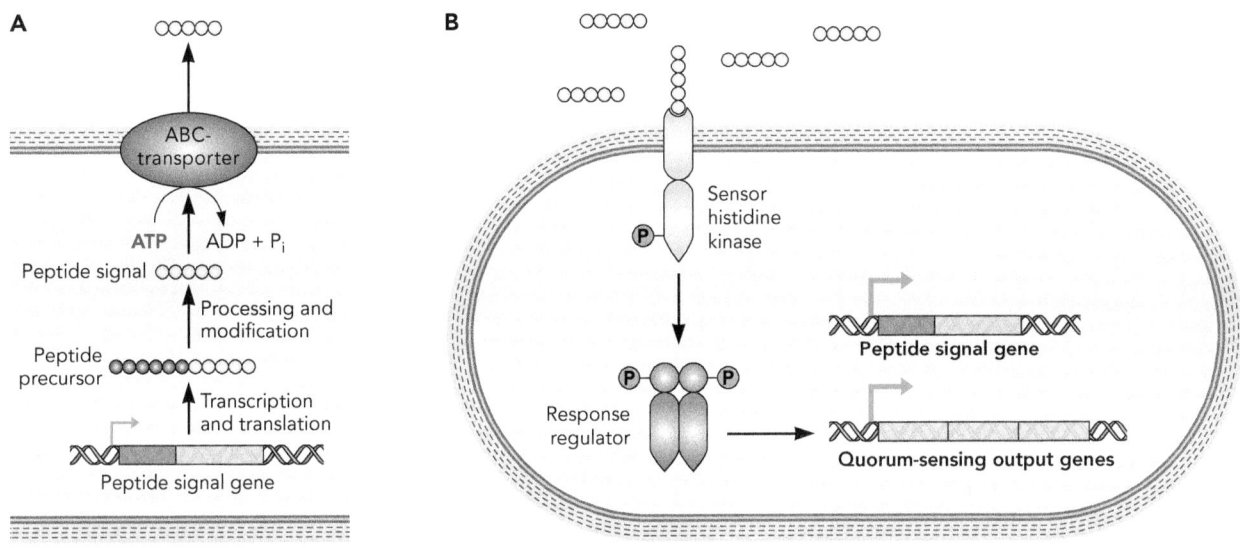

FIGURE 16.7 **Generic model for quorum sensing in Gram-positive bacteria.** (A) The signal molecule is a peptide that is transcribed and translated from a gene in the chromosome and then proteolytically processed and modified prior to active transport across the cell envelope by an ABC transporter. (B) The peptide signal (five yellow circles) is sensed by a sensor histidine kinase protein that first autophosphorylates and then transphosphorylates a response regulator. The response regulator then binds to the DNA to activate expression of target genes, which may include the gene encoding the peptide signal, creating a positive feedback loop.

virulence factors, sporulation (Chapter 17), and **cellular competence**, which allows cells to take up exogenous DNA (Table 16.1).

Bacillus subtilis uses quorum sensing to activate both competence and sporulation, two mutually exclusive processes. When cell density reaches a certain threshold, cells will first induce competence in an attempt to acquire new genes from neighboring cells. At higher cell densities, sporulation can be induced (Chapter 17). There are two peptides, ComX and competence and sporulation factor (CSF), involved in quorum-sensing regulation of cellular competence (Fig. 16.8). ComX is a 10-amino-acid oligopeptide that is processed from a larger 55-amino-acid precursor peptide. One tryptophan residue in ComX is modified by attachment of a lipid. ComX is sensed by the ComP sensor histidine kinase. When the quorum-sensing signal ComX reaches a threshold concentration, ComP autophosphorylates a histidine residue and then transphosphorylates its cognate response regulator ComA. ComA-P serves as an activator for the transcription of *comS*, leading to the activation of a multistep pathway for the development of competence.

ComA-P is regulated by RapC (a phosphatase), which is under control of the CSF signaling system. A 40-amino-acid peptide encoded by the *phrC* gene is partially processed and actively transported out of the cell before it is proteolytically processed into a 5-amino-acid oligopeptide called CSF. CSF accumulates outside the cell in a cell-density-dependent manner and then reenters the cell through an active transporter, Opp. As CSF begins to reenter the cell and is present at low cytoplasmic concentrations (1 to 5 nM), it inhibits the activity of RapC, resulting in higher levels of ComA-P and the activation of competence genes.

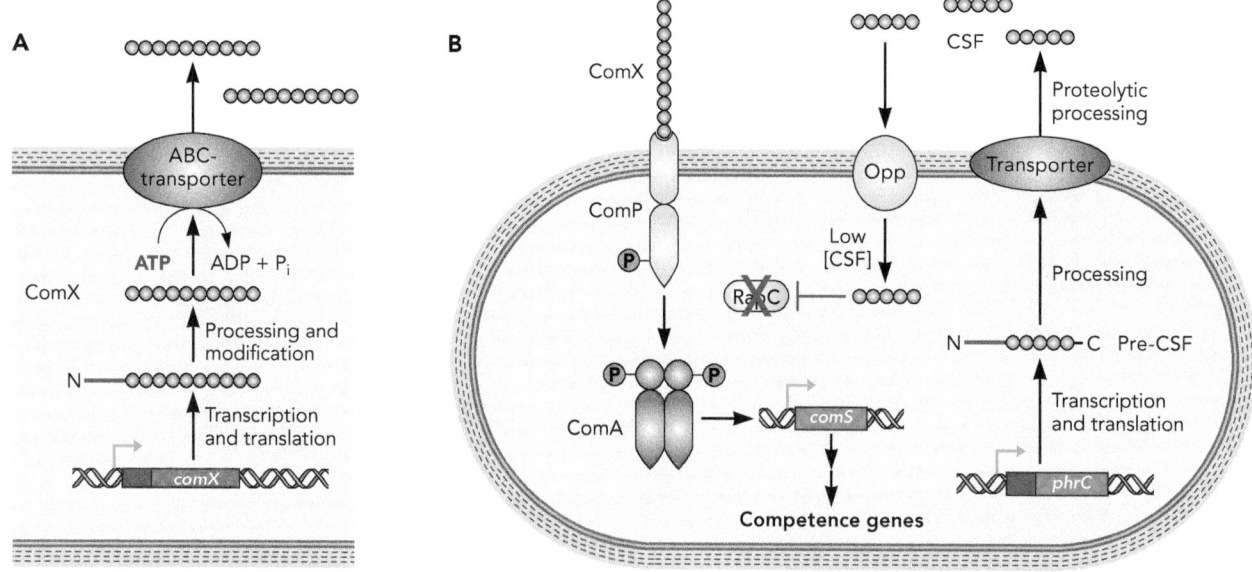

FIGURE 16.8 **Quorum-sensing control of genetic competence in *Bacillus subtilis*.** (A) The peptide transcribed and translated from *comX* undergoes proteolytic processing and modification to convert it to the decapeptide ComX, which is actively transported outside of the cell. (B) ComX is sensed by the sensor histidine kinase ComP, which autophosporylates and then transphosphorylates the response regulator ComA. ComA-P activates transcription of *comS*. ComS then initiates expression of the genes needed for the cell to become competent to take up exogenous DNA. A peptide transcribed and translated from *phrC* is partially processed and actively transported outside of the cell, where it is further proteolytically processed to produce a pentapeptide signal, competence and sporulation factor (CSF). As CSF accumulates outside of the cell, it is transported into the cytoplasm by Opp. At low concentrations, CSF represses the activity of the RapC phosphatase, keeping ComA in a phosphorylated state. This results in continued expression of the competence genes.

Interspecies Communication: the LuxS System

The AHL- and peptide-based quorum-sensing systems mediate self-recognition in that the signal produced by a given bacterial species is also sensed by members of that species. However, evidence of cross talk between different bacteria was discovered in the bioluminescent bacterium *Vibrio harveyi*. In addition to an AHL signal, *V. harveyi* was found to produce another signal that is recognized by phylogenetically unrelated bacteria. This "universal" signal was first called AI-2, as it was the second autoinducer found in *V. harveyi*. Structurally, it was identified as a boronated **furanone** synthesized by the LuxS protein (Fig. 16.9). The LuxS quorum-sensing system is widespread in both Gram-negative (e.g., *Vibrio cholerae*) and Gram-positive bacteria. AI-2 is involved in modulating the expression of virulence genes, biofilm formation, and motility in many bacteria. In particular, the AI-2 system may be important for cross-species communication in bacterial communities where different groups of bacteria cooperate and/ or compete (e.g., the human microbiome and biofilms; Chapter 18).

Regulatory Cascade Controlling Quorum Sensing in *Vibrio cholerae*

V. cholerae is a human pathogen capable of causing global pandemics of the disease cholera (Chapter 7). Here we will use it to highlight the complexity of a quorum-sensing system, where different components are joined together in

A

B

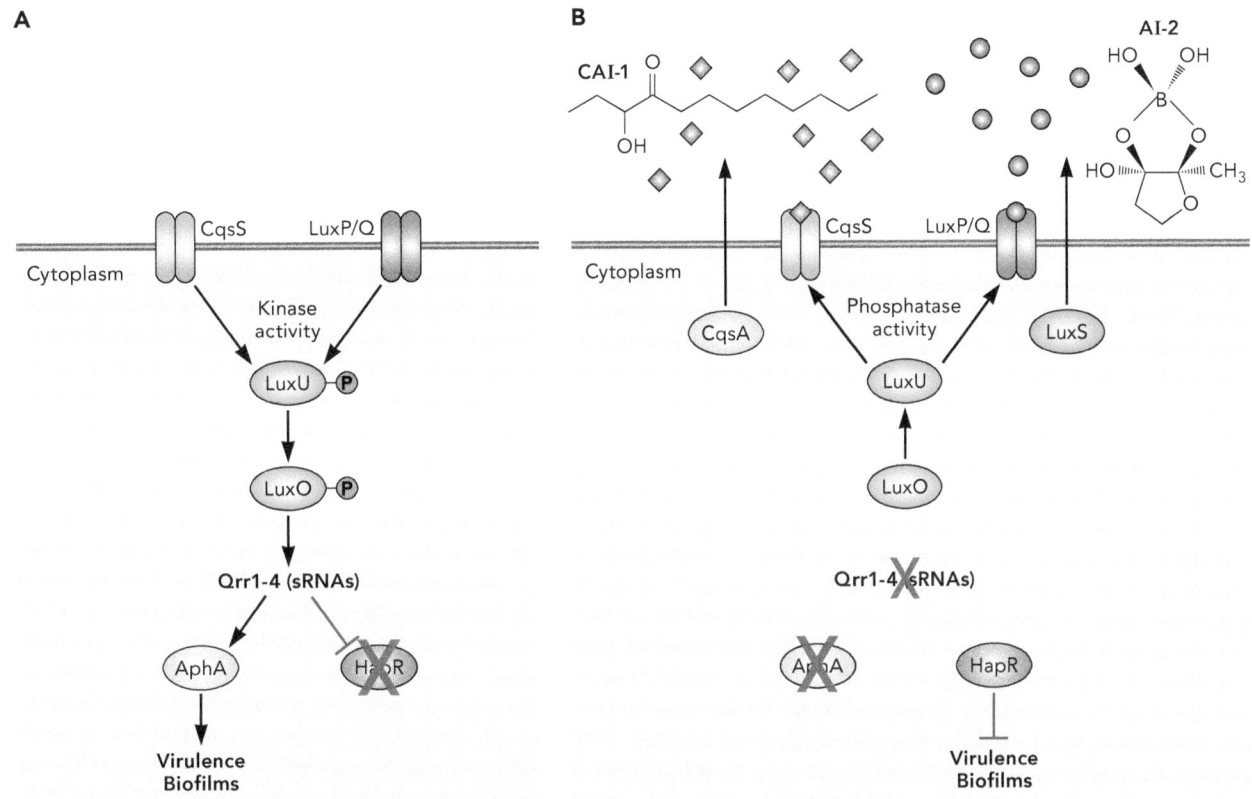

FIGURE 16.9 **Quorum-sensing control of virulence and biofilm formation in *Vibrio cholerae*.** (A) At low cell density, when there are low levels of the signaling molecules in the environment, CqsS and LuxP/Q act as kinases and phosphorylate LuxU. LuxU-P can then phosphorylate LuxO. LuxO-P activates expression of four sRNAs, Qrr1-4, which posttranscriptionally promote production of AphA and inhibit production of HapR. AphA enhances expression of virulence factors (i.e., cholera toxin) and biofilm formation genes at low cell density. (B) At high cell density, the two quorum-sensing signals, CAI-1 (produced by CqsA) and AI-2 (produced by LuxS), are bound by CqsS and LuxP/Q, respectively. The signal receptors (CqsS and LuxP/Q) now function as phosphatases. Under these conditions, AphA is not produced and HapR inhibits transcription of the virulence and biofilm genes.

the signal transduction pathway to control the quorum-sensing-regulated output. These types of signaling cascades are common in *Vibrio* spp., including *V. fischeri, V. harveyi,* and *V. cholerae. V. cholerae* uses two major parallel quorum-sensing systems with two different signals, cholera autoinducer-1 (CAI-1) and autoinducer-2 (AI-2; see above). The autoinducers interact with two cognate inner membrane sensor histidine kinase receptors, CqsS and LuxP/Q, respectively. The receptors transduce information to LuxO, a global response regulator that generates a different response depending on the signal concentration. In *V. cholerae*, at low cell density the quorum-sensing regulator AphA promotes production of virulence factors and biofilm structures, whereas at high cell density regulation by HapR causes rapid dispersion of bacteria from the biofilm, facilitating transmission of the pathogen.

When autoinducers are at a low concentration outside the cell, they are not associated with the CqsS and LuxP/Q sensor histidine kinases, which function to phosphorylate LuxU, a protein that catalyzes phosphotransfer to LuxO. LuxO-P then activates the transcription of a set of small RNA (sRNA) molecules, quorum regulatory RNAs (Qrr1-4; Fig. 16.9A). The sRNAs together with the RNA chaperone Hfq (Chapter 8) posttranscriptionally activate and repress the translation of two master regulators, AphA and HapR, respectively. AphA activates expression of genes for biofilm formation and virulence factors, including cholera toxin (Chapter 7).

Conversely, when the autoinducers are at a high concentration, they bind to the CqsS and LuxP/Q receptors (Fig. 16.9B). The receptors now function as phosphatases, dephosphorylating LuxU-P and LuxO-P, rendering them inactive. LuxO inactivation results in translation of the regulator HapR and downregulation of virulence and biofilm formation. Thus, low concentrations of autoinducers promote infection by *V. cholerae* via biofilm formation and virulence. High concentrations of autoinducers obstruct biofilm formation and virulence, thus permitting dispersion of the bacterium out of the intestine of its host.

Quorum Quenching

The process of blocking quorum sensing has been termed "**quorum quenching.**" The notion that it might be possible to use quorum quenching to interfere with native bacterial quorum sensing actually comes from nature, as several naturally occurring mechanisms that block quorum sensing have been identified. One type of quorum quenching occurs by degrading or modifying the AHL signal produced by Gram-negative bacteria through the activity of lactonase, acylase, or oxidoreductase enzymes (Fig. 16.10). **Lactonases** inactivate AHLs by cleaving the ester bond of the lactone ring to yield acyl-homoserines. On the other hand, **acylases** inactivate AHLs by hydrolyzing the amide bond connecting the lactone ring and the acyl side chain, generating homoserine lactone and a fatty acid, while **oxidoreductases** reduce keto groups to hydroxyl groups on the acyl chain. For example, some Gram-positive bacteria like *Bacillus* species produce a lactonase (AiiA) that cleaves the lactone ring of Gram-negative AHLs, rendering the AHL inactive. In fact, genetically modified plants producing AiiA have been shown to be more resistant to certain pathogens.

Another type of quorum quenching affects AI-2-producing bacteria. Furanones produced by red algae can block formation of bacterial biofilms on their surfaces by interfering with the bacterial boronated furanone signal recognition.

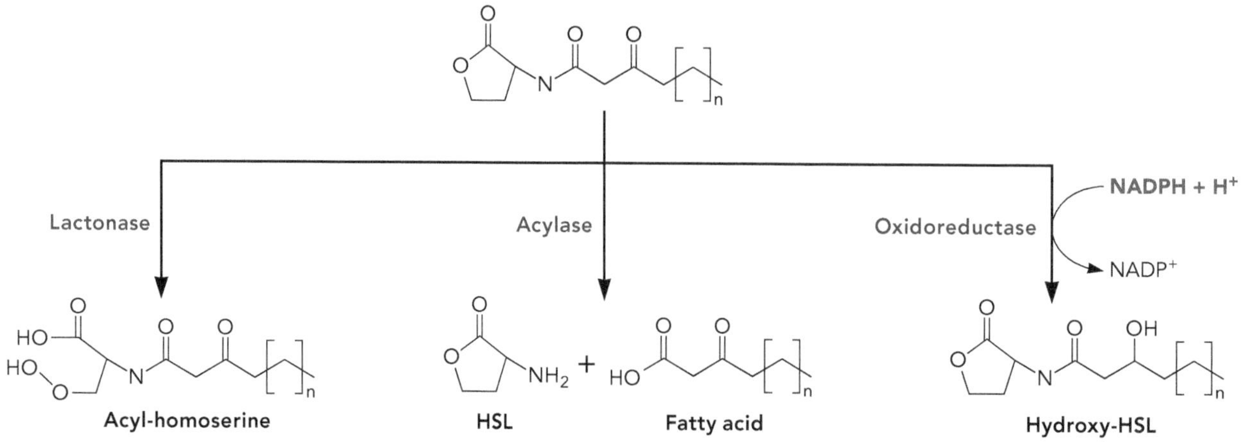

FIGURE 16.10 **Enzymes associated with quorum quenching of acyl-homoserine lactone (AHL).** The AHL signals produced by Gram-negative proteobacteria can be inactivated by lactonase, acylase, or oxidoreductase enzymes, as depicted.

Surface application of furanones has been applied to the treatment of environmental surfaces to prevent biofouling. Generation of synthetic molecules that mimic various quorum-sensing signals and therefore might interfere with normal recognition of the native signal is another approach to quorum quenching. However, there is the complication of possible cross talk between quorum-sensing signals. For example, *Agrobacterium* TraR recognizes a broad spectrum of AHLs when high levels of protein are produced from a multi-copy plasmid and thus it has been used as a reporter to detect AHL production. The native AHL in *Agrobacterium* has eight carbons in its acyl chain, but its LuxR homolog, TraR, can also recognize the 12-carbon AHL of *P. aeruginosa* and the 6-carbon AHL of *V. fischeri*. However, the latter two organisms cannot recognize each other's signals. These factors complicate the development of approaches to manipulate the bacterial quorum-sensing systems and resulting outcomes to the benefit of humans (i.e., promote beneficial bacterial behaviors or interfere with harmful ones). The hope is that studies of quorum quenching will lead to new classes of antimicrobial chemicals that interfere with communication systems to block infection or reduce the virulence of specific pathogens. Unfortunately, because of the immense complexity of natural systems, the mechanisms that turn off a harmful or undesired response for one bacterium could enhance an undesired output in another organism.

Learning Outcomes: After completing the material in this chapter, students should be able to. . .

1. define the concept of quorum sensing

2. describe the symbiotic relationship between *Vibrio fischeri* and squid

3. explain the biochemical mechanism and regulation of bacterial bioluminescence

4. compare and contrast the fundamental components for the quorum-sensing systems of Gram-negative proteobacteria and Gram-positive bacteria

5. apply quorum quenching to manipulate bacterial phenotypes

Check Your Understanding

1. Define this terminology: quorum sensing (QS), autoinducer, threshold level, light organ, luciferase, autoinducer synthase, acyl homoserine lactone (AHL), homologs, peptide signals, cellular competence, furanone, quorum quenching

2. Explain how the concept of a threshold concentration of autoinducer plays a critical role in quorum sensing. (LO1)

3. How does each organism benefit in the symbiotic relationship between the squid *Euprymna scolopes* and the bacterium *Vibrio fischeri*? (LO2)

4. Juvenile squid are not born with bacterial symbionts. How is chemotaxis involved in populating the light organ? (LO2)

5. What is the function of each polypeptide involved in bioluminescence in *Vibrio fischeri*? Why is each required for light production? (LO3)

   a) LuxR:

   b) LuxI:

   c) LuxAB:

   d) LuxCDE:

   e) Flavin oxidoreductase:

6. Luciferase enzyme levels and activity vary at high and low cell density. (LO3)
   a) Fill in the table below to indicate enzyme levels and activity at high and low cell density.

Condition	Are AHL levels inside the cell low or high?	Are LuxI levels inside the cell low or high?	Are LuxAB levels inside the cell low or high?	Is LuxR active or inactive?	Is bioluminescence produced?
Low cell density					
High cell density					

   b) How does AHL get into *Vibrio fischeri* cells?

   c) What activates LuxR?

   d) What does LuxR do when it is activated?

   e) How does this result in bioluminescence?

7. Quorum sensing (QS) in *Pseudomonas aeruginosa* is very complex, as it has a number of QS regulatory systems that interact with each other to control the transcription of hundreds of genes. Two of these QS systems are based on the synthesis and detection of *N*-acyl-homoserine lactone (AHL). (LO4)

   a) Describe the function of the molecules in the LasR/I QS system.

      i. LasR:

      ii. LasI:

iii. 3-oxo-C12-HSL:

iv. Give examples of genes controlled by the LasR/I QS system:

b) Describe the function of the molecules in the RhlR/I QS system.

   i. RhlR:

   ii. RhlI:

   iii. C4-HSL:

   iv. Give examples of genes controlled by the RhlR/RhlI QS system:

c) Order the steps (1 to 8) below that indicate the events leading up to the activation of the LasR/I and RhlR/I QS systems.

   _____ The transcription of *rhlR* is enhanced, and RhlR is produced.

   _____ RhlR binds its target promoters in its regulon.

   _____ RhlR binds C4-HSL, converting RhlR to an active conformation.

   _____ LasR binds its target promoters, including the *rhlR* promoter.

   _____ LasR binds 3-oxo-C12-HSL, converting LasR to an active conformation.

   _____ C4-HSL is synthesized by RhlI, accumulates, and diffuses back into the cell.

   _____ At low cell density, *lasR* and *lasI* are transcribed at low levels.

   _____ At high cell density, 3-oxo-C12-HSL accumulates, reaching an intracellular threshold concentration.

8. *Pseudomonas aeruginosa* has an additional QS system called the *Pseudomonas* quinolone signal (PQS) system, which produces a non-AHL signal. (LO4)

   a) What is the autoinducer signal in the PQS system?

   b) What is the regulatory protein?

   c) Give an example of genes controlled by the PQS system.

9. Quorum-sensing systems in Gram-negative (e.g., proteobacteria) and Gram-positive bacteria ultimately have a similar output (i.e., gene expression controlled by cell density), but the underlying mechanisms are very different. (LO4)

   a) What type of molecule typically acts as an autoinducer in proteobacteria? What type of molecule typically acts as an autoinducer in Gram-positive bacteria?

   b) How do autoinducer signal molecules in Gram-negative proteobacteria and Gram-positive bacteria typically exit the cells that made them? Explain these differences.

   c) How and where in the cell do the regulatory proteins in Gram-negative proteobacteria and Gram-positive bacteria typically sense the presence of high concentrations of autoinducer signal? Explain these differences.

10. Bacterial competence and sporulation are both processes that are controlled by quorum sensing (QS) in *Bacillus subtilis*, a Gram-positive bacterium. (LO4)

   a) At high cell densities, cells will first induce competence in an attempt to acquire new genes from neighboring cells.

      i. Why might it benefit bacteria to become competent only at high cell density?

      ii. Match each entity in the *Bacillus* competence QS system with its corresponding description and/or function.

   _____ ComX
   _____ *comS*
   _____ ComP
   _____ ComA

   A. A 10-amino acid peptide that acts as the autoinducer signal
   B. A membrane-bound sensor histidine kinase that phosphorylates ComA
   C. A regulatory protein that, when phosphorylated, activates transcription of *comS*
   D. The target gene important for competence controlled by ComA

      iii. Briefly describe how a two-component system regulates competence in Gram-positive bacteria.

      iv. Describe the function of each entity in the *Bacillus* competence QS system.

      • The peptide encoded by the *phrC* gene:
      • CSF:
      • RapC:
      • Opp:

v. How does RapC **inhibit** the initiation of cell competence?

vi. How does the low concentration of CSF inside the cell **promote** cell competence?

11. Cholera is caused by the pathogen *Vibrio cholerae*, which colonizes the small intestine of its host. One of the most important *V. cholerae* virulence factors is cholera toxin (CT). Match each quorum-sensing molecule with its corresponding description and/or function. (Note: Some terms can be used more than once.) (LO4)

_____ CAI-1

_____ AI-2

_____ CqsS

_____ LuxPQ

_____ LuxU

_____ LuxO

_____ Qrr1-4

_____ Hfq

_____ AphA

_____ HapR

A. Autoinducer signal

B. Cytoplasmic response regulator

C. Membrane-bound sensor histidine kinase

D. Activator protein for biofilm and toxin genes

E. Repressor protein for biofilm and toxin genes

F. RNA binding protein

G. sRNA

12. Answer the following questions to describe how *Vibrio cholerae* virulence factors are influenced by cell density through a quorum-sensing (QS) system. Note that HapR represses the transcription of *aphA* and Qrr1-4 sRNAs destabilize the *hapR* mRNA. (LO4)

a) Under conditions of **low** *V. cholerae* density:

i. What do the membrane-bound sensor histidine kinases do when the concentration of autoinducer is below the threshold concentration?

ii. Are Qrr1-4 sRNAs transcribed under these conditions?

iii. Is *hapR* transcribed and translated? Why or why not?

iv. How does this impact the transcription of *aphA*?

b) Under conditions of **high** *V. cholerae* density:

i. To which membrane-bound sensor histidine kinases do CAI-1 and AI-2 bind, respectively?

ii. What do the membrane-bound sensor histidine kinases do when the autoinducer is above the threshold concentration?

iii. Are the Qrr1-4 sRNAs transcribed?

iv. Is *hapR* transcribed and translated? Why or why not?

v. How does this impact the transcription of *aphA*?

c) Based on what you know about the control of AphA and HapR production, would you predict that the QS system in *V. cholerae* promotes the production of CT at lower cell density or high cell density?

13. Because quorum sensing (QS) controls so many bacterial cell functions, scientists have been looking for ways to disrupt QS (referred to as quorum quenching) in order to control undesirable bacterial behaviors. One type of quorum quenching occurs by degrading or modifying the AHL signal (structure shown) produced by Gram-negative bacteria through the activity of lactonase, acylase, or oxidoreductase enzymes. Define the function of each enzyme. (LO5)

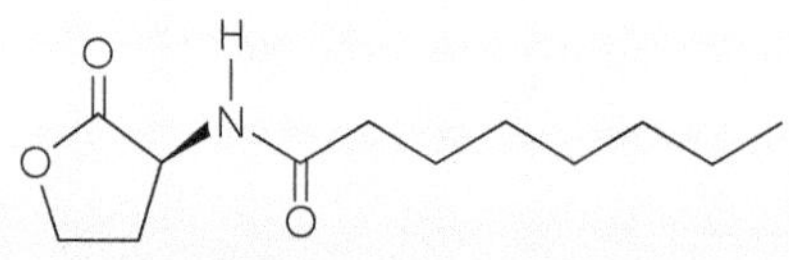

a) AHL lactonases:

b) AHL acylases:

c) AHL oxidoreductases:

Dig Deeper

14. The symbiotic relationship between the squid *Euprymna scolopes* and *Vibrio fischeri* requires that *V. fischeri* colonize the light organ of *E. scolopes*, then grow to a threshold density to induce bioluminescence. (LO4)

   a) The colonization of *E. scolopes* is a very specific process with many interactions between symbiont and host. For example, *V. fischeri* cells form in a biofilm at the opening to the tissues that will form the light organ, which induces *E. scolopes* to release a chemoattractant. Within 1 to 3 hours, a few bacteria detach from the biofilm and migrate into the tissues that will form the light organ, where a few pioneer cells grow into a population of several hundred thousand cells. Previous research has shown that the light organ of *E. scolopes* releases two compounds, *N*-acetylglucosamine (GlcNAc) and *N*-acetylglucosamine dimer (GlcNAc)$_2$, to which *V. fischeri* is attracted. To determine if these compounds are the signals required for colonization, you do the following.

   Step 1. Set up *V. fischeri* cell suspensions in dishes with either seawater or seawater with GlcNAc or (GlcNAc)$_2$ (see diagram).

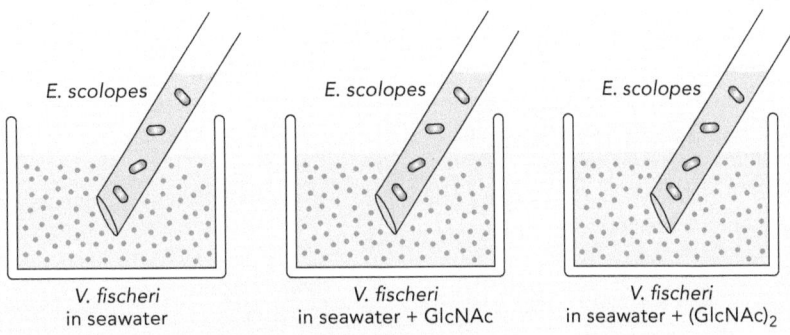

Step 2. Insert capillary tubes containing **E. scolopes** into each dish.

Step 3. Count the number of squid successfully colonized by *V. fischeri* after 24 hours (see graph).

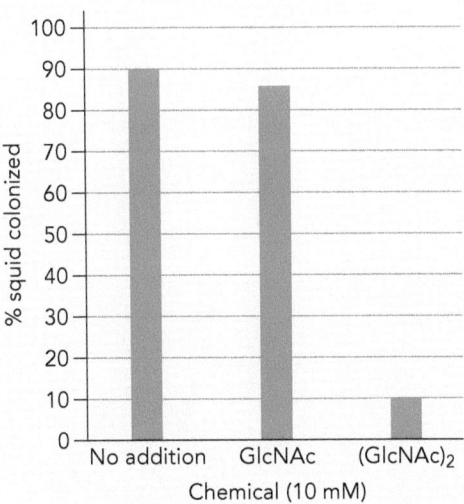

If colonization depends on chemotaxis, then as *E. scolopes* releases a chemical, a gradient will form to which *V. fischeri* can respond. The addition of a high concentration of a given chemical in the seawater will eliminate any chemical gradient that is generated by compounds released from the squid.

   i. The suspension containing seawater only (no addition) serves as a control. What does this control tell you about squid colonization? Explain these results.

   ii. Which compound serves as the signal for colonization? Why do you think colonization was disrupted by the addition of one compound, but not the other?

b) After colonizing its host, *V. fischeri* induces the *lux* operon, which is responsible for bioluminescence. To investigate the regulation of bioluminescence, you grow wild-type *V. fischeri* cells and three mutant strains (Δ*luxA*, Δ*luxI*, and Δ*luxR*) in growth medium with and without added 3-oxo-C6-HSL.

i. What role does 3-oxo-C6-HSL play in the quorum-sensing system that induces bioluminescence?
ii. Given the results below, predict which mutant strain gave each result.

Strain	Luminescence (photons/second/cell)	
	No added 3-oxo-C6-HSL	3-oxo-C6-HSL added
A) Wild type	25	40,000
B)	>0.01	>0.01
C)	5	35,000
D)	5	5

iii. Explain your reasoning for each mutant strain.

15. The initial density-dependent experiments by Nealson et al. explored the regulation of bioluminescence in *Vibrio fischeri*. (LO3)

a) Nealson et al. (see Fig. 16.4) showed that after adding cells to growth medium, the cells grew exponentially (x) but there was a lag in bioluminescence (o). After the lag, did cell density and bioluminescence increase at the same rate? Explain the molecular basis for these results.

b) On the DNA diagram shown, draw in the positions of LuxR, RNAP, and AI at 180 minutes (see Fig. 16.4). Are *luxR* and *luxAB* being transcribed and translated at this time?

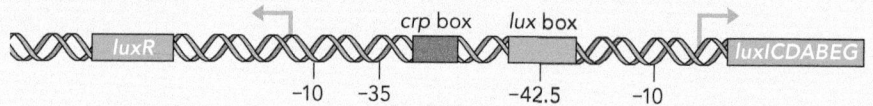

c) At the time, Nealson did not know if the *lux* genes (specifically *luxAB*) are controlled at the level of transcription (i.e., if mRNA production is controlled) or translation (i.e., if the mRNA for luciferase is always made but is not available to be translated). To investigate this, you set up three flasks of *V. fischeri* in growth medium and measure luminescence over time. After 3 hours (at the arrow) you add antibiotics to two separate flasks:

- Chloramphenicol (an antibiotic that inhibits protein synthesis; dotted line)
- Rifampin (an antibiotic that inhibits mRNA synthesis; dashed line)

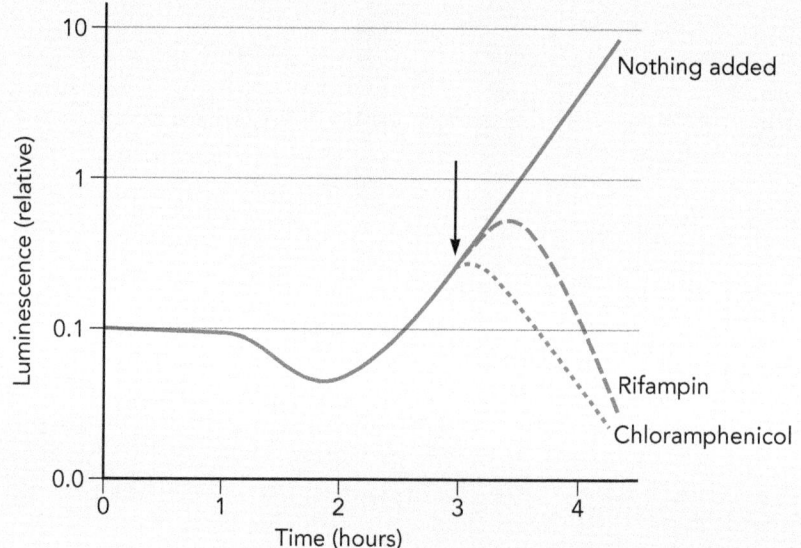

Your results show that in the flask with no added antibiotics, luminescence increases with cell growth. However, when you add either rifampin or chloramphenicol, luminescence decreases.

i. These data support the hypothesis that the synthesis of LuxAB is controlled at the level of transcription. Explain why both rifampin and chloramphenicol cause a decrease in luminescence.

ii. Suppose instead that the synthesis of LuxAB was controlled at the level of translation (i.e., the mRNA was made but not translated). Would you expect rifampin to decrease luminescence? Would you expect chloramphenicol to decrease luminescence? Explain your reasoning.

16. *Pseudomonas aeruginosa* is a Gram-negative pathogen that causes infections in patients with burn wounds, cystic fibrosis, and various immunocompromised conditions. *P. aeruginosa* has a range of virulence factors, including biofilm production, proteases, and toxins, that impede the host immune system. Many of these virulence factors are under the control of the quorum sensing (QS) systems LasI/R and RhlI/R. You have isolated the gene for a new protease (LasX) and want to investigate its regulation. (LO4)

a) To investigate whether the *lasX* gene is under the control of a QS system, you make a *lasX::lacZ* transcriptional fusion and monitor β-galactosidase activity (β-gal; orange line) as a function of growth (blue line). Do your data support the hypothesis that the *lasX* gene is under the control of a QS system? Explain your reasoning.

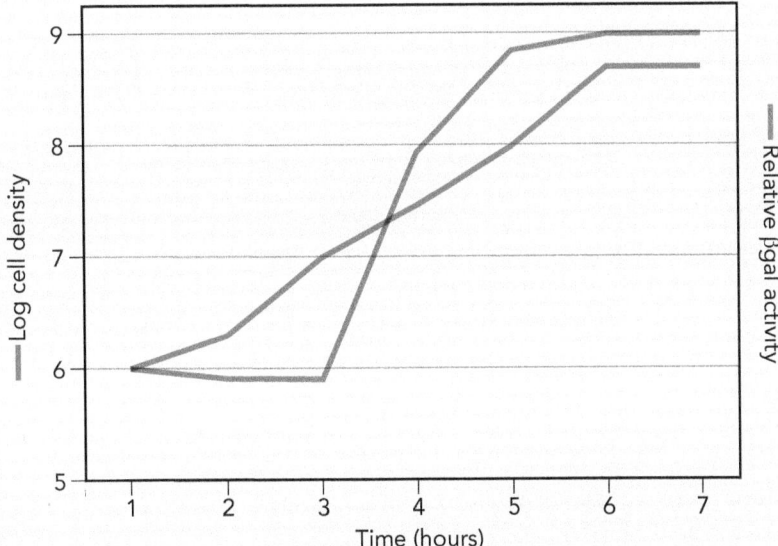

b) To further investigate, you again look at the activity of the *lasX::lacZ* fusion as a function of cell growth, this time in Δ*lasR* and Δ*rhlR* deletion mutant strains. Consider how these two QS systems work together (Fig. 16.6).

i. If *lasX* were under the control of LasR, would you expect to observe β-galactosidase activity in the Δ*lasR* deletion mutant? Explain your reasoning.

ii. If *lasX* were under the control of LasR, would you expect to see β-galactosidase activity in a Δ*rhlR* deletion mutant? Explain your reasoning.

iii. If *lasX* were under the control of RhlR, would you expect to see β-galactosidase activity in a Δ*rhlR* deletion mutant? Explain your reasoning.

iv. If *lasX* were under the control of RhlR, would you expect to see β-galactosidase activity in a Δ*lasR* deletion mutant? Explain your reasoning.

c) The results from your growth experiments suggest that *lasX* is under the control of RhlR. To test this hypothesis, you carry out an electrophoretic mobility shift assay (EMSA; Chapter 8). You purify a 200-bp segment of the *lasX* gene (including the transcription start site and upstream DNA). You add purified LasR and RhlR to separate samples and subject the samples to gel electrophoresis.

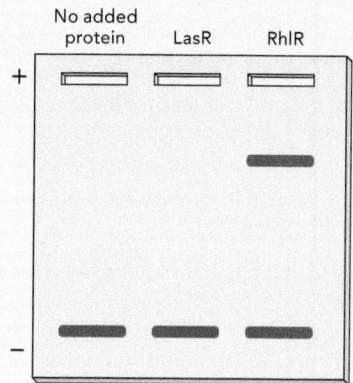

i.   Briefly explain the EMSA experiment. Include in your answer any controls needed and what information the controls can provide.

ii. Your results for the EMSA are shown. What do these data indicate about whether *lasX* is controlled by LasR or RhlR?

17. *Staphylococcus aureus* is a Gram-positive opportunistic pathogen that can cause skin infections, food poisoning, and, most devastating, toxic shock syndrome, which involves the production of the toxic shock syndrome toxin (TSST-1). TSST-1 is encoded by the *tst* gene, which is tightly regulated and expressed at high cell density. The environmental signals and molecular mechanisms that control the production of TSST-1 are complex and not fully understood.

a) Because the *tst* gene is expressed at high cell density, you predict that it might be under the control of a quorum sensing (QS) system. *S. aureus* has a QS system that is similar to the QS system that controls competence in *Bacillus*. Listed below are the critical components of the *S. aureus* QS system (Agr for accessory gene regulator). Fill in the table below to compare the QS system of *S. aureus* to the QS-based control of competence in *Bacillus*. One answer has been provided.

Function	*S. aureus* protein	*Bacillus* protein homolog
Sensor histidine kinase	AgrC	
Response regulator	AgrA	
Autoinducer peptide signal (AIP)	AgrD	
ABC transporter that exports the peptide signal	AgrB	ComQ

b) The response regulator protein AgrA controls the *agrACDB* operon and the gene encoding RNAIII (see diagram). Based on your understanding of the function of the various proteins in the *S. aureus* QS system and the *Bacillus* competence QS system, draw a model of how the system could work in *Staphylococcus*.

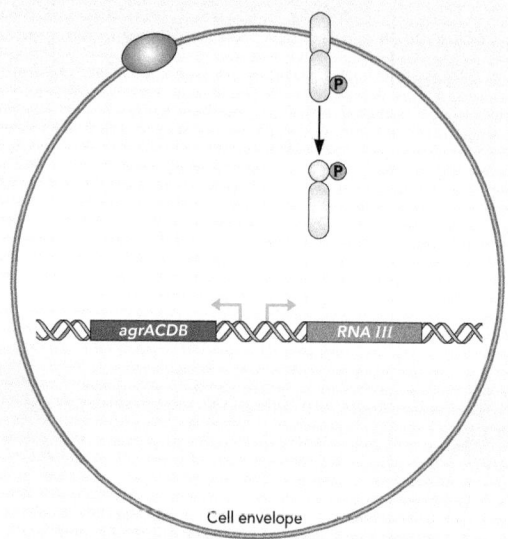

c) RNAIII is a small RNA molecule that can bind to and inhibit the initiation of translation of a target mRNA (for review, see Chapter 8). You want to investigate the role of RNAIII in controlling TSST-1 production. You measure the concentration of TSST-1 in a wild-type strain and a strain with the gene encoding RNAIII deleted.

i. In the graph below, insert bars representing the expected results for a strain with the RNAIII gene deleted if RNAIII had a negative effect on TSST-1 production at the level of transcription (wild-type results are shown in blue; *tst* indicates the gene that codes for TSST-1).

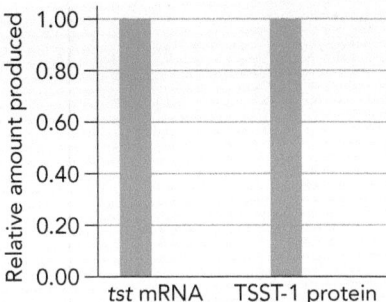

ii. In the graph below, insert bars representing the expected results for a strain with the RNAIII gene deleted if RNAIII had a negative effect on TSST-1 production at the level of translation (wild-type results are shown in blue).

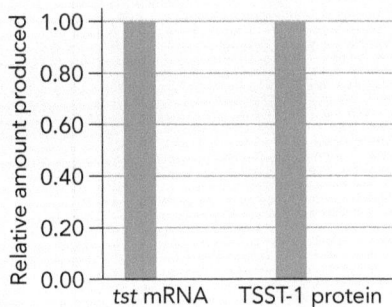

18. You have a strain of *Agrobacterium tumefaciens* that was constructed to detect the presence of different types of AHLs. This biosensor strain does not itself make AHLs. However, AHLs produced by another bacterium can diffuse into the biosensor strain and be detected. The detection system includes a quorum sensing (QS) regulatory protein, the activator TraR. When AHL binds to TraR, TraR takes on its active conformation and binds to target DNA promoters to initiate the transcription of genes in its regulon. This biosensor strain contains a β-galactosidase (*lacZ*) gene under the control of TraR. In the presence of a synthetic indicator compound, X-gal (5-bromo-4-chloro-3-indolyl-β-D-galactopyranoside), β-galactosidase cleaves X-gal, producing a blue product. Therefore, a blue color indicates the presence of AHL. (LO5)

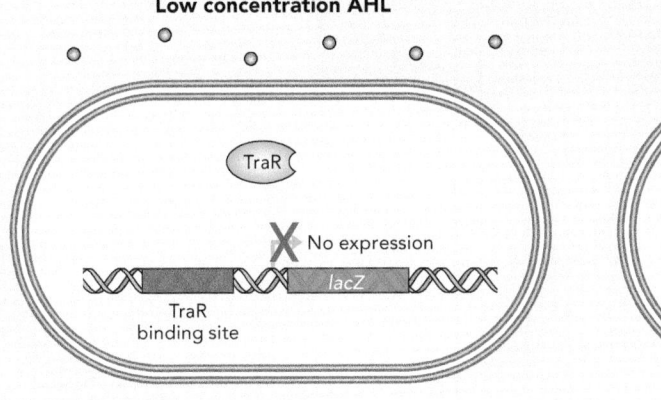
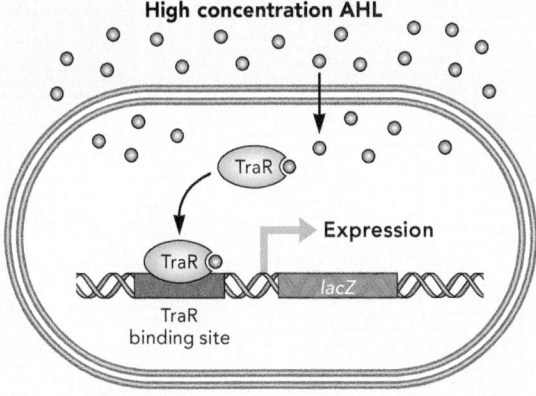

a) To test this system, you set up an agar plate experiment, in which X-gal is added to nutrient agar and the biosensor strain is inoculated onto the plate surface. AHL is added at the point of inoculation. Each sentence below describes a step in this biosensor QS process. Put these events in the correct order from step 1 (shown) to step 6.

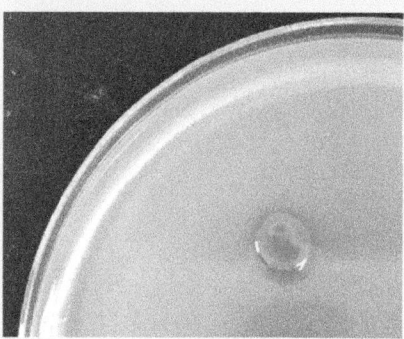

  __1__   AHL is added to the inoculation spot.

  _____   X-gal is cleaved, producing a blue color.

  _____   TraR binds upstream of *lacZ*.

  _____   β-galactosidase is produced.

  _____   AHL diffuses into biosensor cells.

  _____   AHL binds to TraR.

b) Convinced that the biosensor strain is working, you decide to use it to test for quorum quenching by three strains of *Bacillus*. You set up a similar experiment with X-gal added to nutrient agar and the biosensor strain spot inoculated onto the plate surface (spots 1 to 4).

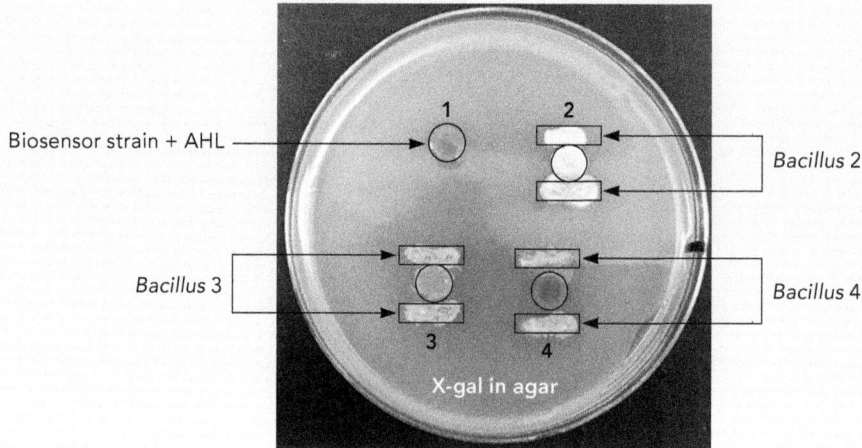

Step 1. You add biosensor strain + AHL to each spot (1-4).

Step 2. To spot 2, you streak *Bacillus* strain 2 on both sides of the biosensor spot.

Step 3. To spot 3, you streak *Bacillus* strain 3 on both sides of the biosensor spot.

Step 4. To spot 4, you streak *Bacillus* strain 4 on both sides of the biosensor spot.

Which strains produce quorum-quenching compounds? For each strain, record your observation and interpret the results.

Bacillus strain	Observation (white or blue?)	Interpretation: Does this strain produce a quorum quencher? Explain how you know.
2		
3		
4		

c) Further investigation reveals that one of your strains is a known lactonase producer. Explain what lactonase does and how it inhibits quorum sensing.

d) Another of your strains makes a molecule that mimics (has a similar structure to) the autoinducer signal. Explain why this mimic molecule would inhibit quorum sensing.

416 Making Connections

416 Oxidative Stress

419 Heat Shock Response

420 Sporulation

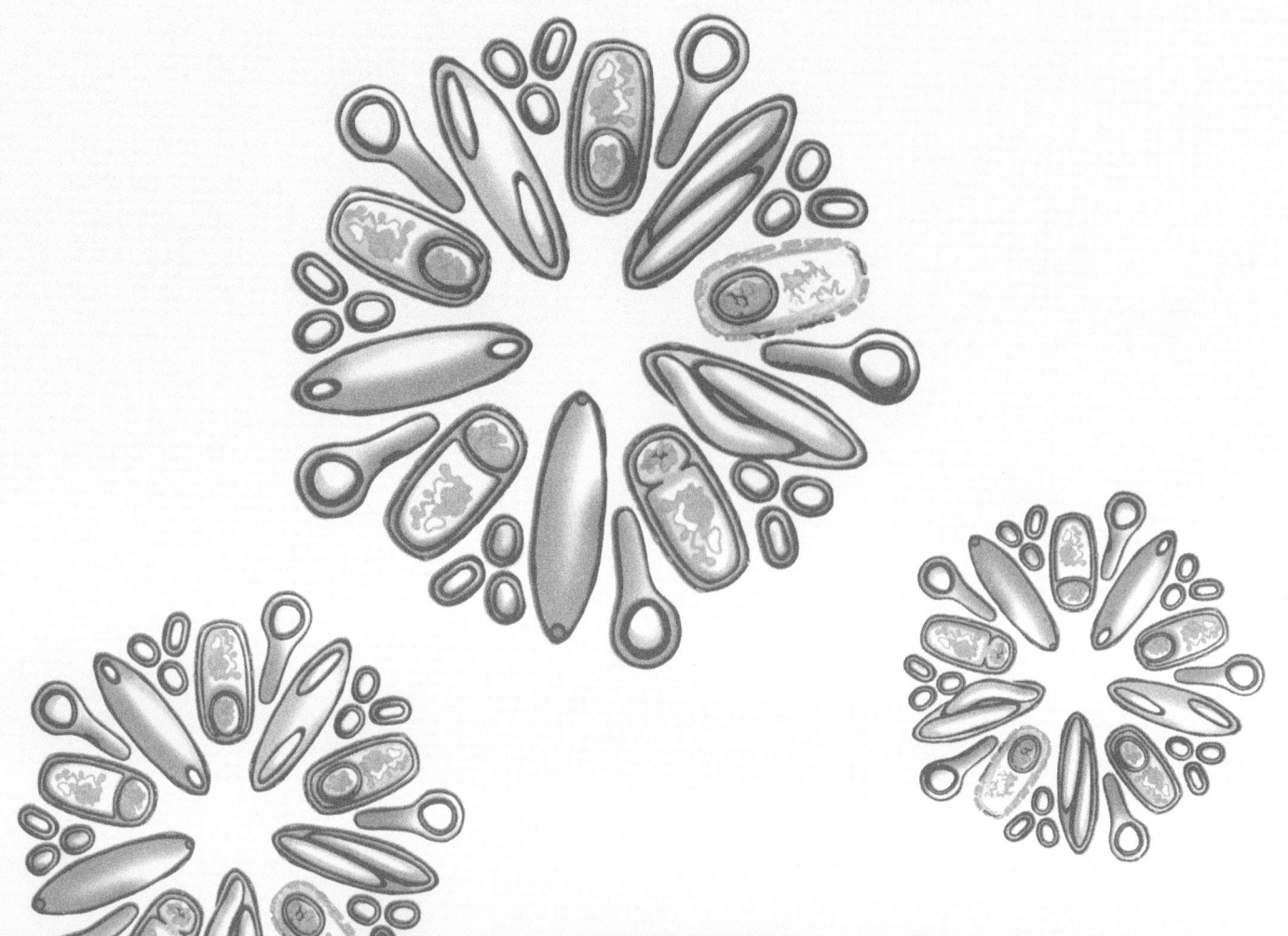

17

STRESS RESPONSES

After completing the material in this chapter, students should be able to . . .

1. list four enzymatic reactions used by bacteria to eliminate oxidative stress

2. outline two regulatory mechanisms (SoxR and OxyR) that control the oxidative stress response in *Escherichia coli*

3. describe the heat shock responses of *Escherichia coli*

4. explain the stages of endospore formation and germination in *Bacillus subtilis*

5. discuss the role of kinases and sigma factors in the control of endospore formation in *Bacillus subtilis*

Microbial Physiology: Unity and Diversity, First Edition. Ann M. Stevens, Jayna L. Ditty, Rebecca E. Parales and Susan M. Merkel.
© 2024 American Society for Microbiology.
Companion website: www.wiley.com/go/stevens/microbialphysiology

Making Connections

Microbes must be able to rapidly adjust to changing environmental conditions. Environmental signals such as nutrient availability or other signals from quorum sensing (Chapter 16) can produce physiological changes that are critical for survival of a microorganism under the new conditions. For example, while some bacteria are motile and have the capacity to move toward nutrients or away from repellents (Chapter 15), nonmotile bacteria must be able to adapt to stressful environmental conditions in a different manner. Oxidative stress and heat shock stress are two conditions to which all living cells respond. In addition, some bacteria have evolved the unique ability to differentiate into dormant resting cells, such as endospores, that permit them to survive harsh environmental conditions. These mechanisms of responding to stress are the focus of this chapter.

Oxidative Stress

All organisms living in the presence of oxygen must have a way to deal with oxidative stress. For aerobic organisms, **reactive oxygen species (ROS)** are formed both spontaneously in air and during the process of aerobic respiration. Anaerobic organisms that can tolerate the levels of O_2 found in air must also deal with oxidative stress. Strictly anaerobic bacteria often don't have these coping mechanisms, so they are even more susceptible to the presence of not just reactive oxygen species, but oxygen itself. The conversion of oxygen into reactive species is summarized below (Table 17.1).

Superoxide radicals are naturally formed as byproducts of aerobic respiration when a molecule of oxygen accepts an extra electron. Superoxide radicals are substrates in the formation of hydroxyl radicals (see below), which can damage proteins, DNA, RNA, and lipids in the cell. Several mechanisms to eliminate superoxide radicals have evolved (Table 17.1). Although spontaneous dismutation chemically converts superoxide radicals to hydrogen peroxide, some aerobic bacteria employ the enzyme **superoxide dismutase (SOD)** to catalyze a more rapid conversion of superoxide radicals to hydrogen peroxide

TABLE 17.1 **Reactive oxygen reactions**

Formation of reactive oxygen species	
$O_2 + e^- \rightarrow O_2^-$	Superoxide radical
$O_2^- + e^- + 2\,H^+ \rightarrow H_2O_2$	Hydrogen peroxide
$H_2O_2 + e^- + H^+ \rightarrow H_2O + OH\cdot$	Hydroxyl radical
$O_2^- + Fe^{3+} \rightarrow O_2 + Fe^{2+}$ $H_2O_2 + Fe^{2+} + H^+ \rightarrow Fe^{3+} + OH\cdot + H_2O$	Haber-Weiss reaction
Enzymatic reactions that eliminate reactive oxygen species	
$2\,O_2^- + 2H^+ \rightarrow H_2O_2 + O_2$	Superoxide dismutase
$O_2^- + 2\,H^+ + cyt\ c_{red} \rightarrow H_2O_2 + cyt\ c_{ox}$	Superoxide reductase
$2\,H_2O_2 \rightarrow 2\,H_2O + O_2$	Catalase
$H_2O_2 + NADH + H^+ \rightarrow 2\,H_2O + NAD^+$	Peroxidase

and oxygen ($2 O_2^- + 2 H^+ \rightarrow H_2O_2 + O_2$). Strict anaerobes must have a different mechanism to eliminate superoxide radicals, since the production of oxygen would be detrimental to their survival. Therefore, these anaerobes use the enzyme **superoxide reductase** to reduce the superoxide radical to hydrogen peroxide at the expense of reduced cytochrome c in the electron transport chain (ETC; $O_2^- + 2 H^+ + cyt\ c_{red} \rightarrow H_2O_2 + cyt\ c_{ox}$).

As discussed above, **hydrogen peroxide** is often formed during the elimination of superoxide radicals. Hydrogen peroxide is a strong oxidizing agent that can also damage cell components. In addition, hydrogen peroxide is produced during the oxidative burst response used by immune system cells to kill pathogens. Therefore, microbes have evolved enzymes to degrade hydrogen peroxide (Table 17.1). In aerobes, the primary means to remove hydrogen peroxide is the enzyme **catalase**, which produces water and oxygen gas from hydrogen peroxide ($H_2O_2 + H_2O_2 \rightarrow 2 H_2O + O_2$). In strict anaerobes, an alternative enzymatic reaction that does not generate oxygen is used. This reaction is catalyzed by **peroxidases**, which use NADH + H$^+$ to reduce hydrogen peroxide to water, at the cost of potential energy for the cell.

Cells must eliminate superoxide radicals and hydrogen peroxide because they can be converted into **hydroxyl radicals** by the two-step Haber-Weiss reaction (Table 17.1). Since there are no known protective enzymes to remove hydroxyl radicals, cells must prevent their formation by eliminating the precursors of hydroxyl radicals. Hydroxyl radicals are the most reactive of all oxygen intermediates. They are very strong oxidizing agents capable of damaging DNA, RNA, protein, and lipid molecules in the cell.

PHYSIOLOGICAL RESPONSE TO OXIDATIVE STRESS IN *ESCHERICHIA COLI*

Bacteria can sense and respond to the presence of ROS. In *Escherichia coli*, superoxide radicals are sensed by the regulatory protein SoxR (Fig. 17.1A). SoxR contains an FeS prosthetic group that is sensitive to changes in the redox state of the cytoplasm. The oxidized form of SoxR serves as a transcription factor that binds DNA and activates transcription of *soxS*. SoxS is a transcription factor that activates the expression of a regulon, including *sodA*, which encodes superoxide dismutase, and *nfo*, which encodes an endonuclease important for DNA excision repair.

Hydrogen peroxide is sensed by the regulatory protein OxyR (Fig. 17.1B). OxyR contains disulfide bonds when it is in its oxidized form. Oxidized OxyR activates transcription of a regulon, including *katG*, which encodes catalase, and *gorA*, which encodes glutathione reductase. KatG degrades hydrogen peroxide as described above. GorA serves to maintain high levels of reduced **glutathione**, an important internal molecule that senses the redox state of the cytoplasm. Glutathione is a tripeptide with the sequence glutamate-cysteine-glycine (Fig. 17.1C) that plays a variety of roles in different bacterial species. The ratio of reduced glutathione (GSH) to oxidized glutathione (GSSG) in the cytoplasm is tightly controlled through the action of GorA, which uses reducing equivalents from NADPH + H$^+$ to maintain high levels of reduced glutathione in the cell. The cytoplasm of all cells is typically a reducing environment. Under conditions of oxidative stress in the cytoplasm, the cysteine residue in glutathione becomes oxidized and can form a disulfide bond with other molecules in the cell. In *E. coli*,

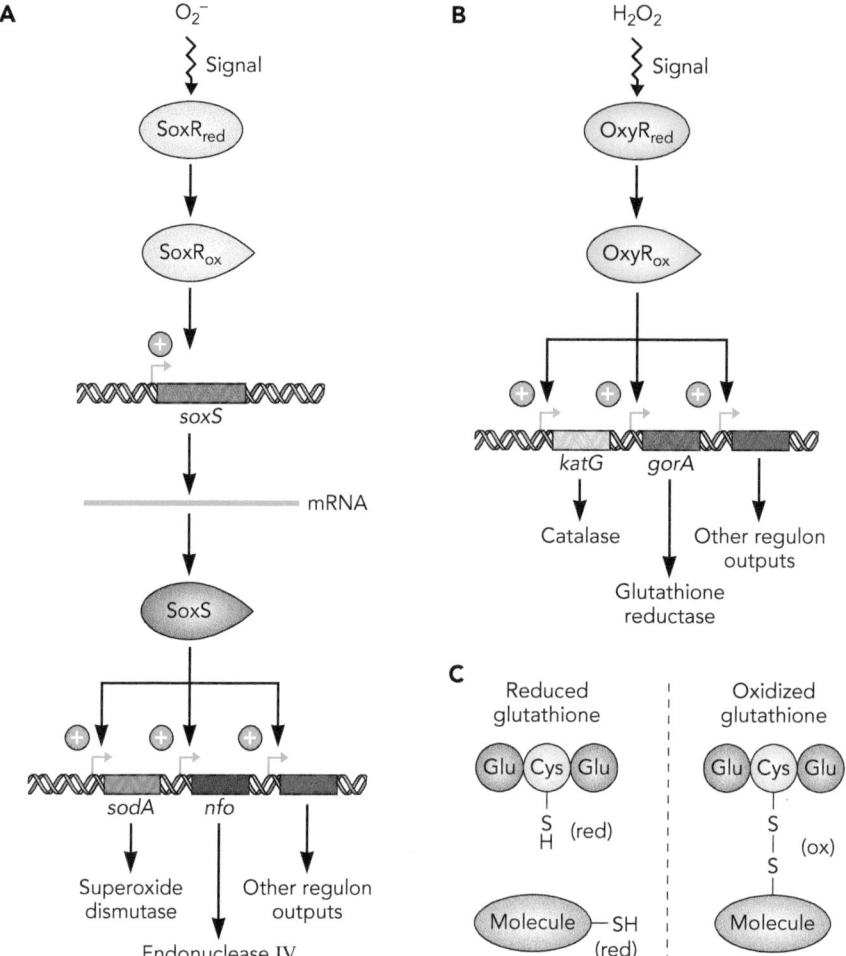

FIGURE 17.1 *Escherichia coli* **response to oxidative stress.** (A) In the presence of superoxide radicals, the FeS cluster in the SoxR sensor protein becomes oxidized. The oxidized form of SoxR binds to the promoter of the *soxS* gene, activating its transcription. The transcriptional activator SoxS turns on expression of the *sox* regulon, which includes genes that encode superoxide dismutase (SodA), an enzyme that catalyzes the elimination of superoxide, and an endonuclease (Nfo) that participates in DNA repair. (B) When the OxyR protein senses hydrogen peroxide, disulfide bonds are formed between cysteine residues in the protein. The oxidized form of OxyR activates expression of a regulon, which includes genes that encode catalase (KatG), an enzyme that breaks down hydrogen peroxide, and glutathione reductase (GorA), which reduces glutathione. (C) Glutathione is a tripeptide that allows a cell to sense the redox state of the cytoplasm. Reduced glutathione has a sulfhydryl group on its cysteine residue, whereas oxidized glutathione forms disulfide bonds (S-S) with other molecules in the cell. See the text for additional details.

upon oxidative stress, the oxidized form of glutathione positively regulates the KefB/C antiporter. Once activated, KefB/C causes a net efflux of potassium ions out of the cell and an influx of protons into the cell. This results in acidification of the cytoplasm, protonating some molecules and minimizing their reactivity with ROS. In other microbes, ROS may be detoxified via direct interaction with GSH or via specific enzymes, which use GSH to catalyze the reduction of peroxides.

Heat Shock Response

Another stress that microbes experience is heat shock. The **heat shock response**, first discovered in *Drosophila*, is a universal stress response found in all living cells. Later research revealed that stressors other than heat can also trigger this physiological response. In addition to elevated temperatures (e.g., 42°C for *E. coli*), chemicals such as ethanol or hydrogen peroxide or changes in the pH that damage proteins can also initiate the heat shock response. Further research has shown that the resulting increase in the concentration of denatured protein in the cytoplasm is the signal that actually triggers this physiological response. Chaperones and chaperonins, capable of folding proteins, are always produced at relatively low levels due to their role in protein secretion (Chapter 14). In addition, they also play a critical role in the heat shock response, as described below.

In *E. coli* cells growing at 30°C, *rpoH* is transcribed using the housekeeping σ^{70} (RpoD) bound to core RNA polymerase (RNAP; Fig. 17.2A). RpoH, also known as σ^{32}, is the sigma factor that is specifically involved in the heat shock response. Although *rpoH* is constitutively expressed, σ^{32} is not translated efficiently due to secondary structure within its mRNA. The small amount of σ^{32} that is translated at 30°C is held in an inactive conformation due to interactions with the general chaperones DnaK, DnaJ, and GrpE. Excess σ^{32} has a short half-life since it is degraded by proteases, further maintaining its cellular concentration at a very low level. However, when temperatures rise to ~42°C, the small amount of σ^{32} that is available can rapidly deal with the stress event without having to wait for transcription and translation of *rpoH* to occur (Fig. 17.2B). When the concentration of denatured proteins in the cytoplasm increases due to high temperature, the chaperones release σ^{32} to initiate the process of repairing the protein damage. After σ^{32} is released from the chaperones, it can join core RNAP and activate the genes in its regulon. Furthermore, the secondary structure within the *rpoH* mRNA is disrupted at the higher temperature, allowing for higher rates of translation. In addition, during a stress event, the stability and activity of σ^{32} are enhanced. The heat shock genes in the σ^{32} regulon encode more than 20 different proteins, including DnaK, DnaJ, and GrpE, which are capable of refolding damaged proteins, and proteases such as Lon and Clp, which degrade damaged proteins that cannot be repaired. When enough chaperones are available to repair the damage and bind σ^{32}, the transient heat shock response terminates.

At even higher temperatures (45 to 50°C), another, more recently discovered extracytoplasmic heat shock response is triggered in *E. coli* (Fig. 17.3). σ^{24} (RpoE) is required to activate genes needed to repair membranes and extracytoplasmic proteins in the periplasm and outer membrane that can be damaged at extreme temperatures. At lower temperatures, σ^{24} remains bound to the cytoplasmic membrane protein RseA (an anti-sigma factor) and is inactive under these conditions. However, when high temperature or osmotic stress denatures outer membrane proteins, RseA is degraded by a specific protease. Degradation of RseA releases σ^{24}, which can then complex with core RNAP and transcribe

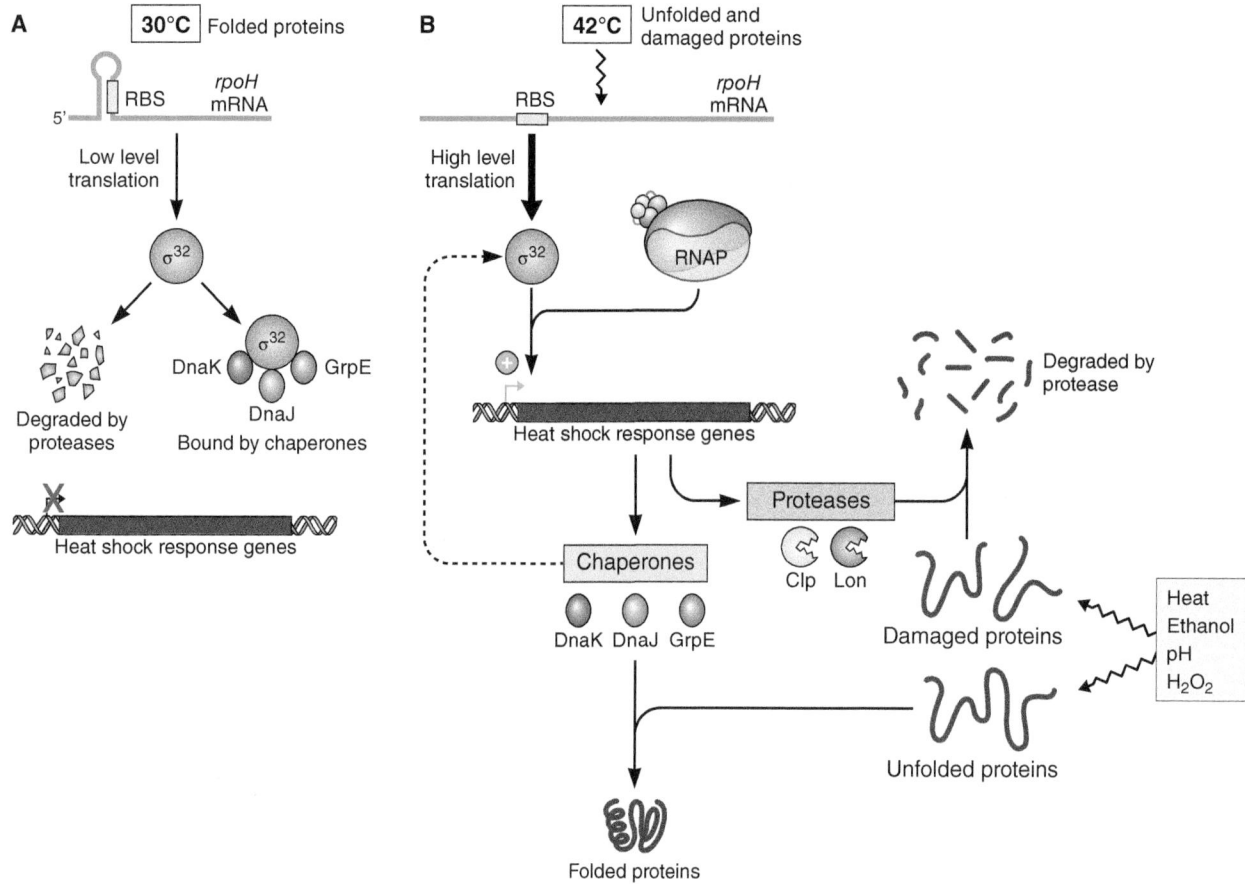

FIGURE 17.2 **Heat shock response in *Escherichia coli*.** The alternative sigma factor σ^{32} (RpoH) is required to transcribe the heat shock response genes. (A) At moderate temperatures (30°C), the ribosome binding site (RBS) upstream of *rpoH* mRNA is sequestered in a hairpin structure, resulting in a low rate of translation. The small amount of σ^{32} that is produced is either degraded by proteases or bound by chaperones (DnaK, DnaJ, and GrpE) that keep σ^{32} in an inactive form. Under these conditions, transcription of the heat shock response genes does not occur. (B) At higher temperatures (42°C), the *rpoH* mRNA hairpin denatures and production of σ^{32} increases. The chaperones that normally sequester σ^{32} are utilized to refold proteins denatured by the higher temperature. Thus, σ^{32} associates with core RNA polymerase (RNAP) and transcription of heat shock response genes occurs. Chaperones and proteases are produced at higher levels, and they function to refold the salvageable proteins and degrade denatured proteins that are beyond repair, respectively. Once protein refolding is complete, the chaperones will again bind to σ^{32} (dashed arrow). Other factors besides heat that damage proteins, including ethanol, pH, and H_2O_2, may also trigger a "heat shock" response.

the cell envelope repair genes (e.g., genes for synthesis of phospholipids and lipopolysaccharide, and chaperones for outer membrane proteins). Once repairs have been made, RseA is again produced and resumes its function as an anti-sigma factor, rendering σ^{24} inactive.

Sporulation

Sporulation is the ultimate stress response for certain bacteria experiencing unfavorable environmental conditions. The low-G+C Gram-positive bacteria *Bacillus* and *Clostridium* (members of the *Firmicutes*), as well as phylogenetically close relatives (e.g., *Heliobacterium*, an anerobic photoheterotroph, and *Desulfotomaculum*, a sulfate reducer) form **endospores**. These metabolically dormant

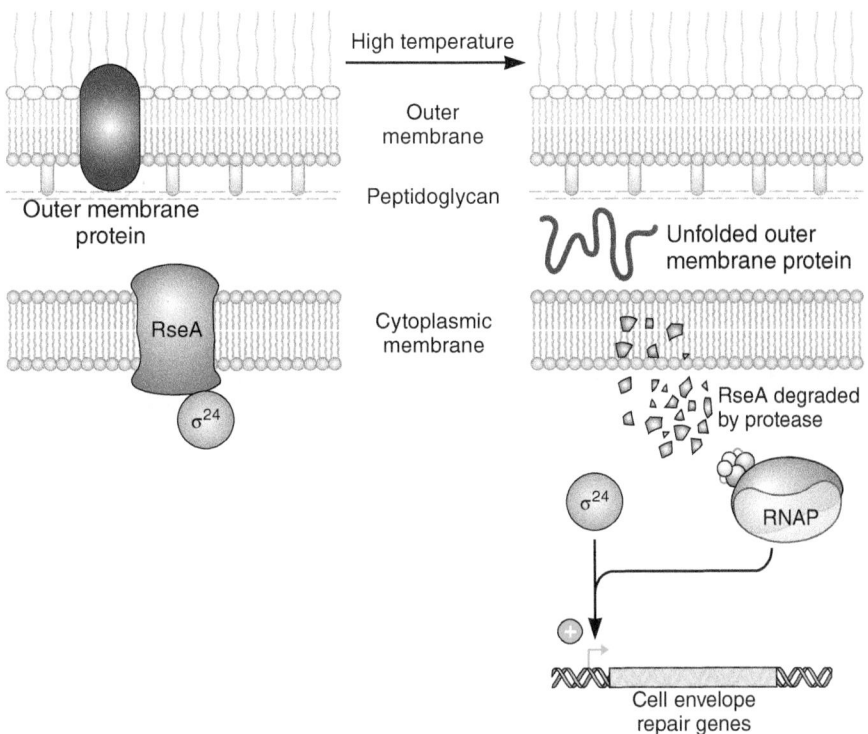

High temperature

Outer membrane

Peptidoglycan

Outer membrane
protein

Unfolded outer
membrane protein

RseA

Cytoplasmic
membrane

σ^{24}

RseA degraded
by protease

σ^{24}

RNAP

+

Cell envelope
repair genes

FIGURE 17.3 **Heat shock response to extreme temperatures in** *Escherichia coli.* In addition to damaging cytoplasmic proteins, temperatures above 45°C damage extracytoplasmic cell envelope components such as membrane lipids and proteins. Repair of these cell envelope components requires core RNA polymerase (RNAP) associated with σ^{24}. Although the gene encoding σ^{24} is constitutively expressed at lower temperatures, the sigma factor is bound by the anti-sigma factor RseA, rendering it inactive. RseA is degraded by a protease when the cell encounters temperatures above 45°C. σ^{24} is then able to associate with core RNAP and transcribe the genes encoding the relevant repair enzymes.

structures are resistant to a range of environmental stresses, including UV light, desiccation, heat, and chemical disinfectants. Thus, understanding sporulation and germination is critical to controlling endospore-forming organisms in industrial and clinical settings. The model organism that has been used to study endospore formation is *Bacillus subtilis*, and although the process of sporulation is generally conserved among endospore formers, some differences have been noted in species other than *Bacillus*. It is also important to note that a few other select microorganisms form different types of spores via alternative differentiation pathways. These include the Gram-negative myxobacteria and high-G+C Gram-positive *Actinobacteria* in the genus *Streptomyces* (Chapter 18). Thus, sporulation has only evolved in a few select bacterial lineages; it is not a universal process in bacteria.

STAGES OF SPORULATION IN *BACILLUS SUBTILIS*

During vegetative growth of *B. subtilis*, two daughter cells are formed from the mother cell through binary fission. During sporulation, however, just one endospore is formed; sporulation is therefore not a reproductive process in *B. subtilis*. The process of sporulation is initiated when carbon or nitrogen sources become limiting. When cells sense nutrient limitation, they initially undergo

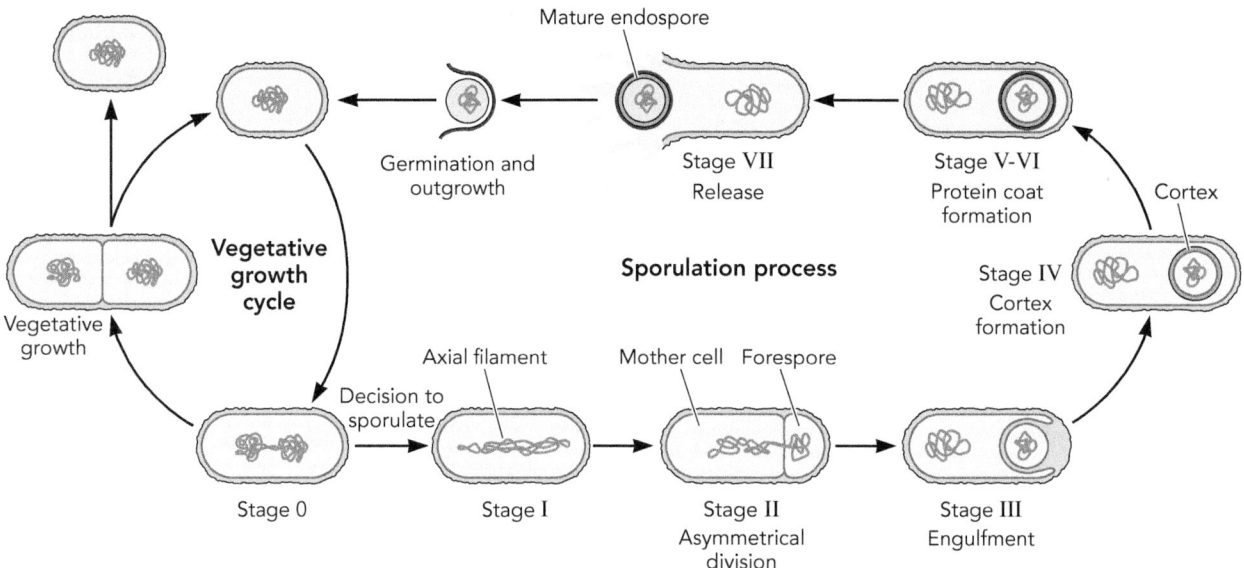

FIGURE 17.4 **Stages in the formation of endospores by *Bacillus subtilis*.** Vegetative cells (Stage 0) continue to replicate until they sense nutrient deprivation. Once the decision is made to sporulate, cells enter Stage I, during which the two copies of the chromosome are aligned along the long axis of the cell, forming an axial filament. In Stage II, asymmetric division occurs, as a septum forms between the mother cell and forespore. One copy of the chromosome is actively transferred into the forespore from the mother cell. In Stage III, the mother cell engulfs the forespore. In Stage IV, a cortex made of peptidoglycan is synthesized between the two membranes surrounding the forespore. In Stages V and VI, additional layers of protein are added to form the endospore coat, and the endospore matures and develops resistance properties. In Stage VII, the mature endospore is released by lysis of the mother cell. When environmental conditions are favorable, the endospore germinates and vegetative growth resumes.

a series of physiological changes as they attempt to acquire needed nutrients. For example, the chemotaxis and motility system is upregulated to allow cells to search for nutrients, antibiotics are produced to inhibit the growth of competing microbes, and various exoenzymes are produced and secreted to break down polymeric carbon sources like proteins and carbohydrates that might be available in the environment. If nutrient limitation continues, cells will become competent and take up DNA from the environment (Chapter 16). Finally, when the population density is high and cells remain unable to grow due to nutrient limitations, the sporulation process begins.

Once a cell commits to the process of sporulation, it takes ~8 to 10 hours under laboratory conditions to form an endospore. The stages of sporulation have been defined based on mutants that cannot progress beyond a particular stage and on the appearance of cells under an electron microscope (Fig. 17.4). The genes associated with the different stages of sporulation are named accordingly. For example, the gene *spo0A* is important in Stage 0 (zero). Spo0A encodes the master regulator of sporulation; mutants lacking *spo0A* never proceed to Stage I and are incapable of sporulating (see below). Stage 0 occurs when a vegetative cell encounters nutrient deprivation and must make the decision about whether or not to sporulate. The cell has two copies of the chromosome at Stage 0, so it could either continue to divide or form an endospore. Once the decision is made to initiate sporulation, Stage I begins. At this point the cell is committed to forming an endospore; therefore, the decision is highly regulated by multiple inputs (see below). While Stage I is poorly characterized, one visible change involves the two nucleoids coalescing to form an axial filament. During Stage II, the cell undergoes an asymmetric division that produces two different cell types. The larger cell is the mother cell and the smaller cell is the **forespore**, the nascent internal structure that will eventually become the mature endospore. The transition from Stage II to Stage III involves engulfment, during which the thin layer of peptidoglycan in the asymmetric septum is degraded and the mother cell membrane expands to surround the forespore. Engulfment is complete when the mother cell membrane fuses. At this point the forespore is surrounded by two membranes and is free inside the mother cell cytoplasm. During Stage IV, a peptidoglycan layer called the **spore cortex** is synthesized between the two membranes of the forespore. The spore cortex has a slightly altered chemical structure and a different degree of cross-linking than the peptidoglycan in vegetative cells. Spore coat proteins are deposited around the forespore during Stage V. In Stage VI, endospore maturation takes place, with the endospore achieving full dormancy and resistance properties. Finally, in Stage VII, the mother cell lyses and the dormant endospore is released. The endospore can remain completely dormant for very long periods of time; viable endospores have been reported in prehistoric amber and Egyptian pyramids. Germination and outgrowth occur when specific nutrients are sensed, and then vegetative growth by binary fission resumes (see below).

RESISTANCE PROPERTIES OF ENDOSPORES

The endospores of *B. subtilis* are metabolically inactive. In addition, they have specialized properties that ensure their survival for extended periods of time until environmental conditions warrant germination of the spore and a return to vegetative growth. For example, spores exhibit resistance to UV light, which can damage DNA by causing the formation of pyrimidine dimers. During sporulation, **small acid-soluble proteins (SASPs)** that coat the DNA and change its conformation from the common B-form helix to a special A-form helix are produced. The germinating endospore has a specialized system to repair specific types of pyrimidine dimers (photoproducts) that form in endospores, and the shape of the A-form helix makes it easier for this system to repair DNA during germination.

Heat resistance is another endospore property that is mediated by a number of physiological changes in the endospore. The endospore core becomes dehydrated as the concentrations of divalent calcium cations and dipicolinic acid increase and bind free water molecules in the endospore. In the absence of water, the endospore is metabolically inactive. Under the conditions of dehydration, the spore coat proteins are in a precipitated state, making them resistant to heat denaturation. In addition to changing the DNA conformation, SASPs reduce the rate of DNA depurination in dehydrated endospores.

Finally, endospores are resistant to chemicals and enzymes. The impermeability of the spore coat layers affords protection against enzymes like lysozyme and chemicals such as hydrogen peroxide by serving as a barrier. The dehydration of the core prevents chemicals from diffusing into the endospore, and thus prevents damage that would otherwise occur. The metabolic dormancy also prevents the action of certain antibiotics that are only effective against actively growing cells (e.g., penicillin, chloramphenicol, and rifampin). In addition, some *Bacillus* cells (e.g., *Bacillus anthracis*) have an exosporium, a semipermeable, proteinaceous layer outside the spore coat that provides additional protection against enzymes and antibodies.

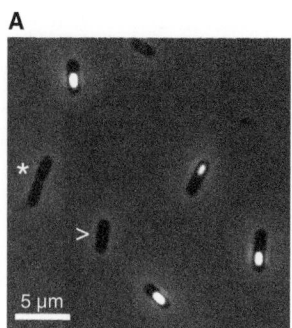

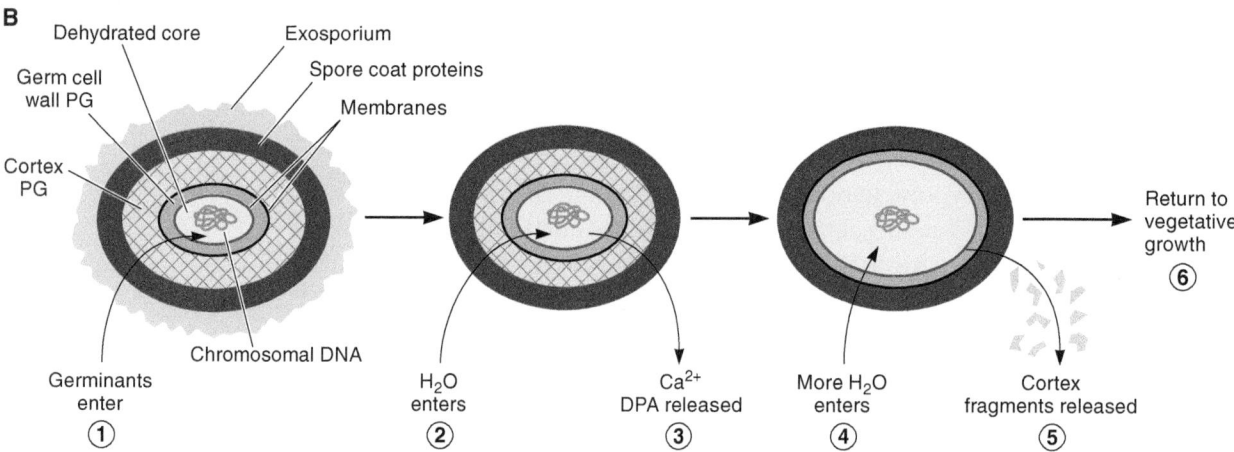

FIGURE 17.5 **Endospore visualization and germination in *Bacillus*.** (A) *B. subtilis* endospores appear phase bright (white) under phase-contrast microscopy, whereas the mother cell, nonsporulating and nondividing vegetative cells (indicated by white >), and cells undergoing binary fission (indicated by white*) appear dark. Image reprinted from Hutchison EA, et al. 2014. *Microbiol Spectr* 2(5). (B) During the process of endospore germination, (1) the dormant, dehydrated endospore senses the presence of specific germinant nutrients and then (2) begins to take up water. (3) Calcium ions and dipicolinic acid (DPA) are released, (4) more water is taken up, and (5) the cortex peptidoglycan (PG) is specifically degraded. During this time, small acid-soluble proteins (SASPs, not shown) in the core are utilized as carbon and energy sources, and (6) the resulting metabolically active cell begins vegetative growth.

PROCESS OF GERMINATION

As described, endospores remain metabolically inactive due to their low concentration of water and high concentration of calcium ions and dipicolinic acid. The dehydrated state of endospores gives them a high refractive index, which makes endospores appear "phase bright" when viewed by phase-contrast microscopy (Fig. 17.5A). The process of germination, which takes several hours, precedes the return to vegetative growth. The first step of germination is the sensing of a **germinant** (typically a nutrient source, e.g., certain amino acids) that must enter the endospore through the protective spore layers and bind to a membrane-associated receptor (Fig. 17.5B). Once the germinant is recognized, water begins to enter the endospore and calcium ions and dipicolinic acid are released. Experimentally, the external concentrations of these chemicals can be used to monitor the progression of germination. The process of germination can also be monitored microscopically, because as water is taken up, the refractive index decreases and the endospores lose their phase bright appearance. At this point the endospore begins to regain some metabolic activity and SASPs in the core are utilized as carbon and energy sources. Additional water rehydrates the endospore, causing it to expand. This is made possible through the degradation of the cortex peptidoglycan, which is structurally distinctive from the vegetative peptidoglycan. As cortex fragments are released, the endospore enlarges and ultimately transitions into an actively growing vegetative cell.

The process of germination is key to killing bacterial endospores, as vegetative cells are much more susceptible to disinfectants, antibiotics, UV light, and heat. Endospore contamination is an ongoing issue in hospital and nursing home settings where the endospores of pathogens like *Clostridium difficile* (recently classified as *Clostridioides*) may be present and could result in a deadly form of gastroenteritis. In addition, the elimination of endospores is important in dealing with *Bacillus anthracis*, a bioterrorism agent of serious concern. Similarly, space exploration vehicles must be decontaminated to remove endospores before they leave Earth.

REGULATION OF *BACILLUS SUBTILIS* SPORULATION INITIATION

A complex regulatory process that takes into account multiple internal and external signals occurs before a cell commits to the sporulation process during Stage 0. As described above, the lack of available nutrients is a primary factor in triggering sporulation. The cell must also be at an appropriate point in the cell cycle with two copies of the genome present. In addition, the cell must not be competent for uptake of exogenous DNA (Chapter 16). The cell can either be competent or undergo sporulation; the two processes are mutually exclusive. During Stage 0, there are both external signals (nutrient levels and high concentrations of the competence and sporulation factor [CSF] peptide) and internal signals (chromosome replication) that collectively determine if initiation of sporulation will occur.

At least five sensor histidine kinases are involved in signal recognition and response during the initiation of sporulation. KinB, which is membrane associated, and KinA, which is cytoplasmic, appear to be the most important (Fig. 17.6). When the sensor histidine kinases are activated and autophosphorylate, they then transphosphorylate Spo0F, a regulatory protein. Spo0F-P can be dephosphorylated by the RapA and RapB phosphatases. However, dephosphorylation of Spo0F-P by RapA and RapB is inhibited by the CSF quorum-sensing signal when it is present at relatively high concentrations (>20 nM), indicating a high population density. Thus, under conditions of nutrient starvation at high cell densities, Spo0F will be highly phosphorylated and it will initiate a regulatory signal transduction cascade triggering endospore formation.

Spo0F has an N-terminal domain typical of a two-component response regulator (Chapter 7). However, Spo0F does not bind DNA, because it lacks the C-terminal domain that normally contains the helix-turn-helix motif. The apparent function of Spo0F is to acquire phosphoryl groups from the sensor histidine kinases and transfer them via the phosphorelay system to the master regulator of sporulation, Spo0A (Fig. 17.6). The phosphoryl group from Spo0F-P is first transferred to Spo0B. Spo0B is a phosphoprotein phosphotransferase that takes the phosphoryl group from Spo0F-P and transfers it to Spo0A. Spo0A is a classic response regulator that binds to promoters and regulates gene expression when in its phosphorylated form. Spo0A-P regulates more than 100 genes; it directly activates the transcription of genes involved in promoting sporulation and represses genes that inhibit sporulation. About 25 of these genes encode

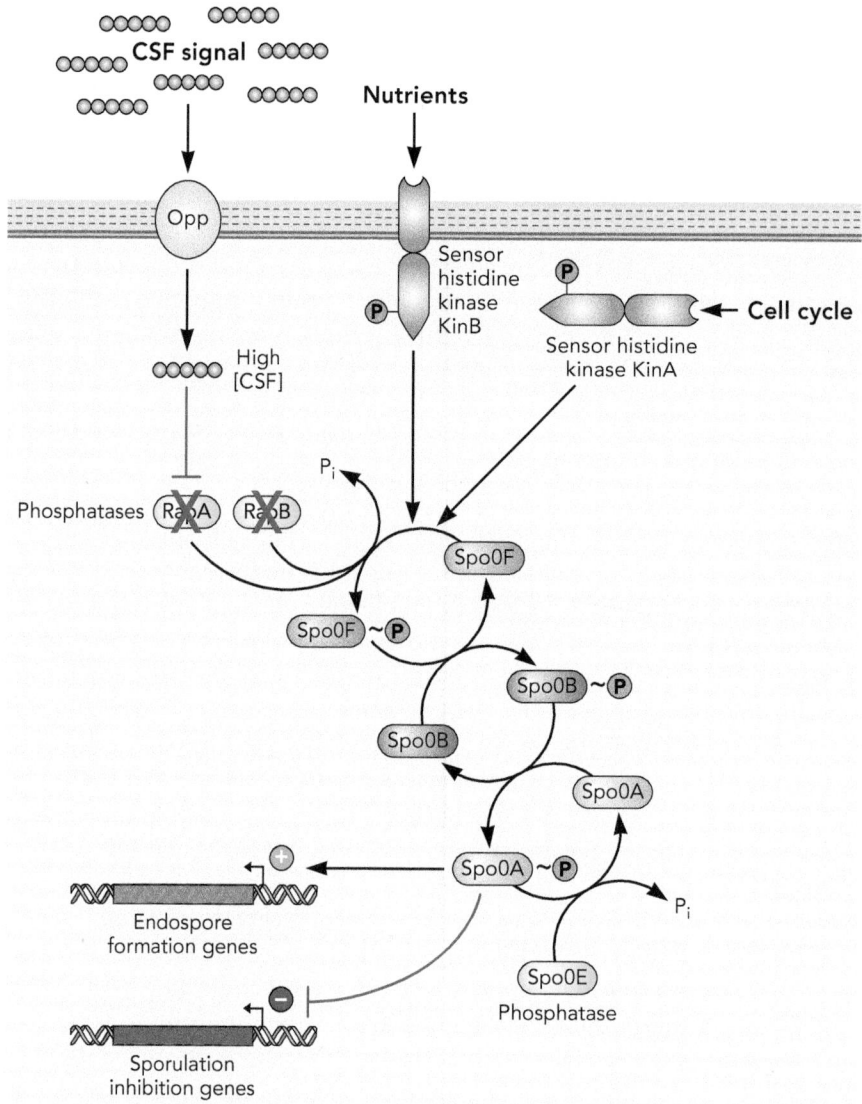

Phosphorylation cascade during the process of sporulation initiation in *Bacillus subtilis.* In this complex two-component signaling process, several histidine kinases monitor environmental signals during the decision to initiate sporulation. Various signals, including nutrient limitation and the appropriate stage of the cell cycle, cause the kinases KinA and KinB to autophosphorylate. A phosphoryl group is then passed to the response regulator Spo0F. The phosphoryl group on Spo0F-P is subsequently passed to a second response regulator, Spo0A, via the phosphotransfer protein Spo0B. Spo0A-P, the master transcriptional regulator of sporulation, activates the expression of sporulation genes and represses the expression of genes that inhibit sporulation. Several phosphatases (RapA, RapB, and Spo0E) can modulate the level of phosphorylation of Spo0F and Spo0A based on additional environmental signals, such as the population density (via competence and sporulation factor [CSF] quorum sensing), to ensure that conditions are appropriate for sporulation to initiate.

activators or repressors that regulate the expression of ~300 additional genes involved in sporulation. In total, the expression of ~10% of the genes in the *B. subtilis* genome changes at the initiation of sporulation. Another check on sporulation initiation is provided by the phosphatase Spo0E, which can dephosphorylate Spo0A-P and stop the initiation of sporulation in response to additional environmental signals. However, when Spo0A-P is phosphorylated, it

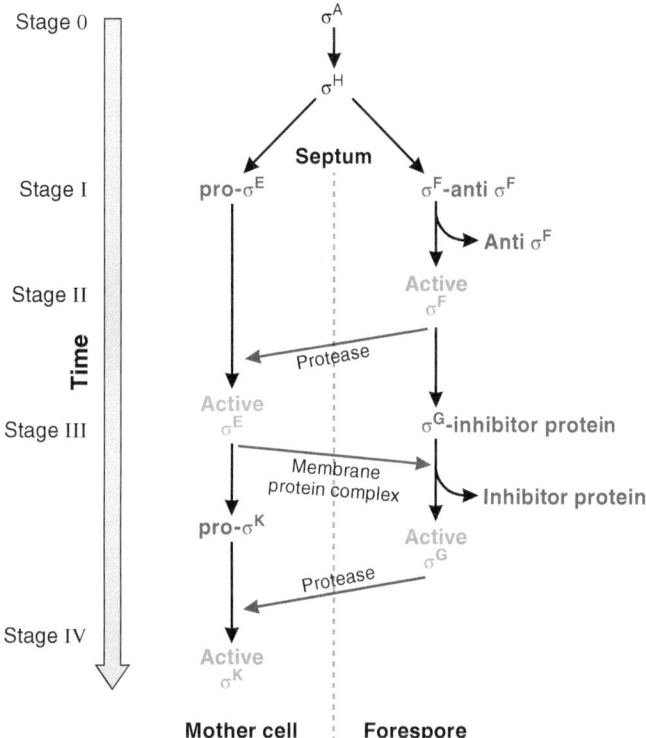

FIGURE 17.7 Sigma factors controlling temporal and spatial differentiation during endospore formation. The timeline (stage of sporulation) and location (mother cell versus forespore) of sigma (σ) factor activation during the stages of sporulation is shown. The mother cell and forespore are separated by the septum (black dotted line). Communication between the mother cell and forespore (crisscross spatial regulation, purple arrows) results in activation of each sigma factor at the appropriate time (temporal cascade regulation). In the mother cell, sigma factors E and K are produced as inactive pro-proteins (red) that are proteolytically processed into their active forms (green). In the forespore, sigma factors F and G are held in an inactive state through the binding of inhibitor proteins (red), but after the inhibitor is removed, the sigma factor will become active (green). See the text for details.

functions as the master regulator controlling the regulon of genes necessary for the cell to progress into Stage I of sporulation (Fig. 17.4). One key gene that is repressed by Spo0A-P is *abrB*, which encodes a repressor of the σ^H gene. When *abrB* is no longer expressed, σ^H levels increase. σ^H, a stationary-phase sigma subunit, is critical for the expression of genes necessary for subsequent stages of spore formation (Fig. 17.7).

TEMPORAL AND SPATIAL GENE REGULATION DURING SPORULATION IN *BACILLUS SUBTILIS*

In Chapter 15, the concept of temporal regulatory control during the synthesis and assembly of a flagellum was introduced. During *B. subtilis* endospore formation, not only is temporal control critical between the different stages of sporulation (i.e., Stages 0 to VII), but spatial control between the mother cell and forespore is also involved (Fig. 17.7). Both temporal and spatial regulation mechanisms occur simultaneously and involve transcriptional and posttranslational regulation (Chapters 7 and 8). Each sigma factor is required for transcription of the subsequent sigma factor in the same cellular compartment in a temporal manner, creating a **cascade of sigma factors** essential for endospore formation. Spatial regulation between the forespore and mother cell is controlled by posttranslational **crisscross regulation**, whereby the sigma activity in one compartment is required for sigma activation in the other compartment.

In vegetative cells facing starvation, the housekeeping sigma factor, σ^A (and Spo0A-P as described above), is needed for the expression of the σ^H gene. σ^H in turn controls expression of the gene encoding σ^F. The gene encoding pro-σ^E is also expressed during this time frame. **Pro-sigma factors**, like pro-σ^E, contain an inhibitory sequence of amino acid residues at the N-terminus rendering them inactive until cleaved. Pro-σ^E is present in the mother cell, while σ^F is present in the forespore. σ^F is bound and held inactive by an anti-σ^F factor protein (SpoIIAB). Subsequently, an anti-anti-sigma factor protein (SpoIIAA) binds and inactivates the anti-sigma factor in the forespore, releasing the active form of σ^F. Active σ^F is required for the expression of a protein (SpoIIR) needed to activate a Stage II protease (SpoIIGA) specifically required in the mother cell. σ^F is also required for the synthesis of σ^G in the forespore. σ^G is initially held in an inactive form by a σ^G-specific inhibitor protein. When the Stage II protease (SpoIIGA) becomes active in the mother cell, it cleaves off the inhibitory sequence of amino acid residues in pro-σ^E, converting it to active σ^E. Active σ^E is required for production of pro-σ^K in the mother cell, as well as a complex of membrane proteins that inactivate the σ^G inhibitor protein in the forespore. As a result, σ^G becomes active in the forespore and stimulates production of a Stage III protease that subsequently processes the inactive pro-σ^K in the mother cell, converting it to active σ^K, which is required for expression of the genes needed for late-stage endospore development.

This cascade of sigma factors involves differential expression of genes in the mother cell and forespore. But how are genes differentially expressed when each compartment presumably carries the same genes and regulatory components? One strategy involves the timing of the transfer of a copy of the chromosome from the mother cell to the forespore, which happens in stage II over approximately 15 minutes and requires ATP hydrolysis. The chromosome is transferred in a precise orientation, always leading with the origin of replication. The location of the *spoIIAB* gene (encoding anti-σ^F) near the replication terminus of the chromosome is critical, as this portion of the chromosome is transferred last. Therefore, for several minutes there is no *spoIIAB* gene in the forespore, which means that no new anti-σ^F is being made in the forespore. During this time, anti-anti-σ^F inactivates the remaining anti-σ^F in the forespore, releasing active σ^F. In contrast, in the mother cell, both the σ^F gene and the anti-σ^F gene are continuously transcribed and translated, maintaining σ^F in its bound inactive form.

In summary, both temporal and spatial regulation are important in the cascade of sigma factors controlling sporulation. Each sigma factor is required for the activity of the next sigma factor in the same cellular compartment in a temporal manner. In the forespore, σ^F (Stage II) is necessary to synthesize σ^G (Stage III), and in the mother cell, σ^E (Stage II) is necessary to synthesize σ^K (Stage IV). Crisscross regulation coordinates the developmental process in the two compartments in a spatial manner (e.g., σ^F [forespore] → σ^E [mother cell] → σ^G [forespore] → σ^K [mother cell]) and ensures proper endospore formation in *B. subtilis*.

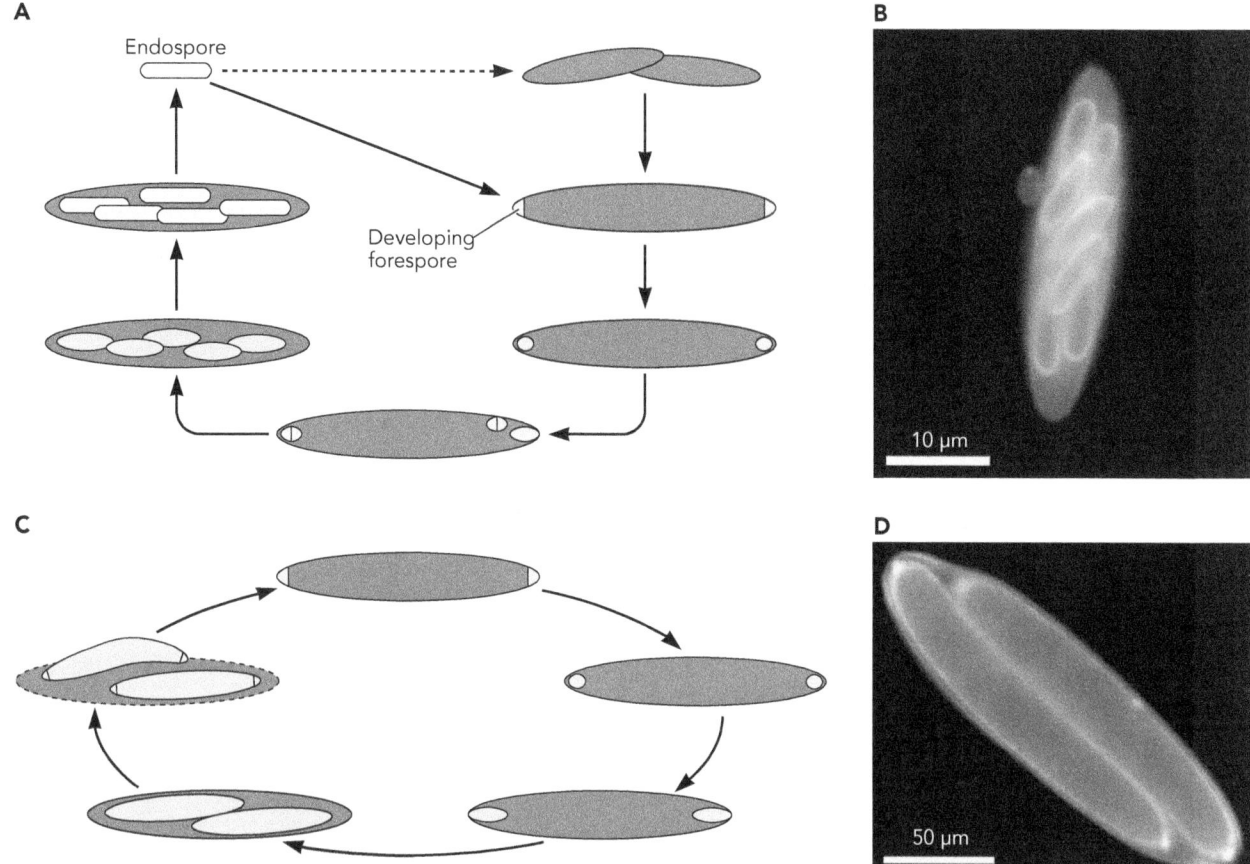

FIGURE 17.8 **Sporulation as the basis of reproduction in some bacteria.** (A) The life cycle of *Metabacterium polyspora* involves endospores that occasionally germinate and grow by binary fission (dashed arrow) or, more frequently (solid arrow), develop forespores (gray) at each pole of the cell. Moving clockwise, these forespores undergo binary fission, then elongate and develop into multiple mature endospores. (B) Fluorescence micrograph of *M. polyspora* showing cell membranes and spore coats (white outlines) of multiple mature endospores stained with FM1-43. Reprinted from Hutchison EA, et al. 2014. *Microbiol Spectr* 2(5), with permission. (C) In *Epulopiscium* spp., twin offspring form by asymmetric division at both cell poles (gray). Starting at the top and moving clockwise, engulfment of offspring cells occurs, followed by offspring cell elongation. The offspring cells initiate the next generation of offspring at their poles before they are released from the original mother cell, restarting the cycle. (D) Fluorescence micrograph of *Epulopiscium* spp. type B with two internal daughter cells. Cellular DNA stained with DAPI (white outlines) is located at the periphery of the cytoplasm of the mother cell and each offspring. Reprinted from Mendell JE, et al. 2008. *Proc Natl Acad Sci U S A* 105(18):6730-6734, with permission.

ALTERNATIVE LIFE CYCLES INVOLVING SPORULATION

It is of interest to note that some other members of the *Firmicutes*, besides *B. subtilis*, have evolved alternative forms of reproduction that are based on the endospore formation process. Although most endospore formers produce one endospore per cell, some can produce multiple endospores per cell. *Metabacterium polyspora*, a symbiont of guinea pigs, produces four or more endospores per cell and appears to use sporulation as its primary form of cell replication (Fig. 17.8A and B). Although pure cultures of *M. polyspora* have not yet been obtained in the lab, current evidence indicates that its cells rarely undergo standard binary fission in which one cell elongates and divides in half to produce two equivalent daughter cells. Unlike *B. subtilis*, which forms a single forespore at one pole, *M. polyspora* produces a forespore at each pole, and each forespore has the capacity to divide into additional forespores that mature into endospores.

A single cell produces multiple endospores that then germinate into vegetative cells after lysis of the mother cell (Fig.17.8A). Therefore, each generation results in the production of multiple offspring.

Another member of the *Firmicutes* is *Epulopiscium*, one of the largest known bacteria, with cells of up to 0.7 mm in length. *Epulopiscium* are symbionts found in the gut of surgeonfish. Like *M. polyspora*, *Epulopiscium* forms forespore-like structures called twin offspring at both cell poles (Fig. 17.8C and D). However, rather than maturing into metabolically inactive endospores, the two offspring elongate into metabolically active vegetative cells that are released when the mother cell lyses. Although no isolates of *Epulopiscium* have been obtained in pure culture, genome sequence analyses indicate that genes encoding the early stages of endospore formation are present, whereas those required for endospore maturation are absent. *Epulopiscium* cells have therefore evolved a reproductive strategy that uses the initial steps in the sporulation process to internally produce metabolically active offspring within the mother cell cytoplasm. These examples illustrate how some bacteria have evolved novel reproductive strategies that are based on the process of endospore formation. These alternatives to binary fission presumably offer a selective advantage in their symbiotic associations with specific eukaryotic hosts.

Learning Outcomes: After completing the material in this chapter, students should be able to . . .

1. list four enzymatic reactions used by bacteria to eliminate oxidative stress

2. outline two regulatory mechanisms (SoxR and OxyR) that control the oxidative stress response in *Escherichia coli*

3. describe the heat shock responses of *Escherichia coli*

4. explain the stages of endospore formation and germination in *Bacillus subtilis*

5. discuss the role of kinases and sigma factors in the control of spore formation in *Bacillus subtilis*

Check Your Understanding

1. Define this terminology: reactive oxygen species (ROS), superoxide radicals, superoxide dismutase (SOD), superoxide reductase, hydrogen peroxide, catalase, peroxidase, hydroxyl radicals, glutathione, heat shock response, endospore, forespore, spore cortex, small acid-soluble proteins (SASPs), germinant, cascade of sigma factors, crisscross regulation, pro-sigma factor

2. Aerobic and anaerobic bacteria often respond differently to various reactive oxygen species (ROS). (LO1)

   a) Fill in the table to indicate the formula and reactions of various ROS.

ROS	ROS chemical formula	Reaction(s) that form the ROS
Superoxide radical		
Hydrogen peroxide		
Hydroxyl radicals		

   b) Fill in the table as indicated to compare the enzymes that aerobic and anaerobic bacteria use to inactivate ROS.

Cell type	ROS	Enzyme	Reaction
Aerobic bacteria	Superoxide radicals		
Strictly anaerobic bacteria	Superoxide radicals		
Aerobic bacteria	Hydrogen peroxide		
Strictly anaerobic bacteria	Hydrogen peroxide		

   c) Anaerobic cells often use different mechanisms to deal with ROS. Why must anaerobic cells eliminate superoxide radicals using a mechanism different from aerobic cells?

   d) Why must anaerobic cells eliminate hydrogen peroxide using a mechanism different from aerobic cells?

3. Hydroxyl radicals are the most reactive of all reactive oxygen species. What strategy is used by bacteria to prevent the formation of hydroxyl radicals? (LO1)

4. Order the statements below (1 to 4) regarding the physiological response of *Escherichia coli* to superoxide radicals. (LO2)

   ____ The oxidized form of SoxR binds DNA and activates transcription of *soxS*.

   ____ The gene for superoxide dismutase is transcribed and translated.

   ____ Superoxide radicals are sensed by the regulatory protein SoxR, which becomes oxidized.

   ____ SoxS binds DNA and activates the expression of the genes in its regulon.

5. Order the statements below (1 to 4) regarding the physiological response of *Escherichia coli* to hydrogen peroxide. (LO2)

_____ The oxidized form of OxyR binds DNA and activates the expression of the genes in its regulon.

_____ *katG* and *gorA* are transcribed and translated.

_____ Hydrogen peroxide oxidizes the reduced form of the regulatory protein OxyR.

_____ Catalase converts H_2O_2 to O_2 and water.

6. Explain how glutathione triggers a chain of events that provides the cell protection against oxidative stress in *Escherichia coli*. (LO2)

7. The heat shock response is common to all living cells. (LO3)

a) "Heat shock response" is in some ways a misnomer for this process—what else can initiate the response? To what signal is the cell actually responding?

b) In *Escherichia coli*, at 30°C, *rpoH* (encodes σ^{32}) is transcribed and translated, but the sigma factor is not active in the cell. How does the presence of inactive σ^{32} benefit *E. coli* when the environment changes from 30 to 42°C?

c) The following three components work to limit the amount of RpoH in *E. coli* at 30°C. How does a change in temperature to 42°C cause each component to release σ^{32} to initiate the heat shock response?

- Chaperones (DnaK, DnaJ, and GrpE)
- Proteases
- Stem-loop structure in the *rpoH* mRNA

d) Once σ^{32} is activated during the heat shock response, the genes encoding chaperones (DnaK, DnaJ, and GrpE) and proteases (Lon and Clp) are overexpressed. What role do these proteins play during the heat shock response at higher temperature?

e) At even higher temperatures (45 to 50°C), another, more recently discovered heat shock response is triggered. At lower temperatures, σ^{24} (RpoE) is bound to the cytoplasmic membrane protein RseA. Order the steps below (1 to 4) to indicate what happens at temperatures above 42°C.

_____ Cell envelope repair genes are transcribed and translated.

_____ Damaged RseA is degraded by proteases.

_____ High temperature denatures periplasmic and outer membrane proteins.

_____ σ^{24} complexes with core RNA polymerase.

8. Some low-G+C Gram-positive bacteria (*Firmicutes*) can respond to unfavorable environmental conditions by forming endospores. (LO4)

a) What triggers sporulation?

b) List four actions cells can take to ameliorate this trigger condition before initiating endospore formation.

c) Label each event listed with the stage of sporulation (Stages I to VII) in which it occurs.

_____ After chromosome replication, the two nucleoids coalesce to form an axial filament.

_____ Spore coat proteins are deposited around the forespore.

_____ The cell undergoes an asymmetric division to produce two different cell types.

_____ The endospore achieves full dormancy and resistance properties.

_____ The mother cell lyses and the dormant endospore is released.

_____ The spore cortex is synthesized between the two membranes of the forespore.

_____ The thin layer of peptidoglycan in the asymmetric septum is degraded and the mother cell engulfs the forespore.

9. *Bacillus* endospores are incredibly resistant to a range of environmental factors. How does each component listed contribute to endospore resistance? (LO4)

   a) Small acid-soluble proteins (SASPs)

   b) Higher concentrations of Ca^{2+} and dipicolinic acid

   c) Spore coat layers

10. Put the following steps in the germination process of endospores in the correct order (1 to 5). (LO4)

   ____ A nutrient such as an amino acid infiltrates the spore coat.

   ____ The cortex peptidoglycan begins to degrade as the cell increases in size.

   ____ The germinant is detected by a membrane receptor.

   ____ The metabolic activity of the cell increases.

   ____ Water enters the cell as Ca^{2+} and dipicolinic acid are released.

11. The initiation of sporulation is highly regulated and dependent on nutrient concentrations, cell density, and the cell cycle. (LO5)

   a) Endospore formation begins when a phosphorylation relay is initiated by kinases KinA and KinB (Stage 0). Briefly describe the function of each protein in this phosphorylation relay system.

   - KinA and KinB

   - Spo0F-P

   - Spo0B-P

   - Spo0A-P

   - Spo0E

   b) To ensure coordination between competence and sporulation, sporulation is also controlled by RapA, RapB, and competence and sporulation factor (CSF).

   i. What is the function of RapA and RapB, and how do they influence sporulation initiation?

   ii. How does CSF work with RapA and RapB to ensure the initiation of sporulation occurs at high CSF concentrations?

12. Further steps in sporulation are controlled by a cascade of sigma factors. (LO5)

   a) Fill in the blanks below with the letter for the appropriate sigma factor.

   i. In vegetative cells, sigma ____ controls the expression of most housekeeping genes.

   ii. Under starvation conditions, sigma ____ (the stationary-phase sigma factor) is transcribed and translated.

   iii. Sigma H initiates the expression of sigma ____ (in the forespore).

   iv. Sigma ____ is bound by an anti-sigma factor, which is released by an anti-anti-sigma factor.

   v. Active sigma F is needed for production of a protease required to activate pro-sigma ____ in the mother cell and for the synthesis of sigma ____ in the forespore.

   vi. Sigma ____ is initially held in an inactive form in the forespore by an inhibitor protein.

   vii. Pro-sigma ____ and ____ are converted to active sigma factors by proteases made in the forespore.

   viii. Active sigma K is required for production of pro-sigma ____ in the mother cell.

   ix. Sigma ____ becomes active in the forespore and stimulates the production of a protease that converts the immature pro-sigma K in the mother cell to a mature sigma factor.

   b) The cascade of sigma factors involves both transcriptional and posttranslational regulation.

   i. Which sigma factors are bound and held inactive by an anti-sigma factor protein? Do these sigma factors reside in the forespore or mother cell?

   ii. Which sigma factors contain an inhibitory sequence of amino acid residues at the N-terminus, rendering them inactive until cleaved? Do these sigma factors reside in the forespore or mother cell?

   c) σ^H is critical for the expression of subsequent sporulation genes. How is the activity of σ^H controlled by Spo0A-P and AbrB?

Dig Deeper

13. Different bacteria respond to reactive oxygen species (ROS) in distinct ways. (LO1, LO2)

 a) *Clostridium* species are Gram-positive, obligate anaerobes that have a strictly fermentative metabolism and are highly sensitive to oxygen. What enzymes would you expect to be present in these organisms to detoxify ROS while maintaining an anoxic environment? Explain your reasoning.

 b) *Streptococcus* species are Gram-positive, aerotolerant organisms that have a strictly fermentative metabolism.

 i. What enzymes might these organisms use to detoxify ROS? Explain your reasoning.

 ii. Transcriptome analysis of *Streptococcus pneumoniae* under oxic conditions (like the upper respiratory tract) revealed the upregulation of several genes. Would you expect the expression of *sodA* to increase or decrease during infection of the upper respiratory tract? Explain your reasoning.

 iii. One upregulated gene of *S. pneumoniae* encodes a glutathione reductase (GorA). In *Escherichia coli*, what role does glutathione reductase play in reducing the effects of oxidative stress?

 c) *E. coli* can either make glutathione or transport it from the environment through an ABC transporter. An analysis of the *S. pneumoniae* genome shows that *S. pneumoniae* does not have the genes for glutathione biosynthesis. However, the genome of *S. pneumoniae* does encode numerous ABC transporters and has been shown to take up glutathione through an as-yet unidentified transport system.

 i. How could you use the *E. coli* gene encoding the glutathione transporter to look for a putative gene for glutathione transport in *S. pneumoniae*?

 ii. How could a mutant strain of *S. pneumoniae* lacking the putative gene for glutathione transport help you to confirm the identity of the transport protein? What would you expect the physiological phenotype to be?

 iii. You find a putative gene for glutathione transport in *S. pneumoniae* and name it *gshT*. You believe GshT is critical to *S. pneumoniae* survival and pathogenicity in the respiratory airways of mice. To test this hypothesis, you infect mice with either wild-type *S. pneumoniae* cells or a *gshT* mutant strain. Assuming your hypothesis is correct, predict how each of the listed parameters would change (increase, decrease, or remain the same) in mutant cells relative to wild type.

Parameter	Result of mutant compared to wild type
The number of *S. pneumoniae* cells in the nasopharyngeal region over time	
The number of hours before onset of symptoms	
Survival rate of infected mice	

14. Bacterial pathogens experience abrupt changes in temperature as they move between their external reservoirs and warm-blooded hosts. A temperature of 37°C often serves as a signal that the bacterium has been ingested and induces a major shift in gene expression and metabolism. Temperature can affect gene expression and protein activity by impacting the secondary structures of proteins, RNA, and DNA. The Gram-negative bacterial pathogens *Yersinia pestis* (causes bubonic plague), *Yersinia enterocolitica* (causes food poisoning), and *Yersinia pseudotuberculosis* (causes fever and intestinal distress) all exhibit this temperature response. You have identified a new toxin, ToxI, in *Y. pseudotuberculosis* that interferes with the host immune response. You want to investigate the regulation of the gene encoding this toxin in response to temperature. (LO3)

 a) To determine if *toxI* expression is influenced by temperature, you make transcriptional and translational fusions of the *toxI* gene to the *luxCDABE* genes (encoding bioluminescence); create different strains of *Y. pseudotuberculosis*, each with a different construct; grow the cells at 25°C (blue) and 37°C (orange); and monitor light production (light emission data at 25°C normalized to 1).

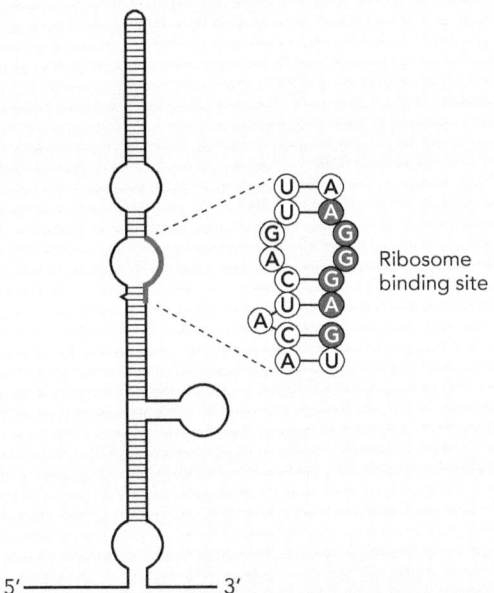

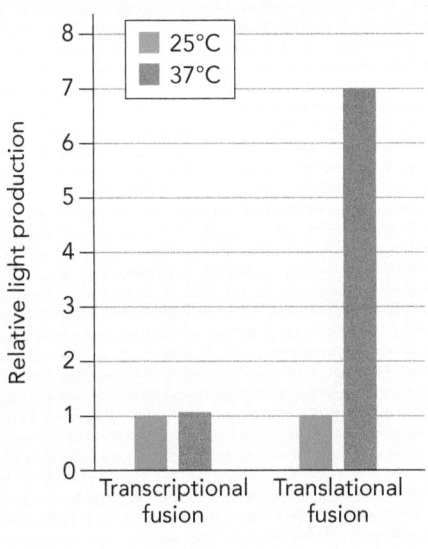

i. Which of the above constructs (1 or 2) shows a transcriptional fusion and which shows a translational fusion? Explain your answer.

ii. Your results are shown in the graph to the right. What do the data indicate about whether ToxI production is controlled by temperature at the level of transcription or translation based on light production from the reporter genes?

b) RNA thermometers often facilitate temperature-induced regulation. To further explore the regulation of *toxI*, you look at the predicted structure (shown below) of the *toxI* mRNA at 25°C.

i. In general, how does temperature influence the translation of mRNA when a stem-loop structure forms? Is the mRNA more likely to be translated at a lower temperature or at a higher temperature?

ii. Given the predicted structure of the *toxI* mRNA at 25°C, do you expect the structure of the *toxI* mRNA to be affected by higher temperatures and potentially impact translation? Explain your reasoning.

iii. You change the GA bases (highlighted in yellow) to UC. Predict whether this sequence change would likely increase or decrease *toxI* mRNA translation at 37°C. Explain your reasoning.

iv. To explore the impact of this sequence change (from GA to UC) on gene regulation, you make translational fusions of the *toxI* gene to *luxCDABE* using your altered gene sequence (with bases GA changed to UC), grow wild-type cells and your UC mutant strain at 25°C (blue) and 37°C (orange) for 4 hours, and monitor light production (light emission data at 25°C normalized to 1). How do results (shown in the graph to the right) for the mutant compare to those for wild-type cells at each temperature? What do these results indicate about the mechanism controlling *toxI* mRNA translation in response to temperature? Explain your reasoning.

v. Your next experiment investigates the impact of the sequence change (from GA to UC) in *toxI* on the pathogenicity of *Y. pseudotuberculosis*. You infect some mice (internal temperature 37°C) with wild-type *Y. pseudotuberculosis* (blue line in the graph below) and some with *Y. pseudotuberculosis* cells carrying the altered *toxI* RBS region (the UC strain; orange line). You monitor the survival of mice over 6 days. How do results for the mutant strain compare to those for wild-type cells? Explain why there is a difference. What do these data indicate about the importance of ToxI for causing disease?

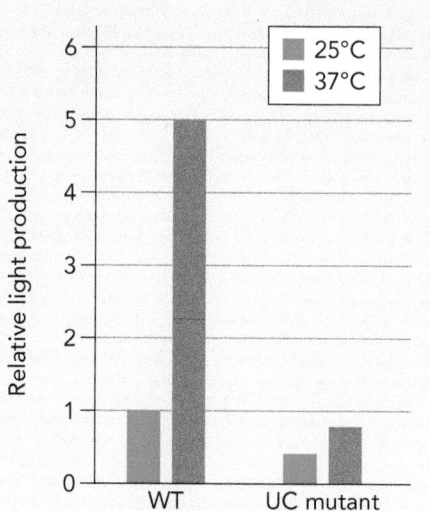

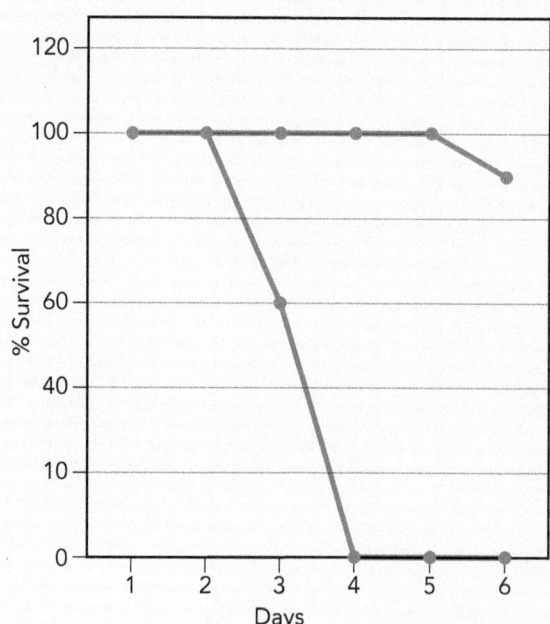

15. During sporulation in *Bacillus subtilis*, the vegetative cell differentiates into a forespore and mother cell. You are curious about the metabolism that takes place in these separate cellular compartments—do the forespore and mother cell both carry out biosynthesis of small molecules like amino acids? Do they both synthesize macromolecules like mRNA and proteins? (LO4)

   a) To investigate the hypothesis that both the forespore and mother cell carry out these metabolic activities equally, you make translational fusions of key metabolic genes (listed) with the green florescent protein (GFP).

   - *citZ*, citrate synthase, involved in the tricarboxylic acid (TCA) cycle

   - *argH*, argininosuccinate lyase, involved in arginine biosynthesis

   - *rpoC*, RNA polymerase subunit involved in transcription

   - *rplL*, ribosomal protein involved in translation

You create different strains of *B. subtilis*, each with a different construct.

| cirZ::GFP | argH::GFP | rpoC::GFP | rplL::GFP |

You then monitor fluorescence in forespores (blue) and mother cells (orange). The results are shown on the graph, with the fluorescence of each construct in the mother cell normalized to 1.

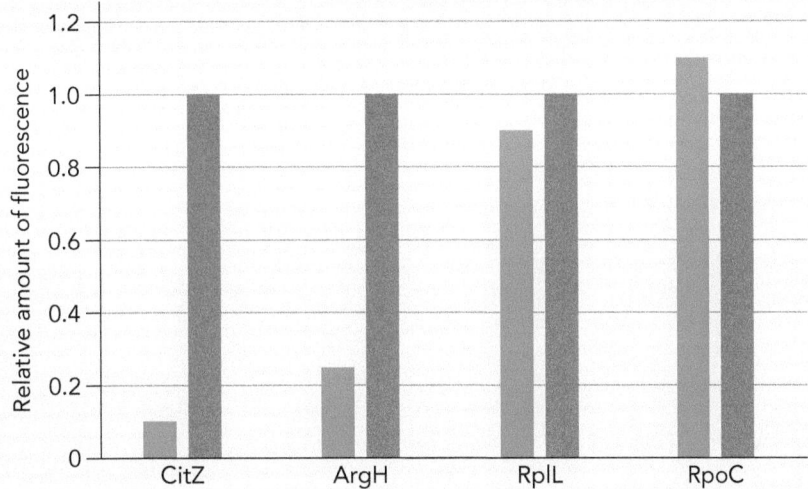

i. Which proteins are involved in enzymatic reactions of central metabolism? How do the results for these translational fusions in the *Bacillus* mother cells (orange) compare with those in the forespores (blue)? Explain these results.

ii. Which proteins are involved in transcription and translation? How do the results for these fusions in the *Bacillus* mother cells (orange) compare with those of the forespores? Explain these results.

iii. Do these data support or refute your hypothesis? Explain your reasoning.

b) Because the volume of the forespore is small (~10% of the mother cell) and the forespores were found to not synthesize arginine (based on the results of the *argH::GFP* fusion above), the forespore has fewer building blocks available from which to make needed macromolecules. You hypothesize that the forespore depends on metabolic precursors (e.g., amino acids) from the mother cell that are transported into the forespore.

There is evidence in the literature for a transport channel connecting the mother cell and forespore made of both mother cell proteins (encoded by the *spoIIIA* operon, called A) and a forespore protein (SpoIIQ, called Q). The A-Q complex could provide a channel-like "feeding tube" through which the mother cell provides small molecules to the forespore. To test this hypothesis, you allow vegetative *Bacillus* cells to take up a small fluorescent dye. You remove the remaining external dye and induce sporulation. You then monitor the movement of fluorescence from the mother cells into forespores over time in wild-type cells (blue) and in cells with the Q gene deleted (orange).

i. What do these results (shown in the graph to the right) tell you about the movement of the fluorescent dye over time in wild-type cells?

ii. What do these results tell you about the movement of the fluorescent dye over time in the strain with the Q gene deleted?

iii. Do these data support or refute your hypothesis? Explain your reasoning.

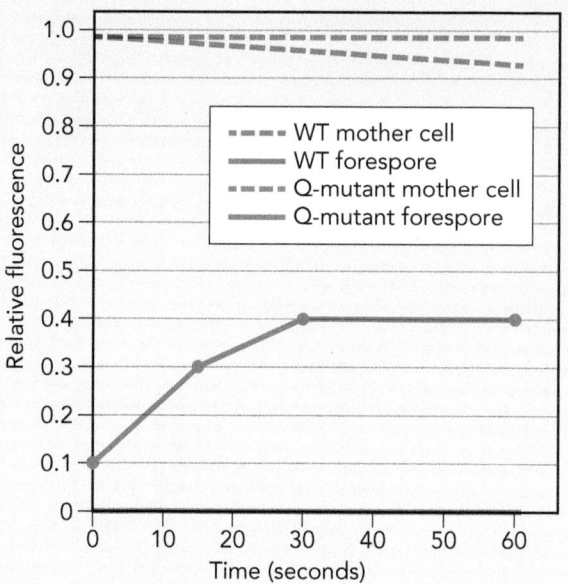

16. *Bacillus anthracis* is a Gram-positive bacterium that causes anthrax, a potentially deadly disease in livestock and humans. In 2001, letters laced with anthrax endospores appeared in the U.S. mail. Five people were killed and 17 were sickened from this intentional release of *B. anthracis* endospores. There is currently much interest in learning how to best combat any future releases, as endospores are resistant to extreme temperature, desiccation, pH, radiation, and other chemical treatments. You believe that endospores could hypothetically be controlled by triggering spore germination, as the vegetative cells could be killed much more easily. Germination begins when endospore receptors sense specific signal molecules (e.g., nutrients). (LO4)

a) Previous research showed that endospore germination in *Bacillus subtilis* can be triggered by an alanine-inosine (Ala-Ino) solution. You set up an experiment to determine if Ala-Ino can also initiate germination in *B. anthracis* spores. You know that germination in endospore-forming bacteria can be monitored by observing the release of calcium dipicolinic acid (DPA) from the core. You grow *B. subtilis* and *B. anthracis* on separate agar plates for 2 weeks and harvest the spores into sterile water, heat the spore solution at 65°C for 30 minutes, add 10 mM Ala-Ino solution, and incubate at 37°C. You then periodically remove samples and measure the absorbance (optical density at 270 nm [OD_{270}]) of the culture supernatant to obtain an estimate of the amount of DPA released into the solution (blue = *B. anthracis*; orange = *B. subtilis*).

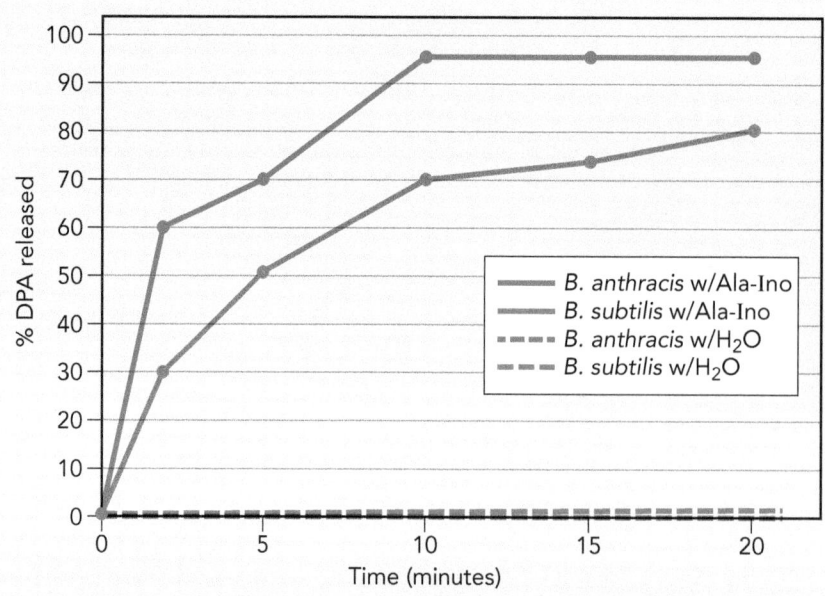

   i. Why did you need to heat the spore solutions at 65°C for 30 minutes before adding Ala-Ino?

   ii. Do your results indicate that Ala-Ino can initiate germination in *B. subtilis* and *B. anthracis* spores? Explain how you know.

b) You are curious as to whether triggering germination can make common decontamination methods more effective in eliminating *B. anthracis*. You again harvest *B. anthracis* spores and heat the spore solution at 65°C for 30 minutes. You dilute the spores in either sterile water (blue) or an Ala-Ino (orange) solution to a final concentration of 10^7 spores/ml. After a 20-minute germination period, you treat the spores with 1% peroxide (H_2O_2). You heat the spore solutions at 65°C for 30 minutes and determine the spore concentration by counting the number of spores using phase-contrast microscopy. What effect does pretreating spores with Ala-Ino have on decontamination with peroxide? Explain these results.

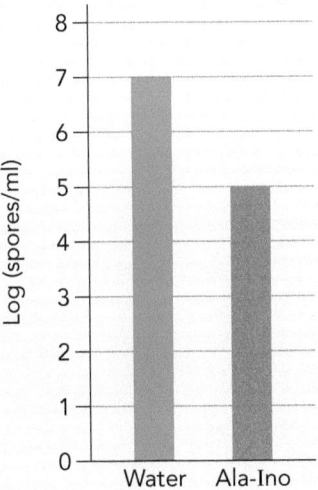

c) In reality, decontamination most often occurs on surfaces. Design an experiment to test the hypothesis that inducing germination could enhance decontamination treatments on a stainless-steel surface.

442 Making Connections

443 A Simple Model for Bacterial Cellular Differentiation: *Caulobacter crescentus*

444 Differentiation in Filamentous Cyanobacterial Species

447 Life Cycle of Filamentous Spore-Forming *Streptomyces*: An Example of Bacterial Multicellularity

449 Life Cycle of Myxobacteria: Predatory Spore-Forming Social Bacteria

452 Biofilms: The Typical State of Microorganisms in the Environment

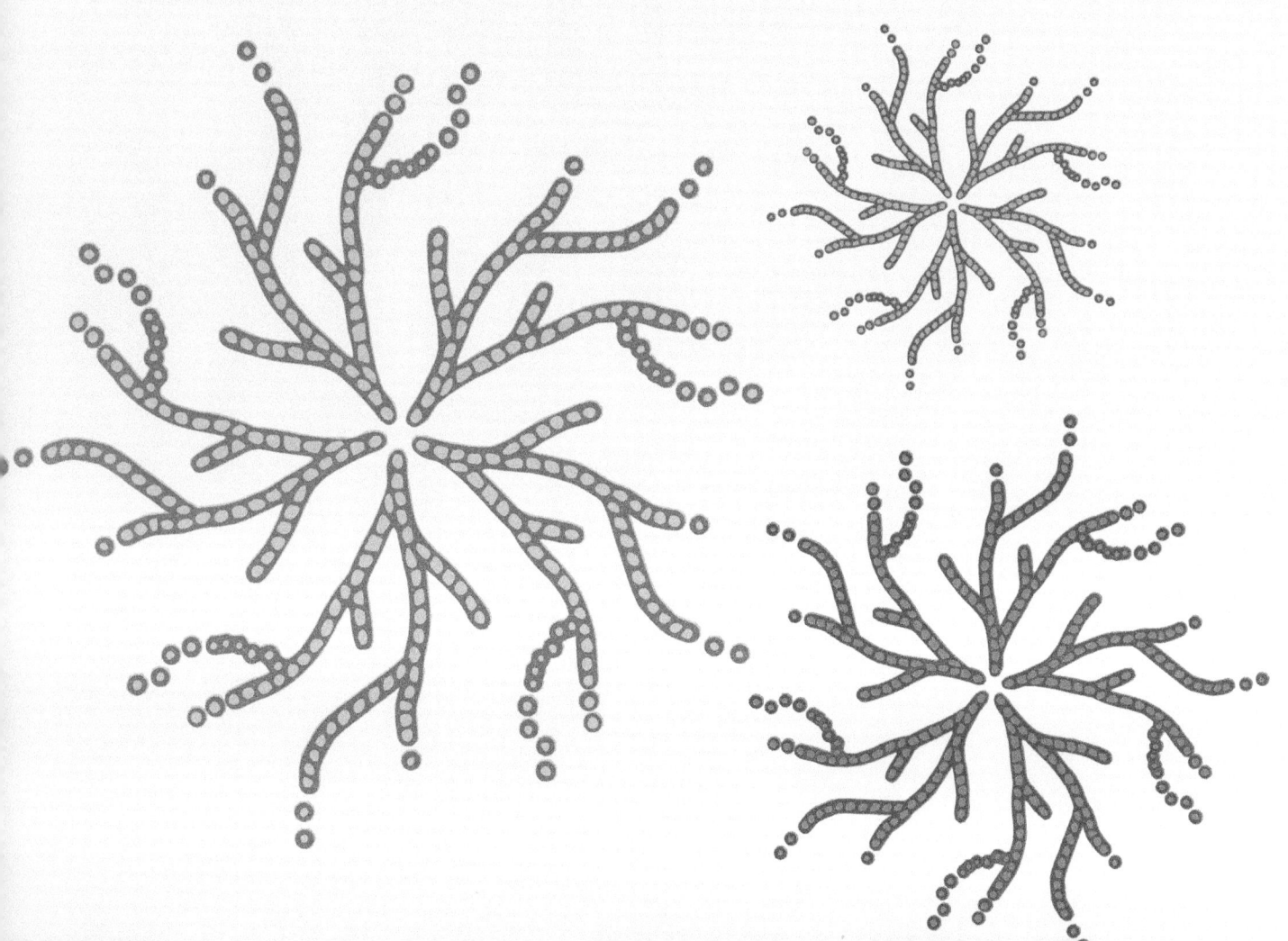

18

LIFESTYLES INVOLVING BACTERIAL DIFFERENTIATION

After completing the material in this chapter, students should be able to . . .

1. compare and contrast stalked and swarmer cell morphologies in *Caulobacter crescentus*

2. describe the process of heterocyst and hormogonium differentiation by filamentous cyanobacterial species

3. explain the reproductive life cycle of *Streptomyces*

4. outline the processes involved in formation of a fruiting body in myxobacteria

5. describe the formation and resulting structural microenvironments of complex biofilms

6. contrast the benefits of growth in biofilms for microbes with the problems biofilms cause in clinical and industrial settings

Microbial Physiology: Unity and Diversity, First Edition. Ann M. Stevens, Jayna L. Ditty, Rebecca E. Parales and Susan M. Merkel.
© 2024 American Society for Microbiology.
Companion website: www.wiley.com/go/stevens/microbialphysiology

Making Connections

In Chapter 16, we explored bacterial quorum sensing as a way that individual bacterial cells communicate with one another to exhibit cooperative group behaviors across a population. In some bacteria, cell-to-cell communication triggers cell differentiation into more than one distinctive cellular physiology within the population, such as competence and endospore formation in *Bacillus* (Chapters 16 and 17). In these cases, the specialized abilities of some cells in the population benefit the survival of the genetic lineage for the entire population. During binary fission (Chapter 4), one mother cell forms two identical daughter cells through vertical genetic transmission. Here we will describe the unique life cycles of several microbes that have been studied as models for cellular differentiation, where cells with identical genetic information have distinctive morphologies. This will include the life cycle of the dimorphic bacterium *Caulobacter*; the differentiation process of the filamentous cyanobacterium *Nostoc*; the formation of distinctive types of spores formed in the streptomycetes and myxobacteria; and the development of complex biofilms, the natural growth state of most microbes (Fig. 18.1).

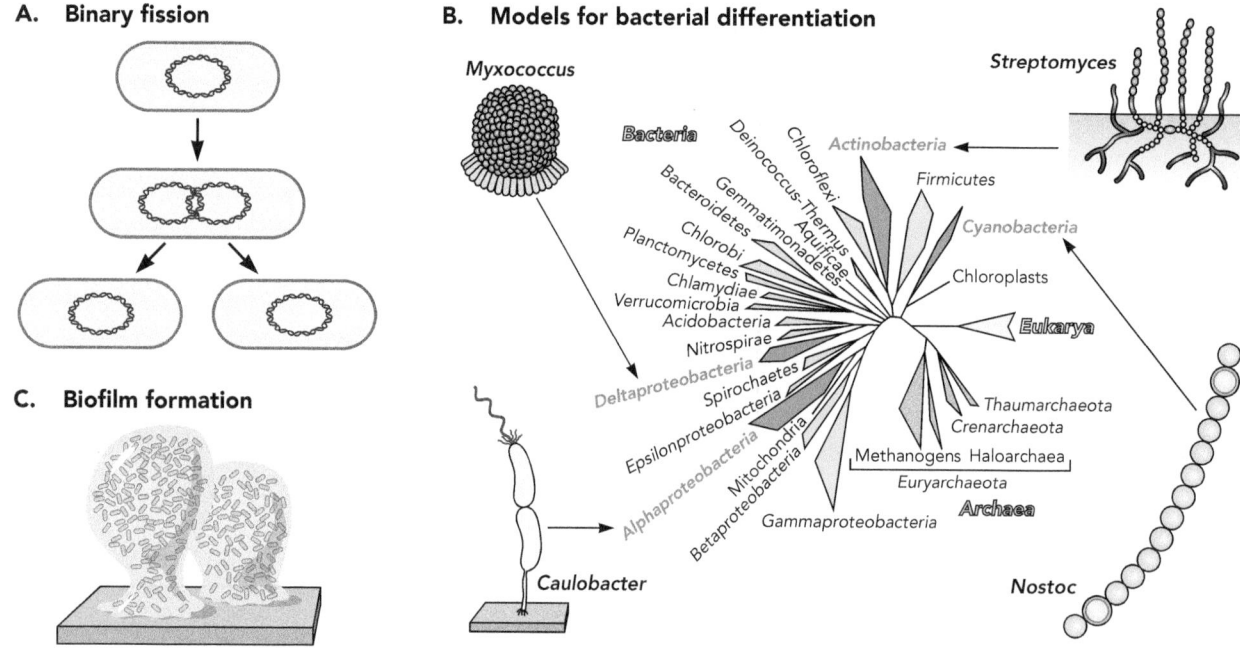

A. Binary fission

B. Models for bacterial differentiation

Myxococcus

Streptomyces

Caulobacter

Nostoc

C. Biofilm formation

FIGURE 18.1 **Introduction to bacterial differentiation.** (A) Most bacteria reproduce via binary fission, whereby a mother cell grows to twice its length and then divides into two equivalent daughter cells. (B) Bacteria with more complex life cycles that serve as models for bacterial differentiation include *Caulobacter*, *Nostoc*, *Streptomyces*, and *Myxococcus*; the phylogenetic locations of these bacterial genera are highlighted on the phylogenetic tree (highlighted in dark coloring and bold green text). See the text for more details about the lifestyles of these microorganisms. (C) Most microbes in nature live in biofilms, which are multicellular communities surrounded by extracellular polymeric substance (EPS) that are typically attached to surfaces.

A Simple Model for Bacterial Cellular Differentiation: *Caulobacter crescentus*

Caulobacter crescentus is a Gram-negative alphaproteobacterium with a distinctive crescent shape due to the presence of a structural protein called crescentin. Caulobacters are commonly found growing on surfaces in low-nutrient aqueous environments. *C. crescentus* serves as a model system for studying asymmetric cell division and bacterial cell differentiation. These bacteria have a dimorphic life cycle with two distinct cell types, **stalked cells** and **swarmer cells** (Fig. 18.2A). The stalked mother cells are the replicative forms, whereas the motile swarmer cells are incapable of DNA replication and cell division. Stalked cells bind to surfaces through an adhesive holdfast at the base of their stalk; they are sessile and incapable of motility. Under high cell density in laboratory cultures, multiple stalked cells may adhere to one another, forming complex rosette structures (Fig. 18.2B). When the stalked cell divides, it first elongates to twice its length, a septum is built at the midpoint, and a new flagellated swarmer cell is released. Swarmer cells have a single polar flagellum; upon being released, they use chemotaxis and motility (Chapter 15) to locate an environment with sufficient

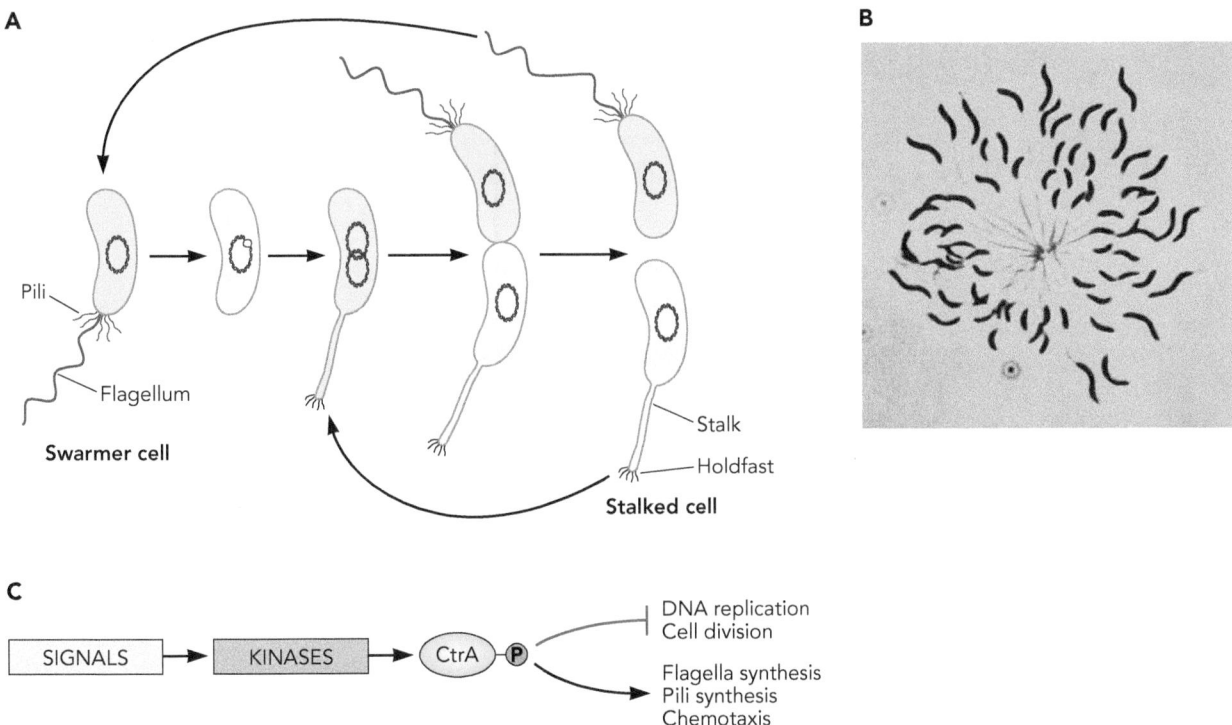

FIGURE 18.2 **Life cycle of *Caulobacter crescentus*.** (A) *C. crescentus* cells divide by unequal binary fission, forming two distinct cell types, nondividing motile swarmer cells with a flagellum and pili at one pole, and nonmotile reproductive stalked cells with a holdfast. Green coloring indicates cells with high concentrations of the key regulatory protein, CtrA-P. (B) Microscopic image showing a rosette of *C. crescentus* cells with multiple cells linked together through their holdfasts. Photo credit: Rebecca E. Parales. (C) The signal transduction cascade controlling CtrA includes poorly defined input signals that trigger several kinases to phosphorylate CtrA. CtrA-P then represses genes associated with DNA replication and cell division and activates genes associated with flagella and pili synthesis and chemotaxis in swarmer cells. See the text for details.

nutrients. The swarmer cell then discards its flagellum and differentiates into a stalked cell, growing a stalk at the same pole. This nascent mother cell attaches to a surface by its adhesive holdfast, replicates its DNA, and divides, continuing the cycle (Fig. 18.2A). Each stalked cell is capable of multiple rounds of replication. The ability of the swarmer cell to move away from the mother cell helps to reduce competition for limited nutrients between the stalked mother cell and the progeny swarmer cell.

The phosphorylated form of the response regulator CtrA (cell cycle transcription regulator A) in *Caulobacter* serves as the master regulator for transcription of genes involved in cellular morphology and division (Fig. 18.2C). The level of phosphorylation of CtrA in the swarmer cell is modulated by a complex signal transduction cascade involving multiple kinases that control the intracellular levels of CtrA-P. The signals that control this process have not been fully elucidated. Swarmer cells have a high concentration of CtrA-P (Fig. 18.2A; indicated by green color), which inhibits DNA replication by binding to the chromosomal origin of replication (*ori*). In addition, CtrA-P represses expression of the essential cell division gene *ftsZ* (Chapter 4), thus preventing cell division. Among other things, CtrA-P also activates genes for the synthesis of flagella, chemotaxis proteins, and pili production (Chapter 15). During the transition from a swarmer cell to a stalked cell, CtrA is degraded. When CtrA-P levels are low, the flagellum is lost, a stalk is formed, and DNA synthesis and cell division occur, completing the conversion of the swarmer cell into a replicative stalked cell. While CtrA-P levels are low in stalked cells due to protease degradation, the levels of other regulatory proteins increase. These regulatory proteins help facilitate the initiation of chromosome replication, cell division, and stalk formation. By the time a new swarmer cell is formed, CtrA levels rise again.

Differentiation in Filamentous Cyanobacterial Species

As described in Chapter 10, the *Cyanobacteria* form a monophyletic group that is characterized by the ability to carry out oxygenic photosynthesis. Many cyanobacterial species also fix nitrogen (Chapter 12), which poses a problem since nitrogenase is irreversibly inactivated by oxygen. Some filamentous cyanobacteria, including members of the genera *Nostoc* and *Anabaena*, solve this problem by producing specialized nitrogen-fixing cells called **heterocysts** as part of their complex life cycle (Fig. 18.3A). Heterocysts are terminally differentiated and do not divide; thus, the price of nitrogen fixation is the loss of reproductive cells in the population. In addition to vegetative cells and heterocysts, these organisms also form cold- and desiccation-resistant resting cells, called **akinetes**, which are analogous to but morphologically distinct from spores (Chapter 17). Some filamentous cyanobacteria also produce smaller, nonreplicating filaments called **hormogonia** that are capable of gliding motility (Chapter 15). Akinetes, heterocysts, and hormogonia represent three cellular differentiation strategies that allow filamentous *Nostoc* to respond to growth-limiting environmental conditions; akinetes assume a quiescent state and wait for the environmental conditions to improve, heterocysts provide fixed nitrogen to adjacent vegetative cells in the filament, and hormogonia can migrate to a more favorable habitat where they can resume their photoautotrophic mode of growth. Heterocyst and

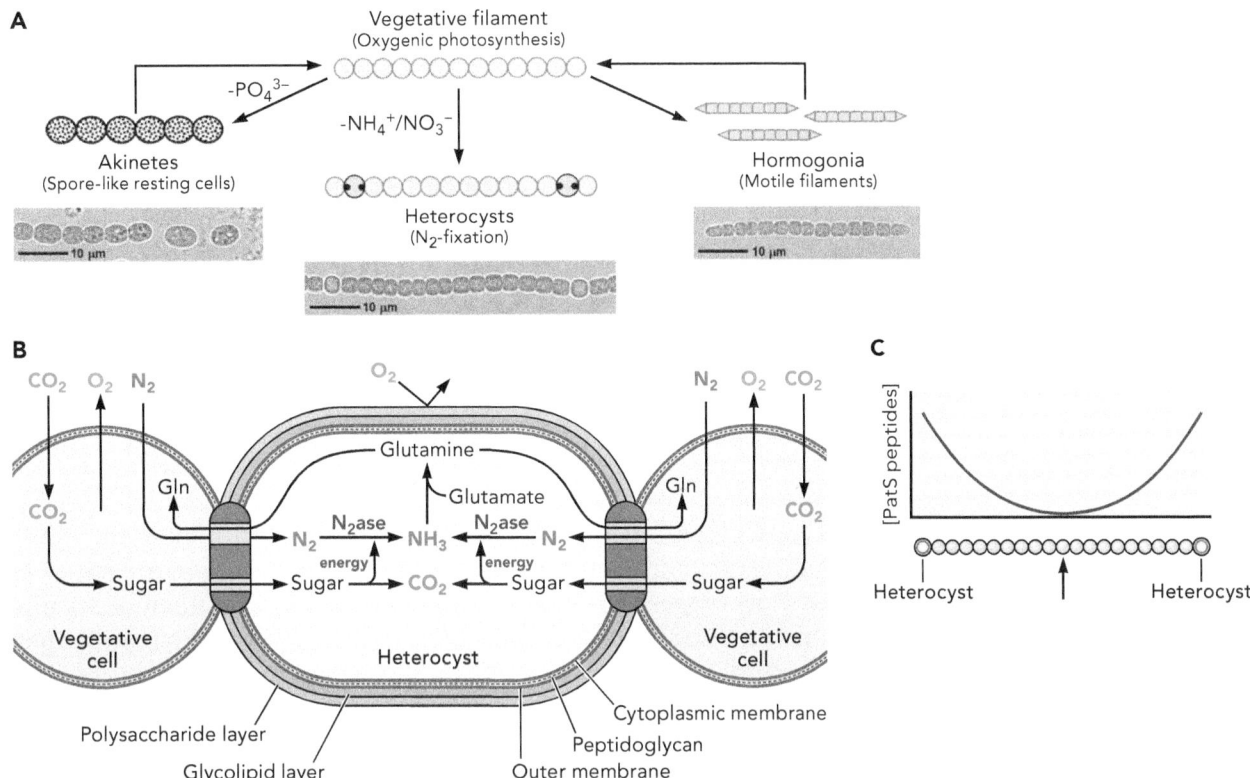

FIGURE 18.3 **Life cycle of the cyanobacterium *Nostoc punctiforme.*** (A) The different types of cells and structures that develop during the life cycle of *Nostoc* are depicted as cartoon models with corresponding microscopy images. Reprinted with permission from Risser D. 2023. *Appl Environ Microbiol* 89:e0039223. (B) The symbiotic relationship between a heterocyst and adjacent vegetative cells within a *Nostoc* filament. Carbon dioxide (CO_2, blue) and nitrogen gas (N_2, orange) enter vegetative cells. Sugars are formed and oxygen gas (O_2, green) is released as the vegetative cells perform oxygenic photosynthesis. Sugars and N_2 are transported from vegetative cells into the heterocyst, whereas O_2 cannot enter. As a result, the oxygen concentration within the heterocyst is sufficiently low to allow nitrogenase to catalyze the reduction of N_2 to ammonia (NH_3). Energy (in the form of ATP and reduced coenzymes) required for N_2 fixation is provided through the catabolism of sugars. The NH_3 is incorporated into glutamate to form glutamine, which is then transported to the neighboring vegetative cells. (C) Graph depicting the relative concentration of PatS peptides in heterocysts (yellow) and vegetative cells (green) along the *Nostoc* filament. A new heterocyst will next form midway between the two existing heterocysts (arrow) where the PatS peptides concentration is sufficiently low; this maintains the pattern of ~1 heterocyst every 10 vegetative cells. See the text for details.

hormogonia formation by *Nostoc* will be discussed in more detail here as these two processes are better understood than the production of akinetes.

Members of the genus *Nostoc* are multicellular cyanobacteria that grow as chains of cells in filaments that can be hundreds of cells long. When biologically available forms of nitrogen (other than N_2 gas) become limiting, about 10% of the vegetative cells differentiate into photosynthetically inactive nitrogen-fixing heterocysts (Fig. 18.3B). During the differentiation process, heterocysts stop making RubisCO, lose accessory photosynthetic pigments, and no longer have a functioning photosystem II. Genes encoding nitrogenase and glutamine synthetase (Chapter 12) are expressed in heterocysts. As a result, heterocysts do not fix CO_2 or produce oxygen, and they metabolize sugar obtained from neighboring photosynthetic cells to generate the reduced coenzyme needed by nitrogenase to fix N_2. Photosystem I, however, remains functional so that heterocysts can produce ATP by photophosphorylation. As discussed in Chapter 12, nitrogenase requires large amounts of ATP to catalyze the reduction of

N_2 gas. This differentiation process results in a division of labor in which two functionally and morphologically distinct but interdependent cell types are located in the same filament. Importantly, the resulting spatial separation of the photosynthetic and nitrogen fixation processes serves to protect the oxygen-sensitive nitrogenase enzyme, as it is housed within cells that no longer carry out oxygenic photosynthesis. To further protect nitrogenase, heterocysts have a high respiration rate and are surrounded by a thick cell wall containing glycolipid and polysaccharide layers that limits the entry of O_2. In addition, some *Nostoc* spp. produce a heme-containing membrane protein called **cyanoglobin,** which reversibly binds O_2, providing additional protection to nitrogenase by sequestering O_2.

The structural and metabolic changes necessary to produce heterocysts result from the establishment of gene expression patterns that are different from those in the vegetative cells. NtcA, a global regulator that belongs to the CRP (cAMP receptor protein) family of transcription factors (Chapter 7), responds to nitrogen limitation when it binds to α-ketoglutarate (also called 2-oxoglutarate), a signal of intracellular nitrogen limitation.

Activated NtcA triggers a regulatory cascade that results in the formation of PatS, which plays a key role in establishing the pattern of heterocyst distribution along the filament. *Nostoc* typically forms 1 heterocyst every ~10 to 15 vegetative cells (Fig. 18.3C). The product of *patS* is a 17-amino-acid peptide that is processed into peptides of 5, 6, or 8 amino acids. These short peptides are transferred from developing heterocysts to neighboring cells to inhibit differentiation. The concentration of PatS peptides is progressively lower in cells that are more distant from the PatS-producing heterocysts. Thus, along the filament, cells with a sufficiently low concentration of PatS peptides have the capacity to differentiate into a heterocyst, generating the cellular pattern.

Many filamentous cyanobacteria are capable of gliding motility on surfaces (Chapter 15). *Nostoc* filaments are capable of transiently differentiating into specialized gliding filaments called hormogonia in response to various environmental signals such as nutrient levels or specific wavelengths of light (Fig. 18.3A). The early stages of hormogonium development appear to be controlled by three alternative sigma factors. SigJ, SigC, and SigF form a hierarchical cascade to control the expression of genes including those for cell division, type IV pilus formation, exopolysaccharide production, chemotaxis, and phototaxis. Gliding hormogonia sense their environment and move toward more favorable locations. For example, specific photoreceptors allow hormogonia to use phototaxis to sense and move toward wavelengths of light that support photosynthesis. *Nostoc* is also capable of forming symbiotic relationships with certain plants, and motility and chemotaxis are important for *Nostoc* to locate and colonize its plant partners in response to unidentified plant signals. The cells in the hormogonia are slightly smaller than those in vegetative filaments because they undergo an additional round of division during the differentiation process. Hormogonia, which glide at a rate of about 1 μm/second, utilize both exopolysaccharides and pili to enable gliding motility. Although details of this unique motility mechanism are still being worked out, it is known that extension and retraction of pili at the leading poles of cells in the filament are coordinated, and hormogonium polysaccharide is secreted from the periplasm through outer membrane pores as the pili extend.

Life Cycle of Filamentous Spore-Forming *Streptomyces*: An Example of Bacterial Multicellularity

Streptomyces spp. are members of the *Actinobacteria*, which is a major lineage of the Gram-positive bacteria with characteristically high genomic G+C content. *Streptomyces* are generally aerobic chemoheterotrophs that are abundant in soil and marine environments. They are responsible for the earthy odor of soil, which results from the production of the volatile secondary metabolite geosmin. **Secondary metabolites** refer to chemicals that are produced via pathways that are not absolutely required for growth and survival; their production typically occurs in the stationary phase. *Streptomyces* spp. collectively make more than 100,000 bioactive compounds with wide-ranging functions (including antimicrobial, antiviral, cytotoxic and antitumor, antihypertensive, immunosuppressive, insecticidal, antioxidative, plant growth-promoting, and herbicidal agents). In fact, many of the antibiotics that are currently in use (e.g., streptomycin, chloramphenicol, erythromycin, kanamycin, tetracycline, and vancomycin) are produced by *Streptomyces* spp.

Streptomyces spp. have a unique life cycle. In contrast to unicellular bacteria like *Escherichia coli* and *Bacillus*, which form surface colonies composed of individual cells that replicate via binary fission (Fig. 18.4A), *Streptomyces* form a filamentous multigenomic network that only replicates by producing spores. The life cycle of *Streptomyces* (Fig. 18.4B) begins with the germination of a spore. At a minimum, spores require aqueous conditions to germinate, but germination is faster and more efficient when nutrients are also present. When spores are viewed by microscopy, the process of germination can be observed to occur in three steps: darkening, swelling, and emergence of the germ tube. Darkening is the result of the loss of light refraction as the dehydrated spore becomes hydrated and the coat is hydrolyzed by enzymes stored within the spore. New peptidoglycan is synthesized, and further influx of water causes swelling and allows the reactivation of additional stored enzymes. During this period, trehalose, a disaccharide of glucose that stabilizes the internal contents of the spore, is cleaved to glucose, which is used as an energy source as metabolic activity resumes. In addition, translation of stored mRNA begins, and DNA replication is initiated as the germ tube emerges from the spore. Vegetative growth of nonmotile branching filaments called **hyphae** occurs from the germ tube, forming the **substrate mycelium**, which grows below the growth surface, anchoring the growing colony to the soil (or agar medium in the lab). Chromosomes continue to replicate during vegetative growth, resulting in multigenomic hyphae with occasional cross walls.

As the vegetative mycelia grow and extend, they produce hydrolytic exoenzymes (e.g., amylases and proteases; Chapter 11) to degrade organic polymers in the soil. As the substrate mycelium depletes the available nutrients and senses nutrient limitation, the *Streptomyces* colony transitions into the reproductive and dispersal phase of its life cycle. During this time, the colony begins to produce aerial hyphae that grow upward, forming the **aerial mycelium**. Nongrowing portions of the vegetative mycelium lyse, providing nutrients to support growth of the aerial mycelium and the production of spores. At the same time, secondary metabolites with antimicrobial properties are produced to inhibit competing microbes from utilizing the released nutrients. Once aerial mycelial growth ceases, the multigenomic aerial filaments synchronously septate, forming chains

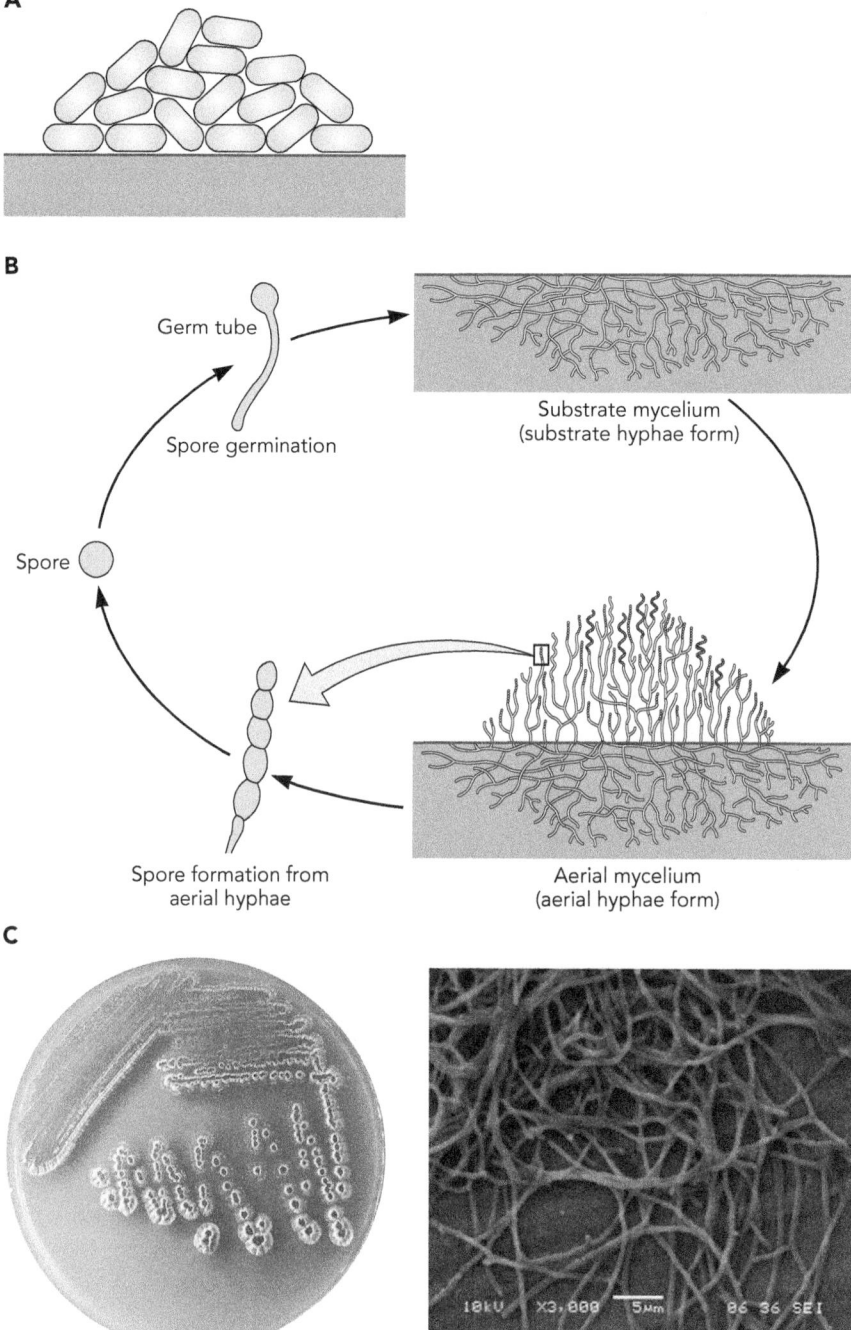

A

B

Germ tube

Spore germination

Substrate mycelium
(substrate hyphae form)

Spore

Aerial mycelium
(aerial hyphae form)

Spore formation from
aerial hyphae

C

FIGURE 18.4 **Colony development in *Escherichia coli* versus *Streptomyces* spp.** (A) *E. coli* (and most unicellular bacteria) grow as individual cells within colonies on the surface of their growth medium. (B) The life cycle of the actinobacterium *Streptomyces* is depicted. Clockwise from the left, under appropriate environmental conditions, a *Streptomyces* spore germinates, forming a germ tube that leads to the development of multigenomic hyphae with occasional cross walls. The hyphae grow down into the substrate, forming the substrate mycelium. The substrate mycelium contains multiple copies of the chromosome. As nutrients become limiting, aerial hyphae grow upward, forming an aerial mycelium above the surface. Spores develop at the tips of the aerial mycelium. As they mature, septa form between individual spores, and the spores are released and the cycle is repeated. See the text for details. (C) *Streptomyces* colonies growing on an agar surface (left) and a scanning electron micrograph (right) of mycelial filaments and hyphae. Reprinted with permission from Tenebro CP et al. 2023. *Microbiol Spectr* 11:e0366122.

of prespores each carrying a single copy of the genome. The prespores undergo maturation, becoming metabolically dormant. The surface of spores is hydrophobic due to a coating of hydrophobic proteins called **chaplins**, and the core of the spore is desiccated. As in *Bacillus* endospore development (Chapter 17), the complex developmental process in *Streptomyces* is controlled by a cascade of sigma factors whose activity is regulated by anti-sigma factors. The production of spores gives the *Streptomyces* colonies a distinctive fuzzy appearance, and in many species, the production of pigmented secondary metabolites results in changes in the color of the colonies. The life cycle is completed when the mature spores are dispersed, often by wind, to new locations where they germinate when conditions are favorable. Although not as resistant to environmental stresses as endospores (Chapter 17), *Streptomyces* spores can lie dormant for many years—the maximum reported survival period is 70 years.

Life Cycle of Myxobacteria: Predatory Spore-Forming Social Bacteria

Myxobacteria are Gram-negative *Deltaproteobacteria* that exhibit a unique life cycle (Fig. 18.5A) in which single cells cooperate to form multicellular structures and generate spores by a completely different mechanism from *Bacillus* (Chapter 17) or *Streptomyces*. Myxobacteria are chemoheterotrophs that can utilize a variety of organic polymers and monomers as carbon and energy sources. As cooperative predators, they work together as a **swarm** to kill and feed on other bacteria and fungi. When swarms contact colonies of other microorganisms, they release exoenzymes to lyse the cells and digest the cellular material. This group feeding behavior requires ~10^5 cells to be successful. When sufficient nutrients are available, individual cells undergo vegetative growth via binary fission. Under starvation conditions, the population of cells undergoes developmental changes resulting in the formation of structures known as **fruiting bodies** (Fig. 18.5B).

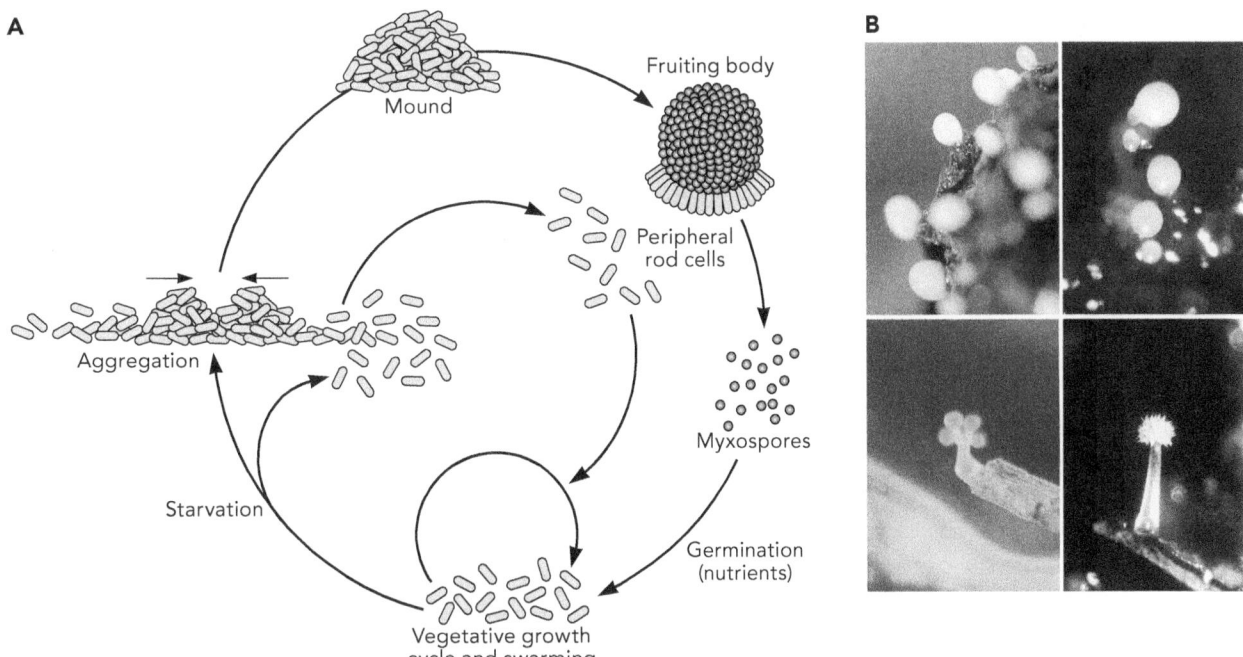

FIGURE 18.5 **Life cycle of the myxobacterium *Myxococcus xanthus*.** (A) The different types of cells and structures that develop during the life cycle of *M. xanthus* are depicted as a cartoon model. Clockwise from the bottom, vegetative cells exhibit swarming behavior as they feed as a group on other microorganisms. Under starvation conditions the cells aggregate, form mounds, and develop into a fruiting body that contains myxospores. Some cells in the population remain outside the fruiting body as peripheral rod cells. See the text for details. (B) Magnified images of simple mounds (top) and more complex (bottom) fruiting body structures of various myxobacteria. Reprinted with permission from Dawid W. 2000. *FEMS Microbiol Rev* 24:403–427.

Fruiting bodies are formed from a mixture of exopolysaccharides, protein, and other macromolecules derived from lysed cells. The fruiting body structure is filled with metabolically quiescent and environmentally resistant **myxospores**. In addition, **peripheral rod cells** form and associate with the outer surface of the fruiting body structure. Peripheral rod cells are thought to provide a reservoir of replicating cells should conditions improve. The fruiting body structure ensures that when conditions do improve, the germination of a sufficient number of myxospores will result in a population of vegetative cells that is capable of predation. Multiple signaling molecules are important to coordinate group feeding behavior and fruiting body formation that enable individual cells to function as a multicellular unit.

The differentiation process of fruiting body formation involves a series of steps that require social behavior and a solid surface where cells can exhibit gliding motility. The signaling pathways for fruiting body formation are very complex and are not well understood; however, it is known that once the process has been initiated, it cannot be reversed. At least five different signals participate in *Myxococcus xanthus* development; this discussion will be limited to two of them. Similar to endospore formation in *Bacillus*, the *M. xanthus* cells begin the sporulation process when they sense nutrient limitation. Starvation triggers an internal stringent response (Chapter 4) generating (p)ppGpp, which initiates a signal transduction cascade resulting in the expression of the genes required for the secretion of proteases (Fig. 18.6A). The secreted proteases produce **A-signal**, which is a mixture of extracellular amino acids (and additional unidentified components) important for the initial stage of fruiting body development. The concentration of A-signal functions as a cell density signal (a form of quorum sensing; Chapter 16). When the amino acid components of A-signal reach concentrations greater than 10 mM (requiring ~10^8 cells), the process of fruiting body formation begins. A-signal is sensed by a membrane-bound receptor protein called SasS, a sensor histidine kinase that phosphorylates the SasR response regulator. SasR-P activates transcription of the regulatory gene *fruA*. FruA is phosphorylated by another kinase. The resulting low level of FruA-P can bind to high-affinity DNA targets, causing expression of the early development genes needed for initiation of fruiting body formation (Fig. 18.6A).

The next stage of *M. xanthus* development is called aggregation, which is also not well understood. What is known is that aggregation requires **C-signal** (Fig. 18.6B). C-signal is a peptide signal processed through an unknown mechanism from CsgA, a protein that is produced and thought to be localized near the cell surface during aggregation. C-signal is proposed to coordinate cell-to-cell contact via connecting as-yet unidentified C-signal receptors on neighboring cells, helping to align cells during aggregation as they move into mounds. It is thought that C-signal binding triggers higher level phosphorylation of FruA by a kinase. High concentrations of FruA-P then activate *frz* and *dev* genes necessary for aggregation and sporulation, respectively.

Myxobacteria are capable of two types of gliding motility (Chapter 15): **adventurous** (A-type) and **social** (S-type). Adventurous motility is a type of single-cell movement currently thought to be activated by cell surface complexes that directly interact with the growth surface. In contrast, social motility involves cell-to-cell contact and type IV pili at the cell poles. After the myxobacteria have aligned end-to-end and aggregated into mounds, some cells within the mound will lyse and provide nutrients to other members of the population. The remainder

A

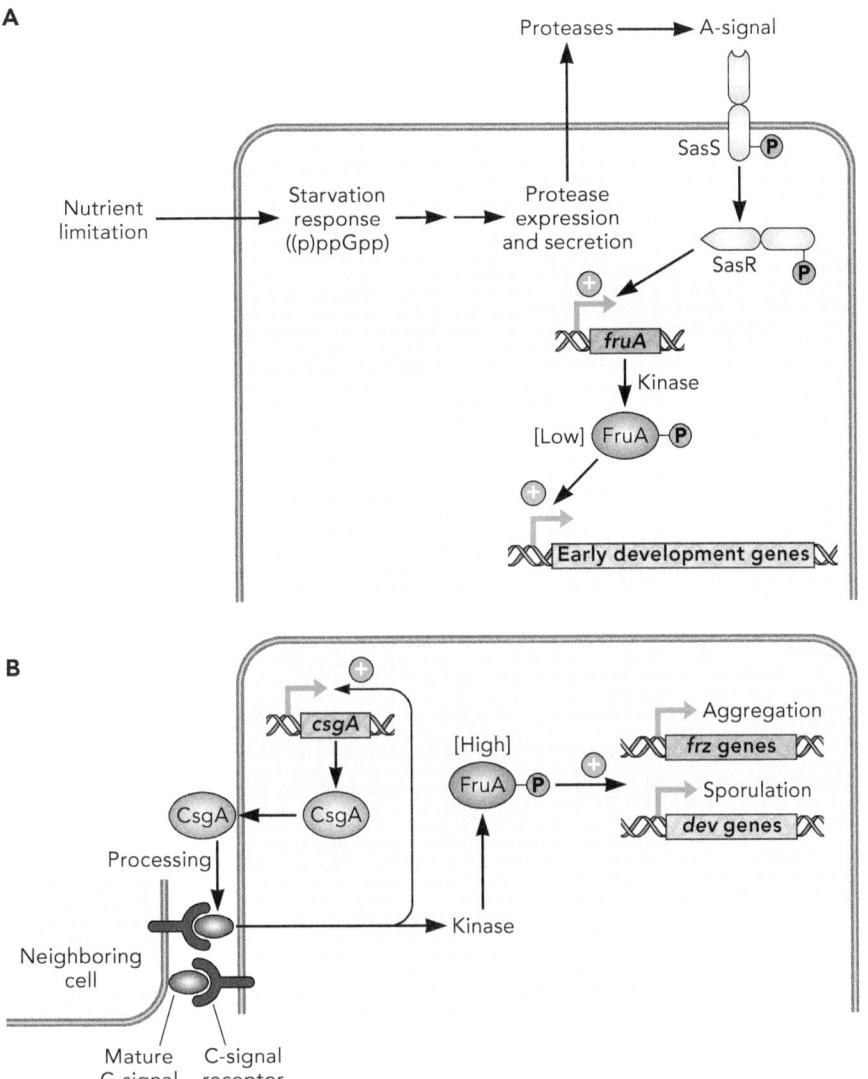

FIGURE 18.6 **Proposed signaling pathways for *Myxococcus xanthus* development.** (A) When nutrients become limiting, vegetative cells produce the internal signal molecule (p)ppGpp, which leads to protease production and secretion resulting in the production of A-signal. A-signal is sensed by the sensor histidine kinase SasS, which autophosphorylates and then transphosphorylates the response regulator SasR. SasR-P activates expression of the gene encoding FruA. Low levels of FruA-P are produced by the activity of a kinase, and FruA-P activates transcription of the early development genes. (B) CsgA associates with the outer membrane and is processed into a mature form known as the C-signal. C-signal is thought to be detected by as-yet unidentified C-signal receptors on neighboring cells. Cell-to-cell contact via C-signal results in a positive feedback loop leading to higher levels of *csgA* expression and the activation of a kinase that phosphorylates FruA. When FruA-P is present at high levels, it activates expression of the *frz* genes required for cell aggregation and the *dev* genes required for sporulation. See the text for details.

of the cells will undergo a differentiation process and become myxospores. Finally, cells that do not become part of the fruiting body differentiate into peripheral rod cells. Peripheral rod cells are morphologically similar to vegetative or stationary-phase cells but have different gene expression patterns. Myxospores have a protein coat and highly cross-linked peptidoglycan that makes them heat and desiccation resistant, but not to the degree of endospores formed by Gram-positive bacteria (Chapter 17). Depending on the genus and species, the fruiting bodies formed by myxobacteria can be as simple as small mounds about the size of a pinhead (referred to as "haystacks" as in *M. xanthus*) or complex structures with many protruding arms (resembling small "trees") (Fig. 18.5B). They are often brightly colored in the yellow-orange-brown range due to the carotenoid pigments that they produce and are visible to the human eye. When conditions change or fruiting bodies are dispersed to a favorable environment (e.g., via animals or rain), the myxospores in the fruiting body can germinate as a group and resume vegetative growth by binary fission. The number of myxospores in the fruiting body will form a sufficient vegetative cell population for predatory feeding behavior upon germination.

Biofilms: The Typical State of Microorganisms in the Environment

In natural environments, bacteria and archaea rarely grow as individual **planktonic cells** suspended in solution. More commonly, microbes grow in **biofilms**, aggregates of cells that are typically attached to surfaces and encased in an extracellular polymeric substance (EPS) layer. EPS is primarily composed of polysaccharide (and is therefore also known as exopolysaccharide), but in addition EPS contains proteins, DNA, and lipids. Biofilms grow in almost every environment, both natural (e.g., on the surface of rocks, soil, plants, and in animal hosts) and anthropogenic (e.g., clinical settings, sewage treatment plants, and many kinds of industrial settings). These aggregates of cells can be **monoclonal** (containing one strain or species) or they can include upwards of thousands of different microorganisms (*Bacteria*, *Archaea*, and *Eukarya*). In most cases, biofilms are not a homogeneous layer of cells attached to a surface, but rather they grow as a heterogeneous community of physiologically distinct subpopulations that respond to and alter their microenvironment.

The formation of a biofilm begins when motile planktonic cells attach to and aggregate on a surface (Fig. 18.7). Thus, chemotaxis and motility (Chapter 15) are important to biofilm formation. The initial cell aggregates will then grow into microcolonies and a mature biofilm will form over time (Fig. 18.8). Biofilms may become multilayered stratified structures consisting of different microbes in distinctive layers, such as those that form on teeth or in microbial mats (Fig. 18.8 B and E). More complex biofilms may have mushroom-shaped towers surrounded by channels that facilitate nutrient delivery and waste removal. **Pellicles** (thin layers of growth on the surface of a liquid) are a type of biofilm, as are **flocs**, which are clumps of aggregated microbes in a liquid suspension. Flocs aid in the process of wastewater treatment by removing nutrients, harmful molecules, and pathogens through metabolism and adhesion. Regardless of the three-dimensional structure of the biofilm, the microorganisms within are encased in a protective EPS matrix as described above.

Once threshold high cell densities are reached during biofilm development, the regulatory process of bacterial quorum sensing (Chapter 16) enables the

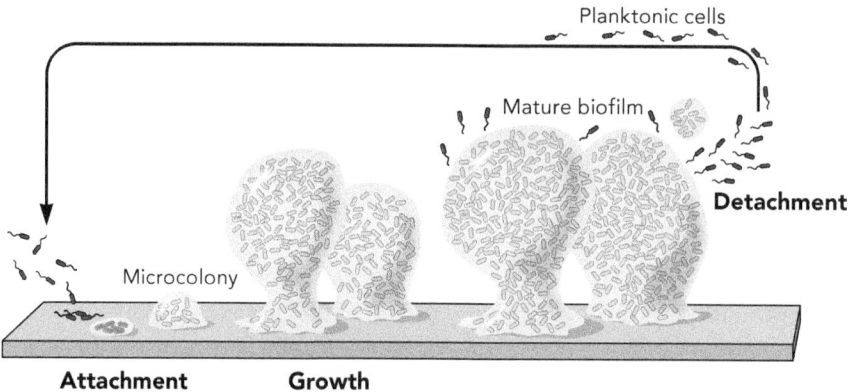

FIGURE 18.7 **Complex biofilm formation.** Left to right, complex biofilms form when motile planktonic cells attach to a surface and develop into a microcolony. Further growth results in the formation of a mature biofilm with a mushroom-like structure. Motile planktonic cells detach from the biofilm matrix and swim to a new environment to start the cycle over again.

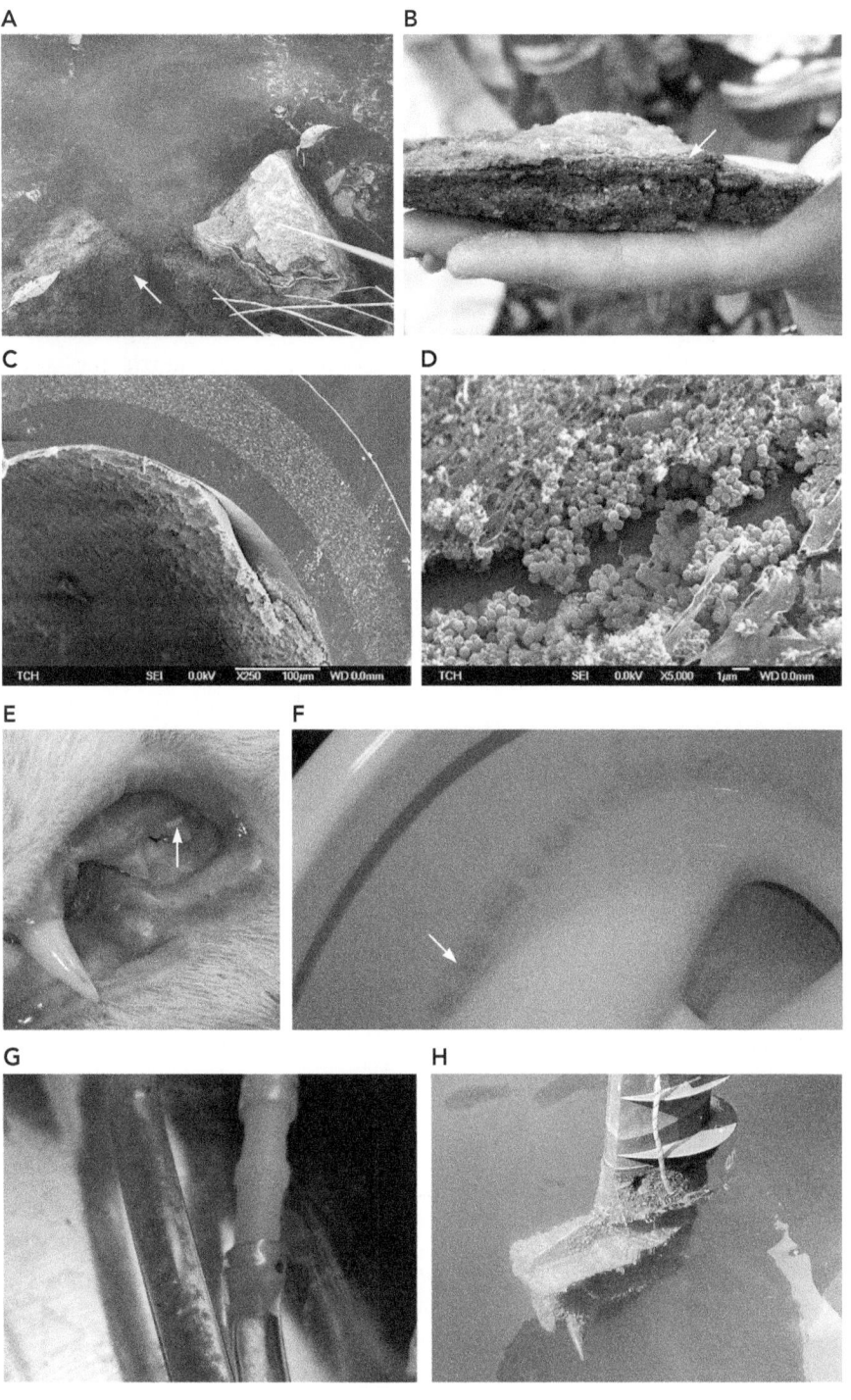

FIGURE 18.8 **Biofilms form on many natural and anthropogenic surfaces.** (A) Biofilms on rocks below the surface of a freshwater stream (photo credit: Ann M. Stevens), (B) a microbial mat (photo credit: Andrew C. Hawkins), (C) an electron micrograph of a biofilm inside of a catheter, (D) an electron micrograph of the catheter microbial community depicted in panel C (reprinted with permission from Pammi M et al. 2013. *BMC Microbiol* 13:257), (E) plaque formation on the tooth of a cat (photo credit: Elena M. Oosterhuis), (F) biofilm growth in a toilet bowl (photo credit: Elena M. Oosterhuis), (G) water flocs passing through a piece of tubing (photo credit: Ayella Maile-Moskowitz), and (H) a biofilm on a boat propeller creating drag (photo credit: Ann M. Stevens). In some images, white arrows indicate the biofilm.

proper expression of the genes necessary for mature biofilm formation. For example, quorum-sensing mutants of *Pseudomonas aeruginosa* that are unable to produce acylhomoserine lactone (AHL) autoinducer signals form flat biofilms instead of the complex mushroom-shaped dome structures made by wild-type cells (Fig. 18.9A). In addition, the internal second messenger signal cyclic diguanylate (di-GMP) is also important for regulating biofilm formation and motility in bacteria (Fig. 18.9B). Cyclic di-GMP is synthesized by diguanylate cyclase enzymes that typically have the conserved five amino acid motif, GGDEF. Phosphodiesterase

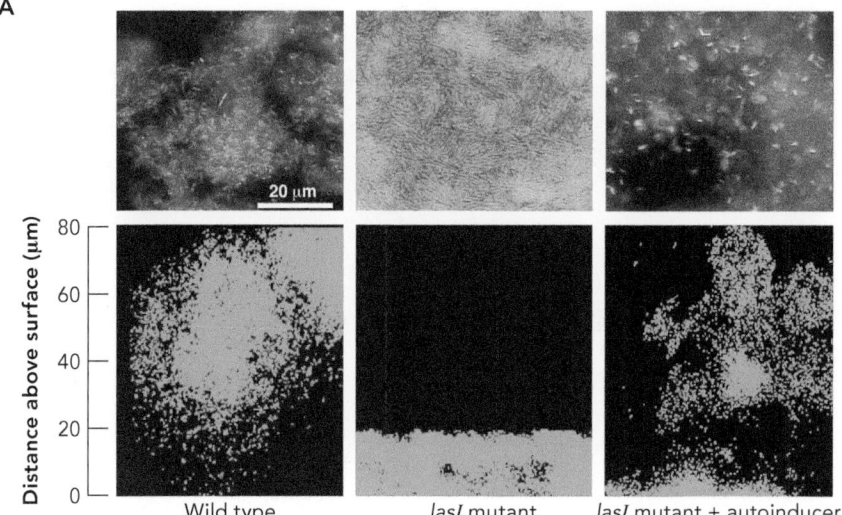

FIGURE 18.9 **External signals and environmental conditions influence biofilm formation.** (A) Quorum-sensing signals such as acyl-homoserine lactones (AHL) are involved in complex biofilm formation. Epifluorescence (top row) and scanning confocal microscopy (bottom row) images are shown. (Left) A wild-type *Pseudomonas aeruginosa* biofilm expressing green fluorescent protein has mushroom-like structures that rise above the surface. (Middle) A quorum-sensing *lasI* mutant strain incapable of synthesizing AHL has a flat dense biofilm structure. (Right) Addition of exogenous AHL restores the wild-type biofilm structure. Reprinted with permission from Davies DG et al. 1998. *Science* 280:295–298. (B) The secondary messenger molecule cyclic-di-GMP (c-di-GMP) is produced from GTP by diguanylate cyclases characterized by the amino acid motif GGDEF and is degraded by phosphodiesterases characterized by the EAL or HD-GYP amino acid motifs. The c-di-GMP signal leads to the repression of motility and virulence gene expression and activation of the genes needed for biofilm formation. pGpG, linear diguanylate.

enzymes, with either an EAL or an HD-GYP amino acid motif, degrade cyclic di-GMP. A single bacterium will utilize multiple diguanylate cyclases and phosphodiesterases to modulate the concentration of cyclic di-GMP in the cytoplasm. In *P. aeruginosa*, contact with surfaces results in higher concentrations of cyclic di-GMP, which leads to the synthesis of higher levels of exopolysaccharides and other EPS matrix components important for biofilm formation. Cells with lower levels of cyclic di-GMP are motile, enabling dispersion of planktonic cells from the surface of the biofilm. This illustrates how even in monoclonal biofilms, gene expression differs in different regions of the biofilm as cells respond to the changes in their microenvironment, giving individual cells drastically different physiological profiles.

Cells on the outside of biofilms have better access to nutrients and oxygen than those on the interior of the community (Fig. 18.10). Therefore, cells on the

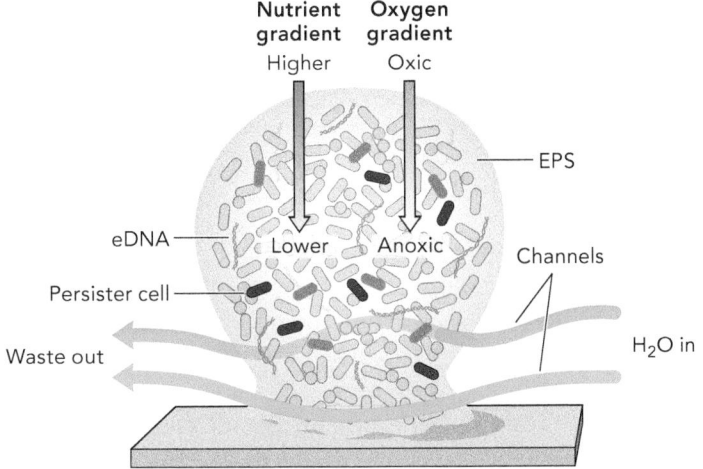

FIGURE 18.10 **The heterogenous nature of biofilms.** A mixed-community biofilm consisting of more than one species of microbes is depicted. The cells are encased in extracellular polymeric substance (EPS) and extracellular enzymes. Extracellular DNA (eDNA) is available for uptake, which enhances horizontal gene transfer. Channels within the biofilm structure enable the delivery of water and nutrients and the removal of waste products. Biofilms have gradients of nutrients and oxygen with higher levels at the surface and lower levels in the interior. As a result, cells in the center of the biofilm have slower growth rates. Slowly growing cells and metabolically inactive persister cells contribute to the biofilm's enhanced resistance to many antimicrobial chemicals. See the text for details.

surface typically grow faster and use aerobic respiration for their metabolism. Inside the biofilm, the EPS can limit the diffusion of oxygen and nutrients, creating microgradients. Under anoxic conditions, cellular metabolism is slowed and growth will occur via anaerobic respiration or fermentation. **Persister cells** with no discernable metabolic activity may even arise within a biofilm. For example, in *Staphylococcus aureus* monoculture biofilms, different regions of the biofilm are composed of cells that are aerobically respiring, fermenting, dormant (e.g., persister cells), or dead depending on the availability of oxygen and/or nutrients.

Cross-feeding through the sharing of metabolic intermediates and end products is a common characteristic of biofilms. Photosynthetic cyanobacterial mats (Fig. 18.8 B), which can grow centimeters thick, contain phototrophs that carry out oxygenic photosynthesis and provide oxygen, hydrogen, and organic carbon to heterotrophs growing in the biofilm. Other species, including anoxygenic phototrophs and sulfur bacteria, drive sulfur oxidation and reduction, respectively, within the microbial mat biofilm (the sulfur cycle; Chapter 11). In *E. coli* biofilms, organic acid waste products of fermenting bacteria in the center of a biofilm were shown to feed aerobically respiring bacteria on the surface, as in anaerobic food webs (Chapter 11). Not only does the structure of a biofilm enable a division of labor that enhances metabolic efficiency across a population or community, but as we will see below, the structure also affords the cells within the biofilm matrix protection from external stresses.

The EPS layer helps by retaining water, thereby preventing dehydration when the biofilm is exposed to air. The matrix also serves as a physical barrier preventing the diffusion of inhibitory chemicals (e.g., antibiotics and disinfectants) toward the cells embedded in the biofilm. In addition, some bacteria make and release exoenzymes that detoxify antibiotics; these enzymes are present in the EPS, further inhibiting the action of antibiotics and benefiting the entire community. Some bacteria make compounds that assist with nutrient sequestration in the nutrient-limited internal biofilm environment. For example, many pseudomonads make siderophores (Chapter 13) that can bind iron. The siderophores are released into the EPS matrix where **cheater cells** (cells that utilize molecules from other cells without having to expend resources to make them) can use the siderophores to gain access to limited iron. Quorum-sensing cheaters also arise when some cells rely on neighboring cells to produce autoinducer signals at high cell density, conserving energy for their own growth at the expense of other cells in the population or community. When bacterial cell density is high, extracellular DNA (eDNA) is released from cells (both living and dead), facilitating horizontal gene transfer (i.e., transduction, transformation, and conjugation), which permits the dissemination of useful genes (e.g., for antibiotic resistance) through the microbial community.

Bacterial biofilms can cause a variety of major health problems. As mentioned, antibiotics are less effective against cells in a biofilm. As shown in Griffith's experiment (Chapter 13), the EPS layer (capsule) also makes it more difficult for white blood cells in the host immune system to detect and destroy encapsulated bacteria, potentially leading to chronic infections. Chronic infections are often characterized by the presence of metabolically inactive persister cells. Since many antibiotics work by interfering with cell growth and biosynthesis, the persister cells are more resistant to the effects of many antibiotics. Biofilms are particularly problematic in medical implants and urinary catheters that become contaminated with bacterial biofilms (Fig. 18.8 C and D), resulting in chronic and recurrent systemic infections that are recalcitrant to antibiotic therapy. If antibiotic treatment is ineffective, the implant must be removed and replaced. In addition, biofilm formation plays a major role in lung infections that affect people with the genetic disorder cystic fibrosis. However, probably the most common biofilms found in animals and humans form on the surface of teeth as plaque (Fig. 18.8 E). Brushing and flossing mechanically remove the biofilm and the chemicals in toothpaste and mouth rinses are used to reduce biofilm levels in an effort to prevent dental caries and gum disease.

Biofilms also create challenges for industry, particularly in the food processing industry, as biofilms are difficult to remove from machinery surfaces. Common foodborne pathogens associated with biofilm formation on inanimate surfaces or the food itself include, but certainly aren't limited to, *Salmonella* spp., *Staphylococcus aureus*, *Listeria monocytogenes*, *Campylobacter jejuni*, and *E. coli* O157:H7. *Legionella pneumophila* outbreaks are commonly associated with water distribution pipes that become lined with biofilms despite chemical treatment (e.g., chlorine or chloramine in drinking water). Reducing conditions and acidic waste products from industrial biofilms can damage machinery and storage tanks (e.g., petroleum industry) and can clog outflow pipes. Biofilms on ships (Fig. 18.8 H) increase drag, reduce speed, and increase fuel consumption, resulting in estimated annual losses of billions of dollars globally.

In summary, biofilms are complex multicellular microbial communities growing in a more natural physiological state than planktonic cultures in the laboratory. It is fundamental basic research on the topics covered in this text, such as metabolism, growth, regulation, structure/function, and motility, that lay the foundation for understanding more complex microbial systems. Some processes are shared across all cells, or certain types of cells, and bring unity to our understanding of microbial life. Other processes enable the great diversity of life that exists in the microbial world; microorganisms have carved out unique and varied niches where they thrive, often growing in biofilm communities.

Learning Objectives: After completing the material in this chapter, students should be able to. . .

1. compare and contrast stalked and swarmer cell morphologies in *Caulobacter crescentus*

2. describe the process of heterocyst and hormogonium differentiation by filamentous cyanobacterial species

3. explain the reproductive life cycle of *Streptomyces*

4. outline the processes involved in formation of a fruiting body in myxobacteria

5. describe the formation and resulting structural microenvironments of complex biofilms

6. contrast the benefits of growth in biofilms for microbes with the problems biofilms cause in clinical and industrial settings

Check Your Understanding

1. Define this terminology: stalked cells, swarmer cells, heterocysts, akinetes, hormogonia, cyanoglobin, secondary metabolites, hyphae, substrate mycelium, aerial mycelium, chaplins, swarm, fruiting bodies, myxospores, peripheral rod cells, A-signal, C-signal, planktonic cells, biofilms, monoclonal, pellicles, flocs, persister cells, cheater cells

2. What are the general functions of stalked and swarmer cells in *Caulobacter crescentus*? Which cell type is motile? Which cell type reproduces? (LO1)

3. What is the function of each protein in the *Caulobacter crescentus* life cycle? (LO1)

   a) CtrA:

   b) FtsZ:

4. Regulation of the life cycle of *Caulobacter crescentus* involves the activity of CtrA. (LO1)

   a) What impact does phosphorylation have on the activity of CtrA?

   b) Which cells, swarmer or stalked, have a higher concentration of CtrA-P?

   c) Do high levels of CtrA-P activate or repress each of the following functions?

   i. DNA synthesis

   ii. Transcription of *ftsZ*

   iii. Flagella and pili synthesis

   iv. Chemotaxis

5. The cyanobacterium *Nostoc* can differentiate into several cell types. (LO2)

   a) Compare *Nostoc* cell types by filling in the table below.

Cell type	Undergo cell division?	Fix nitrogen?	Are cold- and desiccation resistant?	Capable of gliding motility?
Vegetative cells				
Heterocysts				
Hormogonia				
Akinetes				

   b) Under what nutrient conditions is each of the *Nostoc* cell types found? How does each cell type help *Nostoc* survive?

   i. Vegetative cells:

   ii. Heterocysts:

   iii. Hormogonia:

   iv. Akinetes:

6. For each statement, indicate if it applies to vegetative cells (V); heterocysts (H); neither (N), or both (B). (LO2)
These cells:

____ are surrounded by a thick cell wall containing glycolipids and polysaccharides.

____ are desiccation-resistant.

____ produce oxygen as a byproduct of photophosphorylation.

____ use a functional photosystem *I* to produce ATP.

____ can reduce N_2 to NH_3.

7. List four strategies that allow *Nostoc* heterocysts to protect their nitrogenase from oxygen. (LO2)

8. The differentiation of vegetative cells to heterocysts is a complicated process that results in the formation of a heterocyst every ~10 cells. What is the function of each of the following molecules in heterocyst formation? (LO2)

   a) α-Ketoglutarate:

   b) NtcA:

   c) PatS:

9. Explain how hormogonia help *Nostoc* survive under the following growth conditions: (LO2)

   a) Low light:

   b) Forming symbiotic relationships with plants:

10. *Streptomyces* spp. collectively make more than 100,000 bioactive secondary metabolites with a wide range of functions (e.g., antimicrobial, antiviral, plant growth-promoting, and herbicidal). Why would an organism make compounds such as these that are not directly required for growth? (LO3)

11. Unlike *Bacillus*, which make endospores as a survival strategy, *Streptomyces* spp. make spores as the only means of replication. (LO3)

    a) Compare the spore types made by *Streptomyces* and *Bacillus* species. Indicate if each statement applies to *Streptomyces* spores (S), *Bacillus* endospores (E), neither (N), or both (B).

    ____ One cell makes one spore.

    ____ Spores have an outer coating made of chaplins.

    ____ Spores have a desiccated core.

    ____ Differentiation is controlled by a cascade of sigma factors.

    ____ Spores are killed by moderate heat.

    b) Order these events in the *Streptomyces* life cycle (from 1 to 6), beginning with the germination of a spore.

    ____ Vegetative growth of branching filaments called hyphae occurs from the germ tube.

    ____ Trehalose is cleaved to glucose, which fuels transcription, translation, and DNA replication.

    ____ The mycelium forms and grows down into the substrate, anchoring the growing colony.

    ____ The influx of water causes swelling and allows the reactivation of stored enzymes.

    ____ The germ tube emerges from the spore.

    ____ The coat is hydrolyzed by spore enzymes as the desiccated spore becomes hydrated.

12. Match each function (A to E in the box provided) with the appropriate event in the *Streptomyces* life cycle. (LO3)

    ____ Nongrowing portions of the vegetative mycelium lyse

    ____ Secondary metabolites with antimicrobial properties are produced

    ____ Sensing nutrient limitation, the colony begins producing nonbranching aerial hyphae

    ____ The core of the spore is desiccated

    ____ The vegetative mycelia produce hydrolytic exoenzymes

A. to degrade organic compounds in the soil for use as nutrients for growth

B. to provide an aerial mycelium platform for the release of spores

C. to provide nutrients for the aerial hyphae

D. to inhibit competing microbes from utilizing the released nutrients

E. to stabilize DNA and proteins and make the spore resistant to environmental stresses

13. *Myxococcus xanthus* has two life phases that are controlled by the nutrient conditions in the environment: cooperative predation and multicellular development. (LO4)

  a) Cooperative predation allows the swarm of myxobacterial cells to seek out and collectively digest colonies of other microorganisms.

    i. Under what conditions (high or low nutrients) does cooperative predation occur?

    ii. By what process does cell replication happen under these nutrient conditions?

    iii. Explain how each of the following contributes to the ability of myxobacterial cells to prey on other microorganisms in a cooperative way.

      • Gliding motility:

      • Release of exoenzymes:

  b) The early stages of multicellular development in *M. xanthus* are triggered by nutrient depletion, followed by a population-wide sensing of accumulated signals. Define each molecule involved in the life cycle of *M. xanthus* and describe its function.

    • (p)ppGpp:

    • Secreted proteases:

    • A-signal:

    • SasS:

    • SasR:

    • FruA-P (at low concentration):

  c) The later stages of *M. xanthus* development involve the aggregation of cells, which, once started, leads to the formation of mounds and fruiting bodies. While the details are not totally understood, describe the hypothesized function of each molecule involved in the later stages of fruiting body development in *M. xanthus*.

    • CsgA:

    • C-signal:

    • FruA-P (at high concentration):

  d) After the *M. xanthus* cells have aligned end-to-end and aggregated into mounds, cells within the mound will eventually form the fruiting body structure and three different cell types. Describe each and explain their functions.

    • Myxospores (~10% of cell population):

    • Peripheral rod cells (~30% of cell population):

    • Autolysing cells (~60% of cell population):

  e) Adventurous (A-type) and social (S-type) motility are involved in both cooperative predation and multicellular development. Define each.

14. The cellular concentration of cyclic-di-GMP varies in *Pseudomonas aeruginosa* cells as a function of the stage of biofilm development and position within the mature biofilm. (LO5)

  a) What two enzymes control the concentration of cyclic-di-GMP in *P. aeruginosa*? Describe the function of each.

  b) Predict whether the concentration of cyclic-di-GMP in *P. aeruginosa* would be relatively high or low in biofilm cells during each stage of biofilm development listed.

- Planktonic cells:
- Attaching cells:
- Detaching cells:

15. In a mature biofilm, cell differentiation is driven by changes in gene expression in response to different microenvironments. Explain the impact of each environmental gradient on the metabolism of cells in the biofilm. (LO5)

    a) Oxygen gradient:

    b) Nutrient gradient:

16. Biofilms offer microorganisms many benefits. List four ways that the extracellular polymeric substance (EPS) layer can provide survival benefits to the cells living within the EPS matrix. (LO6)

17. Biofilms can be very problematic for humans. (LO6)

    a) Explain how the biofilm allows cells to evade the action of white blood cells of the host immune system.

    b) Explain two ways that biofilms help cells to evade the action of antibiotics.

Dig Deeper

18. *Caulobacter crescentus* has two distinct cell forms: a motile but nonreplicating swarmer cell and a nonmotile but replicating stalked cell. The transition from one form to the other involves a complicated series of regulatory events that is influenced by the availability of nutrients. (LO1)

    a) Like *Escherichia coli*, the swarmer cells of *C. crescentus* undergo a stringent response. Briefly describe the stringent response in *E. coli* (Chapter 4). What triggers the stringent response? What are alarmones, and what end result do they cause?

    b) Under low-nutrient conditions, the stringent response in *C. crescentus* slows the transition from a swarmer cell to a stalked cell. The sensor histidine kinase CckA causes the phosphorylation of the master regulator, CtrA. How do higher levels of CtrA-P in swarmer cells ensure that the cells do not transition to stalked cells in low-nutrient conditions, but stay "locked" in the swarmer phase?

    c) When nutrient conditions improve, the concentration of CtrA-P begins to drop and transition to the stalked phase begins. What causes the concentration of CtrA-P to drop? Why must the cell be able to degrade CtrA?

    d) Other essential enzymes are involved in the *C. crescentus* cell cycle.

       i. DnaA initiates DNA replication and regulates expression of *gcrA*.

       ii. GcrA is a cell cycle regulator that forms a stable complex with RNA polymerase; it regulates components of the DNA replisome and the segregation machinery.

       iii. FtsZ is a membrane-associated protein essential for cell division.

    Considering the characteristics of each cell type (swarmer and stalked), indicate which cells would have high concentrations of each listed protein.

Cell type	CtrA-P	DnaA	GcrA	FtsZ
Swarmer				
Stalked				

19. Ammonium assimilation is highly regulated by the amount of NH_4^+ in the environment. Research suggests that, in the absence of NO_3^-, α-ketoglutarate is the sensor that connects NH_4^+ concentrations in the environment to nitrogen regulation. In *Anabaena*, the link between α-ketoglutarate and NH_4^+ assimilation occurs through glutamate dehydrogenase and the GS/GOGAT (glutamine synthetase / glutamine oxoglutarate aminotransferase) pathway (Chapter 12). (LO2)

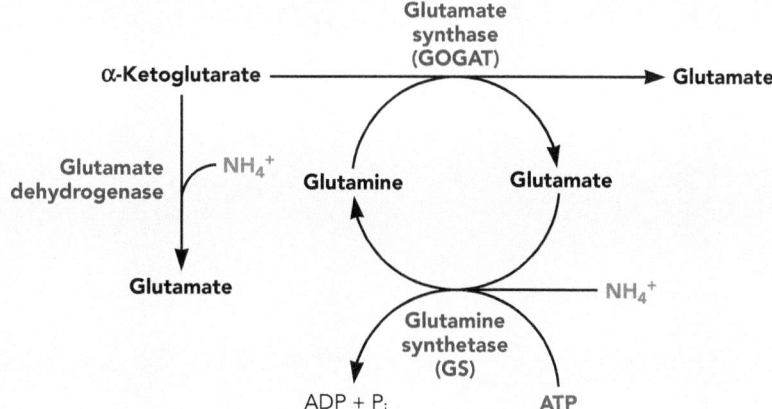

a) You grow *Anabaena* in the presence of either NH_4^+ or NO_3^- and measure the amount of intracellular α-ketoglutarate (see graph to the right). Considering the ammonium assimilation pathways shown above, why would the intracellular concentration of α-ketoglutarate decrease when cells are growing on NH_4^+ compared to NO_3^-?

b) In cyanobacteria, NtcA is a global regulator that controls many genes involved in nitrogen metabolism. α-Ketoglutarate is an allosteric effector of NtcA. When α-ketoglutarate accumulates in cells, it binds to NtcA, enhancing transcriptional activation of the NtcA DNA-binding targets. Therefore, α-ketoglutarate is the signal molecule that connects the concentration of NH_4^+ to the activation of NtcA. Fill in the table below to describe these relationships.

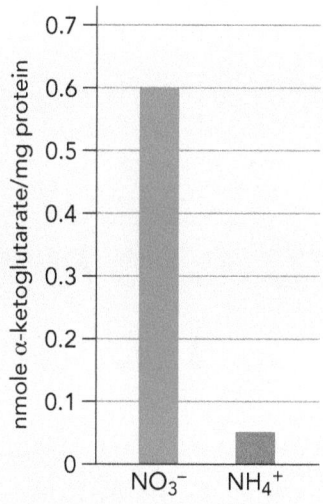

When $[NH_4^+]$ is	Intracellular [α-ketoglutarate] is . . . (relatively high or low)	NtcA is . . . (active or inactive)
High		
Low		

c) The pattern of heterocyst formation in the cyanobacterium *Anabaena* is coordinated by the transcription factor HetR and the PatS peptides. Activated NtcA (NtcA*) triggers a regulatory cascade that initiates heterocyst formation by enhancing the transcription of *hetR*. HetR then activates transcription of *patS*, resulting in relatively high concentrations of these proteins in heterocysts. Draw in the relationships between the proteins NtcA*, HetR, and PatS in the cell diagram.

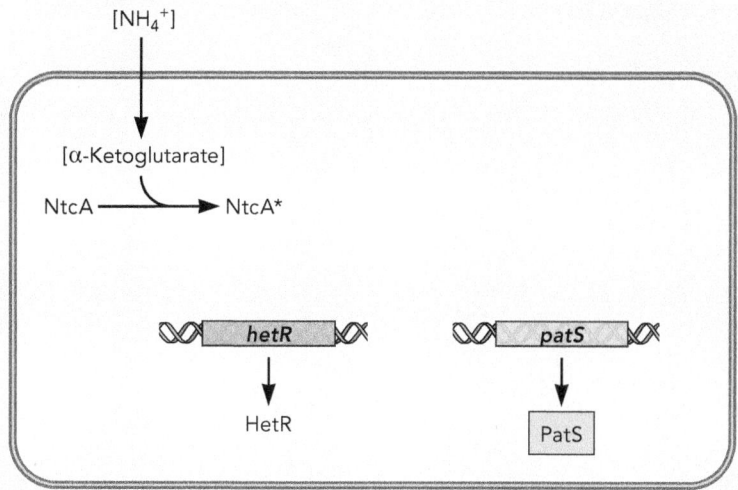

d) PatS is processed into PatS peptides, which are exported from heterocysts to adjacent cells in the filament. Describe how the concentration of PatS peptides plays a key role in determining the spacing of heterocysts.

e) Predict the pattern of heterocyst distribution of each *Anabaena* strain by matching the diagram with the strain.

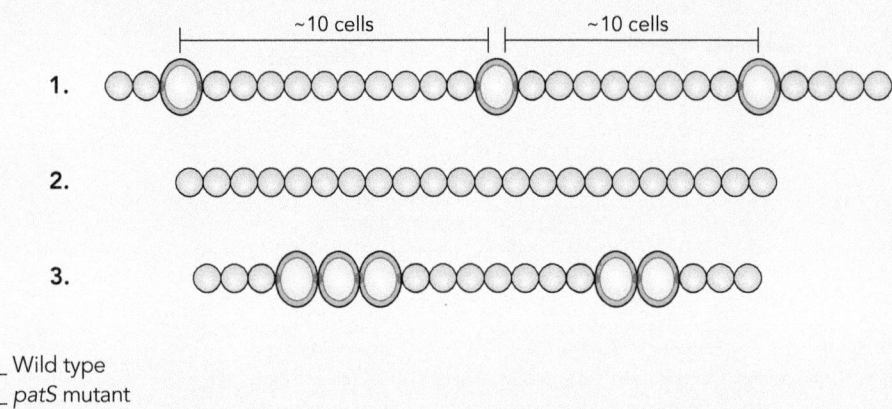

_____ Wild type
_____ *patS* mutant
_____ *hetR* mutant

20. Sporulation and the biosynthesis of streptomycin (an antibiotic that binds the 30S subunit of the ribosome, thus inhibiting translation) in *Streptomyces griseus* are both controlled by a density-dependent mechanism. In the early stages of growth, ArpA (the A-factor receptor) acts as a repressor protein, inhibiting transcription of the gene coding for the global regulator AdpA .

As *S. griseus* grows, A-factor (a signal hormone produced by *S. griseus*) accumulates in the cell until the concentration of A-factor reaches a critical level, typically near stationary growth.

A-factor then binds to the repressor ArpA, releasing it from the *adpA* promoter, thus allowing the transcription and translation of *adpA*. AdpA then transcriptionally activates more than 70 different genes, including those involved in sporulation and antibiotic production. (LO3)

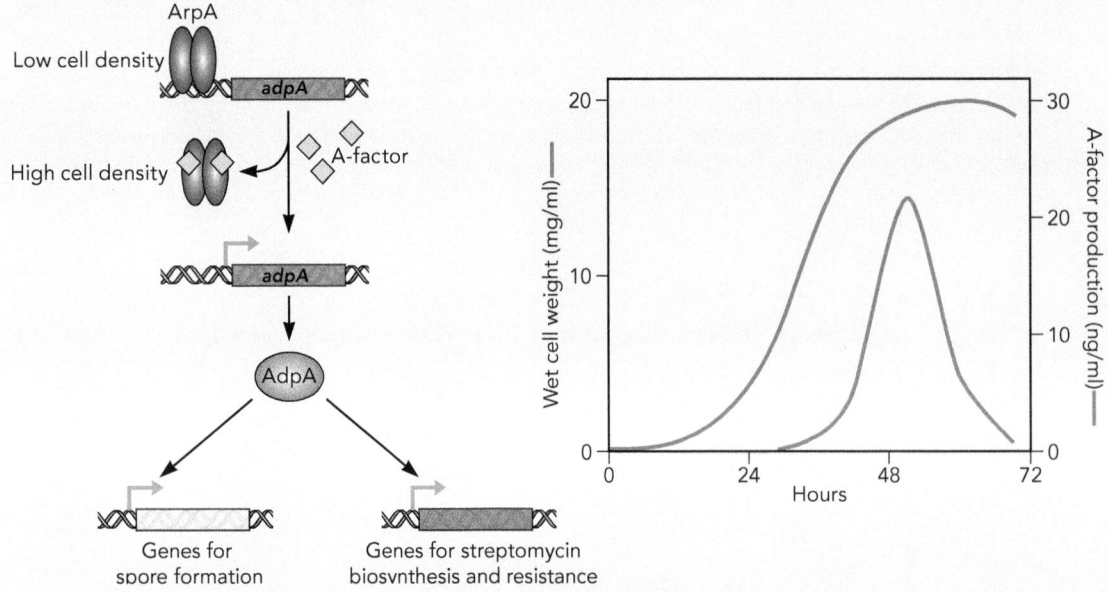

a) How is the control of *adpA* expression by A-factor similar to a quorum-sensing system? What acts as the autoinducer? What responds to the concentration of autoinducer?

b) Considering the life cycle of *S. griseus*, what is the benefit of initiating the sporulation pathway at high cell density?

c) You have a wild-type strain of *S. griseus* and a strain carrying a mutation in *arpA*. You grow each strain in a high-nutrient medium and periodically examine cells from each culture to determine the percentage of cells that have begun to sporulate. You obtain the results shown in the figure. Explain these results.

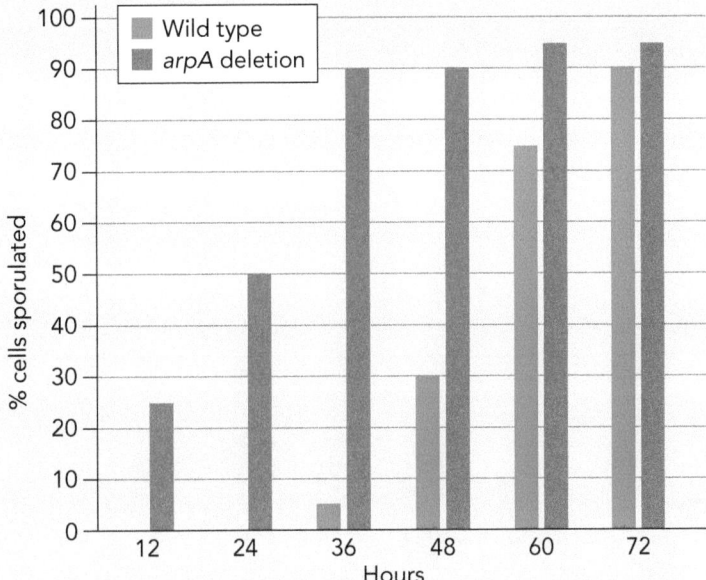

21. AdpA regulates streptomycin biosynthesis by binding to a promoter region upstream of the streptomycin biosynthesis (*strR*) and streptomycin resistance (*strA/B*) genes. (LO3)

   a) *Streptomyces griseus* produces streptomycin as a secondary metabolite during stationary phase.

      i. How does making antibiotics at this point in the *S. griseus* life cycle benefit these cells?

      ii. Why is it critical that the streptomycin resistance genes be controlled simultaneously with the streptomycin biosynthesis genes?

   b) There is much interest in genetically engineering *S. griseus* to control the synthesis, and therefore the production, of antibiotics.

      i. What deletion mutation could allow cells to make streptomycin all the time? Explain your reasoning.

      ii. You have the gene encoding AdpA on a plasmid under constitutive control. From the list of possible phenotypes below, predict the phenotype of a wild-type strain of *S. griseus* that carries this plasmid. Explain your reasoning.

      • would not make streptomycin

      • would make streptomycin only at high cell density

      • would always make streptomycin

22. Cell differentiation in myxobacteria involves three critical processes: (i) sensing of nutrient limitation, (ii) population-wide signal sensing, and (iii) a cascade of transcriptional responses leading to aggregation and sporulation. (LO4)

   a) *Myxococcus xanthus* cells begin to differentiate when they sense nutrient limitation, which triggers an internal stringent response. *M. xanthus* has a gene that is homologous to *relA* in *Escherichia coli*. In the model presented in the text (Fig. 18.6A), (p)ppGpp triggers a regulatory cascade that leads to the production of FruA, a regulatory protein that initiates the transcription of the genes involved in early fruiting body development.

i. Order the steps (from 1 to 6) to describe this process.

_____ A mixture of peptides and amino acids is generated by extracellular proteolysis.

_____ A-signal accumulates to a threshold level outside the cells.

_____ A-signal binds to the sensor histidine kinase SasS, causing SasS to autophosphorylate.

_____ SasR-P initiates expression of early development genes, resulting in low levels of FruA and FruA-P.

_____ (p)ppGpp production ultimately results in the production of proteases.

_____ SasS-P phosphorylates the response regulator SasR.

ii. Fruiting body development is dependent on A-signal accumulating in the environment above a threshold level. Why do you think it is necessary to have a high concentration of cells at this point in the *M. xanthus* cell cycle?

iii. What is the role of low levels of FruA-P at this stage of development?

b) Recent research suggests that (p)ppGpp initiates the production of two critical proteins, AsgA (a sensor histidine kinase, HK) and AsgB (a response regulator, RR), that are required for the expression of key genes in the processes leading to fruiting body formation (see green arrows in the figure below). Given this model, fill in the following table to predict which events would likely occur in an *asgA* deletion mutant strain of *M. xanthus*. Explain your reasoning.

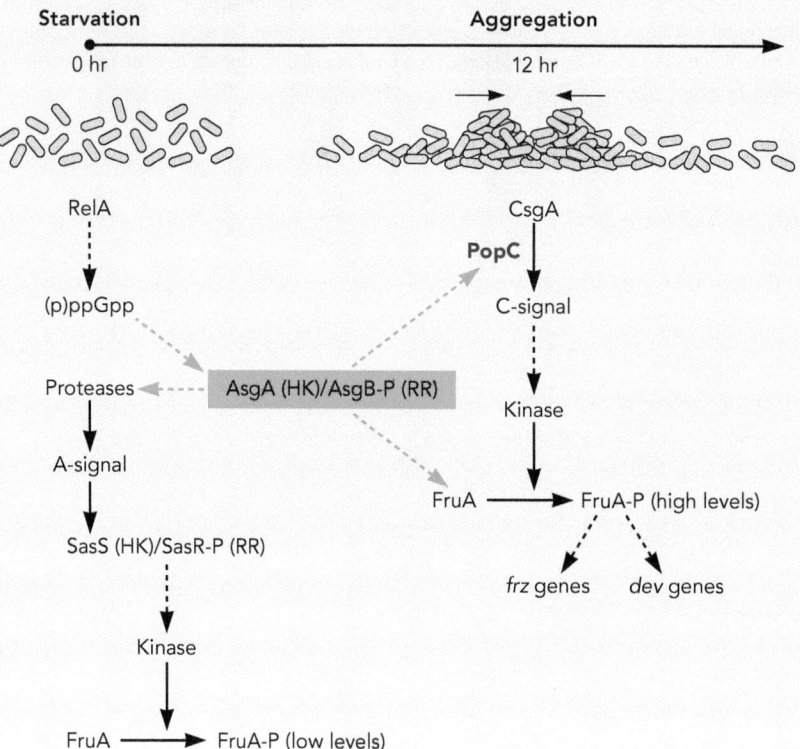

Event	Will event occur in *asgA* deletion mutant?	Reasoning
RelA activity		
(p)ppGpp synthesis		
A-signal production		
SasR phosphorylation		
CsgA protein production		
C-signal production		
FruA phosphorylation		
Cell aggregation		

23. Differentiation within biofilms is influenced by the formation of microenvironments, which drives changes in the microbial physiology and further impacts the microenvironment. An example of this is the role that the quorum-sensing (QS) system LasR/RlhR (Chapter 16) plays in biofilm formation in *Pseudomonas aeruginosa*. (LO5, LO6)

    You are interested in investigating the relationship between QS and biofilm formation in *P. aeruginosa*. You set up two culture chambers with growth media. To one you add wild-type cells of *P. aeruginosa* (WT); to the other you add a Δ*lasI* deletion strain. (Las*I* is the synthase that makes the autoinducer signal, 3-oxo-C12-homoserine lactone.) You place a number of glass slides in each chamber upon which biofilms can form. At each time point you remove a glass slide and determine the cell density in the biofilm. Using microscopy, you are also able to measure biofilm thickness after 96 hours (see Fig. 18.9A in text). Results are shown in the table.

	Cell density at:		Biofilm thickness at:
	8 hours		**96 hours**
WT	3×10^7 cells/ml	5×10^9 cells/ml	100 μm
Δ*lasI*	3×10^7 cells/ml	5×10^7 cells/ml	20 μm

    a) In the first 8 hours, biofilm formation involves attachment to the slide and the beginning of microcolony growth. Why do you think the wild-type and Δ*lasI* deletion strain have about the same amount of growth on the glass slide after 8 hours?

    b) How do the viable cell counts and biofilm thickness for the wild-type and Δ*lasI* deletion strain compare after 96 hours? Explain these results.

24. Dental caries, a ubiquitous disease affecting billions of people globally, is characterized by biofilm formation that leads to acid damage of the tooth enamel. While the bacterium *Streptococcus mutans* (a Gram-positive facultative anaerobe) is considered to be the primary cause of tooth decay, dental biofilms can be highly diverse. One model depicts the dental biofilm as having a core of *S. mutans* (light purple cocci) surrounded by a layer of other types of cells (e.g., *Streptococcus oralis*). (LO5, 6)

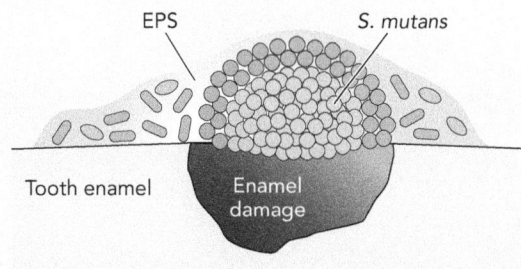

a) To investigate dental biofilm formation, you

- Set up a culture of 10^6 CFU/ml of exponentially growing *S. mutans* and *S. oralis* cells in liquid growth medium.
- Add hydroxyapatite disks (to mimic tooth enamel) and incubate for 48 hours.
- Transfer the biofilm-laden disks to a sterile 10% saliva solution with or without 1% sucrose (a fermentable sugar).
- Measure the total number of cells (by viable cell counting) on replicate disks and determine the pH profiles using microelectrodes on others. You obtain the results shown.

Treatment	Biofilm thickness (mm)	CFU/biofilm	pH at enamel surface
Biofilm, no sucrose	300	10^7	7
Biofilm + sucrose	330	10^7	4.5

Propose a theory that would explain why the pH changed at the enamel surface while the CFU/biofilm was the same with and without sucrose.

b) You then use your biofilm disks incubated with sucrose to test the effectiveness of treating the biofilm with chlorhexidine (CHX), a broad-spectrum antibiotic that disrupts cell membranes and is used in clinical settings to control inflammatory dental conditions. After the incubation in sterile saliva solution containing 1% sucrose, you carefully move some disks into 0.2% CHX for 10 minutes. You then rinse the disks in sterile saliva solution and measure biofilm thickness, cell density, and pH. You obtain the results shown in the table.

Treatment	Biofilm thickness (mm)	CFU/biofilm	pH at enamel surface
Biofilm + sucrose, no CHX	330	10^7	4.5
Biofilm + sucrose, with 10 min CHX treatment	300	10^6	5.7

Based on these data, do you think this treatment was effective enough for clinical use? Did this treatment eliminate the biofilm? How did the pH at the enamel surface compare in samples with and without CHX treatment?

c) You are curious about the role the extracellular polymeric substance (EPS) plays in protecting the biofilm from the action of CHX. You grow some biofilms on disks as described above, treating some disks with dextranase (an enzyme the breaks down EPS) for 60 minutes before transferring disks to sterile saliva or the CHX treatment. You obtain the results shown in the table.

Treatment	CFU/biofilm
Biofilm + sucrose	10^7
Biofilm + sucrose + dextranase	10^7
Biofilm + sucrose + CHX	10^6
Biofilm + sucrose + dextranase + CHX	10^4

i. Do these data support the hypothesis that the EPS provides protection against CHX? Explain your reasoning.

ii. Why did you include a treatment for biofilm + sucrose + dextranase? What does this control tell you?

INDEX

Please note: entries in *italics* refer to figures, whereas **bold** entries refer to tables.

+1sites 148
2C *see* 2-phosphoglycolate
2D SDS-PAGE 59–61
3C *see* 3-phosphoglycerate
16S/18S rRNA gene 7–11
3HP-4HB cycle *see* 3-hydroxypropionate-
    4-hydroxybutyrate cycle
3HP bi-cycle *see*
    3-hydroxypropionate bi-cycle

a
ABC *see* ATP binding cassette
accessory pigments 210, *211*, 213
accessory proteins
    Sec-dependent transport 336–337
    Tat system 340
acetate 253–256
    Cra 158
    fermentation 117, *119*
    glyoxylate cycle 41–42
Acetobacterium spp., acetogenesis
    253–256
acetogens 253–256
    reductive acetyl-CoA pathway
      193, *194*
acetyl-CoA 22, 36–37
    carbon fixation 193–196, *194*,
      196, 197, *199*
    reductive pathway 193, *194*
acetyl-CoA carboxylase 195, *196*
activator proteins 153–155
active transport 308–311
    Gram-negative secretion systems
      341–347
acylases 402
acyl-homoserine lactone (AHL)
    395–396

adaptation, chemotaxis 374, 378
adenosine diphosphate (ADP) 27, *27*
adenosine monophosphate (AMP) 136
adenosine phosphosulfate (APS) 257
adenosine triphosphate (ATP) 27
    fermentation 117–118, *118–119*
    mitochondria 111, *112*
    production 22–23, 26–28,
      37–40, 95–105
    synthesis 95–105
    yield 29, *30*, 111, 118, *118–119*, 241
adenylate cyclase 139
adenylation 136
ADP *see* adenosine diphosphate
ADP-ribosylation 137–138
adventurous motility 450
aerial mycelium 447–448
aerobic anoxygenic phototrophy
    224–225
aerobic respiration 111–117
    bacterial 113–117, *114*, 241
    E. coli 241, *242*
    methanotrophs 252–253, *254*
    methylotrophy 252–253
    mitochondria 111–113, *112*
Agrobacterium tumefaciens 345
AHL *see* acyl-homoserine lactone
akinetes 444
alarmones 84–85
aldolase 25–26
allosteric regulation 29–31, 129–134
    biosynthetic pathways 130–131
    branched pathways 131–134
    concerted feedback 132–133
    cumulative regulation 132–134
    feedback inhibition 129, 132–134
    isozymes 132

    kinetics 129–130, *131*
    positive 129, 134
    precursor induction 134
    sequential feedback 134
amino acids
    biosynthesis 24, 158–161
    disulfide bonds 347, *348*
    phosphorylation targets 134
    starvation responses 84–85
ammonia
    anammox 290–293, *292*
    assimilation 136, 285–287
    nitrogen fixation 277–284
ammonia monooxygenase
    (AMO) 288, *289*
ammonia-oxidizing archaea
    (AOA) 277, 290
ammonia-oxidizing bacteria (AOB) 277,
    287–290, *289*
AMO *see* ammonia monooxygenase
AMP *see* adenosine monophosphate
amphipathic 304
amylases 236
anabolism 23
anaerobic anoxygenic phototrophy
    218–224
    Chlorobi 221–224, *222*
    Chloroflexi 221–224, *222*
    Proteobacteria 218–221, *220*, 224–225
anaerobic food web
    carbon cycle 234–256
    interspecies hydrogen transfer 262–263
    metabolic processes *235*
    sulfur cycle 256–261
    syntrophy 261–263
anaerobic methanotrophs
    (ANME) 262–263

Microbial Physiology: Unity and Diversity, First Edition. Ann M. Stevens, Jayna L. Ditty, Rebecca E. Parales
and Susan M. Merkel.
© 2024 American Society for Microbiology.
Companion website: www.wiley.com/go/stevens/microbialphysiology

anaerobic respiration 115–119
    acetogens 253–256
    carbon cycle 241–256, 261–263
    chemolithoautotrophs 246–253
    E. coli 241–246, *242*
    electron transport chains 115–117
    fermentation 117–119
    sulfur cycle 256–263
    syntrophy 261–263
    tricarboxylic acid cycle 41
analysis
    enzyme assays 55–58
    gene regulation 163–164
    genome 64–65
    motility 381–382
    phylogeny 7–11
    proteome 59–63
    SDS-PAGE 59–61
    transcriptome 64–65
anammox 290–293, *292*
anammoxosomes *292*, 293
anaplerotic reactions 37–40
ANME *see* anaerobic methanotrophs
anoxic *see* anaerobic
anoxic degradation, aromatic acids *238*,
        239–241, *240*
anoxygenic photoheterotrophy 224
anoxygenic phototrophy 218–225
    aerobic 224–225
    anaerobic 218–224
    Chlorobi/Chloroflexi 221–224, *222*
    Firmicutes 224
    Proteobacteria 218–221, *220*, 224–225
antiporters 99, 310, 418
anti-sigma factor 369
AOA *see* ammonia-oxidizing archaea
AOB *see* ammonia-oxidizing bacteria
APS *see* adenosine phosphosulfate
aquaporins 306–308
ArcA/ArcB 244
Archaea 6, 12
    ammonia-oxidizing 277
    carbon dioxide fixation 186, *187*,
        193–196, *194*
    cell walls 315, *316*
    denitrification 293–294
    dicarboxylate/4-hydroxybutyrate
        cycle 197, *199*
    flagella 370
    3-hydroxypropionate-
        4-hydroxybutyrate cycle 195, *199*
    nitrogen fixation 277–284
    reductive acetyl-CoA pathway 193, *194*
    sulfate-reducing 256–259, 261–263
    sulfur-oxidizing 259–261
    syntrophy 261–263
archaellum 370
aromatic acids
    degradation 236–241, *238*
        anoxic *238*, 239–241, *240*
        dioxygenases 237–239, *239*
        monooxygenases 237, *239*
        oxic 236–239, *238–240*
artificial membrane vesicles 101–102, *102*

A-signal 450
aspartate transcarbamoylase 130
assays, enzyme 55–58
assimilation
    ammonia 136, 285–287
    nitrate 284–285
ATP *see* adenosine triphosphate
ATP binding cassette (ABC) transport
        310–311, *311*
ATP sulfurylase 257
ATP synthase *97*, 98–99, 111–113
    bacteria 113, *114*
    mitochondria 111–113, *112*
    yield 111
attenuation 158–161
attenuators 159–161
autoinducers 390, 394–395, 401–402
autoinducer synthase 395
autotrophs 178, 184
autotrophy 184–204
    Calvin cycle 187–191
    dicarboxylate-4-hydroxybutyrate
        cycle 197, *199*
    3-hydroxypropionate-
        4-hydroxybutyrate cycle 195, *199*
    3-hydroxypropionate bi-cycle 195
    reductive acetyl-CoA pathway 193, *194*
    reductive TCA cycle 191, *192*
Avogadro's number 55

b
Bacillus subtilis 13, 399, *400*, 421–428
Bacteria 6, 12–14
    aerobic respiration 115–117, 241, *242*
    ammonia-oxidizing 277
    anaerobic respiration 115–119
    Calvin cycle 187–191, *188*
    carbon dioxide fixation 186, 187–193,
        188, *192*, *194*, 195, *196*
    cell walls 311–315, *312–314*
    cytoplasmic membranes 304–306,
        305, *307*
    denitrification 293–294
    electron transport chains 113–117, *114*
    fermentation 117–119
    flagella 364–370
    3-hydroxypropionate bi-cycle 195, *196*
    nitrite-oxidizing 287–288, 290, *291*
    nitrogen fixation 277–284
    proton motive force 113–115, *114*
    reductive acetyl-CoA pathway 193, *194*
    reductive TCA cycle 191, *192*
    sulfate-reducing 256–259
    sulfur-oxidizing 259–261
    surface motility 371–372, *372*
bacteriochlorophylls (Bchl) 209, *211*
    wavelengths 210
bacteriophage 319, 345–346, *347*
bacteroids 283
Bam complex 343, 351
base pairing, small RNAs 161–163, *162*
batch culture 78–84
    chemostats 83–84
    diauxic growth 80–81

growth phases 78–81
    kinetics 81–82
    requirements 86
    steady-state 78, 83–84
    yield coefficient 82
Bchl *see* bacteriochlorophylls
benzoate catabolism 239, *240*
biased random walk 373
bicarbonate 40, 190
binary fission 78, 85, 85–86, *86*
bioenergetics 95–105
    ATP synthase *97*, 98–99
    chemiosmotic theory 98
biofilms 322
    cellular differentiation 452–456
    cheater cells 455–456
    Gram-negative bacteria 395–398
    Gram-positive bacteria 398–400
    health problems 456
    industrial impacts 456
    persister cells 455
    quorum sensing 388–413, 452–455
    symbiotic bioluminescence 391–395
bioluminescence 391–395
biosynthesis
    allosteric regulation 130–131
    amino acids 24
    anaplerotic reactions 37–40
    cytidine triphosphate 130
    fatty acids 24
    lipids 24
    nucleotides 24
    sugars 24, 31
    tryptophan 131
biosynthetic precursors, essential 23–25
1,3-bisphosphoglycerate 26–27, 117
Bordella pertussis 345
Bradford assay 56–57
branched pathways, allosteric
        regulation 131–134
branched tricarboxylic acid
        pathway 40–41
Braun's lipoprotein 312
buffers 55

c
C1 carriers 249–251, *250*
Calvin cycle 187–191, *188*
cAMP *see* cyclic AMP
cAMP receptor protein (CAP) 139, *140*
CAP *see* cAMP receptor protein; catabolite
        activator protein
capillary assay 381
capsule 321–322, 352
carbon, starvation responses 85
carbon cycle 178, *180*, 234–256, *235*
    acetogens 253–256
    aromatic acid degradation 236–241, *238*
    benzoate catabolism 239, *240*
    chemolithoautotrophs 246–253
    β-ketoadipate pathway 239, *240*
    methanogens 248–251
    methanotrophs 252–253, *254*
    methylotrophy 251–253

overview 234–235, *235*
polymer degradation 236, *237*
syntrophy 261–263
carbon dioxide fixation 187–199
    Calvin cycle 187–191, *188*
    dicarboxylate/4-hydroxybutyrate
        cycle 197, *199*
    diversity of 197–199
    3-hydroxypropionate-
        4-hydroxybutyrate cycle 195, *199*
    3-hydroxypropionate bi-cycle 195, *196*
    reductive acetyl-CoA pathway 193, *194*
    reductive TCA cycle 191, *192*
carbon flow
    glycolysis 29–31, 43–44
    reversing 43–44
    tricarboxylic acid cycle 43–44
carbonic anhydrase 190
carbon monoxide dehydrogenase-
    acetyl-CoA synthase 193
carbon sources 178
    chemolithoautotrophs **246**
carbon storage regulators 162, *163*
carboxysomes 190–191
carotenoids 210, *211*
cascade of sigma factors 427–428
catabolism 23
    aromatic acids 236–241, *238*
    polymer degradation 236, *237*
catabolite activator protein
    (CAP) 139, *140*
catabolite repression 139–140, *140*
catabolite repressor/activator (Cra) 158
catalase 417
Caulobacter crescentus 443–444
*C. diphtheriae see Corynebacterium
    diphtheriae*
cDNA *see* complementary DNA
cell–cell signalling
    quorum sensing 388–413
    rhizobia and plants 282–284
cell division
    chromosome segregation 86–87
    proteins 85, *86*
cell envelopes 13, 302–360
    biofilms 322
    capsule 321–322, 352
    cell walls 311–315, 348–349
    component transport and
        localization 334–348
    cytoplasmic membranes 304–311, *305*,
        307–309, 311
    Gram-negative bacteria *308*, 312–321,
        313–314, 319–320
    Gram-positive bacteria 312, *313–314*
    membrane proteins 334–341, *336*,
        338, 350–351
    outer membranes 315–320, 341–347,
        349–351
    periplasm 320–321
    phospholipids 304–306, *305, 307*
    Sec-dependent transport 334–337,
        340–343, 345–346, 351
    S-layers 322

SRP-dependent transport 338–340, *338*
    structure 302–331
    Tat transport system 339–340, 341–343
cell shape, proteins 85, *86*
cellular competence 399, *400*
cellular components 50–71
    DNA 63–64
    E.coli 53
    enzyme assays 55–59
    genome 59, 63–65
    molecular concentrations 52–55
    proteome 54–55, 59–63
    RNA 61–63
    transcriptome 61–62, 64–65
cellular differentiation 440–466
    biofilms 452–456
    Caulobacter crescentus 443–444
    Cyanobacteria 444–446
    Myxobacteria 449–451
    Nostoc punctiforme 444–445, *444*
    Streptomyces 447–449
cellular growth 73–92
    batch culture 78–84
    binary fission 78
    chemostats 83–84
    chromosome segregation 86–87
    diauxic 80–81, 139–140
    generation time 78
    kinetics 81–82
    monitoring 74–77
    phases 78–81
    requirements 80
    stationary phase 79, 84–85
    steady-state 78, 83–84
    yield coefficient 82
cellular proton levels 100
cellulases 236
cell walls 7, 13, 311–315
    archael 315, *316*
    bacterial 311–315, *312–314*
    Gram-positive/negative 312, *313–314*
    lysis 55
    mycobacteria 313–315
    peptidoglycan transport 348–349
central dogma, regulatory modes 128, *128*
central metabolism 20–49
    anaplerotic reactions 37–40
    coenzymes 28–29
    connections 38
    core concepts 22–23
    Embden–Meyerhof–Parnas pathway
        22, 25–28
    Entner–Doudoroff pathway 33–36
    glyoxylate cycle 41–42
    pentose phosphate pathway 31–33
    transition reaction 36–37
    tricarboxylic acid cycle 37–41
CFUs *see* colony-forming units
chaperones 336
    heat shock response 419–420
    outer membrane insertion
        350–351, *351*
chaperonins 336
chaplins 448

cheater cells 455–456
chemiosmotic theory 98
chemoeffectors 374–375
chemoheterotrophs 179
    sulfur cycle 256–259
chemolithoautotrophs 178, 186, **246**
chemolithoautotrophy 246–256
    acetogens 253–256
    Calvin cycle 187–191, *188*
    carbon dioxide fixation 186–199, *188,
        192, 194, 196–199*
    dicarboxylate-4-hydroxybutyrate
        cycle 197, *199*
    E. coli 241–246, *242*
    3-hydroxypropionate-
        4-hydroxybutyrate cycle 195, *199*
    3-hydroxypropionate bi-cycle 195, *196*
    methanogens 248–251
    methanotrophs 252–253, *254*
    methylotrophy 251–253, 252–253
    reductive acetyl-CoA pathway 193, *194*
    reductive TCA cycle 191, *192*
    sulfate reduction 256–259
    sulfur cycle 256–261
    sulfur oxidation 259–261
    syntrophy 261–263
chemolithotrophs 179, 234
    carbon cycle 246–256
    nitrification 287–293
    sulfur cycle 256–261
chemoorganoheterotrophs 179
chemoorganotrophs 234
chemoorganotrophy
    aromatic acid degradation 236–241
    E. coli 241–246
    polymer degradation 236
chemostats 83–84
chemotaxis 372–382
    adaptation 374, 378, 380
    biased random walk 373
    conservation and variation 380–381
    E. coli 374–380
    excitation 374, 377–378
    laboratory methods 381–382
    repellants 380
    steady-state 378
chemotrophs 178
chemotrophy 232–272
    anaerobic food web 261–263
    carbon cycle 234–256
    sulfur cycle 256–261
    syntrophy 261–263
Che proteins 374–380
Chl *a see* chlorophyll *a*
Chlorobi
    light-harvesting complexes
        213, *214*
    phototrophy *214*, 221–224, *222*
Chloroflexi
    3-hydroxypropionate bi-cycle
        195, *196*
    light-harvesting complexes 213, *214*
    phototrophy *214*, 221–224, *222*
chlorophyll *a* (Chl *a*) 209, *211*

chlorophylls 209–225
  cellular structures 211–215
  Chlorobi/Chloroflexi 221–224
  Cyanobacteria 215–218
  Firmicutes 224
  pigments 210, *211*
  Proteobacteria 218–221, 224–225
  structures *210–211*
  wavelengths 209–211
chlorosomes 213, *214*
cholera toxin 137–138, *138*, 402
chromosomes, segregation 86–87
cis-acting regulatory agents 153–155
cis-*trans* isomerization *226*, 227
citrate lyase 191
citric acid cycle 22, 36–44, *37–41*
  see also Krebs cycle
Class I-activated promoters 155
classification, enzymes 58
Class II-activated promoters 155
Clostridium spp, acetogenesis 253–256
computer-based tracking and modeling
    381–382
coenzymes 28–29, 248–251
colony-forming units (CFUs),
    measurement 76–77
comammox 288
competence and sporulation factor (CSF)
    399, *400*, 425
complementary DNA (cDNA) 163
Complex I 110–111
Complex II 111
Complex III 111
Complex IV 111
complex media 80
concentrations
  Avogadro's number 55
  estimation 52–54
  molarity 55
  physiologically relevant 54–55
concerted feedback 132–133
conditional mutants 85
constitutive expression 148
core polysaccharide 317
core RNA polymerase 149
Corynebacterium diphtheriae (*C.
    diphtheriae*) 137
coupled reactions 310
coupled sites 111–113
covalent modifications 134–140
  adenylation 136
  glycosylation 137–138
  methylation 138
  phosphorylation 134–136
  sugar phosphotransferase system
      139–140, *140*
Cra *see* catabolite repressor/activator
Crenarchaeota, carbon fixation
    195–197, *199*
C ring 366, *367*
crisscross regulation 428
cross-linking
  peptidoglycan 312, *313*
  pseudopeptidoglycan 315, *316*

cross talk 400
CRP 139, *140*, 157–158
crude cell extract 34
CSF *see* competence and sporulation factor
C-signal 450
CsrA/CsrB 162, *163*
CTD *see* C-terminal domains
C-terminal domains (CTD) 135–136
CTP *see* cytidine triphosphate
CtrA 444
cumulative regulation 132–134
Cyanobacteria spp.
  Calvin cycle 187–191, *188*
  cellular differentiation 444–446
  light-harvesting complexes 213, *214*
  photoautotrophy *214*, 215–218, *216*
cyanoglobin 446
cyclic AMP (cAMP) 139
cyclic diguanylate (di-GMP) 453
cyclic electron flow 218
cysteine, double bonds 347, *348*
cytidine triphosphate (CTP) 130
cytochrome oxidases 111, *112*,
    114, *114*, 116
cytochromes 108–116, *112*, *114*
  ammonia-oxidizing bacteria 288, *289*
  anammox *292*, 293
  bacterial respiration 114, *114*, 116, 244
  denitrification 294
  heme 108, *109*
  mitochondrial respiration **110**, *112*, 113
  phototrophy 217, 219–221, 223–224
  sulfate reduction 257, *258*
  sulfur oxidation *260*, 261
cytoplasmic membrane 13
  phospholipids 304–306, *305*, *307*, 348
  protein transport systems 334–348
  transport 306–311, *307–309*, *311*
  variations in 306

d
D *see* daltons
DAHP synthase *see*
    3-deoxy-ᴅ-arabinoheptulosonic
    acid-7-phosphate synthase
daltons (D) 60
DC-4HB cycle *see* dicarboxylate-
    4-hydroxybutyrate cycle
defined media 80
denitrification 293–294
3-deoxy-ᴅ-arabinoheptulosonic acid-
    7-phosphate synthase (DAHP
    synthase) 131
DHAP *see* dihydroxyacetone phosphate
diauxic growth 80–81, 139–140
diazotrophs 277–284
  biochemistry 278–279
  regulation 280–282
  symbiosis 282–284, *284*
dicarboxylate-4-hydroxybutyrate cycle
    (DC-4HB cycle) 197, *199*
differentiation 440–466
  biofilms 452–456
  Caulobacter crescentus 443–444

Cyanobacteria 444–446
  Myxobacteria 449–451
  Nostoc punctiforme 444–445, *444*
  Streptomyces 447–449
diffusion
  facilitated 306, *308*
  simple 306
di-GMP *see* cyclic diguanylate
diguanylate cyclase 453–454
dihydroxyacetone phosphate (DHAP) 25
dinitrogen reductase 278, *279*
dinotrogenase 278, *279*
dioxygenases 237–239, *239*
diphtheria toxin 137
direct count method 75–76
disulfide bonds 347, *348*
divisome 85, *86*
DNA 63–64
  cellular competence 399, *400*
  chromosome segregation 86–87
  operons 148, 156–161, 281–282,
      367–369, *368*
  replication 86, *87*
  transcription 149–161
DnaJ 419–420
DnaK 336, 419–420
DNA–protein interaction tests
    164–168
domains, phlogenetic 6–7
doubling time 78
dry weight measurement 77
DsbA/DsbB 347, *348*
DskA 84–85
DsrA 161

e
E$_I$ *see* enzyme I
E$_{II}$ *see* enzyme II
Earth history 14–15
E. coli see Escherichia coli
eDNA *see* extracellular DNA
ED pathway *see* Entner–Doudoroff
    pathway
EF-2 *see* elongation factor-2
effector molecules, allosteric 129–134
electrochemical gradients 98–102
electrode potentials 109–111, **110**,
    112, 116, 241
electron acceptors 110, 114–117
electron bifurcation 251
electron donors 116
  chemolithoautotrophs 244, **246**, 248
  E.coli 244
  nitrogen fixation 278
  sulfur oxidizers 259–261
electron sources 178
electron transport chains (ETC)
    108–117, *112*, *114*
  ammonia-oxidizing bacteria
      288–290, *289*
  anammox *292*, 293
  bacterial 113–117, *114*
  Chlorobi/Chloroflexi 221–224, *222*
  complexes 108

Cyanobacteria 215–218, *216*
denitrification 294
E. coli 113–115, *114, 116*, 244–246
Firmicutes 224
methanogens 251
mitochondria 110–113, *112*
nitrite-oxidizing bacteria 290, *291*
phototrophy 211, *212*, 215–224
prosthetic groups 108, *109*
Proteobacteria 218–221, *220*, 224–225
proton pumps 111, *112*, 114–115, *114*
Q cycle 111, *112*, 113
Q loop 118
reduction potentials 109–111,
110, *112*, 116
sulfate reduction 257, *258*
sulfur oxidation *260*, 261
terminal electron acceptors
110, 114–117
electrophoresis 164, *165*
electrophoretic mobility shift assay
(EMSA) 164–168, *166*
elongation 151
elongation factor-2 (EF-2) 137
Embden–Meyerhof–Parnas (EMP)
pathway 22, 25–28
carbon flow 29–31, 43–44
fermentation 117
see also glycolysis
EMP pathway *see* Embden–Meyerhof–
Parnas pathway
EMSA *see* electrophoretic mobility
shift assay
endoflagella 365, *366*
endospores 420–425, *422*
formation 421–423
germination 424–425
properties 423
endoymbiotic theory 7
energy sources 178
enterochelin *320*
Entner–Doudoroff (ED) pathway
33–36
environmental impacts, proton motive
force 100–101
EnvZ 318
enzyme assays 55–58
enzyme I (E_I) 139, *140*
enzyme II (E_{II}) 139–140, *140*
enzymes
adenylation 136
allosteric regulation 129–134
assays 55–58
classification 58
covalent modifications 134–140
disulfide bond formation 347, *348*
glycosylation 137–138
kinetics 58–59, 129–130, *131*
methylation 138
phosphorylation 134–136
see also individual enzymes. . .
EPS *see* extracellular polymeric
substance
erythrose-4-phosphate 22–26

Escherichia coli (*E. coli*)
aerobic respiration 113–115, *114,
116*, 241, *242*
anaerobic respiration 115–117,
241–246, *242*
cellular composition 53
chemoorganotrophy 241–246
chemotaxis 374–380
electron transport chains 113–117, *114,
116*, 244–246
fermentation *242, 243*
sigma factors **153**
ETC *see* electron transport chains
ethanol fermentation 117, *118*
Eukarya 6–7
Euryarchaeota, methanogens 248–251
evolution, timeline 14–15
evolutionary distance 8
excitation, chemotaxis 374, 377–378
exoenzymes 236
exponential growth kinetics 81–82
extracellular DNA (eDNA) 456
extracellular polymeric substance
(EPS) 322
extremophiles 12

f
facilitated diffusion 306–308
factor-dependent termination 151, *152*
factor-independent termination 151
FAD/FADH$_2$ *see* flavin adenine
dinucleotide
fatty acids
biosynthesis 24
Cra 158
glyoxylate cycle 41–42
feedback inhibition 29–31, 129, 132–134
concerted 132–133
cumulative 132
sequential 134
fementation 117–119, *118–119*
fermentation *28, 29, 242, 243*
ferredoxin
methanogens 251
nitrogen fixation 278–279
in photosystem I 217–218
ferredoxin-dependent α-ketoglutarate
synthase 191
FeS *see* iron-sulfur
filamentous *Cyanobacteria* 444–446
filamentous *Streptomyces* 447–449
Firmicutes
reductive acetyl-CoA pathway
193, *194*
sporulation 429–430
FISH *see* fluorescent *in situ* hybridization
flagella
archaeal 370
bacterial 364–367
chemotaxis 372–380
classification 364–365
filament synthesis 365–369, *368*
Gram-negative bacteria 366, *367*
structure 365–366, *367*

swarmer cells 443–444
tumbling 369, *370*
flavin adenine dinucleotide (FAD/FADH$_2$)
37, 39–40, 108–111, *109*, 114, 241
bacteria 114
denitrification 294
mitochondria 111, *112*
reduction potential **110**
flavonoid compounds 282
flavoproteins 108–111, *109, 112*, 114
FlhDC 367–369, *368*
FliA/FliM 369
flippases 348, *349*
flocs 452
flow cytometry 76
fluorescence microscopy 74
fluorescent *in situ* hybridization (FISH) 76
FNR *see* fumarate nitrate reductase
folding, proteins 336–338
forespore 423
fructose-1,6-bisphosphate 25–26
fructose-6-phosphate 22–26, 189
fruiting bodies 449–450
FtsY 338
FtsZ *78, 85, 86*
fumarate 116–117
fumarate nitrate reductase (FNR)
regulator 244
fumarate reductase 191
furanones 400

g
G6P *see* glucose-6-phosphate
Gammaproteobacteria 12–13
see also Proteobacteria
Gemmatimonadetes, phototrophy 225
general porins 317–318
generation time 78–79
gene regulation 148–177
attenuation 158–161
global regulators 158
laboratory methods 163–168
nitrogen fixation 280–282
operons 148, 156–161, 281–282
phosphorylation 134–136
phosphotransferase system
139–140, *140*
small RNAs 161–163, 317
transcription 148–161
genes
reporter 164
transcription 148–161
transcription assays 163–164
genome 59, 63–66
germination 424–425
gliding motility 371–372, 444–446,
450–451
global nutrient cycles 179–180
global regulators 148, 158
gluconeogenesis 31, 158, 191
glucose-6-phosphate (G6P) 22–26,
31–32
glucose, phosphotransferase
system 139–140

glutamate, ammonia assimilation 136, 285–287
glutamate dehydrogenase 286
glutamine oxoglutarate aminotransferase (GOGAT) 285
glutamine synthetase (GS) 136, 285
glutathione 417–418
glutathione reductase (GOR) 417
glyceraldehyde-3-phosphate 22–27
glyceraldehyde-3-phosphate dehydrogenase 26
glycerol transport 308
glycocalyx 321–322
glycogen, biosynthesis 162
glycolysis 22, 25–28
  carbon flow 29–31, 43–44
  fermentation 117
  see also Embden–Meyerhof–Parnas pathway
glycosylation 137–138
glyoxylate cycle 41–42, 134
GOGAT *see* glutamine oxoglutarate aminotransferase
GOR *see* glutathione reductase
GorA 417
G proteins, ADP-ribosylation 138
Gram-negative bacteria 12–13
  cell envelope 308, 312, 313–314, 315–320, 319–320
  cell wall 312, 313–314
  flagella 366, 367
  outer membrane 315–320, 341–347, 349–351
  periplasm 320–321, 339–340, 350–351
  quorum sensing 395–398, 395
  secretion systems 335–347
  surface motility 371–372, 372
Gram-positive bacteria 13–14
  cell envelope 312, 313–314
  quorum sensing 398–400, 399–400
  secretion systems 335–340, 345, 346
  surface motility 371–372, 372
gratuitous inducers 158
green nonsulfur bacteria
  3-hydroxypropionate bi-cycle 195, 196
  phototrophy 221–224
green sulfur bacteria
  phototrophy 221–224
  reductive TCA cycle 191, 192
GroEL 336
GroES 336
group translocation 138–140, 140
growth
  batch culture 78–84
  binary fission 78
  chemostats 83–84
  chromosome segregation 86–87
  diauxic 80–81, 139–140
  generation time 78
  kinetics 81–82
  monitoring 74–77
  phases 78–81
  requirements 80
  stationary phase 79, 84–85
  steady-state 78, 83–84

yield coefficient 82
growth rate (k) 81–82
GrpE 419–420
GS *see* glutamine synthetase
GSH *see* reduced glutathione
guanosine pentaphosphate (pppGpp) 84–85
guanosine tetraphosphate (ppGpp) 84–85

h
Halobacterium, retinal-based phototrophy 225–227
HAO *see* hydroxylamine oxidoreductase
health impacts, biofilms 456
heat shock response 419–420
helper domains 342
heme 108, 109
hemolysin 343
heterocysts 444
heterolactic fermentation 117, 118
heterotrophs 178
hexokinase 25, 29
Hfq 161, 162
high-affinity energy-dependent transporters 319–320, 320
histidine kinase 135–136, 135
H$_4$MPT *see* tetrahydromethanopterin
homolactic fermentation 117, 118
horizontal gene transfer 10–11, 456
hormogonia 444
housekeeping functions/genes 148
HPr 139, 140
hybridization, Northern blots 164, 165
hydrazine synthase 293
hydrogen
  methanogenesis 248–251, 250
  nitrogen fixation 278–279
hydrogenases 251
hydrogen peroxide 417
hydrogen transfer, interspecies 262–263
4-hydroxybutyrate, carbon fixation 195, 199
hydroxylamine oxidoreductase (HAO) 288, 289
hydroxyl radicals 417
3-hydroxypropionate-4-hydroxybutyrate cycle (3HP-4HB cycle) 195, 199
3-hydroxypropionate bi-cycle (3HP bi-cycle) 195, 196
hyphae 447

i
IAA *see* indole acrylic acid
incomplete tricarboxylic acid pathway 40–41
indole acrylic acid (IAA) 158–159
inducer exclusion 139
inducers, gratuitous 157
industrial impacts, biofilms 456
infection threads 283
initiation, transcription 149–151
injectisome 343–345, 344
interspecies communication, quorum sensing 400

interspecies hydrogen transfer 262–263
inverted vesicles 101–102, 102
ion-driven transport 309, 310
ionophores 102
IPTG *see* isopropyl-β-D-thiogalactopyranoside
iron-sulfur (FeS) proteins 108, 109, 110–111, 112, 114, 114, 116
isocitrate 37
isoelectric point (pI) 59–60
isomerase 25
isomerization reactions 31–33
isopropyl-β-D-thiogalactopyranoside (IPTG) 157
isozymes 131

k
k *see* growth rate
KatG 417
KDO *see* ketodeoxyoctonate
KDPG *see* 2-keto-3-deoxy-6-phosphogluconate
KefB/C antiporter 418
β-ketoadipate pathway 239, 240
ketodeoxyoctonate (KDO) 317
2-keto-3-deoxy-6-phosphogluconate (KDPG) 33–35
α-ketoglutarate 22–23, 37–41, 287
kinases 134–136
kinetics
  enzymes 58–59, 129–130, 131
  growth 81–82
Krebs cycle 22, 36–44, 37–41
  anaplerotic reactions 37–40
  branched/incomplete 40–41
  carbon flow 43–44
  glyoxylate cycle 41–42
  see also tricarboxylic acid cycle

l
laboratory methods
  gene regulation 163–164
  motility 381–382
lac operon 156–158
Lac permease 310
β-lactam antibiotics 350
lactate fermentation 117, 118–119
lactonases 402
ladderanes 292, 293
lag phase 79
LamB 318–319
last universal common ancestor (LUCA) 8, 14–15
leader sequence/leader peptide 335
  see also signal sequences
leghemoglobin 283–284
light-harvesting complexes 211–215, 213–214
  phycobilisomes 213, 214
  reaction centers 211, 212
  regulation 213
  thylakoids 213, 214
lignin 236, 237
ligninolytic oxoenzymes 236
linear electron flow 218

lipases 236
lipid A 315–317
lipid carriers 348–349, 352
lipids, biosynthesis 24
lipopolysaccharide (LPS) 13, 317, 349–350, *351*
lipoteichoic acid 13
lithotrophs 178
localization
    membrane proteins 337, 340–341, *340*, 350–351, *351*
    phospholipids 348
lophotrichous bacteria 365
Lowry assay 56
LPS *see* lipopolysaccharide
LUCA *see* last universal common ancestor
luciferase 392–394
lux operon 394–396
LuxS system 400
lysis, cell walls 55
lysozyme 55

m
M *see* molarity
malonyl-CoA 195–197
mannitol 139–140
mannose 139–140
maximum parsimony 8
MCPs *see* methyl-accepting chemotaxis proteins
measurement
    growth 74–77
    proton motive force 101–102
    see also monitoring
membrane proteins
    rhodopsins 225–227, *226*
    see also transmembrane proteins
menaquinone (MQ) **110**, 116, *222*, *223*
messenger RNA (mRNA) 62
    assays 163–164
    monocistrionic 148
    polycistrionic 148
    riboswitches 162–163, *163*
    sRNA binding 162, *162–163*
    transcription 149–151, *150*, *152*
    transcriptome 164
metabolism
    autotrophic 184–204
    central 20–49
        anaplerotic reactions 37–40
        coenzymes 28–29
        connections 38
        core concepts 22–23
        Embden–Meyerhof–Parnas pathway 22, 25–28
        Entner–Doudoroff pathway 33–36
        glyoxylate cycle 41–42
        pentose phosphate pathway 31–33
        transition reaction 36–37
        tricarboxylic acid cycle 37–41
    diversity of 178–179
    see also respiration
metabolome 59

metagenomics 66
methanofuran *249*, 251
methanogens
    chemolithoautotrophy 248–251
    reductive acetyl-CoA pathway 193, *194*
    syntrophy 261–263
methanol 252–253
methano monooxygenase (MMO) 252
methanotrophs 252–253, *254*, 262–263
methyl-accepting chemotaxis proteins (MCPs) 375–380
methylation 138
    chemotaxis 374, 378, 380
methylotrophy 251–253
methylreductase complex cofactors *249*, 251
Michaelis–Menten kinetics 58–59, 129–130
microbiomes 66
microRNAs 63
MinCDE proteins 85, *86*
minimal media 80
mitochondria
    ATP synthase 97, 98–99
    chemiosmotic theory 98
    electron transport chain 110–113, *112*
    proton motive force 95–105, 111–113, *112*
MMO *see* methano monooxygenase
MoFe *see* molybdenum-iron-sulfur cofactor
molarity (M) 55
molecular concentrations
    estimation 52–54
    physiologically relevant 54–55
molybdenum-iron-sulfur cofactor (MoFe) proteins 278–280, *279*
monitoring
    growth 74–77
    see also measurement
monocistrionic 148
monooxygenases 237, *239*, 252
monotrichous bacteria 364
MotA/MotB 366–367, *367*
motility 364–387
    biased random walk 373
    chemotaxis 372–381
    classification 364–365
    conservation and variation 380–381
    flagella 443–444
    flagella-driven 364–370
    gliding 371–372, 444–446, 450–451
    laboratory methods 381–382
    random walk 372–373
    regulation of synthesis 367–369
    tumbling 369, *370*, 372–374, *373*
    twitching 371, *372*
MQ *see* menaquinone
M ring 367, *367*
mRNA *see* messenger RNA
murein *see* peptidoglycan
murein lipoprotein 312
mycobacteria, cell walls 313–315
Myxobacteria spp. 449–451
myxospores 450

n
N-acetylglucosamine (NAG) 311, *312*, 348, *350*
N-acetylmuramic acid (NAM) 311, *312*, 348, *350*
NADH
    ammonia-oxidizing bacteria 288–290
    central metabolism 22–23, 27–29, 27, 37–40
    denitrification 294
    fermentation *28*, 29, 117, *118–119*
    mitochondria 110–111, *112*
    nitrogen fixation 278–279
    reduction potential 110–111, 241
NADH dehydrogenase 110–111, *112*, 114, *114*, 116
NADPH
    ammonia assimilation 285–286
    ammonia-oxidizing bacteria 290
    central metabolism 22–23, 27, 28–29, 31, 37–40
    chemolithoautotrophy 247
    nitrite-oxidizing bacteria 290
    phototrophy 215–224, *216*, *220*, *222*
NAG *see* N-acetylglucosamine
NAM *see* N-acetylmuramic acid
Na⁺MF *see* sodium motive force
negative regulation, small RNAs 161–162
nif operons 281–282
nitrate reduction 284–285
nitrification 287–293
    ammonia oxidation 287–290, *289*
    nitrite oxidation 287–288, 290, *291*
nitrite-oxidizing bacteria (NOB) 287–288, 290, *291*
nitrite reductase 285, 293
nitrite reduction, assimilatory 284–285
nitrogenase enzyme 278, *279*
nitrogen assimilation 277, 284–287
nitrogen cycle *180*, 274–301
    ammonia assimilation 136, 285–287
    anammox 290–293, *292*
    assimilation 277, 284–287
    denitrification 293–294
    fixation 277–284
    nitrate reduction 284–285
    nitrification 287–293
    overview 276–277
nitrogen fixation 277–284
    biochemistry 278–279
    regulation 280–282
    symbiosis 282–284, *284*
nitrogen gas 277
NOB *see* nitrite-oxidizing bacteria
nod factors 282
non-phosphotransferase sugars 139
Northern blots 164
Nostoc punctiforme 444–445, *444*
NtcA 446
NTD *see* N-terminal domains
N-terminal domains (NTD) 135–136
NtrB 280
NtrC-P 280–281

nucleotides, biosynthesis 24
nucleotidyl modification 136
nutrient cycles, global 179–180

o

O-antigen 317, 349, *350–351*
OD *see* optical density
omics 53, 64–65
OmpC/OmpF 317–318, *319*
OMV *see* outer membrane vesicles
operator sequences 153
operons 148, 156–161
   flagellar synthesis 367–369, *368*
   lac 156–158
   lux 394–396
   nif 281–282
   trp 158–161
optical density (OD) 74–75, 77
organotrophs 178
origin of replication 86–87
outer membranes
   lipopolysaccharide 349–350, *351*
   protein insertion 350–351, *351*
   protein secretion systems 341–347
outer membrane vesicles (OMV) 398
oxaloacetate 22–23, 40, 191, 197
oxic degradation, aromatic acids
   236–239, *238–240*
oxidases, *E. coli* regulation 244–246, *245*
oxidation-decarboxylation reac-
   tions 31–32
oxidative metabolism, anaplerotic
   reactions 37–40
oxidative phosphorylation 27–28, 96,
   97, 179, 241
oxidative stress 416–418
oxidoreductases 288, *289*, 402
oxygenic photoautotrophy 214,
   215–218, *216*
oxygen sensing, *E. coli* 244–246, *245*

p

P680 217
P840 223
p870 219
partitioning, chromosomes 87
passenger domains 342–343
passive transport 306–308
PC *see* plastocyanin
PCR *see* polymerase chain reaction
pellicles 452
penicillin-binding proteins *350*
pentose phosphate pathway (PPP)
   31–33, 191
PEP *see* phosphoenolpyruvate
peptide signals 398–399, *400*
peptidoglycan (murein) 7, 13,
   311–312, *312–314*
   synthesis 349, *350*
   transport 348–349
peripheral rod cells 450
periplasm 14, 320–321
   endoflagella 365, *366*
   protein transport 339–340, 350–351

peritrichous bacteria 365
peroxidases 417
persister cells 455
Petroff–Hausser chambers 75
pH, proton motive force 100–101
phases, cellular growth 78–81
pheophytin 217
phlogeny
   16S/18S rRNA gene 7–11
   domains 6–7
phosphoenolpyruvate carboxylase 40
phosphoenolpyruvate (PEP) 22–27, 40,
   117, 139, 191
phosphofructokinase 25, 29
6-phosphogluconate 31–35
6-phosphogluconolactone 31–32
3-phosphoglycerate (3C) 22, 26, 27, 189
2-phosphoglycolate (2C) 190
phospholipids 304–306, *305*, *307*, 348
phosphorelay systems 134–136
phosphoribulokinase 189–190
phosphorylases 134–136
phosphorylation
   chemotaxis 374, 377–378
   proteins 134–136, 244
   quorum sensing 399, *400*
   sporulation 425–427
   substrate-level 26–29, 37
phosphotransferase transport system (PTS)
   25, 138–140, *140*
phospodiesterases 453–454
photoautotrophs 178, 186, 208
   Calvin cycle 187–191, *188*
   carbon dioxide fixation 187–191, *188*,
   192, 195, *196*
   3-hydroxypropionate bi-cycle 195, *199*
   reductive TCA cycle 191, *192*
photoheterotrophs 179
photoorganoheterotrophs 179
photophosphorylation 96, *97*, 179
   see also phototrophy
photosynthesis 208
photosystem II (PSII) 217
photosystem I (PSI) 217–218
photosystems 208–227
   Chlorobi/Chloroflexi 214, 221–224, *222*
   chlorophyll-based 209–225
   Cyanobacteria 214, 215–218, *216*
   Firmicutes 224
   Gemmatimonadetes 225
   Halobacterium 225–227
   Proteobacteria 214, 218–221, *220*,
   224–225, *226*
   proton motive force 215, *216*, 219–224,
   220, *222*, 225, *226*
   rhodopsin-based 225–227
phototrophs 178, 208
phototrophy 206–231
   cellular structures 211–215
   chlorophyll-based 209–225
      cellular structures 211–215,
      212, *217–218*
      Chlorobi/Chloroflexi 214, 221–224, *222*
      Cyanobacteria 214, 215–218, *216*

Firmicutes 224
   pigments 210, *211*, 213
   Proteobacteria *214*, 218–221,
   220, 224–225
   retinal-based 225–227
phycobilins 210, *211*, 213
phycobilisomes 213, *214*
phylogenetic tree 4–12
phylogeny 4–17
   Earth history 14–15
   evolutionary distance 8
pI *see* isoelectric point
pigments 210, *211*, 213
Planctomyces, reductive acetyl-CoA
   pathway 193, *194*
planktonic cells 452
plants, rhizobia symbiosis 282–284, *284*
plastocyanin (PC) *216*, 217
plastoquinone (PQ) *216*, 217
PMF *see* proton motive force
polar flagella *345*, 364–370, *370*
polycistrionic 148
polymerase chain reaction (PCR)
   10, 164, *165*
polymers, degradation 236, *237*
polysome 151
poly(U) tails 151
porins 317–319
positive allosteric regulation 129, 134
positive regulation, small RNAs 161
posttranscriptional regulation
   161–163, *162–163*
   overview *129*
   small RNAs 63, 161–163, 317
posttranslational regulation 126–146
   adenylation 136
   allosteric regulation 129–134
   covalent modifications 134–140
   glycosylation 137–138
   methylation 138
   overview 128, *129*
   phosphorylation 134–136
   phosphotransferase system
   138–140, *140*
ppGpp *see* guanosine tetraphosphate
PPP *see* pentose phosphate pathway
pppGpp *see* guanosine pentaphosphate
PQ *see* plastoquinone
precursor induction 134
primary fermentation 117, *118*
primary producers 186
primers 76
P ring 366, *367*
production, ATP 22–23, 26–28,
   37–40, 95–105
prokaryotes 6
promoters 148
promotors 154–155
propionic acid fermentation 117, *119*
propionyl-CoA carboxylase 195, *196*
pro-sigma factors 428
prosthetic groups 108, *109*
proteases *236*, 419–420
protein concentration, measurement 77

protein–DNA interactions, laboratory
    methods 164–168
proteins
    adenylation 136
    cell division 85, *86*
    cell shape 85, *86*
    chaperones 350–351
    covalent modifications 134–140
    disulfide bonds 347, *348*
    folding 336–338
    glycosylation 137–138
    methylation 138
    outer membrane insertion
        350–351, *351*
    periplasmic transport 339–340,
        339, 350–351
    phosphorylation 134–136
    posttranslational regulation 126–146
    secretion systems 341–347
    translation 338–339
    translocases 336–339
    transport systems 334–348
Proteobacteria spp.
    chlorophyll-based phototrophy *214*,
        218–221, *220*, 224–225
    methanotrophs 252–253
    photosynthetic membranes 213, *214*
    retinal-based phototrophy 225–227
    Sec-dependent transport *336*, 337
    see also Escherichia coli
proteome 59–63
proteorhodopsins 225–227
proton levels, cellular 100
proton motive force (PMF) 95–105
    active transport 310
    ammonia-oxidizing bacteria
        288–290, *289*
    anammox *292*, 293
    ATP synthase 97, *98*–99
    bacteria 113–115, *114*
    chemiosmotic theory 98
    chlorophyll-based photosystems
        211–215, *216*, 219–224, *220, 222*
    coupled sites 111–114
    environmental impacts 100–101
    flagella 366, 369
    mathematical equation 100
    measurement 101–102
    methanogens 248–251
    mitochondria 111–113, *112*
    nitrite-oxidizing bacteria 290, *291*
    quantification 99–102
    rhodopsin-based photosystems
        225, *226*
    sulfate reduction 257
    sulfur oxidation 261
proton pumps 111, *112*, 114–115, *114*
pseudogenes 65
Pseudomonas aeruginosa 396–398
pseudomurein *see* pseudopeptidoglycan
pseudopeptidoglycan (pseu-
    domurein) 315
PSI *see* photosystem I
PSII *see* photosystem II

PTS *see* phosphotransferase
    transport system
purple bacteria
    Calvin cycle 187–191, *188*
    phototrophy 218–221
pyruvate 22–27, *36*–37, 40, 117
pyruvate dehydrogenase 36–37
pyruvate kinase 26, 27, 29
pyruvate synthase 191

q
Q cycle 111, *112*, 113, 217, 219–221,
    223–224
Q loop 114, 118
qRT-PCR *see* quantitative reverse tran-
    scription polymerase chain reaction
quantification
    growth 74–77
    proton motive force 99–102
quantitative reverse transcription
    polymerase chain reaction
    (qRT-PCR) 164, *165*
quinones 108, *109*, 110–117
    bacterial metabolism *112*, 114–117
    mitochondrial metabolism
        110–113, *114*
    phototrophy 217, 219–221, 224
    reduction potentials **110**, 116
quorum quenching 402–403
quorum sensing 388–413
    Bacillus subtilis 399, *400*
    biofilms 452–455
    Gram-negative bacteria 395–398, *395*
    Gram-positive bacteria
        398–400, *399–400*
    interspecies 400
    lux operon 394–396, *395*
    Pseudomonas aeruginosa 396–398
    Vibrio cholerae 400–402
    Vibrio fischeri 391–395

r
random walk 372–373
RBS *see* ribosome binding sites
reaction centers 211, *212*
reactive oxygen species (ROS) 416–418
rearragement reactions 31–33
redox *see* reduction/oxidation
reduced coenzymes 22–23, 28–29
    see also NADH; NADPH
reduced glutathione (GSH) 417–418
reduction/oxidation (redox) reactions,
    glycolysis 24
reduction potentials 109–111, **110**,
    112, 116, 241
reductive acetyl-CoA pathway 193, *194*
reductive TCA cycle (rTCA cycle) 191, *192*
regulation
    adenylation 136
    allosteric 129–134
    attentuation 158–161
    biosynthetic pathways 130–131
    branched pathways 131–134
    carbon flow in glycolysis 29–31

catabolite repression 139–140, *140*
cis-acting agents 153–155
concerted feedback 132–133
covalent modification 134–140
Cra 158
CRP 157–158
cumulative 132–134
E. coli terminal ETC complex 244–246
feedback inhibition 129, 132–134
flagellar synthesis 367–369, *368*
glycosylation 137–138
importance of 128–129
inducer exclusion 139
isozymes 131
lac operon 156–158
light-harvesting complexes 213
lux operon 394–396
methylation 138
nitrogen fixation 280–282
phosphorylation 134–136, 244
phosphotransferase system 138–140
porins 318
posttranscriptional 161–163, *162–163*
posttranslational 126–146
precursor induction 134
quorum sensing 388–413
riboswitches 162–163, *163*
sequential feedback 134
small RNAs 161–163, *162–163*, 317
sporulation 425–428
trans-acting agents 153–155, 157–158
transcription 139–140, *140*, 148–161,
    153–161, 244–246, *245*
trp operon 158–161
two-component systems 134–136,
    244–246
regulons 148, 367–369, *368*
RelA 84
repellants, chemotaxis 380
replication, DNA 86, *87*
reporter genes 164
repressor proteins 153–155, 158–161
reproduction
    binary fission 78
    sporulation 420–430, 447–451
requirements, for growth 80
respiration 106–125
    acetogenic 253–256
    aerobic 111–117, 241, *242*
    ammonia-oxidizing bacteria
        288–290, *289*
    anaerobic 115–119, 241–263, *242*
    complexes 108
    electron transport chains 108–117,
        112, 114
    fermentation 117–119, *118–119*,
        242, 243
    methanogenic 248–251, *250*
    methanotrophic 252–253
    methylotrophic 251–253
    mitochondria 110–113
    nitrite-oxidizing bacteria 290, *291*
    prosthetic groups 108, *109*
    proton pumps 111, *112*, 114–115, *114*

respiration (*Continued*)
    Q cycle 111, *112*, 113
    Q loop 114, 118
    reduction potentials 109–111,
        110, *112*, 116
response regulators 135–136, *135*
retinal
    phototrophy 225–227
    structure *210*
reverse electron flow 221, **246**, 247
rhizobia 282–284
rhodopsins 225–227, *226*
ribose-5-phosphate 22–26, 31–32, 189
ribosomal RNA (rRNA) 7–11, 61–62
ribosome binding sites (RBS) 161–163
riboswitches 162–163, *163*
riboswitch regulation 162–163, *163*
ribulose-5-phosphate 31–32
ribulose-1,5-bisphosphate carboxylase/
    oxygenase (RubisCO) 189–191, *190*
ribulose monophosphate (RuMP)
    cycle 252, *254*
rich media 80
right-side-out vesicles 101–102, *102*
ring-cleavage dioxygenases 239
RNA 61–63
    ribosomal 7–12
RNAP *see* RNA polymerase
RNA polymerase holoenzyme 149–151
RNA polymerase (RNAP) 149–151, *150*,
    152, 155, 280–281
RNA-Seq *see* transcriptome sequencing
root nodules 282–283
ROS *see* reactive oxygen species
RpoD/RpoH 419–420
RpoS 84, 161, *162*
rRNA *see* ribosomal RNA
RseA 419–420
rTCA cycle *see* reductive TCA cycle
RubisCO *see* ribulose-1,5-bisphosphate
    carboxylase/oxygenase
RuMP *see* ribulose monophosphate

s
SASPs *see* small acid-soluble proteins
SDS-PAGE 59–61
Sec *see* secretory
SecA/SecB 336
secondary fermentation 117, *119*
secondary metabolites 447
secondary structure, 16S riboso-
    mal RNA 8–9
second messengers 139
secretins 341–342
secretion
    Sec-dependent 336–337, 341–343,
        342, 345, *346*
    Tat system 339–340, *339*, 341–343, *342*
secretory (Sec)-dependent transport
    system 334–337, 340–343,
        345–346, 351
    accessory proteins 336–337
    membrane proteins 334–337, *336*,
        340–341, 351

processes 337
secretion 336–337, 341–343,
    342, 345, *346*
signal sequences 335, 337
SecYEG translocase 338
segregation, chromosomes 86–87
self-recognition 400
semi-quinone (UQ^{dot*}) 113
sequential feedback regulation 134
serine cycle 252, *254*
siderophores 320
sigma (σ) factors 149, 153–155, **153**,
    419, 427–428
signal peptidase cleavage sites
    335, 337–340
signal recognition particle (SRP)-
    dependent transport 338–340,
    338
signal sequences
    membrane integration 340–341
    Sec 334–335, 337
    SRP 338
    Tat 339
simple diffusion 306
Skp 350–351
S-layers 322
slime layer 321–322
small acid-soluble proteins (SASPs) 423
small RNAs (sRNAs) 63, 161–163, 317
social motility 450
SOD *see* superoxide dismutase
sodium motive force (Na$^+$MF)
    248–251, *250*, 255
sonicators 55
SoxR 417
specific activity 57
specific porins 318–319
spectroscopy 74–75
spore cortex 423
sporulation 420–430
    Bacillus subtilis 421–428
    Firmicutes 429–430
    fruiting bodies 449–450
    Myxobacteria 449–451
    regulation 425–428
    stages 421–423
    Streptomyces 447–449
SpoT 85
squid 391–395
S ring 366, *367*
sRNAs *see* small RNAs
SRP *see* signal recognition particle
staining 75–76
stalked cells 443
starvation, sporulation 420–430
stationary phase 79, 84–85
steady-state growth 83–84
stop transfer sequences 341
Streptomyces spp. 447–449
stress responses 414–439
    heat shock 419–420
    oxidative stress 416–418
    sporulation 420–430
stringent response 84

substrate-level phosphorylation 26–29,
    37, 96, 97, 179
substrate mycelium 447
succinate dehydrogenase 111
succinyl-CoA 22–23, 37–41, 195
sugars
    biosynthesis 24, 31
    lac operon 156–158
    phosphotransferase system
        138–140, *140*
sulfate reduction 256–259
sulfhydryl groups 347, *348*
Sulfolobales, 3-hydroxypropionate-
    4-hydroxybutyrate cycle 195, *196*
sulfur cycle *180*, 256–261
    sulfate reduction 256–259, *258*
    sulfur oxidation 259–261, *260*
    syntrophy 261–263
sulfur-oxidizing bacteria 259–261
superoxide dismutase (SOD) 416–417
superoxide radicals 416–417
superoxide reductase 417
SurA 351
surface motility 371–372, *372*
swarmer cells 443–444
swarms 449
swimming motility 364–369
    archael 370
    bacterial 364–370
    classification 364–365
    mechanism 369–370
    regulation of synthesis 367–369
    tumbling 369, *370*
swim plate assay 381
symbiosis
    nitrogen fixation 282–284, *284*
    quorum sensing 391–395
symporters 99, 310
synthesis
    ATP 95–105
    cytidine triphosphate 130
    flagella 367–369, *368*
    lipopolysaccharide 349–350, *351*
    peptidoglycan 349, *350*
    tryptophan 131
synthetic media 80
syntrophy 261–263

t
TA *see* transketolase
Tat *see* twin-arginine translocation
TCA *see* tricarboxylic acid
T-DNA 345
teichoic acid 13
temperature, Calvin cycle 189
terminal electron acceptors 110, 114–117
    chemolithoautotrophs **246**, 248
    sulfate reducers 257–259
    sulfur oxidation 259
terminal oxidase complex 111, *112*
termination, transcription 151
termination sites 86
tethered cell assays 381
tetrahydrofolate (THF) 251

tetrahydromethanopterin
(H₄MPT) *239*, 251
THF *see* tetrahydrofolate
threshold level 390
thylakoids 213, *214*
TK *see* transketolase
T*n*SS *see* type *n* secretion system
TonB-dependent transporter
319–320, *320*
total dry weight measurement 77
total protein concentration 56–57
trans-acting regulatory agents 153–155,
157–158
transcription 148–161
analysis 163–164
attenuation 158–161
cis-acting agents 153–155
Cra 158
CRP 157–158
elongation 151
initiation 149–151
overview *150*
promotors 154–155
regulation 153–161
sigma factors 153–155
termination 151
terminology 148
trans-acting agents 153–155,
157–158
transcriptional fusion 164
transcriptional regulation 153–161
attenuation 158–161
E. coli oxygen sensing 244–246, *245*
lac operon 156–158
nif operons 281–282
overview 128, *129*
phosphorelay systems 134–136
phosphotransferase system
139–140, *140*
trp operon 158–161
transcription assays 163–164
transcription factors *135*, *136*, 139
transcriptome 59, 64–66, 164
transcriptome sequencing
(RNA-Seq) 164, *165*
transfer RNA (tRNA) 8, 62
transition reaction 22, 36–37
transketolase (TA) reaction 31–33
transketolase (TK) reaction 31–33
translation 8

SRP-dependent transport 338–339
translational fusion 164
translocases 336–339
transmembrane proteins
localization 336–341, *336*, *338*,
350–351, *351*
rhodopsins 225–227, *226*
Sec-dependent transport 334–337,
336, 340, 351
SRP-dependent transport 338–340, *338*
translocases 336–339
transport 334–351
cytoplasmic 334–348
outer membrane 350–351, *351*
signal sequences 335, 337–340
transport
cell wall components 348–349
cytoplasmic membrane components 334–348
cytoplasmic membranes 306–311,
307–309, *311*
outer membranes 315–320, *318–320*
periplasmic 339–340, 350–351
phospholipids 348
Sec-dependent 334–337, 340–343,
345–346, 351
SRP-dependent 338–340, *338*
Tat system 339–340, *339*, 341–343
tricarboxylic acid (TCA) cycle 22,
36–44
anaplerotic reactions 37–40
branched/incomplete 40–41
carbon flow 43–44
glyoxylate cycle 41–42
reductive 191, *192*
see also Krebs cycle
tRNA *see* transfer RNA
trp operon 158–161
TrpR 158–161
tryptophan 131, 158–161
tumbling 369, *370*, 372–374, *373*
twin-arginine translocation (Tat) protein
secretion system 339–340, *339*,
341–343, *342*
twitching motility 371, *372*
two-component regulatory systems
134–136, 244, 375–380
type I secretion system (T1SS) 343, *344*
type II secretion system (T2SS)
341–342, *342*

type III secretion system (T3SS)
343–345, *344*
type IV secretion system (T4SS)
345, *346*
type IV topoisomerases 87
type V secretion system (T5SS)
342–343, *342*, 351
type VI secretion system (T6SS)
345–346, *347*

u
U *see* poly
ubiquinone (UQ)
phototrophy 219–221
reduction potential **110**, 116
respiration 108, *109*, 111–116,
112, *114*
UDP-NAM *see* UDP nucleotide-*N*-
acetylmuramic acid
UDP nucleotide-*N*-acetylmuramic acid
(UDP-NAM) 348, *350*
undecaprenyl phosphate cycle
349, *350*
uniporters 309
unrooted Woese phylogenetic tree
9, 12
UQ *see* ubiquinone
UQᵈᵒᵗ⁺ *see* semi-quinone

v
V. cholerae see Vibrio cholerae
vesicles, artificial 101–102, *102*
viable count method 76–77
Vibrio cholerae (*V. cholerae*) 137–138,
345–346
quorum sensing 400–402
Vibrio fischeri 391–395

w
water, transmembrane transport 306–308
wavelengths, chlorophylls 209–211
Woese phylogenetic tree 9, 12
Wood–Ljungdahl pathway 193, *194*

x
xylulose-5-phosphate 31–32

y
Yersinia spp. 345
yield coefficient 82